Graduate Texts in Mathematics 102

Springer
New York
Berlin
Heidelberg
Barcelona
Hong Kong
London
Milan
Paris
Singapore
Tokyo

Graduate Texts in Mathematics

(continued after index)

V. S. Varadarajan

Lie Groups, Lie Algebras, and Their Representations

Springer

V.S. Varadarajan
Department of Mathematics
University of California
Los Angeles, CA 90024
USA

AMS Subject Classifications: 17B05, 17B10, 17B20, 22-01, 22E10, 22E46, 22E60

Library of Congress Cataloging in Publication Data
Varadarajan, V.S.
Lie groups, Lie algebras, and their representations.
(Graduate texts in mathematics; 102)
Bibliography : p.
Includes index.
1. Lie groups. 2. Lie algebras.
3. Representations of groups. 4. Representations of algebras. I. Title. II. Series.
QA387.V35 1984 512′.55 84-1381

Printed on acid-free paper.

This book was originally published in the Prentice-Hall Series in Modern Analysis, 1974

9 8 7 6 5 4 3

ISBN 978-0-387-90969-1 ISBN 978-1-4612-1126-6 (eBook)
DOI 10.1007/978-1-4612-1126-6 SPIN 10721666

Yet here is no confusion: central-ruled
Divergent plungings, run through with a thread
Of pattern never snapping, cleave the tree
Into a dozen stubborn tusslings, yieldings,
That, balancing, bring the whole top alive.
Caught in the wind this night, the full-leaved boughs,
Tied to the trunk and governed by that tie,
Find and hold a center that can rule
With rhythm all the buffeting and flailing,
Till in the end complex resolves to simple.

from *Tree in Night Wind*

ABBIE HUSTON EVANS

PREFACE

ओं तत्सवितुर्वरेण्यम् भर्गो देवस्य धीमहि धियो यो नः प्रचोदयात् ॥

This book has grown out of a set of lecture notes I had prepared for a course on Lie groups in 1966. When I lectured again on the subject in 1972, I revised the notes substantially. It is the revised version that is now appearing in book form.

The theory of Lie groups plays a fundamental role in many areas of mathematics. There are a number of books on the subject currently available —most notably those of Chevalley, Jacobson, and Bourbaki—which present various aspects of the theory in great depth. However, I feel there is a need for a single book in English which develops both the algebraic and analytic aspects of the theory and which goes into the representation theory of semisimple Lie groups and Lie algebras in detail. This book is an attempt to fill this need. It is my hope that this book will introduce the aspiring graduate student as well as the nonspecialist mathematician to the fundamental themes of the subject.

I have made no attempt to discuss infinite-dimensional representations. This is a very active field, and a proper treatment of it would require another volume (if not more) of this size. However, the reader who wants to take up this theory will find that this book prepares him reasonably well for that task.

I have included a large number of exercises. Many of these provide the reader opportunities to test his understanding. In addition I have made a systematic attempt in these exercises to develop many aspects of the subject that could not be treated in the text: homogeneous spaces and their cohomologies, structure of matrix groups, representations in polynomial rings, and complexifications of real groups, to mention a few. In each case the exercises are graded in the form of a succession of (locally simple, I hope) steps, with hints for many. Substantial parts of Chapters 2, 3 and 4, together with a suitable selection from the exercises, could conceivably form the content of a one year graduate course on Lie groups. From the student's point of view the prerequisites for such a course would be a one-semester course on topological groups and one on differentiable manifolds.

The book begins with an introductory chapter on differentiable and analytic manifolds. A Lie group is at the same time a group and a manifold, and the theory of differentiable manifolds is the foundation on which the subject should be built. It was not my intention to be exhaustive, but I have made an effort to treat the main results of manifold theory that are used subsequently, especially the construction of global solutions to involutive systems of differential equations on a manifold. In taking this approach I have followed Chevalley, whose Princeton book was the first to develop the theory of Lie groups globally. My debt to Chevalley is great not only here but throughout the book, and it will be visible to anyone who, like me, learned the subject from his books.

The second chapter deals with the general theory. All the basic results and concepts are discussed: Lie groups and their Lie algebras, the correspondence between subgroups and subalgebras, the exponential map, the Campbell–Hausdorff formula, the theorems known as the fundamental theorems of Lie, and so on.

The third chapter is almost entirely on Lie algebras. The aim is to examine the structure of a Lie algebra in detail. With the exception of the last part of this chapter, where applications are made to the structure of Lie groups, the action takes place over a field of characteristic zero. The main results are the theorems of Lie and Engel on nilpotent and solvable algebras; Cartan's criterion for semisimplicity, namely that a Lie algebra is semisimple if and only if its Cartan–Killing form is nonsingular; Weyl's theorem asserting that all finite-dimensional representations of a semisimple Lie algebra are semisimple; and the theorems of Levi and Mal'čev on the semidirect decompositions of an arbitrary Lie algebra into its radical and a (semisimple) Levi factor. Although the results of Weyl and Levi–Mal'čev are cohomological in their nature (at least from the algebraic point of view), I have resisted the temptation to discuss the general cohomology theory of Lie algebras and have confined myself strictly to what is needed (*ad hoc* low-dimensional cohomology).

The fourth and final chapter is the heart of the book and is a fairly complete treatment of the fine structure and representation theory of semisimple Lie algebras and Lie groups. The root structure and the classification of simple Lie algebras over the field of complex numbers are obtained. As for representation theory, it is examined from both the infinitesimal (Cartan, Weyl, Harish-Chandra, Chevalley) and the global (Weyl) points of view. First I present the algebraic view, in which universal enveloping algebras. left ideals, highest weights, and infinitesimal characters are put in the foreground. I have followed here the treatment of Harish-Chandra given in his early papers and used it to prove the bijective nature of the correspondence between connected Dynkin diagrams and simple Lie algebras over the complexes. This algebraic part is then followed up with the transcendental theory. Here compact Lie groups come to the fore. The existence and conjugacy of their maximal tori are established, and Weyl's classic derivation of his great character formula is given. It is my belief that this dual treatment of representation theory is not only illuminating but even essential and that the infinitesimal and global parts of the theory are complementary facets of a very beautiful and complete picture.

In order not to interrupt the main flow of exposition, I have added an appendix at the end of this chapter where I have discussed the basic results of finite reflection groups and root systems. This appendix is essentially the same as a set of unpublished notes of Professor Robert Steinberg on the subject, and I am very grateful to him for allowing me to use his manuscript.

It only remains to thank all those without whose help this book would have been impossible. I am especially grateful to Professor I. M. Singer for his help at various critical stages. Mrs. Alice Hume typed the entire manuscript, and I cannot describe my indebtedness to the great skill, tempered with great patience, with which she carried out this task. I would like to thank Joel Zeitlin, who helped me prepare the original 1966 notes; and Mohsen Pazirandeh and Peter Trombi, who looked through the entire manuscript and corrected many errors. I would also like to thank Ms. Judy Burke, whose guidance was indispensable in preparing the manuscript for publication.

I would like to end this on a personal note. My first introduction to serious mathematics was from the papers of Harish-Chandra on semisimple Lie groups, and almost everything I know of representation theory goes back either to his papers or the discussions I have had with him over the past years. My debt to him is too immense to be detailed.

V. S. VARADARAJAN

Pacific Palisades

PREFACE TO THE SPRINGER EDITION (1984)

Lie Groups, Lie Algebras, and Their Representations went out of print recently. However, many of my friends told me that it is still very useful as a textbook and that it would be good to have it available in print. So when Springer offered to republish it, I agreed immediately and with enthusiasm. I wish to express my deep gratitude to Springer-Verlag for their promptness and generosity. I am also extremely grateful to Joop Kolk for providing me with a comprehensive list of errata.

CONTENTS

CHAPTER 1

DIFFERENTIABLE AND ANALYTIC MANIFOLDS

1.1. Differentiable Manifolds

We shall devote this chapter to a summary of those concepts and results from the theory of differentiable and analytic manifolds which are needed for our work in the rest of the book. Most of these results are standard and adequately treated in many books (see for example Chevalley [1], Helgason [1], Kobayashi and Nomizu [1], Bishop and Crittenden [1], Narasimhan [1]).

Differentiable structures. For technical reasons we shall permit our differentiable manifolds to have more than one connected component. However, all the manifolds that we shall encounter are assumed to satisfy the second axiom of countability and to have the same dimension at all points. More precisely, let M be a Hausdorff topological space satisfying the second axiom of countability. By a (C^∞) *differentiable structure* on M we mean an assignment

$$\mathfrak{D} : U \mapsto \mathfrak{D}(U) \quad (U \text{ open}, \subseteq M)$$

with the following properties:

(i) for each open $U \subseteq M$, $\mathfrak{D}(U)$ is an algebra of complex-valued functions on U containing 1 (the function identically equal to unity)

(ii) if V, U are open, $V \subseteq U$ and $f \in \mathfrak{D}(U)$, then $f|V \in \mathfrak{D}(V)$;[1] if V_i $(i \in J)$ are open, $V = \cup_i V_i$, and f is a complex-valued function defined on V such that $f|V_i \in \mathfrak{D}(V_i)$ for all $i \in J$, then $f \in \mathfrak{D}(V)$

(iii) there exists an integer $m > 0$ with the following property: for any $x \in M$, one can find an open set U containing x, and m *real* functions $x_1, \ldots, x_m$ from $\mathfrak{D}(U)$ such that (a) the map

$$\xi : y \mapsto (x_1(y), \ldots, x_m(y))$$

is a homeomorphism of U onto an open subset of $\mathbf{R}^m$ (real m-space), and (b)

[1] If F is any function defined on a set A, and $B \subseteq A$, then $F|B$ denotes the restriction of F to B.

for any open set $V \subseteq U$ and any complex-valued function f defined on V, $f \in \mathfrak{D}(V)$ if and only if $f \circ \xi^{-1}$ is a C^∞ function on $\xi[V]$.

Any open set U for which there exist functions $x_1, \ldots, x_m$ having the property described in (iii) is called a *coordinate patch;* $\{x_1, \ldots, x_m\}$ is called a *system of coordinates on* U. Note that for any open $U \subseteq M$, the elements of $\mathfrak{D}(U)$ are continuous on U.

It is not required that M be connected; it is, however, obviously locally connected and metrizable. The integer m in (iii) above, which is the same for all points of M, is called the *dimension* of M. The pair $(M,\mathfrak{D})$ is called *differentiable* (C^∞) *manifold.* By abuse of language, we shall often refer to M itself as a differentiable manifold. It is usual to write $C^\infty(U)$ instead of $\mathfrak{D}(U)$ for any open set $U \subseteq M$ and to refer to its elements as (C^∞) *differentiable functions* on U. If U is any open subset of M, the assignment $V \mapsto C^\infty(V)$ ($V \subseteq U$, open) gives a C^∞ structure on U. U, equipped with this structure, is a C^∞ manifold having the same dimension as M; it is called the *open submanifold* defined by U. The connected components of M are all open submanifolds of M, and there can be at most countably many of these.

Let k be an integer ≥ 0, $U \subseteq M$ any open set. A complex-valued function f defined on U is said to be of *class* C^k on U if, around each point of U, f is a k-times continuously differentiable function of the local coordinates. It is easy to see that this property is independent of the particular set of local coordinates used. The set of all such f is denoted by $C^k(U)$. (We omit k when $k = 0$: $C(U) = C^0(U)$. $C^k(U)$ is an algebra over the field of complex numbers $\mathbf{C}$ and contains $C^\infty(U)$.

Given any complex-valued function f on M, its *support*, supp f, is defined as the complement in M of the largest open set on which f is identically zero. For any open set U and any integer k with $0 \leq k \leq \infty$, we denote by $C_c^k(U)$ the subspace of all $f \in C^k(M)$ for which supp f is a compact subset of U.

There is no difficulty in constructing nontrivial elements of $C^\infty(M)$. We mention the following results, which are often useful.

(i) Let $U \subseteq M$ be open and $K \subseteq U$ be compact; then we can find $\varphi \in C^\infty(M)$ such that $0 \leq \varphi(x) \leq 1$ for all x, with $\varphi = 1$ in an open set containing K, and $\varphi = 0$ outside U.

(ii) Let $\{V_i\}_{i \in J}$ be a locally finite[2] open covering of M with $Cl(V_i)$ (Cl denoting closure) compact for all $i \in J$; then there are $\varphi_i \in C^\infty(M)$ $(i \in J)$ such that

(a) for each $i \in J$ $\varphi_i \geq 0$ and supp φ_i is a (compact) subset of V_i

(b) $\sum_{i \in J} \varphi_i(x) = 1$ for all $x \in M$ (this is a finite sum for each x, since $\{V_i\}_{i \in J}$ is locally finite).

$\{\varphi_i\}_{i \in J}$ is called a *partition of unity subordinate to the covering* $\{V_i\}_{i \in J}$.

[2]A family $\{E_i\}_{i \in J}$ of subsets of a topological space S is called *locally finite* if each point of X has an open neighborhood which meets E_i for only finitely many $i \in J$.

Tangent vectors and differential expressions. Let M be a C^∞ manifold of dimension m, fixed throughout the rest of this section. Let $x \in M$. Two C^∞ functions defined around x are called *equivalent if* they coincide on an open set containing x. The equivalence classes corresponding to this relation are known as *germs of C^∞ functions at x*. For any C^∞ function f defined around x we write $\mathbf{f}_x$ for the corresponding germ at x. The algebraic operations on the set of differentiable functions give rise in a natural and obvious fashion to algebraic operations on the set of germs at x, converting the latter into an algebra over $\mathbf{C}$; we denote this algebra by $\mathbf{D}_x$. A germ is called *real* if it is defined by a real C^∞ function. The real germs form an algebra over $\mathbf{R}$. For any germ $\mathbf{f}$ at x we write $\mathbf{f}(x)$ to denote the common value at x of all the C^∞ functions belonging to $\mathbf{f}$. It is easily seen that any germ at x is determined by a C^∞ function defined on all of M.

Let $\mathbf{D}_x^*$ be the algebraic dual of the complex vector space $\mathbf{D}_x$, i.e., the complex vector space of all linear maps of $\mathbf{D}_x$ into $\mathbf{C}$. An element of $\mathbf{D}_x^*$ is said to be *real* if it is real-valued on the set of real germs. A *tangent vector to M at x* is an element v of $\mathbf{D}_x^*$ such that

$$(1.1.1)\qquad \begin{cases} \text{(i)} & v \text{ is real} \\ \text{(ii)} & v(\mathbf{fg}) = \mathbf{f}(x)v(\mathbf{g}) + \mathbf{g}(x)v(\mathbf{f}) \text{ for all } \mathbf{f}, \mathbf{g} \in \mathbf{D}_x. \end{cases}$$

The set of all tangent vectors to M at x is an $\mathbf{R}$-linear subspace of $\mathbf{D}_x^*$, and is denoted by $T_x(M)$; it is called the *tangent space to M at x*. Its complex linear span $T_{xc}(M)$ is the set of all elements of $\mathbf{D}_x^*$ satisfying (ii) of (1.1.1). Let U be a coordinate patch containing x with $x_1, \ldots, x_m$ a system of coordinates on U, and let

$$\tilde{U} = \{(x_1(y), \ldots, x_m(y)) : y \in U\}.$$

For any $f \in C^\infty(U)$ let $\tilde{f} \in C^\infty(\tilde{U})$ be such that $\tilde{f} \circ (x_1, \ldots, x_m) = f$. Then the maps

$$f \mapsto \left(\frac{\partial \tilde{f}}{\partial t_j}\right)_{t_1 = x_1(x), \ldots, t_m = x_m(x)}$$

for $1 \leq j \leq m$ ($t_1, \ldots, t_m$ being the usual coordinates on $\mathbf{R}^m$) induce linear maps of $\mathbf{D}_x$ into $\mathbf{C}$ which are easily seen to be tangent vectors; we denote these by $(\partial/\partial x_j)_x$. They form a basis for $T_x(M)$ over $\mathbf{R}$ and hence of $T_{xc}(M)$ over $\mathbf{C}$.

Define the element $1_x \in \mathbf{D}_x^*$ by

$$(1.1.2)\qquad 1_x(\mathbf{f}) = \mathbf{f}(x) \quad (\mathbf{f} \in \mathbf{D}_x).$$

1_x is real and linearly independent of $T_x(M)$. It is easy to see that for an element $v \in \mathbf{D}_x^*$ to belong to the complex linear span of 1_x and $T_x(M)$ it is necessary and sufficient that $v(\mathbf{f}_1 \mathbf{f}_2) = 0$ for all $\mathbf{f}_1, \mathbf{f}_2 \in \mathbf{D}_x$ which vanish at x. This leads naturally to the following generalization of the concept of a tangent vector. Let

$$\mathbf{J}_x = \{\mathbf{f} : \mathbf{f} \in \mathbf{D}_x, \mathbf{f}(x) = 0\} \tag{1.1.3}$$

Then $\mathbf{J}_x$ is an ideal in $\mathbf{D}_x$. For any integer $p \geq 1$, $\mathbf{J}_x^p$ is defined to be the linear span of all elements which are products of p elements from $\mathbf{J}_x$; $\mathbf{J}_x^p$ is also an ideal in $\mathbf{D}_x$. For any integer $r \geq 0$ we define a *differential expression of order* $\leq r$ to be any element of $\mathbf{D}_x^*$ which vanishes on $\mathbf{J}_x^{r+1}$; the set of all such is a linear subspace of $\mathbf{D}_x^*$ and is denoted by $T_{xc}^{(r)}(M)$. The real elements in $T_{xc}^{(r)}(M)$ from an $\mathbf{R}$-linear subspace of $T_{xc}^{(r)}(M)$, spanning it (over $\mathbf{C}$), and is denoted by $T_x^{(r)}(M)$. We have $T_x^{(0)}(M) = \mathbf{R}\cdot 1_x$, $T_x^{(1)}(M) = \mathbf{R}\cdot 1_x + T_x(M)$, and $T_x^{(r)}(M)$ increases with increasing r. Put

$$\begin{aligned} T_x^{(\infty)}(M) &= \bigcup_{r\geq 0} T_x^{(r)}(M) \\ T_{xc}^{(\infty)}(M) &= \bigcup_{r\geq 0} T_{xc}^{(r)}(M). \end{aligned} \tag{1.1.4}$$

$T_{xc}^{(\infty)}(M)$ is a linear subspace of $\mathbf{D}_x^*$, and $T_x^{(\infty)}(M)$ is an $\mathbf{R}$-linear subspace spanning it over $\mathbf{C}$.

It is easy to construct natural bases of the $T_x^{(r)}(M)$ in local coordinates. Let U be a coordinate patch containing x and let $\tilde{U}$ and $x_1, \ldots, x_m$ be as in the discussion concerning tangent vectors. Let (α) be any multiindex, i.e., $(\alpha) = (\alpha_1, \ldots, \alpha_m)$ where the α_j are integers ≥ 0; put $|\alpha| = \alpha_1 + \cdots + \alpha_m$. Then the map

$$f \mapsto \left(\frac{\partial^{|\alpha|}\tilde{f}}{\partial t_1^{\alpha_1}\cdots\partial t_m^{\alpha_m}}\right)_{t_1 = x_1(x), \ldots, t_m = x_m(x)} \qquad (f \in C^\infty(U))$$

induces a linear function on $\mathbf{D}_x$ which is real. Let $\partial_x^{(\alpha)}$ denote this (when $(\alpha) = (0)$, $\partial_x^{(\alpha)} = 1_x$). Clearly, $\partial_x^{(\alpha)} \in T_x^{(r)}(M)$ if $|\alpha| \leq r$.

Lemma 1.1.1. *Let $r \geq 0$ be an integer and let $x \in M$. Then the differential expressions $\partial_x^{(\alpha)}$ ($|\alpha| \leq r$) form a basis for $T_x^{(r)}(M)$ over* $\mathbf{R}$ *and for $T_{xc}^{(r)}(M)$ over* $\mathbf{C}$.

Proof. Since this is a purely local result, we may assume that M is the open cube $\{(y_1, \ldots, y_m) : |y_j| < a \text{ for } 1 \leq j \leq m\}$ in $\mathbf{R}^m$ with x as the origin. Let $t_1, \ldots, t_m$ be the usual coordinates, and for any multiindex $(\beta) = (\beta_1, \ldots, \beta_m)$ let $\mathbf{t}^{(\beta)}$ denote the germ at the origin defined by $t_1^{\beta_1} \ldots t_m^{\beta_m}/\beta_1! \cdots \beta_m!$

Let f be a real C^∞ function on M and let $g_{x_1,\ldots,x_m}(t) = f(tx_1, \ldots, tx_m)$ $(-1 \leq t \leq 1, (x_1, \ldots, x_m) \in M)$. By expanding $g_{x_1,\ldots,x_m}$ about $t = 0$ in its Taylor series, we get

$$g_{x_1,\ldots,x_m}(t) = \sum_{0\leq s\leq r} \frac{t^s}{s!} g_{x_1,\ldots,x_m}^{(s)}(0) + \frac{1}{r!}\int_0^t (t-u)^r g_{x_1,\ldots,x_m}^{(r+1)}(u)\,du$$

for $0 \leq t \leq 1$. Putting $t = 1$ and evaluating the t-derivatives of $g_{x_1,\ldots,x_m}$ in terms of the partial derivatives of f, we get, for all $(x_1, \ldots, x_m) \in M$,

$$f(x_1, \ldots, x_m) = \sum_{|\beta| \leq r} \frac{x_1^{\beta_1} \cdots x_m^{\beta_m}}{\beta_1! \cdots \beta_m!} \partial_x^{(\beta)}(\mathbf{f}) + \sum_{|\alpha| = r+1} \frac{x_1^{\alpha_1} \cdots x_m^{\alpha_m}}{\alpha_1! \cdots \alpha_m!} h^{(\alpha)}(x_1, \ldots, x_m),$$

where

$$h^{(\alpha)}(x_1, \ldots, x_m) = (r+1) \int_0^1 (1-u)^r \left(\frac{\partial^{\alpha_1 + \cdots + \alpha_m} f}{\partial t_1^{\alpha_1} \cdots \partial t_m^{\alpha_m}}\right)(ux_1, \ldots, ux_m)\, du.$$

Clearly, the $h^{(\alpha)}$ are real C^∞ functions on M. Passing to the germs at the origin, we get

$$\mathbf{f} = \sum_{|\beta| \leq r} \partial_x^{(\beta)}(f)\mathbf{t}^{(\beta)} + \sum_{|\alpha| = r+1} \mathbf{t}^{(\alpha)}\mathbf{h}^{(\alpha)}.$$

Since $\mathbf{t}^{(\alpha)} \in \mathbf{J}_x^{r+1}$ for any (α) with $|\alpha| = r+1$, we get, for any $\lambda \in T_x^{(r)}(M)$,

$$\lambda = \sum_{|\beta| \leq r} \lambda(\mathbf{t}^{(\beta)})\partial_x^{(\beta)}$$

This shows that the $\partial_x^{(\beta)}(|\beta| \leq r)$ span $T_x^{(r)}(M)$ over $\mathbf{R}$. On the other hand, the $\partial_x^{(\beta)}$ are linearly independent over $\mathbf{R}$ or $\mathbf{C}$, since

$$\partial_x^{(\beta)}(\mathbf{t}^{(\gamma)}) = \begin{cases} 0 & (\gamma) \neq (\beta) \\ 1 & (\gamma) = (\beta) \end{cases}$$

This proves the lemma.

Vector fields. Let $X\,(x \mapsto X_x)$ be any assignment such that $X_x \in T_{xc}(M)$ for all $x \in M$. Then for any function $f \in C^\infty(M)$, the function $Xf: x \mapsto X_x(\mathbf{f}_x)$ is well defined on M, $\mathbf{f}_x$ being the germ at x defined by f. If U is any coordinate patch and $x_1, \ldots, x_m$ are coordinates on U, there are unique complex-valued functions $a_1, \ldots, a_m$ on U such that

$$X_y = \sum_{1 \leq j \leq m} a_j(y)\left(\frac{\partial}{\partial x_j}\right)_y \quad (y \in U).$$

X is called a *vector field on* M if $Xf \in C^\infty(M)$ for all $f \in C^\infty(M)$, or equivalently, if for each $x \in M$ there exist a coordinate patch U containing x and coordinates $x_1, \ldots, x_m$ on U such that the a_j defined above are C^∞ functions on U. A vector field X is said to be *real* if $X_x \in T_x(M)\ \forall\ x \in M$; X is real if and only if Xf is real for all real $f \in C^\infty(M)$. Given a vector field X, the mapping $f \longrightarrow Xf$ is a derivation of the algebra $C^\infty(M)$; i.e., for all f and

$g \in C^\infty(M)$,

$$X(fg) = f \cdot Xg + g \cdot Xf. \tag{1.1.5}$$

This correspondence between vector fields and derivations is one to one and maps the set of all vector fields onto the set of all derivations of $C^\infty(M)$. Denote by $\mathfrak{I}(M)$ the set of all vector fields on M. If $X \in \mathfrak{I}(M)$ and $f \in C^\infty(M)$, $fX : x \mapsto f(x)X_x$ is also a vector field. In this way, $\mathfrak{I}(M)$ becomes a module over $C^\infty(M)$. We make in general no distinction between a vector field and the corresponding derivation of $C^\infty(M)$.

Let X and Y be two vector fields. Then $X \circ Y - Y \circ X$ is an endomorphism of $C^\infty(M)$ which is easily verified to be a derivation. The associated vector field is denoted by $[X, Y]$ and is called the *Lie bracket* of X with Y. The map

$$(X, Y) \mapsto [X, Y]$$

is bilinear and possesses the following easily verified properties:

$$\left\{\begin{array}{ll} \text{(i)} & [X,X] = 0 \\ \text{(ii)} & [X,Y] + [Y,X] = 0 \\ \text{(iii)} & [X,[Y,Z]] + [Y,[Z,X]] + [Z,[X,Y]] = 0 \end{array}\right. \tag{1.1.6}$$

(X, Y, and Z being arbitrary in $\mathfrak{I}(M)$). If X and Y are real, so is $[X, Y]$. The relation (iii) of (1.1.6) is known as the *Jacobi identity*.

Differential operators. Let $r \geq 0$ be an integer and let

$$D : x \mapsto D_x \tag{1.1.7}$$

be an assignment such that $D_x \in T^{(r)}_{xc}(M)$ for all $x \in M$. If $f \in C^\infty(M)$, the function $Df : x \mapsto D_x(\mathbf{f}_x)$ is well defined on M, $\mathbf{f}_x$ being the germ defined by f at x. If U is a coordinate patch and $x_1, \ldots, x_m$ are coordinates on U, then by Lemma 1.1.1 there are unique complex functions $a_{(\alpha)}$ on U such that

$$D_y = \sum_{|\alpha| \leq r} a_{(\alpha)}(y)\partial_y^{(\alpha)} \quad (y \in U).$$

D is called a *differential operator on M* if $Df \in C^\infty(M)$ for all $f \in C^\infty(M)$, or equivalently, if for each $x \in M$ we can find a coordinate patch U containing x with coordinate $x_1, \ldots, x_m$ such that the $a_{(\alpha)}$ defined above are in $C^\infty(U)$. The smallest integer $r \geq 0$ such that $D_x \in T^{(r)}_{xc}(M)$ for all $x \in M$ is called the *order* (ord(D)) or the *degree* (deg(D)) of D. For any differential operator D on M and $x \in M$, D_x is called the *expression* of D at x. If Df is real for any real-valued $f \in C^\infty(M)$, we say that D is *real*. The set of all differential operators on M is denoted by $\mathrm{Diff}(M)$. If $f \in C^\infty(M)$ and $D \in \mathrm{Diff}(M)$, $fD: x \mapsto f(x) D_x$ is again a differential operator; its order cannot exceed the order of D. Thus $\mathrm{Diff}(M)$ is a module over $C^\infty(M)$. A *vector field* is a differential operator of order ≤ 1. If $\{V_i\}_{i \in J}$ is an open covering of M and $D_i (i \in J)$ is a differential operator on V_i such that

(a) $\sup_{i \in J} \operatorname{ord}(D_i) < \infty$
(b) if $V_{i_1} \cap V_{i_2} \neq \phi$, the restrictions of D_{i_1} and D_{i_2} to $V_{i_1} \cap V_{i_2}$ are equal,

then there exists exactly one differential operator D on M such that for any $i \in J$ D_i is the restriction of D to V_i.

Let D $(x \mapsto D_x)$ be a differential operator of order $\leq r$. We also denote by D the endomorphism $f \mapsto Df$ of $C^\infty(M)$. This endomorphism is then easily verified to have the following properties:

(1.1.8)
- (i) it is *local;* i.e., if $f \in C^\infty(M)$ vanishes on an open set U, Df also vanishes on U
- (ii) if $x \in M$, and $f_1, \ldots, f_{r+1}$ are $r + 1$ functions in $C^\infty(M)$ which vanish at x, then
$$(D(f_1 f_2 \cdots f_{r+1}))(x) = 0.$$

Conversely, it is quicky verified that given any endomorphism E of $C^\infty(M)$ satisfying (ii) of (1.1.8) for some integer $r \geq 0$, E is local and there is exactly one differential operator D on M such that $Df = Ef$ for all $f \in C^\infty(M)$; and $\operatorname{ord}(D) \leq r$. In view of this, we make no distinction between a differential operator and the endomorphism of $C^\infty(M)$ induced by it. It follows easily from the expression of a differential operator in local coordinates that if D_1 and D_2 are differential operators of respective orders r_1 and r_2, then $D_1 D_2$ is also a differential operator, and its order is $\leq r_1 + r_2$; moreover, $D_1 D_2 - D_2 D_1$ is a differential operator of order $\leq r_1 + r_2 - 1$. $\mathrm{Diff}(M)$ is thus an algebra (not commutative); if $\mathrm{Diff}(M)_r$ is the set of elements of $\mathrm{Diff}(M)$ of order $\leq r$, $r \mapsto \mathrm{Diff}(M)_r$ converts $\mathrm{Diff}(M)$ into a filtered algebra. A differential operator of order 0 is just the operator of multiplication by a C^∞ function; if u is in $C^\infty(M)$ we denote again by u the operator $f \mapsto uf$ of $C^\infty(M)$.

If $M = \mathbf{R}^m$ and D is a differential operator of order $\leq r$, there are unique C^∞ functions $a_{(\alpha)}$ $(|\alpha| \leq r)$ on M (*coefficients* of D) such that

$$D = \sum_{|\alpha| \leq r} a_{(\alpha)} \frac{\partial^{|\alpha|}}{\partial t_1^{\alpha_1} \cdots \partial t_m^{\alpha_m}},$$

$t_1, \ldots, t_m$ being the linear coordinates on M. It is natural to ask whether such global representations exist on more general manifolds. The following theorem gives one such result.

Theorem 1.1.2. *Let $X_1, \ldots, X_m$ be m vector fields on M such that $(X_1)_x, \ldots, (X_m)_x$ form a basis of $T_{x^c}(M)$ for each $x \in M$. For any multiindex $(\alpha) = (\alpha_1, \ldots, \alpha_m)$ let $X^{(\alpha)}$ be the differential operator*

$$X^{(\alpha)} = X_1^{\alpha_1} X_2^{\alpha_2} \cdots X_m^{\alpha_m} \tag{1.1.9}$$

(when $(\alpha) = (0) X^{(\alpha)} = 1$, the identity operator). Then the $X^{(\alpha)}$ are linearly independent over $C^\infty(M)$. If D is any differential operator of order $\leq r$, we can find unique C^∞ functions $a_{(\alpha)}$ on M such that

$$D = \sum_{|\alpha| \leq r} a_{(\alpha)} X^{(\alpha)}. \tag{1.1.10}$$

If the X_i are real, then for any real differential operator D the $a_{(\alpha)}$ defined by (1.1.10) *are all real.*

Proof. For any integer $r \geq 0$, let $\mathfrak{D}_r$ denote the complex vector space of all differential operators on M of the form $\sum_{|\alpha| \leq r} f_{(\alpha)} X^{(\alpha)}$, the $f_{(\alpha)}$ being C^∞ functions on M. Note that $\mathfrak{D}_1$ contains all vector fields. In fact, if Z is any vector field, we can write $Z = \sum_{1 \leq j \leq m} c_j X_j$ for uniquely defined functions c_j. To see that the c_j are in $C^\infty(M)$, let U be a coordinate patch with coordinates $x_1, \ldots, x_m$. Then there are C^∞ functions d_j, a_{jk} on U $(1 \leq j, k \leq m)$ such that $Z_y = \sum_{1 \leq j \leq m} d_j(y)(\partial/\partial x_j)_y$ and $(X_j)_y = \sum_{1 \leq k \leq m} a_{jk}(y)(\partial/\partial x_k)_y$ for all $y \in U$. Since the $(X_j)_y$ $(1 \leq j \leq m)$ are linearly independent for all y, the matrix (a_{jk}) is invertible. If a^{jk} are the entries of the inverse matrix, they are in $C^\infty(U)$ and $c_j = \sum_{1 \leq j \leq m} d_k a^{kj}$ on U.

We begin the proof of the theorem by showing that if l is an integer ≥ 1 and $Z_1, \ldots, Z_l$ are l vector fields, then the product $Z_1 \cdots Z_l$ belongs to $\mathfrak{D}_l$. For $l = 1$, this is just the remark made in the previous paragraph. Proceed by induction on l. Let $l > 1$, and assume that the result holds for any $l - 1$ vector fields. Let $Z_1, \ldots, Z_l$ be l vector fields, and write $E = Z_1 \cdots Z_l$.

Notice first that if $Y_1, \ldots, Y_l$ are any l vector fields, $F = Y_1 \cdots Y_l$, and F' is the product obtained by interchanging two adjacent Y's, then $F - F'$ is a product of $l - 1$ vector fields. So $F - F' \in \mathfrak{D}_{l-1}$ by the induction hypothesis. Since any permutation is a product of such adjacent interchanges, it follows from the induction hypothesis that $Y_1 \cdots Y_l - Y_{i_1} Y_{i_2} \cdots Y_{i_l} \in \mathfrak{D}_{l-1}$ for any permutation $(i_1, \ldots, i_l)$ of $(1, \ldots, l)$. But if $1 \leq j_1 \leq j_2 \leq \cdots \leq j_l \leq m$, then $X_{j_1} \cdots X_{j_l} = X^{(\alpha)}$ for a suitable (α) with $|\alpha| = l$, so that $X_{j_1} \cdots X_{j_l} \in \mathfrak{D}_l$. Hence, from what we proved above, if $(k_1, \ldots, k_m)$ is any permutation of $(1, \ldots, m)$ and (α) is any multi-index with $|\alpha| \leq l$, then $X_{k_1}^{\alpha_1} \cdots X_{k_m}^{\alpha_m} \in \mathfrak{D}_l$.

Now consider E. By the induction hypothesis, there exist C^∞ functions $b_{(\beta)}$ and c_j on M such that $Z_1 = \sum_{1 \leq j \leq m} c_j X_j$ and $Z_2 \cdots Z_l = \sum_{|\beta| \leq l-1} b_{(\beta)} X^{(\beta)}$. So

$$\begin{aligned} E &= \sum_{1 \leq j \leq m} \sum_{|\beta| \leq l-1} c_j(X_j \circ b_{(\beta)}) X^{(\beta)} \\ &= \sum_{1 \leq j \leq m} \sum_{|\beta| \leq l-1} c_j b_{(\beta)} X_j X^{(\beta)} + \sum_{1 \leq j \leq m} \sum_{|\beta| \leq l-1} c_j (X_j b_{(\beta)}) X^{(\beta)}. \end{aligned}$$

Since, for all (β) with $|\beta| \leq l-1$, $X_j X^{(\beta)} \in \mathfrak{D}_l$ (by what was seen in the preceding paragraph), we have $E \in \mathfrak{D}_l$.

We can now complete the proof of the theorem. Let $r \geq 0$ be any integer. Let U be a coordinate patch with coordinates $x_1, \ldots, x_m$ and let $\partial^{(\alpha)}$ be the differential operators $y \mapsto \partial_y^{(\alpha)}$ on U. By the result of the preceding paragraph (applied to the manifold U), there exist C^∞ functions $a_{(\alpha),(\beta)}$ on U such that

$$\partial^{(\alpha)} = \sum_{|\beta| \leq r} a_{(\alpha),(\beta)} X^{(\beta)} \quad (|\alpha| \leq r) \tag{1.1.11}$$

on U. This shows at once that for any $y \in M$, the $X_y^{(\beta)} (|\beta| \leq r)$ span $T_{yc}^{(r)}(M)$; since their number is exactly the dimension of $T_{yc}^{(r)}(M)$, they must be linearly independent too. Therefore, if D is a differential operator of order $\leq r$, we can find unique functions $a_{(\beta)}$ on M such that

$$D = \sum_{|\beta| \leq r} a_{(\beta)} X^{(\beta)} \tag{1.1.12}$$

To prove that the $a_{(\alpha)}$ are C^∞, we restrict our attention to U and use the above notation. We select C^∞ functions $g_{(\alpha)}$ on U such that $D = \sum_{|\alpha| \leq r} g_{(\alpha)} \partial^{(\alpha)}$ on U. Then by (1.1.11) and (1.1.12) we have, on U,

$$a_{(\beta)} = \sum_{|\alpha| \leq r} g_{(\alpha)} a_{(\alpha),(\beta)} \quad (|\beta| \leq r),$$

proving that the $a_{(\beta)}$ are C^∞. The last statement is obvious. This proves the theorem.

We shall often use Harish-Chandra's notation for denoting the application of differential operators. Thus, if f is a C^∞ function and D a differential operator, $f(x; D)$ denotes the value of Df at $x \in M$.

Exterior differential forms. Let W be a finite-dimensional vector space of dimension m over a field F of characteristic 0. Put $\Lambda_0(W) = F$, and for any integer $k \geq 1$, define $\Lambda_k(W)$ as the vector space of all k-linear skew-symmetric functions on $W \times \cdots \times W$ (k factors) with values in F. $\Lambda_k(W)$ is then 0 if $k > m$, and $\dim \Lambda_k(W) = \binom{m}{k}$, $1 \leq k \leq m$. We write $\Lambda(W)$ for the direct sum of the $\Lambda_k(W)$, $0 \leq k \leq m$ and write $\wedge$ for the operation of exterior multiplication in $\Lambda(W)$ which converts it into an associative algebra over F, its unit being the unit 1 of F. We assume that the reader is, familiar with the defintion of $\wedge$ and the properties of $\Lambda(W)$ (cf. Exercises 9–11). If $\varphi, \varphi' \in \Lambda_1(W)$ (= dual of W), $\varphi \wedge \varphi' = -\varphi' \wedge \varphi$; in particular, $\varphi \wedge \varphi = 0$. More generally, if $\varphi \in \Lambda_r(W)$ and $\varphi' \in \Lambda_{r'}(W)$, then $\varphi \wedge \varphi' \in \Lambda_{r+r'}(W)$, and $\varphi \wedge \varphi' = (-1)^{rr'}\varphi' \wedge \varphi$. If $\{\varphi_1, \ldots, \varphi_m\}$ is a basis for $\Lambda_1(W)$, and $1 \leq k \leq m$, the $\binom{m}{k}$ elements $\varphi_{i_1} \wedge \cdots \wedge \varphi_{i_k}$ $(1 \leq i_1 < \cdots < i_k \leq m)$ form a basis for $\Lambda_k(W)$. Note that dim $\Lambda_m(W) = 1$ and that $\varphi_1 \wedge \cdots \wedge \varphi_m$ is a basis for it. If $\psi_1, \ldots, \psi_m$ is another basis for $\Lambda_1(W)$, where $\psi_i = \sum_{1 \leq j \leq m} a_{ij}\varphi_j$ $(1 \leq i \leq m)$, and if A is the matrix $(a_{ij})_{1 \leq i, j \leq m}$, then

$$\psi_1 \wedge \cdots \wedge \psi_m = \det(A)\cdot\varphi_1 \wedge \cdots \wedge \varphi_m \tag{1.1.13}$$

A 0-*form* is a C^∞ function on M. Let $1 \leq k \leq m$ and let

$$\omega : x \mapsto \omega_x$$

be an assignment such that $\omega_x \in \Lambda_k(T_{xc}(M))$ for all $x \in M$. ω is said to be *real* if ω_x is real-valued on $T_x(M) \times \cdots \times T_x(M)$ for all $x \in M$. Let U be a coordinate patch and let $x_1, \ldots, x_m$ be a system of coordinates on it. For $y \in U$, let $\{(dx_1)_y, \ldots, (dx_m)_y\}$ be the basis of $T_y(M)^*$ dual to $\{(\partial/\partial x_1)_y, \ldots, (\partial/\partial x_m)_y\}$. Then there are unique functions $a_{i_1,\ldots,i_k}$ $(1 \leq i_1 < i_2 < \cdots < i_k \leq m)$ defined on U such that

$$\omega_y = \sum_{1 \leq i_1 < \cdots < i_k \leq m} a_{i_1,\ldots,i_k}(y)(dx_{i_1})_y \wedge \cdots \wedge (dx_{i_k})_y \quad (y \in U).$$

ω is said to be a k-form if all the $a_{i_1,\ldots,i_k}$ are C^∞ functions on U (for all possible choices of U).

Suppose ω $(x \mapsto \omega_x)$ is an assignment such that $\omega_x \in \Lambda_k(T_{xc}(M))$ for all $x \in M$. Let $Z_1, \ldots, Z_k$ be vector fields. Then the function

$$\omega(Z_1, \ldots, Z_k) : x \mapsto \omega_x((Z_1)_x, \ldots, (Z_k)_x)$$

is well defined on M. It is easy to show that ω is a k-form if and only if this function is C^∞ on M for all choices of $Z_1, \ldots, Z_k$. The map

$$(Z_1, \ldots, Z_k) \mapsto \omega(Z_1, \ldots, Z_k)$$

of $\mathfrak{I}(M) \times \cdots \times \mathfrak{I}(M)$ into $C^\infty(M)$ is skew-symmetric and $C^\infty(M)$-multilinear (i.e., **C**-multilinear and respects the module actions of $C^\infty(M)$); the correspondence between such maps and k-forms is a bijection. If ω is a k-form and $f \in C^\infty(M), f\omega : x \mapsto f(x)\omega_x$ is also a k-form. So the vector space of k-forms is also a module over $C^\infty(M)$. If ω is a k-form and ω' is a k'-form, then $x \mapsto \omega_x \wedge \omega'_x$ is usually denoted by $\omega \wedge \omega'$. It is a $(k + k')$-form, and $\omega \wedge \omega' = (-1)^{kk'}$ $\omega' \wedge \omega$.

We write $\mathcal{A}_0(M) = C^\infty(M)$ and $\mathcal{A}_k(M)$ for the $C^\infty(M)$-module of all k-forms. Let $\mathcal{A}(M)$ be the direct sum of all the $\mathcal{A}_k(M)$ $(0 \leq k \leq m)$. Under $\wedge$, $\mathcal{A}(M)$ is an algebra over $C^\infty(M)$.

Suppose $f \in C^\infty(M)$. Then for any vector field Z, $Zf \in C^\infty(M)$, and so there is a unique 1-form, denoted by df, such that

$$(df)(Z) = Zf \quad (Z \in \mathfrak{J}(M)). \tag{1.1.14}$$

If U is a coordinate patch with coordinates $x_1, \ldots, x_m$, then

$$(df)_y = \sum_{1 \leq j \leq m} \left(\frac{\partial f}{\partial x_j}\right)(y)(dx_j)_y \quad (y \in U).$$

In particular, on U, dx_j is the 1-form $y \mapsto (dx_j)_y$. More generally, there is a *unique* endomorphism d $(\omega \mapsto d\omega)$ of the vector space $\mathcal{A}(M)$ with the following properties:

$$\left\{\begin{array}{ll} \text{(i)} & d(d\omega) = 0 \text{ for all } \omega \in \mathcal{A}(M) \\ \text{(ii)} & \text{if } \omega \in \mathcal{A}_r(M),\ \omega' \in \mathcal{A}_{r'}(M), \text{ then } d(\omega \wedge \omega') = (d\omega) \wedge \\ & \omega' + (-1)^r \omega \wedge d\omega' \\ \text{(iii)} & \text{if } f \in \mathcal{A}_0(M),\ df \text{ is the 1-form } Z \mapsto Zf\ (Z \in \mathfrak{J}(M)) \end{array}\right. \tag{1.1.15}$$

Let U be a coordinate patch, let $x_1, \ldots, x_m$ coordinates on it, and let

$$\omega = \sum_{1 \leq i_1 < \cdots < i_k \leq m} a_{i_1, \ldots, i_k} dx_{i_1} \wedge \cdots \wedge dx_{i_k}$$

on U. Then on U

$$d\omega = \sum_{1 \leq i_1 < \cdots < i_k \leq m} da_{i_1, \ldots, i_k} \wedge dx_{i_1} \wedge \cdots \wedge dx_{i_k}. \tag{1.1.16}$$

The elements of $\mathcal{A}(M)$ are called *exterior differential forms* on M. The endomorphism d $(\omega \mapsto d\omega)$ is the operator of *exterior differentiation* on $\mathcal{A}(M)$.

We now discuss briefly some aspects of the theory of integration on manifolds. We confine ourselves to the integration of m-forms on m-dimensional manifolds.

We begin with unoriented or Lebesgue integration. Let M be, as usual, a C^∞ manifold of dimension m, and ω any m-form on M. It is then possible to associate with ω a nonnegative Borel measure on M. To see how this is done, consider a coordinate patch U with coordinates $x_1, \ldots, x_m$, and let $\tilde{U} = \{(x_1(y), \ldots, x_m(y)) : y \in U\}$; for any C^r function f on U, let $\tilde{f} \in C^r(\tilde{U})$ be such that $\tilde{f} \circ (x_1, \ldots, x_m) = f$. Now, we can find a real C^∞ function w_U on U such that $\omega = w_U dx_1 \wedge \cdots \wedge dx_m$ on U. The standard transformation formula for multiple integrals then shows that for any $f \in C_c(U)$, the integral

$$\int_{\tilde{U}} \tilde{f}(t_1, \ldots, t_m) |\tilde{w}_U(t_1, \ldots, t_m)| \, dt_1 \ldots dt_m$$

does not depend on the choice of coordinates $x_1, \ldots, x_m$. In other words, there is a nonnegative Borel measure μ_U on U such that for all $f \in C_c(U)$ and any system $(x_1, \ldots, x_m)$ of coordinates on U

$$\int_U f \, d\mu_U = \int_{\tilde{U}} \tilde{f} \, |\tilde{w}_U| \, dt_1 \cdots dt_m.$$

The measures μ_U are uniquely determined, and this uniqueness implies the existence of a unique nonnegative Borel measure μ on M such that μ_U is the restriction of μ to U for any U. Thus, for any coordinate patch U and any system $(x_1, \ldots, x_m)$ of coordinates on U we have, for all $f \in C_c(U)$,

$$\int_U f \, d\mu = \int_{\tilde{U}} \tilde{f}(t_1, \ldots, t_m) |\tilde{w}_U(t_1, \ldots, t_m)| \, dt_1 \cdots dt_m. \tag{1.1.17}$$

We write $\omega \sim \mu$ and say that μ *corresponds* to ω.

Let M be as above. M is said to be *orientable* if there exists an m-form on M which does not vanish anywhere on M. Two such m-forms, ω_1 and ω_2, are said to be *equivalent* if there exists a positive function g (necessarily C^∞) such that $\omega_2 = g\omega_1$. An *orientation* on M is an equivalence class of nowhere-vanishing m-forms on M. By M being *oriented* we mean that we are given M together with a distinguished orientation; the members of this class are then said to be positive (in symbols, >0).

Suppose now that M is oriented. Let η be any m-form on M with compact support. Select an m-form $\omega > 0$ and write $\eta = g\omega$, where $g \in C_c^\infty(M)$; let μ_ω be the measure corresponding to ω. We then define

$$\int_M \eta = \int_M g \, d\mu_\omega. \tag{1.1.18}$$

It is not difficult to show that this definition is dependent only on η and the orientation of M, and not on the particular choice of ω. Finally, if $\omega > 0$ is as above we often write $\int_M f\omega$ for $\int_M f \, d\mu_\omega$.

Theorem 1.1.3. *Let M be oriented and ω a positive m-form on M. Let μ be the nonnegative Borel measure on M which corresponds to ω. Then, given any differential operator D on M, there exists a unique differential operator $D^\dagger$ on M such that*

$$\int_M Df \cdot g \, d\mu = \int_M f \cdot D^\dagger g \, d\mu \tag{1.1.19}$$

for all $f, g \in C^\infty(M)$ with at least one of f and g having compact support. $D^\dagger$ has the same order as D and $D \mapsto D^\dagger$ is an involutive antiautomorphism of the algebra Diff(M).

Proof. Given $D \in \text{Diff}(M)$ and $g \in C^\infty(M)$, the validity of (1.1.19) for all $f \in C_c^\infty(M)$ determines $D^\dagger g$ uniquely. So if $D^\dagger$ exists, it is unique. It is also clear that if $D^\dagger$ is a differential operator such that (1.1.19) is satisfied whenever f and g are in $C_c^\infty(M)$, then (1.1.19) is satisfied whenever at least one of f and g lies in $C_c^\infty(M)$. The uniqueness implies quickly that the set $\mathfrak{D}_M$ of all $D \in \text{Diff}(M)$ for which $D^\dagger$ exists is a subalgebra, that $\mathfrak{D}_M^\dagger = \mathfrak{D}_M$, and that $D \mapsto D^\dagger$ is an involutive antiautomorphism of $\mathfrak{D}_M$. It remains only to prove that $\mathfrak{D}_M = \text{Diff}(M)$.

Let U be a coordinate patch, and let $(x_1, \ldots, x_m)$ be a coordinate system U with $\omega = w_U dx_1 \wedge \cdots \wedge dx_m$ on U, where $w_U > 0$ on U. Put $\tilde{U} = \{(x_1(y), \ldots, x_m(y)) : y \in U\}$ and for any $h \in C^\infty(U)$ denote by $\tilde{h}$ the element of $C^\infty(\tilde{U})$ such that $\tilde{h} \circ (x_1, \ldots, x_m) = h$. A simple partial integration shows that if $1 \leq j \leq m$, $f, g \in C_c^\infty(U)$,

$$\int_{\tilde{U}} \left(\frac{\partial \tilde{f}}{\partial t_j}\right) \tilde{g}\tilde{w}_U \, dt_1 \cdots dt_m = -\int_{\tilde{U}} \left(\frac{\partial \tilde{g}}{\partial t_j} + \frac{1}{\tilde{w}_U}\frac{\partial \tilde{w}_U}{\partial t_j}\tilde{g}\right) \tilde{f}\tilde{w}_U \, dt_1 \cdots dt_m.$$

If Z_j is the vector field $y \mapsto (\partial/\partial x_j)_y$ on U, and $\varphi_j \in C^\infty(U)$ is defined by $\varphi_j = w_U^{-1} \cdot (Z_j w_U)$, it is clear that $Z_j^\dagger$ exists and is the differential operator of order 1 given by $Z_j^\dagger = -(Z_j + \varphi_j)$. If $h \in C^\infty(U)$, $h^\dagger$ exists and coincides with h. But by Theorem 1.1.2, Diff(U) is algebraically generated by $C^\infty(U)$ and the vector fields Z_j, $1 \leq j \leq m$. Hence $\mathfrak{D}_U = \text{Diff}(U)$. Moreover, the above argument shows that for any $E \in \text{Diff}(U)$ the order of $E^\dagger$ is $\leq$ order of E.

Let D be any differential operator on M. From what we have just proved it is clear that for each coordinate patch U one can find a differential operator $D_U^\dagger$ on U such that $\text{ord}(D_U^\dagger) \leq \text{ord}(D)$ and for all $f, g \in C_c^\infty(U)$

$$\int_M (Df)\cdot g \, d\mu = \int_M f \cdot (D_U^\dagger g) \, d\mu.$$

The uniqueness of $\dagger$ shows that the $D_U^\dagger$ match on overlapping coordinate patches. So there is a differential operator D' on M such that $C_U^\dagger$ is the restriction of D' to U for any arbitrary coordinate patch U. Moreover, if U is any coordinate patch, we have

$$\int_M (Df) g \, d\mu = \int_M f(D'g) \, d\mu$$

for all $f, g \in C_c^\infty(U)$. A simple argument based on partitions of unity shows that this equation is valid for all $f, g \in C_c^\infty(M)$. In other words, $D^\dagger$ exists and coincides with D'. Our construction makes it clear that $\text{ord}(D^\dagger) \leq \text{ord}(D)$

for all $D \in \mathrm{Diff}(M)$. Since $D^{\dagger\dagger} = D$, this shows that $\mathrm{ord}(D) \leq \mathrm{ord}(D^\dagger)$, so that necessarily $\mathrm{ord}(D) = \mathrm{ord}(D^\dagger)$ for all $D \in \mathrm{Diff}(M)$. The theorem is proved.

$D^\dagger$ is called the *formal adjoint of D relative to ω.*

Mappings. Let M, N be C^∞ manifolds. A continuous map

$$\pi : M \longrightarrow N$$

is said to be *differentiable* (C^∞) if for any open set $U \subseteq N$ and any $g \in C^\infty(U)$, $g \circ \pi \in C^\infty(\pi^{-1}(U))$. Suppose π is differentiable, $x \in M$, $y = \pi(x)$. Then with respect to coordinates $x_1, \ldots, x_m$ around x, and $y_1, \ldots, y_n$ around y, π is given by differentiable functions.

If g, g' are C^∞ around y and coincide in an open set containing y, then $g \circ \pi$ and $g' \circ \pi$ coincide in an open set containing x. Thus the map $g \mapsto g \circ \pi$ $(g \in C^\infty(N))$ induces an algebra homomorphism π^* $(\mathbf{u} \mapsto \mathbf{u} \circ \pi)$ of $\mathbf{D}_y$ into $\mathbf{D}_x$. If $X_x \in T_{xc}(M)$, there is a unique $Y_y \in T_{yc}(N)$ such that $Y_y(\mathbf{u}) = X_x(\pi^*(\mathbf{u}))$; we write $Y_y = (d\pi)_x(X_x)$. Thus

$$(d\pi)_x(X_x)(\mathbf{u}) = X_x(\pi^*(\mathbf{u})) \quad (\mathbf{u} \in \mathbf{D}_y).$$

$(d\pi)_x$ is a linear map of $T_{xc}(M)$ into $T_{yc}(N)$, called the *differential of π at x*. It is clear that $(d\pi)_x$ maps the tangent space $T_x(M)$ into the tangent space $T_y(N)$. A special case of this arises when M is an open subset of the real line $\mathbf{R}$. In this case, for any $\tau \in M$, $D_\tau = (d/dt)_{t=\tau}$ is a basis for $T_\tau(M)$, and it is customary to write

$$\dot{\pi}(\tau) = \left(\frac{d}{dt}\right)_{t=\tau} \pi(t) = (d\pi)_\tau(D_\tau). \tag{1.1.20}$$

$\dot{\pi}(\tau)$ is thus an element of $T_{\pi(\tau)}(N)$.

If $p \geq 1$ is any integer, it is obvious that $\pi^*(\mathbf{J}_y^p) \subseteq \mathbf{J}_x^p$, so given any $v \in T_{xc}^{(\infty)}(M)$, there is a unique $v' \in T_{yc}^{(\infty)}(N)$ such that $v'(\mathbf{u}) = v(\pi^*(\mathbf{u}))$ for all $\mathbf{u} \in \mathbf{D}_y$. We write $v' = (d\pi)_x^{(\infty)}(v)$; thus

$$(d\pi)_x^{(\infty)}(v)(\mathbf{u}) = v(\pi^*(\mathbf{u})) \quad (\mathbf{u} \in \mathbf{D}_y). \tag{1.1.21}$$

It is obvious that $(d\pi)_x^{(\infty)}$ maps $T_x^{(r)}(M)$ into $T_y^{(r)}(N)$ for any integer $r \geq 0$ and that $(d\pi)_x^{(\infty)} \mid T_x(M) = (d\pi)_x$. We refer to $(d\pi)_x^{(\infty)}$ as the *complete differential of π at x.* If D is a differential operator on M, there need not in general exist a differential operator D' on N such that $(d\pi)_x^{(\infty)}(D_x) = D'_{\pi(x)}$ for all $x \in M$. If such a D' exists, we shall say (following Chevalley) that D and D' are *π-related.* Given $D \in \mathrm{Diff}(M)$ and $D' \in \mathrm{Diff}(N)$, it is easy to show that D and D' are π-related if and only if $D(u \circ \pi) = (D'u) \circ \pi$ for all $u \in C^\infty(N)$. If

$D_j \in \mathrm{Diff}(M)$ and $D'_j \in \mathrm{Diff}(N)$ are π-related ($j = 1, 2$), then $D_1 \circ D_2$ and $D'_1 \circ D'_2$ are π-related.

Let $\pi : M \longrightarrow N$ be a C^∞ map and ω any r-form on N. For $x \in M$ let $(\pi^*\omega)_x$ be the r-linear form defined by

$$(1.1.22) \qquad (\pi^*\omega)_x(v_1, \ldots, v_r) = \omega_{\pi(x)}((d\pi)_x(v_1), \ldots, (d\pi)_x(v_r)),$$

for $v_1, \ldots, v_r \in T_{xc}(M)$. Then $(\pi^*\omega)_x \in \Lambda_r(T_{xc}(M))$, and $x \mapsto (\pi^*\omega)_x$ is an r-form on M. We denote this form by $\pi^*\omega$. $\pi^* : \omega \mapsto \pi^*\omega$ has the following properties:

$$(1.1.23) \qquad \begin{cases} \text{(i)} & \pi^*(u\omega) = (u \circ \pi)\pi^*\omega \quad (u \in C^\infty(N)) \\ \text{(ii)} & d(\pi^*\omega) = \pi^*(d\omega) \\ \text{(iii)} & \pi^*(\omega_1 \wedge \omega_2) = (\pi^*\omega_1) \wedge (\pi^*\omega_2). \end{cases}$$

($\omega, \omega_1, \omega_2, \in \mathfrak{a}(N)$ are arbitrary).

We consider now the special case where the differentiable map π is a homeomorphism of M onto N and π^{-1} is also a differentiable map. π is then called a *diffeomorphism*. In this case π induces natural isomorphisms between the respective spaces of functions, differential operators, etc. For instance, let $N = M$ and $\alpha : x \mapsto \alpha(x)$ a diffeomorphism of M onto itself. Then α induces the automorphism $u \mapsto u^\alpha$ of $C^\infty(M)$ where $u^\alpha(x) = u(\alpha^{-1}(x))$ for all $x \in M, u \in C^\infty(M)$. This in turn induces the automorphism $D \mapsto D^\alpha$ of the algebra $\mathrm{Diff}(M)$; $D^\alpha(u) = (D(u^{\alpha^{-1}}))^\alpha$, for all $D \in \mathrm{Diff}(M)$ and $u \in C^\infty(M)$. The set of all diffeomorphisms of M is a group under composition. If α, β are diffeomorphisms of M onto itself, the $D^{\alpha\beta} = (D^\beta)^\alpha$ for $D \in \mathrm{Diff}(M)$. Similarly we have the automorphism $\omega \mapsto \omega^\alpha$ of $\mathfrak{a}(M)$.

Let π $(M \mapsto N)$ be a C^∞ map ($m = \dim(M)$, $n = \dim(N)$), $x \in M$, and let $(d\pi)_x$ be surjective. Let $y = \pi(x)$. Then $m \geq n$, and it is well known that in suitable coordinates around x and y, π looks like the projection $(t_1, \ldots, t_m) \mapsto (t_1, \ldots, t_n)$ around the origin in $\mathbf{R}^m$. In fact, let $x_1, \ldots, x_m$ be coordinates around x, and $y_1, \ldots, y_n$ coordinates around y with $x_i(x) = y_j(y) = 0$, $1 \leq i \leq m$, $1 \leq j \leq n$. There are C^∞ functions $F_1, \ldots, F_n$ defined around $\mathbf{0}_m = (0, \ldots, 0) \in \mathbf{R}^m$ such that $y_j \circ \pi = F_j(x_1, \ldots, x_m)$ $(1 \leq j \leq n)$ around x. Since $(d\pi)_x$ is surjective, a standard argument shows that the matrix $(\partial F_j/\partial t_k)_{1 \leq j \leq n, 1 \leq k \leq m}$ has rank n at $\mathbf{0}_m$. By permuting the x_i if necessary, we assume that the $n \times n$ matrix $(\partial F_j/\partial t_k)_{1 \leq j, k \leq n}$ is non-singular at $\mathbf{0}_m$. It is then clear that the functions $y_1 \circ \pi, \ldots, y_n \circ \pi, x_{n+1}, \ldots, x_m$ form a system of coordinates around x; and with respect to these and the y_j, π looks like the projection $(t_1, \ldots, t_m) \mapsto (t_1, \ldots, t_n)$. It follows from this that the set $M_1 = \{z : z \in M, (d\pi)_z \text{ is surjective}\}$ is open in M, that $\pi[M_1]$ is open in N, and that π is an open map of M_1 onto $\pi[M_1]$. π is called a *submersion* if $(d\pi)_x$ is surjective for all $x \in M$. If π is a submersion and $\pi[M] = N$, N is called a *quo-*

tient of M relative to π. It follows from the local description of π given above that if N is a quotient of M relative to π, then for any open set $U \subseteq N$, a function g on U is C^∞ if and only if $g \circ \pi$ is C^∞ on $\pi^{-1}(U)$. In other words, in this case, the differentiable structure of N is completely determined by π and the differentiable structure on M.

We now consider maps with injective differentials. Here it is necessary to exercise somewhat greater care than in the case of a submersion. Let M and N be C^∞ manifolds of dimensions m and n respectively, and $\pi\,(M \longrightarrow N)$ a C^∞ map. Let $x \in M$, $y = \pi(x)$ and suppose that $(d\pi)_x$ is injective. Then $m \leq n$, and in suitable coordinates around x and y, π looks like the injection $(t_1, \ldots, t_m) \mapsto (t_1, \ldots, t_m, 0, \ldots, 0)$ around the origin. More precisely, we can find all of the following: a coordinate patch U containing x with coordinates $x_1, \ldots, x_m$; a coordinate patch V containing y with coordinates $y_1, \ldots, y_n$; and a number $a > 0$ with the following properties:

(1.1.24)
(i) $\xi\,(z \mapsto (x_1(z), \ldots, x_m(z)))$ is a diffeomorphism of U onto I_a^m, with $\xi(x) = \mathbf{0}_m$; $\eta\,(z' \mapsto (y_1(z'), \ldots, y_n(z')))$ is a diffeomorphism of V onto I_a^n, with $\eta(y) = \mathbf{0}_n$.[3]
(ii) $\eta \circ \pi \circ \xi^{-1}$ is the map
$$(t_1, \ldots, t_m) \mapsto (t_1, \ldots, t_m, 0, \ldots, 0)$$
of I_a^m into I_a^n.

To see this, let $x'_1, \ldots, x'_m$ be coordinates around x and let $y'_1, \ldots, y'_n$ be coordinates around $y = \pi(x)$ with $x'_i(x) = y'_j(y) = 0$, $1 \leq i \leq m$, $1 \leq j \leq n$. Let $F_j (1 \leq j \leq n)$ be C^∞ functions around $\mathbf{0}_m$ such that $y'_j \circ \pi = F_j(x'_1, \ldots, x'_m)$ around $x (1 \leq j \leq n)$. Since $(d\pi)_x$ is injective, the matrix $(\partial F_j/\partial t_k)_{1 \leq j \leq n, 1 \leq k \leq m}$ has rank m at $\mathbf{0}_m$. By permuting the y'_j if necessary, we may assume that the $m \times m$ matrix $(\partial F_j/\partial t_k)_{1 \leq j,k \leq m}$ is nonsingular at $\mathbf{0}_m$. It is then clear that the functions $y'_1 \circ \pi, \ldots, y'_m \circ \pi$ form a system of coordinates around x. Let $x_i = y'_i \circ \pi (1 \leq i \leq m)$. Let G_p be C^∞ functions around $\mathbf{0}_m$ such that $y'_p \circ \pi = G_p(x_1, \ldots, x_m)$ around $x\,(m < p \leq n)$. Define $y_i = y'_i\ (i \leq m)$, $y_p = y'_p - G_p(y'_1, \ldots, y'_m)\,(m < p \leq n)$. Then we have (1.1.24) for suitable U, V, $a > 0$. It follows from (1.1.24) that there is a sufficiently small open set U around x such that π is a homeomorphism of U onto $\pi[U]$.

π is called an *immersion* if $(d\pi)_x$ is injective for all $x \in M$; an *imbedding* if it is an one to-one immersion; and a *regular imbedding* if it is an imbedding and if π is a homeomorphism of M onto $\pi[M]$, the latter being given the topology inherited from N. The properties (1.1.24) are not in general strong

[3]For any integer $k \geq 1$ and any $b > 0$, we write I_b^k for the cube in $\mathbf{R}^k$ defined by
$$I_b^k = \{(t_1, \ldots, t_k)\colon -b < t_j < b \quad \text{for} \quad 1 \leq j \leq k\}.$$
The origin of $\mathbf{R}^k$ is denoted by $\mathbf{0}_k$.

enough to ensure that a given imbedding is regular or has other nice properties. Note, however, that if π is an imbedding the equations (1.1.24) completely determine the differentiable structure of M in terms of π and the differentiable structure of N: if $W \subseteq M$ is open and f is a complex-valued function on W, f is C^∞ if and only if for each $x \in W$ one can find an open set U with $x \in U \subseteq W$, an open set V containing $y = \pi(x)$ with $\pi[U] \subseteq V$, and $g \in C^\infty(V)$ such that $f(z) = g(\pi(z))$, $z \in U$.

The next theorem describes some of the nice properties of regular imbeddings. Recall that a subset A of a topological space E is said to be *locally closed* (in E) if it is a relatively closed subset of some open subset of E, or equivalently, if it is open in its closure.

Theorem 1.1.4. *Let π be a regular imbedding of M into N. Then $\pi[M]$ is locally closed in N. For each $x \in M$, we can choose $U, V, x_1, \ldots, x_m, y_1, \ldots, y_n$ such that, in addition to* (1.1.24), *we have*

$$\pi[U] = \pi[M] \cap V. \tag{1.1.25}$$

If P is any C^∞ manifold, and u is any map of P into M, u is C^∞ if and only if $\pi \circ u$ is a C^∞ map of P into N.

Proof. Let $U', V', x_1, \ldots, x_m, y_1, \ldots, y_n$, and $a' > 0$ be such that the relations (1.1.24) are satisfied (with U', V', and a' replacing U, V, and a, respectively). Since π is a homeomorphism onto $\pi[M]$, $\pi[U']$ is open in $\pi[M]$, so there is an open set V'' in N such that $\pi[U'] = V'' \cap \pi[M]$. Let $V_1 = V' \cap V''$. Then V_1 is an open subset of N containing $y = \pi(x)$ and $\pi[U'] = V_1 \cap \pi[M]$. Choose a with $0 < a < a'$ such that $\eta^{-1}(I_a^n) \subseteq V_1$ and $\xi^{-1}(I_a^m) \subseteq U'$. Then if we set $U = \xi^{-1}(I_a^m)$ and $V = \eta^{-1}(I_a^n)$, we have (1.1.25). Note that $\pi[U] = \pi[M] \cap V$ is closed in V by (1.1.24). Now select open sets $V_i (i \in I)$ in N such that $\pi[M] \subseteq \bigcup_{i \in I} V_i$ and $\pi[M] \cap V_i$ is closed in V_i for each $i \in I$. Then it is clear that $\pi[M]$ is closed in $\bigcup_{i \in I} V_i$; thus $\pi[M]$ is locally closed. For the last assertion, let P be a C^∞ manifold, and let u be a map of P into M such that $\pi \circ u$ is a C^∞ map of P into N. Let $p \in P$, $x = u(p)$, $y = \pi(x)$. There is an open set W in P containing p such that $(\pi \circ u)[W] \subseteq V$; then $u[W] \subseteq U$. It follows at once from a consideration of coordinates that u is a C^∞ map of W into U.

The universal property contained in the last assertion of Theorem 1.1.4 is an important consequence of the regularity of an imbedding. However, even some irregular imbeddings possess this property. Let π be an imbedding of M into N. We shall call π *quasi-regular* if the following property is satisfied: if P is any C^∞ manifold and u any map of P into M, u is C^∞ if and only if $\pi \circ u$ is C^∞ from P to N. There are imbeddings which are quasi-regular but not regular (Exercise 1).

Submanifolds. Let M, N be C^∞ manifolds. Then M is called a *submanifold* of N if

$$(1.1.26) \quad \begin{cases} \text{(i)} & M \subseteq N \text{ (set-theoretically)} \\ \text{(ii)} & \text{the identity map of } M \text{ into } N \text{ is an imbedding.} \end{cases}$$

M is said to be a *regular* (resp. *quasi-regular*) submanifold if the identity map of M into N is regular (resp. quasi-regular). If M is a submanifold of N and $x \in M$, we shall identify $T_{xc}^{(\infty)}(M)$ with its image in $T_{xc}^{(\infty)}(N)$ under the complete differential of the identity map of M into N.

As we have observed already, the relations (1.1.24) have the following consequence: given a subset $M \subseteq N$ and a topology on M under which M is a Hausdorff second countable space and which is finer than the one induced from N, there is *at most one* differentiable structure on M so that M becomes a submanifold of N. If such a structure exists, we shall equip M with it and refer to M as a submanifold of N. If the topology on M is the one induced by N, then the differentiable structure described above, if it exists, will convert M into a regular submanifold of N.

Theorem 1.1.5. *Let N be a C^∞ manifold and let $M \subseteq N$. In order that M, equipped with the relative topology, be a (regular) submanifold of N, it is necessary and sufficient that the following be satisfied. There exists an integer m with $1 \leq m \leq n$ such that given any $x \in M$, one can find an open set V of N containing x and $n - m$ real differentiable functions $f_1, \ldots, f_{n-m}$ on V such that*

$$(1.1.27) \quad \begin{cases} (i) & V \cap M = \{z : z \in V,\ f_1(z) = \cdots = f_{n-m}(z) = 0\} \\ (ii) & (df_1)_x, \ldots, (df_{n-m})_x \text{ are linearly independent elements of } T_x(N)^*. \end{cases}$$

If this is the case, $\dim(M) = m$, and M is a locally closed subset of N.

Proof. The only thing that needs to be proved is that if M satisfies the conditions described above, then it becomes a regular submanifold of N; Theorem 1.1.4 implies the remaining assertions. Also if $m = n$, (1.1.27) reduces to the condition $V \subseteq M$, so that in this case M is an open submanifold of N. We may thus assume $1 \leq m < n$.

Fix $x \in M$ and let $V, f_1, \ldots, f_{n-m}$ be as in (1.1.27). It is then clear that we can find a system of coordinates $x_1, \ldots, x_n$ in a neighborhood of x such that $x_j(x) = 0 (1 \leq j \leq n)$ and $x_{m+j} = f_j (1 \leq j \leq n - m)$. By replacing V by a smaller open set, we may assume that the homeomorphism $\xi\ (y \mapsto (x_1(y), \ldots, x_n(y)))$ maps V onto I_a^n for some $a > 0$. Then ξ maps $M \cap V$ onto $I_a^m \times \mathbf{0}_{n-m}$. In other words, $U = M \cap V$ is a regularly imbedded submanifold of V, hence of N. Since $x \in M$ is arbitrary, it follows that we can write $M = \bigcup_{i \in I} U_i$, where each U_i is open in M and is a regular submanifold of N. If $i, j \in I$ are such that $U_{ij} = U_i \cap U_j \neq \phi$, then U_{ij} is open in both U_i and U_j and is a regular submanifold of N under each of the C^∞ structures induced by U_i and U_j. These two structures must be the same, so U_{ij} is an open submanifold of both U_i and U_j. It then follows that there is a unique C^∞ structure for M such that each $U_i(i \in I)$ becomes an open submanifold of M. This structure converts M into a regular submanifold of N.

Product manifolds. Let $M_j(j = 1, 2)$ be a C^∞ manifold of dimension m_j, and let $M = M_1 \times M_2$. Equip M with the product topology; it is then Hausdorff and second countable. Let $U \subseteq M$ be an open subset and f a complex function defined on U. We say f is C^∞ if the following condition is satisfied: for any $(a_1, a_2) \in U$ there are coordinate patches V_j around a_j and coordinates $x_{j1}, \ldots, x_{jm_j}$ on $V_j(j = 1, 2)$ such that (i) $V_1 \times V_2 \subseteq U$, and (ii) if $\tilde{V}_j$ is the image of V_j under the map $z \mapsto (x_{j1}(z), \ldots, x_{jm_j}(z))$, there is a C^∞ function φ on $\tilde{V}_1 \times \tilde{V}_2$ such that

$$f(b_1, b_2) = \varphi(x_{11}(b_1), \ldots, x_{1m_1}(b_1), x_{21}(b_2), \ldots, x_{2m_2}(b_2))$$

for all $(b_1, b_2) \in V_1 \times V_2$. $U \mapsto C^\infty(U)$ is a differentiable structure for M; it is called the *product* of the structures on M_1 and M_2. M is called the *product* of the C^∞ manifolds M_1 and M_2. If π_j is the natural projection of M on M_j, π_j is a submersion. If N is a C^∞ manifold and $u: y \mapsto (u_1(y), u_2(y))$ is a map of N into M, then u is C^∞ if and only if u_1 and u_2 are C^∞.

Suppose $x = (x_1, x_2) \in M$. Given functions $f_j \in C^\infty(U_j)$, where $x_j \in U_j$ $(j = 1, 2)$, we write $f_1 \otimes f_2$ for the element of $C^\infty(U_1 \times U_2)$ given by $(f_1 \otimes f_2)(a_1, a_2) = f_1(a_1)f_2(a_2)(a_j \in U_j)$. The map $f_1, f_2 \mapsto f_1 \otimes f_2$ induces a natural injection of $\mathbf{D}_{x_1} \otimes \mathbf{D}_{x_2}$ into $\mathbf{D}_x$. If X_j is a tangent vector to M_j at x_j $(j = 1, 2)$, there is exactly one tangent vector X to M at x such that for $\mathbf{u}_j \in \mathbf{D}_{x_j}$ $(j = 1, 2)$,

$$X(\mathbf{u}_1 \otimes \mathbf{u}_2) = \mathbf{u}_1(x_1)X_2(\mathbf{u}_2) + \mathbf{u}_2(x_2)X_1(\mathbf{u}_1); \tag{1.1.28}$$

$X_1, X_2 \mapsto X$ is a linear isomorphism of $T_{x_1}(M_1) \times T_{x_2}(M_2)$ with $T_x(M)$. More generally, if $v_j \in T_{x_j}^{(r_j)}(M_j)$ $(j = 1, 2)$ there is exactly one $v \in T_x^{(\infty)}(M)$ such that

$$v(\mathbf{f}_1 \otimes \mathbf{f}_2) = v_1(\mathbf{f}_1)v_2(\mathbf{f}_2) \quad (\mathbf{f}_j \in \mathbf{D}_{x_j}); \tag{1.1.29}$$

$v \in T_x^{(r_1 + r_2)}(M)$ and the map $v_1 \otimes v_2 \mapsto v$ extends uniquely to a linear isomorphism of $T_{x_1c}^{(\infty)}(M_1) \otimes T_{x_2c}^{(\infty)}(M_2)$ onto $T_{xc}^{(\infty)}(M)$. We shall often identify these two spaces and write $v_1 \otimes v_2$ for the element v defined by (1.1.29); in particular, the tangent vector X defined by (1.1.28) is nothing but $X_1 \otimes 1_{x_2} + 1_{x_1} \otimes X_2$.

If D_j is a differential operator on M_j $(j = 1, 2)$, then $D: (x_1, x_2) \mapsto (D_1)_{x_1} \otimes (D_2)_{x_2}$ is a differential operator $M_1 \times M_2$; we write $D_1 \otimes D_2$ for D.

These considerations can be extended easily to products of more than two manifolds.

1.2. Analytic Manifolds

We begin by recalling the definition of an analytic function of m variables, real or complex. Let $U \subseteq \mathbf{R}^m$ be any open set and let f be a function defined on U with values in $\mathbf{C}$. f is said to be *analytic* on U if, given any $(x_1^0, \ldots, x_m^0) \in U$, we can find an $\eta > 0$ and a power series

$$\sum_{r_1, \ldots, r_m \geq 0} c_{r_1, \ldots, r_m}(x_1 - x_1^0)^{r_1} \cdots (x_m - x_m^0)^{r_m} \quad (c_{r_1, \ldots, r_m} \in \mathbf{C})$$

around $(x_1^0, \ldots, x_m^0)$ such that the series converges absolutely and uniformly for all $(x_1, \ldots, x_m)$ with $\max_{1 \leq j \leq m} |x_j - x_j^0| < \eta$, to the sum $f(x_1, \ldots, x_m)$. For an open set $U \subseteq \mathbf{C}^m$, a similar definition of a *complex analytic* or *holomorphic* function on U can be given. The functions which are analytic on U form an algebra under the usual operations. Analytic functions of analytic functions are analytic.

The definition of a *real* analytic manifold is similar to that of a C^∞ manifold. Let M be a Hausdorff space satisfying the second axiom of countability. A *real analytic structure* for M is an assignment

$$\mathfrak{A} : U \mapsto \mathfrak{A}(U) \quad (U \text{ open}, \subseteq M)$$

such that

(i) $\mathfrak{A}$ possesses properties (i) and (ii) of a differentiable structure (cf. §1.1).

(ii) There exists an integer $m > 0$ with the following property: for each $x \in M$, can find an open set U containing x and m *real* functions $x_1, \ldots, x_m$ from $\mathfrak{A}(U)$ such that (a) the map $\xi : y \mapsto (x_1(y), \ldots, x_m(y))$ is a homeomorphism of U with an open subset of $\mathbf{R}^m$, and (b) if W is any open subset of U, $\mathfrak{A}(W)$ is precisely the set of all functions of the form $F \circ \xi$, with F analytic on $\xi[W]$.

The pair $(M, \mathfrak{A})$ (and, by abuse of language, M itself) is said be a *real analytic manifold* of *dimension m*. For an open $U \subseteq M$, the elements of $\mathfrak{A}(U)$ are called the *analytic functions* on U. As before, any open set such as U in (ii) above is called a *coordinate patch;* and $x_1, \ldots, x_m$ are called *analytic coordinates* on U.

Let $U \subseteq M$ be open and let f be a complex-valued function defined on U. We define f to be C^∞ if for each $x \in U$, f is a C^∞ function of the local analytic coordinates around x. The assignment $U \mapsto C^\infty(U)$ is easily seen to be a differentiable structure for M. We shall call this the C^∞ structure underlying the analytic structure. Note that $\mathfrak{A}(U) \subseteq C^\infty(U)$ for all open U. The entire theory of differentiable manifolds now becomes available to M.

Let M and N be analytic manifolds and π a map of M into N. The definition of the analyticity of π is analogous to the C^∞ case. π is called an *analytic isomorphism* or an *analytic diffeomorphism* if it is bijective and if both π and π^{-1} are analytic. It is a consequence of the classical theorem on implicit and inverse functions that if π $(M \longrightarrow N)$ is analytic and bijective and if $(d\pi)_x$ is bijective for all $x \in M$, then π^{-1} is analytic, so that π is an analytic diffeomorphism.

Let M be an analytic manifold and D a differential operator on M. For any open set $U \subseteq M$, let D_U denote the restriction of D to U. D is called *analytic* if for each open U, $D_U: f \mapsto D_U f$ leaves $\mathfrak{A}(U)$ invariant. Let U be a coordinate patch, $x_1, \ldots, x_m$ analytic coordinates on U, and let $D_U = \sum_{|\alpha| \leq r} a_{(\alpha)} \partial^{(\alpha)}$. Then if D is analytic, $a_{(\alpha)} \in \mathfrak{A}(U)$; conversely, if for each $x \in M$ we can find analytic coordinates $x_1, \ldots, x_m$ around x such that $D = \sum_{|\alpha| \leq r} a_{(\alpha)} \partial^{(\alpha)}$ on an open set around x with analytic $a_{(\alpha)}$, then D is an analytic differential operator. Similarly, a definition of analyticity can be given for differential forms. The analytic differential operators form a subalgebra of Diff(M). If ω is an analytic m-form which is real and vanishes nowhere, D an analytic differential operator, and $D^\dagger$ the formal adjoint of D with respect to ω, then it is easy to verify that $D^\dagger$ is analytic. If ω, ω' are analytic r-forms, then $d\omega$ and $\omega \wedge \omega'$ are analytic; if π $(M \longrightarrow N)$ is analytic and ω is an analytic r-form on N, so is $\pi^*\omega$ on M.

The concepts of products and quotients of analytic manifolds as well as submanifolds of analytic manifolds are defined exactly as in the C^∞ case, with analytic functions and coordinate systems replacing the C^∞ ones. The definitions and results of §1.1 concerning maps with surjective and injective differentials remain valid with this modification. In particular, Theorems 1.1.4 and 1.1.5 remain true in the analytic case: if N is an analytic manifold and M a subset of N equipped with the relative topology, then M is a regular analytic submanifold of N of dimension m $(1 \leq m \leq n)$ if and only if for each $x \in M$ we can find an open subset V of N containing x and $n - m$ real-valued analytic functions $f_1, \ldots, f_{n-m}$ on V such that (i) $V \cap M$ is precisely the set of common zeros of $f_1, \ldots, f_{n-m}$ in V, and (ii) $(df_1)_x, \ldots, (df_{n-m})_x$ are linearly independent elements of $T_x(N)^*$.

A *complex analytic* or *holomorphic* manifold of complex dimension m is defined in the same way as a real analytic manifold, with holomorphic functions replacing real analytic functions. Given a complex analytic manifold M of dimension m, there is an underlying real analytic structure for M in which M is a real analytic manifold of dimension $2m$; if $U \subseteq M$ is open and f is a real-valued function on U, f will be analytic in this real analytic structure if and only if the following is satisfied: for each $x \in U$, we can find holomorphic coordinates $z_1, \ldots, z_m$ around x such that f is a real analytic function of the $2m$ real functions $\mathrm{Re}(z_1), \ldots, \mathrm{Re}(z_1), \mathrm{Im}(z_1), \ldots, \mathrm{Im}(z_m)$ in a sufficiently small open neighborhood of x.

Let M be a complex analytic manifold and let $x \in M$. Two functions defined and holomorphic in an open set containing x are called *equivalent* if they coincide in some open neighborhood of x. The equivalence classes are called the *germs of holomorphic functions at* x. In the usual way, they form an algebra over $\mathbf{C}$, denoted by $\mathbf{H}_x$; for any $\mathbf{f} \in \mathbf{H}_x$, write $\mathbf{f}(x)$ for the common value at x of the elements of $\mathbf{f}$. The *holomorphic tangent vectors* to M at x are then the linear functions v on $\mathbf{H}_x$ such that $v(\mathbf{fg}) = \mathbf{f}(x)v(\mathbf{g}) + \mathbf{g}(x)v(\mathbf{f})$ for all $\mathbf{f}, \mathbf{g} \in \mathbf{H}_x$. They form a complex vector space, the so called *holomorphic tangent space to* M *at* x; this vector space is denoted by $T_x(M)$. More generally, let $\mathbf{J}_x$ be the ideal in $\mathbf{H}_x$ of all $\mathbf{u}$ with $\mathbf{u}(x) = 0$; then for any integer $r \geq 0$, a *holomorphic differential expression* at x is a linear function v on $\mathbf{H}_x$ which vanishes on $\mathbf{J}_x^{r+1}$. The set of all such is a vector space denoted by $T_x^{(r)}(M)$. As before, we put $T_x^{(\infty)}(M) = U_{r \geq 0} T_x^{(r)}(M)$. Holomorphic vector fields, differential forms, and differential operators can now be defined as in the real analytic case; no changes are needed.

The same situation provails with respect to the concepts of quotient and submanifolds of complex analytic manifolds. In particular, the analogues of Theorems 1.1.4 and 1.1.5 are true in the complex analytic case also.

Algebraic sets. The version of Theorem 1.1.5 for analytic manifolds is very useful in showing that certain subsets of $\mathbf{R}^n$ or $\mathbf{C}^n$ are regular analytic submanifolds. The simplest examples are obtained when we take M to be the set of zeros of a collection of *polynomials*. For example, let $p \geq 1, q \geq 1$ be integers and let F be the polynomial on $\mathbf{R}^{p+q}$ defined by

$$F(x_1, \ldots, x_{p+q}) = x_1^2 + \cdots + x_p^2 - x_{p+1}^2 - \cdots - x_{p+q}^2$$

Let M be the set of zeros of F and $M_0 = M \backslash \{\mathbf{0}\}$ ($\mathbf{0}$ is the origin in $\mathbf{R}^{p+q}$). Then, for $x \in M$, $(dF)_x \neq 0$ if and only if $x \in M_0$. So M_0 is a regular analytic submanifold of dimension $p + q - 1$. It is not difficult to show that M does not look like a manifold around $\mathbf{0}$. $\mathbf{0}$ is called a singular point, and points of M_0 are called regular; the set of regular points is thus open in M and forms a regular submanifold of $\mathbf{R}^{p+q}$. We now prove a theorem of H. Whitney [1] which asserts that the above example is somewhat typical. We work in $\mathbf{R}^n$; the case of sets of zeros in $\mathbf{C}^n$ of complex polynomials can be handled similarly.

Let $U \subseteq \mathbf{R}^n$ be an open set, fixed throughout this discussion; let $\mathcal{P}$ be the algebra of all polynomial functions on $\mathbf{R}^n$ with real coefficients. For any subset $\mathfrak{F} \subseteq \mathcal{P}$ let

$$Z(\mathfrak{F}) = \{u : u \in U, P(u) = 0 \ \forall\ P \in \mathfrak{F}\}. \tag{1.2.1}$$

Any subset of U which is $Z(\mathfrak{F})$ for some $\mathfrak{F} \subseteq \mathcal{P}$ is called an *algebraic subset of* U. For any subset M of U, let

$$\mathcal{J}(M) = \{P : P \in \mathcal{P}, P(u) = 0 \ \forall \ u \in M\}; \tag{1.2.2}$$

$\mathcal{J}(M)$ is an ideal in $\mathcal{P}$. Note that $Z(\mathfrak{F})$ is also the set of common zeros of the elements of $\mathcal{J}(Z(\mathfrak{F}))$, so that any algebraic subset of U is of the form $Z(\mathcal{J})$ for some ideal $\mathcal{J} \subseteq \mathcal{P}$. Now, if $\mathcal{J}$ is an ideal in $\mathcal{P}$, we can find $P_1, \ldots, P_r \in \mathcal{J}$ such that $\mathcal{J} = \mathcal{P}P_1 + \cdots + \mathcal{P}P_r$ (Hilbert basis theorem); $\{P_1, \ldots, P_r\}$ is called an *ideal basis* for $\mathcal{J}$. So any algebraic set is of the form $Z(\mathfrak{F})$ for a finite subset $\mathfrak{F}$ of $\mathcal{P}$.

Suppose now that M is an algebraic subset of U. For any $u \in M$, let $r_M(u)$ be the dimension of the linear space spanned by the differentials $(dP)_u$, $P \in \mathcal{J}(M)$. $r_M(u)$ is called the *rank* of M at u. The relation

$$d(PQ)_u = P(u)(dQ)_u + Q(u)(dP)_u$$

shows that if $\{Q_1, \ldots, Q_p\}$ is an ideal basis for $\mathcal{J}(M)$, $r_M(u)$ is also the dimension of the linear space spanned by $(dQ_1)_u, \ldots, (dQ_p)_u$. Put

$$r = \max_{u \in M} r_M(u) \tag{1.2.3}$$

$$M_0 = \{u : u \in M, r_M(u) = r\} \tag{1.2.4}$$

The points of M_0 are called *regular;* those of $M \setminus M_0$ are called *singular*. Now, for any $P_1, \ldots, P_s \in \mathcal{J}(M)$, $(dP_1)_u, \ldots, (dP_s)_u$ are linearly independent if and only if the matrix $(\partial P_i/\partial t_j)_{1 \leq i \leq s, 1 \leq j \leq n}$ is of rank s at u. It follows easily from this that M_0 is a nonempty open subset of M, being the set of all $u \in M$ where the rank of the matrix $((\partial Q_i/\partial t_j)_u)$ is maximum.

Theorem 1.2.1. (*Whitney*) *Let notation be as above. Then M_0 is a nonempy open subset of M and is a regular analytic submanifold of* $\mathbf{R}^n$ *of dimension* $n - r$.

Proof. We follow Whitney's proof. It is enough to prove that each point of M_0 can be surrounded by a connected open subset of M_0 which is a regular analytic submanifold of $\mathbf{R}^n$ of dimension $n - r$. Fix $u_0 \in M_0$; we may assume that $u_0 = \mathbf{0}$. We can then select $P_1, \ldots, P_r \in \mathcal{J}(M)$ such that $(dP_1)_0, \ldots, (dP_r)_0$ are linearly independent. The matrix $(\partial P_i/\partial t_j)_{1 \leq i \leq r, 1 \leq j \leq n}$ therefore has rank r at $\mathbf{0}$. By permuting the coordinates if necessary, we may assume that

$$\left(\frac{\partial(P_1, \ldots, P_r)}{\partial(t_1, \ldots, t_r)}\right)_0 \neq 0.$$

It is then obvious that

$$\left(\frac{\partial(P_1,\ldots,P_r,t_{r+1},\ldots,t_n)}{\partial(t_1,\ldots,t_r,t_{r+1},\ldots,t_n)}\right)_0 \neq 0.$$

Write $y_i = P_i$, $1 \leq i \leq r$, $y_i = t_i$, $r+1 \leq i \leq n$. Clearly, we can choose an open set V and an $a > 0$ such that (i) $\mathbf{0} \in V \subseteq U$, and (ii) $v \mapsto (y_1(v), \ldots, y_n(v))$ is an analytic diffeomorphism of V with the cube I_a^n. Let

(1.2.5) $$V_0 = \{v : v \in V, y_1(v) = \cdots = y_r(v) = 0\};$$

then V_0 is a connected regular analytic submanifold of $\mathbf{R}^n$ of dimension $n - r$, $\mathbf{0} \in V_0$, and $V \cap M \subseteq V_0$. It is now enough to prove that $V_0 \subseteq M$. For suppose this proved: then $V_0 = V \cap M$, so V_0 is an open subset of M. Since $(dP_1)_v, \ldots, (dP_r)_v$ are linearly independent for all $v \in V_0$, $V_0 \subseteq M_0$. So V_0 would be an open subset of M_0 containing u_0 and imbedded as a regular analytic submanifold of dimension $n - r$ of $\mathbf{R}^n$.

We now prove that $V_0 \subseteq M$. Let A be the algebra of all real-valued analytic functions on V. Write $\mathfrak{s} = \mathfrak{s}(M)$ and $\bar{\mathfrak{s}} = A\mathfrak{s}$, the ideal in A generated by $\mathfrak{s}$. We claim that $\bar{\mathfrak{s}}$ is invariant under the derivations $\partial/\partial y_j$, $r + 1 \leq j \leq n$. It is enough to prove that $\partial/\partial y_j\, \mathfrak{s} \subseteq \bar{\mathfrak{s}}$ for $r + 1 \leq j \leq n$. Fix j with $r + 1 \leq j \leq n$, $F \in \mathfrak{s}$. Write $P_l = t_l$ if $r + 1 \leq l \leq n$ and $l \neq j$, and $P_j = F$. Then

(1.2.6) $$\frac{\partial(P_1,\ldots,P_n)}{\partial(y_1,\ldots,y_n)} = \frac{\partial(P_1,\ldots,P_n)}{\partial(t_1,\ldots,t_n)} \cdot \frac{\partial(t_1,\ldots,t_n)}{\partial(y_1,\ldots,y_n)}.$$

Now

$$\frac{\partial(P_1,\ldots,P_n)}{\partial(y_1,\ldots,y_n)} = \frac{\partial F}{\partial y_j}.$$

Furthermore,

$$\varphi = \frac{\partial(t_1,\ldots,t_n)}{\partial(y_1,\ldots,y_n)} \in A.$$

On the other hand, consider

$$P = \frac{\partial(P_1,\ldots,P_n)}{\partial(t_1,\ldots,t_n)}.$$

We have

$$P = \frac{\partial(P_1,\ldots,P_r,F)}{\partial(t_1,\ldots,t_r,t_j)}.$$

Since $P_1, \ldots, P_r$, $F \in \mathfrak{s}$, P has to vanish at all points of M, as otherwise there would be points of M where $\mathfrak{s}$ has rank $\geq r + 1$. So $P = 0$ on M. Since P is a *polynomial*, $P \in \mathfrak{s}$. (1.2.6) now shows that

$$\frac{\partial F}{\partial y_j} = \varphi P \in \bar{\mathfrak{s}}.$$

It follows from the above result that for any $F \in \bar{\mathfrak{J}}$ and any multiindex $(\alpha) = (\alpha_{r+1}, \ldots, \alpha_n)$, $(\partial/\partial y_{r+1})^{\alpha_{r+1}} \cdots (\partial/\partial y_n)^{\alpha_n} F \in \bar{\mathfrak{J}}$. Now, it is trivial to see that any element of $\mathfrak{J}$ vanishes at $\mathbf{0}$. So if $F \in \bar{\mathfrak{J}}$, *all* the derivatives $(\partial/\partial y_{r+1})^{\alpha_{r+1}} \cdots (\partial/\partial y_n)^{\alpha_n} F$ vanish at $\mathbf{0}$. Since F is analytic and V_0 is connected, this implies that F vanishes on V_0. In particular, all elements of $\mathfrak{J}$ vanish on V_0. So $V_0 \subseteq M$. As mentioned earlier, this is sufficient to prove the theorem.

1.3. The Frobenius Theorem

The aim of this section is to introduce the concept of involutive systems of tangent spaces on an analytic manifold and to prove that such systems are integrable. At the local level this is just the classical Frobenius theorem. However, for applications to the theory of Lie groups, the local form of the theorem is not adequate, and it becomes necessary to construct global integral manifolds. We shall follow Chevalley's elegant method of doing this. We restrict ourselves to the analytic case; the C^∞ versions of our theorems can be proved by means of analogous arguments.

Let M be an analytic manifold of dimension m. An assignment $\mathfrak{L}: x \mapsto \mathfrak{L}_x (x \in M)$ is called a *system of tangent spaces* (*of rank* p) if $\mathfrak{L}_x$ is a linear subspace of dimension p contained in $T_x(M)$ for all $x \in M$. The system $\mathfrak{L}$ is said to be *nontrivial* if $1 \leq p \leq m - 1$. *We shall consider only nontrivial systems in this section.* Given a system $\mathfrak{L}$ of tangent spaces of rank p, a vector field X is said to *belong to* $\mathfrak{L}$ on an open set U if $X_x \in \mathfrak{L}_x$ for all $x \in U$. $\mathfrak{L}$ is said to be an *analytic system* (*a.s.*) if for each $x \in M$ we can find an open set U containing x and p analytic vector fields ($p =$ rank $\mathfrak{L}$) $X_1, \ldots, X_p$ on U such that $(X_1)_y, \ldots, (X_p)_y$ span $\mathfrak{L}_y$ for all $y \in U$. $\mathfrak{L}$ is said to be an *involutive analytic system* (*i.a.s.*) if it has the additional property: let U be an open subset of M and let X, Y be two analytic vector fields which belong to $\mathfrak{L}$ on U; then $[X, Y]$ belongs to $\mathfrak{L}$ on U.

Given an a.s. $\mathfrak{L}$, an analytic submanifold S of M is said to be an *integral manifold* of $\mathfrak{L}$ if (*a*) S is connected, and (b) for each $y \in S$, $T_y(S) = \mathfrak{L}_y$. We do *not* require that S be a regular analytic submanifold, and so the topology of S could be strictly finer than the one induced from M. $\mathfrak{L}$ is said to be *integrable* if each point of M lies in some integral manifold of $\mathfrak{L}$.

An integrable a.s. $\mathfrak{L}$ is necessarily involutive. To prove this, we need only verify that if $x \in M$ and X and Y are analytic vector fields which belong to $\mathfrak{L}$ in some open neighborhood of x, then $[X, Y]_x \in \mathfrak{L}_x$. Now, there is an integral manifold S of $\mathfrak{L}$ through x. Replacing S by a sufficiently small open subset of it containing x, we may assume that S is a (connected) regular submanifold of M and that $S \subseteq U$, where U is open in M and where X and Y are defined on U and belong to $\mathfrak{L}$ on it. Then X' $(y \mapsto X_y)$ and Y' $(y \mapsto Y_y)$ $(y \in S)$ are analytic vector fields of S; if i is the identity map of S into

M, X', X and Y', Y are i-related. So $[X', Y']$ and $[X, Y]$ are i-related. This implies that $[X, Y]_x \in \mathfrak{L}_x$.

Let M (resp. N) be an analytic manifold and $\mathfrak{M}$ (resp. $\mathfrak{N}$) an a.s. on M (resp. N). $\mathfrak{M}$ and $\mathfrak{N}$ are called *isomorphic* if there is an analytic diffeomorphism $\pi\ (M \longrightarrow N)$ such that $(d\pi)_x(\mathfrak{M}_x) = \mathfrak{N}_{\pi(x)}$ for all $x \in M$. If $\mathfrak{L}(x \mapsto \mathfrak{L}_x)$ is an a.s. on M and $U \subseteq M$ an open set, $\mathfrak{L}$ induces on U an a.s. $\mathfrak{L} \mid U$, by restriction. Let $a > 0$ and let us consider the cube I_a^m in $\mathbf{R}^m$. Let $t_1, \ldots, t_m$ be the usual coordinates in $\mathbf{R}^m$. For any $x \in I_a^m$ let $\mathfrak{L}_x^{p,m,a}$ be the linear span of $(\partial/\partial t_1)_x$, $\ldots, (\partial/\partial t_p)_x$. Then $\mathfrak{L}^{p,m,a}: x \mapsto \mathfrak{L}_x^{p,m,a}$ is an i.a.s. If $a_{p+1}, \ldots, a_m$ are fixed numbers between $-a$ and $+a$, the submanifold

$$\{(t_1, \ldots, t_m) : t_{p+1} = a_{p+1}, \ldots, t_m = a_m\}$$

is an integral manifold of $\mathfrak{L}^{p,m,a}$. $\mathfrak{L}^{p,m,a}$ is called a *canonical* i.a.s. The classical Frobenius theorem asserts that, *locally*, every i.a.s. is isomorphic to a canonical one.

The proof of the local Frobenius theorem depends on the following two lemmas; the first lemma proves the theorem in question for the case $p = 1$.

Lemma 1.3.1. *Let M be an analytic manifold, X any real analytic vector field on M, and $x \in M$ a point such that $X_x \neq 0$. Then there are analytic coordinates $x_1, \ldots, x_m$ around x such that $X_y = (\partial/\partial x_1)_y$ for all y in an open neighborhood of x.*

Proof. Select analytic coordinates $z_1, \ldots, z_m$ around x such that $z_1(x) = \cdots = z_m(x) = 0$ and $X_x(z_1) \neq 0$. Then there are real analytic functions $G_1, \ldots, G_m$, defined on I_a^m (for some $a > 0$) such that $G_1(0, \ldots, 0) \neq 0$ and

$$X_y = \sum_{1 \leq i \leq m} G_i(z_1(y), \ldots, z_m(y)) \left(\frac{\partial}{\partial z_i}\right)_y$$

for all y in an open neighborhood of x. Consider the system of differential equations

$$\frac{du_i}{dt} = G_i(u_1(t), \ldots, u_m(t)) \quad (1 \leq i \leq m). \tag{1.3.1}$$

By the standard existence theorem (cf. Appendix, Theorem 1.4.1), we can select b with $0 < b < a$ and real analytic functions $u_1, \ldots, u_m$ on I_b^m such that

(a) $|u_j(t, y_2, \ldots, y_m)| < a$ for $1 \leq j \leq m$ and $(t, y_2, \ldots, y_m) \in I_b^m$
(b) for fixed $(y_2, \ldots, y_m) \in I_b^{m-1}$, the functions $u_1(\cdot, y_2, \ldots, y_m), \ldots, u_m(\cdot, y_2, \ldots, y_m)$ satisfy (1.3.1) on the open interval $(-b, b)$ with the initial conditions

$$u_1(0, y_2, \ldots, y_m) = 0, \quad u_2(0, y_2, \ldots, y_m) = y_2, \quad \ldots, \quad u_m(0, y_2, \ldots, y_m) = y_m.$$

Then the analytic map

$$\tau : (t,y_2,\ldots,y_m) \mapsto (u_1(t,y_2,\ldots,y_m),\ldots,u_m(t,y_2,\ldots,y_m))$$

has the nonvanishing Jacobian $G_1(0,\ldots,0)$ at $\mathbf{0}$, and $\tau(\mathbf{0}) = \mathbf{0}$. So it is an analytic diffeomorphism on an open set containing $\mathbf{0}$. Therefore, there exist functions $F_1,\ldots,F_m$, defined and analytic around $\mathbf{0}$, vanishing at $\mathbf{0}$, such that the map $(v_1,\ldots,v_m) \mapsto (F_1(v_1,\ldots,v_m),\ldots,F_m(v_1,\ldots,v_m))$ inverts τ around $\mathbf{0}$. Let $x_j = F_j(z_1,\ldots,z_m)$. Then $x_1,\ldots,x_m$ form a system of analytic coordinates around x. It is easy to verify that $X_y = (\partial/\partial x_1)_y$ for all y in some open set containing the point x.

Lemma 1.3.2. *Let M be an analytic manifold, $x \in M$, and let $X_1,\ldots,X_p$ be real analytic vector fields defined on an open set U containing x such that (i) $(X_1)_y,\ldots,(X_p)_y$ are linearly independent for $y \in U$, and (ii) $[X_j,X_k] = 0$, $1 \leq j, k \leq p$. Then we can choose coordinates $x_1,\ldots,x_m$ around x such that, in an open set around x,*

$$X_j = \frac{\partial}{\partial x_j} + \sum_{1 \leq s < j} a_{js} \frac{\partial}{\partial x_s} \quad (1 \leq j \leq p), \tag{1.3.2}$$

where the a_{js} are defined and analytic around x.

Proof. We prove this by induction on p. For $p = 1$ this follows at once from Lemma 1.3.1. Let $1 < p \leq m$, and assume the result for $X_1,\ldots,X_{p-1}$. Then we can choose a connected open set V with $x \in V \subseteq U$ and coordinates $u_1,\ldots,u_m$ on V such that

$$X_j = \frac{\partial}{\partial u_j} + \sum_{1 \leq s < j} b_{js} \frac{\partial}{\partial u_s} \quad (1 \leq j \leq p-1), \tag{1.3.3}$$

where the b_{js} are analytic on V. Write $X_p = \sum_{1 \leq s \leq m} g_s\, \partial/\partial u_s$, where the g_s are analytic on V, and put $X'_p = \sum_{p \leq s \leq m} g_s\, \partial/\partial u_s$. From (1.3.3) and the condition (i) of the lemma we conclude easily that $(X'_p)_y \neq 0$ for all $y \in V$. On the other hand, the conditions $[X_p,X_j] = 0$ yield the relations

$$\sum_{1 \leq s \leq m} g_s \left[\frac{\partial}{\partial u_s}, X_j\right] - \sum_{1 \leq s \leq m} (X_j g_s) \frac{\partial}{\partial u_s} = 0 \tag{1.3.4}$$

on V, for $1 \leq j \leq p-1$. Now (1.3.3) shows that, for $1 \leq s \leq m$ and $1 \leq j \leq p-1$, $[\partial/\partial u_s, X_j]$ is a linear combination of only the $\partial/\partial u_t$ with $1 \leq t \leq p-1$. Hence (1.3.4) implies that $X_j g_s = 0$ on V for $1 \leq j \leq p-1$, $p \leq s \leq m$. A simple argument based on (1.3.3) now shows that $\partial/\partial u_j g_s = 0$ on V for $1 \leq j \leq p-1$, $p \leq s \leq m$. Since V is connected, this implies that, for each s with $p \leq s \leq m$, g_s is a function of $u_p, u_{p+1}, \ldots, u_m$ only. An application of Lemma 1.3.1 now shows that we can replace $u_p, \ldots, u_m$ by analytic functions $v_p, \ldots, v_m$ with the following properties: (a) $u_1, \ldots, u_{p-1}, v_p, \ldots, v_m$ form a system of coordinates around x, and (b) $X'_p = \partial/\partial v_p$ around x. Let $x_j = u_j$ for $1 \leq j \leq p-1$ and $x_j = v_j$ for $p \leq j \leq m$. Then (1.3.2) is satisfied in the coordinate system $(x_1, \ldots, x_m)$.

Theorem 1.3.3. (*Local Frobenius Theorem*) *Let $\mathfrak{L}$ $(x \mapsto \mathfrak{L}_x)$ be an involutive nontrivial analytic system of tangent spaces of rank p on an analytic manifold M of dimension m. Then, for any $x \in M$, we can find an open set U containing x and an $a > 0$ such that $\mathfrak{L} | U$ is isomorphic to the canonical i.a.s. $\mathfrak{L}^{p,m,a}$. In particular, $\mathfrak{L}$ is integrable.*

Proof. The theorem is equivalent to the following: given $x \in M$ we can choose analytic coordinates $x_1, \ldots, x_m$ around x such that $\mathfrak{L}_y$ is spanned by $(\partial/\partial x_1)_y, \ldots, (\partial/\partial x_p)_y$ for all y in an open set containing x. Since the canonical involutive analytic systems $\mathfrak{L}^{p,m,a}$ are integrable, this would imply that $\mathfrak{L}$ is integrable. Fix $x \in M$. Let $z_1, \ldots, z_m$ be analytic coordinates around x and let $Z_1, \ldots, Z_p$ be analytic vector fields such that (i) $Z_1, \ldots, Z_p$ are defined on an open set U containing x and the $z_1, \ldots, z_m$ are coordinates on U, and (ii) $(Z_1)_y, \ldots, (Z_p)_y$ span $\mathfrak{L}_y$ for all $y \in U$. We may then write $Z_j = \sum_{1 \leq r \leq m} a'_{jr} \partial/\partial z_r$, where the a'_{jr} are analytic functions on U. Clearly, some $p \times p$ submatrix of $(a'_{jr})_{1 \leq j \leq p, 1 \leq r \leq m}$ is invertible at x. We may assume without losing generality that $(a'_{jr})_{1 \leq j, r \leq p}$ is invertible at x and that U is so small that this matrix is invertible on U. Let b_{ij} $(1 \leq i, j \leq p)$ be the entries of the inverse matrix. Then the b_{ij} are analytic functions on U. Let $X_j = \sum_{1 \leq s \leq p} b_{js} Z_s$. Then: (i) $(X_1)_y, \ldots, (X_p)_y$ span $\mathfrak{L}_y$ for all $y \in U$, and (ii) $X_j = \partial/\partial z_j + \sum_{p+1 \leq r \leq m} c_{jr} \partial/\partial z_r$, $1 \leq j \leq p$, the c_{jr} being analytic functions on U.

We now claim that $[X_j, X_k] = 0$, $1 \leq j, k \leq p$. Fix such j, k. Since $\mathfrak{L}$ is involutive, $[X_j, X_k]$ belongs to $\mathfrak{L}$ on U. Therefore $[X_j, X_k] = \sum_{1 \leq s \leq p} f_s X_s$, where the f_s are analytic functions on U; in particular, f_s is the coefficient of $\partial/\partial z_s$ in $[X_j, X_k]$ for $1 \leq s \leq p$. On the other hand, the formula (ii) above for the X_r shows that $[X_j, X_k]$ is a linear combination of only the $\partial/\partial z_r$ with $p+1 \leq r \leq m$. This implies that the f_s are all zero, i.e., that $[X_j, X_k] = 0$.

Now use Lemma 1.3.2 to choose analytic coordinates $x_1, \ldots, x_m$ around x such that, for $1 \leq j \leq p$, $X_j = \partial/\partial x_j + \sum_{1 \leq s < j} a_{js} \partial/\partial x_s$, the a_{js} being analytic around x. This representation shows that $(\partial/\partial x_1)_y, \ldots, (\partial/\partial x_p)_y$ span $\mathfrak{L}_y$ for all y in some open set containing x. This completes the proof of the theorem.

Let $U \subseteq M$ be an open set, $x_1, \ldots, x_m$ a system of coordinates on U, and $a > 0$. We say that $(U; x_1, \ldots, x_m; a)$ is *adapted to* $\mathfrak{L}$ if the map $u \mapsto (x_1(u), \ldots, x_m(u))$ is an analytic diffeomorphism of U with I_a^m and if $\mathfrak{L}_u$ is spanned by $(\partial/\partial x_1)_u, \ldots ,(\partial/\partial x_p)_u$ for all $u \in U$. In this case, for any $\mathbf{a} = (a_{p+1}, \ldots ,a_m) \in I_a^{m-p}$, we define $U(\mathbf{a})$ by

$$U(\mathbf{a}) = \{u : u \in U, x_{p+1}(u) = a_{p+1}, \ldots ,x_m(u) = a_m\}. \tag{1.3.5}$$

The $U(\mathbf{a})$ are regularly imbedded integral manifolds of $\mathfrak{L}$.

The local Frobenius theorem is not adequate for applications, since the integral manifolds have been constructed only locally. For full effectiveness it is necessary to obtain them in the large. This was done by Chevalley [1]; we shall follow his method of "piecing together" the local integral manifolds to obtain the global ones. However, this has to be done with some care, because the global manifolds are not always regularly imbedded.

It is easy to see that an arbitrary integral manifold of $\mathfrak{L}$ is a union of open subsets of the form $U(\mathbf{a})$. In fact, let $(U; x_1, \ldots ,x_m; a)$ be adapted to $\mathfrak{L}$ and let S be an integral manifold of $\mathfrak{L}$ with $S \cap U \neq \phi$; then $S \cap U$ is open in S. If $\bar{x}_j = x_j | S \cap U$, we have $d\bar{x}_j = 0$ $(p + 1 \leq j \leq m)$, so that these $\bar{x}_j$ are *locally constant* on $S \cap U$. In other words, each connected component of $S \cap U$ (in the topology induced by S) is contained in some $U(\mathbf{a})$. Since these components are open in S, it follows that $S \cap U(\mathbf{a})$ is open in S for any $\mathbf{a} \in I_a^{m-p}$. But then, for any such $\mathbf{a}$, the identity map of $S \cap U(\mathbf{a})$ into $U(\mathbf{a})$ is analytic with a bijective differential. This shows that $S \cap U(\mathbf{a})$ is open in S, as well as in $U(\mathbf{a})$; both S and $U(\mathbf{a})$ induce the same topology on it.

Lemma 1.3.4. *If S_1 and S_2 are any two integral manifolds of $\mathfrak{L}$, then $S_1 \cap S_2$ is open in S_1 as well as in S_2; both S_1 and S_2 induce the same topology on it. The integral manifolds of $\mathfrak{L}$ are all quasi-regularly imbedded in M.*

Proof. Let $u \in S_1 \cap S_2$. Select an open set U containing u, coordinates $x_1, \ldots ,x_m$ on U, and $a > 0$ such that $(U; x_1, \ldots ,x_m; a)$ is adapted to $\mathfrak{L}$. Let $\mathbf{a} \in I_a^{m-p}$ be such that $u \in U(\mathbf{a})$. It is then clear from what we said above that $S_1 \cap S_2 \cap U(\mathbf{a})$ is open in S_1 as well as in S_2, both of which induce the same topology on it. This leads at once to the first assertion. For the second, let S be any integral manifold of $\mathfrak{L}$, N any analytic manifold, and π an analytic map of N into M such that $\pi[N] \subseteq S$. We shall prove that π is an analytic map of N into the analytic manifold S. Fix $y \in N$ and let $u = \pi(y)$. Choose an open set U containing u, coordinates $x_1, \ldots ,x_m$ on U, and $a > 0$, such that $(U; x_1, \ldots ,x_m; a)$ is adapted to $\mathfrak{L}$. Let $\mathbf{a} \in I_a^{m-p}$ be such that $u \in U(\mathbf{a})$, and let T be the connected component of $S \cap U(\mathbf{a})$ containing u(in the topology of S). We claim that T is also the connected component of $S \cap U$ in the topology of U, which contains u. Indeed, if T' is the component in question, then obviously $T \subseteq T'$. On the other hand, since S is second countable, $S \cap U$ has at most countably many connected components (in the topology of S), so that $S \cap U \subseteq \bigcup_{\mathbf{b} \in F} U(\mathbf{b})$ for some countable set $F \subseteq I_a^{m-p}$. But then the map $u \mapsto (x_{p+1}(u), \ldots, x_m(u))$, which is continuous on $S \cap U$ in the topology induced from U, takes at most countably many values. Therefore, it must be a constant on each connected component of $S \cap U$ in the topology of U, in particular on T'. Thus $T' \subseteq S \cap U(\mathbf{a})$; and, since we have already proved that both S and $U(\mathbf{a})$ induce the same topology on $S \cap U(\mathbf{a})$, we must have $T = T'$. This proves our claim. If W is the connected component of $\pi^{-1}(U)$ containing y, it is then clear that W is open and $\pi[W] \subseteq T$. Since T is open in the regularly imbedded $U(\mathbf{a})$, π is an analytic map of W into T. Hence π is an analytic map of W into S; this leads to the second assertion.

Lemma 1.3.5. *Let A be a connected Hausdorff space which is locally connected. Suppose $A = \bigcup_{n=1}^{\infty} A_n$ where each A_n is open in A and each connected component of A_n is second countable for each n. Then A is itself second countable.*

Proof. Let $\mathfrak{C}_n$ be the class of (open) sets which are connected components of A_n, and $\mathfrak{C} = \bigcup_{n=1}^{\infty} \mathfrak{C}_n$. Since there cannot exist an uncountable family of mutually disjoint nonempty open sets in a second countable space it follows that, given $F \in \mathfrak{C}$, there are only countably many $F' \in \mathfrak{C}$ such that $F \cap F' \neq \phi$. We now define the families $J_0, J_1, \ldots$ of open subsets of A as follows. We select $E \in \mathfrak{C}$ arbitrarily and define $J_0 = \{E\}$; for $s \geq 1$, $J_s = \{F\colon F \in \mathfrak{C}, F \cap F' \neq \phi \text{ for some } F' \in J_{s-1}\}$. The J_s $(s \geq 0)$ are well defined inductively. A simple induction on s shows that they are all countable. Let $B = \bigcup_{s=0}^{\infty} \bigcup_{F \in J_s} F$. Then B is open and second countable. If $v \in Cl(B)$, we can find $F \in \mathfrak{C}$ such that $v \in F$; and as $F \cap B \neq \phi$, there is an $s \geq 0$ and an $F' \in J_s$ such that $F \cap F' \neq \phi$. This shows that $F \in J_{s+1}$ and hence that $v \in B$. B is thus open and closed. Since A is connected, $A = B$. A is thus second countable.

Theorem 1.3.6. (*Global Frobenius Theorem*) *Let M be an analytic manifold $\mathfrak{L}$ $(x \mapsto \mathfrak{L}_x)$ an involutive analytic system of tangent spaces of rank p. Given any point of M, there is one and exactly one maximal integral manifold of $\mathfrak{L}$ containing that point. Any (nonempty) integral manifold of $\mathfrak{L}$ is quasi-regularly imbedded in M and is an open submanifold of precisely one maximal integral manifold of $\mathfrak{L}$.*

Proof. Let $\mathfrak{J}$ be the collection of all subsets of M which are unions of integral manifolds of $\mathfrak{L}$. It follows from Lemma 1.3.4 that $\mathfrak{J}$ is a topology for M finer than its original topology. It is clear that $(M,\mathfrak{J})$ is a Hausdorff locally connected space. Let $\{M_\xi : \xi \in J\}$ be the set of connected components of $(M,\mathfrak{J})$. Each M_ξ is an open subspace of $(M,\mathfrak{J})$ and if S is any integral manifold of $\mathfrak{L}$, the underlying topological space of S is an open subspace of exactly one M_ξ.

We now prove that the M_ξ are second countable. Fix $\xi \in J$. Let U be an open set with coordinates $x_1, \ldots, x_m$ and let $a > 0$ be such that $(U; x_1, \ldots, x_m; a)$ is adapted to $\mathfrak{L}$. Since M_ξ as well as the $U(\mathbf{a})$ are open in $(M,\mathfrak{J})$, it follows that $M_\xi \cap U(\mathbf{a})$ is an open subspace of $U(\mathbf{a})$ for all $a \in I_a^{m-p}$. Now, $M_\xi \cap U$ is open in M_ξ and is the disjoint union of the $M_\xi \cap U(\mathbf{a})$, so each connected component of $M_\xi \cap U$ is an open subspace of some $U(\mathbf{a})$ and is therefore second countable. Since M (and hence M_ξ) can be covered by countably many open sets such as U, Lemma 1.3.5 can be used to conclude that M_ξ is second countable.

It is now obvious that there is a unique analytic structure on M_ξ such that each integral manifold of $\mathfrak{L}$ contained in M_ξ is an open submanifold of M_ξ. With this structure, M_ξ becomes a submanifold of M. It is also obvious that each M_ξ is a maximal integral manifold of $\mathfrak{L}$. Theorem 1.3.6 is completely proved.

It may be remarked that Theorems 1.3.3 and 1.3.6 are valid in the complex analytic case also. No change is necessary either in the formulations or in the proofs.

1.4. Appendix

In this appendix we discuss briefly some elementary results on analytic systems of ordinary differential equations. We work in $\mathbf{R}^m$ or $\mathbf{C}^m$. For any $a > 0$, let

$$I_a^m = \{(t_1, \ldots, t_m) : t_j \in \mathbf{R}, |t_j| < a \quad \text{for} \quad 1 \leq j \leq m\},$$
$$J_a^m = \{(z_1, \ldots, z_m\} : z_j \in \mathbf{C}, |z_j| < a \text{ for } \quad 1 \leq j \leq m\}.$$

Let $a > 0$ and let $G_1, \ldots, G_m$ be m real functions defined and analytic on I_a^m. We consider the system of ordinary differential equations:

$$\frac{du_j}{dt} = G_j(u_1(t), \ldots, u_m(t)) \quad (1 \leq j \leq m). \tag{1.4.1}$$

If the G_j are defined and holomorphic on J_a^m, we consider the system

$$\frac{du_j}{dz} = G_j(u_1(z), \ldots, u_m(z)) \quad (1 \leq j \leq m). \tag{1.4.2}$$

Theorem 1.4.1. *Let $a > 0$ and let $G_1, \ldots, G_m$ be real functions defined and analytic on I_a^m. Then*

(a) if u_j, v_j $(1 \leq j \leq m)$ are analytic functions defined on an open interval Δ containing 0 such that $(u_1, \ldots, u_m)$ and $(v_1, \ldots, v_m)$ are both solutions of (1.4.1) on Δ with $u_j(0) = v_j(0)$ $(1 \leq j \leq m)$, then $u_j = v_j$ on Δ for $1 \leq j \leq m$.

(*b*) *there exists* b *with* $0 < b < a$ *and real analytic functions* u_j *on* I_b^{m+1} $(1 \le j \le m)$ *such that*

(i) $|u_j(t, y_1, \ldots, y_m)| < a$ *for* $(t, y_1, \ldots, y_m) \in I_b^{m+1}$

(ii) $$\frac{\partial u_j(t, y_1, \ldots, y_m)}{\partial t} = G_j(u_1(t, y_1, \ldots, y_m), \ldots, u_m(t, y_1, \ldots, y_m))$$

$$u_j(0, y_1, \ldots, y_m) = y_j$$

for $(t, y_1, \ldots, y_m) \in I_b^{m+1}$, $1 \le j \le m$.

Proof. (a) If $(\varphi_1, \ldots, \varphi_m)$ is solution of (1.4.1), we have, for $1 \le j \le m$,

$$\varphi_j'(0) = G_j(\varphi_1(0), \ldots, \varphi_m(0))$$
$$\varphi_j''(0) = \sum_{1 \le r \le m} \frac{\partial G_j}{\partial t_r}(\varphi_1(0), \ldots, \varphi_m(0))\varphi_r'(0),$$

and so on. A simple induction on $s \ge 1$ shows that the initial vector $(\varphi_1(0), \ldots, \varphi_m(0))$ completely determines the values of all the derivatives $\varphi_j^{(s)}(0)$ $(s \ge 1, 1 \le j \le m)$. So if the φ_j are analytic on an open interval Δ containing 0, they are completely determined on Δ by the initial vector $(\varphi_1(0), \ldots, \varphi_m(0))$. (a) follows at once from this.

(b) Replacing a by a smaller positive number, we may assume that the power series expansions of the G_j around the origin converge absolutely and uniformly in I_a^m. Hence the G_j are restrictions to I_a^m of holomorphic functions on J_a^m. We also denote the latter by G_j. Let $0 < c < a$, and

$$\gamma = \max_{1 \le j \le m} \sup_{(z_1, \ldots, z_m) \in J_c^m} |G_j(z_1, \ldots, z_m)|.$$

Then γ is finite. Choose a constant $L \ge 1$ such that

$$\max_{1 \le j \le m} |G_j(z_1, \ldots, z_m) - G_j(z_1', \ldots, z_m')| \le L \max_{1 \le j \le m} |z_j - z_j'| \tag{1.4.3}$$

for all $(z_1, \ldots, z_m), (z_1', \ldots, z_m') \in J_c^m$. Finally, select b with $0 < b < c(1 + \gamma)^{-1}$ and $Lb < 1$.

Now define a sequence

$$(u_{1,N}, \ldots, u_{m,N}) \quad (N = 0, 1, 2, \ldots)$$

of vector-valued functions as follows. Put

$$u_{j,0} = 0 \quad (1 \le j \le m); \tag{1.4.4}$$

for $N \geq 1$ and $(z, z_1, \ldots, z_m) \in J_b^{m+1}$, put

$$u_{j,N}(z,z_1,\ldots,z_m) = z_j + \int_0^z G_j(u_{1,N-1}(z',z_1,\ldots,z_m),\ldots,u_{m,N-1}(z',z_1,\ldots,z_m))dz', \tag{1.4.5}$$

where the integral is taken along the line segment from 0 to z. We claim that for any $N \geq 0$, the $u_{j,N}(1 \leq j \leq m)$ are well defined and holomorphic on J_b^{m+1}, and that

$$u_{j,N}(z, z_1, \ldots, z_m)| < c$$

for $1 \leq j \leq m$ and $(z,z_1,\ldots,z_m) \in J_b^{m+1}$. We prove this claim by induction on N. For $N = 0$ there is nothing to prove. Let $N \geq 1$ and assume the result for $N - 1$. It is clear from (1.4.5) that $u_{j,N}$ is well defined and holomorphic on J_b^{m+1}. Further, if $(z,z_1,\ldots,z_m) \in J_b^{m+1}$, we have for $1 \leq j \leq m$

$$\begin{aligned} |u_{j,N}(z,z_1,\ldots,z_m)| &\leq b + \gamma\left|\int_0^z dz'\right| \\ &\leq b(1+\gamma) \\ &< c, \end{aligned}$$

carrying forward the induction. Our claim is thus proved.

Now for $N \geq 1$ and $(z,z_1,\ldots,z_m) \in J_b^{m+1}$

$$\begin{aligned} &|u_{j,N+1}(z,z_1,\ldots,z_m) - u_{j,N}(z,z_1,\ldots,z_m)| \\ &\quad \leq Lb \max_{1\leq j\leq m} \sup_{(z,z_1,\ldots,z_m)\in J_b^{m+1}} |u_{j,N}(z,z_1,\ldots,z_m) - u_{j,N-1}(z,z_1,\ldots,z_m)|, \end{aligned}$$

from (1.4.5) and (1.4.3). Applying this estimate in succession and noting that $|u_{j,1}(z,z_1,\ldots,z_m)| < c$ for $1 \leq j \leq m$ and that $(z,z_1,\ldots,z_m) \in J_b^{m+1}$, we get

$$\max_{1\leq j\leq m} \sup_{(z,z_1,\ldots,z_m)\in J_b^{m+1}} |u_{j,N+1}(z,z_1,\ldots,z_m) - u_{j,N}(z,z_1,\ldots,z_m)| \leq c(Lb)^N.$$

Since $Lb < 1$, it follows that the series

$$\sum_{N\geq 0} \{u_{j,N+1}(z,z_1,\ldots,z_m) - u_{j,N}(z,z_1,\ldots,z_m)\}$$

converges uniformly in J_b^{m+1} for $1 \leq j \leq m$. Let $u_j(z,z_1,\ldots,z_m)$ be the sum. Then u_j is holomorphic on J_b^{m+1} and

$$u_j(z,z_1,\ldots,z_m) = \lim_{N\to\infty} u_{j,N}(z,z_1,\ldots,z_m) \quad (1 \leq j \leq m) \tag{1.4.6}$$

for $(z,z_1,\ldots,z_m) \in J_b^{m+1}$. (1.4.6) and (1.4.5) now yield

$$u_j(z,z_1,\ldots,z_m) \tag{1.4.7}$$
$$= z_j + \int_0^z G_j(u_1(z',z_1,\ldots,z_m),\ldots,u_m(z',z_1,\ldots,z_m))\,dz'$$

for $1 \leq j \leq m$ and $(z,z_1,\ldots,z_m) \in J_b^{m+1}$. Restricting to I_b^{m+1} and differentiating (1.4.7) with respect to z, we get

$$\frac{\partial u_j(t,y_1,\ldots,y_m)}{\partial t} = G_j(u_1(t,y_1,\ldots,y_m),\ldots,u_m(t,y_1,\ldots,y_m))$$
$$u_j(0,y_1,\ldots,y_m) = y_j$$

for $1 \leq j \leq m$ and $(t,y_1,\ldots,y_m) \in I_b^{m+1}$. The u_j being analytic on I_b^{m+1}, the theorem is proved.

The holomorphic version of Theorem 1.4.1 with the differential equations (1.4.2) instead of (1.4.1) is proved as above with minor variations. We leave its formulation and proof to the reader.

In applications it often happens that the G_j depend analytically on certain parameters. In this case, the solutions u_j also have the same analytic dependence on these parameters.

Theorem 1.4.2. *Let N be an analytic manifold, $a > 0$, and let the real functions G_j be defined and analytic on $I_a^m \times N$. Fix $x \in N$. Then we can find b with $0 < b < a$, an open subset N_x of N containing x, and real analytic functions $u_1,\ldots,u_m$ on $I_b^{m+1} \times N_x$ such that*

$$\frac{\partial u_j(t,y_1,\ldots,y_m,x')}{\partial t} = G_j(u_1(t,y_1,\ldots,y_m,x'),\ldots,u_m(t,y_1,\ldots,y_m,x'))$$
$$u_j(0,y_1,\ldots,y_m,x') = y_j$$

for $1 \leq j \leq m$, $(t,y_1,\ldots,y_m) \in I_b^{m+1}$ and $x' \in N_x$.

Proof. We may assume that for some $d > 0$, $N = I_d^n$, $x = (0,\ldots,0)$, and that the G_j are the restrictions to $I_a^m \times I_d^n$ of functions (denoted again by G_j) defined and holomorphic on $J_a^m \times J_d^n$. let $0 < c < a, 0 < e < d$, and let $N' = J_e^n$. Define γ by

$$\gamma = \max_{1 \leq j \leq m} \sup_{(z_1,\ldots,z_m) \in J_c{}^m,\, x' \in N'} |G_j(z_1,\ldots,z_m,x')|$$

and let $L \geq 1$ be a constant such that

$$|G_j(z_1,\ldots,z_m,x') - G_j(z'_1,\ldots,z'_m,x')| \leq L \max_{1 \leq j \leq m} |z_j - z'_j|$$

for all $x' \in N'$ $(z_1, \ldots, z_m)$, $(z'_1, \ldots, z'_m) \in J^m_c$. Choose b such that $0 < b < c(1+\gamma)^{-1}$ and $Lb < 1$; we then define the sequence $u_{j,N}$ as follows. For $N = 0$, put $u_{j,0} = 0$ $(1 \leq j \leq m)$; for $N \geq 1$ write

$$\begin{aligned} u_{j,N}(z,z_1, \ldots, z_m,x') \\ = z_j + \int_0^z G_j(u_1(z', z_1, \ldots, z_m, x'), \ldots, u_m(z', z_1, \ldots, z_m, x'), x')\, dz' \end{aligned}$$

for $1 \leq j \leq m$, $(z,z_1, \ldots, z_m) \in J^{m+1}_b$, $x' \in N'$. Theorem 1.4.2 is now proved by arguing exactly as in the previous theorem. We leave the details to the reader.

The same proof also gives the holomorphic version of the above result.

EXERCISES

1. Consider $\mathbf{C}^2$ as a four-dimensional real analytic manifold, and let $\mathbf{T}^2 = \{(z_1,z_2): z_1,z_2 \in \mathbf{C}, |z_1| = |z_2| = 1\}$; show that $\mathbf{T}^2$ is a regularly imbedded compact submanifold. Prove that if $\alpha \in \mathbf{R}$ is irrational, the map $t \mapsto (e^{it},e^{i\alpha t})$ $(t \in \mathbf{R})$ is an imbedding of $\mathbf{R}$ into $\mathbf{T}^2$ which is quasi-regular but not regular.

2. Let $n \geq 2$ and let π be the map of $\mathbf{R}^n$ into $\mathbf{R}^1$ given by

$$\pi(x_1, \ldots, x_n) = x_1^2 + \cdots + x_n^2.$$

Let $M = \mathbf{R}^n \setminus \{0\}$, $N = \{t: t \in R, t > 0\}$. Let $D = \partial^2/\partial x_1^2 + \cdots + \partial^2/\partial x_n^2$. Prove that there is a unique differential operator $\tilde{D}$ on N such that D and $\tilde{D}$ are π-related. Calculate $\tilde{D}$.

3. (a) Let F be a field of characteristic 0; V (resp. W) a vector space over F of finite dimension m (resp. n); and γ a linear map of V onto W with kernel U. Let λ (resp. μ) be a nonzero element of $\Lambda_m(V)$ (resp. $\Lambda_n(W)$). Prove that there is exactly one $\nu \in \Lambda_{m-n}(U)$ with the following property: let $u_1, \ldots, u_{m-n}, v_1, \ldots, v_n$ be a basis for V such that $u_1, \ldots, u_{m-n}$ span U; then

$$\nu(u_1, \ldots, u_{m-n}) = \frac{\lambda(u_1, \ldots, u_{m-n}, v_1, \ldots, v_n)}{\mu(\gamma v_1, \ldots, \gamma v_n)}$$

We write $\nu = (\lambda/\mu)_\gamma$.

(b) Let M and N be analytic manifolds of dimensions m and n respectively. Let $\omega^1 \in \mathfrak{a}_m(M)$ and $\omega^2 \in \mathfrak{a}_n(N)$, and suppose that ω^1 and ω^2 vanish nowhere. Let π be a submersion of M onto N, and for each $y \in N$ let $P_y = \pi^{-1}(\{y\})$. Prove that the P_y are closed regular submanifolds of M. For $y \in N$ and $x \in P_y$ let $\omega^y_x = (\omega^1_x/\omega^2_x)_{(d\pi)_x}$. Prove that $\omega^y : x \mapsto \omega^y_x$ is an element of $\mathfrak{a}_{m-n}(P_y)$ for each $y \in N$ and that $y \mapsto \omega^y$ is analytic in a natural sense.

(c) Prove that if $f \in C_c^\infty(M)$, then

$$\int_M f\omega^1 = \int_N \left(\int_{P_y} (f \mid P_y)\omega^y\right) \omega^2.$$

4. Let M be an analytic manifold and $\mathfrak{L}$ $(x \mapsto \mathfrak{L}_x)$ an analytic system of tangent spaces of rank p. A given 1-form ω is said to belong to $\mathfrak{L}$ on an open set U if $\omega_x \mid \mathfrak{L}_x = 0$ for all $x \in U$. Prove that $\mathfrak{L}$ is involutive if and only if the following condition is satisfied: for any $x \in M$ and any analytic 1-form ω which belongs to $\mathfrak{L}$ in an open neighborhood of x, we can find an open set U containing x and analytic 1-forms $\alpha_1, \ldots, \alpha_q, \omega_1, \ldots, \omega_q$ on U such that $\omega_1, \ldots, \omega_q$ belong to $\mathfrak{L}$ on U, and $d\omega = \sum_{1 \le i \le q} \alpha_i \wedge \omega_i$ on U.

5. Let $\pi : M \longrightarrow N$ be an analytic map. Assume that π is a submersion and that $\pi[M] = N$. Let C_π be the set of all elements of $C^\infty(M)$ of the form $g \circ \pi$ where $g \in C^\infty(N)$. Let D be a differential operator on M. Prove that there exists a differential operator D' on N such that D' is π-related to D if and only if D maps C_π into itself, and that in this case D' uniquely determined by D.

6. (a) Let $t_1, \ldots, t_m$ be the usual coordinates in $\mathbf{R}^m$, $\mathbf{t} = (t_1, \ldots, t_m)$, $r^2 = t_1^2 + \cdots + t_m^2$. Let φ be the function defined by

$$\varphi(\mathbf{t}) = \begin{cases} Ce^{-1/1-r^2} & r^2 < 1 \\ 0 & r^2 \ge 1, \end{cases}$$

where $C > 0$ is a constant such that $\int_{\mathbf{R}^m} \varphi \, dt_1 \cdots dt_m = 1$. Put $\varphi_\epsilon(\mathbf{t}) = \varphi(\epsilon^{-1}\mathbf{t})(\epsilon > 0, \mathbf{t} \in \mathbf{R}^m)$. Prove that if $0 \le k \le \infty$, then for any $f \in C_c^k(\mathbf{R}^m)$, and any $(\beta) = (\beta_1, \ldots, \beta_m)$ with $|\beta| \le k$;

$$\partial^{(\beta)}(f * \varphi_\epsilon) \longrightarrow \partial^{(\beta)} f \qquad (\epsilon \longrightarrow 0+),$$

the convergence being uniform over $\mathbf{R}^m$ (here $*$ denotes convolution and $\partial^{(\beta)} = (\partial/\partial t_1)^{\beta_1} \cdots (\partial/\partial t_m)^{\beta_m}$).

(b) Use a partition of unity argument to deduce from (a) the following result. Let M be a C^∞ manifold, K a compact subset of M and U an open set with $K \subseteq U$. Then, given any $f \in C_c^k(M)$ with $\operatorname{supp} f \subseteq K$, we can find a sequence $\{f_n\}_{n \ge 1}$ of elements of $C_c^\infty(M)$ such that (i) $\operatorname{supp} f_n \subseteq U$ for all $n \ge 1$, and (ii) if D is any differential operator on M of order $\le k$, $Df_n \longrightarrow Df$ uniformly over M.

7. (a) Let V be a finite-dimensional vector space over $\mathbf{R}$, V_c its complexification. We assume that $V \subseteq V_c$. We regard V as an analytic manifold in the usual manner. Let S be the symmetric algebra over V_c. For any $u \in V$ let $\partial(u)$ be the endomorphism of $C^\infty(V)$ given by

$$(\partial(u)f)(x) = \left[\frac{d}{dt} f(x + tu)\right]_{t=0} \qquad (x \in V, f \in C^\infty(V)).$$

Prove that $\partial(u)$ is a vector field and that $u \mapsto \partial(u)$ $(u \in V)$ extends uniquely to an isomorphism (denoted again by ∂) of S onto the subalgebra of Diff (V) consisting of all differential operators which are invariant under all translations of V. If $u_1, \ldots, u_s \in V$, prove that for any $f \in C^\infty(V)$,

$$f(x; \partial(u_1 \cdots u_s)) = \left[\frac{\partial^s}{\partial t_1 \cdots \partial t_s} f(x + t_1 u_1 + \cdots + t_s u_s)\right]_{t_1 = \cdots = t_s = 0} \quad (x \in V).$$

(b) Let $\{u_1, \ldots, u_m\}$ be a basis for V and let dV be the m-form on V such that

$$dV(\partial(u_1), \ldots, \partial(u_m)) = 1.$$

Prove that dV is invariant under all translations and that the corresponding measure is a Lebesgue measure on V.

(c) Prove that there is a unique automorphism $* : a \mapsto a^*$ of S such that $u^* = -u$ for all $u \in V$; and that $\partial(a^*)$ is the formal adjoint of $\partial(a)$ relative to dV for any $a \in S$.

The next exercise examines the geometric significance of the condition for an a.s. to be involutive.

8. (a) Let M be a compact analytic manifold, X an analytic vector field on M. Prove that there is a unique family $\{\xi_t^X : t \in \mathbf{R}\}$ of analytic diffeomorphisms of M such that

(i) $\xi_0^X =$ identity, $\xi_{t+t'}^X = \xi_t^X \xi_{t'}^X (t, t' \in \mathbf{R})$

(ii) $t, x \mapsto \xi_t^X(x)$ is an analytic map of $\mathbf{R} \times M$ into M

(iii) $\left(\frac{d}{dt}\xi_t^X(x)\right)_{t=0} = X_x \quad (x \in M)$.

(b) Let M be as in (a), $\mathfrak{L} : x \mapsto \mathfrak{L}_x$ an i.a.s. on M. Let X be an analytic vector field on M belonging to $\mathfrak{L}$. Prove that the ξ_t^X leave $\mathfrak{L}$ invariant.

(c) Obtain, for noncompact M, local versions of (a) and (b) and deduce a geometric criterion for an a.s. to be involutive.

Exercises 9–11 discuss exterior algebras.

9. Let m be an integer ≥ 1, and F a field of characteristic 0. Let C be a vector space of dimension 2^m over F, and $\{e_A\}$, a basis of C indexed by the collection of all subsets of $\{1, \ldots, m\}$. Write $e_\phi = 1_C$ and $e_A = e_{i_1, \ldots, i_p}$ if $A = \{i_1, \ldots, i_p\}$ with $1 \leq i_1 < \cdots < i_p \leq m$. Prove the existence of a unique bilinear map $u,v \mapsto u \wedge v$ of $C \times C$ into C such that (i) C becomes an associative algebra over F with 1_C as unit, (ii) $e_A \wedge e_B = 0$ if $A \cap B \neq \phi$, and (iii) if $A \cap B = \phi$, $A = \{i_1, \ldots, i_p\}$, $B = \{j_1, \ldots, j_q\}$, and $A \cup B = \{s_1, \ldots, s_r\}$, with $i_1 < \cdots < i_p$, $j_1 < \cdots < j_q$, and $s_1 < \cdots < s_r$, then $e_A \wedge e_B = \epsilon e_{A \cup B}$, where $\epsilon = +1$ or -1 according as the rearrangement $\{s_1, \ldots, s_r\} \longrightarrow \{i_1, \ldots, i_p, j_1, \ldots, j_q\}$ is induced by an even or odd permutation. Prove also that $e_{i_1, \ldots, i_p} = e_{i_1} \wedge \cdots \wedge e_{i_p}$ $(i_1 < \cdots < i_p)$ and deduce that 1_C and the e_i generate C.

10. Let V be a vector space of dimension m over F; $\mathfrak{J}$, the tensor algebra over V; $\mathfrak{J}_0 = F \cdot 1$; and for $r \geq 1$, $\mathfrak{J}_r$, the subspace of $\mathfrak{J}$ spanned by all elements of the form $x_1 \otimes \cdots \otimes x_r$ ($x_i \in V$). Let $\Pi_0 = \{1\}$ and, for $r \geq 1$, let Π_r be the group of all permutations of $\{1, \ldots, r\}$. Let $\mathcal{I} = \sum_{x, x' \in V} \mathfrak{J} \otimes (x \otimes x' + x' \otimes x) \otimes \mathfrak{J}$.

(a) Prove that $\mathcal{I}$ is a proper two-sided ideal of $\mathfrak{J}$. (To check $\mathcal{I} \neq \mathfrak{J}$, note that $\mathcal{I} \subseteq \sum_{r \geq 2} \mathfrak{J}_r$.)

(b) Let E be the quotient algebra $\mathfrak{J}/\mathcal{I}$; let $\wedge$ be the product operation in E; and let $\pi : a \mapsto \bar{a}$ be the natural map of $\mathfrak{J}$ onto E. Prove the following universal property of (E, V): if A is any associative algebra (with unit) over F and λ is a linear map of V into A such that $\lambda(x)^2 = 0$ for all $x \in V$, then there is a unique homomorphism $\bar{\lambda}$ of E into A such that $\lambda(x) = \bar{\lambda}(\bar{x})$ ($x \in V$).

(c) Let $\{x_1, \ldots, x_m\}$ be a basis for V. Prove that the elements $\bar{x}_{i_1} \wedge \bar{x}_{i_2} \wedge \cdots \wedge \bar{x}_{i_m}$ ($1 \leq i_1 < \cdots < i_p \leq m$) and $\bar{1}$ form a basis for E. Deduce that $\dim(E) = 2^m$. (Use the relations $\bar{x}_i \wedge \bar{x}_j + \bar{x}_j \wedge \bar{x}_i = 0$ to prove that these elements span E; thus $\dim(E) \leq 2^m$. If C is as in Exercise 9 and $\lambda : V \longrightarrow C$ the linear map with $\lambda(x_i) = e_i$ ($1 \leq i \leq m$), then by (b), $\bar{\lambda}(E) = C$ so that $\dim(E) \geq 2^m$.)

(d) For any $\sigma \in \Pi_r$, let $\sigma : t \mapsto \sigma(t)$ be the linear automorphism of $\mathfrak{J}_r$ such that $\sigma(v_1 \otimes \cdots \otimes v_r) = v_{\sigma(1)} \otimes \cdots \otimes v_{\sigma(r)}$ ($v_i \in V$); let $\epsilon(\sigma) = \pm 1$ according as σ is even or odd. Let $\mathfrak{A}_r$ be the subspace of all $t \in \mathfrak{J}_r$ such that $\sigma t = \epsilon(\sigma) t$ for all $\sigma \in \Pi_r$. If $P_r = (1/r!) \sum_{\sigma \in \Pi_r} \epsilon(\sigma)\sigma$, prove that P_r is a projection of $\mathfrak{J}_r$ onto $\mathfrak{A}_r$. Deduce that $\mathfrak{A}_r = 0$ if $r > m$ and $\dim(\mathfrak{A}_r) = \binom{m}{r}$ ($0 \leq r \leq m$)

(e) Let $\mathfrak{A} = \sum_r \mathfrak{A}_r$. Prove that $\mathfrak{J}$ is the direct sum of $\mathcal{I}$ and $\mathfrak{A}$. (If $v_1, \ldots, v_r \in V$ and σ is the interchange of an adjacent pair, $v_1 \otimes \cdots \otimes v_r \equiv \epsilon(\sigma)\sigma(v_1 \otimes \cdots \otimes v_r) \bmod \mathcal{I}$; so $t \equiv P_r(t) \bmod \mathcal{I}$ for all $t \in \mathfrak{J}_r$. Now use a dimension argument.)

(f) For $t, t' \in \mathfrak{A}$, let $t \wedge t' \in \mathfrak{A}$ be the unique element such that $\pi(t \wedge t') = \pi(t) \wedge \pi(t')$. Prove that if $t \in \mathfrak{A}_r$ and $t' \in \mathfrak{A}_{r'}$, $t \wedge t' = P_{r+r'}(t \otimes t')$.

$E = E(V)$ is the exterior algebra over V. It is usual to identify V with $\pi(V)$ so that $V \subseteq E(V)$.

11. Let V, F, m be as in Exercise 10. For $r \geq 1$ let M_r be the vector space of all r-linear maps of $V \times \cdots \times V$ (r factors) into k ($M_1 = V^*$). Let M be the direct sum of $M_0 = k \cdot 1$ and the M_r ($r \geq 1$). For $\varphi \in M_r$ and $\varphi' \in M_{r'}$, let $\varphi \otimes \varphi' \in M_{r+r'}$ be defined by $\varphi \otimes \varphi' : x_1, \ldots, x_r, \ldots, x_{r+r'} \mapsto \varphi(x_1, \ldots, x_r) \varphi'(x_{r+1}, \ldots, x_{r+r'})$. Let Π_r be as in Exercise 10. For $\sigma \in \Pi_r$ and $\varphi \in M_r$, $\sigma\varphi \in M_r$ is defined by $\sigma\varphi(x_1, \ldots, x_r) = \varphi(x_{\sigma(1)}, \ldots, x_{\sigma(r)})$. Let $\Lambda_0 = k \cdot 1$ and $\Lambda_r = \{\varphi : \varphi \in M_r, \sigma\varphi = \epsilon(\sigma)\varphi$ for all $\sigma \in \Pi_r\}$ ($r \geq 1$). Put $\Lambda = \sum_{r \geq 0} \Lambda_r$. Use Exercise 10 and the canonical identification of M with the tensor algebra over V^* to get the following results:

(a) $\Lambda_r = 0$ if $r > m$, $\dim(\Lambda_r) = \binom{m}{r}$ if $0 \leq r \leq m$, and $P_r = (1/r!) \sum_{\sigma \in \Pi_r} \epsilon(\sigma)\sigma$ is a projection of M_r onto Λ_r.

(b) For $t \in \Lambda_r$, $t' \in \Lambda_{r'}$, let $t \wedge t' = P_{r+r'}(t \otimes t')$. Extend $\wedge$ bilinearly to $\Lambda \times \Lambda$. Prove that Λ becomes an associative algebra over F with 1 as unit.

(c) Let $\{\varphi_1,\ldots,\varphi_m\}$ be a basis for V^*. Prove that 1 and $\varphi_{i_1}\wedge\cdots\wedge\varphi_{i_p}$ $(1\leq i_1<\cdots<i_p\leq m)$ form a basis for Λ. Deduce the existence of a unique algebra isomorphism of Λ onto the exterior algebra $E(V^*)$ that is the identity on V^*.

(d) Let $\varphi\in\Lambda_r,\varphi'\in\Lambda_{r'}$. Prove that $\varphi\wedge\varphi'=(-1)^{rr'}\varphi'\wedge\varphi$. Deduce that $\sum_{r\text{ even}}\Lambda_r$ is the center of Λ.

(e) Let ϵ be the endomorphism of Λ such that $\epsilon=(-1)^r$ on Λ_r. Prove that ϵ is an involutive automorphism of Λ. Let L be a linear map of Λ_1 into Λ_r for some $r\geq 1$. Prove that there is a unique derivation (resp. antiderivation) D_L of Λ extending L if r is odd (resp. even). Here, an endomorphism D of Λ is called a derivation (resp. antiderivation) if $D(\omega\wedge\omega')=D\omega\wedge\omega'+\omega\wedge D\omega'$ (resp. $D(\omega\wedge\omega')=D\omega\wedge\omega'+\epsilon(\omega)\omega\wedge D\omega'$) for all $\omega,\omega'\in\Lambda$.

12. M is a C^∞ manifold of dimension m; $\mathfrak{a}(M)$ is as in §1.1.

(a) Let d be an endomorphism of $\mathfrak{a}(M)$ satisfying (1.1.15). Prove that d is local in the following sense: if $\omega\in\mathfrak{a}(M)$ and $\omega|U=0$ on some open set $U\subseteq M$, then $d\omega|U=0$.

(b) Let d be as in (a), $U\subseteq M$ an open submanifold. Deduce from (a) the existence of a unique endomorphism d^U of $\mathfrak{a}(U)$ such that (i) $(d\omega)|U=d^U(\omega|U)$ $(\omega\in\mathfrak{a}(M))$, and (ii) d^U is local. Prove further that d^U satisfies (1.1.15).

(c) Let U be as in (b) and such that there are $x_1,\ldots,x_m\in C^\infty(U)$ forming a coordinate system on U. Prove that there is exactly one endomorphism of $\mathfrak{a}(U)$ satisfying (1.1.15), and that it is given by (1.1.16).

(d) Deduce from (a)–(c) that there is exactly one endomorphism d of $\mathfrak{a}(M)$ satisfying (1.1.15), and that for U as in (c), d^U is given by (1.1.16). (To prove existence of d, define d^U by (c) for U as in (c), and patch up the local definitions.)

(e) Let $0\leq p\leq m$ and let ω be any p-form. Prove the following global description of $d\omega$: if $X_1,\ldots,X_{p+1}$ are smooth vector fields on M,

$$(p+1)d\omega(X_1,\ldots,X_{p+1})=\sum_{1\leq i\leq p+1}(-1)^{i-1}X_i\cdot\omega(X_1,\ldots,\hat{X}_i,\ldots,X_{p+1})$$
$$+\sum_{1\leq i<j\leq p+1}(-1^{i+j}\omega([X_i,X_j],\ldots,\hat{X}_i,\ldots,\hat{X}_j,\ldots,X_{p+1}).$$

(Here, $\wedge$ over an X_i indicates it should be omitted. To prove this we may assume $\omega=f(df_1\wedge\cdots\wedge df_p)$ where $f,f_1,\ldots,f_p\in C^\infty(M)$. Then $(p+1)!d\omega(X_1,\ldots,X_{p+1})=\sum_{\sigma\in\Pi_{p+1}}\epsilon(\sigma)(X_{\sigma(1)}f)(X_{\sigma(2)}f_1)\cdots(X_{\sigma(p+1)}f_p)$. The coefficient $\epsilon(\sigma)$ is $X_{\sigma(1)}\{f(X_{\sigma(2)}f_1)\cdots(X_{\sigma(p+1)}f_p)\}-fX_{\sigma(1)}\{(X_{\sigma(2)}f_1)\cdots(X_{\sigma(p+1)}f_p)\}$. Simplify the second expression by the Liebniz formula.)

13. Let M be as in Exercise 12. For $Y\in\mathfrak{J}(M)$ and $\eta\in\mathfrak{a}_q(M)$ $(q\geq 1)$, let η_Y be the $(q-1)$-form $Z_1,\ldots,Z_{q-1}\mapsto\eta(Y,Z_1,\ldots,Z_{q-1})$ $(Z_i\in\mathfrak{J}(M))$; for $\eta\in\mathfrak{a}_0(M)$, put $\eta_Y=0$.

(a) For $\omega\in\mathfrak{a}_p(M)$ and $Y\in\mathfrak{J}(M)$, prove that

$$L_Y\omega:(X_1,\ldots,X_p)\mapsto\sum_{1\leq i\leq p}(-1)^i\omega([Y,X_i],\ldots,\hat{X}_i,\ldots)+Y\cdot\omega(X_1,\ldots,X_p)$$

is a p-form. Extend L_Y to an endomorphism of $\mathfrak{a}(M)$, denoted again by L_Y, and prove that it is a derivation of $\mathfrak{a}(M)$. Prove also that $L_X L_Y - L_Y L_X = L_{[X,Y]}$ $(X, Y \in \mathfrak{J}(M))$.

(b) Prove that $(p+1)(d\omega)_Y = -pd(\omega_Y) + L_Y\omega$ $(\omega \in \mathfrak{a}_p(M),\ Y \in \mathfrak{J}(M))$. Deduce that if $d\omega = 0$, $L_Y\omega = d(p\omega_Y)$. (Use (e) of Exercise 12).

(c) Prove that d commutes with L_Y. (Use (b) to prove that $L_Y d\omega = dL_Y\omega = (p+1)d(d\omega)_Y$ for $\omega \in \mathfrak{a}_p(M)$).

14. Let M be a C^∞ manifold of dimension m. If T is a C^∞ manifold and $\eta_t (t \in T)$ are p-forms on M, η_t is said to be smooth in t if for $X_1, \ldots, X_p \in \mathfrak{J}(M)$, $t,x \mapsto \eta_t(X_1, \ldots, X_p)(x)$ is C^∞ on $T \times M$. If $T = \mathbf{R}$ and η_t is a p-form smooth in t, $d\eta_t/dt$, $\int_0^t \eta_\tau\, d\tau$, etc., are defined in the obvious fashion. If α is any diffeomorphism of M, $\omega \mapsto \omega^\alpha$ (resp. $X \mapsto X^\alpha$) is the induced automorphism of $\mathfrak{a}(M)$ (resp. $\mathfrak{J}(M)$). Let $1 \le p \le m$.

(a) If $\eta_t (t \in \mathbf{R})$ is a p-form smooth in t, so are $d\eta_t/dt$, $(\eta_t)_Y$, $d\eta_t$, etc., and

$$d\left(\frac{d}{dt}\eta_t\right) = \frac{d}{dt}(d\eta_t).$$

(b) Let $\xi_t (t \in \mathbf{R})$ be an one-parameter group of diffeomorphisms of M, i.e., $\xi_{t+t'} = \xi_t\xi_{t'} (t, t' \in \mathbf{R})$ and $t,x \mapsto \xi_t(x)$ is a C^∞ map of $\mathbf{R} \times M$ into M. If $\omega \in \mathfrak{a}_p(M)$, prove that $\omega_t = \omega^{\xi_{-t}}$ is smooth in t.

(c) Let $X \in \mathfrak{J}(M)$ be defined by

$$X_x = \left(\frac{d}{dt}\xi_t(x)\right)_{t=0} \quad (x \in M).$$

Prove that

$$\frac{d}{dt}\omega_t = L_X\omega_t \quad \forall\, t.$$

(It is enough to consider $t = 0$. Note first for $Y \in \mathfrak{J}(M)$ and $f \in C^\infty(M)$, $(d(Y^{\xi_t}f)(x)/dt)_{t=0} = -([X,Y]f)(x)$ $(x \in M)$. Let $X_i \in \mathfrak{J}(M)$ and $X_i^{\xi_t} = \sum_{1\le j\le m} a_{ij}(x:t)\,\partial/\partial x_j$ in local coordinates, $1 \le i \le p$. Observe now that

$$\left(\frac{\partial}{\partial t}a_{ij}(x:t)\right)_{t=0} = -([X,X_i]x_j)(x)$$

and calculate

$$\left(\frac{d}{dt}\omega_t(X_1, \ldots, X_p)(x)\right)_{t=0} = \left(\frac{d}{dt}\omega(X_1^{\xi_t}, \ldots, X_p^{\xi_t})(\xi_t x)\right)_{t=0}.$$

(d) Let X, ω, ω_t be as in (b) and (c) and let $d\omega = 0$. Let $\eta_t = p\int_0^t (\omega_\tau)_X\, d\tau$. Prove that $d\eta_t = \omega_t - \omega$ for all t. (Hint: Let $\Delta_t = d\eta_t - (\omega_t - \omega)$. Then $d\Delta_t/dt = 0$ by (c) and (b) of Exercise 13.)

$L_X\omega$ is called the *Lie derivative* of ω by X; (c) gives its differential interpretation.

15. Let M be oriented. Let ζ be an $(m-1)$-form with compact support. Prove that $\int_M d\zeta = 0$. Deduce that if M is compact and ζ is any $(m-1)$-form, $\int_M d\zeta = 0$.

CHAPTER 2

LIE GROUPS AND LIE ALGEBRAS

2.1. Definition and Examples of Lie Groups

The notion of a Lie group is obtained by imitating the definition of a topological group.

Let G be a topological group. Suppose there is an analytic structure on the set G, compatible with its topology, which converts it into an analytic manifold and for which the maps

$$\left\{ \begin{array}{ll} (x,y) \mapsto xy & (x,y \in G) \\ x \mapsto x^{-1} & (x \in G) \end{array} \right. \tag{2.1.1}$$

of $G \times G$ into G and of G into G, respectively, are both analytic. Then G, together with this analytic structure, is called a *Lie group*. As usual, by abuse of language, we shall refer to G itself as a Lie group. According as the analytic structure is real or complex, G is called a *real* or a *complex* Lie group. A connected Lie group is called an *analytic* group.

The underlying topological group of a Lie group is obviously locally compact and second countable. If G is a complex Lie group, then the underlying topological group, together with the real analytic structure corresponding to the complex analytic structure, forms a real Lie group (cf. §1.2). We shall refer to this as the real Lie group underlying the complex Lie group G.

If G_i, $1 \leq i \leq n$, are Lie groups, then the product group $G_1 \times \cdots \times G_n$, equipped with the product analytic structure, is a Lie group; we shall denote it by $G_1 \times \cdots \times G_n$. If G_1 and G_2 are Lie groups, a map π of G_1 into G_2 is called an *isomorphism* of Lie groups if it is an isomorphism of the underlying groups as well as of the analytic manifolds.

Let G be a Lie group. It is then immediate from the definition that $(x,y) \mapsto xy^{-1}$ is an analytic map of $G \times G$ into G. For any fixed $a \in G$, let l_a and r_a be the left and right translations of G defined by

$$l_a x = ax \qquad r_a x = xa \quad (x \in G). \tag{2.1.2}$$

Then l_a and r_a are analytic diffeomorphisms of the analytic manifold G. If we write

$$i_a x = axa^{-1} = x^a \quad (x \in G), \tag{2.1.3}$$

then i_a is an analytic automorphism of the Lie group G.

It is natural to ask whether the class of Lie groups is enlarged if one replaces the analytic manifolds in the definition of a Lie group by C^k manifolds $(0 \leq k \leq \infty)$. For example, let G be a topological group, and let us assume that for some integer k $(0 \leq k \leq \infty)$ G has the structure of a C^k manifold for which the maps (2.1.1) are of class C^k. It is then natural to call G a C^k group. One may then raise the question whether it is possible to equip G with an analytic structure compatible with its C^k structure, under which it is a Lie group. It is comparatively easy to prove that there cannot be more than one analytic structure with this property; we shall give a proof of this later on. The question of existence, however, is more difficult to settle. When $k = \infty$ or is at least sufficiently large, the existence of such a compatible analytic structure on a C^k group was a classical result, proved by Schmidt and known to Hilbert. The case $k = 0$ is the fifth problem of Hilbert. Its solution for compact groups was by von Neumann in 1933; the general case was settled only in recent times, as a result of the contributions of Gleason, Montgomery–Zippin, and other mathematicians. We refer the interested reader to the book of Montgomery and Zippin [1] for a treatment of this and related questions. The fact that every C^k group underlies a unique Lie group shows that we can restrict ourselves to analytic and Lie groups without any loss of generality.

Given a Lie group G and a subgroup H of G which is not necessarily closed in G, we shall call H a *Lie subgroup* of G if (i) H is a Lie group, and (ii) the identity mapping of H into G is an imbedding of the analytic manifold H into the analytic manifold G, i.e., H is an analytic submanifold of G. If H is regularly imbedded in G, then H is a topological subgroup of G. A connected Lie subgroup is called an *analytic subgroup*.

Theorem 2.1.1 *Let G be a Lie group, real or complex. Suppose H is a subgroup which is at the same time a quasi-regularly imbedded submanifold of G. Then H, together with this analytic structure, is a Lie subgroup of G. If H is a regularly imbedded submanifold of G, then H is closed in G.*

Proof. The map $(x,y) \mapsto xy^{-1}$ of $G \times G$ into G is analytic. Hence, by restriction, $\varphi: (x,y) \mapsto xy^{-1}$ is an analytic map of $H \times H$ into G. Since $\varphi[H \times H] \subseteq H$ and H is quasi-regularly imbedded, φ is an analytic map of $H \times H$ into H. This proves that H is a Lie group. It is obviously a Lie subgroup of G. Suppose now that H is regularly imbedded in G. Then H is locally closed in G; and in particular, H is open in its closure. Let $\bar{H}$ be the closure of H in G. Then $\bar{H}$ is a subgroup of G, and H an open subgroup of $\bar{H}$. But an open subgroup of a topological group is necessarily closed; hence, H is closed in $\bar{H}$. This implies that $H = \bar{H}$; i.e., H is closed in G.

We now discuss a few examples of Lie groups.

(1) $\mathbf{R}^m$, the additive group of m-tuples of real numbers, is a real analytic group. $\mathbf{C}^m$, the additive group of m-tuples of complex numbers, is a complex analytic group.

(2) Let $\mathbf{C}^*$ be the multiplicative group of nonzero complex numbers; the analytic structure of $\mathbf{C}^*$ is that of an open submanifold $\mathbf{C}$. Then $\mathbf{C}^*$ is a complex analytic group. For any integer $m \geq 1$, $\mathbf{C}^{*m} = \mathbf{C}^* \times \cdots \times \mathbf{C}^*$ (m factors) is an abelian complex analytic group of (complex) dimension m.

(3) Let

$$\mathbf{T}^m = \{(z_1, \ldots, z_m) \in \mathbf{C}^{*m} : |z_j| = 1 \text{ for } 1 \leq j \leq m\} \tag{2.1.4}$$

Then $\mathbf{T}^m$ is a connected compact subgroup of $\mathbf{C}^{*m}$. We equip $\mathbf{C}^{*m}$ with the real analytic structure underlying its complex analytic structure and define the functions F_j by $F_j(z_1, \ldots, z_m) = (\operatorname{Re} z_j)^2 + (\operatorname{Im} z_j)^2 - 1$; then F_j $(1 \leq j \leq m)$ are real analytic functions and $\mathbf{T}^m$ is the set of common zeros of $F_1, \ldots, F_m$. It is easy to see that $dF_1, \ldots, dF_m$ are linearly independent at all points of $\mathbf{T}^m$. Thus $\mathbf{T}^m$ is a compact regular analytic submanifold of the real analytic manifold underlying $\mathbf{C}^{*m}$. By Theorem 2.1.1, $\mathbf{T}^m$ is a compact analytic group. It is called the *m-dimensional torus*. If π is the map of $\mathbf{R}^m$ onto $\mathbf{T}^m$ given by

$$\pi(x_1, \ldots, x_m) = (e^{2i\pi x_1}, \ldots, e^{2i\pi x_m}), \tag{2.1.5}$$

then π is a submersion of $\mathbf{R}^m$ onto $\mathbf{T}^m$. Thus $\mathbf{T}^m$ may be regarded as the quotient manifold of $\mathbf{R}^m$ relative to π. Note that π is a homomorphism and its kernel is $\mathbf{Z}^m$, the set of all $(x_1, \ldots, x_m)$ where all the x_j are integers.

(4) Let $n \geq 1$ and let $\mathfrak{M}(n,\mathbf{R})$ be the real vector space of all $n \times n$ real matrices. We denote by a_{ij} $(1 \leq i, j \leq n)$ the linear function which associates with any matrix its ijth entry. Let $GL(n,\mathbf{R})$ be the set of all invertible elements of $\mathfrak{M}(n,\mathbf{R})$. $GL(n,\mathbf{R})$ is open in $\mathfrak{M}(n,\mathbf{R})$, and we regard it as an open submanifold of $\mathfrak{M}(n,\mathbf{R})$. Under matrix multiplication $GL(n,\mathbf{R})$ becomes a real Lie group. In an analogous manner, $GL(n,\mathbf{C})$ becomes a complex Lie group. The Lie subgroups of $GL(n,\mathbf{R})$ and $GL(n,\mathbf{C})$ provide the most important examples of Lie groups. More abstractly, if V is a vector space of finite dimension over $\mathbf{R}$ (resp. $\mathbf{C}$), the group $GL(V)$ of linear automorphisms of V is a real (resp. complex) Lie group.

(5) Let $n \geq 1$ and let $T^u(n,\mathbf{R})$ be the set of all upper triangular $n \times n$ matrices with real entries whose diagonal elements are all equal to 1: $A \in \mathfrak{M}(n,\mathbf{R})$ belongs to $T^u(n,\mathbf{R})$ if and only if $a_{ij}(A) = \delta_{ij}$ for $1 \leq j \leq i \leq n$. $T^u(n,\mathbf{R})$ is a closed subgroup of $GL(n,\mathbf{R})$. Since it is an affine subspace of $\mathfrak{M}(n,\mathbf{R})$, it is a regular analytic submanifold of $GL(n,\mathbf{R})$, hence a Lie sub-

group (Theorem 2.1.1). Similarly one can define the complex Lie group $T^u(n,\mathbf{C})$.

(6) Let $SL(n,\mathbf{R})$ be the closed subgroup of $GL(n,\mathbf{R})$ consisting of all elements of determinant 1. For $A \in \mathfrak{M}(n,\mathbf{R})$ let $f(A) = \det(A) - 1$. Then f is an analytic function, and $SL(n,\mathbf{R})$ is the set of zeros of f. If A_{ij} is the cofactor of a_{ij} in the matrix $(a_{rs})_{1\le r,s\le n}$, a simple calculation shows that

$$df = \sum_{1\le i,j\le n} A_{ij}\,da_{ij}. \tag{2.1.6}$$

It follows from this formula that if df vanishes at some $A_0 \in \mathfrak{M}(n,\mathbf{R})$, then all the cofactors of the elements of A_0 must be zero, so that $\det(A_0)$ must vanish. Consequently df is nonzero at all points of $SL(n,\mathbf{R})$. This proves that $SL(n,\mathbf{R})$ is a closed regular analytic submanifold of $GL(n,\mathbf{R})$ of dimension $n^2 - 1$. Theorem 2.1.1 allows us to conclude that $SL(n,\mathbf{R})$ is a closed Lie subgroup of $GL(n,\mathbf{R})$. An analogous treatment can be given for $SL(n,\mathbf{C})$.

(7) Let $O(n,\mathbf{R})$ be the group of $n \times n$ real orthogonal matrices. It is a compact subgroup of $GL(n,\mathbf{R})$. We shall prove that it is a regular analytic submanifold of $\mathfrak{M}(n,\mathbf{R})$, of dimension $\frac{1}{2}n(n-1)$. Theorem 2.1.1 will then imply that it is a compact Lie subgroup of $GL(n,\mathbf{R})$.

Let q_{ij} be the function on $\mathfrak{M}(n,\mathbf{R})$ defined by

$$q_{ij} = \sum_{1\le s\le n} a_{is}a_{js} - \delta_{ij} \quad (1 \le i,j \le n). \tag{2.1.7}$$

Then $q_{ij} = q_{ji}$, and $O(n,\mathbf{R})$ is the set of common zeros of all the q_{ij}. For any $A \in \mathfrak{M}(n,\mathbf{R})$ let $\delta(A)$ be the dimension of the vector space spanned by the differentials $(dq_{ij})_A$. We now show that if $A \in \mathfrak{M}(n,\mathbf{R})$ is invertible, $\delta(A) = \frac{1}{2}n(n+1)$. Fix an invertible A in $\mathfrak{M}(n,\mathbf{R})$. Denote by Q the $n \times n$ matrix (q_{ij}). Then $Q = AA^t - 1$ and so (the suffix denotes that the derivatives are evaluated at A)

$$\left(\frac{\partial Q}{\partial a_{kl}}\right)_A = AE_{lk} + E_{kl}A^t \tag{2.1.8}$$

where E_{rs} is the $n \times n$ matrix whose uvth entry is $\delta_{ur}\delta_{vs}$. Let S_A be the vector space of all matrices $B = (b_{kl})$ such that $\sum_{1\le k,l\le n} b_{kl}(\delta Q/\delta a_{kl})_A = 0$. It is then obvious that $\delta(A) = n^2 - \dim(S_A)$. On the other hand, using the above expression for $(\partial Q/\partial a_{kl})_A$, we find that

$$\begin{aligned} B \in S_A &\Longleftrightarrow \sum_{1\le k,l\le n} b_{kl}(AE_{lk} + E_{kl}A^t) = 0 \\ &\Longleftrightarrow AB^t + BA^t = 0 \\ &\Longleftrightarrow BA^t \text{ is skew-symmetric.} \end{aligned}$$

But, since A is invertible, $X \mapsto XA^t$ is a linear automorphism of the vector space $\mathfrak{M}(n,\mathbf{R})$. So S_A has the same dimension as the vector space of real $n \times n$ skew-symmetric matrices, which is $\frac{1}{2}n(n-1)$. Thus

$$\delta(A) = \tfrac{1}{2}n(n+1) \quad (A \text{ invertible}).$$

In particular, for $A \in O(n,\mathbf{R})$, $\delta(A) = \frac{1}{2}n(n+1)$. Thus the $\frac{1}{2}n(n+1)$ 1-forms dq_{ij} $(1 \leq i \leq j \leq n)$ are linearly independent at all points of $O(n,\mathbf{R})$. $O(n,\mathbf{R})$ is thus a regular analytic submanifold of dimension $\frac{1}{2}n(n-1)$. An analogous discussion can be given for $O(n,\mathbf{C})$. *Note that $O(n,\mathbf{C})$ is not compact.*

For $A \in O(n,\mathbf{R})$, $\det(A) = \pm 1$. The subgroup $SO(n,\mathbf{R})$ of all real orthogonal matrices of determinant 1 is an open and closed subgroup of $O(n,\mathbf{R})$ of index 2. It is thus also a compact Lie subgroup of $GL(n,\mathbf{R})$. $SO(n,\mathbf{C}) = SL(n,\mathbf{C})$ is analogously open and closed and if index 2 in $O(n,\mathbf{C})$.

(8) The discussion on $SL(n,\mathbf{R})$ and $O(n,\mathbf{R})$ can be generalized to include all algebraic subgroups of $GL(n,\mathbf{R})$. A subgroup $G \subseteq GL(n,\mathbf{R})$ is said to be *real algebraic* if G is an algebraic subset of the open set of invertible elements of $\mathfrak{M}(n,\mathbf{R})$ (cf. §1.2). Similarly, a *complex algebraic group* is a subgroup of $GL(n,\mathbf{C})$ which is at the same time an algebraic subset of it. For instance, the orthogonal group is algebraic. We have the following theorem.

Theorem 2.1.2. *Let G be a real (resp. complex) algebraic group. Then G is a closed real (resp. complex) Lie subgroup of $GL(n,\mathbf{R})$ (resp. $GL(n,\mathbf{C})$).*

Proof. We discuss only the real case; the complex case is treated along the same lines. Let $\mathcal{P}$ be the algebra of all polynomials in the entries a_{ij} with real coefficients. Write $U = GL(n,\mathbf{R})$, and let $G \subseteq U$ be an algebraic group. In view of Theorem 2.1.1, it is enough to prove that G is a regular submanifold of U. Let $\mathfrak{g}$ be the ideal of all elements of $\mathcal{P}$ which vanish on G. For $A \in U$ let R_A be the vector space spanned by the differentials $(df)_A$, $f \in \mathfrak{g}$; let $d_A = \dim(R_A)$.

Suppose $A \in G$, and let l_A be the left translation $B \mapsto AB$ of U. For any $p \in \mathcal{P}$, let p^A be the function $X \mapsto p(AX)$ on $\mathfrak{M}(n,\mathbf{R})$. Then $p^A \in \mathcal{P}$, and $p \mapsto p^A$ is an automorphism of the algebra $\mathcal{P}$ which leaves the ideal $\mathfrak{g}$ invariant. Now, for any $B \in U$, the differential $(dl_A)_B$ is an isomorphism of the tangent space to U at B onto the tangent space to U at AB; the dual of this isomorphism maps $(dp)_{AB}$ onto $(dp^A)_B$ for any $p \in \mathcal{P}$. In particular, the vector space R_{AB} gets mapped onto the vector space R_A under this dual. Thus $d_{AB} = d_B$ for all $B \in U$. We thus see that d_A is constant for $A \in G$.

Now apply Whitney's Theorem 1.2.1. The constancy of d_A for $A \in G$ shows that all points of G are regular and enables us to conclude that G itself is a regular analytic submanifold of U. As mentioned at the beginning, this proves that G is a closed Lie subgroup of $GL(n,\mathbf{R})$.

(9) Let B be the skew-symmetric bilinear form on $\mathbf{C}^{2n} \times \mathbf{C}^{2n}$ given by

$$B(\mathbf{x},\mathbf{y}) = \sum_{1 \leq p \leq n} (x_p y_{n+p} - x_{n+p} y_p), \tag{2.1.9}$$

where $\mathbf{x} = (x_1, \ldots, x_{2n})$ and $\mathbf{y} = (y_1, \ldots, y_{2n})$. The *symplectic group* $Sp(n,\mathbf{C})$ is defined to be the subgroup of $GL(2n,\mathbf{C})$ of all elements which leave B invariant. Let

$$F = \begin{pmatrix} 0 & I_n \\ -I_n & 0 \end{pmatrix}, \tag{2.1.10}$$

where I_n is the $n \times n$ identity matrix. Then it is easy to show that $Sp(n,\mathbf{C})$ is the subgroup of all $A \in GL(2n,\mathbf{C})$ such that $A^t F A = F$. $Sp(n,\mathbf{C})$ is thus a complex algebraic group, hence a closed complex Lie subgroup of $GL(2n,\mathbf{C})$ by Theorem 2.1.2. The analogously defined algebraic subgroup of $GL(2n,\mathbf{R})$ is denoted by $Sp(n,\mathbf{R})$.

It is customary to refer to the groups $GL(n,\mathbf{C})$, $SL(n,\mathbf{C})$, $SO(n,\mathbf{C})$, $O(n,\mathbf{C})$, and $Sp(n,\mathbf{C})$ as the *complex classical groups*.

(10) Let $U(n,\mathbf{C})$ be the unitary group in n dimensions, i.e., the subgroup of all matrices in $GL(n,\mathbf{C})$ that leave the Hermitian form $x_1\bar{x}_1 + \cdots + x_n\bar{x}_n$ invariant. If $\dagger$ denotes adjoints, then $A \in U(n,\mathbf{C})$ if and only if $AA^\dagger = A^\dagger A = 1$. $SU(n,\mathbf{C})$ denotes $SL(n,\mathbf{C}) \cap U(n,\mathbf{C})$. If $Sp(n,\mathbf{C})$ is defined as above, we write $Sp(n) = Sp(n,\mathbf{C}) \cap U(2n,\mathbf{C})$. $U(n,\mathbf{C})$, $SU(n,\mathbf{C})$, and $Sp(n)$ are all compact groups.

Let $\mathbf{e}_p = (\delta_{p1}, \ldots, \delta_{pn})$ $(1 \leq p \leq n)$ be the usual basis of $\mathbf{C}^n$ over $\mathbf{C}$. Then $\mathbf{e}_1, \ldots, \mathbf{e}_n, i\mathbf{e}_1, \ldots, i\mathbf{e}_n$ $(i^2 = -1)$ is a basis of $\mathbf{C}^n$ considered as a vector space over $\mathbf{R}$. This enables us to identify $GL(n,\mathbf{C})$ with a real algebraic subgroup of $GL(2n,\mathbf{R})$. Under this identification, $SU(n,\mathbf{C})$, $U(n,\mathbf{C})$, and $Sp(n)$ are easily seen to be real algebraic subgroups of $GL(2n,\mathbf{R})$. These are therefore all Lie groups.

$U(n,\mathbf{C})$, $SU(n,\mathbf{C})$, $SO(n,\mathbf{R})$, and $Sp(n)$ are usually referred to as the *compact classical groups*.

Note that if G is a Lie group and G^0 is the connected component of G containing 1, G^0 is an open and closed subgroup of G; hence G^0 is an open analytic subgroup of G. We leave it to the reader to verify that if H is a Lie subgroup of G, H^0 is an analytic subgroup of G, and that H^0 is regularly imbedded in G if H is.

If G is a countable discrete group, $G^0 = \{1\}$. We shall regard G^0 as an analytic group (of dimension 0), so G will be a Lie group.

2.2. Lie Algebras

Let k be a field of characteristic 0. A vector space $\mathfrak{g}$ over k is called a *Lie algebra over k* if there is a map

$$(X,Y) \mapsto [X,Y] \quad (X, Y, [X,Y] \in \mathfrak{g})$$

of $\mathfrak{g} \times \mathfrak{g}$ into $\mathfrak{g}$ with the following properties:

$$(2.2.1) \quad \begin{cases} \text{(i)} & (X,Y) \mapsto [X,Y] \text{ is bilinear} \\ \text{(ii)} & [X,Y] + [Y,X] = 0 \quad (X, Y \in \mathfrak{g}) \\ \text{(iii)} & [X,[Y,Z]] + [Y,[Z,X]] + [Z,[X,Y]] = 0 \quad (X, Y, Z \in \mathfrak{g}). \end{cases}$$

For $X, Y \in \mathfrak{g}$, $[X,Y]$ is called the *bracket* of X with Y. The relation (iii) of (2.2.1) is known as the *Jacobi identity. All the Lie algebras considered by us are finite-dimensional unless we explicitly state otherwise.*

Let $\mathfrak{g}$ be a Lie algebra over k and let $\{X_1, \ldots, X_n\}$ be a basis of $\mathfrak{g}$ (as a vector space). Then there are uniquely determined constants $c_{rsp} \in k$ $(1 \leq r, s, p \leq n)$ such that

$$(2.2.2) \qquad [X_r, X_s] = \sum_{1 \leq p \leq n} c_{rsp} X_p$$

The c_{rsp} are called the *structure constants* of $\mathfrak{g}$ relative to the basis $\{X_1, \ldots, X_n\}$. The identities (2.2.1) then lead to the following relations:

$$(2.2.3) \quad \begin{cases} \text{(i)} & c_{rsp} + c_{srp} = 0 \quad (1 \leq r, s, p \leq n) \\ \text{(ii)} & \sum_{1 \leq p \leq n} (c_{rsp}c_{ptu} + c_{stp}c_{pru} + c_{trp}c_{psu}) = 0 \\ & \qquad\qquad (1 \leq r, s, t, u \leq n). \end{cases}$$

If $\mathfrak{g}$ is a Lie algebra over k and K is a field containing k, the K-vector space $\mathfrak{g} \otimes_k K$ has a unique structure of a Lie algebra over K such that

$$(2.2.4) \qquad [X \otimes 1, Y \otimes 1] = [X,Y] \otimes 1 \quad (X, Y \in \mathfrak{g})$$

We denote this Lie algebra by $\mathfrak{g}_K$ and refer to it as the *extension of* $\mathfrak{g}$ *to* K. We shall usually identify $\mathfrak{g}$ with its image in $\mathfrak{g}_K$ under the map $X \mapsto X \otimes 1$ $(X \in \mathfrak{g})$. If $k = \mathbf{R}$ and $K = \mathbf{C}$, we write $\mathfrak{g}_c$ for $\mathfrak{g} \otimes_{\mathbf{R}} \mathbf{C}$ and call it the *complexification* of $\mathfrak{g}$.

Let $\mathfrak{g}$ be a Lie algebra over k. Given two linear subspaces $\mathfrak{a}$ and $\mathfrak{b}$ of $\mathfrak{g}$, we denote by $[\mathfrak{a},\mathfrak{b}]$ the linear space spanned by $[X,Y]$ with $X \in \mathfrak{a}$, $Y \in \mathfrak{b}$. A linear subspace $\mathfrak{h}$ of $\mathfrak{g}$ is called a *subalgebra* if $[\mathfrak{h},\mathfrak{h}] \subseteq \mathfrak{h}$; it is called an *ideal* if $[\mathfrak{g},\mathfrak{h}] \subseteq \mathfrak{h}$. If $\mathfrak{g}$, $\mathfrak{g}'$ are Lie algebras over k, and π $(X \mapsto \pi(X))$ a linear map of $\mathfrak{g}$ into $\mathfrak{g}'$, π is called a *homomorphism* if it preserves the bracket operations, i.e., if

$$(2.2.5) \qquad [\pi(X),\pi(Y)] = \pi([X,Y]) \quad (X, Y \in \mathfrak{g}).$$

If π is a homomorphism, then $\pi[\mathfrak{g}]$ is a subalgebra of $\mathfrak{g}'$, and the kernel of π is an ideal in $\mathfrak{g}$. Conversely, let $\mathfrak{g}$ be a Lie algebra over k and $\mathfrak{h}$ an ideal of $\mathfrak{g}$.

Let $\mathfrak{g}' = \mathfrak{g}/\mathfrak{h}$ be the quotient vector space, and π the canonical linear map of $\mathfrak{g}$ onto $\mathfrak{g}'$. For $X' = \pi(X)$ and $Y' = \pi(Y)$ let

$$[X',Y'] = \pi([X,Y]). \tag{2.2.6}$$

Then it is easy to show that $[X',Y']$ is well defined and that $\mathfrak{g}'$ becomes a Lie algebra over k with this definition of the bracket. π is then a homomorphism of $\mathfrak{g}$ onto $\mathfrak{g}'$ with $\mathfrak{h}$ as its kernel. $\mathfrak{g}'$ is called the *quotient* of $\mathfrak{g}$ by $\mathfrak{h}$; we continue to denote it by $\mathfrak{g}/\mathfrak{h}$.

Let $\mathfrak{g}_i$ $(1 \leq i \leq m)$ be Lie algebras over k. Then $\mathfrak{g} = \mathfrak{g}_1 \times \cdots \times \mathfrak{g}_m$ becomes a Lie algebra over k if we define

$$[(X_1,\ldots,X_m),(Y_1,\ldots,Y_m)] = ([X_1,Y_1],\ldots,[X_m,Y_m]) \tag{2.2.7}$$

for $(X_1,\ldots,X_m)$ and $(Y_1,\ldots,Y_m)$ in $\mathfrak{g}$. $\mathfrak{g}$ is called the *product* of the Lie algebras $\mathfrak{g}_i$ $(1 \leq i \leq m)$.

We now give some examples of Lie algebras.

(1) Let $\mathfrak{g}$ be any finite-dimensional vector space over k. If we define $[X,Y] = 0$ for all $X, Y \in \mathfrak{g}$, then $\mathfrak{g}$ becomes a Lie algebra over k. It is said to be *abelian*.

(2) Let V be a finite-dimensional vector space over k. For any two endomorphisms X and Y of V, define

$$[X,Y] = XY - YX. \tag{2.2.8}$$

With this bracket, the vector space of all endomorphisms of V becomes a Lie algebra over k. It is denoted by $\mathfrak{gl}(V)$. More concretely, the vector space of all $n \times n$ matrices with entries from k becomes a Lie algebra over k if we define the bracket by (2.2.8). This Lie algebra is generally denoted by $\mathfrak{gl}(n,k)$. As important subalgebras of $\mathfrak{gl}(n,k)$ we mention the following: $\mathfrak{sl}(n,k)$, the subalgebra of all matrices of trace 0; $\mathfrak{o}(n,k)$, the subalgebra of all matrices which are skew-symmetric; and when $n = 2m$, $\mathfrak{sp}(m,k)$, the subalgebra of all matrices A such that $A^tF + FA = 0$, F being the matrix defined by

$$F = \begin{pmatrix} 0 & I_m \\ -I_m & 0 \end{pmatrix} \tag{2.2.9}$$

(here I_m is the $m \times m$ identity matrix). The verification that these are in fact subalgebras is elementary and is left to the reader.

(3) Let $\mathfrak{gl}(n,\mathbf{C})_{\mathbf{R}}$ denote $\mathfrak{gl}(n,\mathbf{C})$ considered as a Lie algebra over $\mathbf{R}$. Then $X \mapsto -X^\dagger$ is an involutive automorphism of $\mathfrak{gl}(n,\mathbf{C})_{\mathbf{R}}$, † denoting the operation of taking adjoints. The elements which are fixed by this involution form a subalgebra, denoted by $\mathfrak{u}(n,\mathbf{C})$. More generally, let $p \geq 1, q \geq 1$ be two in-

tegers with $n = p + q$ and let D be the diagonal matrix

$$D = \mathrm{diag}(\underbrace{1,\ldots,1}_{p \text{ elements}},\underbrace{-1,\ldots,-1}_{q \text{ elements}}) \tag{2.2.10}$$

Then $X \mapsto -DX^tD$ is an involutive automorphism of $\mathfrak{gl}(n,\mathbf{C})_{\mathbf{R}}$ whose fixed points form a subalgebra, denoted by $\mathfrak{u}(p,q,\mathbf{C})$. $\mathfrak{su}(p,q,\mathbf{C}) = \mathfrak{sl}(n,\mathbf{C}) \cap \mathfrak{u}(p,q,\mathbf{C})$ is a subalgebra of $\mathfrak{u}(p,q,\mathbf{C})$. Analogously, with D as in (2.2.10), $X \mapsto -DX^tD$ is an involutive automorphism of $\mathfrak{gl}(n,\mathbf{R})$ whose set of fixed points is a subalgebra, denoted by $\mathfrak{o}(p,q)$.

(4) Let $\mathfrak{A}$ be any algebra over k. We assume that the multiplication in $\mathfrak{A}$ is bilinear but not necessarily associative. An endomorphism D of $\mathfrak{A}$ (considered as a vector space) is called a *derivation* if

$$D(ab) = (Da)b + a(Db) \quad (a, b \in \mathfrak{A}). \tag{2.2.11}$$

If D_1 and D_2 are derivations of $\mathfrak{A}$, then it is easy to see that $[D_1,D_2] = D_1D_2 - D_2D_1$ is also a derivation of $\mathfrak{A}$. If $\mathfrak{A}$ is finite-dimensional, the set of derivations of $\mathfrak{A}$ is a subalgebra of $\mathfrak{gl}(\mathfrak{A})$.

(5) Let M be a real analytic manifold. We shall denote by $\mathfrak{J}_a(M)$ the real vector space of all real analytic vector fields on M. It follows from (1.1.6) that $\mathfrak{J}_a(M)$, equipped with the Lie bracket, is a Lie algebra over $\mathbf{R}$. This Lie algebra is in general infinite-dimensional. However, there are many situations where $\mathfrak{J}_a(M)$ admits a variety of finite-dimensional subalgebras. These play an important role in the theory of Lie groups. For instance, we may take $M = \mathbf{R}^n$ and take $\mathfrak{g}$ to be the set of all vector fields of the form $\sum_{1 \leq i \leq n} a_i(\partial/\partial x_i)$, where $x_1, \ldots, x_n$ are the usual coordinates on M and the a_i are possibly inhomogeneous linear functions of the x's. $\mathfrak{g}$ is a finite-dimensional subalgebra of $\mathfrak{J}_a(M)$.

We also introduce at this stage the notion of a representation of a Lie algebra; with later applications in mind, we allow the representation to be infinite-dimensional. Let $\mathfrak{g}$ be a Lie algebra over k, and V a vector space over k, not necessarily finite-dimensional. By a *representation* of $\mathfrak{g}$ in V we mean a map

$$\pi : X \mapsto \pi(X) \quad (X \in \mathfrak{g})$$

of $\mathfrak{g}$ into the vector space of all endomorphisms of V such that

$$\begin{cases} \text{(i)} & \pi \text{ is linear} \\ \text{(ii)} & \pi([X,Y]) = \pi(X)\pi(Y) - \pi(Y)\pi(X) \quad (X, Y \in \mathfrak{g}). \end{cases} \tag{2.2.12}$$

If V is finite-dimensional, (2.2.12) is equivalent to saying that π is a homomorphism of $\mathfrak{g}$ into $\mathfrak{gl}(V)$. The dimension of V is called the *degree* of π. π is said to be the *trivial* representation if $\dim V = 1$ and $\pi(X) = 0$ for all $X \in \mathfrak{g}$. In many cases of interest V is an algebra over k and each $\pi(X)$ a derivation of V.

Suppose $\mathfrak{g}$ is any Lie algebra over k. For any $X \in \mathfrak{g}$, let ad X denote the endomorphism of $\mathfrak{g}$ given by

$$\text{ad } X : Y \mapsto [X,Y] \quad (Y \in \mathfrak{g}). \tag{2.2.13}$$

It follows easily from (2.2.1) that ad X is a derivation of $\mathfrak{g}$ and that $X \mapsto \text{ad } X$ is a representation of $\mathfrak{g}$ in $\mathfrak{g}$. This is called the *adjoint representation* of $\mathfrak{g}$. Note that $\mathfrak{g}$ is abelian if and only if ad $X = 0$ for all $X \in \mathfrak{g}$. The kernel of the adjoint representation is the set of all $X \in \mathfrak{g}$ such that $[X,Y] = 0$ for all $Y \in \mathfrak{g}$; it is called the *center* of $\mathfrak{g}$.

There are two important and interesting operations which can be performed on representations to yield new representations. Let $\mathfrak{g}$ be a Lie algebra over k and let π_i be a representation of $\mathfrak{g}$ in a vector space V_i, $i = 1, \ldots, r$. Let $V = V_1 \times \cdots \times V_r$, and for any $X \in \mathfrak{g}$ let $\pi(X)$ be the endomorphism of V defined by

$$\pi(X)(v_1, \ldots, v_r) = (\pi_1(X)v_1, \ldots, \pi_r(X)v_r) \tag{2.2.14}$$

for all $(v_1, \ldots, v_r) \in V$. It is easy to see that π is a representation of $\mathfrak{g}$ in V; it is called the *direct sum* of the representations π_i $(1 \leq i \leq r)$. Further, let

$$W = V_1 \otimes V_2 \otimes \cdots \otimes V_r$$

For any $X \in \mathfrak{g}$ let $\pi(X)$ be the endomorphism of W given by

$$\left\{ \begin{aligned} \pi(X) = \pi_1(X) \otimes 1 \otimes \cdots \otimes 1 + 1 \otimes \pi_2(X) \otimes 1 \otimes \cdots \otimes 1 \\ + \cdots + 1 \otimes 1 \otimes \cdots \otimes 1 \otimes \pi_r(X). \end{aligned} \right. \tag{2.2.15}$$

We leave it to the reader to verify that π is a representation of $\mathfrak{g}$ in W. It is called the *tensor product* of $\pi_1, \ldots, \pi_r$ and is denoted by $\pi_1 \otimes \cdots \otimes \pi_r$.

It is important not to confuse the notion of tensor products of representations of a Lie algebra with another notion, the so-called outer tensor product. Let $\mathfrak{g}_i$ be a Lie algebra over k, and π_i a representation of $\mathfrak{g}_i$ in V_i. Let $\mathfrak{g} = \mathfrak{g}_1 \times \cdots \times \mathfrak{g}_r$ be the product of the $\mathfrak{g}_i$, and for any $(X_1, \ldots, X_r) \in \mathfrak{g}$ let

$$\begin{aligned} \pi(X_1, \ldots, X_r) = \pi_1(X_1) \otimes 1 \otimes \cdots \otimes 1 + \cdots \\ + 1 \otimes 1 \otimes \cdots \otimes \pi_r(X_r). \end{aligned} \tag{2.2.16}$$

It is easy to verify that π_0 is a representation of $\mathfrak{g}_0$ in W. It is called the *outer tensor product* of $\pi_1, \ldots, \pi_r$, and is denoted by $\pi_1 \times \pi_2 \times \cdots \times \pi_r$. Note that

$$(2.2.17)\qquad (\pi_1 \otimes \cdots \otimes \pi_r)(X) = (\pi_1 \times \pi_2 \times \cdots \times \pi_r)(X, \ldots, X) \quad (X \in \mathfrak{g})$$

if $\mathfrak{g}_1 = \cdots = \mathfrak{g}_r = \mathfrak{g}$.

Representations π_i of a Lie algebra $\mathfrak{g}$ in V_i ($i = 1, 2$) are said to be *equivalent* if there is a linear isomorphism ξ of V_1 onto V_2 such that

$$(2.2.18)\qquad \xi\pi_1(X)\xi^{-1} = \pi_2(X) \quad (X \in \mathfrak{g}).$$

A representation π of a Lie algebra $\mathfrak{g}$ in V is said to be *irreducible* if 0 and V are the only subspaces of V which are invariant under all $\pi(X)$, $X \in \mathfrak{g}$. Let π be non-irreducible and W ($\neq 0$, $\neq V$) a subspace invariant under all $\pi(X)$, $X \in \mathfrak{g}$; for any $X \in \mathfrak{g}$ let $\pi_W(X)$ (resp. $\pi_{V/W}(X)$) be the endomorphism induced by $\pi(X)$ on W (resp. V/W). Then π_W (resp. $\pi_{V/W}$) is a representation of $\mathfrak{g}$ in W (resp. V/W); it is called the *subrepresentation* (resp. *quotient representation*) defined by W.

2.3. The Lie Algebra of a Lie Group

Let G be a real Lie group. We denote its identity element by 1. Then the vector space $\mathfrak{I}_a(G)$ of all analytic real vector fields on G is a Lie algebra over $\mathbf{R}$, the bracket being the usual Lie bracket. For any $b \in G$, l_b ($x \mapsto bx$) is an analytic diffeomorphism of the analytic manifold G. It therefore induces an automorphism $X \mapsto X^{l_b}$ of the Lie algebra $\mathfrak{I}_a(G)$. An element $X \in \mathfrak{I}_a(G)$ is said to be *left-invariant* if $X^{l_b} = X$ for all $b \in G$. It is obvious that the set of all left-invariant real analytic vector fields forms a subalgebra of the Lie algebra $\mathfrak{I}_a(G)$. We denote this Lie algebra by $\mathfrak{g}$ and call it the *Lie algebra of* G. In a similar manner, if G is a complex Lie group, the set of all left-invariant holomorphic vector fields on G is a Lie algebra over $\mathbf{C}$, denoted by $\mathfrak{g}$ and called the *Lie algebra of* G.

Theorem 2.3.1 *Let G be a Lie group and $\mathfrak{g}$ its Lie algebra. Then the map*

$$(2.3.1)\qquad X \mapsto X_1 \quad (X \in \mathfrak{g}),$$

is a linear isomorphism of $\mathfrak{g}$ onto the tangent space $T_1(G)$ to G at 1. In particular

$$(2.3.2)\qquad \dim(G) = \dim(\mathfrak{g}).$$

Proof. We give the proof in the real case. The complex case is handled in a similar fashion.

If $X \in \mathfrak{g}$ and $b \in G$, the left invariance of X implies that

$$X_b = (dl_b)_1(X_1).$$

Hence $X_1 = 0$ implies $X_b = 0$ for all $b \in G$; i.e., $X = 0$. Thus the map (2.3.1) is injective. We now prove that it is surjective. Let $v \in T_1(G)$. For any $b \in G$ we define the tangent vector $X_b \in T_b(G)$ by

$$X_b = (dl_b)_1(v).$$

We prove that X $(b \mapsto X_b)$ is an element of $\mathfrak{g}$. If $x \in G$, the relation $l_{xb} = l_x \circ l_b$ implies that

$$\begin{aligned}(dl_x)_b(X_b) &= (dl_x)_b((dl_b)_1(v))\\ &= (dl_{xb})_1(v)\\ &= X_{xb}.\end{aligned}$$

X is thus invariant under all left translations. We now assert that X is an analytic vector field. In view of the left invariance, it is enough to check that X is analytic around the identity element 1. Select coordinates $x_1, \ldots, x_n$ on an open subset U of G with $1 \in U$ and $x_1(1) = \cdots = x_n(1) = 0$. Since the map $(x,y) \mapsto xy$ of $G \times G$ into G is analytic, there are functions F_i $(1 \leq i \leq n)$ defined and analytic around the origin of $\mathbf{R}^n \times \mathbf{R}^n$ such that

$$x_i(ab) = F_i(x_1(a), \ldots, x_n(a) : x_1(b), \ldots, x_n(b)) \tag{2.3.3}$$

for $1 \leq i \leq n$ and all a,b in some open set V with $1 \in V$ and $VV \subseteq U$. Let $c_1, \ldots, c_n \in \mathbf{R}$ be such that $X_1 = \sum_{1 \leq j \leq n} c_j(\partial/\partial x_j)_1$. If $a \in V$, $X_a x_i = X_1 y_i$, where y_i is the function $b \mapsto x_i(ab)$ on V $(1 \leq i \leq n)$. Hence, from (2.3.3),

$$X_a x_i = \sum_{1 \leq j \leq n} c_j \left(\frac{\partial}{\partial v_j} F_i(x_1(a), \ldots, x_n(a) : v_1, \ldots, v_n)\right)_{v_1 = \cdots = v_n = 0}. \tag{2.3.4}$$

The expression (2.3.4) shows that the functions $a \mapsto X_a x_i$ $(1 \leq i \leq n)$ are analytic on V. This proves that X is an analytic vector field. Thus $X \in \mathfrak{g}$.

The map $X \mapsto X_1$ is therefore a linear bijection of $\mathfrak{g}$ onto $T_1(G)$. The assertions of the theorem follow at once from this.

It follows at once from Theorem 2.3.1 that for any $b \in G$, the map $X \mapsto X_b$ $(X \in \mathfrak{g})$ is a linear isomorphism of $\mathfrak{g}$ with $T_b(G)$. This isomorphism enables us to identify the tangent space to G at b with $\mathfrak{g}$. We refer to this as the *canonical identification.*

It is possible to develop the theory of Lie groups by defining the Lie algebra of a Lie group G to be the set of all right-invariant analytic vector fields. It should be noted, however, that a right-invariant vector field is in general not left-invariant. If G is abelian, it is obvious that left and right invariance coincide.

If G is a Lie group, its Lie algebra $\mathfrak{g}$ is an "invariant" of the Lie group structure of G. The fundamental problem in the theory of Lie groups is that of determining the extent to which G is determined by $\mathfrak{g}$. We shall prove that G is determined up to "local isomorphism" by its Lie algebra. We shall also prove the much deeper result that any abstractly given Lie algebra over **R** (resp. **C**) is isomorphic to the Lie algebra of a real (resp. complex) Lie group. The proofs of these theorems require considerable preparation and will be carried out in subsequent sections. At this stage we content ourselves with the remark that $\mathfrak{g}$ is already determined by the component of 1 of G. In fact, let G^0 be the component of 1 of G and $\mathfrak{g}^0$ its Lie algebra. G^0 is an open submanifold of G. For any $X \in \mathfrak{g}$, let X^0 be its restriction to G^0. Then $X \mapsto X^0$ is an isomorphism of $\mathfrak{g}$ with $\mathfrak{g}^0$ (by Theorem 2.3.1). In rough terms, the structure of G beyond G^0 cannot be obtained from $\mathfrak{g}$.

We now give a few examples of the correspondence $G \mapsto \mathfrak{g}$.

(1) Let V be a vector space over **R** of finite dimension n. It is an analytic manifold in the usual way and a real Lie group under addition. For $v \in V$ denote by $\partial(v)$ the derivation of $C^\infty(V)$ given by

$$(\partial(v)f)(u) = \left\{\frac{d}{dt}f(u + tv)\right\}_{t=0} \qquad (u \in V). \tag{2.3.5}$$

$\partial(v)$ is a left-invariant analytic vector field, and $v \mapsto \partial(v)$ is a linear isomorphism of V with its Lie algebra. For fixed $u \in V$, $f \mapsto (\partial(v)f)(u)$ defines a tangent vector $\partial(v)_u$ to V at u, and $v \mapsto \partial(v)_u$ is a linear isomorphism of V with the tangent space to V at u. This enable us to identify the tangent space to V at any of its points with V itself. We shall always do this and refer to it as the *canonical identification*. Let $x_1, \ldots, x_n$ be a basis of the dual V^* of V, and $e_1, \ldots, e_n$ the corresponding dual basis of V. If $v \in V$, then

$$v = \sum_{1 \le i \le n} x_i(v)e_i,$$

and an easy calculation shows that

$$\partial(v)_u = \sum_{1 \le i \le n} x_i(v)\left(\frac{\partial}{\partial x_i}\right)_u \qquad (u \in V).$$

It follows at once from this that the Lie algebra of V is abelian.

(2) Let V be as in (1). The space $\mathfrak{M}$ of endomorphisms of V is a vector space over **R** of dimension n^2 and is an analytic manifold in the usual manner. Let $G = GL(V)$; then G is an open submanifold of $\mathfrak{M}$. We now "determine" the Lie algebra $\mathfrak{g}$ of G. We canonically identify the tangent space to $\mathfrak{M}$ at any of its points with $\mathfrak{M}$ itself (cf. (1)). Given $X \in \mathfrak{g}$, the tangent vector X_1 at the identity element 1 is, by virtue of our identification, an element of $\mathfrak{M}$ itself. Let us denote this element by X^0. Since $\dim(\mathfrak{M}) = n^2 = \dim(\mathfrak{g})$, it is clear

that $X \mapsto X^0$ is a linear isomorphism of $\mathfrak{g}$ with $\mathfrak{M}$. To complete the "determination" of $\mathfrak{g}$ it only remains to calculate $[X,Y]^0$ in terms of X^0 and Y^0 for any two elements X and Y of $\mathfrak{g}$.

Let $X, Y \in \mathfrak{g}$. Let f be an arbitrary real *linear* function on $\mathfrak{M}$. For $A \in G$ let f^A be the linear function on $\mathfrak{M}$ defined by $f^A(C) = f(AC)$ $(C \in \mathfrak{M})$. Then in the notation for differential operators (cf. §1.1)

$$\begin{aligned} f(A;Y) &= f^A(1;Y^0) \\ &= \left\{\frac{d}{dt}f(A + tAY^0)\right\}_{t=0} \\ &= f(AY^0). \end{aligned}$$

Therefore

$$\begin{aligned} f(1;XY) &= (Yf)(1;X^0) \\ &= \left\{\frac{d}{dt}f(Y^0 + tX^0Y^0)\right\}_{t=0} \\ &= f(X^0Y^0). \end{aligned}$$

It follows from this that

$$f(1;[X,Y]) = f(X^0Y^0 - Y^0X^0). \tag{2.3.6}$$

Since f is an arbitrary real linear function on $\mathfrak{M}$, and since there exist n^2 real linear functions on $\mathfrak{M}$ which form a system of coordinates for $\mathfrak{M}$ (hence for G), the formula (2.3.6) leads us to

$$[X,Y]^0 = X^0Y^0 - Y^0X^0. \tag{2.3.7}$$

In other words, the map $X \mapsto X^0$ is a Lie algebra isomorphism of $\mathfrak{g}$ with $\mathfrak{gl}(V)$. We shall henceforth identify $\mathfrak{g}$ with $\mathfrak{gl}(V)$ via this isomorphism, so that $\mathfrak{gl}(V)$ becomes the Lie algebra of $GL(V)$.

(3) Let A be an associative algebra with unit 1. We assume that A is defined over **R** and that it is of finite dimension; the complex case may be treated along the same lines. Denote by G the group of invertible elements of A. For $x, y \in A$, let $\lambda_x(y) = xy$. The function $f: x \mapsto \det(\lambda_x)$ is a polynomial function on A, and $x \in G$ if and only if $f(x) \neq 0$. G is thus open in A. Equipped with the topology and analytic structure inherited from A, G becomes a Lie group. Let $\mathfrak{g}$ denote its Lie algebra.

Since A is a vector space, we may identify the tangent spaces $T_x(A)$ $(x \in A)$ with A itself in the usual fashion. We then have a linear isomorphism

$$X \mapsto X^0 \quad (X \in \mathfrak{g})$$

of $\mathfrak{g}$ onto A which associates with any $X \in \mathfrak{g}$ the element of A corresponding to the tangent vector defined by X at 1. Proceeding exactly as in the previous example, we find that

$$[X,Y]^0 = X^0Y^0 - Y^0X^0.$$

In other words, if we denote by A_L the Lie algebra whose underlying vector space is that of A and in which the bracket is defined by

$$[u,v] = uv - vu \quad (u, v \in A),$$

then $\mathfrak{g}$ is canonically isomorphic to A_L.

(4) Let $G_1, \ldots, G_n$ be Lie groups and let $\mathfrak{g}_j$ be the Lie algebra of G_j, $1 \leq j \leq n$. Let $G = G_1 \times \cdots \times G_n$ be the product Lie group and $\mathfrak{g}$ its Lie algebra. We canonically identify the tangent space to G at any point $(x_1, \ldots, x_n)$ with $T_{x_1}(G_1) \times \cdots \times T_{x_n}(G_n)$. If $X_j \in \mathfrak{g}_j$ for $1 \leq j \leq n$, the assignment $(x_1, \ldots, x_n) \mapsto ((X_1)_{x_1}, \ldots, (X_n)_{x_n})$ determines an element, say X, of $\mathfrak{g}$. We leave it to the reader to check that the map $(X_1, \ldots, X_n) \mapsto X$ is an isomorphism of the Lie algebra $\mathfrak{g}_1 \times \cdots \times \mathfrak{g}_n$ onto $\mathfrak{g}$. In view of this isomorphism, we shall identify $\mathfrak{g}$ with $\mathfrak{g}_1 \times \cdots \times \mathfrak{g}_n$.

2.4. The Enveloping Algebra of a Lie Group

Let G be a real Lie group. For each $a \in G$, l_a $(x \mapsto ax)$ is an analytic diffeomorphism of G onto itself; it therefore induces an algebra automorphism of the algebra (over **R**) of all analytic real differential operators on G. An analytic real differential operator D is called *left-invariant* if it is invariant under all left translations. Such differential operators form an algebra over **R**, denoted by $\mathfrak{G}$. If G is a complex Lie group, $\mathfrak{G}$ will denote the algebra (over **C**) of all holomorphic differential operators on G invariant under all left translations. In either case $\mathfrak{G}$ is called the *enveloping algebra* of G. If $\mathfrak{g}$ is the Lie algebra of G, then $\mathfrak{g} \subseteq \mathfrak{G}$, and for $X, Y \in \mathfrak{g}$, $[X,Y] = XY - YX$. We shall postpone to the next chapter a closer study of $\mathfrak{G}$, and content ourselves at this stage with the following theorem.

Theorem 2.4.1. *Let G be a Lie group, $\mathfrak{g}$ its Lie algebra, $\mathfrak{G}$ its enveloping algebra. Suppose $\{X_1, \ldots, X_n\}$ is any basis for $\mathfrak{g}$. For any n-tuple $(r_1, \ldots, r_n)$ of integers ≥ 0 let us define the element $X^{(r_1,\ldots,r_n)}$ of $\mathfrak{G}$ by*

$$X^{(r_1,\ldots,r_n)} = X_1^{r_1} \cdots X_n^{r_n} \tag{2.4.1}$$

($X^{(0,\ldots,0)} = 1$, the identity operator). Then, the $X^{(r_1,\ldots,r_n)}$ form a basis for $\mathfrak{G}$. In particular, $\mathfrak{G}$ is algebraically generated by 1 and $\mathfrak{g}$. For any $a \in G$, the map $D \mapsto D_a$ $(D \in \mathfrak{G})$ is a linear isomorphism of $\mathfrak{G}$ with $T_a^{(\infty)}(G)$.

Proof. We work with real Lie groups; the complex case can be treated similarly. Now, $X_1, \ldots, X_n$ are real analytic vector fields on G with the property that $(X_1)_a, \ldots, (X_n)_a$ form a basis for $T_a(G)$ for any $a \in G$; therefore, Theorem 1.1.2 applies. We conclude first from that theorem that the $X^{(r_1,\ldots,r_n)}$ are linearly independent elements of $\mathfrak{G}$. Now suppose $D \in \mathfrak{G}$. The same theorem then implies that for some integer $s \geq 0$ and analytic functions $f_{(r_1,\ldots,r_n)}$ on G $(r_1 + \cdots + r_n \leq s)$,

$$D = \sum_{r_1+\cdots+r_n \leq s} f_{(r_1,\ldots,r_n)} X^{(r_1,\ldots,r_n)}, \tag{2.4.2}$$

the functions $f_{(r_1,\ldots,r_n)}$ being uniquely determined by D. Since the $X^{(r_1,\ldots,r_n)}$ are left-invariant, we have

$$D = D^{l_a} = \sum_{r_1+\cdots+r_n \leq s} (f_{(r_1,\ldots,r_n)})^{l_a} X^{(r_1,\ldots,r_n)} \quad (a \in G). \tag{2.4.3}$$

The relations (2.4.2) and (2.4.3) imply that all the $f_{(r_1,\ldots,r_n)}$ are left-invariant. Hence they are all constant. $\mathfrak{G}$ is thus linearly spanned by the $X^{(r_1,\ldots,r_n)}$. In particular, the X_i generate $\mathfrak{G}$ algebraically.

To prove the last assertion, fix $a \in G$ and write $\tau(D) = D_a$ $(D \in \mathfrak{G})$. The left invariance of the elements of $\mathfrak{G}$ implies at once that τ is injective. On the other hand, if p is any integer ≥ 0 and $\mathfrak{G}^{(p)}$ is the linear span of the $X^{(r_1,\ldots,r_n)}$ with $r_1 + \cdots + r_n \leq p$, τ maps $\mathfrak{G}^{(p)}$ into $T_a^{(p)}(G)$, while $\dim \mathfrak{G}^{(p)} = \dim T_a^{(p)}(G)$. τ thus induces a linear bijection of $\mathfrak{G}^{(p)}$ onto $T_a^{(p)}(G)$ for each $p \geq 0$. Consequently, τ is a linear isomorphism.

As a simple example, let B be a vector space over $\mathbf{R}$ of dimension n. Let $\mathfrak{b}$ be the Lie algebra of the additive group of B, $\mathfrak{B}$ its enveloping algebra. For $X \in B$ let $\partial(X)$ be the derivation of $C^\infty(B)$ defined by (2.3.5). Then $X \mapsto \partial(X)$ is a linear isomorphism of B onto $\mathfrak{b}$. Let S denote the symmetric algebra over B; we assume that $B \subseteq S$. Since $\partial(X)\partial(Y) = \partial(Y)\partial(X)$ for $X, Y \in B$, ∂ extends to a unique homomorphism, denoted again by ∂, of S into $\mathfrak{B}$. Since $\mathfrak{b}$ generates $\mathfrak{B}$, ∂ maps S onto $\mathfrak{B}$. On the other hand, if $X_1, \ldots, X_n$ is a basis of B, the elements $\partial(X_1)^{r_1} \cdots \partial(X_n)^{r_n} = \partial(X_1^{r_1} \cdots X_n^{r_n})$ are linearly independent in $\mathfrak{B}$ by Theorem 2.4.1. It follows from this that ∂ is an isomorphism of the algebra S onto the algebra $\mathfrak{B}$. If $\xi_1, \ldots, \xi_n$ is the basis of B^* dual to the basis $\{X_1, \ldots, X_n\}$ of B, then an easy calculation shows that

$$\partial(X_1^{r_1} \cdots X_n^{r_n})(\xi_1^{s_1} \cdots \xi_n^{s_n}) = \begin{cases} \dfrac{s_1!}{r_1!} \cdots \dfrac{s_n!}{r_n!} \xi_1^{s_1-r_1} \cdots \xi_n^{s_n-r_n} \\ \quad \text{if} \quad r_1 \leq r_n, \ldots, s_1 \leq s_n \\ 0 \qquad \text{otherwise} \end{cases}$$

The linear independence of the $\partial(X_1)^{r_1} \cdots \partial(X_n)^{r_n}$ may also be deduced from these formulae, thereby avoiding the appeal to Theorem 2.4.1.

2.5. Subgroups and Subalgebras

Let G be a Lie group (real or complex) and $\mathfrak{g}$ its Lie algebra. Let H be a Lie subgroup of G and $\bar{\mathfrak{h}}$ its Lie algebra. We denote by i the identity map of H into G. For any $\bar{X} \in \bar{\mathfrak{h}}$, $(di)_1(\bar{X}_1) \in T_1(G)$, so there is a unique element $X \in \mathfrak{g}$ such that $X_1 = (di)_1(\bar{X}_1)$. Write

$$(2.5.1) \qquad X = (di)(\bar{X}) \quad (\bar{X} \in \bar{\mathfrak{h}}).$$

Since $(di)_1$ is a linear injection, it is clear that di is a linear injection of $\bar{\mathfrak{h}}$ into $\mathfrak{g}$. Moreover, $\bar{X}$ and $X = (di)(\bar{X})$ are i-related for any $\bar{X} \in \bar{\mathfrak{h}}$. In fact, if we write λ_x for the left translation of H by an element $x \in H$, then $l_x \circ i = i \circ \lambda_x$; from this we get the relation

$$(2.5.2) \qquad X_x = (di)_x(\bar{X}_x) \quad (x \in H),$$

which asserts the i-relatedness of $\bar{X}$ and X. It follows from this that

$$(2.5.3) \qquad [(di)(\bar{X}),(di)(\bar{Y})] = (di)([\bar{X},\bar{Y}]) \quad (\bar{X},\bar{Y} \in \bar{\mathfrak{h}});$$

di is therefore a Lie algebra injection of $\bar{\mathfrak{h}}$ into $\mathfrak{g}$. If we write

$$(2.5.4) \qquad (di)[\bar{\mathfrak{h}}] = \mathfrak{h},$$

then $\mathfrak{h}$ is a subalgebra of $\mathfrak{g}$, called the *subalgebra of* $\mathfrak{g}$ *defined by* H. Clearly, $\mathfrak{h}$ is also the subalgebra of $\mathfrak{g}$ defined by H^0. Note for later use the following relation, which follows trivially from (2.5.2):

$$(2.5.5) \qquad (di)_x[T_x(H)] = \{X_x : X \in \mathfrak{h}\} \quad (x \in H).$$

The question naturally arises whether one can construct, corresponding to an arbitrary subalgebra $\mathfrak{h}$ of $\mathfrak{g}$, a Lie subgroup H of G which defines $\mathfrak{h}$; and if this is possible, whether H is uniquely determined. It is obvious that we cannot in general expect uniqueness unless H is connected. The aim of this section is to prove that the correspondence $H \mapsto \mathfrak{h}$ is a bijection of the set of all *analytic subgroups* of G onto the set of all subalgebras of $\mathfrak{g}$. The proof depends on the following lemma.

Lemma 2.5.1. *Let G be a Lie group, $\mathfrak{g}$ its Lie algebra, and $\mathfrak{h}$ a subalgebra of $\mathfrak{g}$. For any $x \in G$, let $\mathfrak{L}_x^{\mathfrak{h}}$ be the subspace of $T_x(G)$ consisting of the set of all tangent vectors of the form X_x, $X \in \mathfrak{h}$. Then $(x \mapsto \mathfrak{L}_x^{\mathfrak{h}})$ is an involutive analytic system of tangent subspaces, of rank equal to dim $\mathfrak{h}$, on the manifold G.*

Proof. We work with real Lie groups; the complex case is analogous. Put $\mathfrak{L} = \mathfrak{L}^{\mathfrak{h}}$, let $p = \dim \mathfrak{h}$, and let $X_1, \ldots, X_n$ be a basis for $\mathfrak{g}$ such that

$X_1, \ldots, X_p$ span $\mathfrak{h}$. Since $\mathfrak{h}$ is a subalgebra, there are constants $c_{ijk} \in \mathbf{R}$ such that

$$[X_i,X_j] = \sum_{1 \leq k \leq p} c_{ijk} X_k \quad (1 \leq i,j \leq p).$$

The tangent vector $(X_1)_x, \ldots, (X_p)_x$ span $\mathfrak{L}_x$ for any $x \in G$. Since the X_i are analytic vector fields, we may conclude that $\mathfrak{L}$ is an analytic system of tangent subspaces of rank p. It remains to prove that $\mathfrak{L}$ is involutive.

To prove that $\mathfrak{L}$ is involutive, let U be any open subset of G, and X and Y two analytic vector fields on U which belong to $\mathfrak{L}$ on U; we must prove that $[X,Y]$ belongs to $\mathfrak{L}$ on U. Now, the $(X_1)_x, \ldots, (X_n)_x$ form a basis for $T_x(G)$ for any $x \in G$. Therefore by Theorem 1.1.2 there are analytic functions f_i, g_i $(1 \leq i \leq n)$ on U such that $X = \sum_{1 \leq i \leq n} f_i X_i$ and $Y = \sum_{1 \leq i \leq n} g_i X_i$ on U. Since X and Y belong to $\mathfrak{L}$ on U, $f_i = g_i = 0$ for $p < i \leq n$. Hence on U

$$\begin{aligned}[X,Y] &= \sum_{1 \leq i,j \leq p} f_i(X_i g_j)X_j - \sum_{1 \leq i,j \leq p} g_i(X_i f_j)X_j + \sum_{1 \leq i,j \leq p} f_i g_j [X_i,X_j] \\ &= \sum_{1 \leq k \leq p} h_k X_k,\end{aligned}$$

where for $1 \leq k \leq p$,

$$h_k = \sum_{1 \leq i,j \leq p} c_{ijk} f_i g_j + \sum_{1 \leq i \leq p} (f_i(X_i g_k) - g_i(X_i f_k)).$$

This proves that $[X,Y]$ belongs to $\mathfrak{L}$ on U. $\mathfrak{L}$ is therefore involutive.

Theorem 2.5.2. *The correspondence, which assigns to any analytic subgroup of G the subalgebra of $\mathfrak{g}$ defined by it, is a bijection of the set of all analytic subgroups of G onto the set of all subalgebras of $\mathfrak{g}$. If $\mathfrak{h} \subseteq \mathfrak{g}$ is a subalgebra, the analytic subgroup H that defines $\mathfrak{h}$ is the maximal integral manifold containing 1 of the involutive system $\mathfrak{L}^{\mathfrak{h}}$, while for any $x \in G$, the left coset xH is the maximal integral manifold of $\mathfrak{L}^{\mathfrak{h}}$ passing through x.*

Proof. Let $\mathfrak{L}$ be as in Lemma 2.5.1. We can apply the Chevalley–Frobenius theory of involutive systems (cf. §1.3) to it. Let H be the maximal integral manifold of $\mathfrak{L}$ containing 1.

From the definition of $\mathfrak{L}$ it is clear that

$$(dl_x)_y(\mathfrak{L}_y) = \mathfrak{L}_{xy} \quad (x, y \in G). \tag{2.5.6}$$

It follows from (2.5.6) that if S is an integral manifold of $\mathfrak{L}$ containing y, xS is an integral manifold of $\mathfrak{L}$ containing xy. As a consequence, if S is the maximal integral manifold of $\mathfrak{L}$ containing y, xS is the maximal integral manifold of $\mathfrak{L}$ through xy. Taking $y = 1$, we see that xH is the maximal integral manifold through x for any $x \in G$. Since the maximal integral manifold through any point is unique, we must have $xH = H$ for $x \in H$. It follows easily from this that H is a subgroup of G. We know (Theorem 1.3.7) that H is a connected quasi-regular submanifold of G. Hence, by Theorem 2.1.1, H is an analytic subgroup of G.

We claim that $\mathfrak{h}$ is the subalgebra of $\mathfrak{g}$ defined by H. Suppose $\mathfrak{h}'$ is the subalgebra in question. Then, for any $X \in \mathfrak{g}$, $X \in \mathfrak{h}'$ if and only if $X_1 \in (di)_1[T_1(H)] = \mathfrak{L}_1$, i.e., if and only if $X \in \mathfrak{h}$. This proves that $\mathfrak{h}' = \mathfrak{h}$.

Suppose, finally, that H' is an analytic subgroup of G which defines the same subalgebra $\mathfrak{h}$ of $\mathfrak{g}$. Let i' be the identity mapping of H' into G. By (2.5.5), $(di')_x[T_x(H')] = \mathfrak{L}_x$ for all $x \in H'$. Hence H' is an integral manifold of $\mathfrak{L}$. Since $1 \in H'$, it follows that H' is an open submanifold of H. In particular, H' is an open subgroup of H. But then H' is closed in H, and since H is connected, we must have $H' = H$. Thus H' and H are identical as analytic subgroups of G.

Theorem 2.5.2 is completely proved.

In general, an analytic subgroup is not regularly imbedded, and so is not a topological subgroup. We shall now examine the circumstances under which an analytic subgroup is regularly imbedded.

Lemma 2.5.3. *Let A and B be locally compact second countable groups, and let φ be a continuous homomorphism of A onto B. Then φ is open. If φ is one-to-one, it is a homeomorphism.*

Proof. Let 1_A and 1_B be the respective identities of A and B. Let V be an open set containing 1_A. Select a compact neighborhood V_1 of 1_A such that $V_1V_1^{-1} \subseteq V$. Since A is second countable, we can find a sequence $a_1, a_2, \ldots$ of elements of A such that $A = \bigcup_{n=1}^{\infty} a_nV_1$. Then $B = \bigcup_{n=1}^{\infty} \varphi(a_n)\varphi[V_1]$. The sets $\varphi(a_n)\varphi[V_1]$ being compact, we can apply the Baire Category theorem to infer that for some $n \geq 1$, $\varphi(a_n)\varphi[V_1]$ has nonempty interior. Hence $\varphi[V_1]$ has nonempty interior. Choose $b \in V_1$ such that $\varphi(b)$ is an interior point of $\varphi[V_1]$. Then $1_B = \varphi(b)\varphi(b^{-1})$ is an interior point of $\varphi[V_1]\varphi[V_1^{-1}] = \varphi[V_1V_1^{-1}]$, so that 1_B is an interior point of $\varphi[V]$. If U is an arbitrary open set in A and $a \in U$, 1_B is an interior point of $\varphi[a^{-1}U] = \varphi(a)^{-1}\varphi[U]$ by the above result, showing that $\varphi(a)$ is an interior point of $\varphi[U]$. This proves that $\varphi[U]$ is open. φ is thus an open mapping. The second result is a trivial consequence of the first.

Theorem 2.5.4. *Let G be a Lie group and H a Lie subgroup of G. Then H is quasi-regularly imbedded in G. Moreover, the following conditions on H are equivalent:*

(i) *H is a topological subgroup of G*
(ii) *H is regularly imbedded in G*
(iii) *H is a closed subset of G.*

Proof. Let H^0 be the component of the identity of H. H^0 is quasi-regularly imbedded in G, as we observed in the proof of Theorem 2.5.2. From this the quasi-regularity of H follows easily. For the rest of the theorem (i) $\Rightarrow$ (ii) by definition, while (ii) $\Rightarrow$ (iii) by Theorem 2.1.1. Suppose now that H is a closed subset of G. Then H is locally compact and second countable in the relative topology inherited from G. So, by Lemma 2.5.3, the identity mapping of H into G is a homeomorphism into. This shows that H is a topological subgroup of G. Theorem 2.5.4 is proved.

It is easy to give examples of analytic subgroups which are not topological subgroups. For example, let $G = \mathbf{T}^2$, the two-dimensional torus (cf. (2.1.4)). Let h be the map of $\mathbf{R}^1$ into G given by

$$h(t) = (e^{it}, e^{i\alpha t}) \quad (t \in \mathbf{R}^1), \tag{2.5.7}$$

α being an irrational number. Let $H = h[\mathbf{R}^1]$. Then H is a subgroup of G. It is classical that H is not closed in G—it is actually dense in G. We give to H the analytic structure for which h is an analytic diffeomorphism of $\mathbf{R}^1$ with H. It is then obvious that H is an analytic subgroup of G.

As an example illustrating Theorem 2.5.2, consider $G = GL(n,\mathbf{R})$ and $H = O(n,\mathbf{R})$. H is a closed Lie subgroup of G of dimension $\frac{1}{2}n(n-1)$ (cf. §2.1). We identify canonically the Lie algebra of G with $\mathfrak{g} = \mathfrak{gl}(n,\mathbf{R})$. We now determine the subalgebra $\mathfrak{h}$ of $\mathfrak{g}$ defined by H. We have

$$\dim(\mathfrak{h}) = \tfrac{1}{2}n(n-1). \tag{2.5.8}$$

Let a_{kl} $(1 \leq k, l \leq n)$ be the linear function on $\mathfrak{M}(n,\mathbf{R})$ whose value at any matrix A is its klth entry. Fix $X \in \mathfrak{h}$, and let $X = (b_{ij})_{1 \leq i, j \leq n}$. Then the tangent vector corresponding to X at 1 is

$$X_1 = \sum_{1 \leq k, l \leq n} b_{kl}\left(\frac{\partial}{\partial a_{kl}}\right)_1.$$

Let q_{ij} be the functions defined by (2.1.7) and let Q be the $n \times n$ matrix-valued function whose ijth entry is q_{ij}. Since $q_{ij} = 0$ on H, we have $X_1 q_{ij} = 0$ $(1 \leq i, j \leq n)$, i.e.,

$$\sum_{1 \leq k, l \leq n} b_{kl}\left(\frac{\partial Q}{\partial a_{kl}}\right)_1 = 0. \tag{2.5.9}$$

In view of (2.1.8), we get from (2.5.9)

$$X + X^t = 0, \tag{2.5.10}$$

i.e., X is skew-symmetric. Thus $\mathfrak{h} \subseteq \mathfrak{o}(n,\mathbf{R})$, the Lie algebra of all skew-symmetric matrices. Since dim $\mathfrak{o}(n,\mathbf{R}) = \frac{1}{2}n(n-1)$, we conclude from (2.5.8) that $\mathfrak{h} = \mathfrak{o}(n,\mathbf{R})$.

Given a Lie group G with Lie algebra $\mathfrak{g}$ and an analytic subgroup H of G, Theorem 2.5.2 enables us to identify in a canonical fashion the Lie algebra of H with the subalgebra of $\mathfrak{g}$ defined by H. In particular, the Lie algebras of analytic subgroups of $GL(n,\mathbf{R})$ may be canonically identified with subalgebras of $\mathfrak{gl}(n,\mathbf{R})$. The above example is an illustration of this identification.

2.6. Locally Isomorphic Groups

As we have seen, the Lie algebra of a Lie group G is already determined by the component of the identity G^0 of G. Actually, the converse of this is also true. It turns out that the Lie algebra of G completely determines the local structure of G; i.e., two Lie groups have isomorphic Lie algebras if and only if they are locally isomorphic. These facts constitute what are usually referred to as the first and second fundamental theorems of Sophus Lie. The proof of this theorem will be given in §2.8. We shall devote this section to a brief recapitulation of the basic facts of the theory of a certain class of connected groups and their covering groups. This theory is well known, and we refer the reader to Pontryagin [1] and Spanier [1].

We begin with the notion of paths and their homotopy. Let X be a Hausdorff topological space. A *path* in X is a continuous map f of the unit interval [0,1] into X; if $f(0) = f(1) = x$, f is said to be a *closed path about* x. If f is constant it is said to be a *null* path. X is said to be *arcwise connected* if, given any two points $x, y \in X$, there is a path f in X with $f(0) = x$ and $f(1) = y$. X is said to be *locally arcwise connected* if the collection of all arcwise connected open sets forms a base for the topology of X. Suppose now that X is arcwise connected and locally arcwise connected. Two paths f and g in X are said to be *homotopic* ($f \sim g$ in symbols) if there is a continuous map φ of the unit square $[0,1] \times [0,1]$ into X such that $\varphi(0,t) = f(t)$, $\varphi(1,t) = g(t)$ $(0 \leq t \leq 1)$ and $\varphi(s,0) = f(0) = g(0)$, $\varphi(s,1) = f(1) = g(1)$ $(0 \leq s \leq 1)$. If f and g are paths with $f(1) = g(0)$, their *composition* $h = fg$ is the path given by $h(t) = f(2t)$ $(0 \leq t \leq \frac{1}{2})$ and $h(t) = g(2t-1)$ $(\frac{1}{2} \leq t \leq 1)$, while f^{-1} is the path given by $f^{-1}(t) = f(1-t)$ $(0 \leq t \leq 1)$. It is easy to prove that homotopy is an equivalence relation in the set of paths and that this relation respects formation of compositions and inverses. Let $x \in X$. Then the set of all equivalence classes of closed paths about x becomes a group under composition, the identity of the group being the class of closed paths about x which are homotopic to the null path. The isomorphism class of this group is independent of x. We denote it by $\pi_1(X)$ and call it the *fundamental group* of X. X is said to be *simply connected* if $\pi_1(X)$ is trivial. X is said to be *locally simply connected* if for any

$x \in X$ and any open set U containing x, we can find an open set V such that $x \in V \subseteq U$ and any closed path in V is homotopic in U to a null path. X is called *admissible* if it is arcwise connected, locally arcwise connected and locally simply connected. Any convex open subset of $\mathbf{R}^n$ is simply connected. If $n \geq 3$, the unit sphere in $\mathbf{R}^n$ is simply connected. Any connected manifold is an admissible topological space, and at each of its points there is a basis of simply connected open neighborhoods.

Let X be an admissible topological space. A pair $(\tilde{X},\omega)$ is called a *covering* of X if $\tilde{X}$ is admissible, ω is a continuous map of $\tilde{X}$ onto X, and if the following condition is satisfied: given $x \in X$, there is a connected open set U containing x such that $\omega^{-1}(U)$ is a disjoint union of open sets $\{V_j\}$ and ω is a homeomorphism of V_j onto U for each j; U is said to be *evenly covered* in this situation. $\tilde{X}$ is called the *covering space* and ω the *covering map*. Coverings (X_1,ω_1) and (X_2,ω_2) of X are said to be *equivalent* if there is a homeomorphism ω of X_1 onto X_2 such that $\omega_1 = \omega_2\omega$.

The fundamental theorem of the theory of covering spaces can now be described. Let X be an admissible topological space. Then there is a covering of X with the property that the covering space is simply connected; moreover this covering is determined up to equivalence. It is called a *universal covering* of X because it possesses the following property. Let $(\tilde{X},\omega)$ be a covering of X such that $\tilde{X}$ is simply connected and let (X_1,ω_1) be any covering of X; then there is a continuous map ω' of $\tilde{X}$ onto X_1 such that $\omega = \omega_1\omega'$ and $(\tilde{X},\omega')$ is a covering of X_1. The covering space of a universal covering is called a *universal covering space* of X.

A topological group G is said to be *admissible* if its underlying topological space is admissible in the sense just described. Suppose G is an admissible topological group and $(\tilde{G},\omega)$ a covering of the topological space G. Then a multiplication can be introduced into $\tilde{G}$ under which it becomes a topological group (necessarily admissible) and ω becomes a continuous homomorphism with discrete kernel. Conversely, let $\tilde{G}$ be an admissible topological group, N a discrete normal subgroup, $G = \tilde{G}/N$, and ω the natural homomorphism of $\tilde{G}$ onto G. Then G is admissible, N is contained in the center of $\tilde{G}$, and $(\tilde{G},\omega)$ is a covering of G. Motivated by these results, we introduce the following definition. Let G be an admissible group. An admissible group $\tilde{G}$ is said to be a *covering group* of G if there exists a continuous homomorphism ω with discrete kernel mapping $\tilde{G}$ onto G; ω is then called the *covering homomorphism*. Given any admissible group G there exists a simply connected covering group of G. It is determined up to isomorphism and is known as the *universal covering group* of G. More precisely, let G_j be a simply connected covering group of G, and ω_j the corresponding covering homomorphism $(j = 1, 2)$; then there is an isomorphism ω of G_1 onto G_2 (as topological groups) such that $\omega_1 = \omega_2\omega$. If G_j is a covering group of G with covering homomorphism ω_j $(j = 1, 2)$ and

If G_1 is simply connected, there is a continuous homomorphism ω of G_1 onto G_2 with discrete kernel such that $\omega_1 = \omega_2 \omega$.

One of the most important and useful properties of a simply connected topological group is the possibility of extending local homomorphisms to global ones (*principle of monodromy*). To describe this result precisely we need the concept of a local homomorphism. Let G_j $(j = 1, 2)$ be groups with G_1 arcwise connected. By a *local homomorphism* of G_1 into G_2 we mean a map φ of an open neighborhood U of the identity of G_1 into G_2 with the following property: if a, b, and ab are all in $U, \varphi(ab) = \varphi(a)\varphi(b)$. A homomorphism ψ of G_1 into G_2 is said to *extend* a local homomorphism φ if $\psi(a) = \varphi(a)$ for all a in some open neighborhood of the identity. Since G_1 is connected, it is generated by any arbitrary open neighborhood of its identity, and so there can be at most *one* homomorphism of G_1 into G_2 which extends a given local homomorphism. The *principle of monodromy* asserts that if G_1 is admissible and simply connected, then any local homomorphism of G_1 into G_2 possesses a unique extension to a global homomorphism of G_1 into G_2.

The concept of local homomorphism leads naturally to that of local isomorphism. Let G_j $(j = 1, 2)$ be admissible groups. We shall say that G_1 and G_2 are *locally isomorphic* if there are open neighborhoods U_j of the identity of G_j $(j = 1, 2)$ and a homeomorphism φ of U_1 onto U_2 such that the following condition is satisfied: if $a, b \in U_1$, then $ab \in U_1$ if and only if $\varphi(a)\varphi(b) \in U_2$, and in that case $\varphi(ab) = \varphi(a)\varphi(b)$. Let G be an admissible group and G_1 a covering group of G. Then G and G_1 are locally isomorphic. The relation of local isomorphism is reflexive, symmetric, and transitive. The fundamental result in this context is the one which asserts that two admissible groups G_1 and G_2 are locally isomorphic if and only if there is an admissible group which covers both. In particular, let G be a given admissible group, $\tilde{G}$ the universal covering group of G; then the groups of the form $\tilde{G}/N$, where N is a discrete normal (hence central) subgroup of $\tilde{G}$, are admissible and locally isomorphic to G; moreover, any admissible group locally isomorphic to G is isomorphic to a group of the above type.

Let us now consider the case of an analytic (= connected Lie) group G. Since a connected analytic manifold is necessarily arcwise connected, locally arcwise connected, and locally simply connected, it follows that G is admissible. In this case it turns out that all the admissible groups locally isomorphic to G become analytic groups in a natural manner. The proof of this depends on the following lemma.

Lemma 2.6.1. *Let G be a connected second countable topological group. Suppose U and V are two open neighborhoods of the identity having the following properties:*

(2.6.1) $\begin{cases} \text{(a)} & \textit{there is an analytic structure on } V \textit{ converting it into an analytic manifold} \\ \text{(b)} & UU^{-1} \subseteq V, \textit{ and the map } (u,v) \mapsto uv^{-1} \textit{ is analytic from } U \times U \textit{ into } V. \end{cases}$

Then there exists a unique analytic structure on G such that

(i) *some open neighborhood of* 1 *is an open submanifold of both G and V*
(ii) *for any* $a \in G$, *the left translation* l_a *is an analytic diffeomorphism of G onto itself.*

Moreover, G becomes an analytic group under this analytic structure.

Proof. Let W be an open set containing 1 with $W = W^{-1}$ and $WW \subseteq U$. W is an open submanifold of V. For any $a \in G$ we regard aW as an analytic manifold with the analytic structure for which $x \mapsto ax$ is an analytic diffeomorphism of W onto aW. Suppose $a, a' \in G$ are such that $aW \cap a'W \neq \varnothing$. Then $aW \cap a'W$ is an open submanifold of both aW and $a'W$. In fact, $a'^{-1}a$ and $a^{-1}a'$ belong to $WW \subseteq U$, and (2.6.1) shows that $x \mapsto a'^{-1}ax$ is an analytic diffeomorphism of $W \cap a^{-1}a'W$ onto $W \cap a'^{-1}aW$. This proves that there is an analytic structure on G such that (i) all the aW are open submanifolds of G, and (ii) the left translations l_a $(a \in G)$ are analytic diffeomorphisms of G.

Suppose there exists another analytic structure for G such that the requirements (i) and (ii) of the lemma are satisfied. Let $\bar{G}$ denote the analytic manifold obtained by equipping the topological space G with this analytic structure. The identity map i $(\bar{G} \longrightarrow G)$ is then analytic at 1 with a bijective differential. But then condition (ii) implies that i is analytic at any $a \in G$ and has a bijective differential. i is thus an analytic diffeomorphism.

We now prove that this analytic structure converts G into an analytic group. Write $i_a x = axa^{-1}$, $a, x \in G$. If N is an open neighborhood of 1 such that $N = N^{-1}$ and $NNN \subseteq W$, it follows from (2.6.1) that $(x,y,z) \mapsto xyz$ is an analytic map of $N \times N \times N$ into W. In particular for $a \in N$, i_a is an analytic map of N into W. Using left translations, we conclude easily that i_a is an analytic map of G onto itself for all $a \in N$. Now G, being connected, is generated by the elements of N. Further, $i_{ab} = i_a i_b$ and $i_{a^{-1}} = i_a^{-1}$, $a, b \in G$. Hence i_a is an analytic diffeomorphism of G for all $a \in G$.

We now prove that $(x,y) \mapsto xy^{-1}$ is an analytic map of $G \times G$ into G. Fix $x_0, y_0 \in G$. To prove the analyticity of this map at (x_0,y_0), it is enough to prove that $(x,y) \mapsto (x_0x)(y_0y)^{-1}$ is analytic at $(1,1)$, in view of the fact that left translations are analytic diffeomorphisms. Now $(x_0x)(y_0y)^{-1} = l_{x_0y_0^{-1}} i_{y_0}(xy^{-1})$. So, since $(x,y) \mapsto xy^{-1}$ is analytic at $(1,1)$ by (2.6.1), the required conclusion follows from the analyticity of $l_{x_0y_0^{-1}}$ and i_{y_0}. G is thus an analytic group.

Corollary 2.6.2. *Let G_1 and G_2 be admissible groups satisfying the second axiom of countability. Let*

$$\pi : G_1 \longrightarrow G_2$$

be a continuous homomorphism whose kernel D is a discrete subgroup of G_1 and which maps G_1 onto G_2. Suppose G_1 (resp. G_2) is an analytic group. Then there exists exactly one analytic structure on G_2 (resp. G_1) converting it into an analytic group for which π is an analytic homomorphism.

Proof. For definiteness, let G_1 be an analytic group. Since D is discrete, we can select a compact neighborhood K_1 of the identity 1_1 of G_1 such that $K_1K_1^{-1} \cap D = \{1_1\}$. Let V_1 be the interior of K_1 and $V = \pi[V_1]$. Then V is open in G_2 by Lemma 2.5.3, $1_2 \in V$, and π is a homeomorphism of V_1 onto V. We give to V the analytic structure which makes π a diffeomorphism of V_1 onto V and select an open neighborhood U of 1_2 such that $UU^{-1} \subseteq V$. Then U, V satisfy the conditions of the lemma. So we can convert G_2 into an analytic group in such a way that π is analytic at 1_1. Since π is a homomorphism, π is analytic everywhere.

Now observe that π is an analytic diffeomorphism of an open neighborhood of the identity of G_1 onto an open neighborhood of the identity of G_2. It is clear from Lemma 2.6.1 that this requirement on π determines the analytic structure on G_2 uniquely. On the other hand, it follows from Corollary 2.7.4 (see below) that a surjective analytic homomorphism whose kernel is discrete is necessarily a local diffeomorphism. Hence, the above analytic structure G_2 is the only one with the properties stated in the corollary.

The case when the roles of G_1 and G_2 are interchanged is handled similarly. This completes the proof of the corollary.

Now consider an analytic group G. Let $\tilde{G}$ be a universal covering group of G and ω a covering homomorphism. It follows from Corollary 2.6.2 that we can convert $\tilde{G}$ into an analytic group so that ω becomes an analytic map which is locally diffeomorphic. Since any admissible group locally isomorphic to G is isomorphic to a group $\tilde{G}/D$ where D is a discrete normal subgroup of $\tilde{G}$, Corollary 2.6.2 applies once again and shows that all such admissible groups can be converted into analytic groups in a perfectly natural fashion. This is the fact we mentioned earlier.

We give some examples.

(1) Let $G = \mathbf{R}^m$ and let π be the homomorphism of G onto $\mathbf{T}^m$ given by (2.1.5). Then kernel(π) $= \mathbf{Z}^m$. It is easy to see that the analytic structure defined on $\mathbf{T}^m$ by Corollary 2.6.2 is the same as the one introduced in §2.1.

(2) Let $G = SU(2,\mathbf{C})$. If $x \in GL(2,\mathbf{C})$, then $x \in G$ if and only if

$$x = \begin{pmatrix} \alpha & \beta \\ -\beta^{\mathrm{conj}} & \alpha^{\mathrm{conj}} \end{pmatrix}, \tag{2.6.2}$$

where $\alpha, \beta \in \mathbf{C}$ and $|\alpha|^2 + |\beta|^2 = 1$. Thus the map

$$x \mapsto (\mathrm{Re}\,\alpha, \mathrm{Im}\,\alpha, \mathrm{Re}\,\beta, \mathrm{Im}\,\beta)$$

is a homeomorphism of G onto the unit sphere in $\mathbf{R}^4$. G is thus simply connected. It is easy to see that the center C of G consists of the two matrices $\pm\begin{pmatrix} 1 & 0 \\ 0 & 1 \end{pmatrix}$. Thus, up to isomorphism, G and G/C are the only admissible groups locally isomorphic to G. Actually, $G/C \simeq SO(3,\mathbf{R})$, as we shall see presently.

Let V be the vector space (over $\mathbf{R}$) of all 2×2 Hermitian matrices of trace 0. The map $(x_1,x_2,x_3) \mapsto v(x_1,x_2,x_3)$, where

$$v(x_1,x_2,x_3) = \begin{pmatrix} x_1 & x_2 + ix_3 \\ x_2 - ix_3 & -x_1 \end{pmatrix} \quad ((x_1,x_2,x_3) \in \mathbf{R}^3)$$

is a linear isomorphism of $\mathbf{R}^3$ onto V. For $g \in G$ let $\pi'(g)$ be the linear automorphism $v \mapsto gvg^{-1}$ of V, and let $\pi(g)$ be the linear automorphism of $\mathbf{R}^3$ which corresponds to $\pi'(g)$ under the identification $(x_1,x_2,x_3) \mapsto v(x_1,x_2,x_3)$. Since $\det(v(x_1,x_2,x_3)) = -(x_1^2 + x_2^2 + x_3^2)$ and $\det(v) = \det(\pi'(g)v)$ for $v \in V$, it follows that $\pi(g)$ leaves the form $(x_1,x_2,x_3) \mapsto x_1^2 + x_2^2 + x_3^2$ invariant. So $\pi[G] \subseteq O(3,\mathbf{R})$. Since $\pi(G)$ is connected, $\pi[G] \subseteq SO(3,\mathbf{R})$. An explicit calculation reveals that arbitrary rotations in $\mathbf{R}^3$ around the coordinate axes can be obtained as $\pi(g_t)$ $(t \in \mathbf{R})$, where for g_t we take one of the following three forms:

$$g_t = \begin{pmatrix} e^{it} & 0 \\ 0 & e^{-it} \end{pmatrix}, \quad g_t = \begin{pmatrix} \cos t & \sin t \\ -\sin t & \cos t \end{pmatrix}, \quad g_t = \begin{pmatrix} \cos t & i\sin t \\ i\sin t & \cos t \end{pmatrix}.$$

It follows from this that $\pi[G] = SO(3,\mathbf{R})$. A straightforward argument shows that the kernel of π is C. In particular, $\pi_1(SO(3,\mathbf{R})) = \mathbf{Z}_2$.

(3) $G = SL(2,\mathbf{C})$. It is easily checked that the center C of G is the subgroup consisting of the matrices $\pm\begin{pmatrix} 1 & 0 \\ 0 & 1 \end{pmatrix}$. We show that G is simply connected. This will prove that, up to isomorphism, G and G/C are the only admissible groups locally isomorphic to G.

Let $n \geq 1$ be any integer and let D be the set of all $n \times n$ complex Hermitian matrices which are nonnegative definite. Put $P = D \cap SL(n,\mathbf{C})$. It is a simple consequence of the diagonalization theorem that given any $p \in D$, there is a unique $p' \in D$ such that $p'^2 = p$. It is customary to denote p' by $p^{1/2}$. It is obvious that $\|p^{1/2}\| = \|p\|^{1/2}$, $\|\cdot\|$ being the usual norm of elements of $\mathfrak{M}(n,\mathbf{C})$ regarded as endomorphisms of the complex Hilbert space $\mathbf{C}^n$. Thus the map $p \mapsto p^{1/2}$ of D onto itself is one-to-one, maps compact sets into compact sets, and has a continuous inverse. It is therefore a homeomorphism. Note that it leaves P invariant.

Consider now any $x \in GL(n,\mathbf{C})$. Then there are uniquely defined matrices u and p such that u is unitary, p is Hermitian positive definite, and $x = up$. In fact, we must have $p = (x^\dagger x)^{1/2}$; then $u = xp^{-1}$ and is unitary. This is the classical *polar decomposition* of x. From the formulae for u and p it is clear that if $x \in SL(n,\mathbf{C})$, then $u \in SU(n,\mathbf{C})$ and $p \in P$. From the fact that $p \mapsto p^{1/2}$ is a homeomorphism of P onto itself, it follows that the map $(u,p) \mapsto up$ is a homeomorphism of $SU(n,\mathbf{C}) \times P$ onto $SL(n,\mathbf{C})$; it is also a homeomorphism of $U(n,\mathbf{C}) \times P^+$ onto $GL(n,\mathbf{C})$, where $P^+ = D \cap GL(n,\mathbf{C})$.

Now suppose $n = 2$. We know from (2) that $SU(2,\mathbf{C})$ is simply connected. So it remains to prove that P is simply connected. But P is the set of all matrices p of the form

$$p = \begin{pmatrix} a & \alpha \\ \alpha^{\text{conj}} & \dfrac{1 + |\alpha|^2}{a} \end{pmatrix},$$

where a is real and > 0 and $\alpha \in \mathbf{C}$. Thus P is homeomorphic to $(0,\infty) \times \mathbf{C}$, hence obviously simply connected.

(4) Let G, H be groups with H normal in G and G/H countable. Suppose H is an analytic group and that for each $x \in G$ the map $y \mapsto xyx^{-1}$ of H onto itself is an analytic diffeomorphism. We convert G into a topological group by requiring that H be an open (hence closed) subspace of G and equip it with the unique analytic structure for which aH is an open submanifold for each $a \in G$. We leave it to the reader to verify that G is a Lie group and that $H = G^0$.

2.7. Homomorphisms

We now prove two basic theorems concerning analytic homomorphisms from one Lie group to another. Their proofs depend on the following simple lemmas on certain involutive analytic systems. These are true for both real and complex manifolds; since the complex case needs no new arguments, we give the proofs only for real analytic manifolds.

Lemma 2.7.1. *Let M, N be analytic manifolds and let π be a submersive analytic map of M onto N. Then $\mathfrak{L} : x \mapsto$ kernel $(d\pi)_x$ $(x \in M)$ is an involutive analytic system of tangent spaces of rank equal to $\dim(M) - \dim(N)$. For* *any $x \in M$, let S_x be the connected component containing x of $\pi^{-1}(\{\pi(x)\})$. Then S_x is open and closed in $\pi^{-1}(\{\pi(x)\})$, is a closed regular submanifold of M, and is the maximal integral manifold of $\mathfrak{L}$ passing through x.*

Proof. Let $m = \dim(M)$, $n = \dim(N)$. Clearly, $m = n$ if and only if $(d\pi)_x$ is bijective for all $x \in M$. In this case π is a local diffeomorphism. If $x \in M$, $y = \pi(x)$, and U is an open subset of M containing x on which π is one-to-one, then $\{x\} = U \cap \pi^{-1}(\{y\})$. This proves that $\pi^{-1}(\{y\})$ is a discrete subset of M and establishes the lemma in this special case.

Let $m > n$, $x_0 \in M$, $y_0 = \pi(x_0)$. Choose coordinates $x_1, \ldots, x_m$ (resp. $y_1, \ldots, y_n$) on an open set U containing x_0 (resp. V containing y_0) such that $V = \pi[U]$, $y_i \circ \pi = x_i$ $(1 \leq i \leq n)$, and for some $a > 0$, $(x_1, \ldots, x_m)$ (resp. $(y_1, \ldots, y_n)$) maps U (resp. V) onto I_a^m (resp. I_a^n). It follows from this that for any $z \in U$, $\mathfrak{L}_z$ is spanned by $(\partial/\partial x_{n+1})_z, \ldots, (\partial/\partial x_m)_z$. This proves that $\mathfrak{L}$ is an involutive analytic system of tangent spaces of rank $m - n$. Let $U(x_0) = \{z : z \in U, x_1(z) = \cdots = x_n(z) = 0\}$. Then $U(x_0) = U \cap \pi^{-1}(\{\pi(x_0)\})$; moreover, $U(x_0)$ is a regular submanifold of M and is an integral manifold of $\mathfrak{L}$. Since x_0 is completely arbitrary, it follows from this that (i) for any $x \in M$, $\pi^{-1}(\{\pi(x)\})$ is a closed regular submanifold of M, and (ii) the connected components of $\pi^{-1}(\{\pi(x)\})$, being open and closed in $\pi^{-1}(\{\pi(x)\})$, are closed regular submanifolds of M and integral manifolds of $\mathfrak{L}$. Let $x \in M$, let S_x be the connected component of $\pi^{-1}(\{\pi(x)\})$ containing x, and let S be the maximal integral manifold of $\mathfrak{L}$ passing through x. Since S_x is an integral manifold of $\mathfrak{L}$, S_x is open in S. On the other hand, S_x is closed in M, and therefore closed in S. Since S is connected, $S = S_x$. This proves the lemma.

Lemma 2.7.2. *Let M, N be analytic manifolds and π an analytic map of M into N such that for some integer $p \geq 1$, $\dim(d\pi)_x[T_x(M)] = p$ for all $x \in M$. Then $x \mapsto$* kernel $(d\pi)_x$ *is an involutive analytic system of tangent spaces on M. Moreover, for each $y \in M$ we can find a connected open set U containing y such that $\pi[U]$ is a regular p-dimensional submanifold of N and such that π is a submersion of U onto $\pi[U]$.*

Proof. The entire lemma is local. To prove the first assertion, assume that M and N are open neighborhoods of the origin in $\mathbf{R}^m$ and $\mathbf{R}^n$ respectively, and that π is the map $(t_1, \ldots, t_m) \mapsto (u_1, \ldots, u_n)$, the u_j being analytic functions of $t_1, \ldots, t_m$ which vanish at $t_1 = \cdots = t_m = 0$. Let $\mathfrak{L}_x = \text{kernel}(d\pi)_x$, $x \in M$. By assumption, the matrix $(\partial u_i/\partial t_j)_{1 \leq i \leq n, 1 \leq j \leq m}$ has rank p on M. Without loss of generality we may assume that the submatrix $(\partial u_i/\partial t_j)_{1 \leq i \leq p, 1 \leq j \leq m}$ also has rank p on M. Suppose π_1 is the projection $(x_1, \ldots, x_n) \mapsto (x_1, \ldots, x_p)$ of $\mathbf{R}^n$ onto $\mathbf{R}^p$, and let $\pi' = \pi_1 \circ \pi$. Then $(d\pi')_x$ is surjective for all $x \in M$. In particular, $\pi'[M] = P$ is an open neighborhood of the origin in $\mathbf{R}^p$. From Lemma 2.7.1 it follows that $\mathfrak{L}' : x \mapsto \mathfrak{L}'_x = \text{kernel}(d\pi')_x$ is an involutive analytic system of tangent spaces of rank $m - p$. On the other hand, $\mathfrak{L}_x \subseteq \mathfrak{L}'_x$ and dim $\mathfrak{L}_x = m - p$ for all $x \in M$. Hence $\mathfrak{L}_x = \mathfrak{L}'_x$, $x \in M$.

To prove the second assertion, choose coordinates around any $y \in M$ which are adapted to $\mathfrak{L}$. We assume, in fact, that M is the cube I_a^m for some $a > 0$, that y is the origin of coordinates, and that $\mathfrak{L}_x$ is spanned by $(\partial/\partial t_{p+1})_x, \ldots, (\partial/\partial t_m)_x$ for all $x \in M$; moreover, that N is an open neighborhood of the origin in $\mathbf{R}^n$ and that π is the map $(t_1, \ldots, t_m) \mapsto (u_1, \ldots, u_n)$, the u_j being analytic functions of $t_1, \ldots, t_m$ which vanish at the origin. The fact that the $(\partial/\partial t_j)_x$ $(j > p)$ span kernel$(d\pi)_x$, $x \in M$, implies that $\partial u_r/\partial t_i = 0$ on M for $1 \leq r \leq n$, $p + 1 \leq i \leq m$. Thus the u_r are functions of $t_1, \ldots, t_p$ only. Let $v_r(t_1, \ldots, t_p) = u_r(t_1, \ldots, t_m)$, $1 \leq r \leq n$, $(t_1, \ldots, t_m) \in M$. Then $\pi = \bar{\pi} \circ \pi_2$, where π_2 is the projection $(t_1, \ldots, t_m) \mapsto (t_1, \ldots, t_p)$ of I_a^m onto I_a^p, and $\bar{\pi}$ is the map $(t_1, \ldots, t_p) \mapsto (v_1(t_1, \ldots, t_p), \ldots, v_n(t_1, \ldots, t_p))$ of I_a^p into N. It is obvious that $(d\bar{\pi})_x$ is injective for all $x \in I_a^p$. Hence we can find b with $0 < b < a$ such that $\bar{\pi}[I_b^p]$ is a regular p-dimensional submanifold of N and $\bar{\pi}$ is an analytic diffeomorphism of I_b^p onto it. Let $U = \pi_2^{-1}(I_b^p)$. Then U is a connected open neighborhood of the origin in I_a^m with the required properties.

Theorem 2.7.3. *Let G_j be a Lie group with Lie algebra $\mathfrak{g}_j$ $(j = 1, 2)$, and let π be an analytic homomorphism of G_1 into G_2. Then for any $X \in \mathfrak{g}_1$ there exists exactly one $X' \in \mathfrak{g}_2$ such that X and X' are π-related. Write $X' = (d\pi)(X)$. Then $d\pi$ is a homomorphism of $\mathfrak{g}_1$ into $\mathfrak{g}_2$. Let $\mathfrak{h}_1$ be the kernel of $d\pi$ and $\mathfrak{h}_2 = (d\pi)[\mathfrak{g}_1]$, and denote by H_j the analytic subgroup of G_j corresponding to $\mathfrak{h}_j$ $(j = 1, 2)$. Then* (i) *$\pi[G_1]$ is a Lie subgroup of G_2, π is an analytic map of G_1 onto $\pi[G_1]$, and $H_2 = \pi[G_1^0] = \pi[G_1]^0$; and* (ii) *$H_1$ is a closed analytic subgroup of G_1 and coincides with the component of the identity of the kernel of π.*

Proof. Let 1_j be the identity of G_j $(j = 1, 2)$. It is obvious that given $X \in \mathfrak{g}_1$, there is exactly one $X' \in \mathfrak{g}_2$ such that $X'_{1_2} = (d\pi)_{1_1}(X_{1_1})$. π being a homomorphism, we have

$$l_{\pi(x)} \circ \pi = \pi \circ l_x \quad (x \in G_1). \tag{2.7.1}$$

Consequently,

$$X'_{\pi(x)} = (d\pi)_x(X_x) \quad (x \in G_1). \tag{2.7.2}$$

The relation (2.7.2) shows that X and X' are π-related. The fact that $d\pi$ is a homomorphism of $\mathfrak{g}_1$ into $\mathfrak{g}_2$ follows at once from the π-relatedness proved above.

Define now

$$\begin{cases} \mathfrak{L}_x = \{X_x : X \in \mathfrak{h}_1\} & (x \in G_1) \\ \mathfrak{M}_y = \{X'_y : X' \in \mathfrak{h}_2\} & (y \in G_2). \end{cases} \tag{2.7.3}$$

It follows from (2.7.1) and (2.7.2) that

$$(2.7.4)\qquad \begin{cases} \mathfrak{L}_x = \text{kernel } (d\pi)_x \\ \mathfrak{M}_{\pi(x)} = (d\pi)_x[T_x(G_1)] \end{cases}$$

for all $x \in G_1$. Now, $\mathfrak{h}_2$ is a subalgebra of $\mathfrak{g}_2$, so by Lemma 2.5.1 and Theorem 2.5.2 $\mathfrak{M}$ $(y \mapsto \mathfrak{M}_y)$ is an involutive analytic system of tangent spaces whose maximal integral manifolds are H_2 and its left cosets. On the other hand, by Lemma 2.7.2 we can find a connected open set U containing 1_1 such that $\pi[U] = N$ is a p-dimensional regular submanifold of G_2, where $p = \dim \mathfrak{h}_2$. This shows that for $x \in U$, $\mathfrak{M}_{\pi(x)} = (d\pi)_x[T_x(G_1)]$ is a subspace of $T_{\pi(x)}(N)$. Since $\dim \mathfrak{M}_{\pi(x)} = p$, we conclude that N is an integral manifold of $\mathfrak{M}$. Since $1_2 \in N$, it follows that N is an open submanifold of H_2. Now, U generates G_1^0 and N generates H_2, so we must have $\pi[G_1^0] = H_2$. Since H_2 is quasi-regularly imbedded in G_2, it follows that the inner automorphisms of $\pi[G_1]$ induce analytic automorphisms of H_2. Hence (cf. example (4) of §2.6) we may convert $\pi[G_1]$ into a Lie group in such a way that $H_2 = \pi[G_1]^0$ is an open submanifold of it. Since H_2 is an analytic subgroup of G_2, $\pi[G_1]$ is a Lie subgroup of G_2. Now, H_2 being a quasi-regular submanifold of G_2, π is an analytic map of G_1^0 into H_2. Since π is a homomorphism, we conclude that π is an analytic map of G_1 into $\pi[G_1]$.

It is clear from (2.7.4) that π is a submersion of G_1 onto $\pi[G_1]$. We may therefore conclude from Lemma 2.7.1 that $\mathfrak{L}$ $(x \mapsto \mathfrak{L}_x)$ is an involutive analytic system of tangent spaces on G_1, that $H'_1 = \pi^{-1}(1_2)$ is a closed regular submanifold of G_1, and that the component of H'_1 containing 1_1 is the maximal manifold of $\mathfrak{L}$ through 1_1. This shows that H'_1, the kernel of π, is a closed Lie subgroup of G_1 (Theorem 2.1.1) and that H_1 is the component of the identity of H'_1. This proves everything we wanted.

The following corollary is immediate from Theorem 2.7.3.

Corollary 2.7.4. *Let π be as above. Then $d\pi$ is surjective if and only if $\pi[G_1^0] = G_2^0$, and it is injective if and only if the kernel of π is discrete.*

$d\pi$ is called the *differential of* π. In connection with applications it is often useful to observe the following method of determining $d\pi$ when π is given. Suppose that $X \in \mathfrak{g}_1$ and that x $(t \mapsto x(t))$ is a C^∞ or an analytic map of an open neighborhood of $t = 0$ such that $x(0) = 1_1$ and $(dx/dt)_{t=0} = X_{1_1}$ (cf. (1.1.20)). Then

$$(2.7.5)\qquad (d\pi)(X)_{1_2} = \left(\frac{d}{dt}(\pi \circ x)\right)_{t=0}$$

If G_1, G_2, G_3 are three analytic groups and if π $(G_1 \to G_2)$ and π' $(G_2 \to G_3)$ are analytic homomorphisms, then it is obvious that $d(\pi' \circ \pi) = d\pi' \circ d\pi$.

Suppose $G, G_1, \ldots, G_n$ are analytic groups with respective Lie algebras $\mathfrak{g}$, $\mathfrak{g}_1, \ldots, \mathfrak{g}_n$, and let π_j be an analytic homomorphism of G into G_j, $1 \leq j \leq n$; then $\pi_0 : x \mapsto (\pi_1(x), \ldots, \pi_n(x))$ is an analytic homomorphism of G into $G_1 \times \cdots \times G_n$, and $d\pi_0$ is the homomorphism $X \mapsto (d\pi_1(X), \ldots, d\pi_n(X))$ of $\mathfrak{g}$ into $\mathfrak{g}_1 \times \cdots \times \mathfrak{g}_n$ (here we are canonically identifying the Lie algebra of $G_1 \times \cdots \times G_n$ with $\mathfrak{g}_1 \times \cdots \times \mathfrak{g}_n$).

The converse of Theorem 2.7.3 is not true in general. If λ is a homomorphism of $\mathfrak{g}_1$ into $\mathfrak{g}_2$, there may not exist an analytic homomorphism π of G_1 into G_2 for which $d\pi = \lambda$. The next theorem shows that such a π exists whenever G_1 is simply connected.

Theorem 2.7.5. *Let G_i be an analytic group with Lie algebra $\mathfrak{g}_i$ $(i = 1, 2)$, and let λ be a homomorphism of $\mathfrak{g}_1$ into $\mathfrak{g}_2$. Then there cannot exist more than one analytic homomorphism π of G_1 into G_2 for which $d\pi = \lambda$; if G_1 is simply connected, there is always one such π, and it is unique.*

Proof. Let $G = G_1 \times G_2$ and $\mathfrak{g} = \mathfrak{g}_1 \times \mathfrak{g}_2$. We shall canonically identify $\mathfrak{g}$ with the Lie algebra of G. Define $\mathfrak{h}$ to be the graph of λ, i.e.,

$$\mathfrak{h} = \{(X, \lambda(X)) : X \in \mathfrak{g}_1\}. \tag{2.7.6}$$

As λ is a homomorphism, $\mathfrak{h}$ is a subalgebra of $\mathfrak{g}$. So we can associate with $\mathfrak{h}$ the analytic subgroup of G which defines $\mathfrak{h}$; denote this subgroup by H.

Suppose now that π is an analytic homomorphism of G_1 into G_2 such that $d\pi = \lambda$. Then $\sigma : x \mapsto (x, \pi(x))$ is an analytic homomorphism of G_1 into G, and $d\sigma$ is the homomorphism $X \mapsto (X, \lambda(X))$ of $\mathfrak{g}_1$ into $\mathfrak{g}$. Consequently, $(d\sigma)[\mathfrak{g}_1] = \mathfrak{h}$, and we may conclude from Theorem 2.7.3 that $\sigma[G_1] = H$. In other words, H is the graph of π. Since λ determines H uniquely, we see that π is uniquely determined by λ.

Now assume that G_1 is simply connected. To prove the existence of an analytic homomorphism π of G_1 into G_2 such that $\lambda = d\pi$, let γ be the analytic homomorphism $(x_1, x_2) \mapsto x_1$ of G onto G_1; then $d\gamma$ is the map $(X_1, X_2) \mapsto X_1$ of $\mathfrak{g}$ onto $\mathfrak{g}_1$. Let τ denote the restriction of γ to H; then $d\tau$ is the restriction of $d\gamma$ to $\mathfrak{h}$ (here we are canonically identifying $\mathfrak{h}$ with the Lie algebra of H). Now, the restriction of $d\gamma$ to $\mathfrak{h}$ is obviously an *isomorphism* of $\mathfrak{h}$ onto $\mathfrak{g}_1$. So, by Corollary 2.7.4, τ is a homomorphism of H onto G_1 with a discrete kernel. In other words, H is a covering group of G_1 and τ is a covering homomorphism. Since G_1 is simply connected, τ must be an isomorphism of H onto G_1 (as topological groups). On the other hand, the fact that $d\tau$ is an isomorphism of $\mathfrak{h}$ onto $\mathfrak{g}_1$ implies that $(d\tau)_h$ is a linear bijection of $T_h(H)$ onto $T_{\tau(h)}(G_1)$ for all $h \in H$. τ is therefore an analytic isomorphism. $\tau^{-1} = \sigma$ is then an analytic homomorphism of G_1 onto H. It is clear that

$$\sigma(x) = (x, \pi(x)) \quad (x \in G_1), \tag{2.7.7}$$

π being an analytic homomorphism of G_1 into G_2. From (2.7.7), $d\sigma(X) = (X, d\pi(X))$, $X \in \mathfrak{g}_1$. On the other hand, $d\sigma(X) = (X, \lambda(X))$, $X \in \mathfrak{g}_1$. So $\lambda = d\pi$. The proof of the theorem is complete.

The following example shows that one cannot drop the requirement of simple connectivity. Let $G_1 = SO(3,\mathbf{R})$, $G_2 = SU(2,\mathbf{C})$, and let σ be the covering homomorphism of G_2 onto G_1 constructed in example (2) of §2.6. By Corollary 2.7.4, $d\sigma$ is an isomorphism of $\mathfrak{g}_2$ onto $\mathfrak{g}_1$. Let $\lambda = (d\sigma)^{-1}$. Suppose there is an analytic homomorphism π of G_1 into G_2 such that $\lambda = d\pi$; by Corollary 2.7.4, π must be surjective. Since $\lambda \circ d\sigma = d\sigma \circ \lambda =$ identity, $\sigma \circ \pi = \pi \circ \sigma =$ identity, by the uniqueness part of Theorem 2.7.5. π is thus the inverse of σ. But this is a contradiction, since σ is not one-to-one and so does not possess an inverse.

Corollary 2.7.6. *Let G be a simply connected analytic group, $\mathfrak{g}$ its Lie algebra. Suppose ξ $(X \mapsto X^{\xi})$ is an automorphism of $\mathfrak{g}$. Then there exists exactly one automorphism η $(x \mapsto x^{\eta})$ of G such that $\xi = d\eta$.*

Proof. By Theorem 2.7.5 there is a unique homomorphism η of G into itself such that $\xi = d\eta$. It remains to check that η is an automorphism. By Corollary 2.7.4, η is surjective and its kernel is discrete. G is therefore a covering group of itself, η being the covering homomorphism. Since G is simply connected, η must be a bijection.

The assumption of simple connectedness is essential for the validity of this corollary. Suppose G is not simply connected. Let G^* be a simply connected covering group of G and let π be a covering homomorphism. Given ξ, there is an automorphism η^* of G^* such that $d\eta^* = \xi$. In order that there exist an automorphism η of G such that $d\eta = \xi$, it is necessary and sufficient that η^* leave the kernel of π invariant. This may not always happen.

2.8. The Fundamental Theorem of Lie

We shall now prove the basic theorem of Sophus Lie according to which the Lie algebra of a Lie group is a complete invariant of the local structure of the group.

Let G_1 and G_2 be two Lie groups. We say that they are *locally analytically isomorphic* if there are open neighborhoods U_1 and U_2 of the respective identities and an analytic diffeomorphism ω of U_1 onto U_2 such that the following condition is satisfied: if $x, y \in U_1$, then $xy \in U_1$ if and only if $\omega(x)\omega(y) \in U_2$, and then $\omega(xy) = \omega(x)\omega(y)$. It is obvious that local analytic isomorphism is an equivalence relation and that two Lie groups are locally analytically isomorphic if and only if their components of identity are.

As an example, let us consider two analytic groups G_1 and G_2 and suppose there is an analytic homomorphism ω of G_1 onto G_2 having a discrete kernel. Then G_1 and G_2 are locally analytically isomorphic. In fact, $d\omega$ is an isomorphism of the corresponding Lie algebras by Corollary 2.7.4, so we can find open neighborhoods V_j of the identity of G_j $(j = 1, 2)$ such that ω is an analytic diffeomorphism of V_1 onto V_2; if we select an open neighborhood $U_j = U_j^{-1}$ of the identity of G_j such that $\omega[U_1] = U_2$ and $U_jU_j^{-1} \subseteq V_j$ $(j = 1, 2)$, then U_1, U_2, and ω satisfy the conditions of the definition. The next lemma shows that, when suitably generalized, the above example is typical.

Lemma 2.8.1. *Let G_1 and G_2 be two analytic groups. Then they are locally analytically isomorphic if and only if the following condition is satisfied: there exists a simply connected analytic group G and homomorphisms ω_j of G into G_j such that ω_j is analytic, maps G onto G_j, and has discrete kernel $(j = 1, 2)$.*

Proof. Let G_1 and G_2 be locally analytically isomorphic and suppose that U_1, U_2, and ω are as in the definition. Let G be a simply connected admissible covering group of G_1 with covering homomorphism ω_1. In view of the discussion in §2.6, we may assume that G is an analytic group and ω_1 an analytic homomorphism. By the monodromy principle, we can construct a continuous homomorphism ω_2 of G into G_2 such that $\omega_2(x) = \omega(\omega_1(x))$ for all x in an open neighborhood of the identity of G. From the properties of U_1, U_2, and ω it is clear that $\omega_2[G]$ contains an open set around the identity of G_2. Hence $\omega_2[G] = G_2$. Clearly, ω_2 is analytic at the identity of G. Being a homomorphism, ω_2 is analytic everywhere. On the other hand, both ω and ω_1 are diffeomorphisms around the respective identities, so that ω_2 is also a diffeomorphism around the identity of G. This shows that $d\omega_2$ is an isomorphism. By Corollary 2.7.4, the kernel of ω_2 is discrete. G, G_1, G_2 are thus in the prescribed relation to one other.

Conversely, suppose that G, G_1, G_2, ω_1, ω_2 satisfy the conditions of the lemma. Then G and G_j are locally analytically isomorphic, $j = 1, 2$. This shows that G_1 and G_2 are locally analytically isomorphic. This proves the lemma.

Theorem 2.8.2. *Let G_j be Lie groups and $\mathfrak{g}_j$ the corresponding Lie algebras, $j = 1, 2$. Then $\mathfrak{g}_1$ and $\mathfrak{g}_2$ are isomorphic if and only if G_1 and G_2 are locally analytically isomorphic.*

Proof. We may assume that G_1 and G_2 are analytic groups. Suppose G_1 and G_2 are locally analytically isomorphic; then there is a simply connected analytic group G and analytic homomorphisms ω_j of G onto G_j with discrete kernels. Let $\mathfrak{g}$ be the Lie algebra of G. Then, by Corollary 2.7.4, $d\omega_j$ is an isomorphism of $\mathfrak{g}$ onto $\mathfrak{g}_j$. So $(d\omega_2)(d\omega_1)^{-1}$ is an isomorphism of $\mathfrak{g}_1$ onto $\mathfrak{g}_2$.

Conversely, let λ be an isomorphism of $\mathfrak{g}_1$ onto $\mathfrak{g}_2$. Let G be a simply connected covering group of G_1, with ω_1 as the covering homomorphism. We regard G as an analytic group such that ω_1 is analytic. Let $\mathfrak{g}$ be the Lie algebra of G; then $\lambda \circ d\omega_1$ is an isomorphism of $\mathfrak{g}$ onto $\mathfrak{g}_2$. By Theorem 2.7.5, there is an analytic homomorphism ω_2 of G into G_2 such that $d\omega_2 = \lambda \circ d\omega_1$. Since $\lambda \circ d\omega_1$ is an isomorphism, we conclude by Corollary 2.7.4 that ω_2 is surjective and has discrete kernel. By Lemma 2.8.1, G_1 and G_2 are locally analytically isomorphic. This proves the theorem.

2.9. Closed Lie Subgroups and Homogeneous Spaces. Orbits and Spaces of Orbits

Let G be a Lie group and H a closed Lie subgroup. We show in this section that there exists a natural analytic structure on the left coset space G/H converting it into an analytic manifold on which G "acts analytically."

Let G be a topological group and M a Hausdorff topological space. We say that G *acts* on M if there exists a continuous map

$$(g,x) \mapsto g \cdot x \quad (g \in G, x \in M, g \cdot x \in M) \tag{2.9.1}$$

of $G \times M$ into M such that

$$\begin{cases} \text{(i)} \quad 1 \cdot x = x \quad (x \in M) \\ \text{(ii)} \quad (g_1 g_2) \cdot x = g_1 \cdot (g_2 \cdot x) \quad (g_1, g_2 \in G, x \in M). \end{cases} \tag{2.9.2}$$

If we define

$$t_g(x) = g \cdot x \quad (g \in G, x \in M), \tag{2.9.3}$$

then t_g is a homeomorphism of M onto itself for any $g \in G$, and

$$\begin{cases} t_{g_1 g_2} = t_{g_1} t_{g_2} \quad (g_1, g_2 \in G) \\ t_1 = \text{identity}. \end{cases} \tag{2.9.4}$$

Under these circumstances M is called a *G-space*. G is said to act *transitively* on M if there exists an $x_0 \in M$ such that $g \mapsto g \cdot x_0$ maps G onto M; M is then known as a *transitive G-space* or a *homogeneous space*. In this case, given $x, x' \in M$, there is some $g \in G$ such that $g \cdot x = x'$. If M is a G-space and $x_0 \in M$,

$$G_{x_0} = \{g : g \in G, g \cdot x_0 = x_0\} \tag{2.9.5}$$

is a closed subgroup of G, called the *stability subgroup* at x_0. If M is a transitive G-space and $x, x' \in M$, one has

$$G_{x'} = hG_x h^{-1}, \tag{2.9.6}$$

where h is any element of G with the property that $h \cdot x = x'$; thus the stability subgroups at the various points of M are mutually conjugate.

Suppose now that G is a locally compact second countable group and G_0 a closed subgroup of G. Let $N = G/G_0$ and β the natural map of G onto N. We give to N the quotient topology with respect to β; i.e., a subset $V \subseteq N$ is open if and only if $\beta^{-1}(V)$ is open in G. It is known (and easy to prove) that N is a Hausdorff, locally compact, and second countable space with the above topology. We leave it to the reader to verify that the map

$$(g,\beta(a)) \mapsto g \cdot \beta(a) = \beta(ga) \quad (g, a \in G) \tag{2.9.7}$$

is well-defined and continuous from $G \times N$ into N and defines an action of G on N. It is obvious that G acts transitively on N and that G_0 is the stability subgroup at $\beta(1)$. The following lemma shows that every transitive G-space arises in this manner.

Lemma 2.9.1. *Let M be a G-space, both G and M being locally compact and second countable. Let G act transitively on M, let $x_0 \in M$, and let G_{x_0} be the stability subgroup at x_0. Then the map*

$$\pi : gG_{x_0} \mapsto g \cdot x_0 \quad (g \in G) \tag{2.9.8}$$

is well defined on G/G_{x_0}, is a homeomorphism of G/G_{x_0} onto M, and intertwines the actions of G on the two spaces. In particular, the map $g \mapsto g \cdot x_0$ is an open map of G onto M.

Proof. Let β denote the natural map of G onto $G/G_{x_0} = N$, and let γ be the map $g \mapsto g \cdot x_0$ of G onto M. γ is continuous, and $\gamma = \pi \circ \beta$. Since N has the quotient topology, we conclude that π is continuous. It is obvious that π is one-to-one and intertwines the G-actions on N and M. It remains to prove that π is an open map. Let V be a compact subset of N with non-null interior. Then there is a sequence $a_1, a_2, \ldots$ of elements of G such that $N = \bigcup_n (a_n \cdot V)$, from which we get $M = \bigcup_n (a_n \cdot \pi[V])$. By the Baire category theorem, some $a_n \cdot \pi[V]$, hence $\pi[V]$ itself, has non-null interior. Let $W =$ interior $\pi[V]$ and $V_1 = \pi^{-1}(W)$. Since π is one-to-one and $Cl(V_1)$ is compact, π is a homeomorphism of V_1 onto W. Now, any open subset of N is a union of sets of the form $a \cdot U$, where $a \in G$ and U is an open subset of V_1. Consequently, π is an open map. The proof of the lemma is complete.

Let G, M be locally compact and second countable, G acting on M. We shall say that G acts *analytically* on M if (i) G is a Lie group and M is an analytic manifold, and (ii) the map $(g,x) \mapsto g \cdot x$ is analytic from $G \times M$

into M. In this case each t_g ($g \in G$) is an analytic diffeomorphism of M onto itself. The main purpose of this section is to prove the analytic version of Lemma 2.9.1. This is valid for both real and complex Lie groups and manifolds; the proofs are given only in the real case. We need two lemmas.

Lemma 2.9.2. *Let G be a Lie group acting transitively and analytically on an analytic manifold M and let $x_0 \in M$. Then the stability subgroup G_{x_0} is a closed Lie subgroup of G. If G is transitive, the map*

$$\gamma : g \mapsto g \cdot g_0 \quad (g \in G) \tag{2.9.9}$$

is an analytic submersion of G onto M. In particular, M is the quotient manifold of G relative to the map γ.

Proof. We have

$$\gamma \circ l_g = t_g \circ \gamma \quad (g \in G), \tag{2.9.10}$$

l_g as usual denoting the left translation by g. If we write $V_g = (d\gamma)_g[T_g(G)]$, it follows from (2.9.10) that $V_g = (dt_g)_{x_0}[V_1]$. In particular, $\dim(V_g)$ is constant, $= k$ say, for all $g \in G$. If $y \in G_{x_0}$, we can choose, by Lemma 2.7.2, an open set U containing y, such that $\gamma[U]$ is a k-dimensional regular submanifold of M and such that $\gamma : U \longrightarrow \gamma[U]$ is a submersion. In particular, $\{z : z \in U, \gamma(z) = x_0\} = G_{x_0} \cap U$ is a regularly imbedded submanifold of U, hence of G. Since $y \in G_{x_0}$ is arbitrary, G_{x_0} is a regular submanifold of G, proving the first assertion. Suppose now that G acts transitively on M. Then $\gamma[U]$ is open in M, by Lemma 2.9.1. Hence $k = \dim M$, proving that $(d\gamma)_g$ is surjective for all $g \in G$. This proves the lemma.

Lemma 2.9.3. *Let G be a Lie group with Lie algebra $\mathfrak{g}$, and H a closed Lie subgroup which defines the subalgebra $\mathfrak{h}$ of $\mathfrak{g}$. We denote by $\mathfrak{L}$ ($x \mapsto \mathfrak{L}_x$) the involutive analytic system*[1] *of tangent spaces defined by $\mathfrak{L}_x = \{X_x : X \in \mathfrak{h}\}$, $x \in G$. Write $p = \dim \mathfrak{h}$. Then there exist an $a > 0$, an open subset U of G containing 1, and coordinates $x_1, \ldots, x_m$ on U such that*

$$\begin{cases} \text{(i)} \quad (U; x_1, \ldots, x_m; a) \text{ are adapted to } \mathfrak{L} \\ \text{(ii)} \quad zH \cap U = \{x : x \in U, x_i(x) = x_i(z) \text{ for } p < i \leq m\} \\ \hfill (z \in U). \end{cases} \tag{2.9.11}$$

Proof. Choose coordinates $x_1, \ldots, x_m$ on an open subset V of G containing 1 such that $(\partial/\partial x_1)_y, \ldots, (\partial/\partial x_p)_y$ span $\mathfrak{L}_y$ for all $y \in V$; we may (and do) assume that $y \mapsto (x_1(y), \ldots, x_m(y))$ is an analytic diffeomorphism of V onto

[1]Cf. Lemma 2.5.1.

some cube I_b^m ($b > 0$) and that $x_1(1) = \cdots = x_m(1) = 0$. For $0 < c \leq b$ we write $V_c = \{y : y \in V, (x_1(y), \ldots, x_m(y)) \in I_c^m\}$. For $0 < d \leq b$ and $\mathbf{a} = (a_{p+1}, \ldots, a_m) \in I^{m-p}$, let $V_d(\mathbf{a}) = \{y : y \in V_d, x_i(y) = a_i \text{ for } i > p\}$. For $0 < c \leq b$, $V_c(\mathbf{0})$ is an integral manifold of $\mathfrak{L}$ containing 1, and so it is an open submanifold of H. Moreover, H is regularly imbedded in G. So we can choose b_1 with $0 < b_1 \leq b$ such that $V_{b_1} \cap H = V_{b_1}(\mathbf{0})$; then select b_2, b_3 with $0 < b_j \leq b_1$ ($j = 2, 3$) such that $V_{b_3}^{-1}V_{b_3} \subseteq V_{b_2}$, $V_{b_2}V_{b_2} \subseteq V_{b_1}$. We now prove that (ii) of (2.9.11) is satisfied for $U = V_{b_3}$. This will prove the lemma, with $a = b_3$.

Let $u \in V_{b_3}$ and let $x_i(u) = a_i$, $p < i \leq m$, so that $u \in V_{b_3}(\mathbf{a})$, where $a = (a_{p+1}, \ldots, a_m)$; we must prove that $V_{b_3} \cap uH = V_{b_3}(\mathbf{a})$. Since $V_{b_3}(\mathbf{a})$ is an integral manifold of $\mathfrak{L}$ through u, $V_{b_3}(\mathbf{a}) \subseteq uH$, and we need verify only that $V_{b_3} \cap uH \subseteq V_{b_3}(\mathbf{a})$. Now, $V_{b_2} \cap H = V_{b_2}(\mathbf{0})$, so

$$\begin{aligned} V_{b_3} \cap uH &= u(u^{-1}V_{b_3} \cap H) \\ &\subseteq u(V_{b_2} \cap H) \\ &= uV_{b_2}(\mathbf{0}). \end{aligned}$$

On the other hand, $uV_{b_2}(\mathbf{0})$ is a connected integral manifold of $\mathfrak{L}$ contained in V_{b_1}, so the functions $x_{p+1}, \ldots, x_m$ are constant on $uV_{b_2}(\mathbf{0})$. This proves that $x_{p+1}, \ldots, x_m$ are constant on $V_{b_3} \cap uH$; i.e., $V_{b_3} \cap uH \subseteq V_{b_3}(\mathbf{a})$.

Theorem 2.9.4. *Let G be a Lie group, H a closed Lie subgroup. Then there exists exactly one analytic structure on $G/H = N$ which converts it into an analytic manifold such that the natural action of G on N is analytic. If M is any analytic manifold on which G acts analytically and transitively, $x_0 \in M$, and G_{x_0} is the stability subgroup at x_0, then the map*

$$gG_{x_0} \mapsto g \cdot x_0$$

is an analytic diffeomorphism of G/G_{x_0} onto M.

Proof. Let β be the natural map of G onto N. For each open set $V \subseteq N$ let $\mathfrak{A}(V)$ be the set of all complex-valued functions f defined on V such that $f \circ \beta$ is analytic on $\beta^{-1}(V)$. It is evident that for any $g \in G$, $f \in \mathfrak{A}(g \cdot V)$ if and only if $x \mapsto f(g \cdot x)$ belongs to $\mathfrak{A}(V)$; i.e., $\mathfrak{A}$ is invariant under G. We shall prove that $V \mapsto \mathfrak{A}(V)$ is an analytic structure on N. It is clearly enough to prove the following: for any $x \in N$, we can find an open set A containing x such that the restriction $\mathfrak{A}_A : V \mapsto \mathfrak{A}(V)$ (V open, $\subseteq A$) is an analytic structure on A. Since G is transitive on N and $\mathfrak{A}$ is invariant under G, it is sufficient to prove this for $x_0 = \beta(1)$.

We select an open subset U of G around 1, coordinates $x_1, \ldots, x_m$ on U, and $a > 0$ such that the conditions (2.9.11) are satisfied. Let $A = \beta[U]$.

Then A is an open neighborhood of x_0. Write

$$U_0 = \{x : x \in U, x_1(x) = \cdots = x_p(x) = 0\}. \tag{2.9.12}$$

It is then obvious from (ii) of (2.9.11) that β is an one-to-one map of U_0 onto A. We may therefore, by replacing a with a smaller number if necessary, assume that β is a homeomorphism of U_0 onto A. Clearly, U_0 is a regular submanifold (containing 1) of G of dimension $m - p$. Let ζ be the map of $U_0 \times H$ into G given by

$$\zeta(x,h) = xh \quad (x \in U_0, h \in H). \tag{2.9.13}$$

ζ is obviously analytic while (ii) of (2.9.11) implies that it is one-to-one. We assert that $d\zeta$ is surjective everywhere. For any $x \in U_0$, the image of $\{x\} \times H$ under ζ is xH, and $xH \cap U = \{z : z \in U, x_i(z) = x_i(x), p < i \leq m\}$, from which it follows that the range of $(d\zeta)_{(x,1)}$ contains $(\partial/\partial x_s)_x$, $s = 1, \ldots, p$; the image of $U_0 \times \{1\}$ being U_0, it is obvious that the range of $(d\zeta)_{(x,1)}$ contains $(\partial/\partial x_s)_x$, $p < s \leq m$. $(d\zeta)_{(x,1)}$ is thus surjective. For any $h \in H$, let α_h and β_h be the analytic diffeomorphisms of $U_0 \times H$ and G given respectively by $\alpha_h(x,k) = (x,kh)$ $(x \in U_0, k \in H)$ and $\beta_h(y) = yh (y \in G)$. Then $\beta_h \circ \zeta = \zeta \circ \alpha_h$, so

$$(d\zeta)_{(x,h)} = (d\beta_h)_x \circ (d\zeta)_{(x,1)} \circ (d\alpha_{h^{-1}})_{(x,h)}. \tag{2.9.14}$$

The relation (2.9.14) implies our assertion that $(d\zeta)_{(x,h)}$ is surjective for all $(x,h) \in U_0 \times H$. Since $\dim(U_0 \times H) = \dim G$, we may conclude that $\zeta[U_0 \times H] = U_0H$ is open in G and that ζ is an analytic diffeomorphism of $U_0 \times H$ onto U_0H.

Let γ be the map of A onto U_0 which inverts the restriction of β to U_0. For any open set $V \subseteq A$ and any function f defined on V, let f_0 be the function $x \mapsto f(\beta(x))$ defined on $\gamma[V]$. Then

$$(f \circ \beta \circ \zeta)(x,h) = f_0(x) \quad ((x,h) \in U_0 \times H). \tag{2.9.15}$$

Now, ζ is an analytic diffeomorphism of $U_0 \times H$ onto U_0H. Consequently, $f \mapsto f \circ \beta \circ \zeta$ is an isomorphism of the algebra $\mathfrak{A}(V)$ onto the algebra of all those analytic functions on $\gamma[V] \times H$ which depend only on the first argument. The relation (2.9.15) then implies that $f \mapsto f_0$ is an isomorphism of the algebra $\mathfrak{A}(V)$ onto the algebra of analytic functions on $\gamma[V]$ (considered as an open submanifold of U_0). This proves that $\mathfrak{A}_A$ is an analytic structure on A. Our construction makes it obvious that for any $g \in G$, $\beta(x) \mapsto \beta(gx)$ is an analytic diffeomorphism of N. On the other hand, $d\beta$ is surjective at 1. So, since G is transitive on N, $d\beta$ is surjective everywhere. β is therefore a submersion, and N is the quotient of the analytic manifold G relative to β.

Let η denote the map $(g,\beta(g')) \mapsto \beta(gg')$ $(g, g' \in G)$ of $G \times N$ into N. We denote by π the map $(g,g') \mapsto (g,\beta(g'))$ of $G \times G$ onto $G \times N$. We have

$$(2.9.16) \qquad (\eta \circ \pi)(g,g') = \beta(gg').$$

From (2.9.16) it follows that $\eta \circ \pi$ is an analytic map of $G \times G$ into N. On the other hand, it is obvious that π is a submersion, so $G \times N$ is the quotient of the analytic manifold $G \times G$ relative to π. We may therefore conclude that η is an analytic map of $G \times N$ into N. G thus acts analytically on N.

The uniqueness of the analytic structure on N and the last assertion both follow immediately from Lemmas 2.9.1 and 2.9.2.

The proof of the theorem is complete.

We shall refer to the analytic structure defined above on N as the *analytic structure induced from G.*

Let $V \subseteq N$ be an open set. By a *section for G/H defined on V* we mean an analytic map c of V into G such that

$$(2.9.17) \qquad \beta(c(x)) = x \quad (x \in V).$$

It follows from (2.9.17) that

$$(2.9.18) \qquad (d\beta)_{c(x)} \circ (dc)_x = \text{identity} \quad (x \in V).$$

The relation (2.9.18) shows that $(dc)_x$ is injective for all $x \in V$ and that, for any $x \in V$, the linear spaces $(dc)_x[T_x(N)]$ and $\mathfrak{L}_{c(x)}(= T_{c(x)}(c(x)H))$ are complementary in $T_{c(x)}(G)$. In geometric language, c is an imbedding of V into G with the following property: for any $x \in V$, there is an open neighborhood W of x such that $c[W]$ is a regular submanifold of G transversal to the left cosets of H. If a section is defined on all of N, it is called *global.*

Theorem 2.9.5. *Let G, H be as above and let $N = G/H$. Suppose that $x \in N$ and that $y \in G$ is such that $\beta(y) = x$. Then there is a section c defined in an open neighborhood of x such that $c(x) = y$.*

Proof. Let A and U_0 be as in the proof of Theorem 2.9.4. Then β is an analytic diffeomorphism of U_0 onto A. As before, let γ be the map of A onto U_0 which inverts the restriction of β to U_0. Then γ is a section for G/H, defined on A; and $\gamma(x_0) = 1$, where $x_0 = \beta(1)$. Since $\beta(y) = x$, we must have $y \cdot x_0 = x$. Define c on $y \cdot A$ by

$$(2.9.19) \qquad c(x') = y\gamma(y^{-1} \cdot x') \quad (x' \in y \cdot A).$$

It is easily verified that c is a section with the required properties.

The case when H is a normal subgroup of G is especially important.

Theorem 2.9.6. *Let G be a Lie group, H a closed normal Lie subgroup. Then the analytic structure induced on the topological group $N = G/H$ from G converts N into a Lie group. The natural map β is then an analytic homomorphism of G onto N.*

Proof. Let η be the map $(x,y) \mapsto xy^{-1}$ of $N \times N$ into N; $\bar{\eta}$ the map $(g,g') \mapsto gg'^{-1}$ of $G \times G$ into G; and π the map $(g,g') \mapsto (\beta(g),\beta(g'))$ of $G \times G$ onto $N \times N$. It is obvious that π is a submersion, and hence $N \times N$ is the quotient of the analytic manifold $G \times G$ with respect to π. On the other hand, $\eta \circ \pi = \beta \circ \bar{\eta}$ and so $\eta \circ \pi$ is an analytic map of $G \times G$ into N. So η is analytic from $N \times N$ into N. Theorem 2.9.6 follows at once from this.

When H is a closed normal Lie subgroup of G, the Lie group $N = G/H$ constructed above is called the *quotient of G by H.*

We now give an example where global sections do not exist. Let G be a compact Lie group and H a finite subgroup. We claim that if G is connected, then there exist no global sections for G/H. For if a global section c exists, then the compactness of G implies that $(h,x) \mapsto c(x)h$ is a homeomorphism of $H \times (G/H)$ onto G. This is a contradiction, since $H \times (G/H)$ is not connected. For examples of global sections we refer the reader to the exercises.

We now turn our attention to the case when a Lie group G acts analytically, but not necessarily transitively, on an analytic manifold M. For any $x \in M$, the set

$$G \cdot x = \{g \cdot x : g \in G\} \tag{2.9.20}$$

is called the *orbit* of the point x. Let G_x be the stability subgroup of x. Then the map

$$\pi_x : gG_x \mapsto g \cdot x \tag{2.9.21}$$

is well defined. By Lemma 2.9.2, G_x is a closed Lie subgroup of G. So, since G/G_x is the quotient of G with respect to the map $g \mapsto gG_x$, π_x is analytic. Equip $G \cdot x$ with the analytic structure with respect to which π_x is an analytic diffeomorphism; then we have

Theorem 2.9.7. *$G \cdot x$ is a submanifold of M of dimension $= \dim(G) - \dim(G_x)$. In order that $G \cdot x$ be a regular submanifold of M it is necessary and sufficient that it be locally closed in M.*

Proof. Let i be the inclusion map of $N = G \cdot x$ into M. As in Lemma 2.9.2, $(di)_y[T_y(N)]$ has the same dimension for all $y \in N$. Since i is one-to-one and one-to-one submersions are imbeddings, we can use Lemma 2.7.2 to find, for any $y \in N$, an open neighborhood U of y in N such that i is an imbedding of U in M. This proves that N is a submanifold of M. If N is regular, it has to be locally closed in M, by Theorem 1.1.4. Suppose conversely that N is locally closed in M. Then N is locally compact and second countable in the topology inherited from M. Lemma 2.9.1 now implies that π_x is a homeomorphism of G/G_x with the topological space obtained by equipping N with this relative topology. So N is a regular submanifold of M.

Corollary 2.9.8. *If G is compact, all orbits in M are regular submanifolds.*

When the orbits are regular submanifolds, it is natural to ask whether we can "parametrize" the orbits nicely. Let us write $X = M/G$ for the set of all orbits, and for any $x \in M$ let $\pi(x)$ be the orbit that contains x. Then our question may be formulated as follows: is it possible to convert X into an analytic manifold in such a way that π is an analytic submersion?

Suppose such an analytic structure exists on X. Then π is an open continuous map of M onto X, so a subset Y of X is open if and only if $\pi^{-1}(Y)$ is open in M. In other words, X has the quotient topology. Moreover, X is the quotient of the analytic manifold M with respect to the map π, so the analytic structure on X is uniquely determined. Lemma 2.7.1 implies at once that the orbits of G in M are all closed regular submanifolds of the same dimension $d_0 = \dim(M) - \dim(X)$; in particular, all the stability subgroups have the same dimension $d_s = \dim(G) - d_0$. We now consider the problem of proving the existence of an analytic structure on X with the required properties when the above conditions are satisfied. We limit ourselves to a discussion of a particularly simple situation. We say that G *acts freely* on M if $G_x = \{1\}$ for all $x \in M$. Put

$$(2.9.22) \qquad \begin{cases} \gamma(x,g) = (x,g \cdot x) \\ \Gamma = \gamma[M \times G] = \{(x,g \cdot x) : x \in M, g \in G\}. \end{cases}$$

Observe that if G acts freely on M, γ is one-to-one on $M \times G$.

Lemma 2.9.9. *The quotient topology on X is Hausdorff if and only if Γ is closed in $M \times M$. In this case all the orbits in M are closed regular submanifolds of M.*

Proof. Let Γ be closed, and let $x, y \in M$ be such that $\pi(x) \neq \pi(y)$. Then $(x,y) \notin \Gamma$, so we can find open subsets U, V of M such that $(x,y) \in U \times V \subseteq (M \times M) \setminus \Gamma$. Then $G \cdot U \cap G \cdot V = \varnothing$; hence $\pi[U] \cap \pi[V] = \varnothing$. This proves that X is Hausdorff in the quotient topology. Conversely, suppose X is Hausdorff and $(x,y) \notin \Gamma$. Then $\pi(x) \neq \pi(y)$, so we can find disjoint open subsets Y, Z of X such that $\pi(x) \in Y$ and $\pi(y) \in Z$. Let $U = \pi^{-1}(Y)$, $V = \pi^{-1}(Z)$. Then $(x,y) \in U \times V \subseteq (M \times M) \setminus \Gamma$. This proves that $(M \times M) \setminus \Gamma$ is open. Note that in this case all the orbits are closed in M and are hence regular submanifolds of M by Theorem 2.9.7.

Theorem 2.9.10. *Let G act freely on M. Then the following are equivalent:*

(1) *Γ is closed in $M \times M$, and γ is a homeomorphism of $M \times G$ onto Γ.*

(2) *Γ is closed in $M \times M$; moreover, given any $x \in M$ we can find a regular submanifold N of M passing through x such that (a) $T_x(M)$ is the direct sum of $T_x(N)$ and $T_x(G \cdot x)$, and (b) π is one-to-one on N.*

(3) *There exists an analytic structure on X such that π is an analytic submersion.*

Proof. (1) $\Longrightarrow$ (2). Let $x \in M$. Choose a regular submanifold N_1 of M passing through x such that $T_x(M)$ is the direct sum of $T_x(G \cdot x)$ and $T_x(N_1)$. Let ψ be the map $(g,y) \mapsto g \cdot y$ of $G \times N_1$ into M. It follows from the choice of N_1 that $(d\psi)_{(1,x)}$ maps $T_{(1,x)}(G \times N_1)$ onto $T_x(M)$. But since $G_x = \{1\}$, $\dim(G) = \dim(G \cdot x)$, so $(d\psi)_{(1,x)}$ is actually a bijection. Choosing N_1 small enough, we may therefore ensure that $(d\psi)_{(1,y)}$ is a bijection of $T_{(1,y)}(G \times N_1)$ onto $T_y(M)$ for all $y \in N_1$. Now, if $g \in G$ and λ_g is the analytic diffeomorphism $(h,y) \mapsto (gh,y)$ of $G \times N_1$ onto itself,

$$\psi \circ \lambda_g = t_g \circ \psi,$$

where t_g is given by (2.9.3). Consequently, $(d\psi)_{(g,y)}$ is a bijection of $T_{(g,y)}(G \times N_1)$ onto $T_{g \cdot y}(M)$ for all $g \in G$, $y \in N_1$. In particular, $G \cdot N_1$ is open in M, while ψ is an open map which is even a local homeomorphism. Let U and N_2 be open subsets of G and N_1 respectively with $1 \in U$ and $x \in N_2$, such that ψ is a homeomorphism of $U \times N_2$ onto $U \cdot N_2$.

We claim that for some open neighborhood N of x in N_2, π is one-to-one on N. If this were not so, we could find sequences $\{y_n\}$ in N_2 and $\{g_n\}$ in G such that $y_n \longrightarrow x$ and $y_n' = g_n \cdot y_n \longrightarrow x$ as $n \longrightarrow \infty$, while $y_n \neq y_n'$ for all n. Since γ is a homeomorphism, this implies that $g_n \longrightarrow 1$; hence for some k, $g_k \in U$ and y_k and y_k' are both in N_2. Thus $\psi(1,y_k') = \psi(g_k,y_k)$, so $y_k = y_k'$, a contradiction.

(2) $\Longrightarrow$ (3) We equip X with the quotient topology. Then X is Hausdorff and second countable. For each open subset Y of X we denote by $\mathfrak{A}(Y)$ the set of all complex-valued functions f on Y for which $f \circ \pi$ is analytic on $\pi^{-1}(Y)$.

Let $x \in M$. Select a regular submanifold N of M passing through x and satisfying (a) and (b) of (2). Let ψ be the map $(g,y) \mapsto g \cdot y$ of $G \times N$ into M. Then $(d\psi)_{(1,x)}$ is a bijection of $T_{(1,x)}(G \times N)$ onto $T_x(M)$. Replacing N by an open submanifold of it containing x, we may assume that $(d\psi)_{(1,y)}$ is a bijection of $T_{(1,y)}(G \times N)$ onto $T_y(M)$ for all $y \in N$. As in the previous discussion, we now conclude that $(d\psi)_{(g,y)}$ is a bijection of $T_{(g,y)}(G \times N)$ onto $T_{g \cdot y}(M)$ for all $(g,y) \in G \times N$. On the other hand, it follows from the one-to-one nature of $\pi | N$ that ψ is one-to-one on $G \times N$. Consequently, $G \cdot N$ is open in M and the map ψ is thus an analytic diffeomorphism of $G \times N$ onto $G \cdot N$. If Z is any open subset of N, $G \cdot Z = \psi[G \times Z]$ is then open in M, so $\pi[Z] = \pi[G \cdot Z]$ is open in M; moreover, if f is a complex-valued function defined on $\pi[Z]$, $f \in \mathfrak{A}(\pi[Z])$ if and only if $f \circ \pi \circ \psi$ is an analytic function on $G \times Z$ that depends only on its second argument. In particular, $f \in \mathfrak{A}(\pi[Z])$ if and only if $(f \circ \pi)|Z$ is an analytic function on the open submanifold Z of N_2.

These remarks prove that we have an analytic structure on X and that, for each $x \in M$, there exists a regular submanifold P of M passing through x such that π is an analytic diffeomorphism of P onto an open neighborhood of $\pi(x)$ in X. This proves that π is a submersion of M onto X.

(3) $\Longrightarrow$ (1) Assume now that (3) is satisfied. We have already remarked that X has to be given the quotient topology. By Lemma 2.9.9, Γ is closed in $M \times M$. We now prove that γ is a homeomorphism. Let $\{(x_n,g_n)\}$ be a sequence in $M \times G$ such that $(x_n, g_n \cdot x_n) \longrightarrow (x,y) \in \Gamma$. Then $y = g \cdot x$ for some $g \in G$, $x_n \longrightarrow x$ and $g_n \cdot x_n \longrightarrow y$ as $n \longrightarrow \infty$. Write $h_n = g^{-1}g_n$. We have to prove that $h_n \longrightarrow 1$ as $n \longrightarrow \infty$.

Since π is a submersion, we can find a regular submanifold N of M passing through x such that (a) and (b) of (2) are satisfied. Arguing as in the preceding proof, we may assume that $G \cdot N$ is open in M and that $\psi((g,y) \mapsto g \cdot y)$ is an analytic diffeomorphism of $G \times N$ onto $G \cdot N$. Since both x_n and $h_n \cdot x_n$ converge to x, we may assume that x_n and $h_n \cdot x_n$ belong to $G \cdot N$ for all n. So $x_n = \psi(u_n,y_n)$ and $h_n \cdot x_n = \psi(v_n,z_n)$ for all n. Since ψ is a homeomorphism, we must have $u_n \longrightarrow 1$ and $v_n \longrightarrow 1$ as $n \longrightarrow \infty$. On the other hand, $\psi(h_n u_n, y_n) = \psi(v_n,z_n)$, so $h_n u_n = v_n$ for all n. Hence $h_n \longrightarrow 1$ as $n \longrightarrow \infty$. This completes the proof of the theorem.

Remark. We have actually proved the following stronger result: for each $p \in X$ we can select an open subset Y of X containing p, and an analytic diffeomorphism $\xi_Y = \xi$ of $G \times Y$ onto $\pi^{-1}(Y)$, such that

$$\xi(hg,y) = h \cdot \xi(g,y) \quad (g, h \in G, y \in Y). \tag{2.9.23}$$

In other words, (M,X,π,G) is a *principal G-bundle*.

There are two special cases of this theorem that are worth mentioning.

Corollary 2.9.11. *Let G be a compact Lie group acting freely and analytically on M. Then M/G admits a unique analytic structure for which π is a submersion.*

Proof. In this case it is easily seen that (1) of Theorem 2.9.10. is satisfied.

Corollary 2.9.12. *Let G be a discrete group acting freely and analytically on M, and let Γ be closed in $M \times M$. Then, in order that $X = M/G$ admit an analytic structure such that π is a submersion, it is necessary and sufficient that the following condition be satisfied: for each $x \in M$ we can find an open subset U of M containing x such that $(g \cdot U) \cap U = \varnothing$ for any $g \neq 1$ in G.*

Proof. We shall verify that this condition is equivalent to the fact that γ is a homeomorphism. Suppose γ is a homeomorphism and $x \in M$. If an open neighborhood of x with the required properties does not exist, we can find sequences $\{x_n\}$ and $\{y_n\}$ converging to x as $n \longrightarrow \infty$, with $y_n = g_n \cdot x_n$ for some $g_n \neq 1$ in G for all n. Then $g_n \longrightarrow 1$, so since G is discrete, $g_n = 1$ for all sufficiently large n, a contradiction. In the other direction, let $\{(x_n, g_n)\}$ be a sequence in $M \times G$ such that, for some $(x,g) \in M \times G$, $x_n \longrightarrow x$ and $g_n \cdot x_n \longrightarrow g \cdot x$ as $n \longrightarrow \infty$. Write $h_n = g^{-1}g_n$. Then for some k, x_n and $h_n \cdot x_n$ are in U for all $n \geq k$, U being an open neighborhood of x with the properties described in the statement of this corollary. So $h_n = 1$ for $n \geq k$, proving that $g_n \longrightarrow g$ as $n \longrightarrow \infty$.

2.10. The Exponential Map

Let G be a Lie group and $\mathfrak{g}$ its Lie algebra. In this section we shall introduce the exponential map of $\mathfrak{g}$ into G and study some of its properties. If $G = GL(n,\mathbf{R})$ or an analytic subgroup of it, the exponential map coincides with the usual matrix exponential map—a fact which accounts for its name. The exponential map is probably the most important basic construct associated with G and $\mathfrak{g}$. Many important results in the general theory of Lie groups and Lie algebras depend in some way or the other on the properties of this map.

Let $\mathbf{R}$ be the additive group of real numbers and let t be the usual coordinate on $\mathbf{R}$. $\mathbf{R}$ is an analytic group under addition. The Lie algebra of $\mathbf{R}$ is one-dimensional and is spanned by the vector field $D = d/dt$. For $\tau \in \mathbf{R}$ let $D_\tau = (d/dt)_{t=\tau}$ be the tangent vector defined by D at τ. If M is an analytic manifold and $f: t \mapsto f(t)$ an analytic map of an open neighborhood of τ into M, then we have (cf. (1.1.20))

$$\dot{f}(\tau) = \left(\frac{d}{dt}f\right)_{t=\tau} = (df)_\tau(D_\tau). \tag{2.10.1}$$

Consider now a *real* Lie group G with Lie algebra $\mathfrak{g}$. Let $X \in \mathfrak{g}$. Then the map $tD \mapsto tX$ ($t \in \mathbf{R}$) is a homomorphism of the Lie algebra of $\mathbf{R}$ into $\mathfrak{g}$. Since $\mathbf{R}$ is simply connected, there is a unique analytic homomorphism ξ_X of $\mathbf{R}$ into G such that $d\xi_X(D) = X$ (Theorem 2.7.5). We have

(2.10.2)
$$\begin{cases} \dot{\xi}_X(\tau) = X_{\xi_X(\tau)}, \ \dot{\xi}_X(0) = X_1 \\ d\xi_X(D) = X. \end{cases}$$

Conversely, let η be any analytic homomorphism of $\mathbf{R}$ into G. If we write $X = (d\eta)(D)$, it is obvious that $\eta = \xi_X$. In other words, the correspondence

$$X \mapsto \xi_X$$

is a bijection of $\mathfrak{g}$ onto the set of all analytic homomorphisms of $\mathbf{R}$ into G, such that $d\xi_X(D) = X$ for all $X \in \mathfrak{g}$. Note that ξ_0 is the trivial homomorphism $t \mapsto 1$. If $\tau \in \mathbf{R}$ is fixed, then for any $X \in \mathfrak{g}$, $\eta : t \mapsto \xi_X(\tau t)$ is again an analytic homomorphism of $\mathbf{R}$ into G. Since $(d\eta)_0(D_0) = \tau X_1$, it follows that $\eta = \xi_{\tau X}$; i.e.,

(2.10.3)
$$\xi_{\tau X}(t) = \xi_X(\tau t) \quad (t, \tau \in \mathbf{R}, X \in \mathfrak{g}).$$

It is customary to write

(2.10.4)
$$\xi_X(1) = \exp X \quad (X \in \mathfrak{g}).$$

$\exp : X \mapsto \exp X$ is thus a map of $\mathfrak{g}$ into G, called the *exponential* map. From (2.10.3) we get

(2.10.5)
$$\xi_X(t) = \exp tX \quad (t \in \mathbf{R}, X \in \mathfrak{g}).$$

Also

(2.10.6)
$$\exp 0 = 1.$$

Fix $X \in \mathfrak{g}$. Then for any $x \in G$, the map

(2.10.7)
$$\zeta_x : t \mapsto x \exp tX \quad (t \in \mathbf{R})$$

is an analytic map of $\mathbf{R}$ into G, and since $\zeta_x(t) = l_x(\exp tX)$,

(2.10.8)
$$\dot{\zeta}_x(0) = X_x.$$

In other words, the analytic curve ζ_x is the unique integral curve of the vector field X through the point x. From the definition (2.10.1) we see that for any function f defined and C^∞ around x

(2.10.9)
$$X_x f = f(x;X) = \left(\frac{d}{dt} f(x \exp tX)\right)_{t=0} \quad (x \in G, X \in \mathfrak{g}).$$

Suppose G is a complex Lie group, $\mathfrak{g}$ its Lie algebra. To define the exponential map we proceed as follows. Let $\mathbf{C}$ be the complex analytic group of the complex numbers under addition and let z be the usual coordinate. We denote by E the differential operator d/dz and by E_τ the holomorphic tangent vector $(d/dz)_{z=\tau}$ $(\tau \in \mathbf{C})$. As in the real case, given any $X \in \mathfrak{g}$ there is a unique complex analytic homomorphism ξ_X of $\mathbf{C}$ into G such that

$$d\xi_X(E) = X. \tag{2.10.10}$$

The correspondence

$$X \mapsto \xi_X$$

is a bijection of $\mathfrak{g}$ onto the set of all complex analytic homomorphisms of $\mathbf{C}$ into G. We have

$$\xi_{\tau X}(t) = \xi_X(\tau t) \quad (t, \tau \in \mathbf{C},\ X \in \mathfrak{g}). \tag{2.10.11}$$

The exponential map of $\mathfrak{g}$ into G is then defined by

$$\exp X = \xi_X(1) \quad (X \in \mathfrak{g}). \tag{2.10.12}$$

For $x \in G$, $X \in \mathfrak{g}$, and any function f defined and holomorphic around x, we have

$$X_x f = f(x;X) = \left(\frac{d}{dt} f(x \exp tX)\right)_{t=0}. \tag{2.10.13}$$

Theorem 2.10.1. *Let G be a Lie group, $\mathfrak{g}$ its Lie algebra. Then the exponential map is analytic. Further, it is an analytic diffeomorphism on an open neighborhood of the origin of $\mathfrak{g}$. More generally, let $\mathfrak{g}$ be the direct sum of linear subspaces $\mathfrak{h}_1, \ldots, \mathfrak{h}_s$ $(s \geq 1)$; then there are open neighborhoods B_i of 0 in $\mathfrak{h}_i$ $(1 \leq i \leq s)$ and U of 1 in G, such that the map*

$$\psi : (Z_1, \ldots, Z_s) \mapsto \exp Z_1 \cdots \exp Z_s \tag{2.10.14}$$

is an analytic diffeomorphism of $B_1 \times \cdots \times B_s$ onto U.

Proof. The proofs are essentially the same in both the real and the complex cases. We give the proof in the real case.

We prove first that exp is analytic around $X = 0$. Select coordinates $x_1, \ldots, x_m$ on an open subset U of G containing 1 such that $x_1(1) = \cdots = x_m(1) = 0$, and let $X_i \in \mathfrak{g}$ be such that $(X_i)_1 = (\partial/\partial x_i)_1$, $1 \leq i \leq m$. Then $\{X_1, \ldots, X_m\}$ is a basis for $\mathfrak{g}$. There are functions F_{ki} $(1 \leq k, i \leq m)$, defined and analytic on the open set $\tilde{U} = \{(x_1(y), \ldots, x_m(y)) : y \in U\}$ such that

$$(X_i)_y = \sum_{1 \leq k \leq m} F_{ki}(x_1(y), \ldots, x_m(y)) \left(\frac{\partial}{\partial x_k}\right)_y \quad (y \in U).$$

Denote by F the $m \times m$ matrix $(F_{ki})_{1 \leq k,i \leq m}$. For $1 \leq i \leq m$, write

$$f(t : a_1, \ldots, a_m) = \exp(t(a_1X_1 + \cdots + a_mX_m)) \quad (t, a_1, \ldots, a_m \in \mathbf{R}).$$

It follows from the definition of the exponential map that for fixed $a_1, \ldots, a_m \in \mathbf{R}$, $t \mapsto f(t : a_1, \ldots, a_m)$ is an integral curve of the vector field $a_1X_1 + \cdots + a_mX_m$ with $f(0 : a_1, \ldots, a_m) = 1$. For any fixed $a_1, \ldots, a_m \in \mathbf{R}$, consider the system of differential equations

$$\frac{d\mathbf{y}}{dt} = F(y_1, \ldots, y_m)\mathbf{a}, \tag{2.10.15}$$

where $\mathbf{y}$ and $\mathbf{a}$ are column vectors with respective components $y_1, \ldots, y_m$ and $a_1, \ldots, a_m$. By Theorem 1.4.2 there is an $a > 0$ and an analytic map

$$\mathbf{f} : (t, a_1, \ldots, a_m) \mapsto \mathbf{f}(t : a_1, \ldots, a_m)$$

defined on the cube I_a^{m+1} with values in $\tilde{U}$ such that for fixed $(a_1, \ldots, a_m) \in I_a^m$, the function $t \mapsto \mathbf{f}(t : a_1, \ldots, a_m)$ $(|t| < a)$ is a solution of (2.10.15) with $\mathbf{f}(0 : a_1, \ldots, a_m) = \mathbf{0}$. Going back to U, this means that there is an analytic map

$$\bar{f} : (t, a_1, \ldots, a_m) \mapsto \bar{f}(t : a_1, \ldots, a_m)$$

of the cube I_a^{m+1} into U such that for fixed $(a_1, \ldots, a_m) \in I_a^m$, the curve $t \mapsto \bar{f}(t : a_1, \ldots, a_m)$ $(|t| < a)$ is an integral curve of the vector field $a_1X_1 + \cdots + a_mX_m$ with $\bar{f}(0 : a_1, \ldots, a_m) = 1$. By the uniqueness property of integral curves of vector fields, we must have

$$\bar{f}(t : a_1, \ldots, a_m) = \exp(t(a_1X_1 + \cdots + a_mX_m)) \quad ((t, a_1, \ldots, a_m) \in I_a^{m+1}).$$

It follows from this equation and the analyticity of $\bar{f}$ that the map

$$(a_1, \ldots, a_m) \mapsto \exp(a_1X_1 + \cdots + a_mX_m)$$

is analytic on some cube I_b^m. This proves that exp is analytic on some open neighborhood of $X = 0$.

Now, for any integer $k \geq 1$,

$$\exp X = \left(\exp \frac{1}{k} X\right)^k \quad (X \in \mathfrak{g}).$$

Since the maps $x \mapsto x^k$ of G into G are analytic $(k = 1, 2, \ldots)$, it follows from the proceding result that exp is analytic everywhere.

It remains to prove the last result. Note that for $s = 1$, $\mathfrak{h}_1 = \mathfrak{g}$, and so we obtain as a corollary that exp is an analytic diffeomorphism on an open set around $X = 0$. We shall canonically identify the tangent space to G at any of its points with $\mathfrak{g}$ and, similarly, the tangent space to the vector space $\mathfrak{h}_i$ at any of its points with $\mathfrak{h}_i$ $(1 \leq i \leq s)$; so for any element $(Z_1, \ldots, Z_s) \in \mathfrak{h}_1 \times \cdots \times \mathfrak{h}_s$, $(d\psi)_{(Z_1, \ldots, Z_s)}$ becomes a linear map of $\mathfrak{h}_1 \times \cdots \times \mathfrak{h}_s$ into $\mathfrak{g}$. Clearly, it is enough to prove that $(d\psi)_{(0, \ldots, 0)}$ is a bijection. As $\dim(\mathfrak{h}_1 \times \cdots \times \mathfrak{h}_s) = \dim(\mathfrak{g})$, it is enough to verify surjectivity. Let L be the range of $(d\psi)_{(0, \ldots, 0)}$. If $X \in \mathfrak{h}_i$ and f_i is the map $t \mapsto (0, \ldots, tX, \ldots, 0)$ (zeros in all places except the ith) of $\mathbf{R}$ into $\mathfrak{h}_1 \times \cdots \times \mathfrak{h}_s$, it is clear that $[(d/dt)(\psi \circ f_i)(t)]_{t=0} \in L$. But $(\psi \circ f_i)(t) = \exp tX$, so we can conclude that $\mathfrak{h}_i \subseteq L$. Since this is true for $i = 1, \ldots, s$, we have $L = \mathfrak{g}$. As mentioned earlier, this is sufficient to complete the proof.

Remarks 1. $(d\exp)_X$ is an endomorphism of $\mathfrak{g}$ for each $X \in \mathfrak{g}$ with our identification. We write

$$D(X : Y) = D_X(Y) = (d\exp)_X(Y) \quad (X, Y \in \mathfrak{g}). \tag{2.10.16}$$

It follows from the reasoning above that

$$(d\exp)_0(Y) = Y \quad (Y \in \mathfrak{g}). \tag{2.10.17}$$

2. Since $\mathfrak{g}$ is connected, so is $\exp[\mathfrak{g}]$. Hence

$$\exp[\mathfrak{g}] \subseteq G^0.$$

We shall give examples where $\exp[\mathfrak{g}] \neq G^0$. It is also possible that $(d\exp)_X$ is not surjective for some $X \in \mathfrak{g}$ (cf. §2.14). Since $\exp[\mathfrak{g}]$ contains an open neighborhood of the identity, it follows that the subgroup generated by $\exp[\mathfrak{g}]$ concides with G^0.

3. Let $\{X_1, \ldots, X_m\}$ be a basis for $\mathfrak{g}$. It follows from the above theorem that for some $a > 0$, the map

$$\varphi : (a_1, \ldots, a_m) \mapsto \exp(a_1 X_1 + \cdots + a_m X_m)$$

is an analytic diffeomorphism of the cube I_a^m onto an open subset U_1 of G containing 1. Suppose $x_1, \ldots, x_m$ are the analytic functions on U_1 such that $y \mapsto (x_1(y), \ldots, x_m(y))$ is the map of U_1 onto I_a^m which inverts φ. Then for $1 \leq i \leq m$,

$$x_i(\exp(a_1 X_1 + \cdots + a_m X_m)) = a_i \quad ((a_1, \ldots, a_m) \in I_a^m). \tag{2.10.18}$$

$x_1, \ldots, x_m$ are called the *canonical coordinates of the first kind* around 1, relative to the basis $\{X_1, \ldots, X_m\}$.

Let $\{X_1, \ldots, X_m\}$ be, as before, a basis for $\mathfrak{g}$ and let us consider the map

$$\psi : (a_1, \ldots, a_m) \mapsto \exp(a_1X_1)\exp(a_2X_2)\cdots\exp(a_mX_m)$$

of $\mathbf{R}^m$ into G. ψ is obviously analytic. It follows from the theorem proved above that for some $a > 0$, ψ is an analytic diffeomorphism of the cube I_a^m onto an open subset U_2 of G containing 1. Let $x_1, \ldots, x_m$ be the analytic functions on U_2 such that the map $y \mapsto (x_1(y), \ldots, x_m(y))$ inverts ψ. Then for $1 \leq i \leq m$,

$$x_i(\exp a_1X_1 \exp a_2X_2 \cdots \exp a_mX_m) = a_i \quad ((a_1, \ldots, a_m) \in I_a^m). \tag{2.10.19}$$

$x_1, \ldots, x_m$ are called the *canonical coordinates of the second kind* around 1 with respect to the basis $\{X_1, \ldots, X_m\}$.

We now discuss some examples.

(1) Let V be a vector space (over $\mathbf{R}$ or $\mathbf{C}$) of finite dimension m. Let G be the additive group of V. We have seen earlier that the Lie algebra of G can be canonically identified with V itself. With this identification, $\exp X = X$ for $X \in V$.

(2) Let V be a vector space of finite dimension m over $\mathbf{R}$ or $\mathbf{C}$. For $G = GL(V)$ and $\mathfrak{g} = \mathfrak{gl}(V)$, the exponential map coincides with the usual matrix exponential. The proof of this depends on the following lemma.

Lemma 2.10.2. *Let V be as above and let $\mathcal{E}$ be the algebra of endomorphisms of V. For $A \in \mathcal{E}$ let*

$$e^A = 1 + A + \frac{A^2}{2!} + \cdots + \frac{A^n}{n!} + \cdots. \tag{2.10.20}$$

Then e^A is well defined and $A \mapsto e^A$ is an analytic map of $\mathcal{E}$ into $GL(V)$. If $\lambda_1, \ldots, \lambda_r$ are the eigenvalues of A with respective multiplicities $m_1, \ldots, m_r$, then $e^{\lambda_1}, \ldots, e^{\lambda_r}$ are the eigenvalues of e^A with respective multiplicities $m_1, \ldots, m_r$. In particular,

$$\det(e^A) = e^{\operatorname{tr} A} \quad (A \in \mathcal{E}). \tag{2.10.21}$$

If A and B are in $\mathcal{E}$ and commute, then

$$e^{A+B} = e^A e^B. \tag{2.10.22}$$

Proof. It is enough to prove everything when V is a complex vector space.

Suppose $\{v_1, \ldots, v_m\}$ is a basis for V and $c > 0$ is a constant such that all the matrix entries of an $A \in \mathcal{E}$ relative to this basis are $\leq c$. Then an easy induction shows that for any integer $n \geq 1$, the entries of the matrix of A^n are all $\leq (mc)^n$. It follows from this that the series defining e^A converges over all of $\mathcal{E}$, the convergence being uniform over compact sets of $\mathcal{E}$. The classical theorem on uniformly convergent sequences of holomorphic functions now implies that $A \mapsto e^A$ is a holomorphic map of $\mathcal{E}$ into $\mathcal{E}$.

Suppose $A, B \in \mathcal{E}$ and $AB = BA$. Then an easy induction on n shows that

$$\frac{(A+B)^n}{n!} = \sum_{0 \leq r \leq n} \frac{A^r}{r!} \frac{B^{n-r}}{(n-r)!}.$$

The formula (2.10.22) follows from this by multiplication of the series for e^A and e^B. Since $e^0 = 1$, $e^A e^{-A} = 1$; therefore e^A is invertible for all $A \in \mathcal{E}$, the inverse of e^A being e^{-A}.

It remains to prove the result concerning eigenvalues. Fix $A \in \mathcal{E}$. By reduction theory we can choose a basis $\{v_1, \ldots, v_m\}$ for V such that the matrix (a_{ij}) of A relative to this basis is upper triangular, i.e., $a_{ij} = 0$ $(i > j)$. It is easily seen that the matrix of A^n has the same property for all $n \geq 1$, and the diagonal entries of the matrix of A^n are a_{ii}^n, $1 \leq i \leq m$. A simple calculation now shows that the matrix of e^A is upper triangular with diagonal entries $e^{a_{ii}}$, $1 \leq i \leq m$. Since the characteristic polynomial of A (resp. e^A) is $\prod_{1 \leq i \leq m} (z - a_{ii})$ (resp. $\prod_{1 \leq i \leq m} (z - e^{a_{ii}})$), we are through.

It follows from the above lemma that for any $X \in \mathfrak{gl}(V)$, the map $t \mapsto e^{tX}$ is an analytic homomorphism of $\mathbf{R}$ into $GL(V)$. Writing $A = tX$ in (2.10.20) and differentiating termwise, we get

$$\left(\frac{d}{dt} e^{tX}\right)_{t=0} = X. \tag{2.10.23}$$

We may therefore conclude that

$$\xi_X(t) = e^{tX} \quad (t \in \mathbf{R}, X \in \mathfrak{gl}(V)). \tag{2.10.24}$$

The formula (2.10.24) shows that

$$\exp X = e^X \quad (X \in \mathfrak{gl}(V)). \tag{2.10.25}$$

(3) Let $G = GL(2,\mathbf{R})$, $\mathfrak{g} = \mathfrak{gl}(2,\mathbf{R})$. We now prove that

$$x = \begin{pmatrix} -1 & 1 \\ 0 & -1 \end{pmatrix} \in G^0 \quad \text{but} \quad x \notin \exp[\mathfrak{g}].$$

Suppose to the contrary and let $X \in \mathfrak{g}$ be such that $x = e^X$. Since $tr\ X$ is real and $\det(x) = 1$, (2.10.21) shows that $tr\ X = 0$. So there is a $c \in \mathbf{C}$ such that

c and $-c$ are the eigenvalues of X. The eigenvalues of $x = e^X$ are then e^c and e^{-c}, showing that $e^c = -1$ and $c \neq 0$. Therefore X has distinct eigenvalues. Hence we can find an invertible *complex* 2×2 matrix u such that

$$uXu^{-1} = \begin{pmatrix} c & 0 \\ 0 & -c \end{pmatrix}.$$

Then

$$uxu^{-1} = ue^X u^{-1} = e^{uXu^{-1}} = \begin{pmatrix} -1 & 0 \\ 0 & -1 \end{pmatrix}.$$

Consequently, $(uxu^{-1})^2 = ux^2u^{-1} = 1$. But then $x^2 = 1$, a contradiction. So $x \notin \exp[\mathfrak{g}]$. To prove that $x \in G^0$, note that the curve

$$t \mapsto \begin{pmatrix} \cos t & \sin t \\ -\sin t & \cos t \end{pmatrix} \quad (0 \leq t \leq \pi)$$

joins 1 to -1, and the curve

$$t \mapsto \begin{pmatrix} -1 & t \\ 0 & -1 \end{pmatrix} \quad (0 \leq t \leq 1)$$

joins -1 to x.

We conclude this section with a useful result connecting the exponential map with subgroups and homomorphisms.

Theorem 2.10.3. (1) *Let G_i be a real (resp. complex) Lie group with Lie algebra $\mathfrak{g}_i$ $(i = 1, 2)$, and let π $(G_1 \longrightarrow G_2)$ be an analytic homomorphism. Then*

(2.10.26) $$\pi(\exp X) = \exp(d\pi)(X) \quad (X \in \mathfrak{g}_1).$$

In particular, $(d\pi)(X) = 0$ if and only if $\exp tX$ lies in the kernel of π for all $t \in \mathbf{R}$ (resp. $\mathbf{C}$).

(2) *Let G be a real (resp. complex) Lie group with Lie algebra $\mathfrak{g}$, H an arbitrary Lie subgroup of G, and $\mathfrak{h}$ the subalgebra of $\mathfrak{g}$ defined by H. Suppose $X \in \mathfrak{g}$. Then for X to be in $\mathfrak{h}$ it is necessary and sufficient that $\exp tX \in H$ for all $t \in \mathbf{R}$ (resp. $\mathbf{C}$).*

Proof. We prove both assertions in the real case. Let 1_i be the identity of G_i $(i = 1, 2)$.

To prove (1), let $X \in \mathfrak{g}_1$ and $\eta_X(t) = \pi(\exp tX)$, $(t \in \mathbf{R})$. Then $\dot{\eta}_X(0) = (d\pi)_{1_1}(X_{1_1}) = (d\pi)(X)_{1_2}$, so since η_X is a homomorphism,

(2.10.27) $$\eta_X = \xi_{(d\pi)(X)}.$$

This proves that $\pi(\exp tX) = \exp t(d\pi)(X)$. Putting $t = 1$ we obtain (2.10.26).

It is clear from (2.10.27) that $(d\pi)(X) = 0$ if and only if η_X is trivial, i.e., if and only if $\exp tX$ lies in the kernel of π for all $t \in \mathbf{R}$.

To prove (2), let $\bar{\mathfrak{h}}$ be the Lie algebra of H and i the identity map of H into G. By (1), $i(\exp t\bar{X}) = \exp t(di)(\bar{X})$ for all $\bar{X} \in \bar{\mathfrak{h}}$ and $t \in \mathbf{R}$. So if $X \in \mathfrak{h}$, $\exp tX \in H$ for all $t \in \mathbf{R}$. Conversely, let $X \in \mathfrak{g}$ and $\exp tX \in H$ for all $t \in \mathbf{R}$. Since H is quasi-regularly imbedded in G, $t \mapsto \exp tX$ is an analytic homomorphism of $\mathbf{R}$ into H, so $\exp tX = i(\exp t\bar{X})$ for all $t \in \mathbf{R}$ and some $\bar{X} \in \bar{\mathfrak{h}}$. Therefore, $X = (di)(\bar{X}) \in \mathfrak{h}$. This proves the theorem.

One can describe (2) in less pedantic terms in the following way. Let the Lie algebra of H be canonically identified with $\mathfrak{h}$. Then the exponential map of $\mathfrak{h}$ into H is the restriction to $\mathfrak{h}$ of the exponential map of $\mathfrak{g}$ into G.

2.11. The Uniqueness of the Real Analytic Structure of a Real Lie Group

One of the most important applications of the exponential map is the theorem which asserts that a topological group can be converted into a real Lie group in at most one way. The proof is based on the following lemma.

Lemma 2.11.1. *Let G be a real Lie group and let α be a continuous homomorphism of $\mathbf{R}$ into G. Then α is analytic.*

Proof. Let $\mathfrak{g}$ be the Lie algebra of G. By Theorem 2.10.1 we can choose an open set B around 0 in $\mathfrak{g}$ such that (i) $X \in B, |t| \leq 1 \Rightarrow tX \in B$, and (ii) exp is an analytic diffeomorphism of B onto an open subset U of G containing 1. Let B_1 be an open neighborhood of 0 on B such that (i) $X \in B_1, |t| \leq 1 \Rightarrow tX \in B_1$, and (ii) $B_1 + B_1 \subseteq B$. Since α is continuous, we can find a $b > 0$ such that $\alpha(t) \in \exp B_1$ for $-b < t < b$. We can then write $\alpha(t) = \exp \beta(t)$, $-b < t < b$, β being obviously a continuous map of $(-b,b)$ into B_1.

We now assert that for any $t \in (-b,b)$ and $k = 1, 2, \ldots,$ $\beta(rt/k) = r\beta(t/k)$ for $r = 1, 2, \ldots, k$. For $k = 1$ this is obvious. Fix $t \in (-b,b)$ and an integer $k > 1$. Since $|rt/k| < b$ for $1 \leq r \leq k$, $\beta(rt/k) \in B_1$ for all such r. Suppose that, for some r with $1 \leq r \leq k$, $r\beta(t/k) \in B$; then $\exp r\,\beta(t/k) = (\exp \beta(t/k))^r = (\alpha(t/k))^r = \alpha(rt/k) = \exp \beta(rt/k)$, so $r\beta(t/k) = \beta(rt/k)$. In this case we can even conclude that $r\beta(t/k) \in B_1$. Suppose now that $r\beta(t/k) \notin B$ for some r with $1 \leq r \leq k$, and that s is the smallest such r. Obviously, $1 < s \leq k$. Then $(s-1)\beta(t/k) \in B$, so $(s-1)\beta(t/k) \in B_1$ by the observation made above. But then $s\beta(t/k) = (s-1)\beta(t/k) + \beta(t/k) \in B_1 + B_1 \subseteq B$, contradicting the definition of s. So $r\beta(t/k) \in B$, $1 \leq r \leq k$. Then $r\beta(t/k) = \beta(rt/k)$, $1 \leq r \leq k$, as we saw above.

We thus have $\beta(pt) = p\beta(t)$ for $-b < t < b$ and all rational p with $0 < p < 1$. Since β is continuous, we have $\beta(ct) = c\beta(t)$ for $-b < t < b$ and $0 \leq c \leq 1$. Since $\beta(-t) = -\beta(t)$ for $t \in (-b,b)$, we have $\beta(ct) = c\beta(t)$ for

$t \in (-b,b)$ and $-1 \leq c \leq 1$. Let $X = (2/b)\beta(b/2)$. Then, for $t \in [-b/2,b/2]$,

$$\beta(t) = \beta\left(\frac{2t}{b} \cdot \frac{b}{2}\right)$$
$$= tX.$$

In other words, β is the restriction to $[-b/2,b/2]$ of a linear map of **R** into $\mathfrak{g}$. Hence β is analytic around $t = 0$. So α is analytic around $t = 0$. Since α is a homomorphism, it is analytic everywhere.

Theorem 2.11.2. *Let G_1, G_2 be real Lie groups and π a homomorphism of G_1 into G_2. In order that π be analytic it is necessary and sufficient that for every continuous homomorphism α of* **R** *into G_1, $\pi \circ \alpha$ is continuous. In particular, if π is continuous, π is analytic.*

Proof. The necessity of the conditions is obvious. We now prove their sufficiency. Let $\mathfrak{g}_i$ be the Lie algebra of G_i, $i = 1, 2$. Let $X_1, \ldots, X_m$ be a basis for $\mathfrak{g}_1$. By our assumption, for $1 \leq i \leq m$, $\eta_i : t \mapsto \pi(\exp tX_i)$ is a continuous homomorphism of **R** into G_2, hence, by the preceding lemma, analytic. So the map

$$\varphi : (t_1, \ldots, t_m) \mapsto \eta_1(t_1) \cdots \eta_m(t_m)$$

of $\mathbf{R}^m$ into G_2 is analytic. On the other hand, we saw during the course of our discussion of the canonical coordinates of the second kind in §2.10 that for some $a > 0$, the map

$$\psi : (t_1, \ldots, t_m) \mapsto \exp(t_1X_1) \cdots \exp(t_mX_m)$$

is an analytic diffeomorphism of the cube I_a^m onto an open subset U of G_1 containing 1. Then for $u \in U$ we have

$$\pi(u) = \varphi(\psi^{-1}(u)).$$

This shows that π is analytic on U. Since π is a homomorphism, it is analytic everywhere. This proves the theorem.

As an immediate consequence we obtain

Theorem 2.11.3. *Let G_1, G_2 be real Lie groups and φ a continuous one-to-one homomorphism of G_1 onto G_2. Then φ is an analytic isomorphism of G_1 onto G_2. In particular, a (second countable) topological group can admit at most one real analytic structure compatible with its topology under which it is a real Lie group.*

Proof. By Lemma 2.5.3, φ is a homeomorphism. So both φ and φ^{-1} are analytic by the previous theorem. For the second assertion, let G be a real Lie group and G^* another real Lie group having the same underlying topological group as G. If we apply the first result to the identity map, we get $G = G^*$ as real Lie groups. This proves the theorem.

The uniqueness of the analytic structure does not persist for complex Lie groups. For example, let G be the additive group of complex numbers with the usual complex structure. Then the map $x \mapsto x^{\text{conj}}$ is an automorphism of the underlying topological groups but not an automorphism of the complex structure.

It follows from this theorem and the discussion in §2.6 that given a real Lie group G, any second countable topological group locally (topologically) isomorphic to G can be regarded as a Lie group in a unique fashion. From this we obtain the following significant refinement of Theorem 2.8.2.

Theorem 2.11.4. *Let G_j be real Lie groups and $\mathfrak{g}_j$ the corresponding Lie algebras ($j = 1, 2$). Then the following statements are equivalent:*

(i) *G_1 and G_2 are locally isomorphic as topological groups*
(ii) *G_1 and G_2 are locally isomorphic as Lie groups*
(iii) *$\mathfrak{g}_1$ and $\mathfrak{g}_2$ are isomorphic as Lie algebras*

2.12. Taylor Series Expansions on a Lie Group

Suppose G is a Lie group with Lie algebra $\mathfrak{g}$. Let $x \in G$, $X \in \mathfrak{g}$. We have then seen that $t \mapsto x \exp tX$ is the integral curve of the vector field X through the point x and that, if f is any function defined and analytic around x,

$$(Xf)(x) = f(x;X) = \left(\frac{d}{dt} f(x \exp tX)\right)_{t=0}. \tag{2.12.1}$$

We propose to obtain similar formulae involving higher derivatives. Such formulae lead to expansions analogous to the Taylor series expansions in Euclidean spaces. Throughout this section we shall work with a fixed real Lie group G whose Lie algebra will be denoted by $\mathfrak{g}$. We shall leave to the reader the task of making the necessary changes in the complex case. We shall denote by $\mathfrak{G}$ the enveloping algebra of G, introduced in §2.4.

Lemma 2.12.1. *Let $x \in G$, $X \in \mathfrak{g}$. Then for any integer $k \geq 0$ and any function f defined and C^∞ around x,*

$$(X^k f)(x) = f(x;X^k) = \left(\frac{d^k}{dt^k} f(x \exp tX)\right)_{t=0}. \tag{2.12.2}$$

If f is analytic around x, we have, for all sufficiently small $|t|$,

$$f(x \exp tX) = \sum_{n=0}^{\infty} f(x;X^n)\frac{t^n}{n!}. \tag{2.12.3}$$

Proof. We shall prove by induction on k the more general formula

$$f(x \exp tX;X^k) = \frac{d^k}{dt^k} f(x \exp tX) \tag{2.12.4}$$

for all $t \in \mathbf{R}$ and all f defined and C^∞ around x. For $k = 0$ this is obvious. Assume (2.12.4) for some $k \geq 0$ and all f. Since $X^{k+1} = X \cdot X^k$ and $\exp sX \exp tX = \exp(s + t)X$, we have

$$\begin{aligned} f(x \exp tX;X^{k+1}) &= \left\{\frac{d}{ds}(X^k f)(x \exp(s+t)X)\right\}_{s=0} \\ &= \frac{d}{dt}(X^k f)(x \exp tX) \\ &= \frac{d^{k+1}}{dt^{k+1}} f(x \exp tX) \end{aligned}$$

proving (2.12.4) for $k + 1$. For $t = 0$ we obtain (2.12.2). The relation (2.12.3) follows from the analyticity of the function $t \mapsto f(x \exp tX)$.

Lemma 2.12.2. *Let $x \in G$, $X_1, \ldots, X_s \in \mathfrak{g}$. If f is a function defined and C^∞ around x, then*

$$\left\{ \begin{aligned} &(X_1 \cdots X_s f)(x) = f(x;X_1 \cdots X_s) \\ &\quad = \left(\frac{\partial^s}{\partial t_1 \cdots \partial t_s} f(x \exp t_1 X_1 \cdots \exp t_s X_s)\right)_{t_1 = \cdots = t_s = 0}. \end{aligned} \right. \tag{2.12.5}$$

Proof. Let F be the function $(t_1, \ldots, t_s) \mapsto f(x \exp t_1 X_1 \cdots \exp t_s X_s)$, defined in a neighborhood of the origin in $\mathbf{R}^s$. Then for all sufficiently small $|t_1|, \ldots, |t_{s-1}|$,

$$\left(\frac{\partial}{\partial t_s} F(t_1, \ldots, t_{s-1}, t_s)\right)_{t_s = 0} = (X_s f)(x \exp t_1 X_1 \exp t_{s-1} X_{s-1}).$$

(2.12.5) now follows by induction on s.

We shall now obtain a general expansion formula for functions on G. When suitably specialized, these go over to the usual Taylor series expansions in Euclidean spaces. To formulate the results precisely we need some notation. Fix an integer $s \geq 1$ and elements $X_1, \ldots, X_s \in \mathfrak{g}$. For any ordered s-tuple $\mathbf{n} = (n_1, \ldots, n_s)$ of integers ≥ 0, write $X(\mathbf{n})$ for the coefficient of

$t_1^{n_1} \cdots t_s^{n_s}$ in the formal polynomial

$$\frac{n_1!n_2! \cdots n_s!}{(n_1 + \cdots + n_s)!}(t_1X_1 + \cdots + t_sX_s)^{n_1+\cdots+n_s}.$$

Note that $X_1, \ldots ,X_s$ do not in general commute. When $n_1 = \cdots = n_s = 0$, we put $X(\mathbf{n}) = 1$. The $X(\mathbf{n})$ are elements of $\mathfrak{G}$, and for any $\mathbf{n} = (n_1, \ldots ,n_s)$, the order of the differential operator $X(\mathbf{n})$ is $\leq n_1 + \cdots + n_s$. For example, let $s = 2$. Then

$$X(1,1) = \tfrac{1}{2}(X_1X_2 + X_2X_1)$$
$$X(2,1) = \tfrac{1}{3}(X_1^2X_2 + X_1X_2X_1 + X_2X_1^2).$$

It is also possible to describe $X(\mathbf{n})$ in another manner. Let $\mathbf{n} = (n_1, \ldots ,n_s)$ and let $n = n_1 + \cdots + n_s$. Define the elements $Z_1, \ldots ,Z_n$ of $\mathfrak{g}$ by

$$\begin{cases} Z_j = X_1 & 1 \leq j \leq n_1 \\ Z_{n_1+\cdots+n_k+j} = X_{k+1} & 1 \leq j \leq n_{k+1},\ 1 \leq k \leq s-1. \end{cases}$$

Then

$$X(\mathbf{n}) = \frac{1}{n!}\sum_{(i_1,\ldots,i_n)} Z_{i_1}Z_{i_2} \cdots Z_{i_n}, \tag{2.12.6}$$

where the sum extends over all permutations $(i_1, \ldots ,i_n)$ of $(1,2, \ldots ,n)$. We leave it to the reader to verify (2.12.6).

Theorem 2.12.3. *Let $x \in G$ and let f be a function defined and analytic around x. Let $X_1, \ldots ,X_s \in \mathfrak{g}$. Then there is an $a > 0$ such that*

$$f(x \exp(t_1X_1 + \cdots + t_sX_s)) = \sum_{n_1,\ldots,n_s \geq 0} \frac{t_1^{n_1} \cdots t_s^{n_s}}{n_1! \cdots n_s!} f(x;X(\mathbf{n})), \tag{2.12.7}$$

the series converging absolutely and uniformly in the cube I_a^s.

Proof. Let F be the function $(t_1, \ldots ,t_s) \mapsto f(x \exp(t_1X_1 + \cdots + t_sX_s)$, defined and analytic around the origin in $\mathbf{R}^s$. Write $D_j = \partial/\partial t_j$, $1 \leq j \leq s$. Then for some $a > 0$, we have the following expansion absolutely and uniformly convergent for all $(t_1, \ldots ,t_s) \in I_a^s$:

$$F(t_1, \ldots ,t_s) = \sum_{n_1,\ldots,n_s \geq 0} \frac{t_1^{n_1} \cdots t_s^{n_s}}{n_1! \cdots n_s!}(D_1^{n_1} \cdots D_s^{n_s}F)_0,$$

where the suffix 0 indicates that the derivatives are taken when $t_1 = \cdots = t_s = 0$. Now fix $(t_1, \ldots ,t_s) \in I_a^s$ and let u be the function $t \mapsto F(tt_1, \ldots ,tt_s)$

defined around $t = 0$. Obviously,

$$\frac{u^{(k)}(0)}{k!} = \sum_{\substack{n_1,\ldots,n_s\geq 0\\ n_1+\cdots+n_s=k}} \frac{t_1^{n_1}\cdots t_s^{n_s}}{n_1!\cdots n_s!}(D_1^{n_1}\cdots D_s^{n_s}F)_0$$

for $k = 0, 1, \ldots$. On the other hand, it is clear from (2.12.2) that we have, for $k = 0, 1, \ldots$,

$$u^{(k)}(0) = f(x;(t_1X_1 + \cdots + t_sX_s)^k).$$

The equality of these two expressions for $u^{(k)}(0)$ for all $(t_1, \ldots, t_s) \in I_a^s$ implies that the corresponding coefficients of $t_1^{n_1} \cdots t_s^{n_s}$ must be the same. We therefore obtain

$$f(x;X(\mathbf{n})) = (D_1^{n_1} \cdots D_s^{n_s}F)_0. \tag{2.12.8}$$

(2.12.7) follows at once from this.

As an application of these formulae, we now derive expressions for products and commutators in the group in canonical coordinates. In the following we shall use the symbol $O(t^3)$ to denote any function of the form $t \mapsto t^3u(t)$ where u is defined and analytic around $t = 0$ with values in some finite-dimensional vector space over the reals.

Theorem 2.12.4. *Let $s \geq 1$ and $X_1, \ldots, X_s \in \mathfrak{g}$. Then*

$$\exp tX_1 \cdots \exp tX_s = \exp\left\{t \sum_{1\leq i\leq s} X_i + \frac{t^2}{2} \sum_{1\leq i<j\leq s} [X_i,X_j] + O(t^3)\right\} \tag{2.12.9}$$

for all sufficiently small $|t|$. In particular, for $X, Y \in \mathfrak{g}$ and for all sufficiently small $|t|$,

$$\text{(2.12.10)} \quad \begin{cases} \text{(i)} \quad \exp tX \exp tY = \exp\left\{t(X + Y) + \frac{t^2}{2}[X,Y] + O(t^3)\right\} \\ \text{(ii)} \quad \exp tX \exp tY \exp(-tX) = \exp\{tY + t^2[X,Y] + O(t^3)\} \\ \text{(iii)} \quad \exp tX \exp tY \exp(-tX) \exp(-tY) \\ \qquad\qquad = \exp\{t^2[X,Y] + O(t^3)\}. \end{cases}$$

Proof. Fix $s \geq 1$, $X_1, \ldots, X_s \in \mathfrak{g}$. Let f be a function defined and analytic around 1 and let F be the function $(t_1, \ldots, t_s) \mapsto f(\exp t_1X_1 \cdots \exp t_sX_s)$. Then for sufficiently small $|t|$,

$$F(t, \ldots, t) = f(1) + t \sum_{1\leq i\leq s} (D_iF)_0 + \tfrac{1}{2}t^2 \sum_{1\leq i,j\leq s} (D_iD_jF)_0 + O(t^3);$$

here $D_i = \partial/\partial t_i$, and the suffix zero indicates that the derivatives are taken for $t_1 = \cdots = t_s = 0$. Now, by (2.12.5),

$$(D_iF)_0 = f(1;X_i), \quad (D_iD_jF)_0 = (D_jD_iF)_0 = f(1;X_iX_j) \quad (1 \le i < j \le s).$$

Therefore, for all sufficiently small $|t|$,

$$\text{(2.12.11)} \qquad \begin{cases} F(t,\ldots,t) = f(1) + t \sum_{1 \le i \le s} f(1;X_i) \\ \qquad + \dfrac{t^2}{2}\{\sum_{1 \le i \le s} f(1;X_i^2) + 2 \sum_{1 \le i < j \le s} f(1;X_iX_j)\} + O(t^3). \end{cases}$$

Select a basis $\{\bar{X}_1, \ldots, \bar{X}_m\}$ for $\mathfrak{g}$, and let $x_1, \ldots, x_m$ be the corresponding canonical coordinates of the first kind. If $Z = c_1\bar{X}_1 + \cdots + c_m\bar{X}_m \in \mathfrak{g}$, then $x_k(\exp tZ) = tc_k$ and hence, by (2.12.2),

$$\text{(2.12.12)} \qquad x_k(1;Z^n) = \begin{cases} c_k & \text{if } n = 1 \\ 0 & \text{if } n \neq 1 \end{cases} \quad (1 \le k \le m).$$

Let

$$X_i = \sum_{1 \le k \le m} c_{ik}\bar{X}_k, \qquad [X_i,X_j] = \sum_{1 \le k \le m} d_{ijk}\bar{X}_k.$$

We now apply (2.12.11) to the case when $f = x_k$. Since

$$\sum_{1 \le i \le s} X_i^2 + 2 \sum_{1 \le i < j \le s} X_iX_j = (X_1 + \cdots + X_s)^2 + \sum_{1 \le i < j \le s} [X_i,X_j],$$

we conclude from (2.12.11), on taking into account (2.12.12), that

$$\begin{aligned} x_k(\exp tX_1 \cdots \exp tX_s) &= t \sum_{1 \le i \le s} x_k(1;X_i) \\ &\quad + \frac{t^2}{2} \sum_{1 \le i < j \le k} x_k(1;[X_i,X_j]) + O(t^3) \\ &= t \sum_{1 \le i \le s} c_{ik} + \frac{t^2}{2} \sum_{1 \le i < j \le s} d_{ijk} + O(t^3). \end{aligned}$$

On the other hand, if we define $Z(t)$ for sufficiently small $|t|$ by $\exp Z(t) = \exp tX_1 \cdots \exp tX_s$, then

$$Z(t) = \sum_{1 \le k \le m} x_k(\exp tX_1 \cdots \exp tX_s)\bar{X}_k.$$

It then follows from the above that

$$Z(t) = t \sum_{1 \le i \le s} X_i + \frac{t^2}{2} \sum_{1 \le i < j \le s} [X_i,X_j] + O(t^3).$$

This proves (2.12.9). The relations (2.12.10) then follow by specializing suitably. This proves the theorem.

One can now obtain a direct interpretation of the Lie bracket in $\mathfrak{g}$ in terms of commutators in G. In fact, it follows easily from (iii) of (2.12.10) that the map

$$s \mapsto \exp(s^{1/2}X)\exp(s^{1/2}Y)\exp(-s^{1/2}X)\exp(-s^{1/2}Y) \quad (s \geq 0) \tag{2.12.13}$$

is of class C^1 near $s = 0$, and its derivative at $s = 0$ is $[X,Y]_1$. This was the classical (local) way of introducing the Lie bracket.

Corollary 2.12.5. *Let $X, Y \in \mathfrak{g}$ and let $\{X_n\}$, $\{Y_n\}$ be sequences in $\mathfrak{g}$ such that $X_n \longrightarrow X$ and $Y_n \longrightarrow Y$ as $n \longrightarrow \infty$. Then*

$$\begin{cases} \exp(X+Y) = \lim_{n\to\infty}\left(\exp\frac{X_n}{n}\exp\frac{Y_n}{n}\right)^n \\ \exp[X,Y] = \lim_{n\to\infty}\left(\exp\frac{X_n}{n}\exp\frac{Y_n}{n}\exp\frac{-X_n}{n}\exp\frac{-Y_n}{n}\right)^{n^2}. \end{cases} \tag{2.12.14}$$

This corollary would follow if we showed that the $O(t^3)$ estimates in (2.12.10) are uniform when X and Y vary over compact subsets of $\mathfrak{g}$; and for this it would be enough to verify that the $O(t^3)$ estimates in (2.12.11) are uniform when $X_1, \ldots, X_s$ vary over compact subsets of $\mathfrak{g}$. Fix a norm $\|\cdot\|$ over $\mathfrak{g}$, and for any function f defined and analytic around 1, consider the function

$$g(Y_1, \ldots, Y_s) = f(\exp Y_1 \cdots \exp Y_s) \quad (\|Y_j\| < a,\ 1 \leq j \leq s);$$

g is analytic around $(0, \ldots, 0)$. Let H be the difference between g and its Taylor expansion about $(0, \ldots, 0)$ containing only derivatives of order ≤ 2. Then we can find $C' > 0$, $b' > 0$ such that

$$\|H(Y_1, \ldots, Y_s)\| \leq C' \sum_{1 \leq j \leq s} \|Y_j\|^3 \quad (\|Y_j\| < b',\ 1 \leq j \leq s).$$

It follows from this that if $M > 0$, then with $C = C'sM^3$ and $b = b'/M$,

$$\|H(tX_1, \ldots, tX_s)\| \leq Ct^3 \quad (\|X_j\| < M \text{ for } 1 \leq j \leq s,\ |t| < b).$$

This estimate implies the required uniformity in (2.12.11).

We conclude this section with another application of the exponential map, expecially the formulae (2.12.14). Let G be a real Lie group. It was proved by von Neumann that when $G = GL(n,\mathbf{R})$, all closed subgroups of G are Lie subgroups. This result was later extended by E. Cartan to arbitrary G. We now give a proof of Cartan's result.

Theorem 2.12.6. *Let G be a real Lie group and H a closed subgroup. Then H is a Lie subgroup of G.*

Proof. Let H^0 denote, as usual, the component of the identity of H. The theorem will follow if we prove that H^0 is an analytic subgroup of G and is open in H. For, assuming that this has been done, then by Lemma 2.6.1 there is an analytic structure on H which converts it into a Lie group and for which H^0 is an open submanifold of H; since H^0 is regularly imbedded in G and H is the union of (countably many) disjoint left cosets of H^0, it is clear that H is regularly imbedded in G. H is thus a Lie subgroup of G.

The proof that H^0 is open in H and is an analytic subgroup of G consists of two steps. Let $\mathfrak{g}$ be the Lie algebra of G. We first introduce the set

$$\mathfrak{h} = \{X : X \in \mathfrak{g}, \exp tX \in H \text{ for all } t \in \mathbf{R}\}; \tag{2.12.15}$$

if H were a Lie subgroup, $\mathfrak{h}$ would be the subalgebra of $\mathfrak{g}$ defined by H. The first step consists in proving that $\mathfrak{h}$ is a subalgebra of $\mathfrak{g}$. This would allow us to introduce the analytic subgroup H' of G defined by $\mathfrak{h}$. The second step consists in proving that $H' = H^0$ and that it is open in H.

We prove first that $\mathfrak{h}$ is a subalgebra. Obviously, $0 \in \mathfrak{h}$. If $X \in \mathfrak{h}$ and c, $t \in \mathbf{R}$, $\exp ctX = \exp t(cX)$, so $cX \in \mathfrak{h}$ for all $c \in \mathbf{R}$. Suppose $X, Y \in \mathfrak{g}$ and $t \in R$. It follows from (2.12.14) that as $n \longrightarrow \infty$

$$\exp t(X + Y) = \lim\left(\exp \frac{tX}{n} \exp \frac{tY}{n}\right)^n$$

$$\exp t^2[X,Y] = \lim\left(\exp \frac{tX}{n} \exp \frac{tY}{n} \exp \frac{-tX}{n} \exp \frac{-tY}{n}\right)^{n^2}.$$

Since H is closed in G, these relations imply that $\exp t(X + Y) \in H$ for all $t \in \mathbf{R}$ and $\exp t[X,Y] \in H$ for all $t \geq 0$ in $\mathbf{R}$. If we note that $\exp -t[X,Y] = (\exp t[X,Y])^{-1}$, we can conclude that $X + Y$ and $[X,Y]$ belong to $\mathfrak{h}$. $\mathfrak{h}$ is thus a subalgebra of $\mathfrak{g}$.

Let H' be the analytic subgroup of G defined by $\mathfrak{h}$. Since $\exp \mathfrak{h}$ generates H', we see that $H' \subseteq H^0$. In order to prove that $H' = H^0$ and is open in H, it is obviously sufficient to prove that H' contains an open neighborhood of 1 in H. This will then complete the proof of the theorem.

Select a linear subspace $\mathfrak{b}$ of $\mathfrak{g}$ complementary to $\mathfrak{h}$. By Theorem 2.10.1, we can select open neighborhoods A and B of 0 in $\mathfrak{h}$ and $\mathfrak{b}$ respectively such that (i) the closures of A and B are compact, and (ii) $(X,Y) \mapsto \exp X \exp Y$ is an analytic diffeomorphism of $A \times B$ onto an open neighborhood of 1 in G. Suppose now that H' does not contain an open neighborhood of 1 in H. Then there is a sequence $x_k \in H \setminus H'$ such that $x_k \longrightarrow 1$ as $k \longrightarrow \infty$. We may assume that $x_k = \exp X_k \exp Y_k$ where $X_k \in A$, $Y_k \in B$, and that both X_k and Y_k tend to zero when $k \longrightarrow \infty$. Put $y_k = \exp Y_k$. Since $\exp -X_k \in H'$ and $X_k \longrightarrow 0$, it is clear that $y_k \in H \setminus H'$ and $y_k \longrightarrow 1$; moreover, $Y_k \neq 0$ for any k. So, since $Cl(B)$ is compact, for each $k \geq 1$ we can find an integer

$r_k \geq 1$ such that

$$r_k Y_k \in B, \qquad (r_k + 1)Y_k \notin B. \tag{2.12.16}$$

By passing to a subsequence if necessary, we may assume that $Z = \lim_{k \to \infty} r_k Y_k$ exists. Obviously, $Z \in \mathfrak{b}$. Further, Z cannot be zero. For if $Z = 0$, then $(r_k + 1)Y_k = r_k Y_k + Y_k \longrightarrow 0$ as $k \longrightarrow \infty$, and hence $(r_k + 1)Y_k \in B$ for sufficiently large k, contradicting (2.2.16). Thus $Z \neq 0$, $Z \in \mathfrak{b}$.

We claim that $\exp tZ \in H$ for all $t \in \mathbf{R}$. Since H is a closed subgroup, it is enough to prove that $\exp tZ \in H$ for all t rational and > 0. Since $\exp(m/n)Z = (\exp(1/n)Z)^m$ for integers $m, n \geq 1$, we need only prove $\exp(1/p)Z \in H$ for all integers $p \geq 1$. Fix $p \geq 1$, and write $r_k = s_k p + t_k$, where s_k and t_k are integers, $s_k \geq 0$, $0 \leq t_k < p$. Then

$$\exp \frac{1}{p} r_k Y_k = \exp s_k Y_k \cdot \exp \frac{t_k}{p} Y_k.$$

Now, $Y_k \longrightarrow 0$ and $0 \leq t_k < p$, so $\exp(t_k/p)Y_k \longrightarrow 1$. So, as $\exp(1/p)r_k Y_k \longrightarrow \exp(1/p)Z$, we have $\exp s_k Y_k \longrightarrow \exp(1/p)Z$. On the other hand, $\exp s_k Y_k = y_k^{s_k} \in H$ for all k. So, using the fact that H is closed once again, we conclude that $\exp(1/p)Z \in H$. This proves our claim.

It now follows from (2.12.15) that $Z \in \mathfrak{h}$. Since $Z \neq 0$ and $Z \in \mathfrak{b}$, we reach a contradiction. This proves the theorem.

We remark that the analogue of this theorem for complex groups is false. For instance, let $G = \mathbf{C}^*$ and $H = \mathbf{T}^1$, the one-dimensional torus. Then G is complex analytic and H is closed in G, but H is not a complex Lie subgroup of G.

2.13. The Adjoint Representations of $\mathfrak{g}$ and G

Let $\mathfrak{g}$ be a Lie algebra over a field k of characteristic zero. For $X \in \mathfrak{g}$, let ad X be the endomorphism of $\mathfrak{g}$ defined by

$$(\text{ad } X)(Y) = [X, Y] \quad (Y \in \mathfrak{g}). \tag{2.13.1}$$

We have seen that ad X is a derivation of $\mathfrak{g}$ for all $X \in \mathfrak{g}$ and that $X \mapsto \text{ad } X$ is a representation of $\mathfrak{g}$, the so-called adjoint representation of $\mathfrak{g}$ (cf. §2.2). We write

$$\text{ad } \mathfrak{g} = \{\text{ad } X : X \in \mathfrak{g}\}; \tag{2.13.2}$$

ad $\mathfrak{g}$ is a subalgebra of $\mathfrak{gl}(\mathfrak{g})$.

Now let us consider the case when $k = \mathbf{R}$ (resp. $\mathbf{C}$). Let G be a real (resp. complex) Lie group with Lie algebra $\mathfrak{g}$. Suppose V is a finite-dimensional vector space over $\mathbf{R}$ (resp. $\mathbf{C}$). Then by a *representation* of G in V we mean an analytic homomorphism of G into $GL(V)$. It is obvious that if π is a representation of G in V, then $x,v \mapsto \pi(x)v$ is an analytic map of $G \times V$ into V.

In the theory of group representations it is customary to reserve the term "representation" to denote a somewhat more general type of object. Let G be a real Lie group and V a finite-dimensional vector space over $\mathbf{C}$. By a representation of G in V is then meant an analytic homomorphism of G into $GL(V)_{\mathbf{R}}$ where $GL(V)_{\mathbf{R}}$ is the real Lie group underlying the complex Lie group $GL(V)$. In view of Theorem 2.11.2, a map π of G into $GL(V)$ is a representation in this sense if and only if

$$\text{(2.13.3)} \qquad \begin{cases} \text{(i)} & \pi(1) = 1,\ \pi(xy) = \pi(x)\pi(y) \quad (x, y \in G) \\ \text{(ii)} & \text{for any } v \in V, \text{ the map } x \mapsto \pi(x)v \text{ is continuous from} \\ & G \text{ to } V. \end{cases}$$

In this section we use the term "representation" only in the stricter earlier sense. If G is a complex Lie group and π is a representation of G in a complex vector space V (in our strict sense), then the functions $x \mapsto \pi(x)v$ occuring in (2.13.3) are holomorphic; in the theory of group representations π would be called a *complex analytic or holomorphic representation.*

Let G be a Lie group with Lie algebra $\mathfrak{g}$. The main aim of this section is to show that there is a natural representation of G in $\mathfrak{g}$, the so-called adjoint representation, and that its differential is none other than the adjoint representation of $\mathfrak{g}$. As usual, the proofs are given in the real case; the complex case needs only minor changes.

Before formally introducing the adjoint representation, it is covenient to begin with a more general situation.

Lemma 2.13.1. *Let G be a Lie group acting analytically on an analytic manifold M. Let $x_0 \in M$, and let G_0 be the stability subgroup of G at x_0. For each $g \in G$, let t_g denote the diffeomorphism $x \mapsto g \cdot x$ of M. Then for each $g \in G_0$, $L_g = (dt_g)_{x_0}$ is a linear automorphism of $T_{x_0}(M)$, and L $(g \mapsto L_g)$ is a representation of G_0 in $T_{x_0}(M)$.*

Proof. By Lemma 2.9.2, G_0 is a closed Lie subgroup of G. If $g \in G_0$, $t_g \cdot x_0 = t_{g^{-1}} \cdot x_0 = x_0$, so the linear map $(dt_g)_{x_0}$ is a well-defined endomorphism of the tangent space $T_{x_0}(M)$ having $(dt_{g^{-1}})_{x_0}$ as its inverse. So L_g is an automorphism of $T_{x_0}(M)$ for each $g \in G_0$, $L_1 = 1$, and the composition formula for differentials implies that L is a homomorphism of G_0 into $GL(T_{x_0}(M))$. It remains to check that L is analytic, and it is enough to verify analyticity at the identity.

Let $x_1, \ldots, x_p$ be coordinates on an open subset U of G_0 containing 1, and $y_1, \ldots, y_n$ coordinates on an open subset A of M containing x_0. We may suppose that $x_i(1) = y_j(x_0) = 0$ $(1 \leq i \leq p,\ 1 \leq j \leq n)$ and that for a suitable open set B with $x_0 \in B \subseteq A$, $g \cdot b \in A$ for all $g \in U$ and $b \in B$. Since G_0 is a Lie subgroup of G, G_0 acts analytically on M, so there are functions F_i defined and analytic around $(\mathbf{0},\mathbf{0}) \in \mathbf{R}^p \times \mathbf{R}^n$ such that

$$y_i(g \cdot x) = F_i(x_1(g), \ldots, x_p(g) : y_1(x), \ldots, y_n(x))$$

for $1 \leq i \leq n$, $g \in U$, $x \in B$. Then for fixed $g \in U$,

$$(dt_g)_{x_0}\left(\left(\frac{\partial}{\partial y_s}\right)_{x_0}\right) = \sum_{1 \leq r \leq n} \left(\frac{\partial}{\partial t_s} F_r(x_1(g), \ldots, x_p(g) : t_1, \ldots, t_n)\right)_{t_1 = \cdots = t_n = 0} \left(\frac{\partial}{\partial y_r}\right)_{x_0}$$

for $1 \leq s \leq n$. This shows that the matrix of $(dt_g)_{x_0}$ with respect to the basis $\{(\partial/\partial y_1)_{x_0}, \ldots, (\partial/\partial y_n)_{x_0}\}$ of $T_{x_0}(M)$ has entries which are analytic functions of g on U. L is thus analytic at 1. As mentioned earlier, this is sufficient to complete the proof of the lemma.

Consider now a Lie group G with Lie algebra $\mathfrak{g}$. For each $y \in G$, $i_y : x \mapsto yxy^{-1}$ is an automorphism of G and consequently induces in a natural way an automorphism $X \mapsto X^y$ of the Lie algebra of all analytic vector fields on the analytic manifold G. By transport of structure we have, for any analytic vector field X on G,

(2.13.4) $$X^{yy'} = (X^{y'})^y \quad (y,y' \in G),$$

It follows from the identity $l_x i_y = i_y l_{y^{-1}xy}$ $(x, y \in G)$ that if $X \in \mathfrak{g}$, then $X^y \in \mathfrak{g}$ for $y \in G$. We put

(2.13.5) $$\mathrm{Ad}\,(y)\,X = X^y \quad (y \in G,\ X \in \mathfrak{g}).$$

It is then clear that $\mathrm{Ad}\,(y)$ is an automorphism of the Lie algebra $\mathfrak{g}$ for each $y \in G$ and that $y \mapsto \mathrm{Ad}\,y$ is a homomorphism of G into $GL(\mathfrak{g})$. For any subset $\mathfrak{b}$ of $\mathfrak{g}$ and $y \in G$, write $\mathfrak{b}^y = \{X^y : x \in \mathfrak{b}\}$.

It is possible to introduce the linear transformation $\mathrm{Ad}\,(y)$ of $\mathfrak{g}$ in another manner. Since the map $(y,x) \longrightarrow yxy^{-1}$ is analytic from $G \times G$ into G, the natural action of G on itself by inner automorphisms is analytic. Moreover, $i_y 1 = 1$ for all $y \in G$. So by Lemma 2.13.1, $(di_y)_1$ is an automorphism of $T_1(G)$ for each $y \in G$, and $y \mapsto (di_y)_1$ is a representation of G in $T_1(G)$. On the other hand, $X \mapsto X_1$ is a linear isomorphism of $\mathfrak{g}$ onto $T_1(G)$, and we have as an immediate consequence of the definitions

(2.13.6) $$(X^y)_1 = (di_y)_1(X_1) \quad (X \in \mathfrak{g},\ y \in G).$$

In other words, the map $X \mapsto X_1$ intertwines the linear transformations Ad (y) and $(di_y)_1$ for all $y \in G$. We may thus conclude that $y \mapsto$ Ad (y) is a representation of G in $\mathfrak{g}$. At the same time, it follows from (2.13.6) that for fixed $y \in G$, Ad (y) is the automorphism of $\mathfrak{g}$ which is the differential of the automorphism i_y of G (cf. §2.7). In particular, taking $\pi = i_y$ in (2.10.26), we get the important relation

$$\exp X^y = y \exp X y^{-1} \quad (X \in \mathfrak{g}, y \in G); \tag{2.13.7}$$

i.e., the exponential map intertwines the actions of Ad(y) (on $\mathfrak{g}$) and i_y (on G) for all $y \in G$. Ad : $y \mapsto$ Ad (y) is called the *adjoint representation* of G. It follows from Theorem 2.7.3 that Ad$[G]$ is a Lie subgroup of $GL(\mathfrak{g})$ whose component of identity coincides with Ad$[G^0]$. The basic result concerning the adjoint representation of G is the following.

Theorem 2.13.2. *Let G be a Lie group with Lie algebra $\mathfrak{g}$. Then the differential of the adjoint representation of G is the adjoint representation of $\mathfrak{g}$. In particular, its kernel is the centralizer of G^0 in G, and the subalgebra of $\mathfrak{g}$ defined by this kernel is the center of $\mathfrak{g}$. Moreover,*

$$\text{Ad}(\exp X) = e^{\text{ad } X} \quad (X \in \mathfrak{g}). \tag{2.13.8}$$

Proof. Let λ denote the differential of the adjoint representation of G. Then by Theorem 2.10.3 and the fact the that exponential map of $\mathfrak{gl}(\mathfrak{g})$ is the usual matrix exponential, we have Ad$(\exp tX) = e^{t\lambda(X)}$ for all $X \in \mathfrak{g}$, $t \in \mathbf{R}$. Fix $X \in \mathfrak{g}$ and write $y_t = \exp tX$. Then for any $Z \in \mathfrak{g}$,

$$Z^{y_t} = Z + t\lambda(X)Z + O(t^2) \quad (t \longrightarrow 0). \tag{2.13.9}$$

On the other hand, it follows from (2.13.7) and (ii) of (2.12.10) that

$$\exp tZ^{y_t} = \exp\{tZ + t^2[X,Z] + O(t^3)\} \quad (t \longrightarrow 0). \tag{2.13.10}$$

A comparison of (2.13.9) and (2.13.10) yields at once the conclusion $\lambda(X)Z = [X,Z]$. Thus $\lambda(X) = \text{ad } X$ for $X \in \mathfrak{g}$. This proves the first assertion and implies (2.13.8), as already mentioned above. It remains to determine the kernel of Ad. If $y \in G$, Ad$(y) = 1$ if and only if $X'^y = X'$ for all $X' \in \mathfrak{g}$. In view of (2.13.7), this can happen if and only if $y \exp tX' y^{-1} = \exp tX'$ for all $t \in \mathbf{R}$ and $X' \in \mathfrak{g}$, i.e., if and only if y commutes with exp$[\mathfrak{g}]$. Now the subgroup of G generated by exp$[\mathfrak{g}]$ coincides with G^0 (cf. Remark 2 following Theorem 2.10.1), and hence we can conclude that y is an element of the kernel of Ad if and only if it commutes with G^0. Let $\mathfrak{z}$ be the subalgebra of $\mathfrak{g}$ defined by the kernel of Ad. Then by (2) of Theorem 2.10.3, $X \in \mathfrak{z}$ if and only if Ad$(\exp tX) = e^{t \text{ ad } X} = 1$ for all $t \in \mathbf{R}$, i.e., if and only if ad $X = 0$. So $\mathfrak{z}$ is the center of $\mathfrak{g}$. This proves the theorem.

Corollary 2.13.3. *If $X, Y \in \mathfrak{g}$ and $[X,Y] = 0$, then*

$$\exp(X + Y) = \exp X \exp Y. \tag{2.13.11}$$

G^0 is abelian if and only $\mathfrak{g}$ is abelian; in this case the exponential map is a covering homomorphism of the additive group of $\mathfrak{g}$ onto G_0. A simply connected abelian analytic group is isomorphic to the additive group of a vector space.

Proof. Let Z be the component of the identity of the kernel of Ad, and $\mathfrak{z}$ the corresponding subalgebra of $\mathfrak{g}$. By the theorem just proved $\mathfrak{z}$ is the center of $\mathfrak{g}$. Since $\mathfrak{g}$ is abelian if and only if $\mathfrak{z} = \mathfrak{g}$, it follows that $\mathfrak{g}$ is abelian if and only if $Z = G^0$. Applying the theorem once again, we see that this is true if and only if G^0 is contained in its centralizer, i.e., if and only if G^0 is abelian. Suppose now that $X, Y \in \mathfrak{g}$ are such that $[X,Y] = 0$, G being arbitrary. Let $\mathfrak{h}$ be the subspace spanned by X and Y. Then $\mathfrak{h}$ is a subalgebra of $\mathfrak{g}$ and is abelian. So, by the previous result, the analytic subgroup of G corresponding to $\mathfrak{h}$ is abelian. Since $\exp[\mathfrak{h}]$ is contained in this subgroup, we see that $\eta : t \mapsto \exp tX \exp tY$ is an analytic *homomorphism* of $\mathbf{R}$ into G. On the other hand, we see from (i) of (2.12.10) that

$$\begin{aligned}\dot{\eta}(0) &= (d\exp)_0(X + Y)\\ &= X + Y \quad \text{(by (2.10.17))}\end{aligned}$$

Consequently $\eta = \xi_{X+Y}$, proving that $\exp tX \exp tY = \exp t(X + Y)$ for all $t \in \mathbf{R}$. For $t = 1$ we obtain (2.13.11).

It follows from this that if $\mathfrak{g}$ is abelian, exp is a homomorphism of the additive group of $\mathfrak{g}$. But then $\exp[\mathfrak{g}]$ is a subgroup of G and hence must coincide with G^0. Since exp is an analytic diffeomorphism around $X = 0$, the kernel of exp must be discrete. Because $\mathfrak{g}$ is simply connected, exp must be a covering homomorphism. In particular, if G is a simply connected abelian analytic group, exp must be an isomorphism of $\mathfrak{g}$ onto G. This proves all the statements of the corollary.

It follows from this corollary that the abelian analytic groups are precisely of the form V/D, where V is the additive group of a vector space and D a discrete subgroup. A complete description of the discrete subgroups of the additive vector groups would then lead to a complete description of all abelian analytic groups.

Next we formulate a theorem which describes the connection between normal subgroups of G and ideals in $\mathfrak{g}$. For simplicity of formulation we work with connected groups.

Theorem 2.13.4. *Let G be an analytic group with Lie algebra $\mathfrak{g}$, H an analytic subgroup of G, and $\mathfrak{h}$ the corresponding subalgebra of $\mathfrak{g}$. Then H is normal in G if and only if $\mathfrak{h}$ is an ideal in $\mathfrak{g}$; in this case $\mathfrak{h}^y = \mathfrak{h}$ for all $y \in G$.*

Proof. Suppose H is normal in G and let $X \in \mathfrak{h}$, $y \in G$. Then $\exp tX^y = y \exp tX y^{-1} \in H$ for all $t \in \mathbf{R}$. By (2) of Theorem 2.10.3, $X^y \in \mathfrak{h}$. Thus $\mathfrak{h}^y = \mathfrak{h}$. In particular, if $X \in \mathfrak{h}$ and $Z \in \mathfrak{g}$, $X^{\exp tZ} = e^{t \operatorname{ad} Z} \cdot X \in \mathfrak{h}$ for all $t \in \mathbf{R}$. But then $[Z,X] = ((d/dt)e^{t \operatorname{ad} Z}X)_{t=0} \in \mathfrak{h}$ too. So $\mathfrak{h}$ is an ideal in $\mathfrak{g}$. Conversely, let $\mathfrak{h}$ be an ideal in $\mathfrak{g}$. Then for $X \in \mathfrak{h}$ and $Z \in \mathfrak{g}$, $(\operatorname{ad} Z)^n(X) \in \mathfrak{h}$ for $n = 0,1,\ldots$. This shows that $X^{\exp Z} = e^{\operatorname{ad} Z} \cdot X \in \mathfrak{h}$. So $\mathfrak{h}^y = \mathfrak{h}$ for all $y \in \exp[\mathfrak{g}]$. Since the set of all $y \in G$ such that $\mathfrak{h}^y = \mathfrak{h}$ is a subgroup of G, we can conclude that $\mathfrak{h}^y = \mathfrak{h}$ for all $y \in G$. In view of (2.13.7), we may conclude that $yHy^{-1} = H$ for all $y \in G$; i.e., H is normal. This proves the theorem.

We shall conclude this section with some remarks.

(1) Let G_i be an analytic group with Lie algebra $\mathfrak{g}_i$ $(i = 1, 2)$, and let π be an analytic homomorphism of G_1 into G_2. Then it follows from (2.13.7), (2.10.26), and the definition of Ad that

$$\operatorname{Ad}(\pi(y)) \circ d\pi = d\pi \circ \operatorname{Ad}(y) \quad (y \in G_1). \tag{2.13.12}$$

In particular, let G be any Lie group, H a Lie subgroup. Let $\mathfrak{g}$ be the Lie algebra of G and let us canonically identify the Lie algebra of H with the subalgebra $\mathfrak{h}$ defined by H. If we denote by Ad_G and Ad_H the respective adjoint representations of G and H, then

$$\operatorname{Ad}_H(y) = \operatorname{Ad}_G(y)|\mathfrak{h} \quad (y \in H). \tag{2.13.13}$$

We leave it to the reader to verify that $\mathfrak{h}^y = \mathfrak{h}$ for any $y \in H$ and that (2.13.13) is satisfied for all $y \in H$.

(2) Let $G = GL(V)$, V being a finite-dimensional vector space over $\mathbf{R}$ (or $\mathbf{C}$). We write $\mathfrak{g} = \mathfrak{gl}(V)$. A trivial calculation shows that for $y \in G$ and $X \in \mathfrak{g}$,

$$ye^Xy^{-1} = e^{yXy^{-1}}. \tag{2.13.14}$$

It follows from this that

$$X^y = yXy^{-1} \quad (X \in \mathfrak{g}, y \in G). \tag{2.13.15}$$

In view of the remark (1), (2.13.15) remains valid even if G is only a Lie subgroup of $GL(V)$, provided that we identify its Lie algebra canonically with the subalgebra of $\mathfrak{gl}(V)$ that it defines.

(3) Let G be arbitrary. Since $\operatorname{Ad}(y)$ is an automorphism of $\mathfrak{g}$ for $y \in G$, we have $[X^y,Y^y] = [X,Y]^y$, X, $Y \in \mathfrak{g}$. This shows that

$$\operatorname{ad} X^y = \operatorname{Ad}(y) \cdot \operatorname{ad} X \cdot \operatorname{Ad}(y)^{-1} \quad (y \in G, X \in \mathfrak{g}). \tag{2.13.16}$$

2.14. The Differential of the Exponential Map

In this section we make a closer study of the exponential map. Let G be a Lie group and $\mathfrak{g}$ its Lie algebra. As usual, we identify the tangent spaces to G and $\mathfrak{g}$ at any of their points with $\mathfrak{g}$ itself. With this identification, for any $X \in \mathfrak{g}$, $(d\exp)_X$ becomes an endomorphism of the underlying vector space of $\mathfrak{g}$. We write

$$D(X:Y) = (d\exp)_X(Y) \quad (X, Y \in \mathfrak{g}). \tag{2.14.1}$$

We propose to obtain an explicit formula for $D(X:Y)$. As usual, everything will be proved for real groups; the changes to be made in the complex case are minor and left to the reader.

For $X, Y \in \mathfrak{g}$, the map

$$f: t \mapsto \exp(-X)\exp(X + tY) \quad (t \in \mathbf{R})$$

is analytic from $\mathbf{R}$ into G with $f(0) = 1$. Since $l_{\exp X} f(t) = \exp(X + tY)$, it follows from our identification of the tangent space $T_{\exp X}(G)$ with $\mathfrak{g}$ that $\dot{f}(0)$ is precisely $(d\exp)_X(Y)$. Thus

$$(d\exp)_X(Y) = \left(\frac{d}{dt}\exp(-X)\exp(X + tY)\right)_{t=0}. \tag{2.14.2}$$

Further, the map

$$(t,X,Y) \mapsto \exp(-X)\exp(X + tY)$$

is analytic from $\mathbf{R} \times \mathfrak{g} \times \mathfrak{g}$ into G. Consequently, we conclude from (2.14.2) that

$$D: (X,Y) \mapsto (d\exp)_X(Y)$$

is an analytic map of $\mathfrak{g} \times \mathfrak{g}$ into $\mathfrak{g}$.

Lemma 2.14.1. *Let l, n be integers with $0 \le l \le n$. Then*

$$\sum_{0 \le k \le l} (-1)^k \binom{n+1}{k} = (-1)^l \binom{n}{l} \tag{2.14.3}$$

(here, for integers a, b with $0 \le b \le a$, $\binom{a}{b}$ denotes the binomial coefficient $a!/b!(a-b)!$).

Proof. Follows trivially by induction on l.

Lemma 2.14.2. *Let $\mathfrak{A}$ be an associative algebra over a field k of characteristic zero. For any $a \in \mathfrak{A}$ let d_a be the endomorphism $b \mapsto ab - ba$ of $\mathfrak{A}$. Then for any integer $n \geq 0$,*

$$d_a^n(b) = (-1)^n \sum_{0 \leq p \leq n} (-1)^p \binom{n}{p} a^p b a^{n-p} \quad (b \in \mathfrak{A}). \tag{2.14.4}$$

Proof. Let l_a (resp. r_a) denote the endomorphism $b \mapsto ab$ (resp. $b \mapsto ba$) of $\mathfrak{A}$. It is obvious that l_a commutes with r_a and that $d_a = l_a - r_a$. So for any integer $n \geq 0$,

$$d_a^n = (-1)^n \sum_{0 \leq p \leq n} (-1)^p \binom{n}{p} l_a^p r_a^{n-p}.$$

If we apply both sides to an element $b \in \mathfrak{A}$, we get (2.14.4).

Theorem 2.14.3. *Let G be a Lie group, $\mathfrak{g}$ its Lie algebra. For any $X \in \mathfrak{g}$ let $(d \exp)_X$ denote the differential of the exponential map at X. Then*

$$(d \exp)_X = \sum_{n=0}^{\infty} \frac{(-1)^n}{(n+1)!} (\operatorname{ad} X)^n. \tag{2.14.5}$$

In particular, $(d \exp)_X$ is bijective if and only if no eigenvalue of the endomorphism ad X *is of the form $(-1)^{1/2} 2k\pi$ for a nonzero integer k.*

Proof. Let $X, Y \in \mathfrak{g}$ and let U be an open subset of G containing 1. We choose an $a > 0$ such that $\exp uX \exp(vX + wY) \in U$ for all real u, v, w with $|u| < a$, $|v| < a$, and $|w| < a$. Suppose f is a function defined and analytic on U. Then the function

$$F: u,v,w \mapsto f(\exp uX \exp(vX + wY))$$

is analytic on the cube I_a^3. By choosing a sufficiently small, we may assume that the power series expansion of F about the origin is absolutely and uniformly convergent in I_a^3. We write, for integers $p, q, r \geq 0$,

$$F_{p,q,r} = \left(\frac{\partial^{p+q+r}}{\partial u^p \partial v^q \partial w^r F}\right)(0,0,0).$$

Then

$$F(u,v,w) = \sum_{p,q,r \geq 0} \frac{F_{p,q,r}}{p!q!r!} u^p v^q w^r \quad ((u,v,w) \in I_a^3),$$

and hence

$$\left(\frac{\partial F}{\partial w}\right)(u,v,0) = \sum_{p,q \geq 0} \frac{F_{p,q,1}}{p!q!} u^p v^q \quad ((u,v) \in I_a^2).$$

In particular, taking $-u = v = t$, we have

$$\left(\frac{\partial F}{\partial w}\right)(-t,t,0) = \sum_{n=0}^{\infty} \frac{c_n}{n!} t^n \quad (|t| < a), \tag{2.14.6}$$

where

$$c_n = \sum_{0 \le k \le n} (-1)^k \binom{n}{k} F_{k,n-k,1}. \tag{2.14.7}$$

Now, it follows from the relation (2.14.2) that

$$\left(\frac{\partial F}{\partial w}\right)(-t,t,0) = f(1;(d\exp)_{tX}(Y)) \quad (|t| < a). \tag{2.14.8}$$

On the other hand, we see from (2.12.8) that for any u with $|u| < a$,

$$\left(\frac{\partial^{q+1} F}{\partial v^q \partial w}\right)(u,0,0) = f(\exp uX; \xi),$$

where ξ is the element of the enveloping algebra $\mathfrak{G}$ of G defined as the coefficient of $v^q w$ in the expansion of $(vX + wY)^q/(q + 1)$ as a polynomial in v and w. Thus for $|u| < a$,

$$\left(\frac{\partial^{q+1} F}{\partial v^q \partial w}\right)(u,0,0) = \frac{1}{q+1} \sum_{0 \le s \le q} f(\exp uX; X^s Y X^{q-s}).$$

Differentiating this p times with respect to u at $u = 0$ we obtain, on using (2.12.2), for all integers $p, q \ge 0$,

$$F_{p,q,1} = \frac{1}{q+1} \sum_{0 \le s \le q} f(1; X^{p+s} Y X^{q-s}). \tag{2.14.9}$$

Now substitute this expression for $F_{p,q,1}$ in the formula (2.14.7) for c_n. We then get

$$\begin{aligned} c_n &= \sum_{0 \le k \le n} (-1)^k \binom{n}{k} \cdot \frac{1}{n-k+1} \sum_{0 \le s \le n-k} f(1; X^{k+s} Y X^{n-k-s}) \\ &= \frac{1}{n+1} \sum_{0 \le k \le n} (-1)^k \binom{n+1}{k} \sum_{k \le l \le n} f(1; X^l Y X^{n-l}) \\ &= \frac{1}{n+1} \sum_{0 \le l \le n} f(1; X^l Y X^{n-l}) \sum_{0 \le k \le l} (-1)^k \binom{n+1}{k} \\ &= \frac{1}{n+1} \sum_{0 \le l \le n} (-1)^l \binom{n}{l} f(1; X^l Y X^{n-l}), \end{aligned}$$

by Lemma 2.14.1. Moreover, by Lemma 2.14.2 we have the identity

$$(\mathrm{ad}\, X)^n(Y) = (-1)^n \sum_{0 \leq l \leq n} (-1)^l \binom{n}{l} X^l Y X^{n-l}$$

valid in the associative algebra $\mathfrak{G}$. Hence we have

$$c_n = \frac{(-1)^n}{n+1} f(1; (\mathrm{ad}\, X)^n(Y)).$$

From (2.14.6) and (2.14.8) we then obtain the formula

$$f(1; (d\exp)_{tX}(Y)) = \sum_{n=0}^{\infty} \frac{(-1)^n}{(n+1)!} t^n f(1; (\mathrm{ad}\, X)^n(Y)) \quad (|t| < a). \tag{2.14.10}$$

The open set U and the function f have been arbitrary so far. If we choose U to be a coordinate open set around 1 and f to be an arbitrary member of a system of coordinates on U, we can conclude from (2.14.10) that

$$D(tX : Y) = \sum_{n=0}^{\infty} \frac{(-1)^n}{(n+1)!} t^n (\mathrm{ad}\, X)^n(Y) \quad (|t| < a).$$

Observe that both sides of this relation are analytic functions on **R**; hence they must be equal for all t. Putting $t = 1$ and observing that Y was arbitrary, we get (2.14.5).

For the second assertion, let g be the entire function on **C** defined by

$$g(z) = \sum_{n=0}^{\infty} \frac{(-1)^n}{(n+1)!} z^n \quad (z \in \mathbf{C}).$$

Then $zg(z) = 1 - e^{-z}$, and we easily see that $g(z) = 0$ if and only if $z = (-1)^{1/2} 2k\pi$ for some nonzero integer k. Fix $X \in \mathfrak{g}$. Then $(d\exp)_X$ is bijective if and only if no eigenvalue of $(d\exp)_X$ is zero. But $(d\exp)_X = g(\mathrm{ad}\, X)$, so its eigenvalues are $g(z_1), \ldots, g(z_s)$, where $z_1, \ldots, z_s$ are the eigenvalues of ad X.[2] This leads to the second assertion. The proof of the theorem is complete.

Globally, the exponential map is seldom one-to-one, even when we restrict it to the open set of all $X \in \mathfrak{g}$ such that $(d\exp)_X$ is bijective. We now prove a theorem which explicitly exhibits an open neighborhood of 0 on which exp is an analytic diffeomorphism (cf. Harish-Chandra [6]). We need two lemmas. For any endomorphism L of a finite-dimensional vector space (over **R** or **C**),

[2] This is seen by arguing as in Lemma 2.10.2. In fact, let $Z_1, \ldots, Z_m$ be a basis for the complexification $\mathfrak{g}_c$ of $\mathfrak{g}$ such that the matrix (a_{ij}) of ad X is upper triangular, i.e., $a_{ij} = 0$ $(i > j)$. Then the matrix of g (ad X) is also upper triangular and its diagonal entries are $g(a_{ii})$ $(1 \leq i \leq m)$.

write $\sigma(L)$ for the set of all eigenvalues of L. As usual, for any $z \in \mathbf{C}$, we write $\mathrm{Re}(z)$ and $\mathrm{Im}(z)$ for the real and imaginary parts of z respectively.

Lemma 2.14.4. *Let V be a finite-dimensional vector space over $\mathbf{R}$ or $\mathbf{C}$. Let $\mathfrak{m}$ be the set of all endomorphisms L of V with the property that $|\mathrm{Im}\,\lambda| < \pi$ for each eigenvalue λ of L. Then $\mathfrak{m}$ is an open subset of $\mathfrak{gl}(V)$ containing 0, and the exponential map from $\mathfrak{gl}(V)$ into $GL(V)$ is one-to-one on $\mathfrak{m}$.*

Proof. It is enough to consider the case when V is defined over $\mathbf{C}$, since the real case can be reduced to this by complexification. $\mathfrak{m}$ is well defined and contains 0. We write $n = \dim V$. If $n = 1$, the lemma simply asserts the elementary fact that the exponential function is one-to-one on the subset $\{z : z \in \mathbf{C}, |\mathrm{Im}(z)| < \pi\}$ of $\mathbf{C}$. We propose to show that $\mathfrak{m}$ is open and that if $L_1, L_2 \in \mathfrak{m}$ are such that $e^{L_1} = e^{L_2}$, then $L_1 = L_2$.

We prove first that $\mathfrak{m}$ is open in $\mathfrak{gl}(V)$. We shall deduce this from the following more general fact: if A is an open subset of $\mathbf{C}$ and $\mathfrak{u}_A$ is the set of all $L \in \mathfrak{gl}(V)$ such that $\sigma(L) \subseteq A$, then $\mathfrak{u}_A$ is open in $\mathfrak{gl}(V)$. Suppose this is false. Then we can find $L_0 \in \mathfrak{u}_A$ and a sequence $\{L_n\}$ from $\mathfrak{gl}(V)\backslash\mathfrak{u}_A$ such that $L_n \longrightarrow L_0$ as $n \longrightarrow \infty$. For any $M \in \mathfrak{gl}(V)$ let

$$P(M : t) \equiv \det(t1 - M) \equiv t^n + \sum_{0 \le s < n} c_s(M)t^s$$

be the characteristic polynomial of M. Then $c_s(L_n) \longrightarrow c_s(L_0)$ as $n \longrightarrow \infty$ $(0 \le s < n)$. On the other hand, if λ is any root of the equation $P(M : t) = 0$, we have[3]

$$|\lambda| \le 1 + \sum_{0 \le s < n} |c_s(M)|.$$

Consequently, we can find a constant $\Lambda > 0$ such that all the eigenvalues of all the L_n lie in the disc $\{z : |z| < \Lambda\}$ in $\mathbf{C}$. Now choose a subsequence $\{n_k\}$ and eigenvalues $\lambda_{n_k} \in \sigma(L_{n_k})$ such that $\lambda_{n_k} \notin A$ for all k and $\lambda_{n_k} \longrightarrow \lambda_0$ as $k \longrightarrow \infty$. Then $\lambda_0 \notin A$. But since $P(L_{n_k}, \lambda_{n_k}) = 0$ for all k, $P(L_0, \lambda_0) = 0$, showing that $\lambda_0 \in \sigma(L_0) \subseteq A$. This contradiction proves that $\mathfrak{u}_A$ is open.

Now consider the case when $\sigma(L_1) = \sigma(L_2) = \{0\}$, i.e., when L_1 and L_2 are nilpotent. Then $L_1^n = L_2^n = 0$. For any nilpotent endomorphism K of V, $e^K - 1 = KS$, where S commutes with K, showing that $K' = e^K - 1$ is nilpotent. Now, the relation $x = \log(1 + (e^x - 1))$ (x real and sufficiently small) implies the formal power series identity

$$\sum_{s=1}^{\infty} \frac{(-1)^{s-1}}{s}\left(\sum_{k=1}^{\infty} \frac{T^k}{k!}\right)^s = T, \tag{2.14.11}$$

[3]This may be seen as follows. If $|\lambda| \le 1$, there is nothing to prove. If $|\lambda| > 1$, $\lambda = \sum_{0 \le s \le n} c_s(M)\lambda^{s-(n-1)}$, so $|\lambda| \le \sum_{0 \le s \le n} |c_s(M)|$.

T being an indeterminate. If we replace T by the nilpotent endomorphism K of V, we see at once that

$$K = \sum_{s=1}^{\infty} \frac{(-1)^{s-1}}{s}(e^K - 1)^s \quad (K \text{ nilpotent}). \tag{2.14.12}$$

It follows at once from this that $L_1 = L_2$.

We take up next the case when L_1 and L_2 are both semisimple. Let λ_j be the distinct eigenvalues of L_1 $(1 \leq j \leq s)$, and let V_j be the eigenspace corresponding to λ_j. Then $e^{\lambda_1}, \ldots, e^{\lambda_s}$ are the distinct eigenvalues of e^{L_1}, and V_j is the eigenspace corresponding to e^{λ_j}. Since L_1 is semisimple, V is the direct sum of $V_1, \ldots, V_s$. Now L_2 commutes with $e^{L_2} = e^{L_1}$ and so leaves each V_j invariant. If μ is an eigenvalue of the restriction of L_2 to V_j (j fixed), the $|\text{Im}(\mu)| < \pi$ by assumption; on the other hand, $e^{L_2}v = e^{\lambda_j}v$ for all $v \in V_j$. So $e^{\mu} = e^{\lambda_j}$, implying that $\mu = \lambda_j$. λ_j is thus the sole eigenvalue of L_2 restricted to V_j. Since this restriction is also semisimple, $L_2 v = \lambda_j v$ for all $v \in V_j$. Thus L_1 and L_2 coincide on V_j. Since j was arbitrary, $L_1 = L_2$.

We now come to the general case. Let $L_j = S_j + N_j$ be the Jordan decomposition[4] of L_j into its semisimple part S_j and nilpotent part N_j $(j = 1, 2)$. Then e^{S_j} is semisimple, $e^{N_j} - 1$ is nilpotent, and since S_j and N_j commute, we have in addition that $e^{L_j} = e^{S_j}e^{N_j} = e^{S_j} + e^{S_j}(e^{N_j} - 1)$ $(j = 1, 2;$ cf. Lemma 2.10.2). Moreover, e^{S_j} and $e^{N_j} - 1$ commute, implying in particular that $e^{S_j}(e^{N_j} - 1)$ is nilpotent. So e^{S_j} and $e^{S}(e^{N_j} - 1)$ are the semisimple and nilpotent parts of e^{L_j}, respectively. The uniqueness of Jordan decompositions implies now that $e^{S_1} = e^{S_2}$ and $e^{N_1} = e^{N_2}$. On the other hand, because S_j commutes with N_j and N_j is nilpotent, $\sigma(S_j) = \sigma(L_j)$. So we conclude from the special cases proved earlier that $S_1 = S_2$ and $N_1 = N_2$. Thus $L_1 = L_2$. The proof of the lemma is complete.

From now on, we fix a Lie group G with Lie algebra $\mathfrak{g}$. Let

$$\mathfrak{v} = \{X : X \in \mathfrak{g}, |\text{Im}\,\lambda| < \pi \text{ for each eigenvalue of ad}\,X\}. \tag{2.14.13}$$

Lemma 2.14.5. *$\mathfrak{v}$ is an open connected subset of $\mathfrak{g}$ which is invariant under* Ad$[G]$. *If $\mathfrak{z}$ is the center of $\mathfrak{g}$,*

$$\mathfrak{v} + \mathfrak{z} = \mathfrak{v}., \tag{2.14.14}$$

Proof. For $X \in \mathfrak{g}$ and $y \in G$, the eigenvalues of ad X and ad X^y are the same, by (2.13.16). So $\mathfrak{v}$ is invariant under Ad. If $X \in \mathfrak{v}$, $Z \in \mathfrak{z}$, then

[4]Cf. §3.1.

$\mathrm{ad}(X + Z) = \mathrm{ad}\, X$, because $\mathrm{ad}\, Z = 0$. This proves (2.14.14). For $X \in \mathfrak{v}$ and $t \in \mathbf{R}$ with $0 \leq t \leq 1$, we have $tX \in \mathfrak{b}$. Thus $\mathfrak{v}$ is connected. Suppose $\mathfrak{m}$ is the set of all endomorphisms L of $\mathfrak{g}$ such that $|\mathrm{Im}\, \lambda| < \pi$ for any eigenvalue λ of L; then $\mathfrak{v} = \{X : X \in \mathfrak{g},\ \mathrm{ad}\, X \in \mathfrak{m}\}$. Since $\mathfrak{m}$ is open (Lemma 2.14.4) and ad is continuous, $\mathfrak{v}$ is open. This proves the lemma.

The center $\mathfrak{z}$ of $\mathfrak{g}$ is abelian. We denote by Z the analytic subgroup of G cooresponding to $\mathfrak{z}$. By Theorem 2.13.2, Z is the component of identity of the centralizer of G^0 in G, while Corollary 2.13.3 implies that exp is a covering homomorphism of the additive group of $\mathfrak{z}$ onto Z. Let

$$\Gamma = \{X : X \in \mathfrak{z},\ \exp X = 1\}. \tag{2.14.15}$$

Then Γ is a discrete additive subgroup of $\mathfrak{z}$.

Theorem 2.14.6. *Let notation be as above. Then* $\exp \mathfrak{v} = U$ *is a connected open neighborhood of* 1 *in* G *which is invariant under the inner automorphisms of* G. *The exponential map has bijective differential at all points of* $\mathfrak{v}$, *and for* $X, X' \in \mathfrak{v}$, $\exp X = \exp X'$ *if and only if* $X - X' \in \Gamma$, *where* Γ *is defined by* (2.14.15). *In particular, exp is a covering map of* $\mathfrak{v}$ *onto* U, *and* U *is the quotient of* $\mathfrak{v}$ *with respect to exp (as an analytic manifold). If* Z *is simply connected, exp is an analytic diffeomorphism of* $\mathfrak{v}$ *onto* U; *this is always the case if* G *is a simply connected analytic group.*

Proof. U is obviously connected, and (2.13.7) implies that it is invariant under the inner automorphisms of G. It is immediate from Theorem 2.14.3 that $(d\exp)_X$ is bijective for all $X \in \mathfrak{v}$. This enables us to conclude that U is open in G. Suppose now $X, X' \in \mathfrak{v}$ and $\exp X = \exp X'$. Then $e^{\mathrm{ad}\, X} = e^{\mathrm{ad}\, X'}$ by (2.13.8), and hence $\mathrm{ad}\, X = \mathrm{ad}\, X'$ by Lemma 2.14.5. This proves that $X - X' = Y \in \mathfrak{z}$. But then $\exp X = \exp(X' + Y) = \exp X' \exp Y$ (cf. (2.13.11)) $= \exp X \exp Y$, so $\exp Y = 1$, proving that $X - X' \in \Gamma$. Conversely, if $X \in \mathfrak{v}, Y \in \Gamma$, and $X' = X + Y$, then $X' \in \mathfrak{v}$ by (2.14.14) and $\exp X = \exp X'$. This leads to all the assertions of the theorem except for the last pair. If Z is simply connected, the covering property of $\exp|\mathfrak{z}$ implies that $\Gamma = \{0\}$. So in this case exp is one-to-one on $\mathfrak{v}$, hence an analytic diffeomorphism of $\mathfrak{v}$ onto U. The last assertion follows from a theorem of Mal'čev according to which closed normal analytic subgroups of simply connected analytic groups are also simply connected (cf. Theorem 3.18.2).

We shall conclude this section with an example. Let $G = SO(3,\mathbf{R})$. We may then identify its Lie algebra $\mathfrak{g}$ with the Lie algebra of all 3×3 skew-symmetric real matrices; the exponential map is then the usual matrix exponential. Let

$$X_1 = \begin{pmatrix} 0 & 0 & 0 \\ 0 & 0 & 1 \\ 0 & -1 & 0 \end{pmatrix}, \quad X_2 = \begin{pmatrix} 0 & 0 & 1 \\ 0 & 0 & 0 \\ -1 & 0 & 0 \end{pmatrix}, \quad X_3 = \begin{pmatrix} 0 & 1 & 0 \\ -1 & 0 & 0 \\ 0 & 0 & 0 \end{pmatrix}.$$

Then $\{X_1,X_2,X_3\}$ is a basis for $\mathfrak{g}$, and the commutation rules are

$$[X_1,X_2] = X_3, \quad [X_2,X_3] = X_1, \quad [X_3,X_1] = X_2.$$

It follows easily from this that if $X = x_1X_1 + x_2X_2 + x_3X_3$ $(x_1, x_2, x_3 \in \mathbf{R})$, then the eigenvalues of ad X are 0 and $\pm(-1)^{1/2}(x_1^2 + x_2^2 + x_3^2)^{1/2}$. Theorem 2.14.3 shows that $X = x_1X_1 + x_2X_2 + x_3X_3$ is a singular point of the map exp if and only if $x_1^2 + x_2^2 + x_3^2 = 4k^2\pi^2$ for some nonzero integer k. Given any $y \in G$, we can find a continuous homomorphism $t \mapsto y(t)$ of $\mathbf{R}$ into G such that $y(1) = y$; this follows from the well-known fact that any element of G is a rotation around some axis. So we can write $y = \exp Y$ for some $Y \in \mathfrak{g}$. The exponential map is therefore surjective in this case.

2.15. The Baker–Campbell–Hausdorff Formula

Let G be a Lie group, $\mathfrak{g}$ its Lie algebra. We have seen that if $X, Y \in \mathfrak{g}$ and $[X,Y] = 0$, then $\exp X \exp Y = \exp(X + Y)$. This is in general not true if $[X,Y] \neq 0$, and the question arises naturally whether one can obtain an explicit formula for $\exp X \exp Y$ for arbitrary $X, Y \in \mathfrak{g}$. Now, the exponential map is an analytic diffeomorphism around $0 \in \mathfrak{g}$, so we can find an open neighborhood $\mathfrak{a}$ of 0 and an analytic map $A : (X,Y) \mapsto A(X : Y)$ of $\mathfrak{a} \times \mathfrak{a}$ into $\mathfrak{g}$ such that

$$\exp X \exp Y = \exp A(X : Y) \quad (X, Y \in \mathfrak{a}).$$

The problem raised above may then be regarded as that of determining A explicitly. It is our aim in this section to obtain an expression for A. The formula we obtain is substantially equivalent to what is known as the *Baker–Campbell–Hausdorff formula.*

We begin with an auxiliary lemma.

Lemma 2.15.1. *Let V be a finite-dimensional vector space* (*over* $\mathbf{R}$ *or* $\mathbf{C}$) *equipped with a norm* $|\cdot|$. *Let E be the algebra of all endomorphisms on V and let* $|\cdot|$ *denote the standard*[5] *operator norm in E. Let $a > 0$ and let $\mathbf{F}_a$ be the*

[5]If T is an endomorphism of V,

$$|T| = \sup_{v \in V, |v| \leq 1} |Tv|.$$

algebra of all functions of a complex variable z which are analytic in the disc $\{z : |z| < a\}$; *for* $\varphi \in \mathbf{F}_a$ *we denote by* $\varphi(z) = \sum_{n=0}^{\infty} a_n(\varphi)z^n$ *its power series expansion around* $z = 0$. *Then for any* $L \in E$ *with* $|L| < a$ *and any* $\varphi \in \mathbf{F}_a$,

$$\varphi(L) = \sum_{n=0} a_n(\varphi)L^n$$

is absolutely convergent in E, and the map $\varphi \mapsto \varphi(L)$ *is a homomorphism of* $\mathbf{F}_a$ *into E.*

Proof. Elementary.

We recall that a series $\sum_n x_n$ of vectors in a Banach space is *absolutely convergent* if $\sum_n |x_n| < \infty$; in this case it is also necessarily convergent. It follows from the lemma that if $\varphi \in \mathbf{F}_a$ does not vanish in $\{z : |z| < a\}$, then $\varphi(L)$ is invertible and $\varphi(L)^{-1} = (1/\varphi)(L)$.

Let G be a Lie group and $\mathfrak{g}$ its Lie algebra. We work in the real case, but all our considerations go over to the complex case with only minor changes. We shall equip $\mathfrak{g}$ with a suitable norm $|\cdot|$ so that it becomes a Banach space. Let A be as defined at the beginning of this section with $A(0 : 0) = 0$. For any $\epsilon > 0$ let $\mathfrak{g}_\epsilon = \{Z : Z \in \mathfrak{g}, |Z| < \epsilon\}$ and let η, ζ be two numbers with $0 < \eta < \zeta$ such that exp is an analytic diffeomorphism on $\mathfrak{g}_\zeta$ and $(\exp[\mathfrak{g}_\eta])^2 \subseteq \exp[\mathfrak{g}_\zeta]$; then A is an analytic map of $\mathfrak{g}_\eta \times \mathfrak{g}_\eta$ into $\mathfrak{g}_\zeta$. For $X, Y \in \mathfrak{g}$, put

$$Z(u : v : X : Y) = A(uX : vY); \tag{2.15.1}$$

then $Z(\cdot : \cdot : X : Y)$ is analytic in a neighborhood of (0,0) in $\mathbf{R}^2$, certainly for $|uX| < \eta$, $|vY| < \eta$. Then

$$\exp uX \exp vY = \exp Z(u : v : X : Y). \tag{2.15.2}$$

Let

$$F(t : X : Y) = Z(t : t : X : Y). \tag{2.15.3}$$

F is analytic around $t = 0$, certainly if $|tX| < \eta$, $|tY| < \eta$. If

$$c_n(X : Y) = \frac{1}{n!}\left(\frac{d^n}{dt^n}F(t : X : Y)\right)_{t=0} \quad (n \geq 0), \tag{2.15.4}$$

then for all sufficiently small t

$$F(t : X : Y) = \sum_{n=0}^{\infty} t^n c_n(X : Y), \tag{2.15.5}$$

the series being absolutely convergent. It follows from (i) of (2.12.10) that

(2.15.6)
$$\begin{cases} c_0(X:Y) = 0 \\ c_1(X:Y) = X + Y \\ c_2(X:Y) = \frac{1}{2}[X,Y]. \end{cases}$$

We shall now determine the coefficients c_n. This will be done[6] by deriving a differential equation for F and obtaining its solution as a power series; the coefficients c_n will then be determined by recursion formulae. To this end, let g be the entire function on $\mathbf{C}$ given by

(2.15.7)
$$g(z) = \sum_{n=0}^{\infty} \frac{(-1)^n}{(n+1)!} z^n = \frac{1-e^{-z}}{z}.$$

Put $h = 1/g$. Then h is analytic around $z = 0$, and $h(0) = 1$. A simple calculation shows that $h(-z) = h(z) - z$. So if we put

(2.15.8)
$$f(z) = h(z) - \frac{1}{2}z = \frac{z}{1-e^{-z}} - \frac{1}{2}z.$$

then f is defined and analytic around $z = 0$, $f(0) = 1$, and f is even. We write

(2.15.9)
$$f(z) = f(-z) = 1 + \sum_{p=1}^{\infty} K_{2p} z^{2p}.$$

It is a straightforward verification that the K_{2p}'s are all rational numbers; we leave it to the reader. From Theorem 2.14.3 and Lemma 2.15.1 we see that for all $X \in \mathfrak{g}_c$ the endomorphism $(d \exp)_X = g(\text{ad } X)$ is invertible and

(2.15.10)
$$\begin{cases} g(\text{ad } X)^{-1} = f(\text{ad } X) + \frac{1}{2} \text{ad } X \\ f(\text{ad } X) = 1 + \sum_{p=1}^{\infty} K_{2p}(\text{ad } X)^{2p}. \end{cases}$$

Lemma 2.15.2. *Let $X, Y \in \mathfrak{g}$ and F be as in* (2.15.3). *Let $a > 0$ be such that $a|X| < \eta$, $a|Y| < \eta$. Then F is a solution to the equation*

(2.15.11)
$$\frac{dF}{dt} = f(\text{ad } F)(X + Y) + \frac{1}{2}[X - Y, F]$$

in $(-a,a)$ with the initial condition

(2.15.12)
$$F(0:X:Y) = 0.$$

[6]The development through Theorem 2.15.4 is an adaptation of a treatment of this question by Professor V. Bargmann of Princeton, given in a course many years ago, and is closely related to the treatment of Baker [1] and Hausdorff [1].

Proof. For brevity we denote $Z(u:v:X:Y)$ by $Z(u:v)$ and $F(t:X:Y)$ by $F(t)$. We shall also make the usual identification of the tangent spaces to G and $\mathfrak{g}$ at each of their points with $\mathfrak{g}$. We have

$$\exp uX \exp vY = \exp Z(u:v) \quad (|u| < a, |v| < a).$$

Equating the differentials of the maps $v \mapsto \exp uX \exp vY$ and $v \mapsto \exp Z(u:v)$ we get, in view of our identification of tangent spaces,

$$\begin{aligned} Y &= (d\exp)_{Z(u:v)}\left(\frac{\partial Z}{\partial v}\right) \\ &= g(\operatorname{ad} Z)\left(\frac{\partial Z}{\partial v}\right) \end{aligned}$$

So by (2.15.10), since $Z(u:v) \in \mathfrak{g}_\zeta$,

$$\frac{\partial Z}{\partial v} = f(\operatorname{ad} Z)(Y) + \frac{1}{2}[Z,Y] \quad (|u| < a, |v| < a). \tag{2.15.13}$$

On the other hand, proceeding with the equation

$$\exp(-vY)\exp(-uX) = \exp(-Z(u:v))$$

in an analogous manner but taking differentials with respect to u, we obtain

$$\begin{aligned} -X &= (d\exp)_{(-Z(u:v))}\left(\frac{-\partial Z}{\partial u}\right) \\ &= g(-\operatorname{ad} Z)\left(\frac{-\partial Z}{\partial u}\right), \end{aligned}$$

giving for $|u| < a$ and $|v| < a$

$$\frac{\partial Z}{\partial u} = g(-\operatorname{ad} Z)^{-1}(X).$$

Using (2.15.10) and the fact that f is an even function, we get

$$\frac{\partial Z}{\partial u} = f(\operatorname{ad} Z)(X) - \frac{1}{2}[Z,X] \quad (|u| < a, |v| < a). \tag{2.15.14}$$

Now

$$\frac{dF}{dt} = \left(\frac{\partial Z}{\partial u} + \frac{\partial Z}{\partial v}\right)_{u=v=t}.$$

Consequently, we get (2.15.11) from (2.15.13) and (2.15.14).

Lemma 2.15.3. *Let* $c_n(X:Y)$ *be defined* (2.15.4) *for* $X, Y \in \mathfrak{g}$. *Then they are uniquely determined by the recursion formula*

$$(2.15.15)\qquad \begin{aligned}(n+1)c_{n+1}(X:Y) &= \tfrac{1}{2}[X-Y, c_n(X:Y)]\\ &+ \sum_{p\geq 1, 2p\leq n} K_{2p} \sum_{\substack{k_1,\ldots,k_{2p}>0\\ k_1+\cdots+k_{2p}=n}} [c_{k_1}(X:Y),[\cdots[c_{k_{2p}}(X:Y),X+Y]\cdots]\end{aligned}$$

($n \geq 1$, $X, Y \in \mathfrak{g}$) *and by the condition* $c_1(X:Y) = X + Y$.

Proof. The relations (2.15.15) obviously determine all the c_n uniquely if c_1 is known. We now prove (2.15.15). Fix $X, Y \in \mathfrak{g}$ and write c_n for $c_n(X:Y)$. In what follows, if k is any integer ≥ 1, denote by $O(t^k)$ any function of the form $t \mapsto t^k g(t)$, where g is defined and analytic around $t = 0$ and takes values in an appropriate finite-dimensional real vector space. Fix an integer $n \geq 1$. Then with the above convention, we have

$$\frac{dF}{dt} = c_1 + 2tc_2 + \cdots + (n+1)t^n c_{n+1} + O(t^{n+1})$$

and

$$\operatorname{ad} F(t) = t \operatorname{ad} c_1 + t^2 \operatorname{ad} c_2 + \cdots + t^n \operatorname{ad} c_n + O(t^{n+1}).$$

Hence for any integer $p \geq 1$ with $2p \leq n$,

$$(\operatorname{ad} F(t))^{2p} = \sum_{2p\leq s\leq n} t^s \sum_{\substack{k_1\geq 1,\ldots,k_{2p}\geq 1\\ k_1+\cdots+k_{2p}=s}} \operatorname{ad} c_{k_1} \cdots \operatorname{ad} c_{k_{2p}} + O(t^{n+1}).$$

On the other hand, $\operatorname{ad} F(t) = O(t)$, so

$$\begin{aligned} f(\operatorname{ad} F(t)) &= 1 + \sum_{p\geq 1, 2p\leq n} K_{2p}(\operatorname{ad} F(t))^{2p} + O(t^{n+1})\\ &= 1 + \sum_{1\leq s\leq n} t^s \sum_{p\geq 1, 2p\leq s} K_{2p} \sum_{\substack{k_1,\ldots,k_{2p}\geq 1\\ k_1+\cdots+k_{2p}=s}} \operatorname{ad} c_{k_1} \cdots \operatorname{ad} c_{k_{2p}} + O(t^{n+1}).\end{aligned}$$

If we now substitute these expressions in (2.15.11) and identify the coefficients t^n on both sides, we obtain (2.15.15) without difficulty. This proves the lemma.

The c_n may be calculated from (2.15.15) in succession; unfortunately, the calculations become complicated very rapidly. However, the first few of the c_n may be calculated without too much difficulty. On doing this, one finds the following expressions for c_3 and c_4:

$$(2.15.17)\qquad \begin{cases} c_3(X:Y) = \frac{1}{12}[[X,Y],Y] - \frac{1}{12}[[X,Y],X]\\ c_4(X:Y) = -\frac{1}{48}[Y,[X,[X,Y]]] - \frac{1}{48}[X,[Y,[X,Y]]]\end{cases}$$

for $X, Y \in \mathfrak{g}$. It is clear from the relations (2.15.15) that for any $n \geq 1$, c_n is a polynomial map of $\mathfrak{g} \times \mathfrak{g}$ into $\mathfrak{g}$ whose degree is n.

Let $M \geq 1$ be a constant such that

$$|[X,Y]| \leq M|X||Y| \quad (X, Y \in \mathfrak{g}), \tag{2.15.18}$$

and let H be the function of the complex variable z defined by

$$H(z) = 1 + \sum_{p \geq 1} |K_{2p}| z^{2p}. \tag{2.15.19}$$

Then H is defined and analytic around $z = 0$—certainly for $|z| < 2\pi$. Consider the differential equation

$$\begin{cases} \dfrac{dy}{dz} = \dfrac{1}{2}y + H(y) \\ y(0) = 0. \end{cases} \tag{2.15.20}$$

From the general theory of differential equations, we know that for some constant $\delta > 0$ there is a solution y to (2.15.20) which is holomorphic in the disc $\{z : |z| < \delta\}$. Note that $\delta > 0$ is a *universal constant.* We are now in a position to formulate and prove our first main result in this section. Put

$$\mathfrak{a} = \mathfrak{g}_{\delta/2M} = \left\{\{X : X \in \mathfrak{g}, |X| < \frac{\delta}{2M}\right\}. \tag{2.15.21}$$

Theorem 2.15.4. *Let G be a Lie group, $\mathfrak{g}$ its Lie algebra. For $X, Y \in \mathfrak{g}$, define $c_1(X : Y) = X + Y$ and $c_n(X : Y)$ $(n > 1)$ by the recursion formulae* (2.15.15). *Then for each $n \geq 1$, c_n is a polynomial map of $\mathfrak{g} \times \mathfrak{g}$ into $\mathfrak{g}$ of degree n. Moreover, if $\mathfrak{a}$ is as in* (2.15.21), *the series*

$$\sum_{n=1}^{\infty} c_n(X : Y) = C(X : Y) \tag{2.15.22}$$

converges absolutely for all $X, Y \in \mathfrak{a}$, its sum C defines an analytic map of $\mathfrak{a} \times \mathfrak{a}$ into $\mathfrak{g}$, and

$$\exp X \exp Y = \exp C(X : Y) \quad (X, Y \in \mathfrak{a}). \tag{2.15.23}$$

Proof. It remains to prove only the last group of assertions. Write $c_n = c_n(X : Y)$ and $\alpha = \max(|X|, |Y|)$, X, Y being fixed in $\mathfrak{g}$. From (2.15.15) we find that $|c_1| \leq 2\alpha$, and for $n \geq 1$,

$$\begin{aligned}(n+1)|c_{n+1}| &\leq M\alpha|c_n| \\ &\quad + 2\alpha \sum_{p \geq 1, 2p \leq n} |K_{2p}| M^{2p} \sum_{\substack{k_1 > 0, \ldots, k_{2p} > 0 \\ k_1 + \cdots + k_{2p} = n}} |c_{k_1}| \cdots |c_{k_{2p}}|.\end{aligned} \tag{2.15.24}$$

Let y be the solution of (2.15.20) which is holomorphic in $\{z : |z| < \delta\}$, and let

$$y(z) = \sum_{n \geq 1} \gamma_n z^n \quad (|z| < \delta). \tag{2.15.25}$$

Substitution of (2.15.25) in (2.15.20) then yields

$$\begin{cases} (n+1)\gamma_{n+1} = \frac{1}{2}\gamma_n + \sum_{p \geq 1, 2p \leq n} |K_p| \sum_{\substack{k_1 > 0, \ldots, k_{2p} > 0 \\ k_1 + \cdots + k_{2p} = n}} \gamma_{k_1} \cdots \gamma_{k_{2p}} \\ \gamma_1 = 1. \end{cases} \tag{2.15.26}$$

It follows from (2.15.26) that $\gamma_n \geq 0$ for all n.

We now claim that for $n \geq 1$

$$|c_n| \leq M^{n-1}(2\alpha)^n \gamma_n. \tag{2.15.27}$$

Since $|c_1| \leq 2\alpha$, this is true for $n = 1$. Suppose (2.15.27) is true for c_n with $1 \leq n \leq m$. Then from (2.15.24) we get

$$\begin{aligned} (m+1)|c_{m+1}| &\leq M^m \alpha (2\alpha)^m \gamma_m \\ &\quad + 2\alpha \sum_{p \geq 1, 2p \leq m} |K_{2p}| M^{2p} \sum_{\substack{k_1 > 0, \ldots, k_{2p} > 0 \\ k_1 + \cdots + k_{2p} = m}} M^{m-2p}(2\alpha)^m \gamma_{k_1} \cdots \gamma_{k_{2p}} \\ &= M^m (2\alpha)^{m+1}(m+1)\gamma_{m+1}, \end{aligned}$$

in view of (2.15.26). Thus (2.15.27) is true for all $n \geq 1$. Since the series (2.15.25) converges absolutely if $|z| < \delta$, we see from (2.15.27) that $\sum_{n \geq 1} |c_n|$ converges if $2M\alpha < \delta$, i.e., if X and Y are in $\mathfrak{a}$.

For fixed $X, Y \in \mathfrak{g}$, $c_n(tX : tY) = t^n c_n(X : Y)$ for $t \in \mathbf{R}$. Hence we conclude from (2.15.5) that for fixed X, Y

$$C(tX : tY) = A(tX : tY)$$

for all sufficiently small $|t|$. This implies, in view of the analyticity of both C and A around (0,0), that $C = A$ in a neighborhood of (0,0) in $\mathfrak{g} \times \mathfrak{g}$. In particular,

$$\exp X \exp Y = \exp C(X : Y)$$

whenever X and Y are sufficiently near 0. Once again we use analytic continuation to conclude the validity of this relation on $\mathfrak{a} \times \mathfrak{a}$. This completes the proof of the theorem.

Remarks 1. This theorem shows that the multiplication law in the group is determined uniquely and very explicitly by the Lie algebra structure, at least in a neighborhood of the identity. We thus obtain an alternative proof of the second fundamental theorem of Lie in the following form: if G_i are Lie groups with Lie algebras $\mathfrak{g}_i$ ($i = 1, 2,$), and if φ is a Lie algebra isomorphism of $\mathfrak{g}_1$ onto $\mathfrak{g}_2$, there is a (unique) local analytic isomorphism φ of a neighborhood of the identity in G_1 onto a neighborhood of the identity in G_2 such that $\varphi(\exp_{G_1} X) = \exp_{G_2} \varphi(X)$ for all $X \in \mathfrak{g}_1$ sufficiently near the origin. We leave the details of proving this to the reader.

2. When we introduced the concept of a Lie group we mentioned the problem of showing that any C^k group can be converted into an analytic group, the analytic structure being of course compatible with the C^k structure. Theorem 2.15.4 can be used to prove this, at least when k is sufficiently large. We indicate briefly how this may be done: Suppose, for instance, that G is a C^∞ group; introduce its Lie algebra $\mathfrak{g}$ and the exponential map as in the analytic case, prove the formula (2.14.5) for the differential of the exponential map (cf. ex. 41), and then, as in this chapter, derive the differential equation (2.15.11) together with the initial condition (2.15.12). Let ϵ be such that $0 < \epsilon < \min((\eta/3), (\delta/4M))$. Then for $X, Y \in \mathfrak{g}_\epsilon$, (2.15.11) is valid for $|t| < 2$. From the general theory of ordinary differential equations we conclude that $F(\cdot : X : Y)$ is analytic for $|t| < 2$, and hence

$$F(t : X : Y) = \sum_1^\infty t^n c_n(X : Y) = C(tX : tY)$$

for all sufficiently small t. Since both sides are analytic for $|t| < 2$, the above relation is true for all t with $|t| < 2$. Setting $t = 1$,

$$A(X : Y) = C(X : Y) \quad (X, Y \in \mathfrak{g}_\epsilon).$$

In other words, A is analytic around $(0,0) \in \mathfrak{g} \times \mathfrak{g}$. But this is equivalent to saying that in canonical coordinates of the first kind, multiplication in the group near the identity is given by analytic functions of the coordinates.

3. When one calculates the $c_n(X : Y)$ by means of the recursion formulae (2.15.15), one finds that each $c_n(X : Y)$ is a linear combination of the commutators of the form $[Z_1,[Z_2,[\cdots[Z_{n-1},Z_n]\cdots]]]$ with $Z_i \in \{X,Y\}$ for $1 \leq i \leq n$, the coefficients being universal rational constants. This suggests that there is a formal algebraic theory of the exponential series underlying the analytical theory. This was already implicit in the work of Baker [1], but results explicity emphasizing the formal aspects were first obtained by Hausdorff [1]. It was Dynkin [2] who returned to this question in 1947 and obtained the decisive results, including an exact formula for $c_n(X : Y)$. We refer the reader to the exercises at the end of this chapter for these results.

2.16. Lie's Theory of Transformation Groups

In this section we give a brief treatment of the theory of Lie transformation groups. The local theory was conceived and developed by Sophus Lie himself, and it marked the beginning of the entire theory of Lie groups and Lie algebras. The development of the global aspects of the theory is, however, a relatively recent accomplishment. We refer the reader to Palais [2] for a detailed treatment of the questions that grew out of Lie's work, and to Montgomery–Zippin [1] for the topological aspects of the theory of transformation groups (see also Bourbaki [5]).

Our main concern is with the *infinitesimal description* of the action of an analytic group G on an analytic manifold M. Let $\mathfrak{g}$ be the Lie algebra of G. Then for any $X \in \mathfrak{g}$, the one-parameter group $t \mapsto \exp tX$ acts on G; and therefore one can, following Lie, introduce the vector field $\tau(X)$ on M whose integral curves are of the form $t \mapsto \exp(-tX) \cdot x (x \in M)$. The first fundamental theorem of Lie asserts that $X \mapsto \tau(X)$ is a homomorphism of $\mathfrak{g}$ into the Lie algebra $\mathfrak{J}_a(M)$ of all analytic vector fields on M. Now, the action of G on M is described by a homomorphism, say α, of G into the group of all analytic diffeomorphisms of M; so if for heuristic purposes we regard $\mathfrak{J}_a(M)$ as the "Lie algebra" of the group of all analytic diffeomorphisms of M, we see that τ is the "differential" of α. It is therefore natural to call any homomorphism of $\mathfrak{g}$ into $\mathfrak{J}_a(M)$ an infinitesimal G-transformation group on M, and to refer to τ as the infinitesimal G-transformation group determined by the action of G on M. The second fundamental theorem of Lie asserts that given an arbitrary infinitesimal G-transformation group τ on M, one can construct at least a local action of G on M which determines τ, such a local action being essentially unique.

While Lie himself did not consider the global problem it is obvious that infinitesimal transformation groups do not always generate global transformation groups. To see this, let τ be the infinitesimal G-transformation group determined by a global action of G on M; let M' be an open submanifold of M; and let $\tau'(X) = \tau(X) | M'$ $(X \in \mathfrak{g})$. Then unless M' is invariant in M, τ' will not be determined by a global G-action on M'. The main problem studied in this section thus divides itself into two parts. In the first and classical part we establish the one-to-one correspondence between local and infinitesimal transformation groups; in the second part we investigate the conditions under which a local transformation group "extends" to a global transformation group.

We now begin the formal development. Let G be an analytic group with Lie algebra $\mathfrak{g}$, M an analytic manifold. Let $n = \dim(G)$, $m = \dim(M)$. We treat the real case, leaving the complex case to the reader. By a *local action of G on M* or a *local G-transformation group on M* we mean a map φ of a subset D of $G \times M$ into M such that

(i) D is an open enighborhood of $\{1\} \times M$, and φ is analytic on D

(ii) $\varphi(1 : x) = x \quad (x \in M)$

(iii) if E is the set of all $(g,h,x) \in G \times G \times M$ such that (h,x), (gh,x), $(g,\varphi(h : x))$ are all in D and $\varphi(gh : x) = \varphi(g : \varphi(h : x))$, then E is a neighborhood of $\{1\} \times \{1\} \times M$.

If $D = G \times M$ in (i) and $E = G \times G \times M$ in (iii), φ is called a *global G-action on M* or a *global G-transformation group on M; M* is then a *G*-space (cf. §2.9). By an *infinitesimal G-action on M* or an *infinitesimal G-transformation group on M* we mean a homomorphism of $\mathfrak{g}$ into the Lie algebra $\mathfrak{J}_a(M)$ of all analytic vector fields on M.

Let $\mathfrak{G}$ be the enveloping algebra of G. As usual, for any $g \in G$, we identify $T_g^\infty(G)$ with $\mathfrak{G}$. Suppose now that φ is a local G-transformation group on M. For any $x \in M$, the map

$$\varphi_x : g \mapsto \varphi(g^{-1} : x) \tag{2.16.1}$$

is defined and analytic around 1, so we may introduce its complete differential $(d\varphi_x)_1^\infty$, which is a linear map of $\mathfrak{G}$ into $T_x^{(\infty)}(M)$. Put

$$\tau_\varphi^\infty(a)_x = (d\varphi_x)_1^\infty(a) \quad (a \in \mathfrak{G}). \tag{2.16.2}$$

If we write

$$\varphi(g : x) = g \cdot x, \tag{2.16.3}$$

then for any $x \in M$, and $a = X_1X_2 \cdots X_r$ ($X_i \in \mathfrak{g}$ for all i), and any function f defined and C^∞ around x, the element $\tau_\varphi^\infty(a)_x$ of $T_x^{(\infty)}(M)$ is determined by

$$\tau_\varphi^\infty(a)_x(\mathbf{f}) = \left(\frac{\partial}{\partial t_1} \cdots \frac{\partial}{\partial t_r} f(\exp(-t_rX_r) \cdots \exp(-t_1X_1) \cdot x)\right)_0; \tag{2.16.4}$$

here $\mathbf{f}$ is the germ defined by f at x, and the suffix 0 indicates that the derivatives are taken when $t_1 = \cdots = t_r = 0$.

Lemma 2.16.1. *For any $a \in \mathfrak{G}$, $\tau_\varphi^\infty(a)(x \mapsto \tau_\varphi^\infty(a)_x)$ is an analytic differential operator on M, and the map $\tau_\varphi^\infty(a \mapsto \tau_\varphi^\infty(a))$ is a homomorphism of $\mathfrak{G}$ into the algebra of analytic differential operators on M.*

Proof. The first assertion is immediate from (2.16.4) on using local coordinates. To prove the second it is enough to show that $\tau_\varphi^\infty(ab) = \tau_\varphi^\infty(a)\tau_\varphi^\infty(b)$ when $a = X_1 \cdots X_r$ and $b = Y_1 \cdots Y_s$ ($X_i, Y_j \in \mathfrak{g}$ for all i, j). Let $D = \partial/\partial t_1 \cdots \partial/\partial t_r$, $E = \partial/\partial u_1 \cdots \partial/\partial u_s$, and let a suffix 0 mean that the derivatives are taken when all the variable are 0. Then for any function f defined and C^∞ around x,

$$\begin{aligned}
&(\tau_\varphi^\infty(ab)f)(x)\\
&\quad = (DEf(\exp(-u_sY_s) \cdots \exp(u_1Y_1)\exp(-t_rX_r) \cdots \exp(-t_1X_1) \cdot x))_0\\
&\quad = (D(\tau_\varphi^\infty(b)f)(\exp(-t_rX_r) \cdots \exp(-t_1X_1) \cdot x))_0\\
&\quad = (\tau_\varphi^\infty(a)(\tau_\varphi^\infty(b)f))(x).
\end{aligned}$$

Corollary 2.16.2. *Let $\tau_\varphi = \tau_\varphi^\infty | \mathfrak{g}$. τ_φ is a homomorphism of $\mathfrak{g}$ into the Lie algebra $\mathfrak{J}_a(M)$.*

The infinitesimal G-transformation group τ_φ is said to be *determined by φ*.

Let τ $(X \mapsto \tau(X))$ be any homomorphism of $\mathfrak{g}$ into $\mathfrak{J}_a(M)$. For any $X \in \mathfrak{g}$, the assignment

$$\bar{\tau}(X) : (g,x) \mapsto (X_g, \tau(X)_x) \tag{2.16.5}$$

is an analytic vector field on $G \times M$, and it is obvious that $\bar{\tau} : X \mapsto \tau(X)$ is a Lie algebra injection of $\mathfrak{g}$ into $\mathfrak{J}_a(G \times M)$. Let

$$\mathfrak{L}^\tau_{(g,x)} = \{\bar{\tau}(X)_{(g,x)} : X \in \mathfrak{g}\}; \tag{2.16.6}$$

it is then immediate that

$$\mathfrak{L}^\tau : (g,x) \mapsto \mathfrak{L}^\tau_{(g,x)} \tag{2.16.7}$$

is an involutive analytic system of tangent spaces on $G \times M$, of rank equal to $\dim(G) = n$. The theory of §1.3 is applicable to $\mathfrak{L}^\tau$. For $(g,x) \in G \times M$, denote by $S_{(g,x)}$ the maximal integral manifold of $\mathfrak{L}^\tau$ passing through (g,x).

Lemma 2.16.3. *Let notation be as above. Then*

(1) *if λ_h denotes, for any $h \in G$, the analytic diffeomorphism $(g,x) \mapsto (hg,x)$ of $G \times M$ onto itself, then*

$$\lambda_h[S_{(g,x)}] = S_{(hg,x)}. \tag{2.16.8}$$

(2) *Suppose φ is a global G-transformation group on M such that $\tau_\varphi = \tau$. Then for any $x \in M$, the map*

$$\alpha_{\varphi,x} : g \mapsto (g, \varphi(g^{-1} : x)) \tag{2.16.9}$$

is an analytic diffeomorphism of G onto $S_{(1,x)}$.

(3) *Suppose φ is a local G-transformation group on M such that $\tau_\varphi = \tau$. Let $x_0 \in M$. Then there exists an open connected neighborhood $V = V^{-1}$ of 1 (resp. U of x_0) such that for each $x \in U$, the map $\alpha_{\varphi,x}$ is an analytic diffeomorphism of V onto an open submanifold of $S_{(1,x)}$ that passes through $(1,x)$.*

Proof. The proof of (1) is an immediate consequence of the fact that $\mathfrak{L}^\tau$ is invariant under the λ_h $(h \in G)$. Let φ be a global G-transformation group on M with $\tau_\varphi = \tau$. Then for any $X \in \mathfrak{g}$, $x \in M$,

$$\begin{aligned}(d\alpha_{\varphi,x})_g(X_g) &= (X_g, \tau(X)_{\varphi(g^{-1}:x)}) \\ &= \bar{\tau}(X)_{\alpha_{\varphi,x}(g)}.\end{aligned}$$

So if we write $A_x = \alpha_{\varphi,x}[G]$ and give to A_x the analytic structure that makes $\alpha_{\varphi,x}$ an analytic diffeomorphism, it follows that A_x is a submanifold of $G \times M$ which is an integral manifold of $\mathfrak{L}^\tau$ through $(1,x)$. Thus A_x is an open submanifold of $S_{(1,x)}$. On the other hand, A_x is closed in $G \times M$, hence closed in $S_{(1,x)}$. So $A_x = S_{(1,x)}$ and $\alpha_{\varphi,x}$ is an analytic diffeomorphism of G onto $S_{(1,x)}$. This proves (2). To prove (3), let V (resp. U) be an open connected neighborhood of 1 (resp. x_0) such that $V = V^{-1}$ and the following condition is satisfied: if $g, h \in V$ and $x \in U$, then $\varphi(h : x)$, $\varphi(gh : x)$, and $\varphi(g : \varphi(h : x))$ are all defined, and $\varphi(gh : x) = \varphi(g : \varphi(h : x))$. Then, for any $X \in \mathfrak{g}$, $g \in V$, $x \in U$,

$$\begin{aligned}(d\alpha_{\varphi,x})(X_g) &= (X_g, \tau(X)_{\varphi(g^{-1} : x)}) \\ &= \bar{\tau}(X)_{\alpha_{\varphi,x}(g)}.\end{aligned}$$

This shows, as in the previous instance, that $\alpha_{\varphi,x}$ is an analytic diffeomorphism of V onto an open submanifold of $S_{(1,x)}$ passing through $(1,x)$.

Corollary 2.16.4. *Let τ be a homomorphism of $\mathfrak{g}$ into $\mathfrak{J}_a(M)$. If φ_1 and φ_2 are global G-transformation groups on M with $\tau_{\varphi_1} = \tau_{\varphi_2} = \tau$, then $\varphi_1 = \varphi_2$. If φ_1 and φ_2 are local G-transformation groups on M with $\tau_{\varphi_1} = \tau_{\varphi_2} = \tau$, then $\varphi_1 = \varphi_2$ in a neighborhood of $\{1\} \times M$.*

Proof. First assume that φ_1 and φ_2 are global G-transformation groups on M with $\tau_{\varphi_1} = \tau_{\varphi_2} = \tau$. Let p_G (resp. p_M) be the projection of $G \times M$ onto G (resp. onto M), and let $p_{G,x} = p_G | S_{(1,x)}$ $(x \in M)$. It follows from (2) of the preceding lemma that $p_{G,x}$ is a bijection of $S_{(1,x)}$ onto G and that $\varphi_1(g : x) = \varphi_2(g : x) = p_M \circ p_{G,x}^{-1}(g^{-1})$. Suppose that φ_1 and φ_2 are only local. Fix $x_0 \in M$ and let V, U be as in lemma above such that (3) is satisfied with respect to both φ_1 and φ_2. Fix $x \in U$ and let $A_{x,i} = \alpha_{\varphi_i,x}[V]$ $(i = 1, 2)$. Then $A_{x,1}$ and $A_{x,2}$ are open submanifolds of $S_{(1,x)}$ through $(1,x)$. So $A_{x,1} \cap A_{x,2}$ is a nonempty open submanifold of both $A_{x,1}$ and $A_{x,2}$. It follows from this that the set

$$W_x = \{g : g \in V, \varphi_1(g : x) = \varphi_2(g : x)\}$$

is open in V. Since it is nonempty and closed in V, $W_x = V$. So $\varphi_1 = \varphi_2$ on $V \times U$. Since $x_0 \in M$ was arbitrary, $\varphi_1 = \varphi_2$ in a neighborhood of $\{1\} \times M$.

The above results suggest that given an arbitrary homomorphism τ of $\mathfrak{g}$ into $\mathfrak{J}_a(M)$, one may be able to construct the global transformation group φ which determines τ by defining $\varphi(g : x) = p_M \circ p_{G,x}^{-1}(g^{-1})$. However, this is not always possible, because in general $p_{G,x}$ is not one-to-one on $S_{(1,x)}$, nor does it map $S_{(1,x)}$ onto G. On the other hand, such a definition will certainly work locally and lead to a local G-transformation group. We now turn to the details of this construction.

Lemma 2.16.5. *Let τ be a homomorphism of $\mathfrak{g}$ into $\mathfrak{J}_a(M)$, and let $\mathfrak{L}^\tau$ be defined by* (2.16.6) *and* (2.16.7). *Given $x_0 \in M$, we can find a connected open neighborhood $V = V^{-1}$ of 1 (resp. U of x_0), and an analytic map ψ of $V \times U$ into M such that*

(i) *$\psi(1 : x) = x$ for all $x \in U$*

(ii) *$(g,x) \mapsto (g,\psi(g^{-1} : x))$ is an analytic diffeomorphism of $V \times U$ onto an open neighborhood W of $(1,x_0)$*

(iii) *for each $x \in U$, the map*

$$g \mapsto (g,\psi(g^{-1} : x))$$

is an analytic diffeomorphism of V onto a connected open submanifold of $S_{(1,x)}$ that contains $(1,x)$.

Proof. Select an open neighborhood V_1 (resp. U_1) of 1 (resp. x_0) and functions $z_1, \ldots, z_n$ on V_1, $x_1, \ldots, x_{n+m}$ on $V_1 \times U_1$ such that the following conditions are satisfied: (i) $(z_1, \ldots, z_n)$ form a system of coordinates on V_1 and $(x_1, \ldots, x_{n+m})$ on $V_1 \times U_1$, and (ii) for all $(g,x) \in V_1 \times U_1$, $(\partial/\partial x_1)_{(g,x)}, \ldots, (\partial/\partial x_n)_{(g,x)}$ span $\mathfrak{L}^\tau_{(g,x)}$ or, equivalently $\mathfrak{L}^\tau_{(g,x)}$ is precisely the interection of the null spaces of $(dx_{n+1})_{(g,x)}, \ldots, (dx_{n+m})_{(g,x)}$. For $(g,x) \in V_1 \times U_1$ put

$$\bar{z}_i(g,x) = z_i(g), \quad \bar{w}_j(g,x) = w_j(x) = x_{n+j}(1,x)$$

$(1 \leq i \leq n,\ 1 \leq j \leq m)$.

Now observe that $\mathfrak{L}^\tau_{(1,x_0)}$ does not contain any tangent vector of the form $(0,Y)$ where Y is a nonzero element in $T_{x_0}(M)$. It follows from this that the functions $\bar{z}_1, \ldots, \bar{z}_n, x_{n+1}, \ldots, x_{n+m}$ form a system of coordinates around $(1,x_0)$ and, furthermore, that the functions $w_1, \ldots, w_m$ form a system of coordinates around x_0. Consequently the functions $\bar{z}_1, \ldots, \bar{z}_n, \bar{w}_1, \ldots, \bar{w}_m$ form a system of coordinates around $(1,x_0)$. We can therefore select open neighborhoods $V = V^{-1}$, U, W $(\subseteq V_1 \times U_1)$ of 1, x_0, and $(1,x_0)$ respectively, and an analytic diffeomorphism σ of $V \times U$ onto W such that

$$\bar{z}_i = \bar{z}_i \circ \sigma \quad (1 \leq i \leq n) \qquad \bar{w}_j = x_{n+j} \circ \sigma \quad (1 \leq j \leq m). \tag{2.16.10}$$

The first set of the relations (2.16.10) shows that for any $(g,x) \in V \times U$, the first member of $\sigma(g,x)$ is g. Hence there exists an analytic map ψ of $V \times U$ into M such that

$$\sigma(g,x) = (g,\psi(g^{-1}:x)) \quad ((g,x \in V \times U).$$

The second set of the relations (2.16.10) may obviously be rewritten as

$$x_{n+j}(g,\psi(g^{-1} : x)) = x_{n+j}(1,x) \quad ((g,x) \in V \times U,\ 1 \leq j \leq m). \tag{2.16.11}$$

The equations (2.16.11) show that for fixed $x \in U$, the image V^x of $V \times \{x\}$ under σ is a connected n-dimensional regular submanifold of W passing through $(1,x)$, on which the functions $x_{n+1}, \ldots, x_{n+m}$ are constant. So V^x is an integral manifold of $\mathfrak{L}^\tau$ through $(1,x)$, proving that it is an open connected submanifold of $S_{(1,x)}$. The proof of the lemma is complete.

We are now in a position to state and prove the basic theorem of Lie.

Theorem 2.16.6. *Let τ $(X \mapsto \tau(X))$ be an infinitesimal G-transformation group on M. Then given any $x_0 \in M$, we can find an open neighborhood U of x_0 and a local G-transformation group φ on U such that for any $X \in \mathfrak{g}$, $\tau_\varphi(X) = \tau(X) | U$.*

Proof. Fix $x_0 \in M$ and select V, U and ψ as in the preceding lemma. Let D be the set of all $(g,x) \in V \times U$ such that $\psi(g:x) \in U$. Obviously, D is an open neighborhood of $\{1\} \times U$ in $G \times U$. Let $\varphi = \psi | D$. We wish to prove that φ is a local G-transformation group on U with $\tau_\varphi(X) = \tau(X) | U$ for all $X \in \mathfrak{g}$.

Fix $x \in U$. Since the map $g \mapsto (g,\varphi(g^{-1}:x))$ is analytic from V into $S_{(1,x)}$, it follows that for any $X \in \mathfrak{g}$, $(d\varphi_x)_1(X_1) = \tau(X)_x$, φ_x being defined by (2.16.1). Now select connected open neighborhoods $V_j = V_j^{-1}$ of 1 (resp. U_j of x) $(j = 1, 2)$ such that $V_2^2 \subseteq V_1$, $V_1^2 \subseteq V$, $\psi[V_2 \times U_2] \subseteq U_1$, $\psi[V_1 \times U_1] \subseteq U$. Clearly, for $(g,h,y) \in V_2 \times V_2 \times U_2$, $\varphi(h:y)$, $\varphi(gh:y)$, $\varphi(g:\varphi(h:y))$ are all defined. We shall prove that $\varphi(gh:y) = \varphi(g:\varphi(h:y))$ for all $(g,h,y) \in V_2 \times V_2 \times U_2$ or, what is the same thing, that $\psi(h^{-1}g^{-1}:y) = \psi(h^{-1}:\psi(g^{-1}:y))$ for all $(g,h,y) \in V_2 \times V_2 \times U_2$. Fix $g \in V_2$, $y \in U_2$, write $z = \psi(g^{-1}:y)$, and define

$$\alpha(h) = (gh,\psi(h^{-1}g^{-1}:y)), \qquad \beta(h) = (gh,\psi(h^{-1}:z)) \quad (h \in V_2).$$

It follows from the previous lemma that α (resp. β) is an analytic diffeomorphism of V_2 onto an open submanifold A (resp. B) of $S_{(1,y)}$ (resp. $S_{(g,z)}$). A and B are thus integral manifolds of $\mathfrak{L}^\tau$, and as $(g,z) \in A \cap B$, $A \cap B$ is a nonempty open submanifold of both A and B. The fact that $A \cap B$ is open in A implies that $W = \{h : h \in V_2, \alpha(h) = \beta(h)\}$ is open in V_2. Since W is closed in V_2, $W = V_2$. So $\alpha = \beta$, proving that $\psi(h^{-1}g^{-1}:y) = \psi(h^{-1}:z)$. Since $x \in U$ was arbitrary, this completes the proof that φ is a local G-transformation group on U with $\tau_\varphi(X) = \tau(X) | U$ $(X \in \mathfrak{g})$.

The above theorem is local in both the group and space variables. It is natural to raise the question to what extent the global version of this theorem remains true. This was studied in considerable detail by Palais [2]. We confine ourselves to proving two important results (Theorem 2.16.8 and 2.16.13). We consider first the question of globalization with respect to M. We need a lemma.

Lemma 2.16.7. *Let $\{U_i : i \in I\}$ be a locally finite open covering of M. For each $(i,j) \in I \times I$ and any $x \in U_i \cap U_j$, let U_{ijx} be a neighborhood of x contained in $U_i \cap U_j$. Then for each $x \in M$ we can select a neighborhood U_x of x such that*

(i) *if $x \in U_i \cap U_j$, then $U_x \subseteq U_{ijx}$*
(ii) *if $U_x \cap U_y \neq \varnothing$, there is an $i \in I$ such that $U_x \cup U_y \subseteq U_i$.*

Proof. Let $\{W_i : i \in I\}$ be a locally finite open covering of M with $Cl(V_i) \subseteq U_i$ $(i \in I)$. For $x \in M$, denote by U'_x the intersection of all the U_i, V_j, U_{pqx} that contain x. Since there are only finitely many of these, U'_x is a neighborhood of x. So U'_x meets $Cl(V_i)$ for only finitely many i, from which it follows that

$$U_x = U'_x \cap \bigcap_{i \in I : x \notin Cl(V_i)} (M \backslash Cl(V_i))$$

is a neighborhood of x. With this choice of U_x, (i) is obvious. Suppose $U_x \cap U_y \neq \varnothing$. Then $x \in U_i$ for some $i \in I$, so $U_x \subseteq V_i$. Then $U_y \cap Cl(V_i) \neq \varnothing$, which means that $y \in Cl(V_i)$. So $y \in U_i$, implying $U_y \subseteq U_i$. Thus $U_x \cup U_y \subseteq U_i$.

Theorem 2.16.8. *Let τ be an infinitesimal G-transformation group on M. Then there is a local G-transformation group φ on M such that $\tau_\varphi = \tau$.*

Proof. By Theorem 2.16.6 we can select a locally finite open covering $\{U_i : i \in I\}$ of M and local G-transformation groups φ_i on U_i $(i \in I)$, such that $\tau_{\varphi_i}(X) = \tau(X) | U_i$ $(X \in \mathfrak{g}, i \in I)$. For each $(i,j) \in I \times I$ and each $x \in U_i \cap U_j$, select a neighborhood U_{ijk} of x such that $U_{ijk} \subseteq U_i \cap U_j$, and φ_i and φ_j are defined and equal on a neighborhood of $\{1\} \times U_{ijx}$. By the lemma above we can find for each $x \in M$ an open neighborhood U_x of x satisfying the conditions (i) and (ii) therein. For each $x \in M$ let $I(x) = \{i : i \in I, x \in U_i\}$, and let W_x be the set of all $(g,y) \in G \times U_x$ such that all the φ_i $(i \in I(x))$ are defined and take the same value at (g,y). Since $I(x)$ is finite, W_x is a neighborhood of $\{1\} \times U_x$. Let $\varphi_x = \varphi_i | W_x$, $i \in I(x)$. Suppose $x, y \in M$ and $W_x \cap W_y \neq \varnothing$. Then $U_x \cap U_y \neq \varnothing$, so for some $i \in I$, $U_x \cup U_y \subseteq U_i$. This means that $i \in I(x) \cap I(y)$, so $\varphi_x | W_x \cap W_y = \varphi_y | W_x \cap W_y = \varphi_i | W_x \cap W_y$. Define φ on $\bigcup_x W_x$ by setting $\varphi | W_x = \varphi_x | W_x$; φ is well defined It is now clear that φ is a local G-transformation group on M with $\tau_\varphi = \tau$. In fact, $\bigcup_x W_x$ is a neighborhood of $\{1\} \times M$, while for each $x \in M$ we can find an open neighborhood $V'_x \times U'_x$ of $(1,x)$ in $G \times U_x$ such that $\varphi = \varphi_i$ on $V'_x \times U'_x$ $(i \in I(x))$; this leads to our assertion.

We now consider the more interesting but more difficult problem of globalization in the group variable. First we have the following elementary consequence of Theorem 2.16.8.

Theorem 2.16.9. *Let M be compact and G simply connected. Then for any infinitesimal G-transformation group τ on M there is a unique global G-transformation group φ on M such that $\tau_\varphi = \tau$.*

Proof. The proof depends on the following elementary result whose verification is left to the reader: if Z is any Hausdorff space, $z \in Z$, and A is a neighborhood of $\{z\} \times M$, there is an open neighborhood B of z such that $B \times M \subseteq A$. This said, let ψ be a local G-transformation group on M with $\tau_\psi = \tau$. By the observation made just now we can find open neighborhoods $V_j = V_j^{-1}$ of 1 in G $(j = 1, 2)$ such that (i) ψ is defined on $V_1 \times M$, and (ii) $V_2^2 \subseteq V_1$, and for $(g,h,x) \in V_2 \times V_2 \times M$, $\psi(gh : x) = \psi(g : \psi(h : x))$. For $h \in V_2$ let $t_h(x) = \psi(h : x)$ $(x \in M)$. Then t_h is an analytic map of M into itself, $t_1 =$ identity and $t_{hh'} = t_h \circ t_{h'}$ $(h, h' \in V_2)$. In particular, $t_h \circ t_{h^{-1}} = t_{h^{-1}} \circ t_h =$ identity for $h \in V_2$, so each t_h $(h \in V_2)$ is an analytic diffeomorphism of M. Since G is simply connected, we can find a homomorphism θ $(h \longrightarrow \theta_h)$ of G into the group of all analytic diffeomorphisms of M such that $t_h = \theta_h$ for all h in a neighborhood of 1 in V_2. Write $\varphi(h : x) = \theta_h(x)$ $(h \in G, x \in M)$. It is easily verified that φ is analytic. Hence φ is a global G-transformation group on M with $\tau_\varphi = \tau$. We have already proved the uniqueness of φ.

If M is not compact, it is not always possible to construct global G-transformation groups corresponding to arbitrary infinitesimal transformation groups, even when G is simply connected. For instance, let us consider the case $G = \mathbf{R}$. Then $\mathfrak{g}$ is spanned by d/dt (t being the usual coordinate on $\mathbf{R}$); and if Z is any analytic vector field on M, one knows from classical analysis that it is not always possible to find a global $\mathbf{R}$-transformation group ψ on M such that $\tau_\psi(d/dt) = Z$. If we can do this, we call Z *global*; in this case

$$Z_x = \left(\frac{d}{dt}\psi(-t : x)\right)_{t=0} \qquad (x \in M). \tag{2.16.12}$$

If $\zeta(t)$ is the map $x \mapsto \psi(t : x)$ of M, $\zeta(t)$ is an analytic diffeomorphism of M, and ζ $(t \mapsto \zeta(t))$ is a homomorphism of $\mathbf{R}$ into the group of analytic diffeomorphisms of M. We refer to ζ as the *one-parameter group of analytic diffeomorphisms generated by* Z, and to Z as the *infinitesimal generator of* ζ. Coming back to the case of arbitrary G, suppose φ is a global G-transformation group on M. Then $\tau_\varphi(X)$ is global for each $X \in \mathfrak{g}$, because $(t,x) \mapsto \varphi(\exp tX : x)$ is an $\mathbf{R}$-transformation group on M and

$$\tau(X)_x = \left(\frac{d}{dt}\varphi(\exp(-tX) : x)\right)_{t=0}.$$

Our main result in this context is Theorem 2.16.13 below, which asserts that if τ is an infinitesimal G-transformation group on M such that $\tau(X)$ is global for all $X \in \mathfrak{g}$, then $\tau = \tau_\varphi$ for some global G-transformation group on M, provided G is simply connected. This is a special case of a more general result of Palais [2] (see Exercise 47). In what follows we fix an infinitesimal G-transformation group τ on M and follow the treatment of Palais [2].

Lemma 2.16.10. *Let ζ be an one-parameter group of analytic diffeomorphisms of M, and Z the infinitesimal generator of ζ. If $X \in \mathfrak{g}$, then $\tau(X) = Z$ and only if for each $x \in M$, $(\exp(tX), \zeta(-t \cdot x)) \in S_{(1,x)}$ for all $t \in \mathbf{R}$.*

Proof. Write $f(t) = (\exp(tX), \zeta(-t) \cdot x)$ $(t \in \mathbf{R})$, $x \in M$ being fixed. Suppose $f(t) \in S_{(1,x)}$ for all t. Then f is an analytic map of $\mathbf{R}$ into $S_{(1,x)}$, so $(d/dt\, f(t))_{t=0} \in \mathfrak{L}^{\tau}_{(1,x)}$. Since $(d/dt\, f(t))_{t=0} = (X_1, Z_x)$, we have $Z_x = \tau(X)_x$. Suppose conversely that $\tau(X) = Z$. Then $\bar{Z} : (g,y) \mapsto (X_g, Z_y)$ is an analytic vector field on $S_{(1,x)}$. Suppose $t_0 \in \mathbf{R}$ is such that $f(t_0) \in S_{(1,x)}$. We can find $\epsilon > 0$ and an analytic map $\bar{f} : t \mapsto \bar{f}(t)$ of $(-\epsilon,\epsilon)$ into $S_{(1,x)}$ such that $\bar{f}(0) = f(t_0)$ and $d/dt\, \bar{f}(t) = \bar{Z}_{\bar{f}(t)}$ $(|t| < \epsilon)$. By the uniqueness of integral curves of vector field, $\bar{f}(t) = f(t + t_0)$, $|t| < \epsilon$. So $\{t : t \in \mathbf{R}, f(t) \in S_{(1,x)}\}$ is a nonempty open subset of $\mathbf{R}$. Since this set is obviously closed, it must be $\mathbf{R}$.

Lemma 2.16.11. *Suppose $X_1, \ldots, X_p \in \mathfrak{g}$ are such that $\tau(X_i)$ is global for all $i = 1, \ldots, p$. Let ζ_i be the one-parameter group of analytic diffeomorphisms of M generated by $\tau(X_i)$ $(1 \leq i \leq p)$. Define $\zeta(t_1, \ldots, t_p) = \zeta_1(t_1) \circ \cdots \circ \zeta_p(t_p)$ $((t_1, \ldots, t_p) \in \mathbf{R}^p)$. Then*

$$\Phi : (t_1, \ldots, t_p, x) \mapsto \exp(t_1 X_1) \cdots \exp(t_p X_p), \zeta(t_1, \ldots, t_p)^{-1} \cdot x)$$

is an analytic map of $\mathbf{R}^p \times M$ into $G \times M$ such that, for each $x \in M$, $\Phi(t_1, \ldots, t_p, x) \in S_{(1,x)}$ for all $(t_1, \ldots, t_p) \in \mathbf{R}^p$.

Proof. The analyticity of Φ is obvious. We prove the second assertion by induction on p. The case $p = 1$ is precisely the preceding lemma. Suppose $p > 1$, and assume the result when the number of elements considered from $\mathfrak{g}$ is $p - 1$. Then

$$(\exp(t_2 X_2) \cdots \exp(t_p X_p), \zeta(t_1, \ldots, t_p)^{-1} \epsilon S_{(1, \exp(-t_1 X_1) \cdot x)}$$

for all $x \in M$, $(t_1, \ldots, t_p) \in \mathbf{R}^p$. By (2.16.8),

$$\Phi(t_1, \ldots, t_p, x) \in S_{(\exp(t_1 X_1), \exp(-t_1 X_1) \cdot x)}$$

for all $(t_1, \ldots, t_p) \in \mathbf{R}^p$, $x \in M$. But since $(\exp uX, \exp(-uX) \cdot x)$ lies in $S_{(1,x)}$ for $u \in \mathbf{R}$,

$$S_{(\exp uX,\, \exp(-uX) \cdot x)} = S_{(1,x)} \quad (x \in M,\, u \in \mathbf{R}).$$

This proves what we want.

Lemma 2.16.12. *Let τ be as above and let us suppose that $\tau(X)$ is global for all $X \in \mathfrak{g}$. Let p_G be the canonical projection of $G \times M$ onto G and let $p_{G,x} = p_G | S_{(1,x)}$ $(x \in M)$. Then for each $x \in M$, $p_{G,x}$ is a covering map of $S_{(1,x)}$ onto G.*

Proof. Let $\{X_1, \ldots, X_n\}$ be a basis for $\mathfrak{g}$. Then the map

$$(t_1, \ldots, t_n) \mapsto \exp(t_1 X_1) \cdots \exp(t_n X_n)$$

is an analytic diffeomorphism of a neighborhood of the origin in $\mathbf{R}^n$ onto a neighborhood of 1 in G. Combining this observation with the previous lemma, we establish the existence of a connected open neighborhood $V = V^{-1}$ of 1 in G and an analytic map ψ of $V \times M$ into M with the following properties: (i) $\psi(1 : x) = x$ for all $x \in M$, and (ii) $(g, \psi(g^{-1} : x)) \in S_{(1,x)}$ for all $(g,x) \in V \times M$. For $x \in M$ let

$$\alpha_x(g) = (g, \psi(g^{-1} : x)) \quad (g \in V)$$
$$V^x = \alpha_x[V].$$

It is then easily seen that V^x is a connected open submanifold of $S_{(1,x)}$ containing $(1,x)$ and that α_x is an analytic diffeomorphism of V onto V^x. In particular, α_x is a homeomorphism of V onto V^x, from which it follows that V^x is closed in $p_{G,x}^{-1}(V)$. This shows that V^x is the connected component of $p_{G,x}^{-1}(V)$ that contains $(1,x)$ and that $p_{G,x}$ is a homeomorphism of V^x onto V.

We now claim that $p_{G,x}[S_{(1,x)}] = G$ for all $x \in M$. Since V generates G, it is enough to prove that for each integer $k \geq 1$ and each $y \in M$, $p_{G,y}[S_{(1,y)}]$ contains V^k. We do this by induction on k. This has been shown for $k = 1$ in the previous paragraph. Assume this has been proved for some k, and let $g \in V^{k+1}$. Then $g = g'h$, where $g' \in V$, $h \in V^k$. Fix $x \in M$ and select $y \in M$ such that $(g',y) \in S_{(1,x)}$. Then $S_{(g',y)} = S_{(1,x)}$. On the other hand, by the induction hypothesis, we can find $z \in M$ such that $(h,z) \in S_{(1,y)}$. So $(g,z) = (g'h,z) \in S_{(g',y)} = S_{(1,x)}$ (cf. (2.16.8)).

In order to prove that the $p_{G,x}$ are covering maps it is enough to show that for any $g \in G$ and $x \in G$ and $x \in M$, $p_{G,x}$ maps each connect component of $p_{G,x}^{-1}(gV)$ homeomorphically onto gV. Fix $g \in G$ and a connected component C of $p_{G,x}^{-1}(gV)$; let (g,y) be some element of C. Then $S_{(1,x)} = S_{(g,y)} = \lambda_g[S_{(1,y)}]$. It follows from this that λ_g is a homeomorphism of $p_{G,y}^{-1}(V)$ onto $p_{G,x}^{-1}(gV)$ and that the following diagram commutes:

$$\begin{array}{ccc} gV & \xleftarrow{\quad p_{G,x} \quad} & p_{G,x}^{-1}(gV) \\ \uparrow{\scriptstyle l_g} & & \uparrow{\scriptstyle \lambda_g} \\ V & \xleftarrow[\quad p_{G,y} \quad]{} & p_{G,y}^{-1}(V) \end{array}$$

On the other hand, we have seen above that V^y is the connected component containing $(1,y)$ of $p_{G,y}^{-1}(V)$ and that $p_{G,y}$ is a homeomorphism of V^y onto V. So $C = \lambda_g[V^y]$, and $p_{G,x}$ is a homeomorphism of C onto gV. This completes the proof of the lemma.

Theorem 2.16.13. *Let τ be an infinitesimal G-transformation group on M. Suppose that G is simply connected and that $\tau(X)$ is a global vector field on M for all $X \in \mathfrak{g}$. Then there is a unique global G-transformation group φ on M such that $\tau_\varphi = \tau$.*

Proof. By the preceding lemma and the simple connectedness of G, we obtain the result that $p_{G,x}$ is a homeomorphism of $S_{(1,x)}$ onto G for each $x \in M$. For $(g,x) \in G \times M$, define $\varphi(g^{-1}:x)$ as the unique element of M such that $(g,\varphi(g^{-1}:x)) \in S_{(1,x)}$. We prove that φ is a global G-transformation group on M with $\tau_\varphi = \tau$.

Clearly, $\varphi(1:x) = x$ for all $x \in M$. Suppose $g, h \in G$, $x \in M$. Let $y = \varphi(g^{-1}:x)$. Then $(g,y) \in S_{(1,x)}$, so $S_{(1,x)} = S_{(g,y)} = \lambda_g[S_{(1,y)}]$. Consequently, since $(gh,\varphi(h^{-1}g^{-1}:x)) \in S_{(1,x)}$, $(h,\varphi(h^{-1}g^{-1}:x)) \in S_{(1,y)}$. In other words, $\varphi(h^{-1}g^{-1}:x) = \varphi(h^{-1}:\varphi(g^{-1}:x))$ or, what is the same thing,

$$\varphi(gh:x) = \varphi(g:\varphi(h:x)) \quad ((g,h,x) \in G \times G \times M). \tag{2.16.13}$$

It only remains to verify the analyticity of φ and the equation $\tau_\varphi = \tau$. Suppose V and ψ are as in the proof of the preceding lemma. It is then clear that $\varphi(g:x) = \psi(g:x)$ for $(g,x) \in V \times M$. So φ is analytic on $V \times M$. An easy induction based on (2.16.13) establishes that φ is analytic on $V^k \times M$ for all $k \geq 1$. Since V generates G, we get the analyticity of φ on $G \times M$. Let $x \in M$, $X \in \mathfrak{g}$, and let $Y = (d\varphi_x)_1(X_1)$, where φ_x is the map $g \mapsto \varphi(g^{-1}:x)$. Since α_x $(g \mapsto (g,\varphi(g^{-1}:x)))$ is an analytic map of G into $S_{(1,x)}$, $(d\alpha_x)_1(X_1) \in \mathfrak{L}^\tau_{(1,x)}$. Hence $(X_1,Y) \in \mathfrak{L}^\tau_{(1,x)}$, proving that $Y = \tau(X)_x$. This completes the proof of the theorem.

Remark. It turns out that it is not necessary to require that $\tau(X)$ be global for all $X \in \mathfrak{g}$. Let us say that a subset A of $\mathfrak{g}$ *generates* $\mathfrak{g}$ if $\mathfrak{g}$ is the smallest subalgebra containing A. It can then be shown that the above theorem remains true provided we assume that G is simply connected and that $\tau(X)$ is global for all X belonging to a subset of $\mathfrak{g}$ which generates $\mathfrak{g}$. This is the theorem of Palais referred to earlier. For a proof the reader is referred to Exercise 47.

EXERCISES

1. Determine the Lie algebras of all the matrix Lie groups which have been considered in this chapter.

2. Let k be a field of characteristic zero and $\mathfrak{g}$ a Lie algebra of dimension 2 over k. Prove that either $\mathfrak{g}$ is abelian or there is a basis $\{X, Y\}$ for $\mathfrak{g}$ with $[X, Y] = X$.

3. (a) For any Lie algebra $\mathfrak{g}$ over a field k of characteristic 0, we write $\mathfrak{D}\mathfrak{g}$ for $[\mathfrak{g},\mathfrak{g}]$. Prove that $\mathfrak{D}\mathfrak{g}$ is an ideal in $\mathfrak{g}$ and that $\mathfrak{g}/\mathfrak{D}\mathfrak{g}$ is abelian.

(b) Let $\mathfrak{g}$ be the subalgebra of $\mathfrak{gl}(n,k)$ consisting of all $n \times n$ matrices (a_{ij}) with $a_{ij} = 0$ for $1 \leq j \leq i \leq n$. Determine $\mathfrak{D}\mathfrak{g}$, $\mathfrak{D}(\mathfrak{D}\mathfrak{g}) = \mathfrak{D}^2\mathfrak{g}$, $\mathfrak{D}(\mathfrak{D}^2\mathfrak{g}) = \mathfrak{D}^3\mathfrak{g}$, etc.

4. Let k be a field of characteristic zero and $\mathfrak{g}$ a Lie algebra of dimension 3 over k. The following exercises lead to the classification of all such Lie algebras up to isomorphism (cf. Jacobson [1], pp. 11–14).

(a) If $\dim \mathfrak{D}\mathfrak{g} = 1$ and $\mathfrak{D}\mathfrak{g} \subseteq$ center $\mathfrak{g}$, there is a basis $\{X, Y, Z\}$ for $\mathfrak{g}$ with $[X, Y] = [X,Z] = 0$, $[Y,Z] = X$.

(b) If $\dim \mathfrak{D}\mathfrak{g} = 1$ but $\mathfrak{D}\mathfrak{g} \nsubseteq$ center $\mathfrak{g}$, there is a basis $\{X, Y, Z\}$ for $\mathfrak{g}$ with $[X, Y] = X$, $[X,Z] = [Y,Z] = 0$.

(c) If $\dim \mathfrak{D}\mathfrak{g} = 2$, then $\mathfrak{D}\mathfrak{g}$ is abelian, and there is a basis $\{X, Y, Z\}$ for $\mathfrak{g}$ with $[X, Y] = 0$, $[X,Z] = aX + bY$, $[Y,Z] = cX + dY$ where $\begin{pmatrix} a & b \\ c & d \end{pmatrix} \in GL(2,k)$. Prove further that the isomorphism classes of such $\mathfrak{g}$ are in one-to-one correspondence with the conjugacy classes in $PGL(2,k) = GL(2,k)/k^\times \cdot 1$. Examine the case when k is algebraically closed.

(d) If $\dim \mathfrak{D}\mathfrak{g} = 3$, prove that $\mathfrak{g}$ is simple and that there is a basis $\{X, Y, Z\}$ for $\mathfrak{g}$ such that $[X, Y] = Z$, $[Y,Z] = aX$, $[Z,X] = bY$ where $a, b \in k^\times$. If $k = \mathbf{R}$, prove that such $\mathfrak{g}$ form the two isomorphism classes obtained by taking $a = b = 1$ and $-a = b = 1$ in the above.

(e) If k is algebraically closed, prove that $\mathfrak{sl}(2,k)$ is the only simple Lie algebra of dimension 3 over k (up to isomorphism).

5. Let $\mathfrak{g}_c$ be a Lie algebra over $\mathbf{C}$. By a *real form* of $\mathfrak{g}_c$ is meant a Lie algebra $\mathfrak{g}$ over $\mathbf{R}$ such that (i) $\mathfrak{g}$ is a subalgebra of the real Lie algebra underlying $\mathfrak{g}_c$, and (ii) $\dim_{\mathbf{R}} \mathfrak{g} = \dim_{\mathbf{C}} \mathfrak{g}_c$.

(a) Determine all real forms of $\mathfrak{g}_c = \mathfrak{sl}(n,\mathbf{C})$.

(b) Let $\mathfrak{g}_c$ be the Lie algebra of dimension 3 over $\mathbf{C}$ with a basis $\{X, Y, Z\}$ for which $[X, Y] = 0$, $[X,Z] = X$, and $[Y,Z] = aY$, a being a nonzero complex number. Prove that $\mathfrak{g}_c$ has a real form if and only if either a is real or $|a| = 1$. (Hint: If $\mathfrak{g}$ is a real form of $\mathfrak{g}_c$, then $\mathfrak{D}\mathfrak{g}$ is a real form of $\mathfrak{D}\mathfrak{g}_c = \mathbf{C}\cdot X + \mathbf{C}\cdot Y$, so there is a basis $\{X', Y', Z'\}$ for $\mathfrak{g}_c$ with real structure constants such that X', Y' span $\mathfrak{D}\mathfrak{g}_c$ and $Z' \equiv \rho Z \bmod \mathfrak{D}\mathfrak{g}_c$ for some $\rho \neq 0$. The condition for this is the existence of $M \in GL(2,\mathbf{C})$ such that $\rho M\begin{pmatrix} 1 & 0 \\ 0 & a \end{pmatrix}M^{-1} \in GL(2,\mathbf{R})$. This is so if and only if either $a \in \mathbf{R}$ or $|a| = 1$).

6. Let k be an algebraically closed field of characteristic 0. Prove that the classical Lie algebras $\mathfrak{sl}(n,k)$ $(n \geq 2)$, $\mathfrak{o}(n,k)$ $(n \geq 5)$, and $\mathfrak{sp}(n,k)$ $(n \geq 1$, even) are simple, i.e., do not possess proper nonzero ideals. Prove also that $\mathfrak{sl}(2,k)$ and $\mathfrak{o}(3,k)$ are isomorphic and that $\mathfrak{o}(4,k)$ is the direct sum of two ideals each of which is isomorphic to $\mathfrak{o}(3,k)$.

7. Give an example of a closed subgroup of $GL(2,\mathbf{R})$ which is not algebraic.

8. Let $G = SL(n,C)$. Let $\mathcal{P}$ be the algebra of all polynomials in the matrix entries a_{ij} $(1 \leq i, j \leq n)$, and $\mathcal{I}$ the ideal of those which vanish on G. Let f be the polynomial function $(a_{ij}) \mapsto \det(a_{ij}) - 1$. Prove that $\mathcal{I} = f\mathcal{P}$.

9. (a) Let k be a field of characteristic zero and V a finite-dimensional vector space over k. Let V^* be the dual of V and $\mathcal{P}_V$ the algebra of all polynomial functions on V. For any endomorphism L of V, write L^t for the endomorphism of V^* given by $(L^t f)(v) = f(Lv)$ $(v \in V, f \in V^*)$. Prove that corresponding to any endomorphism L of V there is a unique derivation $\delta(L)$ of $\mathcal{P}_V$ such that $\delta(L)f = -L^t f$ for all $f \in V^*$. Prove also that $L \mapsto \delta(L)$ is a representation of $\mathfrak{gl}(V)$ in $\mathcal{P}_V$.

(b) Take $k = \mathbf{R}$ or $\mathbf{C}$ in (a), and let $G \subseteq GL(V)$ be an algebraic group. Let E be the algebra of endomorphisms of V and $\mathcal{P}$ the algebra of polynomial functions on the vector space E. Write $\mathcal{I}$ for the ideal of all those elements of $\mathcal{P}$ which vanish on G. For each $X \in \mathfrak{gl}(V)$ let R_X be the endomorphism $A \mapsto AX$ of E and let $d(X) = \delta(R_X)$. Prove that the subalgebra $\mathfrak{g}$ of $\mathfrak{gl}(V)$ corresponding to G consists of precisely all those $X \in \mathfrak{gl}(V)$ such that derivation $d(X)$ of $\mathcal{P}$ maps the ideal $\mathcal{I}$ into itself.

10. Let $\mathbf{H}$ be the division algebra of quaternions. We write the elements of $\mathbf{H}$ in the form $q = a + bi + cj + dk$, where $a, b, c, d \in \mathbf{R}$, and i, j, k satisfy the following relations: $i^2 = j^2 = k^2 = -1$, $ij = -ji = k$, $jk = -kj = i$, $ki = -ik = j$. For q as above, define $q^\dagger = a - bi - cj - dk$.

(a) Show that $q \mapsto q^\dagger$ is an involutive antiautomorphism of $\mathbf{H}$, that $qq^\dagger$ is real and ≥ 0 for all $q \in \mathbf{H}$, and that $q \mapsto |q| = (qq^\dagger)^{1/2}$ is a multiplicative norm on $\mathbf{H}$.

(b) Identify $\mathbf{C}$ with $\{q : q \in \mathbf{H}, q = a + bi\}$, and for $q = a + bi + cj + dk \in \mathbf{H}$ write $z_1(q) = a + bi$, $z_2(q) = c - di$. Prove that $q \mapsto (z_1(q), z_2(q))$ is a $\mathbf{C}$-isomorphism of $\mathbf{H}$ with $\mathbf{C}^2$, $\mathbf{H}$ being considered as a right vector space over $\mathbf{C}$.

(c) For $q \in \mathbf{H}$ let $l_q q' = qq'$ $(q' \in \mathbf{H})$, and let σ_q be the endomorphism of $\mathbf{C}^2$ that corresponds to l_q with respect to the isomorphism considered in (b). Let $\mathbf{H}_1$ be the multiplicative group of quaternions of norm 1. Prove that $\mathbf{H}_1$ is a real analytic group, and that $q \mapsto \sigma_q$ is an isomorphism of it with $SU(2,\mathbf{C})$.

(d) For $q_1, q_2 \in \mathbf{H}$ let $\tau(q_1,q_2)$ be the $\mathbf{R}$-linear endomorphism $q' \mapsto q_1 q' q_2^{-1}$ of $\mathbf{H}$. Prove that $\tau : (q_1,q_2) \mapsto \tau(q_1,q_2)$ is an isomorphism of $\mathbf{H}_1 \times \mathbf{H}_1$ with $SO(4,\mathbf{R})$. Verify that the kernel of τ is $\{\pm(1,1)\}$. Deduce that $SO(4,\mathbf{R})$ and $SO(3,\mathbf{R}) \times SO(3,\mathbf{R})$ are locally isomorphic.

(e) Let $\mathbf{H}^n$ be considered in the obvious way as a right vector space over $\mathbf{H}$. For $\mathbf{q} = (q_1, \ldots, q_n)$, $\mathbf{q}' = (q'_1, \ldots, q'_n)$ in $\mathbf{H}^n$, define $\langle \mathbf{q}, \mathbf{q}' \rangle =$

$\sum_{1 \leq r \leq n} q_r(q'_r)^{\dagger}$, $\|\mathbf{q}\| = (\sum_{1 \leq r \leq n} q_r q_r^{\dagger})^{1/2}$. Let G_n denote the group of all automorphisms L of the additive group of $\mathbf{H}^n$ such that $L(\mathbf{q}q') = L(\mathbf{q})q'$ ($\mathbf{q} \in \mathbf{H}^n, q' \in \mathbf{H}$) and $\|L(\mathbf{q})\| = \|\mathbf{q}\|$ ($\mathbf{q} \in \mathbf{H}^n$). Prove that G_n acts transitively on the unit sphere in $\mathbf{H}^n$ with respect to the norm $\|\cdot\|$ and that the stability subgroup of the point $(0, \ldots, 0, 1)$ is canonically isomorphic to G_{n-1}.

(f) For $g \in G_n$ let $\sigma(g)$ denote the element of $GL(2n,\mathbf{C})$ that corresponds to g under the $\mathbf{C}$-isomorphism

$$(q_1, \ldots, q_n) \mapsto (z_1(q_1), z_1(q_2), \ldots, z_1(q_n), z_2(q_1), \ldots, z_2(q_n))$$

of $\mathbf{H}^n$ with $\mathbf{C}^{2n}$, the former being considered as a right vector space over $\mathbf{C}$. Prove that $g \mapsto \sigma(g)$ is an isomorphism of G_n with $Sp(n)$ (as analytic groups) (see Chevalley [1]).

11. (a) Let G be a real Lie group, H a closed Lie subgroup. If H and G/H are connected, prove that G is connected. If G is connected and G/H is simply connected, prove that H is connected and that $\pi_1(G)$ is a quotient group of $\pi_1(H)$.

(b) Prove that $SO(n,\mathbf{R})$, $SU(n,\mathbf{C})$ and $Sp(n)$ act transitively on the respective unit spheres of $\mathbf{R}^n$, $\mathbf{C}^n$, and $\mathbf{C}^{2n}$. Deduce the homeomorphisms $SO(n,\mathbf{R})/SO(n-1,\mathbf{R}) \approx S^{n-1}$, $SU(n,\mathbf{C})/SU(n-1,\mathbf{C}) \approx S^{2n-1}$, and $Sp(n)/Sp(n-1) \approx S^{4n-1}$ (S^k is the k-dimensional sphere).

(c) Prove that $SO(n,\mathbf{R})$ ($n \geq 2$), $SU(n,\mathbf{C})$ ($n \geq 1$), and $Sp(n)$ ($n \geq 1$) are connected while $SU(n,\mathbf{C})$ ($n \geq 2$) and $Sp(n)$ ($n \geq 1$) are simply connected. Prove also that $\pi_1(SO(n,\mathbf{R})) = \mathbf{Z}_2$ for $n \geq 3$, and $= \mathbf{Z}$ for $n = 2$.

(d) Prove that $U(n,\mathbf{C})$ is connected for all $n \geq 1$, and $\pi_1(U(n,\mathbf{C})) = \mathbf{Z}$.

(e) Show that $SU(2,\mathbf{C})$ and $Sp(1)$ are isomorphic.

12. (a) Let V be a vector space over $\mathbf{R}$ or $\mathbf{C}$, and let $X \in \mathfrak{gl}(V)$. Prove that the eigenvalues of ad X are of the form $\lambda - \mu$, where λ and μ are eigenvalues of X. Prove also that if X is semisimple, so is ad X.

(b) Let G_0 (resp $\mathfrak{g}_0$) denote the real Lie group (resp. real Lie algebra) underlying $GL(n,\mathbf{C})$ (resp. underlying $\mathfrak{gl}(n,\mathbf{C})$). Write $U = U(n,\mathbf{C})$, and let $\mathfrak{u}$ be the subalgebra of $\mathfrak{g}_0$ defined by U. Prove that $\mathfrak{u}$ is the Lie algebra of all skew Hermitian elements of $\mathfrak{g}_0$. If $\mathfrak{p}_0$ is the $\mathbf{R}$-linear subspace of all Hermitian elements in $\mathfrak{g}_0$, prove that $\mathfrak{g}_0$ is the direct sum $\mathfrak{u} + \mathfrak{p}_0$ and that

$$[\mathfrak{u},\mathfrak{p}_0] \subseteq \mathfrak{p}_0, \qquad [\mathfrak{p}_0,\mathfrak{p}_0] \subseteq \mathfrak{u}, \qquad [\mathfrak{u},\mathfrak{u}] \subseteq \mathfrak{u}.$$

(c) Prove that the map

$$\psi : (k,X) \mapsto k \exp X \quad (k \in U, X \in \mathfrak{p}_0)$$

is a homeomorphism of $U \times \mathfrak{p}_0$ onto G_0. Deduce that G_0 is connected and that $\pi_1(G_0) = \mathbf{Z}$.

(d) Calculate $(d\psi)_{(k,X)}$ using (2.14.5) and deduce that ψ is an analytic diffeomorphism. (Hint: use (a)).

13. Let G_0 and $\mathfrak{g}_0$ be as in Exercise 12, and let G be a closed Lie subgroup of G_0, with $\mathfrak{g}$ as the corresponding subalgebra of $\mathfrak{g}_0$. For $X \in \mathfrak{g}_0$ let $X^\dagger$ denote the matrix adjoint to X. Let G^0 be the component of 1 in G.

(a) Prove that $G^0 = (G^0)^\dagger$ if and only if $\mathfrak{g} = \mathfrak{g}^\dagger$.

(b) Let $\mathfrak{g} = \mathfrak{g}^\dagger$ and let $\mathfrak{k} = \mathfrak{u} \cap \mathfrak{g}$, $\mathfrak{p} = \mathfrak{p}_0 \cap \mathfrak{g}$. Prove that $\mathfrak{g}$ is the direct sum $\mathfrak{k} + \mathfrak{p}$ and that

$$[\mathfrak{k},\mathfrak{p}] \subseteq \mathfrak{p}, \qquad [\mathfrak{p},\mathfrak{p}] \subseteq \mathfrak{k}, \qquad [\mathfrak{k},\mathfrak{k}] \subseteq \mathfrak{k}.$$

(c) Let K^0 be the analytic subgroup of G corresponding to the subalgebra $\mathfrak{k}$. Prove that K^0 is compact and that the map

$$\psi : (k,X) \mapsto k \exp X \quad (k \in K^0, X \in \mathfrak{p})$$

is an analytic diffeomorphism of $K^0 \times \mathfrak{p}$ onto G^0. Deduce that $\pi_1(G^0) = \pi_1(K^0)$ and that global analytic sections exist on G^0/K^0. (Hint: Use Exercise 12 to prove that $(d\psi)_{(k,X)}$ is bijective. Observe now that $K^0 \exp[\mathfrak{p}]$ is both open and closed in G^0.)

(d) Prove that K^0 is a maximal compact subgroup of G^0 and deduce that $K^0 = U \cap G^0$.

(e) Suppose now that $G = G^\dagger$ and that G is an algebraic subset of the real vector space of all complex $n \times n$ matrices. If $p = \exp X$, where $X \in \mathfrak{p}_0$, show that $p^2 \in G$ if and only if $X \in \mathfrak{p}$. Deduce that in this case, $\exp tX \in G$ for all $t \in \mathbf{R}$. (Hint: For some $u \in U$, $upu^{-1} = \operatorname{diag}(e^{x_1}, \ldots, e^{x_n})$ $(x_i \in \mathbf{R})$. Observe now that if F is a polynomial in n variables and $F(e^{2kx_1}, \ldots, e^{2kx_n}) = 0$ for $k = 1, 2, \ldots$, then $F(e^{tx_1}, \ldots, e^{tx_n}) = 0$ for all $t \in \mathbf{R}$).

(f) Let $K = G \cap U$. Then prove that

$$\psi : (k,X) \mapsto k \exp X$$

is an analytic diffeomorphism of $K \times \mathfrak{p}$ onto G. Deduce that K is a maximal compact subgroup of G.

(g) Prove that $SL(n,\mathbf{C})$ $(n \geq 2)$ and $Sp(n,\mathbf{C})$ $(n \geq 1)$ are connected and simply connected and that $SO(n,\mathbf{C})$ $(n \geq 3)$ is connected with fundamental group $\mathbf{Z}_2$.

14. Let M be a real or complex analytic manifold. Suppose G is a topological group acting transitively and continuously on M with compact stability groups, and let D be a discrete subgroup of G such that (i) each element of D induces an analytic diffeomorphism of M, and (ii) $G_x \cap D = \{1\}$ for all $x \in M$. Then prove that M/D admits an analytic structure with respect to which the natural map of M onto M/D is a submersion.

15. (a) Let G_c be a simply connected complex analytic group with Lie algebra $\mathfrak{g}_c$, let $\mathfrak{g}$ be a real form of $\mathfrak{g}_c$, and let G be the analytic subgroup of the underlying real analytic group of G_c defined by $\mathfrak{g}$ (when $\mathfrak{g}$ is considered as a subalgebra of the real Lie algebra underlying $\mathfrak{g}_c$). Let V be a finite-dimensional vector space over $\mathbf{R}$ and ρ a homomorphism of $\mathfrak{g}$ into $\mathfrak{gl}(V)$. Prove that there is a (unique) analytic homomorphism π of G into $GL(V)$ such that $d\pi = \rho$. (Hint: If ρ_c, V_c is the complexification of ρ, V and π_c the corresponding homomorphism of G_c into $GL(V_c)$, take $\pi = \pi_c | G$).

(b) Taking $G = SL(n,\mathbf{R})$, $G_c = SL(n,\mathbf{C})$ $(n \geq 2)$, show that any Lie group which is a nontrivial covering group of $G = SL(n,\mathbf{R})$ cannot be isomorphic to a matrix Lie group.

16. Prove that if G is a simply connected analytic group, then any normal analytic subgroup of G is necessarily closed in G. (Hint: We may work over $\mathbf{R}$. This depends on the fact that given any Lie algebra over $\mathbf{R}$ there is a real analytic group whose Lie algebra is isomorphic to the given one. Let $\mathfrak{g}$ be the Lie algebra of G, H the normal analytic subgroup, and $\mathfrak{h} \subseteq \mathfrak{g}$ the corresponding ideal. If B is an analytic group whose Lie algebra is isomorphic to $\mathfrak{g}/\mathfrak{h}$, prove that there is a continuous homomorphism of G onto B, the component of identity of whose kernel is H).

17. Let G_i be a real Lie group, V_i a finite-dimensional real vector space, and π_i an analytic homomorphism of G_i into $GL(V_i)$ $(i = 1, 2)$. Write $H = G_1 \times G_2$, $V = V_1 \otimes V_2$, and let $\pi(x_1,x_2) = \pi_1(x_1) \otimes \pi_2(x_2)$, $(x_1,x_2) \in H$. Prove that $d\pi = d\pi_1 \times d\pi_2$. If $G_1 = G_2 = G$ and $\pi(x) = \pi_1(x) \otimes \pi_2(x)$, $x \in G$, prove that $d\pi = d\pi_1 \otimes d\pi_2$.

18. Prove that the exponential map is surjective when $G = SO(n,\mathbf{R})$, $SU(n,\mathbf{C})$, or $GL(n,\mathbf{C})$.

19. Let $G = GL(n,\mathbf{C})$, $\mathfrak{g} = \mathfrak{gl}(n,\mathbf{C})$. Prove that there is an open set U containing 1 in G with the following properties: (i) for each $x \in U$, the series $\log x = \sum_{r=1}^{\infty} (-1)^{r-1} \cdot (x - 1)^r/r$ converges absolutely to an element of $\mathfrak{g}$, (ii) the map $x \mapsto \log x$ is an analytic diffeomorphism of U onto an open neighborhood $\mathfrak{u}$ of 0 in $\mathfrak{g}$, and (iii) $x = e^{\log x}$, $X = \log e^X$ for $x \in U$, $X \in \mathfrak{u}$.

20. (a) Let G be the group of all $n \times n$ real matrices (a_{ij}) with $a_{ij} = \delta_{ij}$ for $i \geq j$. Let $\mathfrak{g}$ be the subalgebra of $\mathfrak{gl}(n,\mathbf{R})$ defined by G. We identify $\mathfrak{g}$ with the Lie algebra of G. Prove that for any $X \in \mathfrak{g}$, ad X is a nilpotent endomorphism of $\mathfrak{g}$. Deduce that exp is an analytic diffeomorphism of $\mathfrak{g}$ onto G, with inverse log given by $\log x = \sum_{r=1}^{n} (-1)^{r-1}(x - 1)^r/r$ $(x \in G)$.

(b) Prove that all analytic subgroups of G are closed and simply connected.

(c) For $X, Y \in \mathfrak{g}$, let $A(X:Y)$ be the unique element of $\mathfrak{g}$ such that $\exp X \exp Y = \exp A(X:Y)$. Prove that A is a polynomial map of $\mathfrak{g} \times \mathfrak{g}$ into $\mathfrak{g}$.

(d) Denote by Γ the subgroup of all matrices in G whose entries are integers. Prove that G/Γ is compact.

21. Let A be a finite-dimensional algebra over $\mathbf{R}$. We assume that the multiplication is bilinear but not necessarily associative.

(a) Prove that if L is an endomorphism of A, L is a derivation if and only if e^{tL} is an automorphism of A for all $t \in \mathbf{R}$.

(b) Show that the group $Aut(A)$ of the automorphisms of A is an algebraic subgroup of $GL(A)$ and that its corresponding subalgebra in $\mathfrak{gl}(A)$ is the Lie algebra of all derivations of A.

(c) Take A to be the Lie algebra $\mathfrak{sl}(n,\mathbf{R})$, and prove that the derivations ad X ($X \in A$) are precisely all the derivations of A. Deduce that $(Aut(A))^0$ is the image under the adjoint representation Ad of $SL(n,\mathbf{R})$.

(d) Show that if $n > 2$, the map $X \mapsto -X^t$ ($X \in A$) is an automorphism of A which does not lie in $(Aut(A))^0$.

22. Let $G = SL(2,\mathbf{C})$ act on itself by inner automorphisms. Determine all the orbits.

23. Let G be an analytic group, $\mathfrak{g}$ its Lie algebra. Let $\mathfrak{M}$ be the vector space of all left-invariant 1-forms on G.

(a) Prove that if φ is an analytic diffeomorphism of G onto itself such that each member of $\mathfrak{M}$ is invariant under φ, there is an element $a \in G$ such that $\varphi = l_a$; i.e., $\varphi(x) = ax$ for $x \in G$.

(b) Suppose there is an analytic diffeomorphism ψ of $\mathfrak{g}$ onto G. Let $\mathfrak{M}_\psi$ be the set of 1-forms on $\mathfrak{g}$ that correspond to the members of $\mathfrak{M}$ under ψ. For $a \in G$, let λ_a be the analytic diffeomorphism of $\mathfrak{g}$ onto itself defined by $\lambda_a = \psi^{-1} \circ l_a \circ \psi$. Deduce from (a) that the map $a \mapsto \lambda_a$ is an isomorphism of G onto the group of all analytic diffeomorphisms of $\mathfrak{g}$ onto itself that leave the members of $\mathfrak{M}_\psi$ invariant.

24. Let G be an analytic group, $\mathfrak{g}$ the Lie algebra of G, and $\{X_1, \ldots, X_m\}$ a basis for $\mathfrak{g}$. Let $k = \mathbf{R}$ or $\mathbf{C}$ according as G is real or complex, and let ψ be the map $(t_1, \ldots, t_m) \mapsto \exp t_1X_1 \cdots \exp t_mX_m$ of k^m into G. Let $\omega_1, \ldots, \omega_m$ be the left-invariant 1-forms on G such that $\omega_i(X_j) = \delta_{ij}$, $1 \leq i, j \leq m$. We identify the tangent space to G at any of its points with $\mathfrak{g}$.

(a) Prove that $(d\psi)_{(t_1,\ldots,t_m)}(\partial/\partial t_s) = X_s^{(\exp t_{s+1}X_{s+1}\cdots\exp t_mX_m)^{-1}}$ for $1 \leq s \leq m$ (when $s = m$, the exponent of X_s is regarded as 1).

(b) Let $\bar{\omega}_j$ be the 1-form on k^m that corresponds to ω_j under ψ, $1 \leq j \leq m$. Let $a_{ij}(x)$ denote ijth entry of the matrix of $\text{Ad}(x)$ relative to $\{X_1, \ldots, X_m\}$ ($x \in G$). Then deduce from (a) that

$$\bar{\omega}_r = \sum_{1 \leq s \leq m} a_{rs}((\exp t_{s+1}X_{s+1} \cdots \exp t_mX_m)^{-1})\, dt_s \quad (1 \leq r \leq m).$$

25. Let G be a Lie group with Lie algebra $\mathfrak{g}$. An analytic 1-form on G which is invariant under all left translations is called a *Maurer–Cartan form.* Let $\mathfrak{m}$ denote the vector space of all such forms (cf. Chevalley [1].

(a) Let $\mathfrak{g}^*$ be the dual of $\mathfrak{g}$. Prove that for any $\omega \in \mathfrak{g}^*$ there is a unique $\bar{\omega} \in \mathfrak{m}$ such that $\bar{\omega}_1(X_1) = \omega(X)$ for all $X \in \mathfrak{g}$ and that the map $\omega \mapsto \bar{\omega}$ is a linear isomorphism of $\mathfrak{g}^*$ onto $\mathfrak{m}$.

(b) Let $\{X_1, \ldots, X_m\}$ be a basis for $\mathfrak{g}$, $\{\omega_1, \ldots, \omega_m\}$ the dual basis of $\mathfrak{g}^*$; for any $\omega \in \mathfrak{g}^*$, let $\bar{\omega} \in \mathfrak{m}$ be defined by (a) and let $\bar{\omega}^*$ be the analytic 1-form on $\mathfrak{g}$ which corresponds to $\bar{\omega}$ under exp. Put $\hat{\mathfrak{m}} = \{\bar{\omega}^* : \omega \in \mathfrak{g}^*\}$. If $x_1, \ldots, x_m$ are the linear coordinates on $\mathfrak{g}$ corresponding to the basis $\{X_1, \ldots, X_m\}$, prove that there are analytic functions a_{ij} of $x_1, \ldots, x_m$ such that

$$\bar{\omega}_i^* = \sum_{1 \leq j \leq m} a_{ij}(x_1, \ldots, x_m)\, dx_j \quad (1 \leq i \leq m).$$

(c) Let $\mathbf{A}$ denote the $m \times m$ matrix whose ijth entry is the function $a_{ij}(x_1, \ldots, x_m)$. Define the structure constants c_{rjk} by $[X_r, X_j] = \sum_{1 \le k \le m} c_{rjk} X_k$. Then prove that

$$d\omega_k = -\frac{1}{2} \sum_{i,j=1}^{m} c_{ijk} \omega_i \wedge \omega_j.$$

(d) Obtain the differential equations

$$\frac{\partial}{\partial t}(t\mathbf{A}(tx_1, \ldots, tx_m)) = I_m - L(x_1, \ldots, x_m) t\mathbf{A}(tx_1, \ldots, tx_m),$$

where $L(x_1, \ldots, x_m)$ is the $m \times m$ matrix whose ijth entry is $\sum_{1 \le r \le m} c_{rji} x_r$, while I_m is the $m \times m$ unit matrix. Hence show that

$$\mathbf{A} = \sum_{n=0}^{\infty} \frac{(-1)^n}{(n+1)!} L^n.$$

(Hint: For fixed $(x_1, \ldots, x_m)$, the map $t \mapsto \exp t(x_1 X_1 + \cdots + x_m X_m)$ is the composition of $(y_1, \ldots, y_m) \mapsto \exp(y_1 Y_1 + \cdots + y_m Y_m)$ followed by $t \mapsto (tx_1, \ldots, tx_m)$. Deduce from this the relation

$$\sum_{1 \le j \le m} a_{kj}(tx_1, \ldots, tx_m) x_j = x_k \quad (1 \le k \le m).$$

Let $\omega'_1, \ldots, \omega'_m$ be the 1-forms on $\mathbf{R}^{n+1}$ that correspond to $\bar{\omega}^*_1, \ldots, \bar{\omega}^*_m$ respectively under the map $(t, x_1, \ldots, x_m) \mapsto (tx_1, \ldots, tx_m)$. Deduce from the previous relation the formula

$$\omega'_k = x_k\, dt + t \sum_{1 \le j \le m} a_{kj}(tx_1, \ldots, tx_m)\, dx_j.$$

Use the relations

$$d\omega'_k = -\sum_{i,j=1}^{m} c_{ij} \omega'_i \wedge \omega'_j$$

to obtain the required differential equations.)

(e) Prove that for any $\omega \in \mathfrak{g}^*$, $(\bar{\omega}^*)_X(Y) = \omega((d\exp)_X(Y))$ $(X, Y \in \mathfrak{g})$, and hence obtain the above expression for $\mathbf{A}$ as a consequence of (2.14.5). Conversely, deduce (2.14.5) from the expression for $\mathbf{A}$.

26. (a) Let G be a real Lie group of dimension m, and let $\mathfrak{q}$ be the real vector space of all analytic m-forms on G which are invariant under all left translations. Prove that $\dim \mathfrak{q} = 1$ and that for any $\omega \in \mathfrak{q}$, the corresponding Borel measure on G is a left Haar measure.

(b) Let μ be a left Haar measure for G and Δ the modular function; i.e., Δ is the continuous homomorphism of G into the positive reals such that $\Delta(y) \int_G f(xy)\, d\mu(x) = \int_G f(x)\, d\mu(x)$ for all $f \in C_c(G)$. Prove that $\Delta(y) = \det \mathrm{Ad}(y)^{-1}$ for all $y \in G$. Hence prove that G is unimodular (i.e., left Haar measures are also right-invariant) if and only if tr ad $X = 0$ for all $X \in \mathfrak{g}$. Deduce that the classical algebras are unimodular. Give an example of a group which is not unimodular.

(c) Let V be a real vector space of dimension $\mathfrak{m}$, $G = GL(V)$. Let λ be a Lebesgue measure on the vector space $\mathfrak{gl}(V)$. Prove that the measure μ on G defined by $d\mu(X) = (\det X)^{-m} d\lambda(X)$ is a left Haar measure on G. Obtain the corresponding formula when V is a complex vector space.

(d) Let G be a Lie group, $\mathfrak{g}$ its Lie algebra. Let $\mathfrak{v}$ be as in Theorem 2.14.6. Assume that exp is one-to-one on $\mathfrak{v}$. Denote by dx a left Haar measure on G and by dX a Lebesgue measure on $\mathfrak{g}$. Prove that there is a constant $c > 0$ such that $\int_{\mathfrak{v}} f(\exp X) |\det (d \exp)_X| \, dX = c \int_{\exp \mathfrak{v}} f(x) \, dx$ for all $f \in C_c(\exp \mathfrak{v})$.

27. (a) Let V be a real vector space of finite dimension n. If D is a discrete subgroup of V, prove that D is isomorphic to $\mathbf{Z}^m$ for some m and that $\dim (\mathbf{R} \cdot D) = m$.

(b) Let G be a connected abelian Lie group. Prove the existence of closed analytic subgroups K, H such that (i) K is compact and contains every compact subgroup of G, (ii) H is isomorphic to $\mathbf{R}^q$ and K is isomorphic to $\mathbf{T}^q$ for some $p, q \geq 0$, and (iii) $G = KH$, $K \cap H = \{1\}$. Prove further that K is uniquely determined by (1), while H is determined up to an automorphism of G.

28. (a) Let D be a discrete subgroup of $\mathbf{C}^n$. If D is of rank $2n$, i.e., D is isomorphic to $\mathbf{Z}^{2n}$, prove that $\mathbf{C}^n/D$ is compact.

(b) Let G be a compact complex analytic group of dimension n. Prove that G is abelian and is isomorphic to $\mathbf{C}^n/D$, where D is a discrete subgroup of $\mathbf{C}^n$ of rank $2n$. [Observe that the adjoint representation of G is trivial.]

(c) Let $G = \mathbf{C}^n/D$ where D is a discrete subgroup of $\mathbf{C}^n$ of rank $2n$. Prove that any holomorphic exterior differential form on G is invariant under translations.

(d) Let $G_1 = \mathbf{C}^n/D_1$ and $G_2 = \mathbf{C}^n/D_2$, where D_1 and D_2 are discrete subgroups of $\mathbf{C}^n$ of rank $2n$. If $\gamma : G_1 \longrightarrow G_2$ is a holomorphic isomorphism of the underlying complex manifolds that takes the identity of G_1 to that of G_2, prove that there is a linear automorphism of $\mathbf{C}^n$ that maps D_1 onto D_2 and induces γ. (Hint: Let π_i be the canonical map $\mathbf{C}^n \longrightarrow G_i$. Prove, using covering space arguments, that there is a complex analytic diffeomorphism $\varphi : \mathbf{C}^n \longrightarrow \mathbf{C}^n$ such that $\pi_2 \circ \varphi = \gamma \circ \pi_1$. Use (c) to prove that the space of Maurer–Cartan forms of $\mathbf{C}^n$ is stable under φ^*, and deduce from Exercise 23 that φ is an affine map.)

In Exercises 29–34, G is an analytic group with Lie algebra $\mathfrak{g}$; M is a C^∞ manifold; $(x,y) \mapsto x \cdot y$ $(x \in G, y \in M)$ a smooth action of G on M; $\dim(M) = m$, $\dim(G) = n$. All functions, forms, etc., considered by us are real. $\mathfrak{A}(M)$ (resp. $\mathfrak{A}(G)$) is the algebra of exterior differential forms on M (resp. G). For $\omega \in \mathfrak{A}(M)$, $x \in G$, let $\omega_x = \omega^{t_x^{-1}}$, where t_x is the diffeomorphism $y \mapsto x \cdot y$ of M. $\omega \in \mathfrak{A}(M)$ is said to be *closed* if $d\omega = 0$; it is said to be *exact* if $\omega \in d(\mathfrak{A}(M))$. $\mathfrak{C}(M)$ is the algebra of all closed forms, $\mathcal{E}(M)$ the ideal of all exact forms, $H(M) = \mathcal{C}(M)/\mathcal{E}(M)$ (the De Rham cohomology algebra) the quotient algebra. If $\mathfrak{D}(M)$ is a subspace of $\mathfrak{a}(M)$, then $\mathfrak{D}_p(M) = \mathfrak{D}(M) \cap \mathfrak{a}_p(M)$. $H^p(M) = \mathcal{C}_p(M)/\mathcal{E}_p(M)$. $\bar{\mathfrak{a}}(M) = \{\omega : \omega \in \mathfrak{a}(M), \omega_x = \omega \text{ for all } x \in G\}$; $\bar{\mathcal{C}}(M) = \mathcal{C}(M) \cap \bar{\mathfrak{a}}(M)$, $\bar{\mathcal{E}}(M) = d(\bar{\mathfrak{a}}(M))$. $\bar{\mathfrak{a}}(G)$ is the algebra of left-invariant elements of $\mathfrak{a}(G)$. $\Lambda(\mathfrak{g})$ is, as usual, the algebra of multilinear forms on $\mathfrak{g}$; $\Lambda_p(\mathfrak{g})$ its subspace of p-forms.

29. Let $1 \leq p \leq m$, $\omega \in \mathfrak{a}_p(M)$.
(a) Prove that ω_x is smooth in $x \in G$ (cf. Exercise 14, Chapter 1).
(b) Suppose ω is closed. Prove that $\omega_x - \omega$ is exact for all $x \in G$. (Use Exercise 14, Chapter 1.)
(c) Let ω be closed, $z \in G$. Prove the existence of an open neighborhood U of z and, for each $u \in U$, of a $(p-1)$-form η_u on M such that (i) η_u is smooth in u, and (ii) $d\eta_u = \omega_u - \omega$ $(u \in U)$. (Hint: We may assume that $z = 1$. For $X \in \mathfrak{g}$ let $\tilde{X} \in \mathfrak{I}(M)$ be defined by $\tilde{X}_x = ((d/dt)(\exp tX \cdot x))_{t=0}$ $(x \in M)$. For $X \in \mathfrak{g}$ let $\eta_{\exp X} = p\int_0^1 (\omega_{\exp sX})_{\tilde{X}}\, ds$. By Exercise 14 of Chapter 1, $d\eta_{\exp X} = \omega_{\exp X} - \omega$. Verify that $\eta_{\exp X}$ is smooth in X.)
(d) Let ω be as above, dx a left Haar measure on G. Prove that for any $g \in C_c^\infty(G)$, $\int_G g(x)\omega_x\, dx - \left(\int_G g(x)\, dx\right)\cdot\omega$ is exact. (Use (c) and a partition of unity argument.)

30. Let G be compact, $1 \leq p \leq m$.
(a) Let $\omega \in \mathcal{C}_p(M)$ and $\bar{\omega} = \int_G \omega_x\, dx$. Prove that $\bar{\omega} - \omega$ is exact.
(b) Suppose $\omega \in \mathcal{E}_p(M) \cap \bar{\mathfrak{a}}(M)$. Prove that $\omega \in \bar{\mathcal{E}}_p(M)$. (If $\omega = d\eta$, where $\eta \in \mathfrak{a}_{p-1}(M)$, then $\omega = d\bar{\eta}$, where $\bar{\eta} = \int_G \eta_x\, dx$.)
(c) Prove that the natural map of $\bar{\mathcal{C}}(M)/\bar{\mathcal{E}}(M)$ (resp. $\bar{\mathcal{C}}_p(M)/\bar{\mathcal{E}}_p(M)$) into $H(M)$ (resp. $H^p(M)$) is an algebra (resp. linear) isomorphism.

31. Let G be arbitrary.
(a) Prove that d leaves $\bar{\mathfrak{a}}(G)$ invariant
(b) For $\omega \in \bar{\mathfrak{a}}_p(G)$ and $X_1, \ldots, X_p \in \mathfrak{g}$, let $\bar{\omega}(X_1, \ldots, X_p) = \omega(X_1, \ldots, X_p)$. Prove that the maps $\omega \mapsto \bar{\omega}$ extend to an algebra isomorphism of $\bar{\mathfrak{a}}(G)$ with $\Lambda(\mathfrak{g})$.
(c) Let d be the endomorphism of $\Lambda(\mathfrak{g})$ defined as follows: $d1 = 0$, and for $\eta \in \Lambda_p(\mathfrak{g})$ $(p \geq 1)$, $X_1, \ldots, X_{p+1} \in \mathfrak{g}$, $(p+1)(d\eta)(X_1, \ldots, X_{p+1}) = \sum_{1 \leq i < j \leq p+1} (-1)^{i+j}\eta([X_i, X_j], \ldots, \hat{X}_i, \ldots, \hat{X}_j, \ldots, X_{p+1})$. Prove that $d\bar{\omega} = \overline{(d\omega)}$ for $\omega \in \bar{\mathfrak{a}}(G)$. Deduce that (i) $d^2 = 0$, and (ii) $d(\omega \wedge \omega') = d\omega \wedge \omega' + \epsilon(\omega)\omega \wedge d\omega'$ $(\omega, \omega' \in \Lambda(\mathfrak{g})$, ϵ as in (e) of Exercise 11, Chapter 1).
(d) For $X \in \mathfrak{g}$, let L_X be the endomorphism of $\Lambda(\mathfrak{g})$ such that $L_X 1 = 0$ and, for $p \geq 1$ and $\omega \in \Lambda_p(\mathfrak{g})$, $(L_X\omega)(X_1, \ldots, X_p) = \sum_{1 \leq i \leq p} (-1)^i \omega([X, X_i], \ldots, \hat{X}_i, \ldots, X_p)$ $(X_1, \ldots, X_p \in \mathfrak{g})$. Prove that each L_X is a derivation of $\Lambda(\mathfrak{g})$ and that $X \mapsto L_X$ is a representation of $\mathfrak{g}$ in $\Lambda(\mathfrak{g})$.
(e) Prove that d commutes with all L_X.

(f) Prove that

$$2(p+1)(d\omega)(X_1, \ldots, X_{p+1}) = \sum_{1 \le i \le p+1} (-1)^{i-1}(L_{X_i}\omega)(X_1, \ldots, \hat{X}_i, \ldots, X_{p+1}) \quad (X_i \in \mathfrak{g}, \omega \in \Lambda_p(\mathfrak{g})).$$

32. Let G be transitive on M. Fix $y_0 \in M$; let H be the stabilizer of y_0 in G, $\mathfrak{h}$ the subalgebra of $\mathfrak{g}$ defined by H, and γ the map $x \mapsto x \cdot y_0$ of G onto M. For $h \in H$, $\sigma(h)$ (resp. $\rho(h)$) is the linear transformation of $\mathfrak{g}/\mathfrak{h}$ (resp. the tangent space $T_{y_0}(M)$ to M at y_0) induced by Ad(h) (resp. h). σ and ρ are representations of H.

(a) Prove that the differential $(d\gamma)_1$ induces an isomorphism of $\mathfrak{g}/\mathfrak{h}$ with $T_{y_0}(M)$ that intertwines the actions of σ and ρ. Let $\underset{\sim}{\gamma}$ be the induced isomorphism of $\Lambda(\mathfrak{g}/\mathfrak{h})$ with $\Lambda(T_{y_0}(M))$.

(b) Let $\xi(\varphi \mapsto \xi\varphi)$ be the injection of $\Lambda(\mathfrak{g}/\mathfrak{h})$ into $\Lambda(\mathfrak{g})$ induced by the natural map of $\mathfrak{g}$ onto $\mathfrak{g}/\mathfrak{h}$. Let $\Lambda(\mathfrak{g}/\mathfrak{h})_\sigma$ be the algebra of all elements of $\Lambda(\mathfrak{g}/\mathfrak{h})$ invariant under all $\sigma(h)$ $(h \in H)$. For any $\omega \in \bar{\mathfrak{A}}(M)$ let φ_ω be the element of $\Lambda(\mathfrak{g}/\mathfrak{h})$ that corresponds to ω_{y_0} via $\underset{\sim}{\gamma}$. Prove that $\varphi_\omega \in \Lambda(\mathfrak{g}/\mathfrak{h})_\sigma$ and that $\omega \mapsto \varphi_\omega$ is a degree-preserving algebra isomorphism of $\bar{\mathfrak{A}}(M)$ with $\Lambda(\mathfrak{g}/\mathfrak{h})_\sigma$. Prove also that $\omega \in \bar{\mathfrak{C}}(M)$ (resp. $\bar{\mathfrak{E}}(M)$) if and only if $d(\xi\varphi_\omega) = 0$ (resp. $\xi\varphi_\omega \in d(\xi(\Lambda(\mathfrak{g}/\mathfrak{h})_\sigma))$). (For $\omega \in \bar{\mathfrak{A}}(M)$, $\gamma^*\omega \in \bar{\mathfrak{A}}(G)$, and (cf. Exercise 31) $\overline{\gamma^*\omega} = \xi\varphi_\omega$, so $\xi\varphi_{d\omega} = d(\xi\varphi_\omega)$.)

33. Let G, M be compact, and G be transitive on M. Let notation be as in Exercise 32. Following E. Cartan, we call (G, M) a *symmetric pair* if there is an involutive automorphism θ of G such that, if G_θ is the subgroup of fixed points of θ and G_θ^0 is the connected component of G_θ containing 1, then $G_\theta^0 \subseteq H \subseteq G_\theta$. We write θ again for the induced involution of $\mathfrak{g}$. We assume that (G,M) is a symmetric pair and that θ has the above significance.

(a) Let $\mathfrak{s} = \{X: X \in \mathfrak{g}, \theta X = -X\}$. Prove that $\mathfrak{g} = \mathfrak{h} + \mathfrak{s}$ is a direct sum and that $\mathfrak{h} = \{X: X \in \mathfrak{g}, \theta X = X\}$.

(b) Prove that $[\mathfrak{h},\mathfrak{h}] \subseteq \mathfrak{h}$, that $[\mathfrak{h},\mathfrak{s}] \subseteq \mathfrak{s}$, that $[\mathfrak{s},\mathfrak{s}] \subseteq \mathfrak{h}$, and that Ad$(h)$ leaves $\mathfrak{s}$ invariant for all $h \in H$.

(c) Let d be the endomorphism of $\Lambda(\mathfrak{g})$ defined in Exercise 31, and ξ as in Exercise 32. Prove that $d(\xi\varphi) = 0$ for all $\varphi \in \Lambda(\mathfrak{g}/\mathfrak{h})$. (Hint: $d(\xi\omega_\omega) = \xi\varphi_{d\omega}$ for $\omega \in \bar{\mathfrak{A}}(M)$, so $d(\xi(\Lambda(\mathfrak{g}/\mathfrak{h})_\sigma)) \subseteq \xi(\Lambda(\mathfrak{g}/\mathfrak{h})_\sigma)$; since $[\mathfrak{s},\mathfrak{s}] \subseteq \mathfrak{h}$, it is immediate that $d(\xi\varphi)(X_1, \ldots, X_{p+1}) = 0$ whenever $X_i \in \mathfrak{s}$ for all i and $\varphi \in \Lambda_p(\mathfrak{g}/\mathfrak{h})$.)

(d) Prove that all members of $\bar{\mathfrak{A}}(M)$ are closed and that $\bar{\mathfrak{E}}(M) = \bar{\mathfrak{A}}(M) \cap \mathfrak{E}(M) = \{0\}$. Deduce that $\mathfrak{C}(M)$ is the direct sum of $\mathfrak{E}(M)$ and $\bar{\mathfrak{A}}(M)$.

(e) Let $\underset{\sim}{\xi}$ be the isomorphism of $\Lambda(\mathfrak{g}/\mathfrak{h})$ with $\Lambda(\mathfrak{s})$ induced by the natural map of $\mathfrak{s}$ onto $\mathfrak{g}/\mathfrak{h}$. Let $\bar{\Lambda}(\mathfrak{s}) = \underset{\sim}{\xi}(\Lambda(\mathfrak{g}/\mathfrak{h})_\sigma)$. Prove that $\bar{\Lambda}(\mathfrak{s})$ is the algebra of all elements of $\Lambda(\mathfrak{s})$ invariant under the action of Ad(h) for all $h \in H$.

(f) For $\omega \in \bar{\mathfrak{A}}(M)$ let $[\omega]$ be the corresponding class in $H(M)$. Prove that $[\omega] \mapsto [\underset{\sim}{\omega}] = \underset{\sim}{\xi}(\varphi_\omega)$ is a degree-preserving algebra isomorphism of $H(M)$ onto $\bar{\Lambda}(\mathfrak{s})$. Deduce that dim $H^p(M) = $ dim $\Lambda_p(\mathfrak{s}) < \infty$ $(0 \le p \le m)$.

34. Let G be compact. Let $\bar{\bar{\mathfrak{a}}}(G)$ be the algebra of all two-sided invariant exterior differential forms on G. Let $\bar{\Lambda}(\mathfrak{g})$ be the subalgebra of all elements of $\Lambda(\mathfrak{g})$ invariant under the adjoint representation of G.

(a) Prove that the map $\omega \mapsto \bar{\omega}$ induces an algebra isomorphism of $\bar{\bar{\mathfrak{a}}}(G)$ with $\bar{\Lambda}(\mathfrak{g})$.

(b) Let $G \times G$ act on G by $((a,b),x) \longrightarrow axb^{-1}$. Prove that $(G \times G,G)$ is a symmetric pair.

(c) Use (b) or a direct argument to prove that $\mathfrak{C}(G)$ is the direct sum of $\mathcal{E}(G)$ and $\bar{\bar{\mathfrak{a}}}(G)$ and that $[\omega] \mapsto \bar{\omega}$ $(\omega \in \bar{\bar{\mathfrak{a}}}(G))$ is a degree-preserving isomorphism of $H(G)$ onto $\bar{\Lambda}(\mathfrak{g})$.

(d) Deduce that $H(G_1)$ and $H(G_2)$ are canonically isomorphic if G_1 and G_2 are locally isomorphic compact analytic groups.

(For Exercises. 29–34 cf. E. Cartan [3]; for the theory of symmetric pairs see the papers of E. Cartan and also Helgason [1]; for an account of the topological properties of Lie groups, see Samelson [1]; see also Borel [1].)

35. (a) Let $V = \mathbf{R}^n$, $H = SO(n)$, and let H act on V in the usual way. Let φ_i be the linear function $(x_1, \ldots, x_n) \mapsto x_i$, and let $\eta = \varphi_1 \wedge \cdots \wedge \varphi_n$. If $1 \leq p \leq n$ and $\omega \in \Lambda_p(V)$ is H-invariant, prove that $\omega = 0$ for $p < n$, and $\omega \in \mathbf{R} \cdot \eta$ if $p = n$.

(b) Let M be the unit sphere in $\mathbf{R}^{n+1}$, $G = SO(n+1)$. Take $y_0 = (0, \ldots, 0,1)$ in the notation of Exercise 32 and verify that (G,M) is a symmetric pair.

(c) Prove that $H^p(M) = 0$ for $1 \leq p < n$ and that $\dim H^n(M) = 1$.

(d) Let $\bar{M}$ be the space obtained by identifying x with $-x$ for all $x \in M$ (projective real n-space). Prove that if n is even, $H^p(\bar{M}) = 0$, $1 \leq p \leq n$.

36. Let $G = \mathbf{T}^n$, the n-torus. Prove that $\dim H^p(G) = \binom{n}{p}$, $0 \leq p \leq n$.

37. Let $G = SO(3)$. Write $b_p = \dim H^p(G)$. Prove that $b_1 = b_2 = 0$, $b_3 = 1$.

38. Let M be the set of all k-dimensional linear subspaces of $\mathbf{C}^{n+1}$ $(1 \leq k \leq n)$.

(a) Prove that the natural actions of both $G = GL(n+1,\mathbf{C})$ and $U = U(n+1,\mathbf{C})$ on M are transitive.

(b) Let y_0 be the element of M defined by the equations $x_{k+1} = 0, \ldots, x_{n+1} = 0$ (x_i are the usual coordinates on $\mathbf{C}^{n+1}$). Determine the stabilizers of y_0 in G and U.

(c) Use (b) to show that M can be equipped with a natural complex structure under which it is compact and of (complex) dimension $k(n+1-k)$, and on which the action of G is complex analytic.

(d) Verify that (U,M) is a symmetric pair.

39. Let G be a real Lie group. Prove that it does not have small subgroups; i.e., prove the existence of an open neighborhood U of 1 such that $\{1\}$ is the only subgroup of G that is entirely contained in U.

40. (a) Let G be a real analytic group, $\mathfrak{g}$ its Lie algebra. Suppose $\mathfrak{g}$ can be given the structure of a complex vector space such that $\mathfrak{g}$ becomes a complex Lie algebra, denoted by $\tilde{\mathfrak{g}}$. Prove that there exists exactly one complex structure on G under which G becomes a complex Lie group, say $\bar{G}$, having the following two properties: (i) G is the real analytic group underlying $\bar{G}$, and (ii) $\bar{\mathfrak{g}}$ is the Lie algebra of $\bar{G}$.

(b) Prove that $\exp_{\bar{G}} = \exp_G$.

(c) Let G be a complex analytic group. If G is compact, prove that all its finite-dimensional (holomorphic) representations are trivial. Deduce that G is commutative.

41. Let G be a real C^∞ group, $\mathfrak{g}$ its Lie algebra, and $\mathfrak{G}$ its enveloping algebra. As in the analytic case we identify $T_x(G)$ and $T_x^{(\infty)}(G)$ with $\mathfrak{g}$ and $\mathfrak{G}$ respectively ($x \in G$). Define D by (2.14.1).

(a) Let A ($t \mapsto A(t)$) be a C^∞ map of some open interval Δ of $\mathbf{R}$ into $\mathfrak{g}$ and let $w(s, t) = \exp(sA(t))$ ($s \in \mathbf{R}$, $t \in \Delta$). Calculate the image of $\partial^2/\partial s \partial t$ in $\mathfrak{G}$ under $(dw)_{(s,t)}^{(\infty)}$ in two different ways and deduce the following equation (here $\dot{A}$ denotes dA/dt):

$$\dot{A} + sD(sA : \dot{A})A = \frac{\partial}{\partial s}(sD(sA : \dot{A}) + sA\,D(sA : \dot{A}).$$

(b) Let $X, Y \in \mathfrak{g}$. Take $A(t) = X + tY$ in (a) and obtain the following differential equation for the function Ξ ($s \mapsto sD(sX : Y)$) ($s \in \mathbf{R}$):

$$\frac{d}{ds}\Xi = -[X,\Xi] + Y$$

(c) Deduce from (b) the formula (2.14.5) for D.

42. Let $\mathfrak{J}$ be an associative algebra with unit 1 over a field k of characteristic 0. $\mathfrak{J}^{(0)} = \mathfrak{J}, \mathfrak{J}^{(1)}, \mathfrak{J}^{(2)}, \ldots$ are k-linear subspaces of $\mathfrak{J}$ such that $\mathfrak{J}^{(m)} \supseteq \mathfrak{J}^{(n)}$ if $0 \leq m \leq n$ and $\mathfrak{J}^{(m)}\mathfrak{J}^{(n)} \subseteq \mathfrak{J}^{(m+n)}$ if $m, n \geq 0$. We also write $\mathfrak{J}^+ = \mathfrak{J}^{(1)}$. Assume that $1 \in \mathfrak{J}^{(0)}$, $1 \notin \mathfrak{J}^{(m)}$ for $m > 0$, and $\bigcap_{n \geq 0} \mathfrak{J}^{(n)} = \{0\}$. Clearly, each $\mathfrak{J}^{(m)}$ ($m \geq 1$) is a two-sided ideal in $\mathfrak{J}$.

(a) For any $x \in \mathfrak{J}$ let ord (x) be the supremum of all integers $m \geq 0$ for which $x \in \mathfrak{J}^{(m)}$, and let $|x| = 2^{-\text{ord}(x)}$. Prove the following properties for the function $|\cdot|$: (i) $0 \leq |x| = |cx| \leq 1\ \forall\ x \in \mathfrak{J}$, $c \in k^\times$; $|x| = 0$ if and only if $x = 0$, and $|1| = 1$; (ii) $|x + y| \leq \max(|x|,|y|)\ \forall\ x, y \in \mathfrak{J}$; (iii) $|xy| \leq |x|\,|y|\ \forall\ x, y \in \mathfrak{J}$. We shall say that $\mathfrak{J}$ is *complete* if it is a complete metric space for the distance function $d(x,y) = |x - y|$ ($x, y \in \mathfrak{J}$). From now on we assume that $\mathfrak{J}$ is complete.

(b) Let $\{x_\lambda : \lambda \in \Lambda\}$ be an indexed family in $\mathfrak{J}$. We say that it is *summable* if there exists $x \in \mathfrak{J}$ such that for each $\varepsilon > 0$ one can find a finite subset $F_\epsilon \subseteq \Lambda$ for which $|x - \sum_{\lambda \in F} x_\lambda| < \varepsilon$ whenever F is a finite set $\supseteq F_\epsilon$. In this case we write $x = \sum_{\lambda \in \Lambda} x_\lambda$. x is, of course, uniquely determined. Prove the following criterion for summability: $\{x_\lambda : \lambda \in \Lambda\}$ is summable if and only if $|x_\lambda| \longrightarrow 0$ in the sense that given any $\epsilon > 0$, we can find a finite subset $F_\epsilon \subseteq \Lambda$ such that $|x_\lambda| < \epsilon$ whenever $\lambda \notin F_\epsilon$.

(c) If $x \in \mathfrak{J}^+$, show that $\{x^n/n! : n \geq 0\}$ and $\{(-1)^{n-1}x^n/n : n \geq 1\}$ are summable. Write $\exp x = \sum_{n \geq 0} (x^n/n!)$, $\log y = \sum_{n \geq 1} (-1)^{n-1}(y - 1)^n/n$ ($x \in \mathfrak{J}^+, y \in 1 + \mathfrak{J}^+$).

(d) Prove that exp is a homeomorphism of $\mathfrak{J}^+$ onto $1 + \mathfrak{J}^+$ and that log is its inverse.

(e) If $x, y \in \mathfrak{J}^+$, prove that $\{x^m y^n/m!n! : m, n \geq 0\}$ is summable and that $\sum_{m,n\geq 0} x^m y^n/m!n! = \exp x \exp y$. Prove that if $xy = yx$, then $\exp(x + y) = \exp x \cdot \exp y = \exp y \cdot \exp x$. Deduce that $\exp x$ is invertible and $\exp(-x) = (\exp x)^{-1}$.

(f) For any $x \in \mathfrak{J}$ let $\theta(x)(y) = xy - yx = [x,y]$ $(y \in \mathfrak{J})$. Prove that $\theta(x)$ is a continuous derivation of $\mathfrak{J}$. If $x \in \mathfrak{J}^+$ and D is a continuous derivation of $\mathfrak{J}$, prove that

$$\exp(-x)D(\exp x) = \sum_{n\geq 0} \frac{(-1)^n}{(n+1)!}\theta(x)^n(D(x)).$$

[Hint: Use Lemmas 2.14.1 and 2.14.2.]

(g) Let T be an indeterminate and $k[[T]]$ the algebra over k of all formal power series $\psi = \sum_{n\geq 0} a_n T^n$ $(a_n \in k)$ in T with coefficients from k. If L is any endomorphism such that $L[\mathfrak{J}^{(m)}] \subseteq \mathfrak{J}^{(m+1)}$ for all $m \geq 0$, prove that L is continuous, that $\psi(L) : y \mapsto \sum_{n\geq 0} a_n L^n(y)$ $(y \in \mathfrak{J})$ is a well-defined continuous endomorphism of $\mathfrak{J}$, and that $\psi \mapsto \psi(L)$ is a homomorphism of $k[[T]]$.

(h) Let $x, y \in \mathfrak{J}^+$ and let $\exp x \exp y = \exp z$ where $z = z(x,y) \in \mathfrak{J}^+$ is uniquely determined by x and y. Let D_1 (resp. D_2) be a continuous derivation of $\mathfrak{J}$ such that $D_1(y) = 0$ and $[x, D_1(x)] = 0$ (resp. $D_2(x) = 0$ and $[y, D_2(y)] = 0$). Let K_{2p} $(p \geq 1)$ be the rational constants defined in §2.15. Prove that

$$D_1(z) = -\frac{1}{2}[z, D_1(x)] + D_1(x) + \sum_{p\geq 1} K_{2p}\theta(z)^{2p}(D_1(x))$$

$$D_2(z) = \frac{1}{2}[z, D_2(y)] + D_2(y) + \sum_{p\geq 1} K_{2p}\theta(z)^{2p}(D_2(y)).$$

[Hint: Imitate proof of Lemma 2.15.2.]

43. Let k be a field of characteristic 0, E a set having two elements U and V. W is the set consisting of all words in E; treat 1 as a word of zero length. $\mathfrak{J}$ is the set of all formal infinite sums $x = \sum_{w\in W} c_w(x)w$, where $c_w(x) \in k$ for all $w \in W$. We refer to elements of $\mathfrak{J}$ as *formal power series in two non-commuting indeterminates U and V*; the $c_w(x)$ are called the *coefficients* of x. Equality, addition, and multiplication by elements of k are coefficient-wise; multiplication is Cauchy. Thus, if $x, y \in \mathfrak{J}$, then

$$c_w(xy) = \sum_{\substack{w', w'' \in W \\ w = w'w''}} c_{w'}(x)c_{w''}(y).$$

For integers $r, s \geq 0$, define $W_{r,s}$ as the set of all words with r letters U and s letters V and put $\mathfrak{J}_{r,s} = \sum_{w\in W_{r,s}} k \cdot w$; for any integer $n \geq 0$, write $\mathfrak{J}_n = \sum_{r+s=n} \mathfrak{J}_{r,s}$, and define $\mathfrak{J}^{(n)}$ to be the set of all x such that $c_w(x) = 0$ whenever length of w is $< n$. For any $x \in \mathfrak{J}$, write $x_{r,s} = \sum_{w\in W_{r,s}} c_w(x)w$, $x_n = \sum_{r+s=n} x_{r,s}$. It is easy to see that $\{\mathfrak{J}^{(m)} : m \geq 0\}$ satisfies the conditions described in Exercise 42. We may thus define $|\cdot|$ on $\mathfrak{J}$ and regard $\mathfrak{J}$ as a topological ring.

(a) Prove that $\mathfrak{J}$ is complete. Prove also that for any $x \in \mathfrak{J}$, $x = \sum_{r,s\geq 0} x_{r,s} = \sum_{n\geq 0} x_n$.

(b) Prove that there exists a unique continuous endomorphism $\partial(U)$ (resp. $\partial(V)$) of $\mathfrak{J}$ such that $\partial(U)x = rx$ if $x \in \mathfrak{J}_{r,s}$ (resp. $\partial(V)x = sx$ if $x \in \mathfrak{J}_{r,s}$). Prove also that $\partial(U)$ and $\partial(V)$ are derivations of $\mathfrak{J}$.

(c) Let $C = C(U:V)$ be the unique element of $\mathfrak{J}^+$ such that $\exp U \exp V = \exp C$. Prove that

$$\partial(U)(C) = -\frac{1}{2}[C,U] + U + \sum_{p\geq 1} K_{2p}\theta(C)^{2p}(U)$$

$$\partial(V)(C) = \frac{1}{2}[C,V] + V + \sum_{p\geq 1} K_{2p}\theta(C)^{2p}(V).$$

From these equations deduce recursion formulae for the $C_{r,s}$ and C_n. Prove that the C_n satisfy the recursion relations (2.15.15) with $C_1 = U + V$.

44. Let notation be as in Exercise 43. For $x, y \in \mathfrak{J}$ put $[x,y] = xy - yx$. $\mathfrak{J}$, equipped with $[\cdot,\cdot]$, is a Lie algebra over k. Denote by $\mathfrak{F}$ the Lie subalgebra of $\mathfrak{J}$ generated by $\{U,V\}$. Let $\mathfrak{F}_{r,s} = \mathfrak{F} \cap \mathfrak{J}_{r,s}$, $\mathfrak{F}_n = \mathfrak{F} \cap \mathfrak{J}_n$. Let ψ be the unique continuous endomorphism of $\mathfrak{J}$ such that $\psi(1) = 0$, $\psi(U) = U$, $\psi(V) = V$, and for any word $w = Z_1 \cdots Z_n$ of length ≥ 2 (i.e., $n \geq 2$ and $Z_i \in \{U,V\}$ for all i), $\psi(w) = [Z_1,[Z_2,[\cdots [Z_{n-1},Z_n]]\cdots]$.

(a) Prove that $\mathfrak{F}$ is spanned by the $\psi(w)$ ($w \in W$) and that $\mathfrak{F}_{r,s}$ (resp. $\mathfrak{F}_n$) is spanned by the $\psi(w)$ with $w \in W_{r,s}$ (resp. $w \in W_n$). Deduce that $\mathfrak{F}$ is the algebraic direct sum of $\mathfrak{F}_{r,s}$ and that $\mathfrak{F}_n = \sum_{r+s=n} \mathfrak{F}_{r,s}$.

(b) Prove that $\psi(w) = nw$ for $w \in \mathfrak{F}_n$. [Hint: This needs the theory of free Lie algebras. See Chapter 3.]

(c) Let $C = C(U:V)$ be as in Exercise 42. Prove that $C_{r,s} \in \mathfrak{F}_{r,s}$ for all r, $s \geq 0$. [Hint: Use Exercise 43 (c)]

(d) Prove that if $r + s > 0$, then

$$(r+s)C_{r,s} = \sum_{m\geq 1} \frac{(-1)^{m-1}}{m} \sum_{\substack{p_1+q_1\geq 1,\ldots,p_m+q_m\geq 1 \\ p_1+\cdots+p_m=r,\, q_1+\cdots+q_m=s}} \frac{\psi(U^{p_1}V^{q_1}\cdots U^{p_m}V^{q_m})}{p_1!q_1!\cdots p_m!q_m!}.$$

(Hint: $C = \log(1+u)$, where $u = \sum_{p+q\geq 1} U^pV^q/p!q!$. So $C_{r,s}$ can be determined by expanding the log. Observe, by (c), that $\psi(C_{r,s}) = (r+s)C_{r,s}$. This formula is due to Dynkin [2]).

45. Use the notation of Exercises 43 and 44. Assume now that $k = \mathbf{R}$ or $\mathbf{C}$. Let $\mathfrak{g}$ be a Lie algebra over k. Suppose that $\|\cdot\|$ is a norm over $\mathfrak{g}$ and that there is a constant $M > 0$ such that $\|[Z,Z']\| \leq M\|Z\|\,\|Z'\|$ for all $Z, Z' \in \mathfrak{g}$. For $X, Y \in \mathfrak{g}$, $c_1(X:Y) = X + Y$, and $c_{n+1}(X:Y)$ ($n \geq 1$) is determined by the recursion relations (2.15.15). Let α be such that $\|X\| < \alpha$, $\|Y\| < \alpha$.

(a) Let $X, Y \in \mathfrak{g}$. Prove the existence of a unique homomorphism $\pi_{X,Y}$ of the Lie algebra $\mathfrak{F}$ into $\mathfrak{g}$ such that $\pi_{X,Y}(U) = X$, $\pi_{X,Y}(V) = Y$.

(b) Prove that $\pi_{X,Y}(C_n(U:V)) = c_n(X:Y)$.

(c) Prove that the series $\sum_{n\geq 0} \|c_n(X:Y)\|$ converges whenever $\|X\| < 1/2M \log 2$ and $\|Y\| < 1/2M \log 2$. (Hint: Use (b) and Exercise 44 (d) to obtain the majorant $\|nc_n(X:Y)\| \leq M^{n-1}\alpha^n\beta_n$, where β_n is the coefficient of z^n in the expansion of $-\log(1-(e^z e^z - 1))$ in powers of z. Observe that this function of z is analytic if $|z| < \frac{1}{2}\log 2$).

(d) Prove that if $\mathfrak{g}$ is finite-dimensional, $c(X:Y) = \sum_{n\geq 0} c_n(X:Y)$ is well defined for $(X,Y) \in \mathfrak{b} \times \mathfrak{b}$, where $\mathfrak{b} = \{Z : Z \in \mathfrak{g}, \|Z\| < 1/2M \log 2\}$, and that $c : (X,Y) \mapsto c(X:Y)$ is an analytic map of $\mathfrak{b} \times \mathfrak{b}$ into $\mathfrak{g}$.

46. Let $k = \mathbf{R}$ or $\mathbf{C}$ and let A be a finite-dimensional associative algebra over k with unit 1. Let $|\cdot|$ be some norm over A. For $x, y \in A$ let $l_x(y) = xy$, $[x,y] = xy - yx$. Write $\mathfrak{g}$ for the Lie algebra thus obtained whose underlying vector space is that of A. Let G be the Lie group of all invertible elements of A. We identify $\mathfrak{g}$ canonically with the Lie algebra of G.

(a) For $x \in A$, let $\|x\| = \sup_{|y|\leq 1} |l_x(y)|$. Prove that $\|\cdot\|$ is an equivalent norm on A and $\|xy\| \leq \|x\|\,\|y\|$ for all $x, y \in A$.

(b) Prove that the exponential map of $\mathfrak{g}$ into G is given by $\exp x = \sum_{n\geq 0} x^n/n!$ $(x \in A)$.

(c) For any $\epsilon > 0$, let $G(\epsilon) = \{y : y \in A, \|y - 1\| < \epsilon\}$ and $\mathfrak{g}(\epsilon) = \{x : x \in A, \|x\| < \epsilon\}$. For $y \in G(1)$, let $\sum_{n\geq 1} (-1)^{n-1}(y-1)^n/n$. Prove that $G(1) \subseteq G$ and that $y \mapsto \log y$ is a well-defined analytic map of $G(1)$ into $\mathfrak{g}$.

(d) Prove that $\exp(\log y) = y$ for $y \in G(1)$; prove further that if $\|x\| < \log 2$, then $\exp x \in G(1)$ and $\log(\exp x) = x$.

(e) Show that $\log[G(1)] = \mathfrak{u}$ is an open neighborhood of 0 in A, that log is an analytic diffeomorphism of $G(1)$ onto $\mathfrak{u}$ with exp as its inverse, and that $\mathfrak{u}$ is the connected component containing 0 of the open set $\exp^{-1}(G(1))$.

(f) For $x, y \in \mathfrak{g}$, define $c_n(x:y)$ $(n \geq 1)$ by the recursion formulae (2.15.15). Prove that the series $\sum_{n\geq 1} c_n(x:y)$ converges absolutely for $(x,y) \in \mathfrak{c} = \mathfrak{g}(\frac{1}{4}\log 2) \times \mathfrak{g}(\frac{1}{4}\log 2)$; if $c(x,y)$ is its sum, prove further that c is analytic on $\mathfrak{c}$ and that $\exp x \exp y = \exp c(x:y)$ for all $(x,y) \in \mathfrak{c}$.

47. Let G be an analytic group with Lie algebra $\mathfrak{g}$, M an analytic manifold, and τ $(X \mapsto \tau(X))$ an infinitesimal G-transformation group on M. Let A be the set of all $X \in \mathfrak{g}$ for which $\tau(X)$ is global.

(a) Prove that if $X \in A$ and $t \in \mathbf{R}$, then $tX \in A$.

(b) Let $X, Y \in A$ and let ξ (resp. η) be the one-parameter group of diffeomorphisms of M that is generated by $\tau(X)$ (resp. $\tau(Y)$). If $s \in \mathbf{R}$ and $\zeta(t) = \eta(s) \circ \eta(t) \circ \xi(-s)$ $(t \in \mathbf{R})$, prove that the one-parameter group ζ is generated by $\tau(Y^{\exp sX})$. Deduce that $Y^{\exp X} \in A$. (Hint: Use Lemmas 2.16.10 and 2.16.11.)

(c) Let $\mathfrak{h}$ be the linear subspace of $\mathfrak{g}$ spanned by A. Prove that $\mathfrak{h}$ is a subalgebra. (Hint: If $X, Y \in A$, $[X,Y] = ((d/ds)Y^{\exp sX})_{s=0} \in \mathfrak{h}$.)

(d) Suppose that A generates $\mathfrak{g}$. Prove that there exists a global G-transformation group φ on M such that $\tau_\varphi = \tau$. (Hint: By (c), we can choose a basis $\{X_1, \ldots, X_n\}$ for $\mathfrak{g}$ such that $X_i \in A$ for all i. From this point the proofs of Lemma 2.16.12 and Theorem 2.16.13 need no change.)

(e) Suppose that $Z_1, \ldots, Z_k$ are analytic vector fields on M. If each Z_i is global and if the Z_i generate a finite-dimensional subalgebra $\mathfrak{a}$ of $\mathfrak{Z}_a(M)$, show that each vector field in $\mathfrak{a}$ is global.

(f) Let $M = \mathbf{R}^2$ with the usual coordinates x and y. Write

$$Z_1 = y\frac{\partial}{\partial x}, \qquad Z_2 = \frac{x^2}{2}\frac{\partial}{\partial y}.$$

Show that Z_1 and Z_2 are global but that neither $Z_1 + Z_2$ nor $[Z_1,Z_2]$ is global.

48. (a) Let G be an n-dimensional analytic group, H a discrete subgroup, and $M = G/H$. Prove the existence of analytic vector fields $X_1, \ldots, X_n$ on M such that (i) the X_i are linearly independent at each point of M, and (ii) there are constants c_{ijk} such that $[X_i,X_j] = \sum_k c_{ijk}X_k$ for all i, j.

(b) Suppose M is a compact n-dimensional connected analytic manifold on which there exist analytic vector fields $X_1, \ldots, X_n$ having properties (i) and (ii) of (a). Prove that M is analytically diffeomorphic to a quotient manifold of the form G/H, where G is an n-dimensional analytic group and H is a discrete subgroup of it. (Hint: Use the global form of the third fundamental theorem of Lie to construct an analytic Lie group G and a basis $\bar{X}_1, \ldots, \bar{X}_m$ of its Lie algebra $\mathfrak{g}$ such that $[\bar{X}_i,\bar{X}_j] = \sum_k c_{ijk}\bar{X}_k$ for all i, j. Let τ be the infinitesimal G-transformation group on M for which $\tau(\bar{X}_i) = X_i$ for all i. Now use Theorem 2.16.9.)

(c) Suppose M is a compact, connected, complex analytic manifold of dimension n admitting n holomorphic vector fields that are linearly independent at each point. Prove that M is holomorphically diffeomorphic to a quotient manifold of the form G/D, where G is a complex analytic group of dimension n and D is a discrete subgroup of it.

49. (a) Let G be a Lie group and H a closed Lie subgroup. Let $\mathfrak{g}$ be the Lie algebra of G, and let $\mathfrak{h}$ be the subalgebra of $\mathfrak{g}$ defined by H. Let $\mathfrak{a}$ be a linear subspace of $\mathfrak{g}$ complementary to $\mathfrak{h}$. Prove the existence of an open neighborhood $\mathfrak{u}$ of 0 in $\mathfrak{a}$ such that (i) $U = \exp[\mathfrak{u}]$ is a regular submanifold of G and exp is an analytic diffeomorphism of $\mathfrak{u}$ onto U, and (ii) U meets each left H-coset at most once. Use this to obtain an alternative proof of Theorem 2.9.4.

(b) Assume that H is connected; prove that for H to be an isolated fixed point for the action of H on G/H it is necessary and sufficient that $\mathfrak{h}$ be its own normalizer in $\mathfrak{g}$, i.e., that $\mathfrak{h} = \{X : X \in \mathfrak{g}, [X,\mathfrak{h}] \subseteq \mathfrak{h}\}$. (Hint: Let N be the normalizer of H. Prove that H is open in N.)

50. Let G be an analytic group over k ($= \mathbf{R}$ or $\mathbf{C}$), $\mathfrak{g}$, the Lie algebra of G. Identify $T_g(G)$ and $T_X(\mathfrak{g})$ with $\mathfrak{g}$ for any $g \in G$, $X \in \mathfrak{g}$ in the usual manner. Suppose $\mathfrak{u}$ is an open neighborhood of 0 in $\mathfrak{g}$ and φ is a C^∞ map of $\mathfrak{u}$ into G such that (i) $\varphi(0) = 1$ and $(d\varphi)_0(X) = X \;\forall\; X \in \mathfrak{g}$, and (ii) if $X \in \mathfrak{g}$ and $t, t' \in k$ are such that tX, $t'X$, and $(t + t')X \in \mathfrak{u}$, then $\varphi((t + t')X) = \varphi(tX)\varphi(t'X)$. Then prove that φ coincides with the exponential map in a neighborhood of 0.

CHAPTER 3

STRUCTURE THEORY

This chapter will be devoted to the development of the general structure theory of Lie algebras and its group-theoretic consequences. Among other things, we shall prove the theorems of Levi-Mal'čev, Weyl, and Ado, as well as the global version of the third fundamental theorem of Lie, namely, that there is an analytic group corresponding to every real or complex Lie algebra. In addition, we shall develop the theory of the universal enveloping algebra in a systematic fashion; this will be of great use in our subsequent discussion of the theory of semisimple Lie groups and their representations.

3.1. Review of Linear Algebra

The aim of this section is to give a brief review of certain concepts and results from linear algebra which will be needed later. The reader who is interested in detailed proofs is referred to standard texts (Jacobson [1], Bourbaki [1, 2], Chevalley [2]).

Throughout this section, k will denote a field of characteristic zero. We recall that if k' is either an algebraically closed extension or a Galois extension[1] of k and Γ the corresponding Galois group, then $k = \{c : c \in k', c^s = c$ for all $s \in \Gamma\}$. Let V be a vector space of finite dimension over k, k' an extension of k; then we write $V^{k'}$ for the canonical extension of V to a vector space over k'. We always identify V with a subset of $V^{k'}$. If k'' is an extension of k with $k \subseteq k'' \subseteq k'$, we may regard $V^{k'}$ as the canonical extension of $V^{k''}$ to a vector space over k'. For any subspace U (resp. endomorphism L) of V, we write $U^{k'}$ (resp. $L^{k'}$) for the subspace of $V^{k'}$ spanned by U (resp. endo-

[1] k' is not necessarily a finite extension of k. So k' Galois over k means: (i) k' is algebraic over k, and (ii) if $\bar{k}$ is an algebraic closure of k' with $k' \subseteq \bar{k}$, then any k-isomorphism of k' into $\bar{k}$ is a k-automorphism of k''. If V is a vector space over k and $s(c \mapsto c^s)$ is an automorphism of k, then by an s-linear automorphism of V is meant an automorphism L of the additive group of V such that $L(cv) = c^s Lv$ for $v \in V$, $c \in k$.

morphism of $V^{k'}$ extending L). Such subspaces and endomorphisms of $V^{k'}$ are said to be *defined over* k. If k' is either an algebraically closed extension or a Galois extension of k and Γ is the Galois group of k' over k, then for any $s \in \Gamma$ there is a unique (s-linear) automorphism s of the additive group of $V^{k'}$ such that $s \cdot v = v$ for all $v \in V$. If U' is any subspace of $V^{k'}$ (resp. L' is any endomorphism of $V^{k'}$), then U' (resp. L') is defined over k if and only if $s \cdot U' = U'$ (resp. $s \cdot L' \cdot s^{-1} = L'$) for all $s \in \Gamma$; in this case, $U' = U^{k'}$, where $U = U' \cap V$ (resp. $L' = L^{k'}$, where $L = L' \mid V$).

Let V be of finite dimension m over k. An endomorphism L of V is called *nilpotent* if for each $v \in V$ there is an integer $n = n(v) > 0$ such that $L^n v = 0$. In this case there is an integer $r > 0$ such that $L^r = 0$. If L is nilpotent, the subspace V_1 of vectors v such that $Lv = 0$ is nonzero, and L induces a nilpotent endomorphism on V/V_1. It follows that we can find subspace $V_0 = 0$, $V_1, \ldots, V_s = V$ $(1 \leq s \leq m)$ such that (i) $V_0 \subseteq V_1 \subseteq \cdots \subseteq V_s$, $\dim V_i < \dim V_{i+1}$ $(0 \leq i \leq s-1)$, and (ii) $V_{i+1} = \{v : v \in V, Lv \in V_i\}$ $(0 \leq i \leq s-1)$. In particular, $L^m = 0$, and there is a basis for V with respect to which the matrix of L has zeros on and below the main diagonal. If k' is an extension of k, an endomorphism L is nilpotent if and only if $L^{k'}$ is. If L is nilpotent, $\operatorname{ad} L : M \mapsto [L,M]$ is a nilpotent endomorphism of $\mathfrak{gl}(V)$ (cf. (2.14.4)).

Let T be an indeterminate and $k[T]$ the algebra of all polynomials in T with coefficients from k. If L is any endomorphism of V and $p = a_0 T^N + a_1 T^{N-1} + \cdots + a_N \in k[T]$, write $p(L) = a_0 L^N + a_1 L^{N-1} + \cdots + a_N 1$. The action $(p,v) \mapsto p \cdot v = p(L)v$ $(p \in k[T], v \in V)$ then converts V into a module over the principal ideal domain $k[T]$. The theory of modules over such rings leads quickly to the main theorems concerning L and its action on V.

The first theorem is the so-called decomposition into primary components. For any $q \in k[T]$ let $V(q)$ be the subspace $\{v : v \in V, q(L)v = 0\}$, and let V_q denote the subspace $\bigcup_{n \geq 1} V(q^n)$.

Theorem 3.1.1. *Let J be the ideal in $k[T]$ of all q such that $q(L) = 0$, and let p be the monic*[2] *polynomial which generates J. Write $p = q_1^{m_1} \cdots q_r^{m_r}$, where $m_1, \ldots, m_r$ are integers >0 and $q_1, \ldots, q_r$ are distinct, monic, irreducible polynomials in T. Then $V = \sum_{1 \leq i \leq r} V(q_i^{m_i})$, the sum being direct, and $V_{q_i} = V(q_i^{m_i})$, $1 \leq i \leq r$. Moreover, there are elements $p_i \in k[T]$ such that $p_i(L)$ is the projection of V on V_{q_i} corresponding to the above direct sum.*

This theorem takes a simpler form when k is algebraically closed. Let us assume this and write $\sigma(L)$ for the set of eigenvalues of L. For $\lambda \in \sigma(L)$ let $V_{L,\lambda} = V_{T-\lambda}$, i.e.,

$$V_{L,\lambda} = \{v : v \in V, (L - \lambda \cdot 1)^s v = 0 \text{ for some integer } s > 0\}. \tag{3.1.1}$$

[2]This means that the highest power of T in p has coefficient 1. p is called the *minimal polynomial* of the endomorphism L.

Then the following theorem follows at once from Theorem 3.1.1.

Theorem 3.1.2. *Let k be algebraically closed. Then V is the direct sum of $V_{L,\lambda}$ ($\lambda \in \sigma(L)$), and there are $p_\lambda \in k[T]$ such that $p_\lambda(L)$ is the projection of V onto $V_{L,\lambda}$ corresponding to the direct sum. The dimension of $V_{L,\lambda}$ is equal to the multiplicity of λ as a root of the characteristic polynomial of L. The restriction of $L - \lambda \cdot 1$ to $V_{L,\lambda}$ is nilpotent.*

The second main result is the construction of the so-called invariant factors. For any $v \in V$, let $J_v = \{p : p \in k[T], p(L)v = 0\}$ and $[v] = \{p(L)v : p \in k[T]\}$.

Theorem 3.1.3. *There are ideals $J_1, \ldots, J_n$ of $k[T]$ such that* (i) $0 \neq J_1 \subseteq J_2 \cdots \subseteq J_n \neq k[T]$, *and* (ii) *for suitable vectors $v_1, \ldots, v_n$ of V, V is the direct sum of the subspaces $[v_1], \ldots, [v_n]$, and $J_{v_i} = J_i$ for $1 \leq i \leq n$. If $I_1, \ldots, I_r$ are ideals in $k[T]$ satisfying* (i) *and* (ii) (*for suitable vectors $w_1, \ldots, w_r$*), *then $r = n$ and $J_s = I_s$ for $1 \leq s \leq n$.*

It is obvious that $1 \leq n \leq m$. Let p_i $(1 \leq i \leq n)$ be monic polynomials generating the ideal J_i. Then $p_1, \ldots, p_n$ form a complete set of invariants for the action of L on V.

Let $\bar{k}$ be an algebraic closure of k, with $k \subseteq \bar{k}$. An endomorphism L of V is said to be *semisimple* if $V^{\bar{k}}$ is the direct sum of eigensubspaces of $L^{\bar{k}}$, or equivalently, if there is a basis of $V^{\bar{k}}$ with respect to which $L^{\bar{k}}$ has a diagonal matrix. It is known that if the characteristic equation of L has $m = \dim(V)$ distinct roots in $\bar{k}$, then L is semisimple. Moreover, we have the following very useful theorem.

Theorem 3.1.4. *Let L be an endomorphism of V. Then $L = S + N$, where* (i) *S is semisimple and N is nilpotent, and* (ii) *S and N commute. S and N are uniquely determined by these requirements, and we can find elements p, q of $k[T]$ whose constant terms are zero, such that $S = p(L)$, $N = q(L)$. In particular, L is semisimple if and only if given any subspace of V invariant under L, one can find a complementary subspace that is also invariant under L.*

We remark that if k is algebraically closed and the P_λ are the projections $V \longrightarrow V_{L,\lambda}$, then $S = \sum_{\lambda \in \sigma(L)} \lambda P_\lambda$, and $N = L - S$. It is obvious that $\sigma(L) = \sigma(S)$. For arbitrary k, the decomposition $L = S + N$ is known as the *Jordan decomposition* of L, and S and N are known respectively as the *semisimple* and *nilpotent parts* of L.

If k is not algebraically closed, the decomposition of Theorem 3.1.2 is available for $L^{\bar{k}}$ in $V^{\bar{k}}$, $\bar{k} \supseteq k$ being an algebraic closure of k. Moreover, for L itself the following theorem (the so-called Fitting decomposition) gives a partial substitute.

Theorem 3.1.5. *Let k be arbitrary. For any endomorphism M of V, let $R(M)$ be the range of M, and $N(M)$ the null space of M. Let L be an endomorphism of V and let*

$$N^+ = \bigcup_{n \geq 1} N(L^n)$$
$$R^- = \bigcap_{n \geq 1} R(L^n). \tag{3.1.2}$$

Then V is the direct sum of N^+ and R^-, both of which are invariant subspaces for any endomorphism M with the property that $(\mathrm{ad}\, L)^s(M) = 0$ for some integer $s > 0$. If L is semisimple, $N^+ = N(L)$ and $R^- = R(L)$.

We conclude this review of the theory of a single endomorphism with two results. The first of these follows simply from Theorem 3.1.2; the second comes out of Theorem 3.1.3.

Theorem 3.1.6. *Let k be algebraically closed, L an endomorphism of V. In order that an endomorphism M leave the subspaces $V_{L,\lambda}$ ($\lambda \in \sigma(L)$) invariant, it is necessary and sufficient that there exist an integer $s > 0$ with $(\mathrm{ad}\, L)^s$ $(M) = 0$.*

Theorem 3.1.7. *Let k be arbitrary, L an endomorphism of V. Denote by A the algebra of all endomorphisms of V of the form $p(L)$, $p \in k[T]$. Then an endomorphism M of V belongs to A if and only if it satisfies the following condition: M commutes with every endomorphism of V that commutes with L.*

Some of the above concepts admit far-reaching generalizations to sets of (possibly noncommuting) endomorphisms of V. We now turn to a description of a few results of this type. It is convenient to formulate these in the context representation of associative algebras.

Let $\mathfrak{A}$ be an associative algebra[3] over k. By a *representation* ρ of $\mathfrak{A}$ we mean a homomorphism of $\mathfrak{A}$ into the associative algebra of endomorphisms of a vector space W over k with $\rho(1) = 1$; ρ is said to *act on* W. Unless we state otherwise, we shall be concerned only with finite-dimensional representations, i.e., those that act on finite-dimensional vector spaces. Let ρ be a representation of $\mathfrak{A}$ acting on W and k' an extension field of k. Denoting by $\mathfrak{A}^{k'}$ the extension of $\mathfrak{A}$ to an algebra over k', it is obvious that there is a unique representation ρ' of $\mathfrak{A}^{k'}$ acting on $W^{k'}$ such that $\rho'(a) = \rho(a)^{k'}$ for all $a \in \mathfrak{A}$; ρ' is said to be the *extension of ρ to k'* and is denoted by $\rho^{k'}$. If ρ and ρ' are representations of $\mathfrak{A}$ acting on W and W' respectively, we say that ρ and ρ' are *equivalent* if there is a linear isomorphism ξ of W onto W' such that

[3] We assume that $\mathfrak{A}$ has a unit, denoted by 1. Any subalgebra of $\mathfrak{A}$ contains 1 by definition. This convention is in force throughout this chapter. If $B \subset \mathfrak{A}$, the *subalgebra* $\mathfrak{B}$ *generated by* B is the smallest subalgebra of $\mathfrak{A}$ containing B; B is said to *generate* $\mathfrak{A}$ if $\mathfrak{B} = \mathfrak{A}$.

$\xi\rho(a) = \rho'(a)\xi$ for all $a \in \mathfrak{A}$; ξ is then said to *intertwine* ρ and ρ'. For any representation ρ of $\mathfrak{A}$, define the *character* of ρ to be the linear function $\chi_\rho : a \mapsto tr\ \rho(a)$ $(a \in \mathfrak{A})$ on $\mathfrak{A}$ with values in k. It is obvious that $\chi_\rho(ab) = \chi_\rho(ba)$ for all $a, b \in \mathfrak{A}$ and that equivalent representations have the same characters.

Given a representation ρ of $\mathfrak{A}$ on W, a subspace W' of W is said to be *invariant* (under ρ) if $\rho(a)[W'] \subseteq W'$ for all $a \in \mathfrak{A}$. Let W' be an invariant subspace, and let us write $\rho_{W'}(a)$ (resp. $\rho_{W/W'}(a)$) for the endomorphism induced by $\rho(a)$ on W' (resp. W/W'), for any $a \in \mathfrak{A}$. Then $\rho_{W'} : a \mapsto \rho_{W'}(a)$ (resp. $\rho_{W/W'} : a \mapsto \rho_{W/W'}(a)$) is a representation of $\mathfrak{A}$ in W' (resp. W/W'); it is called the *subrepresentation* (resp. *quotient representation*) defined by W'. A representation ρ on W is said to be *irreducible* if 0 and W are the only subspaces of W that are invariant under ρ; ρ is called *semisimple* if given any subspace of W invariant under ρ, one can find a subspace complementary to it and invariant under ρ.

Suppose ρ is a representation of $\mathfrak{A}$ in W. Then there exist subspaces $W_0 = 0, W_1, \ldots, W_s = W$ such that (i) $W_i \subseteq W_{i+1}$ and $\dim W_i < \dim W_{i+1}$ $(0 \leq i \leq s-1)$, and (ii) each W_i is ρ-invariant and the representations $\rho_i = \rho_{W_{i+1}/W_i}$ are irreducible $(0 \leq i \leq s-1)$. The integer s is the same for all such chains. If $\{\rho_i\}$ and $\{\sigma_i\}$ are the irreducible representations associated with two chains, there is a permutation $i \mapsto i'$ of $\{1, \ldots, s\}$ such that ρ_i and $\sigma_{i'}$ are equivalent for all i. If ρ is semisimple we can write W as the direct sum of subspaces $W_1, \ldots, W_s$, where the W_i are invariant under ρ and the representations ρ_{W_i} are irreducible, $1 \leq i \leq s$. In this case, for any equivalence class $\mathfrak{v}$ of irreducible representations of $\mathfrak{A}$, the number of i's such that $\rho_{W_i} \in \mathfrak{v}$ is called the *multiplicity* of $\mathfrak{v}$ in ρ and is denoted by $(\rho : \mathfrak{v})$. Even though the W_i are in general not unique, for any equivalence class $\mathfrak{v}$ of irreducible representations, the number $(\rho : \mathfrak{v})$ and the subspace $W_\mathfrak{v} = \sum_{i:\rho_{W_i} \in \mathfrak{v}} W_i$ are uniquely determined once ρ is given.

Let ρ be an irreducible representation of $\mathfrak{A}$ on W. Denote by $\mathfrak{B}$ the algebra of all endomorphisms B of W such that $\rho(a)B = B\rho(a)$ for all $a \in \mathfrak{B}$. It is easily seen that if $B \in \mathfrak{B}$ and $B \neq 0$, B is invertible and $B^{-1} \in \mathfrak{B}$. $\mathfrak{B}$ is thus a division algebra over k. If k is algebraically closed, the eigenspaces of elements of $\mathfrak{B}$ are invariant under ρ, leading to the conclusion that $\mathfrak{B} = k \cdot 1$. These results constitute what is known as *Schur's lemma*. For arbitrary k, the action $(B,v) \mapsto Bv$ $(B \in \mathfrak{B}, v \in W)$ converts W into a left vector space over $\mathfrak{B}$. Let $W^\mathfrak{B}$ denote this left vector space. Then for any $a \in \mathfrak{A}$, $\rho(a)$ is an endomorphism of $W^\mathfrak{B}$.

Theorem 3.1.8. *Let notation be as above. Then $\rho[\mathfrak{A}]$ is the set of all endomorphisms of $W^\mathfrak{B}$. In particular, if k is algebraically closed, then $W^\mathfrak{B} = W$ and $\rho[\mathfrak{A}]$ is the set of all endomorphisms of W.*

There is a generalization of this theorem to the case when ρ is assumed only to be semisimple. In order to formulate it we need some notation. Let D be a set of endomorphisms of a vector space W. Denote by D' the set of all endomorphisms B of W such that $AB = BA$ for all $A \in D$. We write $D'' = (D')'$. D' and D'' are both algebras over k, and $D \subseteq D''$. We then have the following theorem.

Theorem 3.1.9. *Let ρ be a semisimple representation of $\mathfrak{A}$ acting on W. Then $\rho[\mathfrak{A}] = \rho[\mathfrak{A}]''$. In particular, let $\mathfrak{v}$ be an equivalence class of irreducible representations of $\mathfrak{A}$, and let $W_{\mathfrak{v}}$ be the linear span of all invariant subspaces W' of W such that $\rho_{W'} \in \mathfrak{v}$; then $W_{\mathfrak{v}}$ admits exactly one complementary subspace invariant under ρ, and there is an element $z_{\mathfrak{v}} \in \mathfrak{A}$ such that $\rho(z_{\mathfrak{v}})$ is the projection of W onto $W_{\mathfrak{v}}$ parallel to this complementary subspace.*

We remark that if ρ is a representation of $\mathfrak{A}$ in W, and k' is an extension field of k, then ρ is semisimple if and only if $\rho^{k'}$ is semisimple. However, it might happen that ρ is irreducible without $\rho^{k'}$ being so. If ρ is irreducible, then in order that $\rho^{k'}$ remain irreducible for all extension fields k' of k, it is necessary and sufficient that $\rho[\mathfrak{A}]' = k \cdot 1$.

In the sequel we shall on occasion deal with infinite-dimensional representations of $\mathfrak{A}$. For such representations, the concepts of irreducibility and equivalence are defined as in the finite-dimensional case. However, care must be exercised in handling them.

Let ρ be a representation of $\mathfrak{A}$ in a vector space V, not necessarily of finite dimension. Let $v \in V$ be a nonzero cyclic vector for ρ, i.e., $V = \rho[\mathfrak{A}]v$. If

$$\mathfrak{M}_v = \{a : a \in \mathfrak{A}, \rho(a)v = 0\},$$

it is clear that $\mathfrak{M}_v$ is a left ideal of $\mathfrak{A}$. Let $\bar{\mathfrak{A}}_v = \mathfrak{A}/\mathfrak{M}_v$ be the quotient vector space and let $a \mapsto \bar{a}$ be the canonical map of $\mathfrak{A}$ onto $\bar{\mathfrak{A}}_v$; for $a,a' \in \mathfrak{A}$, let $\rho_v(a)\cdot\bar{a}' = \overline{aa'}$. Then $\rho_v(a)$ is an endomorphism of $\bar{\mathfrak{A}}_v$, and $a \mapsto \rho_v(a)$ is a representation of $\mathfrak{A}$. It is obvious that the map

$$\xi : \bar{a} \mapsto \rho(a)v \quad (a \in \mathfrak{A})$$

is a well-defined linear isomorphism of $\bar{\mathfrak{A}}_v$ onto V that intertwines the representations ρ_v and ρ; i.e.,

$$\xi \circ \rho_v(a) \circ \xi^{-1} = \rho(a) \quad (a \in \mathfrak{A}).$$

If W is any subspace $\neq V$ of V invariant under ρ, then the set

$$\mathfrak{M}^W = \{a : a \in \mathfrak{A}, \rho(a)v \in W\}$$

is a left ideal of $\mathfrak{A}$ containing $\mathfrak{M}_v$, and the correspondence $W \mapsto \mathfrak{M}^W$ is a bijection of the set of all ρ-invariant subspaces $\neq V$ of V with the set of all left ideals of $\mathfrak{A}$ containing $\mathfrak{M}_v$; obviously, if $W_1 \subseteq W_2$, then $\mathfrak{M}^{W_1} \subseteq \mathfrak{M}^{W_2}$. In particular, ρ is irreducible if and only if $\mathfrak{M}_v$ is a maximal left ideal. In this case, every vector of V is cyclic, and

$$\bigcap_{w \in V} \mathfrak{M}_w = \text{kernel}(\rho).$$

Conversely, let $\mathfrak{M}$ be a left ideal of $\mathfrak{A}$ ($1 \notin \mathfrak{M}$); let $\bar{\mathfrak{A}} = \mathfrak{A}/\mathfrak{M}$ and let $a \mapsto \bar{a}$ be the canonical map of $\mathfrak{A}$ onto $\bar{\mathfrak{A}}$, for $a, b \in \mathfrak{A}$, let $\rho(a)\cdot\bar{b} = \overline{ab}$. Then $a \mapsto \rho(a)$ is a representation of $\mathfrak{A}$ in $\bar{\mathfrak{A}}$, and $\bar{1}$ is cyclic for ρ. ρ is called the *natural representation* of $\mathfrak{A}$ in $\bar{\mathfrak{A}}$. ρ is irreducible if and only if $\mathfrak{M}$ is maximal. Clearly, $\mathfrak{M} = \mathfrak{M}_{\bar{1}}$ in the earlier notation.

We now give a brief account of Chevalley's theory of replicas of endomorphisms. This plays an important role in Chevalley's theory of algebraic groups over fields of characteristic zero (cf. Chevalley [2,5] Lazard [1]).

Let V be a vector space over k of finite dimension m, V^* its dual. For integers $r, s \geq 0$ with $r + s > 0$, let $V_{r,s}$ be given by

$$V_{r,s} = \underbrace{V \otimes \cdots \otimes V}_{r \text{ times}} \otimes \underbrace{V^* \otimes \cdots \otimes V^*}_{s \text{ times}}. \tag{3.1.3}$$

Given any $x \in GL(V)$, let x^0 be the element of $GL(V^*)$ defined by

$$(x^0 v^*)(v) = v^*(x^{-1}v) \quad (v \in V, v^* \in V^*). \tag{3.1.4}$$

Then it is obvious that $x \mapsto x^0$ is an isomorphism of $GL(V)$ onto $GL(V^*)$. For r, s as above and $x \in GL(V)$, let

$$x^{(r,s)} = \underbrace{x \otimes \cdots \otimes x}_{r \text{ times}} \otimes \underbrace{x^0 \otimes \cdots \otimes x^0}_{s \text{ times}} \tag{3.1.5}$$

Obviously, $x \mapsto x^{(r,s)}$ is a homomorphism of $GL(V)$ into $GL(V_{r,s})$. If $k = \mathbf{R}$ or $\mathbf{C}$, then $x \mapsto x^{(r,s)}$ is a representation of $GL(V)$ acting in $V_{r,s}$, and as such admits a differential, which is a representation of $\mathfrak{gl}(V)$ in $V_{r,s}$. A simple calculation reveals that the resulting representation of $\mathfrak{gl}(V)$ makes sense in an arbitrary field of characteristic zero. We now define it directly. For any $L \in \mathfrak{gl}(V)$, let L be the element of $\mathfrak{gl}(V^*)$ given by

$$(\hat{L}v^*)(v) = -v^*(Lv) \quad (v \in V, v^* \in V^*). \tag{3.1.6}$$

Then it is easily seen that $L \mapsto \hat{L}$ is an isomorphism of $\mathfrak{gl}(V)$ onto $\mathfrak{gl}(V^*)$. For any integers r, s, both >0, and any endomorphism L of V, let

$$\begin{aligned} L_{r,0} &= L \otimes 1 \otimes 1 \otimes \cdots \otimes 1 \\ &\quad + 1 \otimes L \otimes 1 \otimes \cdots \otimes 1 + \cdots + 1 \otimes 1 \otimes \cdots \otimes 1 \otimes L, \end{aligned} \tag{3.1.7}$$

where each tensor product has r factors, and let

$$L_{0,s} = (\hat{L})_{s,0}. \tag{3.1.8}$$

Then $L \mapsto L_{r,0}$ and $L \mapsto L_{0,s}$ are representation of $\mathfrak{gl}(V)$ in $V_{r,0}$ and $V_{0,s}$, respectively (cf. §2.2). Define, for integers $r, s \geq 0$ with $r + s > 0$,

$$L_{r,s} = L_{r,0} \otimes 1 + 1 \otimes L_{0,s} \quad (L \in \mathfrak{gl}(V)). \tag{3.1.9}$$

Write $V_{0,0} = k \cdot 1$, and define $L_{0,0} = 0$ for all $L \in \mathfrak{gl}(V)$, $x^{(0,0)} = 1$ for all $x \in GL(V)$. We leave it to the reader to verify that $L \mapsto L_{r,s}$ is a representation of $\mathfrak{gl}(V)$ in $V_{r,s}$, and that if $k = \mathbf{R}$ or C, this representation is the differential of the representation $x \mapsto x^{(r,s)}$ of $GL(V)$ in $V_{r,s}$.

Note finally that if k' is an extension field of k, $(V_{r,s})^{k'}$ is canonically isomorphic to $(V^{k'})_{r,s}$ for all $r, s \geq 0$. If L is an endomorphism of V, then the endomorphisms $(L_{r,s})^{k'}$ and $(L^{k'})_{r,s}$ correspond under this isomorphism. We shall generally identify $(V_{r,s})^{k'}$ (resp. $(L_{r,s}))^{k'}$ with $(V^{k'})_{r,s}$ (resp. $(L^{k'})_{r,s}$) without comment.

The spaces $V_{1,1}$ and $V_{1,2}$ have useful and interesting interpretations. Given $v \in V$, $v^* \in V^*$, let us define $z(v,v^*)$ to be the endomorphism of V given by

$$z(v,v^*)(w) = v^*(w)v \quad (w \in V). \tag{3.1.10}$$

From the properties of tensor products it follows that there is a linear map ζ of $V \otimes V^*$ into $\mathfrak{gl}(V)$ such that

$$\zeta(v \otimes v^*) = z(v,v^*) \quad (v \in V, v^* \in V^*). \tag{3.1.11}$$

If $\{e_1, \ldots, e_m\}$ is a basis for V and $\{e_1^*, \ldots, e_m^*\}$ the dual basis in V^*, it follows from (3.1.11) that $\zeta(e_l \otimes e_p^*)$ is the endomorphism of V whose matrix relative to $\{e_1, \ldots, e_m\}$ is $(\delta_{il}\delta_{pj})_{1 \leq i,j \leq m}$. This shows that ζ is a linear isomorphism of $V_{1,1}$ with the underlying vector space of $\mathfrak{gl}(V)$. In an analogous manner, let $v \in V$, $v_1^*, v_2^* \in V^*$, and let $y(v,v_1^*,v_2^*)$ be the bilinear map of $V \times V$ into V defined by

$$y(v,v_1^*,v_2^*)(u_1,u_2) = v_1^*(u_1)v_2^*(u_2)v \quad (u_1,u_2) \in V. \tag{3.1.12}$$

We write η for the linear map of $V_{1,2}$ into the vector space B of bilinear maps of $V \times V$ into V such that

$$\eta(v \otimes v_1^* \otimes v_2^*) = y(v,v_1^*,v_2^*) \quad (v \in V, v_1^*,v_2^* \in V). \tag{3.1.13}$$

A simple calculation, analogous to the one performed earlier, shows that η is a linear isomorphism of $V_{1,2}$ onto B. We now have the following lemma.

Lemma 3.1.10. *Let notation be as above. Then for any endomorphism L of V, ζ intertwines the actions of $L_{1,1}$ and ad L; i.e.,*

$$\zeta \circ L_{1,1} \circ \zeta^{-1} = \text{ad } L. \tag{3.1.14}$$

Further, let $\beta \in B$, and let V_β be the algebra whose underlying vector space is V and whose multiplication law is given by $x \cdot y = \beta(x,y)$, $x, y \in V$. Then an endomorphism L is a derivation of V_β if and only if

$$L_{1,2}(\eta^{-1}(\beta)) = 0. \tag{3.1.15}$$

Proof. Let $v, w \in V$, $v^* \in V^*$. Then $[L,\zeta(v \otimes v^*)](w) = v^*(w)Lv - v^*(Lw)v$, so $(\text{ad } L)(\zeta(v \otimes v^*)) = (\zeta \circ L_{1,1} \circ \zeta^{-1})(\zeta(v \otimes v^*))$. From this we get (3.1.14). For the second assertion it is enough to prove that if $t \in V_{1,2}$ and $\beta = \eta(t)$, then for any endomorphism L of V, the element $\eta(L_{1,2}(t))$ of B is given by

$$\eta(L_{1,2}(t))(x,y) = L\beta(x,y) - \beta(Lx,y) - \beta(x,Ly) \tag{3.1.16}$$

for all $x, y \in V$. A trivial calculation shows that this is true for all $t = v \otimes v_1^* \otimes v_2^*$, with $v \in V$, $v_1^*, v_2^* \in V^*$. By linearity, the validity of (3.1.16) for all $v \in V_{1,2}$ follows at once.

Let L be any endomorphism of V. Following Chevalley, we say that an endomorphism M of V is a *replica* of L if for all integers $r, s \geq 0$

$$u \in V_{r,s}, L_{r,s}u = 0 \Longrightarrow M_{r,s}u = 0. \tag{3.1.17}$$

We then define

$$\mathfrak{g}(L) = \{M : M \in \mathfrak{gl}(V), M \text{ is a replica of } L\}. \tag{3.1.18}$$

We begin with two elementary properties of replicas of endomorphisms.

Lemma 3.1.11. *Let L be semisimple (resp. nilpotent). Then $L_{r,s}$ is semisimple (resp. nilpotent) for all $r, s \geq 0$. In particular, ad L is semisimple (resp. nilpotent).*

Proof. If M_1, M_2 are nilpotent endomorphisms on a vector space of dimension n, and $[M_1,M_2] = 0$, then $(M_1 + M_2)^{2n} = 0$. In particular, if N_i are nilpotent endomorphism of a vector space W_i, $i = 1, 2$, then $N_1 \otimes 1 + 1 \otimes N_2$ is also nilpotent. We conclude easily from this that $L_{r,s}$ is nilpotent if L is, for $r, s \geq 0$. We now consider the semisimple case. By replacing k with an extension field we may assume that k is algebraically closed. Since L is semisimple, we can select a basis $\{v_1, \ldots, v_m\}$ for V such that $Lv_j = c_jv_j$, $i \leq j \leq m$, for suitable $c_j \in k$. If $\{v_1^*, \ldots, v_m^*\}$ is the dual basis for V^*, then $L^*v_i^* =$

$-c_i v_i^*$, $1 \le i \le m$. If $r, s \ge 0$ and $r + s > 0$, the tensors

$$v_{j_1} \otimes \cdots \otimes v_{j_r} \otimes v_{i_1}^* \otimes \cdots \otimes v_{i_s}^* = t_{j_1,\ldots,j_r;\, i_1,\ldots,i_s}$$
$$1 \le j_1, \ldots, j_r, i_1, \ldots, i_s \le m$$

form a basis for $V_{r,s}$. Since we have

$$L_{r,s} t_{j_1,\ldots,j_r;\, i_1,\ldots,i_s} = (c_{j_1} + \cdots + c_{j_r} - c_{i_1} - \cdots - c_{i_s}) t_{j_1,\ldots,j_r;\, i_1,\ldots,i_s},$$

the semisimplicity of $L_{r,s}$ follows at once. The second assertion is immediate from (3.1.14).

Corollary 3.1.12. *Let $L = S + N$ be the Jordan decomposition of L, and let $r, s \ge 0$. Then $L_{r,s} = S_{r,s} + N_{r,s}$ is the Jordan decomposition of $L_{r,s}$.*

Proof. By the lemma, $S_{r,s}$ is semisimple and $N_{r,s}$ is nilpotent. Since $[S,N] = 0$, we have $[S_{r,s}, N_{r,s}] = 0$. Hence, $L_{r,s} = S_{r,s} + N_{r,s}$ is the Jordan decomposition of $L_{r,s}$.

Theorem 3.1.13. *Let notation be as above. Then:*

(i) *For any $L \in \mathfrak{gl}(V)$, $\mathfrak{g}(L)$ is an abelian Lie algebra whose elements are of the form $p(L)$ for $p \in k[T]$ with zero constant term.*

(ii) *Let k' be an extension field of k. Then*

$$\mathfrak{g}(L^{k'}) = k' \cdot \{M^{k'} : M \in \mathfrak{g}(L)\},$$
$$\mathfrak{g}(L) = \{M \in \mathfrak{gl}(V) : M^{k'} \in \mathfrak{g}(L^{k'})\}.$$

(iii) *Let $M \in \mathfrak{g}(L)$ and $\bar{V}$ a subspace of V invariant under L. Then $\bar{V}$ is invariant under M. Let $\bar{L}$ and $\bar{M}$ (resp. $\underline{L}$ and $\underline{M}$) be the endomorphisms induced on $\bar{V}$ (resp. $V/\bar{V}$) by L and M respectively. Then $\bar{M} \in \mathfrak{g}(\bar{L})$ (resp. $\underline{M} \in \mathfrak{g}(\underline{L})$).*

(iv) *If $M \in \mathfrak{g}(L)$, $N \in \mathfrak{g}(M)$, then $N \in \mathfrak{g}(L)$.*

(v) *If $M \in \mathfrak{g}(L)$, then $M_{r,s} \in \mathfrak{g}(L_{r,s})$ for all $r, s \ge 0$. In particular, ad M is a replica of ad L.*

(vi) *Let L, M be endomorphisms of V. Then $M \in \mathfrak{g}(L)$ if and only if for each $r, s \ge 0$ one can find $p_{r,s} \in k[T]$ with zero constant term such that $M_{r,s} = p_{r,s}(L_{r,s})$.*

(vii) *Let L, L' be endomorphisms of V, $L = S + N$, $L' = S' + N'$ their Jordan decompositions. Then S and N are replicas of L, and $L' \in \mathfrak{g}(L)$ if and only if $S' \in \mathfrak{g}(S)$ and $N' \in \mathfrak{g}(N)$.*

Proof. Since $[M,N]_{r,s} = [M_{r,s}, N_{r,s}]$, it is clear that $\mathfrak{g}(L)$ is a Lie subalgebra of $\mathfrak{gl}(V)$. Fix $M \in \mathfrak{g}(L)$. By (3.1.14), $M \in \{L\}''$, so by Theorem 3.1.7, $M = p(L)$ for some $p \in k[T]$ with constant term a. If $Lv = 0$ for some nonzero $v \in V$, $Mv = av = 0$, so $a = 0$. If L is invertible, $ML^{-1} \in \{L\}''$, so

$ML^{-1} = p_1(L)$ for some $p_1 \in k[T]$; then $M = q(L)$, where $q(T) = p_1(T)T$ has zero constant term. Since all endomorphisms of the form $p(L)$ ($p \in k[T]$) mutually commute, we have (i).

Let $W_{r,s}$ (resp. $W'_{r,s}$) be the null space of $L_{r,s}$ ($L^{k'}_{r,s}$). Obviously, $W'_{r,s} = k' \cdot W_{r,s}$, so $W'_{r,s}$ is a subspace of $(V^{k'})_{r,s}$ defined over k. For $M' \in \mathfrak{gl}(V^{k'})$, $M' \in \mathfrak{g}(L^{k'})$ if and only if $M'_{r,s} | W'_{r,s} = 0$. It follows easily from this that $\mathfrak{g}(L^{k'})$, as a subspace of $\mathfrak{gl}(V^{k'}) \simeq \mathfrak{gl}(V)^{k'}$, is defined over k. Furthermore, if $M \in \mathfrak{gl}(V)$, then $M^{k'} \in \mathfrak{g}(L^{k'})$ if and only if $M^{k'}_{r,s} | W'_{r,s} = 0$ for all r, s, i.e., if and only if $M_{r,s} | W_{r,s} = 0$ for all r, s, i.e., $M \in \mathfrak{g}(L)$. These two observations lead to (ii).

To prove (iii), we may assume that $\bar{V} \neq 0$, $\bar{V} \neq V$. Since M is a polynomial in L, M leaves $\bar{V}$ invariant. Hence $\bar{M}$ is well-defined. Let β be the restriction map which maps V^* linearly onto $\bar{V}^*$. Fix integers $r, s \geq 0$. We prove that if $u \in \bar{V}_{r,s}$ and $\bar{L}_{r,s}u = 0$, then $\bar{M}_{r,s}u = 0$. Write $V_{[r,s]} = V_{r,0} \otimes \bar{V}_{0,s}$. For any endomorphism N of V which leaves $\bar{V}$ invariant, let $\bar{N} = N | \bar{V}$ and $N_{[r,s]} = N_{r,0} \otimes 1 + 1 \otimes \bar{N}_{0,s}$. It is then obvious that $N_{[r,s]}$ leaves $\bar{V}_{r,s}$ invariant, and $N_{[r,s]} | \bar{V}_{r,s} = \bar{N}_{r,s}$. So it is enough to prove that if $u \in V_{[r,s]}$ and $L_{[r,s]}u = 0$, then $M_{[r,s]}u = 0$. Let $\beta_{r,s}$ be the unique linear map of $V_{r,s}$ onto $V_{[r,s]}$ such that

$$(3.1.19) \qquad \begin{aligned} &\beta_{r,s}(v_1 \otimes \cdots \otimes v_r \otimes v_1^* \otimes \cdots \otimes v_s^*) \\ &\quad = v_1 \otimes \cdots \otimes v_r \otimes \beta(v_1^*) \otimes \cdots \otimes \beta(v_s^*) \end{aligned}$$

for $v_1, \ldots, v_r \in V$ and $v_1^*, \ldots, v_s^* \in V^*$. It is then clear that for any endomorphism N that leaves $\bar{V}$ invariant, $\beta_{r,s}$ intertwines the actions of $N_{r,s}$ on $V_{r,s}$ and $N_{[r,s]}$ on $V_{[r,s]}$. It is therefore enough to prove the following: if $w \in V_{r,s}$ and if $L_{r,s}w$ belongs to the kernel of $\beta_{r,s}$, then $M_{r,s}w$ also belongs to the kernel of $\beta_{r,s}$. Since $M_{[r,s]}\beta_{r,s} = \beta_{r,s}M_{r,s}$, it is clear that $M_{r,s}$ leaves the kernel of $\beta_{r,s}$ invariant. On the other hand, it is a consequence of the assertion (vi) to be proved presently, that $M_{r,s} = p_{r,s}(L_{r,s})$ for some $p_{r,s} \in k[T]$ whose constant term is zero. From these two observations we can deduce the assertion.

Let $\bar{V}^0$ be the annihilator of $\bar{V}$ in V^*. Then both $\hat{L}$ and $\hat{M}$ leave $\bar{V}^0$ invariant. It follows from the above result that $\hat{M} | \bar{V}^0$ is a replica of $\hat{L} | \bar{V}^0$. Now $\bar{V}^0$ is canonically isomorphic to $(V/\bar{V})^*$, and this canonical isomorphism intertwines $\hat{L} | \bar{V}^0$ and $\hat{\underline{L}}$ (resp. $\hat{M} | \bar{V}^*$ and $\hat{\underline{M}}$). So $\hat{\underline{M}}$ is a replica of $\hat{\underline{L}}$. This proves that $\underline{M} \in \mathfrak{g}(\underline{L})$. (iii) is completely proved.

(iv) is trivial.

We now prove (v). Let L, M be endomorphisms of V with $M \in \mathfrak{g}(L)$. Fix integers $r, s \geq 0$ with $r + s > 0$. Suppose r', s' are integers ≥ 0. It is then easy to see that there is a natural isomorphism ζ of $(V_{r,s})_{r',s'}$ onto $V_{rr'+ss',rs'+sr'}$, and that for any endomorphism N of V,

$$\zeta \circ (N_{r,s})_{r',s'} \circ \zeta^{-1} = N_{rr'+ss',rs'+sr'}.$$

Taking $N = L, M$ we get the first assertion in (v). The second follows at once from (3.1.14).

We now come to (vi). Suppose $M \in \mathfrak{g}(L)$. Then $M_{r,s} \in \mathfrak{g}(L_{r,s})$ by (v) for integers $r, s \geq 0$, so by (i), we can find polynomials $p_{r,s} \in k[T]$ with zero constant terms such that $M_{r,s} = p_{r,s}(L_{r,s})$, $r, s \geq 0$. Suppose conversely that $M_{r,s} = p_{r,s}(L_{r,s})$ for such polynomials $p_{r,s}$. Then for any $u \in V_{r,s}$ such that $L_{r,s}u = 0$, $L^i_{r,s} = 0$ for $i = 1,2, \ldots$, so $M_{r,s}u = 0$. From our definition of $\mathfrak{g}(L)$, we can conclude that $M \in \mathfrak{g}(L)$. This proves (vi).

To prove (vii), let $L = S + N$ be the Jordan decomposition of L. Then $L_{r,s} = S_{r,s} + N_{r,s}$ is the Jordan decomposition of $L_{r,s}$ for $r, s \geq 0$ (Corollary 3.1.12), and there are $p_{r,s}, q_{r,s} \in k[T]$ with zero constant terms such that $S_{r,s} = p_{r,s}(L_{r,s})$, $N_{r,s} = q_{r,s}(L_{r,s})$ (Theorem 3.1.4). So by (vi) above, S and N are replicas of L. Let L' be another endomorphism and $L' = S' + N'$ its Jordan decomposition. If $S' \in \mathfrak{g}(S)$ and $N' \in \mathfrak{g}(N)$, then in view of (iii) above and the inclusions $S, N \in \mathfrak{g}(L)$, we have $S', N' \in \mathfrak{g}(L)$. So $L' \in \mathfrak{g}(L)$. We now prove the converse. Let $L' \in \mathfrak{g}(L)$. Since S' and N' belong to $\mathfrak{g}(L')$, we have $S', N' \in \mathfrak{g}(L)$. So for any $r, s \geq 0$, there is $p_{r,s} \in k[T]$ with zero constant term, such that $S'_{r,s} = p_{r,s}(S_{r,s} + N_{r,s})$. Fix $r, s \geq 0$. It is then obvious that $p_{r,s}(S_{r,s} + N_{r,s}) = p_{r,s}(S_{r,s}) + N_{r,s}R$, where R is an endomorphism which is a polynomial in $S_{r,s}$ and $N_{r,s}$, and hence commutes with both. Now $p_{r,s}(S_{r,s})$ is semisimple, $N_{r,s}R$ is nilpotent, and the two endomorphisms commute with each other. As their sum is the semisimple endomorphism $S'_{r,s}$, we must have $S'_{r,s} = p_{r,s}(S_{r,s})$. Since $r, s \geq 0$ are arbitrary, we see that $S' \in \mathfrak{g}(S)$. We now take up the proof that $N' \in \mathfrak{g}(N)$. Clearly, we may assume that k is algebraically closed. We claim that if $v \in V$ and $Nv = 0$, then $N'v = 0$. Write V as the direct sum of the $V_{L,\lambda}(\lambda \in \sigma(L))$. Then $Su = \lambda u$ for $u \in V_{L,\lambda}$, so for $u \in V_{L,\lambda}$, $Nu = 0$ if and only if $Lu = \lambda u$. So if $v \in V$ and $Nv = 0$, we can write $v = \sum_{\lambda \in \sigma(L)} v_\lambda$, where $v_\lambda \in V_{L,\lambda}$ and $Lv_\lambda = \lambda v_\lambda$. But since $N' \in \mathfrak{g}(L)$, there is a $p \in k[T]$ such that $N' = p(L)$. Consequently, $N'v_\lambda = p(\lambda)v_\lambda$, $\lambda \in \sigma(L)$. Since the eigenvalues of N' are all zero, we must have $N'v_\lambda = 0$, $\lambda \in \sigma(L)$. Hence $N'v = 0$. If $r, s \geq 0$, $N'_{r,s} \in \mathfrak{g}(L_{r,s})$, so by the same argument as above, we can conclude that $N'_{r,s}v = 0$ for any $v \in V_{r,s}$ for which $N_{r,s}v = 0$. Thus N' is a replica of N. This completes the proof of (vii). The theorem is proved.

We now have the following simple corollary.

Corollary 3.1.14. *Let A be an algebra over k. Assume that $\dim A < \infty$. If L is a derivation of A, then any replica of L is also a derivation of A. In particular, the semisimple and nilpotent parts of A are also derivations of A.*

Proof. Let $L = S + N$ be the Jordan decomposition of L. By (vii) above, S and N are replicas of L. The present corollary is then an immediate consequence of Lemma 3.1.10 (cf. especially (3.1.15)).

Our aim now is the determination of $\mathfrak{g}(L)$ for arbitrary L. In view of (vi) of the preceding theorem it is enough to do this when L is either semisimple or nilpotent. Moreover, in the semisimple case it is sufficient to confine ourselves to the case when k is algebraically closed, or more generally, when L is diagonalizable in k itself.

Theorem 3.1.15. *Let N be a nilpotent endomorphism of V. Then*

$$\mathfrak{g}(N) = \{cN : c \in k\}. \tag{3.1.20}$$

Let $\{v_1, \ldots, v_m\}$ be a basis for V. For any $\boldsymbol{\lambda} = (\lambda_1, \ldots, \lambda_m) \in k^m$, write $S(\boldsymbol{\lambda})$ for the endomorphism for which $S(\boldsymbol{\lambda}) v_j = \lambda_j v_j$, $1 \leq j \leq m$, and $\langle \boldsymbol{\lambda} \rangle$ for the subset of k^m of all $\boldsymbol{\mu} = (\mu_1, \ldots, \mu_m)$ satisfying the following condition: if $c_1, \ldots, c_m$ are rational integers and $c_1\lambda_1 + \cdots + c_m\lambda_m = 0$, then $c_1\mu_1 + \cdots + c_m\mu_m = 0$. Then, for any $\boldsymbol{\lambda} \in k^m$,

$$\mathfrak{g}(S(\boldsymbol{\lambda})) = \{S(\boldsymbol{\mu}) : \boldsymbol{\mu} \in \langle \boldsymbol{\lambda} \rangle\}. \tag{3.1.21}$$

Proof. We prove that $\mathfrak{g}(N) = k \cdot N$ by induction on $\dim V$. For $\dim V = 1$, $N = 0$, so $\mathfrak{g}(N) = 0$. Let $\dim V \geq 2$. Let $s (1 \leq s \leq \dim V)$ be the integer such that T^s is the minimal polynomial of N. We may assume that $s > 1$. Fix $N' \in \mathfrak{g}(N)$. Then by (iii) of Theorem 3.1.13, N' leaves the range $R(N)$ of N invariant, and $N' | R(N)$ is a replica of $N | R(N)$. Since $\dim R(N) < \dim V$, there is, by the induction hypothesis, a constant $c \in k$ such that $N' = cN$ on $R(N)$, i.e., $N'N = cN^2$. Now there are unique constants $c_1, \ldots, c_{s-1}$ such that $N' = c_1 N + \cdots + c_{s-1}N^{s-1}$; the relation $N'N = cN^2$ then gives $N' = cN$ when $s = 2$ and $N' = cN + c_{s-1}N^{s-1}$ when $s > 2$. Suppose $s > 2$ and $c_{s-1} \neq 0$. Then $N^{s-1} \in \mathfrak{g}(N)$. We shall show that this leads to a contradiction. There is a $p \in k[T]$ with zero constant term such that $(N^{s-1})_{2,0} = p(N_{2,0})$; i.e., $N^{s-1} \otimes 1 + 1 \otimes N^{s-1} = p(N \otimes 1 + 1 \otimes N)$ in $V \otimes V$. Now $N^s = 0$, and hence $(N \otimes 1 + 1 \otimes N)^r = 0$ for $r \geq 2s - 1$. So there are constants $d_r \in k$ $(1 \leq r \leq 2s - 2)$ such that

$$N^{s-1} \otimes 1 + 1 \otimes N^{s-1} = \sum_{1 \leq r \leq 2s-2} d_r (N \otimes 1 + 1 \otimes N)^r.$$

We now expand the powers in the right side. Remembering that $N^p = 0$ for $p \geq s$ and writing $c_{p,q} = d_{p+q}\binom{p+q}{q}$, we get

$$N^{s-1} \otimes 1 + 1 \otimes N^{s-1} = \sum_{\substack{0 \leq p,q \leq s-1 \\ p+q > 0}} c_{p,q} N^p \otimes N^q. \tag{3.1.22}$$

Since the endomorphisms $1, N, N^2, \ldots, N^{s-1}$ are linearly independent, so are the endomorphisms $N^p \otimes N^q$, $0 \leq p, q \leq s - 1$. We may therefore conclude from (3.1.22) that $c_{p,q} = 0$ unless $(p,q) = (0,s-1)$ or $(s-1,0)$; in particular, $d_r = 0$ if $r \neq s - 1$. But since $s > 2$, we can find $p, q > 0$ with $r = p + q =$

$s - 1$. So we have $d_r = 0$, $1 \leq r \leq 2s - 2$. This implies that $N^{s-1} \otimes 1 + 1 \otimes N^{s-1} = 0$, a contradiction.

We now take up the proof of (3.1.21). Let $\{v_1^*, \ldots, v_m^*\}$ be the dual basis of V^*. Fix $\boldsymbol{\lambda} \in k^m$ and write $S = S(\boldsymbol{\lambda})$. We have $\hat{S}v_j^* = -\lambda_j v_j^*$, $1 \leq j \leq m$. Suppose $S' = S(\boldsymbol{\mu})$ for some $\boldsymbol{\mu} \in \langle\boldsymbol{\lambda}\rangle$. Let r, s be integers ≥ 0 with $r + s > 0$. Then the null space of $S_{r,s}$ is spanned by the tensors $v_{i_1} \otimes \cdots \otimes v_{i_r} \otimes v_{j_1}^* \otimes \cdots \otimes v_{j_s}^*$, with $\lambda_{i_1} + \cdots + \lambda_{i_r} - \lambda_{j_1} - \cdots - \lambda_{j_s} = 0$. So, in order to prove that $S' \in \mathfrak{g}(S)$, it is sufficient to show that for any such $i_1, \ldots, i_r$, $j_1, \ldots, j_s$, $\mu_{i_1} + \cdots + \mu_{i_r} - \mu_{j_1} - \cdots - \mu_{j_s} = 0$. This is, however, an obvious implication of the assumption $\boldsymbol{\mu} \in \langle\boldsymbol{\lambda}\rangle$. Conversely, let $S' \in \mathfrak{g}(S)$. As $S' = p(S)$ for some $p \in k[T]$, we have $S' = S(\boldsymbol{\mu})$, where $\mu_j = p(\lambda_j)$ for $1 \leq j \leq m$. In particular, $\lambda_i = \lambda_j$ implies $\mu_i = \mu_j$. For integers $r, s \geq 0$ with $r + s > 0$, we use the reasoning given above to deduce that for any $i_1, \ldots, i_r, j_1, \ldots, j_s$ $(1 \leq i_\mu, j_\nu \leq m)$ such that $\lambda_{i_1} + \cdots + \lambda_{i_r} - \lambda_{j_1} - \cdots - \lambda_{j_s} = 0$, we must have $\mu_{i_1} + \cdots + \mu_{i_r} - \mu_{j_1} - \cdots - \mu_{j_s} = 0$. These relations taken for all $r, s \geq 0$ with $r + s > 0$ imply that $\boldsymbol{\mu} \in \langle\boldsymbol{\lambda}\rangle$. This proves (3.1.21). The proof of the theorem is complete.

Remarks 1. For a given $\boldsymbol{\lambda} \in k^m$, the set $\langle\boldsymbol{\lambda}\rangle$ can be given an alternative description. We regard k as a vector space over $\mathbf{Q}$, the prime subfield of rational numbers. Let C be the $\mathbf{Q}$-subspace of k generated by $\lambda_1, \ldots, \lambda_m$. We leave it to the reader to verify that $\langle\boldsymbol{\lambda}\rangle$ is the set of all $\boldsymbol{\mu} = (\mu_1, \ldots, \mu_m) \in k^m$ with the following property: there is a $\mathbf{Q}$-linear map ψ of C into k such that $\mu_j = \psi(\lambda_j)$, $1 \leq j \leq m$.

2. The relation (3.1.21) determines $\mathfrak{g}(S)$ very explicitly for arbitrary semisimple S when k is algebraically closed. Moreover, if S is semisimple but k is not algebraically closed, one can pass to the algebraic closure $\bar{k}$ in which $S^{\bar{k}}$ is diagonalizable and determine $\mathfrak{g}(S)$ from (3.1.21) and the relation

$$\mathfrak{g}(S) = \{L : L \in \mathfrak{gl}(V), L^{\bar{k}} \in \mathfrak{g}(S^{\bar{k}})\}. \tag{3.1.23}$$

If L is an arbitrary endomorphism of V, and $L = S + N$ is its Jordan decomposition, it follows from the above theorems that

$$\mathfrak{g}(L) = \{S' + cN : S' \in \mathfrak{g}(S), c \in k\}. \tag{3.1.24}$$

We shall now establish the criterion (due to Chevalley) for the nilpotency of an endomorphism. If L is an endomorphism of V, it is easy to show that L is nilpotent if and only if $tr(L^s) = 0$ for $s = 1, 2, \ldots$. The following result improves this significantly.

Theorem 3.1.16. *Let L be an endomorphism of V. Then L is nilpotent if and only if*

$$tr(LL') = 0 \quad \forall\, L' \in \mathfrak{g}(L). \tag{3.1.25}$$

Proof. If L is nilpotent, then (3.1.25) is obvious. Conversely, let L be arbitrary but satisfying (3.1.25). Let $L = S + N$ be the Jordan decomposition of L. We must prove that $S = 0$. We may (and shall) assume that k is algebraically closed. Since $\mathfrak{g}(S) \subseteq \mathfrak{g}(L)$, $tr(LS') = 0$ for all $S' \in \mathfrak{g}(S)$. On the other hand, the elements of $\mathfrak{g}(S)$ commute with N, so $S'N$ is nilpotent for all $S' \in \mathfrak{g}(S)$. Hence $tr(SS') = 0$ for all $S' \in \mathfrak{g}(S)$. Now select a basis $\{v_1, \ldots, v_m\}$ for V such that $Sv_j = \lambda_j v_j$, $1 \leq j \leq m$, for suitable numbers $\lambda_j \in k$. Then it follows from Theorem 3.1.15 that $\sum_{1 \leq j \leq m} \mu_j \lambda_j = 0$ for all $\boldsymbol{\mu} = (\mu_1, \ldots, \mu_m) \in \langle \boldsymbol{\lambda} \rangle$. Let C denote the $\mathbf{Q}$-subspace of k spanned by $\lambda_1, \ldots, \lambda_m$. We may then conclude from the remark following the preceding theorem that $\sum_{1 \leq j \leq m} \psi(\lambda_j)\lambda_j = 0$ for all $\mathbf{Q}$-linear maps ψ of C into k.

Suppose now that $C \neq 0$. It is then obvious that there are nonzero $\mathbf{Q}$-linear maps of C into $\mathbf{Q}$ itself. Let ψ be any one such map. Then applying ψ to the relation $\sum_{1 \leq j \leq m} \psi(\lambda_j)\lambda_j = 0$, we get $\sum_{1 \leq j \leq m} (\psi(\lambda_j))^2 = 0$. Since all the $\psi(\lambda_j)$ are in $\mathbf{Q}$, all of them must vanish. Hence $\psi = 0$, a contradiction. C must therefore be zero. But this implies at once that $S = 0$. The proof of the theorem is complete.

Corollary 3.1.17. *Let A be a finite-dimensional algebra over k, and let L be a derivation of A. Then L is nilpotent if and only if $tr(LL') = 0$ for all $L' \in k[L]$ that are derivations of A.*

For by Corollary 3.1.14, any replica of L is a derivation of A and belongs to $k[L]$.

We shall conclude this section with a discussion of tensor algebras. Let V be a vector space over k, not necessarily of finite dimension. We denote by $\mathfrak{T}_r$ the tensor product $V \otimes \cdots \otimes V$ (r terms), r being ≥ 1, we put $\mathfrak{T}_0 = k \cdot 1$. Elements of $\mathfrak{T}_r$ are called *tensors of degree r*. Let $\mathfrak{T}$ be the direct sum of all $\mathfrak{T}_r$ $(r \geq 0)$. Then there is a unique bilinear map $(u,v) \mapsto u \otimes v$ of $\mathfrak{T} \times \mathfrak{T}$ into $\mathfrak{T}$ such that (i) $c1 \otimes v = cv = v \otimes c1$ for $c \in k$ and $v \in \mathfrak{T}$, and (ii) if r, s are integers ≥ 1, $x_1, \ldots, x_r$ and $y_1, \ldots, y_s \in V$, then $(x_1 \otimes \cdots \otimes x_r) \otimes (y_1 \otimes \cdots \otimes y_s) = x_1 \otimes \cdots \otimes x_r \otimes y_1 \otimes \cdots \otimes y_s$. $\mathfrak{T}$, under $\otimes$, becomes an associative algebra with 1 as its unit. It is called the *tensor algebra over V*. V may identified with a subspace of $\mathfrak{T}$.

$\mathfrak{T}$ is generated by V and possesses the following universal property: if $\mathfrak{A}$ is an associative algebra, and γ a linear map of V into $\mathfrak{A}$, there is a unique homomorphism of $\mathfrak{T}$ into $\mathfrak{A}$ that extends γ. This, in fact, characterizes $\mathfrak{T}$. In other words let $\mathfrak{T}'$ be an associative algebra, π a linear map of V into $\mathfrak{T}'$ such that (i) $\mathfrak{T}'$ is generated by $\pi[V]$, and (ii) if $\mathfrak{A}$ is an associative algebra and γ a linear map of V into $\mathfrak{A}$, there is a (unique) homomorphism γ' of $\mathfrak{T}'$ into $\mathfrak{A}$ with $\gamma = \gamma' \circ \pi$; then π is an injection and there is a unique isomorphism ξ of the algebra $\mathfrak{T}$ onto the algebra $\mathfrak{T}'$ such that $\xi v = \pi v$ for all $v \in V$. We follow the usual practice and call a tensor t *homogeneous* if t belongs to $\mathfrak{T}_r$ for some $r \geq 0$. Obviously, $\mathfrak{T}_r \otimes \mathfrak{T}_s \subseteq \mathfrak{T}_{r+s}$ $(r, s \geq 0)$.

Let X_i $(i \in J)$ be indeterminates. By a *free associative algebra over k generated by the X_i* we mean an associative algebra $\mathfrak{A}$ over k containing the X_i such that (i) the X_i $(i \in J)$ generate $\mathfrak{A}$, and (ii) if $\mathfrak{B}$ is an associative algebra over k and x_i $(i \in J)$ are elements of $\mathfrak{B}$, there is a (unique) homomorphism ξ of $\mathfrak{A}$ into $\mathfrak{B}$ such that $\xi(X_i) = x_i$ for all $i \in J$. If $\mathfrak{A}$ and $\mathfrak{A}'$ are free associative algebras over k generated by the X_i, a standard argument based on the above universal property proves the existence of an algebra isomorphism ζ of $\mathfrak{A}$ onto $\mathfrak{A}'$ such that $\zeta(X_i) = X_i$ for all $i \in J$. To prove the existence of the free associative algebra generated by the X_i, we proceed as follows. Let V be a vector space over k for which the X_i form a basis and let $\mathfrak{J}$ be the tensor algebra over V. It is then obvious that $\mathfrak{J}$ is a free associative algebra over k generated by X_i $(i \in J)$. The elements $X_{i_1} \cdots X_{i_n}$ $(i_1, \ldots, i_n \in J)$ form a basis for $\mathfrak{J}_n$ for $n \geq 0$.

We now return to the earlier context of a vector space V and its tensor algebra $\mathfrak{J}$. If x is a linear bijection of V, there is a unique automorphism $\tilde{x}$ (resp. antiautomorphism $\tilde{x}$) of $\mathfrak{J}$ that extends x; if $v_1, \ldots, v_r \in V$, $\tilde{x}(v_1 \otimes \cdots \otimes v_r) = x(v_1) \otimes \cdots \otimes x(v_r)$ (resp. $\tilde{x}(v_1 \otimes \cdots \otimes v_r) = x(v_r) \otimes \cdots \otimes x(v_1)$). We write $x^{\otimes r} = \tilde{x} | \mathfrak{J}_r$. If x and y are two linear bijections of V, then $\widetilde{xy} = \tilde{x}\tilde{y}$ and $\widetilde{x^{-1}} = \tilde{x}^{-1}$. Let L be a linear map of V into $\mathfrak{J}$. Then there is a unique derivation $\tilde{L}$ of $\mathfrak{J}$ such that $\tilde{L}v = Lv$ for all $v \in V$; in fact, if $\{v_i : i \in J\}$ is a basis for V and if one defines $\tilde{L}$ as the unique endomorphism of $\mathfrak{J}$ such that $\tilde{L}(1) = 0$ and

$$\begin{aligned} L(v_{i_1} \otimes \cdots \otimes v_{i_r}) &= Lv_{i_1} \otimes v_{i_2} \otimes \cdots \otimes v_{i_r} \\ &\quad + v_{i_1} \otimes Lv_{i_2} \otimes \cdots \otimes v_{i_r} + \cdots + v_{i_1} \otimes \cdots \otimes v_{i_{r-1}} \otimes Lv_{i_r} \end{aligned} \tag{3.1.26}$$

$(r \geq 1, i_1, \ldots, i_r \in J)$, then it is an easy verification that $\tilde{L}$ is a derivation of $\mathfrak{J}$ extending L, and that it is the only derivation of $\mathfrak{J}$ with this property. An important and extremely useful special case arises when we take L to be an endomorphism of V. Then it is obvious that $\tilde{L}$ leaves each $\mathfrak{J}_r$ invariant and that

$$\begin{aligned} \tilde{L} | \mathfrak{J}_r &= L \otimes 1 \otimes \cdots \otimes 1 + 1 \otimes L \otimes 1 \otimes \cdots \otimes 1 \\ &\quad + \cdots + 1 \otimes \cdots \otimes 1 \otimes L. \end{aligned} \tag{3.1.27}$$

The map $L \mapsto \tilde{L}$ is a homomorphism of the Lie algebra of all endomorphisms of V into the Lie algebra of all derivations of $\mathfrak{J}$.

For any $r \geq 1$ let Π_r be the group of all permutations of $\{1, \ldots, r\}$. Let ϵ be defined by

$$\epsilon(s) = \begin{cases} +1 & \text{if } s \in \Pi_r \text{ is an even permutation} \\ -1 & \text{if } s \in \Pi_r \text{ is an odd permutation.} \end{cases} \tag{3.1.28}$$

Given $s \in \Pi_r$, there is a unique linear automorphism $s^{(r)}$ of $\mathfrak{J}_r$ such that

$$s^{(r)}(v_1 \otimes \cdots \otimes v_r) = v_{s^{-1}(1)} \otimes \cdots \otimes v_{s^{-1}(r)}$$

for all $v_1, \ldots, v_r \in V$. The correspondence $s \mapsto s^{(r)}$ is a representation of Π_r in $\mathfrak{J}_r$. An element $t \in \mathfrak{J}_r$ is called a *symmetric* (resp. *skew-symmetric*) tensor of rank r if $s^{(r)}(t) = t$ for all $s \in \Pi_r$ (resp. $s^{(r)}(t) = \epsilon(s)t$ for all $s \in \Pi_r$). An element $t \in \mathfrak{J}$ is called symmetric (resp. skew-symmetric) if its component in $\mathfrak{J}_r$ is symmetric (resp. skew-symmetric) for all $r \geq 1$. Suppose $\dim V = m < \infty$. Then if $r > m$, 0 is the only skew-symmetric tensor of degree r; for $1 \leq r \leq m$, the vector space of skew-symmetric tensors of degree r has dimension $\binom{m}{r}$. For any $r \geq 1$, the vector space of all symmetric tensors of degree r has dimension $\binom{m+r-1}{r}$.

A two sided ideal $\mathfrak{a}$ of $\mathfrak{J}$ is said to be *homogeneous* if

$$\mathfrak{a} = \sum_{r \geq 0} (\mathfrak{a} \cap \mathfrak{J}_r). \tag{3.1.29}$$

If J is a set of homogeneous tensors of degree $\geq p$ where p is an integer ≥ 2, then the two-sided ideal $\mathfrak{a}$ generated by J is homogeneous, and $\mathfrak{a} \cap \mathfrak{J}_r = 0$ for $r = 0, \ldots, p-1$. In particular,

$$\mathfrak{a} \cap \mathfrak{J}_r = 0 \quad (r = 0, 1). \tag{3.1.30}$$

Let A be the algebra $\mathfrak{J}/\mathfrak{a}$. Let π be the natural map of $\mathfrak{J}$ onto A. If we put

$$A_r = \pi[\mathfrak{J}_r] \quad (r = 0, 1, \ldots), \tag{3.1.31}$$

then A is the direct sum of the A_r, and we have $A_rA_s \subseteq A_{r+s}$ $(r, s \geq 0)$. In view of (3.1.30) it is clear that π is one-to-one on V, so it is natural to identify V with its image in A under π. Obviously, V generates A. Suppose L is an endomorphism (resp. x is an automorphism) of V such that L (resp. x) maps J into $\mathfrak{a}$. Then the derivation $\tilde{L}$ (resp. automorphism $\tilde{x}$) of $\mathfrak{J}$ maps $\mathfrak{a}$ into $\mathfrak{a}$ and so induces a derivation of A (resp. automorphism of A). Denoting this derivation by $\tilde{L}_A$ (resp. automorphism by $\tilde{x}_A$), it is clear that $[L,M]^{\sim}_A = \tilde{L}_A\tilde{M}_A - \tilde{M}_A\tilde{L}_A$ (resp. $(xy)^{\sim}_A = \tilde{x}_A\tilde{y}_A$) for any two endomorphisms (resp. automorphisms) L, M (resp. x, y) of V. As important examples we mention (i) $J = \{x \otimes y - y \otimes x : x, y \in V\}$, and (ii) $J = \{x \otimes y + y \otimes x : x, y \in V\}$. In example (i) A is the symmetric algebra over V; in example (ii) A is the exterior algebra over V.

The symmetric algebra over V is generally denoted by $S(V)$. It is commutative, graded, and its subspace of homogeneous elements of degree n is denoted by $S_n(V)$ $(n = 0,1,2,\ldots)$. We customarily identify V with its image in $S(V)$. If X_i $(i \in J)$ form a basis for V, the monomials $\Pi_{i \in J} X_i^{m_i}$ $(m_i \geq 0$

integers, $m_i = 0$ for all but finitely many i) form a basis for $S(V)$. If $\mathfrak{A}$ is any commutative associative algebra and φ is a linear map of V into $\mathfrak{A}$, φ extends to a unique homomorphism of $S(V)$ into $\mathfrak{A}$. If $\dim(V) = m < \infty$, $\dim S_n(V) = \binom{m+n-1}{n}$.

Let $\dim(V)$ be finite. Denote by V^* the dual of V. We then have a natural isomorphism ξ of V with V^{**}. Let $P(V^*)$ be the algebra of polynomial functions on V^*, i.e., the algebra of functions on V^* with values in k generated by the linear functions on V^*. The map ξ then extends to an algebra isomorphism of $S(V)$ onto $P(V^*)$. The image of $S_n(V)$ under this isomorphism is the subspace of $P(V^*)$ consisting of homogeneous polynomials of degree n.

The exterior algebra over V is generally denoted by $E(V)$. Let $\dim(V) = m < \infty$. We identify V with its image in $E(V)$, and denote the product operation in $E(V)$ by $\wedge$. If $v, w \in V$, then $v \wedge v = 0$ and $v \wedge w + w \wedge v = 0$. If $\{v_1, \ldots, v_m\}$ is a basis for V, then for any r with $1 \leq r \leq m$, the elements $v_{i_1} \wedge v_{i_2} \wedge \cdots \wedge v_{i_r}$ $(1 \leq i_1 < i_2 < \cdots < i_r \leq n)$ are linearly independent and span $E_r(V)$, the subspace of $E(V)$ consisting of homogeneous elements of degree r. In particular, $\dim(E_r(V)) = \binom{m}{r}$. Of course, $E_0(V) = k \cdot 1$.

3.2. The Universal Enveloping Algebra of a Lie Algebra

In this and in the next section, k will denote a field of characteristic 0, $\mathfrak{g}$ a Lie algebra over k. Our aim is to introduce the universal enveloping algebra of $\mathfrak{g}$ and describe some of its properties. If $k = \mathbf{R}$ or $\mathbf{C}$ and $\mathfrak{g}$ is the Lie algebra of a Lie group G, we shall see in §3.4 that there is a canonical isomorphism of the universal enveloping algebra of $\mathfrak{g}$ with what we have earlier called the enveloping algebra of G, namely the algebra of all analytic left-invariant differential operators on G. The main results are the existence and uniqueness of the universal enveloping algebra and the Poincare–Birkhoff–Witt theorem. In proving these, the restriction to finite-dimensional Lie algebras is rather artificial; we therefore work with a Lie algebra $\mathfrak{g}$ of arbitrary dimension.

Let $\mathfrak{J}$ be the tensor algebra over the underlying vector space of $\mathfrak{g}$. We denote multiplication in $\mathfrak{J}$ by $\otimes$. $\mathfrak{J}_0 = k \cdot 1$, and for any integer $m \geq 1$, $\mathfrak{J}_m$ is the subspace of $\mathfrak{J}$ of all homogeneous tensors of degree m. For $X, Y \in \mathfrak{g}$, let

$$u_{X,Y} = X \otimes Y - Y \otimes X - [X,Y]. \tag{3.2.1}$$

We denote by $\mathfrak{L}$ the subspace of $\mathfrak{J}$ spanned by all elements of the form $t \otimes u_{X,Y} \otimes t'$ $(t, t' \in \mathfrak{J}, X, Y \in \mathfrak{g})$:

$$\mathfrak{L} = \sum_{X,Y \in \mathfrak{g}} \mathfrak{J} \otimes u_{X,Y} \otimes \mathfrak{J}. \tag{3.2.2}$$

Since $u_{X,Y} \in \mathfrak{J}_1 + \mathfrak{J}_2$ for all $X, Y \in \mathfrak{g}$, it is clear that $\mathfrak{L} \subseteq \sum_{m\geq 1}\mathfrak{J}_m$. $\mathfrak{L}$ is thus a proper two-sided ideal in $\mathfrak{J}$. We many thus introduce the quotient algebra $\mathfrak{J}/\mathfrak{L}$. Let

$$\mathfrak{G} = \mathfrak{J}/\mathfrak{L},$$

and let γ be the natural homomorphism of $\mathfrak{J}$ onto $\mathfrak{G}$. Since $\mathfrak{g}$ generates $\mathfrak{J}$, $\gamma[\mathfrak{g}]$ generates $\mathfrak{G}$. We denote the unit of $\mathfrak{G}$ by 1, and for $a, b \in \mathfrak{G}$, write ab for their product.

A pair $(\mathfrak{C},\pi)$, where $\mathfrak{C}$ is an associative algebra[3] over k and π is a linear mapping of $\mathfrak{g}$ into $\mathfrak{C}$, is called a *universal enveloping algebra* of $\mathfrak{g}$ if the following conditions are satisfied: (i) $\pi[\mathfrak{g}]$ generates $\mathfrak{C}$, (ii) $\pi([X, Y]) = \pi(X)\pi(Y) - \pi(Y)\pi(X)$ for all $X, Y \in \mathfrak{g}$, and (iii) if $\mathfrak{A}$ is any associative algebra and ξ is any linear map of $\mathfrak{g}$ into $\mathfrak{A}$ such that $\xi([X, Y]) = \xi(X)\xi(Y) - \xi(Y)\xi(X)$ for all $X, Y \in \mathfrak{g}$, there is a homomorphism ξ' (necessarily unique in view of (i)) of $\mathfrak{C}$ into $\mathfrak{A}$ such that $\xi(X) = \xi'(\pi(X))$ for all $X \in \mathfrak{g}$. We now have

Theorem 3.2.1. *Let $\mathfrak{g}$, $\mathfrak{G}$, and γ be as defined above. Then $(\mathfrak{G}, \gamma)$ is a universal enveloping algebra of $\mathfrak{g}$. If $(\mathfrak{G}, \gamma')$ is another universal enveloping algebra of $\mathfrak{g}$, there exists a unique isomorphism ζ of $\mathfrak{G}$ onto $\mathfrak{G}'$ such that $\zeta(\gamma(X)) = \gamma'(X)$ for all $X \in \mathfrak{g}$.*

Proof. We have already observed that $\gamma[\mathfrak{g}]$ generates $\mathfrak{G}$. Since $\gamma(u_{X,Y}) = 0$, we have

$$\gamma[X,Y]) = \gamma(X)\gamma(Y) - \gamma(Y)\gamma(X)$$

for all $X, Y \in \mathfrak{g}$. Suppose $\mathfrak{A}$ is an associative algebra and ξ is a linear map of $\mathfrak{g}$ into $\mathfrak{A}$ such that $\xi([X, Y]) = \xi(X)\xi(Y) - \xi(Y)\xi(X)$ for all $X, Y \in \mathfrak{g}$. Let $\bar{\xi}$ be the homomorphism of $\mathfrak{J}$ into $\mathfrak{A}$ such that $\bar{\xi}(X) = \xi(X)$ for all $X \in \mathfrak{g}$. Since $\bar{\xi}(u_{X,Y}) = 0$ for all $X, Y \in \mathfrak{g}$, $\bar{\xi} = 0$ on $\mathfrak{L}$. Passing to the quotient algebra $\mathfrak{G}$, we may obtain a homomorphism ξ' of $\mathfrak{G}$ into $\mathfrak{A}$ such that $\bar{\xi} = \xi' \circ \gamma$. We have thus proved that $(\mathfrak{G},\gamma)$ is a universal enveloping algebra of $\mathfrak{g}$.

Suppose $(\mathfrak{G}',\gamma)$ is another universal enveloping algebra of $\mathfrak{g}$. By the extension property there is a homomorphism ζ of $\mathfrak{G}$ into $\mathfrak{G}'$ such that $\zeta(\gamma(X)) = \gamma'(X)$ for all $X \in \mathfrak{g}$; since $\gamma'[\mathfrak{g}]$ generates $\mathfrak{G}'$, ζ is surjective. Interchanging the roles of $(\mathfrak{G},\gamma)$ and $(\mathfrak{G}',\gamma')$, we obtain a homomorphism ζ' of $\mathfrak{G}'$ onto $\mathfrak{G}$ such that $\zeta'(\gamma'(X)) = \gamma(X)$ for all $X \in \mathfrak{g}$. Thus $\zeta \circ \zeta'$ and $\zeta' \circ \zeta$ are the respective identities on $\gamma'[\mathfrak{g}]$ and $\gamma[\mathfrak{g}]$. Hence $\zeta \circ \zeta' = \zeta' \circ \zeta = 1$. In other words, ζ is an isomorphism of $\mathfrak{G}$ onto $\mathfrak{G}'$ and $\zeta(\gamma(X)) = \gamma'(X)$ for all $X \in \mathfrak{g}$.

Let J be a *linearly ordered* set, and $\{X_i : i \in J\}$ a basis for $\mathfrak{g}$. Then it is obvious that $\mathfrak{G}$ is spanned by 1 and the products $\gamma(X_{i_1}) \cdots \gamma(X_{i_s})$ $(i_1,\ldots,i_s \in J,$ $s \geq 1)$. Since

$$\gamma(X_j)\gamma(X_i) = \gamma(X_i)\gamma(X_j) + \gamma([X_j,X_i]),$$

it is almost obvious that $\mathfrak{G}$ is spanned by 1 and the products $\gamma(X_{i_1}) \cdots \gamma(X_{i_s})$

with $i_1 \leq i_2 \leq \cdots \leq i_s$, $s \geq 1$; the *Poincaré–Birkhoff–Witt* theorem asserts that these elements actually constitute a basis for $\mathfrak{G}$ (Theorem 3.2.2). We begin the proof of this theorem with some preparation.

Fix the linearly ordered set J and consider a basis $\{X_i : i \in J\}$ of $\mathfrak{g}$. By a *monomial* (in $\mathfrak{I}$) we mean any tensor which is either 1 or is of the form $X_{i_1} \otimes \cdots \otimes X_{i_p}$ $(p \geq 1, i_1, \ldots, i_p \in J)$. A *standard* monomial is a tensor which is either 1 or is of the form $X_{i_1} \otimes \cdots \otimes X_{i_p}$ $(p \geq 1, i_1 \leq \cdots \leq i_p)$. For any $p \geq 0$, let $\mathfrak{I}_p^0$ be the linear span of the standard monomials of degree p. Obviously, $\mathfrak{I}_0^0 = \mathfrak{I}_0$, $\mathfrak{I}_1^0 = \mathfrak{I}_1$. We write

$$\mathfrak{I}^0 = \sum_{p \geq 0} \mathfrak{I}_p^0$$

More generally, if $p \geq 2$ and $t = X_{i_1} \otimes \cdots \otimes X_{i_p}$, define the *index* $\operatorname{ind}(t)$ of t to be the number of pairs (r,s), $1 \leq r, s \leq p$, for which $r < s$ but $i_r > i_s$; $\operatorname{ind}(t) = 0$ if and only if t is standard. We write $\mathfrak{I}_p^d$ for the linear span of all monomials of degree p and index d. Obviously, $\mathfrak{I}_p = \sum_{d \geq 0} \mathfrak{I}_p^d$, the sum being direct.

Theorem 3.2.2. *Let $\mathfrak{g}$ be a Lie algebra over k, J a linearly ordered set, $\{X_i : i \in J\}$ a basis for $\mathfrak{g}$, and $(\mathfrak{G},\gamma)$ the universal enveloping algebra defined above. Then the elements 1 and $\gamma(X_{i_1}) \cdots \gamma(X_{i_s})$ $(s \geq 1, i_1 \leq \cdots \leq i_s)$ form a basis for $\mathfrak{G}$. In particular, γ is an injection on $\mathfrak{g}$.*

Proof. The theorem is equivalent to proving that $\mathfrak{G}$ is the direct sum of $\mathfrak{L}$ and $\mathfrak{I}^0$. We must therefore prove that $\mathfrak{L} + \mathfrak{I}^0 = \mathfrak{I}$ and $\mathfrak{L} \cap \mathfrak{I}^0 = 0$.

To prove that $\mathfrak{L} + \mathfrak{I}^0 = \mathfrak{I}$, it is clearly more than sufficient to prove that for any $r \geq 0$

$$\mathfrak{I}_r \subseteq \mathfrak{L} + \sum_{0 \leq q \leq r} \mathfrak{I}_q^0. \tag{3.2.4}$$

This is clear for $r = 0, 1$. We prove (3.2.4) for $r \geq 2$ by induction on r. Fix $p \geq 2$ and assume (3.2.4) for $0 \leq r \leq p - 1$. Since $\mathfrak{I}_p = \sum_{d \geq 0} \mathfrak{I}_p^d$, it is enough to prove that $\mathfrak{I}_p^d \subseteq \mathfrak{L} + \sum_{0 \leq q \leq p} \mathfrak{I}_q^0$ for all $d \geq 0$. We shall do this by induction on d. For $d = 0$ this is obvious. Let $d \geq 1$, and assume that $\mathfrak{I}_p^e \subseteq \mathfrak{L} + \sum_{0 \leq q \leq p} \mathfrak{I}_q^0$ for $0 \leq e \leq d - 1$. Let $t = X_{i_1} \otimes \cdots \otimes X_{i_p} \in \mathfrak{I}_p^d$. Since $d \geq 1$, it is clear that we can choose an integer r with $1 \leq r \leq p - 1$ such that $i_r > i_{r+1}$. Define t' as the tensor $X_{j_1} \otimes \cdots \otimes X_{j_p}$, where $j_l = i_l$ for $l \neq r$, $\neq r + 1$, $j_{r+1} = i_r, j_r = i_{r+1}$. Then $t' \in \mathfrak{I}_p^{d-1} \subseteq \mathfrak{L} + \mathfrak{I}^0$. But since $X_{i_r} \otimes X_{i_{r+1}} - X_{i_{r+1}} \otimes X_{i_r} \equiv [X_{i_r}, X_{i_{r+1}}]$ modulo $\mathfrak{L}$, it is easily seen that $t - t' \in \mathfrak{L} + \mathfrak{I}_{p-1} \subseteq \mathfrak{L} + \sum_{0 \leq q \leq p-1} \mathfrak{I}_q^0$, by the induction hypothesis. Hence $t \in \mathfrak{L} + \sum_{0 \leq q \leq p} \mathfrak{I}_q^0$. This proves that $\mathfrak{I}_p^d \subseteq \mathfrak{L} + \sum_{0 \leq q \leq p} \mathfrak{I}_q^0$, and carries the induction forward.

The proof that $\mathfrak{L} \cap \mathfrak{I}^0 = 0$ is more complicated. We shall achieve this by constructing an endomorphism L of $\mathfrak{J}$ such that

$$(3.2.5)\quad \begin{cases} \text{(i)} & L(t) = t \text{ for all standard monomials } t \\ \text{(ii)} & \text{if } p \geq 2,\ 1 \leq s \leq p-1, \text{ and } i_s > i_{s+1}, \text{ then} \\ & L(X_{i_1} \otimes \cdots \otimes X_{i_s} \otimes X_{i_{s+1}} \otimes \cdots \otimes X_{i_p}) \\ & \quad = L(\cdots \otimes X_{i_{s+1}} \otimes X_{i_s} \otimes \cdots) + L(\cdots \otimes [X_{i_s}, X_{i_{s+1}}] \otimes \cdots). \end{cases}$$

Indeed, suppose that such an endomorphism L of $\mathfrak{J}$ has been constructed. It follows easily from (ii) above that $L(t_1 \otimes u_{X_i, X_j} \otimes t_2) = 0$ for all $t_1, t_2 \in \mathfrak{J}$, $i, j \in J$ and, consequently, that L is zero on $\mathfrak{L}$. On the other hand, L is the identity on $\mathfrak{J}^0$. So $\mathfrak{L} \cap \mathfrak{J}^0$ has to be 0.

Define L to be the identity on $\mathfrak{J}_0 + \mathfrak{J}_1$. Suppose that $p \geq 2$ and that L is an endomorphism of $\sum_{0 \leq q \leq p-1} \mathfrak{J}_q$ satisfying (3.2.5) for all monomials of degree $\leq p-1$. We wish to extend L to an endomorphism of $\sum_{0 \leq q \leq p} \mathfrak{J}_q$ which will satisfy (3.2.5) for all monomials of degree $\leq p$. It is clearly enough to define $L(t)$ for all $t = X_{i_1} \otimes \cdots \otimes X_{i_p}$ in such a way that (i) and (ii) of (3.2.5) are satisfied. We do this by induction on $d = \operatorname{ind}(t)$. For $d = 0$, we put $L(t) = t$. Suppose that $d \geq 1$ and that L has been defined so as to satisfy (3.2.5) for all monomials of degree p and index $\leq d-1$. Let $t = X_{i_1} \otimes \cdots \otimes X_{i_p} \in \mathfrak{J}_p^d$. Select an integer r, $1 \leq r \leq p-1$, such that $i_r > i_{r+1}$, and define $L(t)$ by the right side of part (ii) of (3.2.5), with r replacing s. It is not immediately obvious that $L(t)$ is well defined, since the integer r is in general not unique. However, if we can ensure that $L(t)$ is well defined, (i) and (ii) of (3.2.5) would follow at once, and we would have an endomorphism of $\sum_{0 \leq q \leq p-1} \mathfrak{J}_q + \sum_{0 \leq e \leq d} \mathfrak{J}_p^e$ such that (3.2.5) is satisfied for all monomials of degree $\leq p-1$ and of degree p but of index $\leq d$. The inductions on d and p then lead to the existence of L. To show that $L(t)$ is well defined, let l be another integer, $1 \leq l \leq p-1$, with $i_l > i_{l+1}$. Let u and v denote the respective expressions obtained by replacing s with r and with l in the right side of (ii) in (3.2.5). We must show, using the induction hypothesis, that $u = v$. In the following, for brevity we write Y_t for X_{i_t}, $1 \leq t \leq p$. Two cases arise.

Case 1. $|r - l| \geq 2$. We may assume without losing generality that $r \geq l + 2$. Then $p \geq 4$. Since both u and v are in $\sum_{0 \leq q \leq p-1} \mathfrak{J}_q + \sum_{0 \leq e \leq d-1} \mathfrak{J}_p^e$, we can use the induction hypothesis to simplify them both. A simple calculation then shows that both u and v are equal to

$$\begin{aligned} & L(\cdots \otimes Y_{l+1} \otimes Y_l \otimes \cdots \otimes Y_{r+1} \otimes Y_r \otimes \cdots) \\ & + L(\cdots \otimes [Y_l, Y_{l+1}] \otimes \cdots \otimes Y_{r+1} \otimes Y_r \otimes \cdots) \\ & + L(\cdots \otimes Y_{l+1} \otimes Y_l \otimes \cdots \otimes [Y_r, Y_{r+1}] \otimes \cdots) \\ & + L(\cdots \otimes [Y_l, Y_{l+1}] \otimes \cdots \otimes [Y_r, Y_{r+1}] \otimes \cdots). \end{aligned}$$

Case 2. $|r - l| = 1$. We may assume that $r = l + 1$. Then $i_l > i_{l+1} >$

i_{l+2}, and $p \geq 3$. Using the induction hypothesis we obtain, after a simple calculation,

$$\begin{aligned} u &= L(\cdots \otimes Y_{l+2} \otimes Y_{l+1} \otimes Y_l \otimes \cdots) + L(\cdots \otimes Y_l \otimes [Y_{l+1}, Y_{l+2}] \otimes \cdots) \\ &\quad + L(\cdots \otimes [Y_l, Y_{l+2}] \otimes Y_{l+1} \otimes \cdots) + L(\cdots \otimes Y_{l+2} \otimes [Y_l, Y_{l+1}] \otimes \cdots) \\ v &= L(\cdots \otimes Y_{l+2} \otimes Y_{l+1} \otimes Y_l \otimes \cdots) + L(\cdots \otimes [Y_{l+1}, Y_{l+2}] \otimes Y_l \otimes \cdots) \\ &\quad + L(\cdots \otimes Y_{l+1} \otimes [Y_l, Y_{l+2}] \otimes \cdots) + L(\cdots \otimes [Y_l, Y_{l+1}] \otimes Y_{l+2} \otimes \cdots). \end{aligned}$$

On the other hand, it follows from the induction hypothesis that for any $X, Y \in \mathfrak{g}$, $t_1 \in \mathfrak{J}_a$, $t_2 \in \mathfrak{J}_b$ with $a \geq 0$, $b \geq 0$ and $a + b = p - 3$, we have

$$L(t_1 \otimes X \otimes Y \otimes t_2) - L(t_1 \otimes Y \otimes X \otimes t_2) = L(t_1 \otimes [X,Y] \otimes t_2).$$

If we use this in the expression for u and compare the result with the expression for v, we see that the relation $u = v$ would follow provided we show that L annihilates the element

$$\begin{aligned} &\cdots \otimes [Y_l, [Y_{l+1}, Y_{l+2}]] \otimes \cdots + \cdots \otimes [Y_{l+1}, [Y_{l+2}, Y_l]] \otimes \cdots \\ &+ \cdots \otimes [Y_{l+2}, [Y_l, Y_{l+1}]] \otimes \cdots \end{aligned}$$

of $\mathfrak{J}_{p-2}$. This is, however, an immediate consequence of the Jacobi identity for $\mathfrak{g}$.

We have thus proved that $u = v$ in both cases. $L(t)$ is thus well defined. As observed earlier, this proves the entire theorem.

Remarks 1. The Poincaré–Birkhoff–Witt theorem proved above is one of the most fundamental results in the theory of Lie algebras. Together with the theorem on the existence and uniqueness of the universal enveloping algebra, it constitutes the principal device for converting Lie algebra problems into associative algebra problems. As such it occupies a central place in the theory.

2. Since γ is injective on $\mathfrak{g}$, it is possible to identify $\mathfrak{g}$ with its image in $\mathfrak{G}$ under γ. With this identification, $\mathfrak{G}$ will be called *the universal enveloping algebra of* $\mathfrak{g}$. $\mathfrak{g} \subseteq \mathfrak{G}$, $\mathfrak{g}$ generates $\mathfrak{G}$, and

$$[X,Y] = XY - YX \quad (X, Y \in \mathfrak{g}). \tag{3.2.6}$$

We shall henceforth make this identification without explicit comment. If $\dim(\mathfrak{g}) < \infty$ and $\{X_1, \ldots, X_m\}$ is a basis for $\mathfrak{g}$, then the elements $X_1^{r_1} \cdots X_m^{r_m}$ ($r_1, \ldots, r_m$ nonnegative integers), which we call the *standard monomials of the basis* $\{X_1, \ldots, X_m\}$, constitute a basis for $\mathfrak{G}$ by Theorem 3.2.2.

3. Let $\dim(\mathfrak{g}) < \infty$, and let $\{X_1, \ldots, X_m\}$ be a basis for $\mathfrak{g}$; let c_{ijr} be the

corresponding structure constants. Then $\mathfrak{G}$ is generated by the X_i with the relations

$$X_iX_j - X_jX_i = \sum_{1 \leq r \leq m} c_{ijr}X_r \quad (1 \leq i, j \leq m). \tag{3.2.7}$$

The relations (3.2.7) actually give a presentation of $\mathfrak{G}$; in fact, if $\mathfrak{A}$ is any associative algebra generated by elements $x_1, \ldots, x_m$ satisfying the relations

$$x_ix_j - x_jx_i = \sum_{1 \leq r \leq m} c_{ijr}x_r \quad (1 \leq i, j \leq m),$$

then Theorem 3.2.1 leads at once to the existence of a unique homomorphism ζ of $\mathfrak{G}$ onto $\mathfrak{A}$ such that $\zeta(X_i) = x_i$ for $1 \leq i \leq m$.

Theorem 3.2.3. *Let $\mathfrak{g}$ be a Lie algebra k, $\mathfrak{G}$ its universal enveloping algebra.*

(i) *Suppose that V is a vector space and π is a representation of $\mathfrak{g}$ in V. Then there exists a representation π' of the associative algebra $\mathfrak{G}$ in V such that $\pi(X) = \pi'(X)$ for all $X \in \mathfrak{g}$; π' is uniquely determined by π. In other words, π "extends" uniquely to a representation of $\mathfrak{G}$ in V.*

(ii) *Suppose α is an automorphism (resp. antiautomorphism) of $\mathfrak{g}$. Then there is a unique automorphism (resp. antiautomorphism) $\tilde{\alpha}$ of $\mathfrak{G}$ that extends α, and $\widetilde{\alpha^{-1}} = \tilde{\alpha}^{-1}$. If α, β are two automorphisms (resp. antiautomorphisms), then $\widetilde{\alpha\beta} = \tilde{\alpha}\tilde{\beta}$.*

(iii) *If D is a derivation of $\mathfrak{g}$, there is a unique derivation $\tilde{D}$ of $\mathfrak{G}$ that extends D; if D_1, D_2 are two such, then $[\widetilde{D_1, D_2}] = [\tilde{D}_1, \tilde{D}_2]$. In particular, for any $X \in \mathfrak{g}$, ad X is the derivation $a \mapsto Xa - aX$ of $\mathfrak{G}$.*

Proof. (i) If in the definition of the universal enveloping algebra we take $\mathfrak{A}$ to be the associative algebra of endomorphisms of V and ξ as the map $X \mapsto \pi(X)$, we get (i). To prove (ii), let α be an automorphism (resp. antiautomorphism) of $\mathfrak{g}$. Then α is a linear bijection of the underlying vector space of $\mathfrak{g}$ and so can be extended to an automorphism (resp. antiautomorphism) $\bar{\alpha}$ of $\mathfrak{I}$. Since $\bar{\alpha}(u_{X,Y}) = u_{\alpha(X),\alpha(Y)}$ (resp. $\bar{\alpha}(u_{X,Y}) = u_{\alpha(Y),\alpha(X)}$) for X, $Y \in \mathfrak{g}$, we have $\bar{\alpha}[\mathfrak{L}] \subseteq \mathfrak{L}$. On the other hand, if β is another automorphism (resp. antiautomorphism) of $\mathfrak{g}$, $\overline{\alpha\beta} = \bar{\alpha}\bar{\beta}$. In particular, $(\overline{\alpha^{-1}}) = (\bar{\alpha})^{-1}$, and hence α^{-1} leaves $\mathfrak{L}$ invariant. So $\bar{\alpha}$ induces an automorphism (resp. antiautomorphism) of $\mathfrak{G}$, say $\tilde{\alpha}$. Since $\mathfrak{g}$ generates $\mathfrak{G}$, the uniqueness of $\tilde{\alpha}$ is clear. The relation $\overline{\alpha\beta} = \bar{\alpha}\bar{\beta}$ implies that $\widetilde{\alpha\beta} = \tilde{\alpha}\tilde{\beta}$ and $(\widetilde{\alpha^{-1}}) = (\tilde{\alpha})^{-1}$. This proves (ii). Suppose D is a derivation of $\mathfrak{g}$. Then there is a derivation $\bar{D}$ of $\mathfrak{I}$ that extends D. Since $\bar{D}(u_{X,Y}) = u_{DX,Y} + u_{X,DY}$ $(X, Y \in \mathfrak{g})$, $\bar{D}$ leaves $\mathfrak{L}$ invariant. Hence $\bar{D}$ induces a derivation $\tilde{D}$ of $\mathfrak{G}$. If D_1 and D_2 are derivations of $\mathfrak{g}$, $[\overline{D_1, D_2}] = [\bar{D}_1, \bar{D}_2]$. This implies that $[\widetilde{D_1, D_2}] = [\tilde{D}_1, \tilde{D}_2]$. The uniqueness of $\tilde{D}$ follows, as before, from the fact that $\mathfrak{g}$ generates $\mathfrak{G}$. If $D = \mathrm{ad}\,X (X \in \mathfrak{g})$, then $a \mapsto Xa - aX$ is a derivation of $\mathfrak{G}$ that coincides with D on $\mathfrak{g}$. Hence by the uniqueness of $\tilde{D}$, $\tilde{D}a = Xa - aX, a \in \mathfrak{G}$.

Corollary 3.2.4. *There is a unique antiautomorphism $a \mapsto a^t$ of $\mathfrak{G}$ such that*

$$X^t = -X \quad (X \in \mathfrak{g}), \tag{3.2.8}$$

and this antiautomorphism is involutive.

Proof. Since $X \mapsto -X$ is an antiautomorphism of $\mathfrak{g}$, the first assertion follows from the theorem above. Since $a \mapsto (a^t)^t$ is an automorphism of $\mathfrak{G}$ which is the identity on $\mathfrak{g}$, $(a^t)^t = a$ for all $a \in \mathfrak{G}$.

For any $a \in \mathfrak{G}$, a^t is called its *formal transpose*. If $Y_1, \ldots, Y_s \in \mathfrak{g}$, then

$$(Y_1 \cdots Y_s)^t = (-1)^s Y_s Y_{s-1} \cdots Y_1 \quad (s \geq 1). \tag{3.2.9}$$

For $X \in \mathfrak{g}$ and $a \in \mathfrak{G}$, write

$$(\mathrm{ad}\,X)(a) = Xa - aX. \tag{3.2.10}$$

It is usual to refer to the representation $X \mapsto \mathrm{ad}\,X$ of $\mathfrak{g}$ in $\mathfrak{G}$ as the *adjoint representation* of $\mathfrak{g}$ in $\mathfrak{G}$.

Theorem 3.2.5. *Let $\mathfrak{g}$ be a Lie algebra over k, $\mathfrak{G}$ its universal enveloping algebra. If $\mathfrak{a}$ is a subalgebra of $\mathfrak{g}$ and $\mathfrak{A}$ the subalgebra of $\mathfrak{G}$ generated by $\mathfrak{a}$, then $\mathfrak{A}$, together with the identity map of $\mathfrak{a}$ into it, is the universal enveloping algebra of $\mathfrak{a}$. If $\mathfrak{a}$ is an ideal of $\mathfrak{g}$ and $\mathfrak{S} = \mathfrak{G}\mathfrak{a}\mathfrak{G}$, then $\mathfrak{S}$ is a proper (two-sided) ideal of $\mathfrak{G}$, and $\mathfrak{G}/\mathfrak{S}$, taken with the natural map of $\mathfrak{g}/\mathfrak{a}$ into it, is a universal enveloping algebra of $\mathfrak{g}/\mathfrak{a}$. Let $\mathfrak{g}_i$ $(1 \leq i \leq s)$ be Lie algebras over k, $\mathfrak{G}_i$ the universal enveloping algebra of $\mathfrak{g}_i$; $\mathfrak{g} = \mathfrak{g}_1 \times \cdots \times \mathfrak{g}_s$, and $\mathfrak{G}$ the universal enveloping algebra of $\mathfrak{g}$. Then the map*

$$\begin{aligned}(X_1, \ldots, X_s) \mapsto\ & X_1 \otimes 1 \otimes \cdots \otimes 1 \\ & + 1 \otimes X_2 \otimes 1 \otimes \cdots \otimes 1 + \cdots + 1 \otimes 1 \otimes \cdots \otimes 1 \otimes X_s\end{aligned} \tag{3.2.11}$$

extends uniquely to an isomorphism of $\mathfrak{G}$ onto the tensor product $\mathfrak{G}_1 \otimes \cdots \otimes \mathfrak{G}_s$ of the algebras $\mathfrak{G}_i$.

Proof. Let $\mathfrak{A}'$ be the universal enveloping algebra of the subalgebra $\mathfrak{a}$ of $\mathfrak{g}$. We denote the product operation in $\mathfrak{A}'$ by $\cdot$. Then there is a homomorphism ξ of $\mathfrak{A}'$ into $\mathfrak{G}$ such that $\xi(X) = X$ for $X \in \mathfrak{a}$. Since $\mathfrak{a}$ generates $\mathfrak{A}'$ as well as $\mathfrak{A}$, it is clear that ξ maps $\mathfrak{A}'$ onto $\mathfrak{A}$. Let $\{X_j : j \in J\}$ be an ordered basis of $\mathfrak{a}$. Since this can be enlarged to an ordered basis of $\mathfrak{g}$, the monomials

$X_{r_1} \cdots X_{r_s} (r_1 \leq \cdots \leq r_s)$ are linearly independent, so they form a basis for $\mathfrak{A}$. On the other hand, the monomials $X_{r_1} \cdot X_{r_2} \cdot \cdots \cdot X_{r_s}$ $(r_1 \leq r_2 \leq \cdots \leq r_s)$ form a basis for $\mathfrak{A}'$, and we have $\xi(X_{r_1} \cdot X_{r_2} \cdot \cdots \cdot X_{r_s}$. So ξ is an isomorphism. This prove the first assertion.

Let $\mathfrak{a}$ be an ideal, $\mathfrak{b} = \mathfrak{g}/\mathfrak{a}$ the quotient Lie algebra, and γ the natural map of $\mathfrak{g}$ onto $\mathfrak{b}$. Suppose $\mathfrak{C}$ is an associative algebra and π is a linear map of $\mathfrak{b}$ into $\mathfrak{C}$ such that $\pi([X', Y']) = \pi(X')\pi(Y') - \pi(Y')\pi(X')$ for all $X', Y' \in \mathfrak{b}$. It is then obvious that there is a homomorphism γ' of $\mathfrak{G}$ into $\mathfrak{C}$ such that $\gamma'(X) = \pi(\gamma(X))$ for all $X \in \mathfrak{g}$. If $a, b \in \mathfrak{G}$ and $Y \in \mathfrak{a}$, $\gamma'(aYb) = \gamma'(a)$ $\pi(\gamma(Y))\gamma'(b) = 0$; hence $\gamma' = 0$ on $\mathfrak{G}$. So $\mathfrak{S}$ is proper, and γ' induces a homomorphism π' of $\mathfrak{G}/\mathfrak{S}$ into $\mathfrak{C}$. Let η be the natural map of $\mathfrak{G}$ onto $\mathfrak{G}/\mathfrak{S}$. Since $\eta = 0$ on $\mathfrak{a}$, we have a unique linear map $\bar{\eta}$ of $\mathfrak{b}$ into $\mathfrak{G}/\mathfrak{S}$ such that $\bar{\eta}(\gamma(X)) = \eta(X)$, $X \in \mathfrak{g}$. It is then trivial to verify that $\bar{\eta}([X', Y']) = \bar{\eta}(X')\bar{\eta}(Y') - \bar{\eta}(Y')\bar{\eta}(X')$ for all $X', Y' \in \mathfrak{b}$ and that $\pi'(\bar{\eta}(X')) = \pi(X')$ for all $X' \in \mathfrak{b}$. This proves that $(\mathfrak{G}/\mathfrak{S}, \bar{\eta})$ is a universal enveloping algebra of $\mathfrak{b}$.

For the last assertion, we identify the $\mathfrak{g}_i$ with subspaces of $\mathfrak{g}$, so that $[\mathfrak{g}_i, \mathfrak{g}_j] = 0$, $i \neq j$ and $\mathfrak{g} = \mathfrak{g}_1 + \cdots + \mathfrak{g}_s$. Let δ denote the map (3.2.11). Then $\delta([X, Y]) = \delta(X)\delta(Y) - \delta(Y)\delta(X)$ for $X, Y \in \mathfrak{g}$, so δ extends to a homomorphism $\bar{\delta}$ of $\mathfrak{G}$ into $\mathfrak{G}_1 \otimes \cdots \otimes \mathfrak{G}_s$. We prove that $\bar{\delta}$ is a bijection. Let J_i be an ordered set, and let $\{Y_j : j \in J_i\}$ be a basis for $\mathfrak{g}_i$ $(1 \leq i \leq s)$. Let $J = \{(i,j) : 1 \leq i \leq s, j \in J_i\}$, and order J by the following rule: $(i,j) < (i',j')$ if either (a) $i < i'$, or (b) $i = i'$ and $j < j'$ in J_i. Write $Y_j = X_{i,j}, j \in J_i$. Then $\{X_{i,j} : (i,j) \in J\}$ is a basis for $\mathfrak{g}$. If $j_{ir} \in J_i$ $(1 \leq r \leq t_i)$ and $j_{i1} < \cdots < j_{it_i}$, an easy calculation shows that

$$\begin{aligned} \bar{\delta}(X_{1,j_{11}} \cdots X_{1,j_{1t_1}} \cdots X_{s,j_{s1}} \cdots X_{s,j_{st_s}}) \\ = (X_{1,j_{11}} \cdots X_{1,j_{1t_1}}) \otimes \cdots \otimes (X_{s,j_{s1}} \cdots X_{s,j_{st_s}}). \end{aligned} \tag{3.2.12}$$

(3.2.12) shows that $\bar{\delta}$ is a linear isomorphism of $\mathfrak{G}$ onto $\mathfrak{G}_1 \otimes \cdots \otimes \mathfrak{G}_s$.

Corollary 3.2.6. *Let $\mathfrak{g}$ be a Lie algebra over k, $\mathfrak{G}$ its universal enveloping algebra. Then $\mathfrak{G}\mathfrak{g}\mathfrak{G} = \mathfrak{G}\mathfrak{g} = \mathfrak{g}\mathfrak{G}$ $(= \mathfrak{G}^+$, say); and $\mathfrak{G}$ is the direct sum of $k \cdot 1$ and $\mathfrak{G}^+$.*

Proof. Let $\mathfrak{G}^+ = \mathfrak{G}\mathfrak{g}\mathfrak{G}$. Then $\mathfrak{G}^+$ is a proper two-sided ideal in $\mathfrak{G}$ and $\mathfrak{G}/\mathfrak{G}^+$ is the universal enveloping algebra of $\mathfrak{g}/\mathfrak{g} = 0$. So $\dim(\mathfrak{G}/\mathfrak{G}^+) = 1$, proving that $\mathfrak{G}$ is the direct sum of $k \cdot 1$ and $\mathfrak{G}^+$. On the other hand, if J is an ordered set and $\{X_j : j \in J\}$ is a basis for $\mathfrak{g}$, any standard monomial other than 1 lies in both $\mathfrak{G}\mathfrak{g}$ and $\mathfrak{g}\mathfrak{G}$. Hence, by the Poincaré–Birkhoff–Witt theorem $\mathfrak{G} = k \cdot l + \mathfrak{G}\mathfrak{g} = k \cdot 1 + \mathfrak{g}\mathfrak{G}$. Since $\mathfrak{G}\mathfrak{g}$ and $\mathfrak{g}\mathfrak{G}$ are both contained in $\mathfrak{G}^+$, the corollary follows at once.

Corollary 3.2.7. *Let $\mathfrak{a}$ and $\mathfrak{b}$ be subalgebras of $\mathfrak{g}$ such that $\mathfrak{g}$ is their vectorial direct sum. Let $\mathfrak{A}$ (resp. $\mathfrak{B}$) be the subalgebra of $\mathfrak{G}$ generated by $\mathfrak{a}$*

(*resp.* $\mathfrak{b}$). *Then the left* (*resp. right*) *ideal* $\mathfrak{G}\mathfrak{a}$ (*resp.* $\mathfrak{a}\mathfrak{G}$) *is proper, and* $\mathfrak{G}$ *is the vectorial direct sum of* $\mathfrak{B}$ *and* $\mathfrak{G}\mathfrak{a}$ (*resp.* $\mathfrak{a}\mathfrak{G}$).

Proof. Considering an ordered basis for $\mathfrak{g}$ which consists of an ordered basis for $\mathfrak{b}$ followed by an ordered basis for $\mathfrak{a}$, and using the Poincaré–Birkhoff–Witt theorem, we conclude that the map $(b,a) \mapsto ba$ of $\mathfrak{B} \times \mathfrak{A}$ into $\mathfrak{G}$ extends to a linear isomorphism ξ of $\mathfrak{B} \otimes \mathfrak{A}$ onto $\mathfrak{G}$. Let $\mathfrak{A}^+ = \mathfrak{A}\mathfrak{a}$. Since we can canonically identify $\mathfrak{A}$ with the universal enveloping algebra of $\mathfrak{a}$, we deduce from the previous corollary that $\mathfrak{A}$ is the direct sum of $\mathfrak{A}^+$ and $k \cdot 1$. So $\mathfrak{G}$ is the direct sum of $\mathfrak{B}\mathfrak{A}^+ = \xi[\mathfrak{B} \otimes \mathfrak{A}^+]$ and $\mathfrak{B} = \xi[\mathfrak{B} \otimes k \cdot 1]$. But $\mathfrak{B}\mathfrak{A}^+ = \mathfrak{B}\mathfrak{A}\mathfrak{a} = \mathfrak{G}\mathfrak{a}$, so $\mathfrak{G}\mathfrak{a}$ is proper, and $\mathfrak{G} = \mathfrak{G}\mathfrak{a} + \mathfrak{B}$ is a direct sum. The argument for the right ideal is similar.

Remarks 1. Let $\mathfrak{g}$ be any finite-dimensional vector space over k. We may then regard $\mathfrak{g}$ as an abelian Lie algebra by defining $[X, Y] = 0$ for all X, $Y \in \mathfrak{g}$. For X, $Y \in \mathfrak{g}$, $u_{X,Y} = X \otimes Y - Y \otimes X$. $\mathfrak{J}/\mathfrak{L}$ is thus the symmetric algebra $S(\mathfrak{g})$ over $\mathfrak{g}$.

2. Let k' be an extension field of k, $\mathfrak{g}$ a Lie algebra over k, and $\mathfrak{g}^{k'}$ its extension to k'. Let $\mathfrak{G}$ be the universal enveloping algebra of $\mathfrak{g}$ and $\mathfrak{G}^{k'}$ its extension to k'. Then there is a natural isomorphism of $\mathfrak{G}^{k'}$ with the enveloping algebra of $\mathfrak{g}^{k'}$.

One of the most interesting applications of the preceding development is to the theory of free Lie algebras. Let X_i $(i \in I)$ be arbitrary distinct elements. By a *free Lie algebra over k generated by the* X_i we mean a Lie algebra $\mathfrak{g}$ over k such that (i) $X_i \in \mathfrak{g}$ for all $i \in I$, and the X_i generate $\mathfrak{g}$ (i.e., the smallest subalgebra of $\mathfrak{g}$ containing all the X_i is $\mathfrak{g}$ itself, and (ii) if $\mathfrak{h}$ is a Lie algebra over k and X'_i $(i \in I)$ are elements of $\mathfrak{h}$, then there is a Lie algebra homomorphism π of $\mathfrak{g}$ into $\mathfrak{h}$ (necessarily unique) such that $\pi(X_i) = X'_i$ for all $i \in I$. If $\mathfrak{g}$ and $\mathfrak{g}'$ are free Lie algebras generated by the X_i, it is immediate from the definition that there is a (unique) isomorphism π of $\mathfrak{g}$ onto $\mathfrak{g}'$ such that $\pi(X_i) = X_i$ $(i \in I)$. To establish the existence of a free Lie algebra generated by X_i $(i \in I)$, we proceed as follows. Let $\mathfrak{G}$ be the free associative algebra over k generated by X_i $(i \in I)$. For $u, v \in \mathfrak{G}$, let $[u,v] = uv - vu$. Then $\mathfrak{G}$ equipped with $[\cdot,\cdot]$ becomes a Lie algebra; we denote it by $\mathfrak{G}_L$. Let $\mathfrak{g}$ be the smallest subalgebra of $\mathfrak{G}_L$ containing all the X_i $(i \in I)$.

Theorem 3.2.8. $\mathfrak{g}$ *is a free Lie algebra generated by* X_i $(i \in I)$. *Moreover* $\mathfrak{G}$, *together with the identity map of* $\mathfrak{g}$ *into it, is the universal enveloping algebra of* $\mathfrak{g}$.

Proof. Let $\mathfrak{h}$ be a Lie algebra over k, $\mathfrak{H} \supseteq \mathfrak{h}$ the universal enveloping algebra of $\mathfrak{h}$. Suppose X'_i $(i \in I)$ are elements of $\mathfrak{h}$. We denote by ξ the homomorphism of $\mathfrak{G}$ into $\mathfrak{H}$ such that $\xi(X_i) = X'_i$ for all $i \in I$. Let $\pi = \xi \mid \mathfrak{g}$, and let $\mathfrak{g}' = \{u : u \in \mathfrak{G}, \xi(u) \in \mathfrak{h}\}$. Clearly, $\mathfrak{g}'$ is a Lie subalgebra of $\mathfrak{G}_L$, so since $X_i \in \mathfrak{g}'$ for all $i \in I$, $\mathfrak{g} \subseteq \mathfrak{g}'$. So π maps $\mathfrak{g}$ into $\mathfrak{h}$, hence is a homomorphism of $\mathfrak{g}$ into $\mathfrak{h}$. This proves that $\mathfrak{g}$ is a free Lie algebra generated by the X_i. Suppose now that $\mathfrak{A}$ is an associative algebra over k and π is a linear map of $\mathfrak{g}$ into $\mathfrak{A}$ such that $\pi([X, Y]) = \pi(X)\pi(Y) - \pi(Y)\pi(X)$ for $X, Y \in \mathfrak{g}$. Denote by ξ the homomorphism of $\mathfrak{G}$ into $\mathfrak{A}$ such that $\xi(X_i) = \pi(X_i)$, $i \in I$. If $\mathfrak{A}_L$ denotes the Lie algebra whose underlying vector space is that of $\mathfrak{A}$ and for which $[u,v] = uv - vu$ $(u, v \in \mathfrak{A})$, then $\xi|\mathfrak{g}$ and π are both homomorphisms of $\mathfrak{g}$ into $\mathfrak{A}_L$ coinciding on the set $\{X_i : i \in I\}$. Since this set generates $\mathfrak{g}$, $\xi|\mathfrak{g} = \pi$. This proves that $\mathfrak{G}$ is the universal enveloping algebra of $\mathfrak{g}$.

Consider the adjoint representation of $\mathfrak{g}$ in $\mathfrak{G}$. By (i) of Theorem 3.2.3, we may extend this to a representation θ of $\mathfrak{G}$ in $\mathfrak{G}$. We have

$$\begin{aligned} \theta(uv) &= \theta(u)\theta(v) && (u, v \in \mathfrak{G}) \\ \theta(u)(v) &= (\operatorname{ad} u)(v) = [u,v] && (u \in \mathfrak{g}, v \in \mathfrak{G}). \end{aligned} \tag{3.2.13}$$

The representation θ is closely related to the endomorphism ψ of $\mathfrak{G}$ defined by

$$\begin{aligned} \psi(1) &= 0, \qquad \psi(X_i) = X_i \quad (i \in I) \\ \psi(X_{i_1} \cdots X_{i_n}) &= \theta(X_{i_1}) \cdots \theta(X_{i_{n-1}})(X_{i_n}) \\ &= [X_{i_1},[X_{i_2}, \ldots ,[X_{i_{n-1}},X_{i_n}] \ldots] \quad (n \geq 2, i_1, \ldots , i_n \in I). \end{aligned} \tag{3.2.14}$$

It follows easily from this that

$$\psi(uv) = \theta(u)\psi(v)) \quad (u, v \in \mathfrak{G}). \tag{3.2.15}$$

Theorem 3.2.9. (i) *For any integer $n \geq 0$ let $\mathfrak{G}_n$ be the subspace of homogeneous elements*† *of $\mathfrak{G}$ of degree n, and let $\mathfrak{g}_n = \mathfrak{G}_n \cap \mathfrak{g}$. Then $\mathfrak{g}_n$ is the linear span of $\psi(X_{i_1} \cdots X_{i_n})$ $(i_1, \ldots ,i_n \in I)$, and $\psi[\mathfrak{G}_n] = \mathfrak{g}_n$.*

(ii) $\mathfrak{g} = \sum_{n \geq 0} \mathfrak{g}_n$, *and the sum is direct.*

(iii) *Let $u \in \mathfrak{G}_n$. Then $u \in \mathfrak{g}_n$ if and only if $\psi(u) = nu$.*

Proof. $\mathfrak{G}_n$ is the linear span of the $X_{i_1} \cdots X_{i_n}$ $(i_1, \ldots ,i_n \in I)$, and we have

$$\begin{aligned} \psi(X_{i_1} \cdots X_{i_n}) &= [X_{i_1},\psi(X_{i_2} \cdots X_{i_n})] \\ &= X_{i_1}\psi(X_{i_2} \cdots X_{i_{n-1}}) - \psi(X_{i_2} \cdots X_{i_{n-1}})X_{i_1}. \end{aligned} \tag{3.2.16}$$

Using this relation and an induction on n, we find that ψ maps $\mathfrak{G}_n$ into $\mathfrak{G}_n \cap \mathfrak{g} = \mathfrak{g}_n$ for all $n \geq 0$. In order to complete the proofs of (i) and (ii) it is

†Recall that $\mathfrak{G}$ is canonically isomorphic to the tensor algebra over the vector space spanned by the X_i and so is graded. The homogeneous subspace $\mathfrak{G}_n$ of degree n spanned by the $X_{i_1} \ldots X_{i_n}$.

therefore sufficient to prove that $\mathfrak{g}$ is the linear span of the $\psi(X_{i_1} \cdots X_{i_n})$. Let $\mathfrak{m}$ be this linear span. Clearly, $\mathfrak{m} \subseteq \mathfrak{g}$. Now $X_i \in \mathfrak{m}$ for all i, and (3.2.16) shows that $\mathfrak{m}$ is invariant under all $\theta(X_i)$. Hence $\mathfrak{m}$ is invariant under $\theta(u)$, $u \in \mathfrak{G}$. Since $\theta(u)(v) = [u,v]$ for $u,v \in \mathfrak{g}$, this implies that $[\mathfrak{g},\mathfrak{m}] \subseteq \mathfrak{m}$, and in particular that $[\mathfrak{m},\mathfrak{m}] \subseteq \mathfrak{m}$. $\mathfrak{m}$ is thus a Lie subalgebra of $\mathfrak{G}_L$ containing all the X_i, proving that $\mathfrak{g} \subseteq \mathfrak{m}$.

We now prove (iii). If $n \geq 1$, $u \in \mathfrak{G}_n$, and $\psi(u) = nu$, then $u = (1/n)\psi(u) \in \mathfrak{g}_n$ by (i). For the converse, we use induction on n. Assume that $\psi(v) = mv$ for $v \in \mathfrak{g}_m$ and $m < n$. Suppose that $u = \psi(X_{i_1} \cdots X_{i_n})$. We prove that $\psi(u) = nu$. We may assume that $n \geq 2$. Then, writing $v = \psi(X_{i_2} \cdots X_{i_n})$, we have $u = [X_{i_1},v] = X_{i_1}v - vX_{i_1}$. So

$$\begin{aligned}
\psi(u) &= \psi(X_{i_1}v) - \psi(vX_{i_1}) \\
&= [X_{i_1},\psi(v)] - \theta(v)(X_{i_1}) \qquad \text{(by (3.2.15))} \\
&= (n-1)[X_{i_1},v] - [v,X_{i_1}] \quad \text{(by (3.2.13), as } v \in \mathfrak{g}) \\
&= nu.
\end{aligned}$$

Since the $\psi(X_{i_1} \cdots X_{i_n})$ span $\mathfrak{g}_n$, the induction goes forward.

3.3. The Universal Enveloping Algebra as a Filtered Algebra

Let A be an associative algebra over a field k of characteristic 0. A is said to be *graded* if for each integer $n \geq 0$ there is a subspace A_n of A such that (i) $1 \in A_0$ and A is the direct sum of the A_n, and (ii) $A_m A_n \subseteq A_{m+n}$ for $m, n \geq 0$. In this case the elements of $\bigcup_{m=0}^{\infty} A_m$ are called *homogeneous*, and those of A_n are called *homogeneous of degree n;* if $a = \sum_{n \geq 0} a_n$ ($a_n \in A_n$, $a \in A$), then a_n is called the *homogeneous component* of a of degree n.

A is said to be *filtered* if for each integer $n \geq 0$ there is a subspace $A^{(n)}$ of A such that (i) $1 \in A^{(0)}$, $A^{(0)} \subseteq A^{(1)} \subseteq \cdots$, $\bigcup_{n=0}^{\infty} A^{(n)} = A$, and (ii) $A^{(m)}A^{(n)} \subseteq A^{(m+n)}$ for all $m, n \geq 0$. It is convenient to use the convention that $A^{(-1)} = \varnothing$. For $a \in A$, the integer $s \geq 0$ such that $a \in A^{(s)}$ but $\notin A^{(s-1)}$ is called the *degree* of a, and written $\deg(a)$. For $n \geq 0$, $A^{(n)}$ is then the set of all $a \in A$ with $\deg(a) \leq n$. Clearly $\deg(1) = 0$, $\deg(a + b) \leq \max(\deg(a), \deg(b))$, and $\deg(ab) \leq \deg(a) + \deg(b)$ ($a, b \in A$).

Let A be a graded algebra, $A_n (n \geq 0)$ the homogeneous space of degree n. If we put $A^{(n)} = \sum_{0 \leq m \leq n} A_m$ ($n \geq 0$), then it is easily verified that A becomes a filtered algebra. We call it the filtered algebra *associated with* the graded algebra A. However, not every filtered algebra arises in this manner from a graded algebra. Suppose A is a filtered algebra, $A^{(n)}$ the subspace of all elements of degree $\leq n$. We associate a graded algebra with A in the following manner. Let $B_n = A^{(n)}/A^{(n-1)}$ be the quotient vector space and π_n be the natural map of $A^{(n)}$ onto B_n ($n \geq 0$). Define B as the direct sum of the B_n.

Given $\bar{a} \in B_m$ and $\bar{b} \in B_n$, choose $a \in A^{(m)}$ and $b \in A^{(n)}$ such that $\pi_m(a) = \bar{a}$ and $\pi_n(b) = \bar{b}$, and define $\bar{a}\bar{b} = \pi_{m+n}(ab)$. It is easy to verify that $\bar{a}\bar{b}$ is well defined and independent of the choices of a and b. The map $(\bar{a},\bar{b}) \mapsto \bar{a}\bar{b}$ is bilinear from $B_m \times B_n$ into B_{m+n}. These bilinear maps extend to a bilinear map $(\bar{a},\bar{b}) \mapsto \bar{a}\bar{b}$ of $B \times B$ into B. With this as the operation of multiplication, B becomes an associative algebra, and $B_m B_n \subseteq B_{m+n}$ for all $m,n \geq 0$. B is thus a graded algebra and B_n is the subspace of homogeneous elements of degree n. We call it the graded algebra *associated with* the filtered algebra V. Note that $\dim(A^{(n)}) = \sum_{0 \leq m \leq n} \dim(B_m)$, $n \geq 0$.

Let V be a vector space over k, $\mathfrak{J}$ the tensor algebra over V. If we write $\mathfrak{J}_m$ for the space of homogeneous tensors of degree m ($m \geq 0$), $\mathfrak{J}$ becomes a graded algebra. Let $\mathfrak{J}^{(n)} = \sum_{0 \leq m \leq n} \mathfrak{J}_m$. If $\mathfrak{a}$ is any proper two-sided ideal in $\mathfrak{J}$, we can form the quotient algebra $A = \mathfrak{J}/\mathfrak{a}$; if π is the natural map of $\mathfrak{J}$ onto A and we define

$$A^{(n)} = \pi[\mathfrak{J}^{(n)}] \quad (n \geq 0), \tag{3.3.1}$$

it is obvious that A becomes a filtered algebra with $A^{(n)}$ as the subspace of elements of degree $\leq n$. Suppose $\mathfrak{a}$ is a homogeneous ideal, i.e., $\mathfrak{a} = \sum_{m \geq 1} \mathfrak{a} \cap \mathfrak{J}_m$. Then if we define

$$A_n = \pi[\mathfrak{J}_n] \quad (n \geq 0), \tag{3.3.2}$$

it is easily seen that A is a graded algebra with A_n as the homogeneous subspace of degree n, and that the associated filtered algebra is the one defined by (3.3.1). As examples of this we mention the symmetric and exterior algebras over V.

Let $\mathfrak{g}$ be a Lie algebra over k. As usual, we assume that $\mathfrak{g}$ is finite-dimensional unless the contrary is specified. Put $m = \dim(\mathfrak{g})$. Denote by $\mathfrak{J}$ the tensor algebra over $\mathfrak{g}$, by $\mathfrak{G}$ the universal enveloping of $\mathfrak{g}$, and by γ the homomorphism of $\mathfrak{J}$ onto $\mathfrak{G}$ such that $\gamma(X) = X$ for all $\gamma(X) \in \mathfrak{g}$. The kernel of γ is the ideal $\mathfrak{L}$ generated by all elements of the form $u_{X,Y} = X \otimes Y - Y \otimes X - [X,Y]$ ($X, Y \in \mathfrak{g}$). Since $\mathfrak{L}$ is not a homogeneous ideal of $\mathfrak{J}$, we cannot expect $\mathfrak{G}$ to be a graded algebra. It is, however, a filtered algebra with

$$\mathfrak{G}^{(n)} = \gamma[\mathfrak{J}^{(n)}] \quad (n \geq 0). \tag{3.3.3}$$

Theorem 3.3.1. (i) *$\mathfrak{G}^{(0)} = k1$; $\mathfrak{G}^{(1)}$ is the direct sum of $\mathfrak{G}^{(0)}$ and $\mathfrak{g}$; and for any $n \geq 0$, $\mathfrak{G}^{(n)}$ is the linear span of* 1 *and all elements of the form $Z_1 \cdots Z_s$ ($1 \leq s \leq n$, $Z_i \in \mathfrak{g}$ for all i).*

(ii) *If $\{X_1, \ldots, X_m\}$ is a basis for $\mathfrak{g}$, the standard monomials $X_1^{r_1} \cdots X_m^{r_m}$ with $r_1 + \cdots + r_m \leq n$ form a basis for $\mathfrak{G}^{(n)}$.*

(iii) *Let $\bar{\mathfrak{G}}$ be the graded algebra associated with $\mathfrak{G}$. Then $\mathfrak{G}$ is commuta-*

tive. Moreover, the natural map $X \mapsto \bar{X}$ of $\mathfrak{g} \subseteq \mathfrak{G}^{(1)}$ into $\bar{\mathfrak{G}}$ extends to an algebra isomorphism of the symmetric algebra $\mathfrak{S}$ over $\mathfrak{g}$ onto $\bar{\mathfrak{G}}$.

Proof. The relation $\mathfrak{G}^{(0)} = k \cdot 1$ is obvious. Let $n \geq 1$ be arbitrary. Then $\mathfrak{J}^{(n)}$ is spanned by 1 and all tensors of the form $Z_1 \otimes \cdots \otimes Z_s$ with $1 \leq s \leq n$ and $Z_1, \ldots, Z_s \in \mathfrak{g}$. Hence $\mathfrak{G}^{(n)}$ is the linear span of 1 and the $Z_1, \ldots, Z_s$ $(1 \leq s \leq n, Z_i \in \mathfrak{g}$ for all $i)$. For $n = 1$, this gives us $\mathfrak{G}^{(1)} = k \cdot 1 + \mathfrak{g}$. Since $k \cdot 1 \cap \mathfrak{g} = 0$, all statements of (i) are proved.

Let $X_1, \ldots, X_m$ be a basis for $\mathfrak{g}$. Since the standard monomials are linearly independent, (ii) will be proved if we show that $\mathfrak{G}^{(n)}$ is spanned by the $X_1^{r_1} \cdots X_m^{r_m}$ with $r_1 + \cdots + r_m \leq n$. But by (3.2.4), $\mathfrak{J}^{(n)} \subseteq \mathfrak{L} + \sum_{0 \leq q \leq n} \mathfrak{J}_q^0$, and hence $\mathfrak{G}^{(n)} \subseteq \sum_{0 \leq q \leq n} \gamma[\mathfrak{J}_q^0]$. Since the right side of the last inclusion is the linear span of the $X_1^{r_1} \cdots X_m^{r_m}$ with $r_1 + \cdots + r_m \leq n$, (ii) is proved.

We come now to the proof of (iii). Let $\mathfrak{S}$ be the symmetric algebra over $\mathfrak{g}$. We show first that $\bar{\mathfrak{G}}$ is commutative. Suppose $s \geq 2$ and $Z_1, \ldots, Z_s \in \mathfrak{g}$. Since $Z_r Z_{r+1} - Z_{r+1} Z_r = [Z_r, Z_{r+1}] \in \mathfrak{g}$ for $1 \leq r \leq s-1$, it follows that $Z_1 \cdots Z_s - Z_{\sigma(1)} \cdots Z_{\sigma(s)} \in \mathfrak{G}^{(s-1)}$, where σ is the permutation of $\{1, \ldots, s\}$ that interchanges r and $r+1$ while leaving the others fixed. Now, any permutation of $\{1, \ldots, s\}$ is a product of adjacent interchanges. Consequently, we have

$$Z_1 \cdots Z_s - Z_{\sigma(1)} \cdots Z_{\sigma(s)} \in \mathfrak{G}^{(s-1)} \tag{3.3.4}$$

for any permutation σ of $\{1, \ldots, s\}$. In particular, if $X_1, \ldots, X_p, Y_1, \ldots, Y_n$ are arbitrary elements of $\mathfrak{g}$, then $X_1 \cdots X_p Y_1, \ldots, Y_n \equiv Y_1 \cdots Y_n X_1 \cdots Y_p \pmod{\mathfrak{G}^{(n+p-1)}}$ for $n, p \geq 1$. We conclude from this that

$$ab \equiv ba \pmod{\mathfrak{G}^{(n+p-1)}} \quad (a \in \mathfrak{G}^{(n)}, b \in \mathfrak{G}^{(p)}, n, p \geq 0). \tag{3.3.5}$$

The relation (3.3.5) shows that $\bar{\mathfrak{G}}$ is commutative.

Since $\mathfrak{G}^{(1)}$ is the direct sum of $k \cdot 1$ and $\mathfrak{g}$, the natural map of $\mathfrak{G}^{(1)}$ into $\bar{\mathfrak{G}}$ induces an injection of $\mathfrak{g}$ into $\bar{\mathfrak{G}}$. Let $\bar{X}$ denote the image of X under this map. Since $\bar{\mathfrak{G}}$ is commutative, this linear map extends to a homomorphism ξ of $\mathfrak{S}$ into $\bar{\mathfrak{G}}$. We prove that ξ is an isomorphism. Let $\{X_1, \ldots, X_m\}$ be a basis for $\mathfrak{g}$. Denote by $\bar{\mathfrak{G}}_n$ the homogeneous subspace of $\bar{\mathfrak{G}}$ of degree n, and let π_n be the linear map of $\mathfrak{G}^{(n)}$ onto $\bar{\mathfrak{G}}_n$ with kernel $\mathfrak{G}^{(n-1)}$. If $n \geq 1$, the elements $X_1^{r_1} \cdots X_m^{r_m}$ with $r_1 + \cdots + r_m = n$ are linearly independent modulo $\mathfrak{G}^{(n-1)}$. So the elements $\bar{X}_1^{r_1} \cdots \bar{X}_m^{r_m} = \pi_n(X_1^{r_1} \cdots X_m^{r_m})$ $(r_1 + \cdots + r_m = n)$ form a basis for $\bar{\mathfrak{G}}_n$. In other words, the monomials $\bar{X}_1^{r_1} \cdots \bar{X}_m^{r_m}$ $(r_1, \ldots, r_m \geq 0)$ form a basis for $\mathfrak{G}$. The fact that ξ is an isomorphism of $\mathfrak{S}$ onto $\bar{\mathfrak{G}}$ follows immediately.

Corollary 3.3.2. *Let α (resp. D) be an automorphism (resp. derivation) of $\mathfrak{g}$ and let $\tilde{\alpha}$ (resp. $\tilde{D}$) the corresponding automorphism (resp. derivation) of $\mathfrak{G}$. Then $\tilde{\alpha}$ (resp. $\tilde{D}$) maps $\mathfrak{G}^{(n)}$ into itself for all $n \geq 0$.*

Theorems 3.2.2 and 3.3.1 summarize the essential features of $\mathfrak{G}$ as a filtered algebra and are of great use in many applications. We now indicate one of these, namely to the construction of the so-called *symmetrizer map* of $\mathfrak{S}$ onto $\mathfrak{G}$.

For any integer $p \geq 1$ we denote by Π_p the permutation group of $\{1, \ldots, p\}$ Let $s \mapsto s^{(p)}$ be the natural representation of Π_p in $\mathfrak{I}_p$. Let the endomorphisms Q_p of $\mathfrak{I}_p$ be defined by

$$Q_0 = 1,$$
$$Q_p = \frac{1}{p!} \sum_{s \in \Pi_p} s^{(p)} \quad (p \geq 1). \tag{3.3.6}$$

Then $Q_1 = 1$, each Q_p is a projection, and the range of Q_p is the space of homogeneous symmetric tensors of degree p. We write $\bar{\mathfrak{I}}$ for the space of all symmetric tensors and $\bar{\mathfrak{I}}_p = \bar{\mathfrak{I}} \cap \mathfrak{I}_p$ (p > 0). We have

$$\dim \bar{\mathfrak{I}}_p = \binom{m+p-1}{p}. \tag{3.3.7}$$

Lemma 3.3.3. *Let notation be as above. Then $\mathfrak{I}$ is the direct sum of $\mathfrak{L}$ and $\bar{\mathfrak{I}}$. Moreover, if γ is the natural map of $\mathfrak{I}$ onto $\mathfrak{G}$, γ is a linear isomorphism of $\sum_{0 \leq q \leq p} \bar{\mathfrak{I}}_q$ onto $\mathfrak{G}^{(p)}$ $(p \geq 0)$, and of $\bar{\mathfrak{I}}$ onto $\mathfrak{G}$.*

Proof. γ is an isomorphism of $\bar{\mathfrak{I}}_0 = \mathfrak{I}_0$ onto $\mathfrak{G}^{(0)}$. Let $p \geq 1$ be arbitrary. If $X_1, \ldots, X_p \in \mathfrak{g}$, $1 \leq r \leq p-1$, and s is the permutation of $\{1, \ldots, p\}$ that interchanges r and $r+1$ while leaving the others fixed, it is clear that

$$X_1 \otimes \cdots \otimes X_p - s^{(p)}(X_1 \otimes \cdots \otimes X_p) \in \mathfrak{I}_{p-1} + \mathfrak{L}.$$

We easily conclude from this that

$$t - s^{(p)}(t) \in \mathfrak{I}_{p-1} + \mathfrak{L} \quad (s \in \Pi_p, t \in \mathfrak{I}_p). \tag{3.3.8}$$

Averaging over Π_p, we deduce from this the relation

$$t \equiv Q_p(t) \pmod{\mathfrak{I}_{p-1} + \mathfrak{L}}, t \in \mathfrak{I}_p). \tag{3.3.9}$$

Applying γ to (3.3.9), we finally obtain the inclusion

$$\mathfrak{G}^{(p)} \subseteq \mathfrak{G}^{(p-1)} + \gamma[\bar{\mathfrak{I}}_p].$$

A simple induction on p now shows that γ maps $\sum_{0 \leq q \leq p} \bar{\mathfrak{I}}_q$ onto $\mathfrak{G}^{(p)}$ for all $p \geq 0$. On the other hand, it follows from (3.3.7) and (ii) of Theorem 3.3.1 that $\mathfrak{G}^{(p)}$ has the same dimension as $\sum_{0 \leq q \leq p} \bar{\mathfrak{I}}_q$. So γ is a linear isomorphism of $\sum_{0 \leq q \leq p} \bar{\mathfrak{I}}_q$ onto $\mathfrak{G}^{(p)}$ for $p \geq 0$. In particular, γ is a linear isomorphism of $\bar{\mathfrak{I}}$ onto $\mathfrak{G}$. This implies at once that $\mathfrak{I}$ is the direct sum of $\mathfrak{L}$ and $\bar{\mathfrak{I}}$.

Let $\bar{\mathfrak{g}}$ be the abelian Lie algebra whose underlying vector space is the same as that of $\mathfrak{g}$. Then the universal enveloping algebra of $\bar{\mathfrak{g}}$ can be identified with the symmetric algebra $\mathfrak{S}$ over $\mathfrak{g}$. Let $\bar{\gamma}$ be the natural homomorphism of $\mathfrak{J}$ onto $\mathfrak{S}$. Then the kernel of $\bar{\gamma}$ is the two-sided ideal $\mathfrak{M}$ of $\mathfrak{J}$ generated by the elements of the form $X \otimes Y - Y \otimes X$ $(X, Y \in \mathfrak{g})$. Let $\mathfrak{S}_p$ be the homogeneous subspace of $\mathfrak{S}$ of degree p, and let

$$\mathfrak{S}^{(p)} = \sum_{0 \le q \le p} \mathfrak{S}_q \quad (p \ge 0). \tag{3.3.10}$$

We may then conclude from Lemma 3.3.3 that $\bar{\gamma}$ is a linear isomorphism of $\sum_{0 \le q \le p} \bar{\mathfrak{J}}_q$ onto $\mathfrak{S}^{(p)}$ for every $p \ge 0$ and of $\bar{\mathfrak{J}}$ onto $\mathfrak{S}$. In particular, $\mathfrak{J}$ is the direct sum of $\mathfrak{M}$ and $\bar{\mathfrak{J}}$.

We now define λ to be the unique linear map of $\mathfrak{S}$ into $\mathfrak{G}$ such that the diagram

(3.3.11)

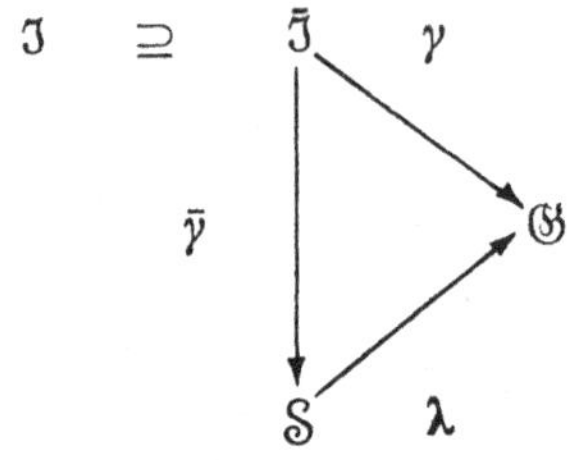

is commutative. It is obvious from the definition that λ is a linear isomorphism of $\mathfrak{S}$ onto $\mathfrak{G}$. It is also clear that for any $p \ge 0$, the diagram

(3.3.12)

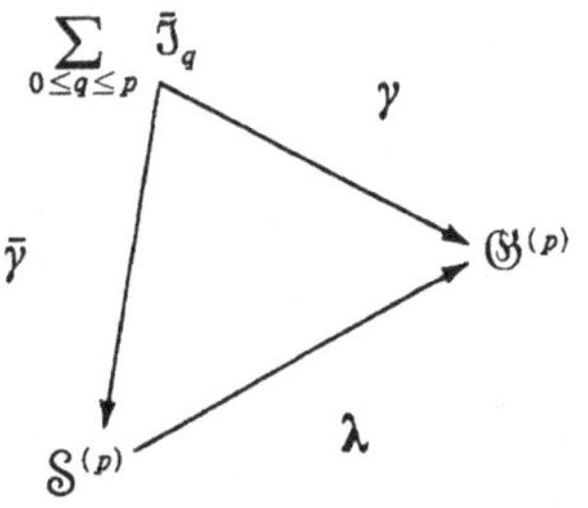

is commutative. The map λ is called the *symmetrizer* of $\mathfrak{S}$ onto $\mathfrak{G}$ (cf. (3.3.13)) and has been used systematically by Harish-Chandra in his work on the representation theory of semisimple Lie groups and Lie algebras.

Theorem 3.3.4 (i) *λ is a linear isomorphism of $\mathfrak{S}$ onto $\mathfrak{G}$ and of $\mathfrak{S}^{(p)}$ onto $\mathfrak{G}^{(p)}$ for $p \ge 0$. If $X_1, \ldots, X_p \in \mathfrak{g}$,*

$$\lambda(\bar{\gamma}(X_1) \cdots \bar{\gamma}(X_p)) = \frac{1}{p!} \sum_{s \in \Pi_p} X_{s(1)} \cdots X_{s(p)}. \tag{3.3.13}$$

(ii) *if* $u \in \mathfrak{S}_n$, $v \in \mathfrak{S}_p$, *then*

$$\lambda(uv) \equiv \lambda(u)\lambda(v) \pmod{\mathfrak{G}^{(n+p-1)}} \tag{3.3.14}$$

(iii) *if* α *(resp. D) is an automorphism (resp. derivation) of* $\mathfrak{g}$, *and* $\tilde{\alpha}$ *and* $\bar{\alpha}$ *(resp.* $\tilde{D}$, $\bar{D}$*) the respective corresponding automorphisms (resp. derivations) of* $\mathfrak{G}$ *and* $\mathfrak{S}$, *then*

$$\lambda \circ \bar{\alpha} = \tilde{\alpha} \circ \lambda, \qquad \lambda \circ \bar{D} = \tilde{D} \circ \lambda. \tag{3.3.15}$$

Proof. Only (3.3.13) remains to be proved in (i). We may assume that $p \geq 1$. Let $X_1, \ldots, X_p \in \mathfrak{g}$ and let $t = X_1 \otimes \cdots \otimes X_p$ Then from (3.3.6), $\gamma(Q_p(t)) = (1/p!) \sum_{s \in \Pi} X_{s(1)} \cdots X_{s(p)}$. On the other hand, since $\mathfrak{S}$ is commutative, $\bar{\gamma}(Q_p(t)) = \bar{\gamma}(X_1) \cdots \bar{\gamma}(X_p)$. We thus have (3.3.13).

We now prove (3.3.14). It is clear from (3.3.4) and (3.3.13) that for arbitrary $X_1, \ldots, X_p \in \mathfrak{g}$,

$$\lambda(\bar{\gamma}(X_1) \cdots \bar{\gamma}(X_p)) \equiv X_1 \cdots X_p \pmod{\mathfrak{G}^{(p-1)}}.$$

Consequently, if $Y_1, \ldots, Y_n, X_1, \ldots, X_p \in \mathfrak{g}$, then, writing $u = \bar{\gamma}(Y_1) \cdots \bar{\gamma}(Y_n)$ and $v = \bar{\gamma}(X_1) \cdots \bar{\gamma}(X_p)$, we have the congruences $\lambda(u) \equiv Y_1 \cdots Y_n \pmod{\mathfrak{G}^{(n-1)}}$, $\lambda(v) \equiv X_1 \cdots X_p \pmod{\mathfrak{G}^{(p-1)}}$, and $\lambda(uv) \equiv Y_1 \cdots Y_n X_1 \cdots X_p \pmod{\mathfrak{G}^{(n+p-1)}}$. It follows from this that $\lambda(uv) \equiv \lambda(u)\lambda(v) \pmod{\mathfrak{G}^{(n+p-1)}}$. (3.3.14) follows from this.

Let α (resp. D) be an automorphism (resp. derivation) of $\mathfrak{g}$ and let $\hat{\alpha}$ (resp. $\hat{D}$) be the automorphism (resp. derivation) of $\mathfrak{J}$ that extends α (resp. D). It is then easy to verify that $\hat{\alpha}$ and $\hat{D}$ both leave $\bar{\mathfrak{J}}$ invariant. On the other hand,

$$\tilde{\alpha} \circ \gamma = \gamma \circ \hat{\alpha}, \qquad \tilde{D} \circ \gamma = \gamma \circ \hat{D}$$
$$\bar{\alpha} \circ \bar{\gamma} = \bar{\gamma} \circ \hat{\alpha}, \qquad \bar{D} \circ \bar{\gamma} = \bar{\gamma} \circ \hat{D}.$$

The relation (3.3.15) now follows from the commutativity of (3.3.11).

Corollary 3.3.5. *For* $X \in \mathfrak{g}$, *let* D_X *be the derivation of* $\mathfrak{S}$ *that extends the endomorphism* ad X *of* $\mathfrak{g}$; *then*

$$\lambda(D_X(u)) = X\lambda(u) - \lambda(u)X \quad (u \in \mathfrak{S}). \tag{3.3.16}$$

Proof. Take $D = \operatorname{ad} X$ in (3.3.15).

Corollary 3.3.6. *Let* $\mathfrak{g}$ *be the direct sum of the subspaces* $\mathfrak{a}_1, \ldots, \mathfrak{a}_r$. *Let* $\mathfrak{S}_i$ *be the subalgebra of* $\mathfrak{S}$ *generated by* 1 *and* $\mathfrak{a}_i$, $\mathfrak{S}_{i,d}$ *the homogeneous subspace of* $\mathfrak{S}_i$ *of degree d*, $\mathfrak{G}_{i,d} = \lambda[\mathfrak{S}_{i,d}]$, *and for integers* $d_1, \ldots, d_r \geq 0$, $\mathfrak{G}_{d_1, \ldots, d_r} = \mathfrak{G}_{1,d_1}\mathfrak{G}_{2,d_2} \cdots \mathfrak{G}_{r,d_r}$. *Then the map* $(u_1, \ldots, u_r) \mapsto \lambda(u_1) \cdots \lambda(u_r)$ *of* $\mathfrak{S}_1 \times \cdots$

$\times\ \mathcal{S}_r$ *into* $\mathfrak{G}$ *extends to a unique linear isomorphism of* $\mathcal{S}_1 \otimes \cdots \otimes \mathcal{S}_r$ *onto* $\mathfrak{G}$. *Moreover, the subspaces* $\mathfrak{S}_{d_1,\ldots,d_r}$ *are all linearly independent, and for* $p \geq 0$,

$$\mathfrak{G}^{(p)} = \sum_{d_1+\cdots+d_r \leq p} \mathfrak{S}_{d_1,\ldots,d_r}. \tag{3.3.17}$$

Proof. We begin by proving (3.3.17) using induction on p. We may assume $p \geq 1$. Now, for $u_i \in \mathcal{S}_{i,d_i}$ with $d_1 + \cdots + d_r \leq p$, we have from (3.3.14)

$$\lambda(u_1 \cdots u_r) \equiv \lambda(u_1) \cdots \lambda(u_r) \pmod{\mathfrak{G}^{(p-1)}}. \tag{3.3.18}$$

On the other hand, it is obvious that

$$\mathcal{S}^{(p)} = \sum_{d_1+\cdots+d_r \leq p} \mathcal{S}_{1,d_1} \cdots \mathcal{S}_{r,d_r} \quad \text{(direct sum)}. \tag{3.3.19}$$

So we obtain from (3.3.18) and (3.3.19) the inclusion

$$\mathfrak{G}^{(p)} \subseteq \mathfrak{G}^{(p-1)} + \sum_{d_1+\cdots+d_r \leq p} \mathfrak{S}_{d_1,\ldots,d_r}.$$

By the induction hypothesis, $\mathfrak{G}^{(p)}$ is contained in the sum of the right of (3.3.17). Since the reverse inclusion is obvious, we obtain (3.3.17). Furthermore, $\dim(\mathfrak{S}_{d_1,\ldots,d_r}) \leq \dim(\mathcal{S}_{1,d_1}) \cdots \dim(\mathcal{S}_{r,d_r})$ for all $d_1, \ldots, d_r \geq 0$, so we conclude from (3.3.19) that

$$\dim(\mathfrak{G}^{(p)}) \geq \sum_{d_1+\cdots+d_r \leq p} \dim(\mathfrak{S}_{d_1,\ldots,d_r}).$$

It follows easily from this estimate that (3.3.17) is a direct sum and that $\dim(\mathfrak{S}_{d_1,\ldots,d_r}) = \dim(\mathcal{S}_{1,d_1}) \cdots \dim(\mathcal{S}_{r,d_r})$ for $d_1 + \cdots + d_r \leq p$. Since $p \geq 0$ is arbitrary, we see that $\mathfrak{G}$ is the direct sum of the $\mathfrak{S}_{d_1,\ldots,d_r}$ and that $\dim(\mathfrak{S}_{d_1,\ldots,d_r}) = \dim(\mathcal{S}_{r,d_1}) \cdots \dim(\mathcal{S}_{r,d_r})$ for all $d_1, \ldots, d_r \geq 0$. If ξ is the unique linear map of $\mathcal{S}_1 \otimes \cdots \otimes \mathcal{S}_r$ into $\mathfrak{G}$ such that $\xi(u_1 \otimes \cdots \otimes u_r) = \lambda(u_1) \cdots \lambda(u_r) (u_i \in \mathcal{S}_i$ for $i = 1, \ldots, r)$, the foregoing conclusions imply at once that ξ is a bijection.

Corollary 3.3.7. *Let* $\mathfrak{a}$ *be a subalgebra of* $\mathfrak{g}$ *and* $\mathfrak{b}$ *a subspace such that* $\mathfrak{g}$ *is the direct sum of* $\mathfrak{a}$ *and* $\mathfrak{b}$. *Let* $S(\mathfrak{b})$ *be the subalgebra of* $\mathcal{S}$ *generated by* 1 *and* $\mathfrak{b}$. *Then*

$$\mathfrak{G} = \mathfrak{G}\mathfrak{a} + \lambda[S(\mathfrak{b})], \tag{3.3.20}$$

the sum being direct.

Proof. Let $S(\mathfrak{a})$ and $\mathfrak{A}$ be the respective subalgebras of $\mathcal{S}$ and $\mathfrak{G}$ generated by 1 and $\mathfrak{a}$. Since we can canonically identify $S(\mathfrak{a})$ with the symmetric algebra of $\mathfrak{a}$, and $\mathfrak{A}$ with the universal enveloping algebra of $\mathfrak{a}$, we must have $\lambda[S(\mathfrak{a})] = \mathfrak{A}$. Let $\mathfrak{A}^+ = \mathfrak{A}\mathfrak{a}$. By Corollary 3.3.6, there is a linear isomorphism η of $S(\mathfrak{b}) \otimes \mathfrak{A}$ onto $\mathfrak{G}$ such that $\eta(b \otimes a) = \lambda(b)a$ for $b \in S(\mathfrak{b})$, $a \in \mathfrak{A}$. Using Corollary 3.2.6, we conclude that $\mathfrak{G}$ is the direct sum of $S(\mathfrak{b})$ and $S(\mathfrak{b})\mathfrak{A}^+$. But $S(\mathfrak{b})\mathfrak{A}^+ = S(\mathfrak{b})\mathfrak{A}\mathfrak{a} = \mathfrak{G}\mathfrak{a}$.

We conclude this section with a theorem which is often useful in finding the center of the universal enveloping algebra of a Lie algebra.

Theorem 3.3.8. *Let* Z *(resp.* $\mathfrak{Z}$*) be the set of all elements u of* S *(resp. a of* $\mathfrak{G}$*) such that* $D_X u = 0$ *for all* $X \in \mathfrak{g}$ *(resp.* $Xa - aX = 0$ *for* $X \in \mathfrak{g}$*),* $D_X(X \in \mathfrak{g})$ *being the derivation of* S *that extends ad X. Then*

(i) Z *and* $\mathfrak{Z}$, *are algebras,* $\lambda[\mathsf{Z}] = \mathfrak{Z}$, *and* $\mathfrak{Z}$ *is the center of* $\mathfrak{G}$

(ii) *if* $u_1 = 1, \ldots, u_r$ *are homogeneous elements of* Z *generating* Z*, then* $\lambda(u_1), \ldots, \lambda(u_r)$ *generate* $\mathfrak{Z}$*; if the* u_i *are algebraically independent, so are the* $\lambda(u_i)$.

Proof. Since each D_X is a derivation, Z is an algebra. For the same reason $\mathfrak{Z}$ is an algebra. Moreover, since $\mathfrak{g}$ generates $\mathfrak{G}$, $\mathfrak{Z}$ is also the center of $\mathfrak{G}$. That $\lambda[\mathsf{Z}] = \mathfrak{Z}$ follows from (3.3.16). This proves (i). We now come to the proof of (ii). Let $\mathsf{Z}_n = \mathsf{Z} \cap \mathsf{S}_n$ and $\mathfrak{Z}_n = \lambda[\mathsf{Z}_n]$ $(n \geq 0)$. Since the derivations D_X leave the homogeneous subspaces invariant, it is easily seen that Z is the direct sum of the Z_n. Hence $\mathfrak{Z}$ is the direct sum of the $\mathfrak{Z}_n$. Obviously, $\mathfrak{Z}_0 = k \cdot 1$.

Suppose that $u_1 = 1, u_2, \ldots, u_r$ are homogeneous and generate Z. We may assume that $d_i = \deg(u_i) > 0$ for all $i \geq 2$. Let $v_i = \lambda(u_i)$, $1 \leq i \leq r$. Denote by $\mathfrak{Z}'$ the algebra generated by $v_1 = 1, v_2, \ldots, v_r$. Then $\mathfrak{Z}' \subseteq \mathfrak{Z}$, and it is enough to prove that $\mathfrak{Z}_n \subseteq \mathfrak{Z}'$ for all $n \geq 0$. This is obvious for $n = 0$. Assume $n \geq 1$ and $\mathfrak{Z}_s \subseteq \mathfrak{Z}'$ for $0 \leq s \leq n-1$. We shall prove that $\mathfrak{Z}_n \subseteq \mathfrak{Z}'$; an induction on n will then complete the argument. Let $b \in \mathfrak{Z}_n$. Then we can write $b = \lambda(a)$, where $a \in \mathsf{Z}_n$. There are constants $c_{n_1,\ldots,n_r} (n_1, \ldots, n_r \geq 0)$ such that

$$a = \sum c_{n_1,\ldots,n_r} u_1^{n_1} \cdots u_r^{n_r},$$

all but finitely many of the $c_{n_1,\ldots,n_r}$ being 0. Since $a \in \mathsf{S}_n$ and $u_i \in \mathsf{S}_{d_i}$, we may assume that the summation is over all $n_1, \ldots, n_r$ with $n_1 d_1 + \cdots + n_r d_r = n$. Define the elements $b' \in \mathfrak{Z}'$ by

$$b' = \sum_{n_1 d_1 + \cdots + n_r d_r = n} c_{n_1,\ldots,n_r} v_1^{n_1} \cdots v_r^{n_r}.$$

It is then clear from (3.3.14) that $b - b' \in \mathfrak{G}^{(n-1)} \cap \mathfrak{Z}$. On the other hand, for any $p \geq 0$, $\mathfrak{Z}_p = \lambda[\mathsf{Z} \cap \mathsf{S}_p] = \mathfrak{Z} \cap \lambda[\mathsf{S}_p]$, so $\mathfrak{Z}_0 + \cdots + \mathfrak{Z}_{n-1} = \mathfrak{G}^{(n-1)} \cap \mathfrak{Z}$. Consequently, $b - b' \in \sum_{0 \leq s \leq n-1} \mathfrak{Z}_s$, so by the induction assumption, $b - b' \in \mathfrak{Z}'$. This proves that $b \in \mathfrak{Z}'$.

Suppose that the u_i are algebraically independent but that the v_i are not. Then there are constants $c_{n_1,\dots,n_r}$ not all zero such that $\sum c_{n_1,\dots,n_r} v_1^{n_1}\cdots v_r^{n_r} = 0$. Let p be the maximum value of $n_1d_1 + \cdots + n_rd_r$ over all $n_1, \dots, n_r$ with $c_{n_1,\dots,n_r} \neq 0$, and let $u = \sum_{n_1d_1+\cdots+n_rd_r=p} c_{n_1,\dots,n_r} u_1^{n_1} \cdots u_r^{n_r}$. Then $u \in S_p \cap \mathfrak{Z}$ and is nonzero. Moreover, we conclude from (3.3.14) and the relation $\sum_{n_1d_1+\cdots+n_rd_r\leq p} c_{n_1,\dots,n_r} v_1^{n_1} \cdots v_r^{n_r} = 0$ that $\boldsymbol{\lambda}(u) \in \mathfrak{G}^{(p-1)} \cap \mathfrak{Z}$. On the other hand, we saw in the previous paragraph that $\mathfrak{G}^{(p-1)} \cap \mathfrak{Z} = \boldsymbol{\lambda}[\mathfrak{S}^{(p-1)} \cap \mathfrak{Z}]$, so we must have $u \in \mathfrak{S}^{(p-1)}$. This is a contradiction. The proof of the theorem is complete.

As an illustrative example, let $\mathfrak{g} = \mathfrak{sl}(2,k)$. Let

$$(3.3.21) \qquad H = \begin{pmatrix} 1 & 0 \\ 0 & -1 \end{pmatrix}, \qquad X = \begin{pmatrix} 0 & 1 \\ 0 & 0 \end{pmatrix}, \qquad Y = \begin{pmatrix} 0 & 0 \\ 1 & 0 \end{pmatrix}.$$

Then $\{H,X,Y\}$ is a basis for $\mathfrak{g}$, and

$$[H,X] = 2X, \qquad [H,Y] = -2Y, \qquad [X,Y] = H.$$

Define the element ω of $\mathfrak{G}$ by

$$(3.3.22) \qquad \omega = H^2 + 2H + 4YX.$$

It is then easy to see that $\omega \in \mathfrak{Z}$ and is the image under $\boldsymbol{\lambda}$ of the element $H^2 + 4XY$ of $\mathfrak{S}$. It can be shown that the algebra $\mathfrak{Z}$ is generated by the homogeneous element $H^2 + 4XY$. Hence $\mathfrak{Z} = k[\omega]$.

Let $\mathfrak{g}$ be arbitrary, π an irreducible representation of $\mathfrak{g}$ in a finite-dimensional vector space V. We extend π to a representation of $\mathfrak{G}$ in V and denote this extension by π again. Suppose now that k is algebraically closed. Then by Schur's lemma, $\pi(z)$ is a scalar multiple of the identity for each $z \in \mathfrak{Z}$. So there exists a homomorphism χ_π of $\mathfrak{Z}$ into k such that $\pi(z) = \chi_\pi(z)\cdot 1$ for all $z \in \mathfrak{Z}$. We call χ_π the *infinitesimal character* of π. When $\mathfrak{g}$ is reductive, χ_π determines the equivalence class of π, as will be proved later on. If the structure of $\mathfrak{Z}$ as an algebra is known, we may then parametrize the finite-dimensional representations of $\mathfrak{g}$ by a subset of the spectrum of $\mathfrak{Z}$.

3.4. The Enveloping Algebra of a Lie Group

In this section we examine the analytic significance of the universal enveloping algebra of the Lie algebra of a Lie group. We work with real groups, leaving it to the reader to make the necessary changes in the proofs for the complex case.

Let G be a real Lie group, $\mathfrak{g}$ its Lie algebra. Denote by $\mathfrak{g}_c$ the complexification of $\mathfrak{g}$, and regard $\mathfrak{g}$ as a subset of $\mathfrak{g}_c$. The elements of $\mathfrak{g}_c$ may be identified with left-invariant analytic vector fields on G. Let $\mathfrak{G}_c$ be the universal enveloping algebra of $\mathfrak{g}_c$ and $\mathfrak{G}$ the subalgebra over $\mathbf{R}$ generated by 1 and $\mathfrak{g}$. We obviously may identify $\mathfrak{G}$ with the universal enveloping algebra of $\mathfrak{g}$.

Theorem 3.4.1. *For any $X \in \mathfrak{g}_c$, let $\partial(X)$ be the differential operator $f \mapsto Xf$ $(f \in C^\infty(G))$. Then the map $X \mapsto \partial(X)$ extends uniquely to an isomorphism ∂ $(a \mapsto \partial(a))$ of $\mathfrak{G}_c$ with the algebra of all left-invariant analytic differential operators on G. Moreover, for $a \in \mathfrak{G}_c$, $\partial(a)$ is real if and only if $a \in \mathfrak{G}$.*

Proof. We have, for all $X, Y \in \mathfrak{G}_c$,

$$\partial([X,Y]) = \partial(X)\partial(Y) - \partial(Y)\partial(X).$$

Let $\mathfrak{D}$ be the algebra (over $\mathbf{C}$) of all left-invariant analytic differential operators on G. Then the map $X \mapsto \partial(X)$ extends to a unique homomorphism ∂ $(a \mapsto \partial(a))$ of $\mathfrak{G}_c$ into $\mathfrak{D}$. Suppose that $\{X_1, \ldots, X_m\}$ is a basis for $\mathfrak{g}$ over $\mathbf{R}$. Then by Theorem 2.4.1, the differential operators $\{\partial(X_1)^{r_1} \cdots \partial(X_m)^{r_m}\}$ $(r_1, \ldots, r_m \geq 0)$ form a basis for $\mathfrak{D}$ over $\mathbf{C}$, and an element of $\mathfrak{D}$ is real if and only if it is in the real span of these. On the other hand, the elements $\{X_1^{r_1} \cdots X_m^{r_m} : r_1, \ldots, r_m \geq 0\}$ form a basis for $\mathfrak{G}_c$ over $\mathbf{C}$, and an element of $\mathfrak{G}_c$ belongs to $\mathfrak{G}$ if and only if it is in the real span of these monomials. Since $\partial(X_1^{r_1} \cdots X_m^{r_m}) = \partial(X_1)^{r_1} \cdots \partial(X_m)^{r_m}$, we have the theorem at once.

In view of the above theorem it is natural to identify $\mathfrak{G}_c$ with $\mathfrak{D}$ via ∂. We shall do so from now on and refer to $\mathfrak{G}$ as the universal enveloping algebra of G. The elements of $\mathfrak{G}_c$ act as differential operators on G; if $y \in G$ and if f is C^∞ around y, then for $Y_1, \ldots, Y_s \in \mathfrak{g}$,

$$f(y; Y_1 \cdots Y_s) = \left(\frac{\partial^s}{\partial t_1 \cdots \partial t_s} f(y \exp t_1 Y_1 \cdots \exp t_s Y_s)\right)_{t_1 = \cdots = t_s = 0} \tag{3.4.1}$$

If $f \in C^\infty(G)$ and $a \in \mathfrak{G}_c$, we write af for the function $y \mapsto f(y; a)$; we have $a(bf) = (ab)f$ $(a, b \in \mathfrak{G}_c)$.

Theorem 3.4.2. *Fix $y \in G$. For each $a \in \mathfrak{G}_c$, let τ_a be the element of $T_{yc}^{(\infty)}(G)$ induced by the linear function $f \mapsto f(y; a)$ on $C^\infty(G)$. Then τ $(a \mapsto \tau_a)$ a linear bijection of $\mathfrak{G}_c$ with $T_{yc}^{(\infty)}(G)$ that maps $\mathfrak{G}_y^{(p)}$ onto $T_y^{(p)}(G)$ for each $p \geq 0$.*

Proof. Clearly, $\tau_1 = 1_y$. If $a = Y_1 \cdots Y_s$ $(1 \leq s \leq r,\ Y_1, \ldots, Y_s \in \mathfrak{G})$, and $f = f_1 \cdots f_{r+1}$, where $f_1, \ldots, f_{r+1}$ are C^∞ functions on G that vanish at y, then $f(y;a) = 0$. This shows that τ maps $\mathfrak{G}^{(r)}$ into $T_y^{(r)}(G)$ for each $r \geq 0$. Suppose now that $a \in \mathfrak{G}_c$ is such that $\tau_a = 0$. Then $f(y;a) = 0$ for all $f \in C^\infty(G)$. Since a is left-invariant, $af = 0$ for all $f \in C^\infty(G)$. Hence $a = 0$, by Theorem 3.4.1. τ is thus an injection. Since $\dim_{\mathbf{C}} \mathfrak{G}_c^{(r)} = \dim_{\mathbf{C}} T_{yc}^{(r)}(G)$ for each $r \geq 0$, τ must map $\mathfrak{G}^{(r)}$ onto $T_y^{(r)}(G)$ for each $r \geq 0$. τ is thus surjective. This proves the theorem.

Let U be an open subset of G. It is then clear from the above theorem that if E is any analytic differential operator on U of order $\leq r$, we can find analytic functions f_i on U and elements $a_i \in \mathfrak{G}_c^{(r)}$ such that

$$Ef = \sum_{1 \leq i \leq s} f_i a_i f \tag{3.4.2}$$

for all $f \in C^\infty(U)$. We abbreviate this as

$$E = \sum_{1 \leq i \leq s} f_i \circ a_i. \tag{3.4.3}$$

The f_i and a_i are not uniquely determined by E. If $\{a_1, \ldots, a_s\}$ is a basis for $\mathfrak{G}^{(r)}$ over $\mathbf{R}$, then we can find unique analytic functions $f_1, \ldots, f_s$ on U such that (3.4.2) is satisfied; in this case, if E is real, the f_i are real.

Suppose E is as above and we have (3.4.3) for suitable a_i, f_i. Let

$$E_y = \sum_{1 \leq i \leq s} f_i(y) a_i \quad (y \in U). \tag{3.4.4}$$

Then $E_y \in \mathfrak{G}_c$, and

$$f(y;E) = f(y;E_y) \tag{3.4.5}$$

for all $f \in C^\infty(U)$ and $y \in U$. Even though f_i and a_i are not uniquely determined by E, the equation (3.4.5) shows that E_y, for any $y \in U$, is determined as the unique element of $\mathfrak{G}_c$ whose image under the isomorphism τ of Theorem 3.4.2 is the element of $T_{yc}^{(\infty)}(G)$ induced by the linear function $f \mapsto f(y;E)$ on $C^\infty(G)$. E_y is called the *local expression* of E at y. It is an easy verification that E is real if and only if $E_y \in \mathfrak{G}$ for each $y \in U$. It is obvious from (3.4.5) that $E = 0$ if and only if $E_y = 0$ for all $y \in U$. In particular, E is uniquely determined by the map $y \mapsto E_y$ $(y \in U)$.

The formula (3.4.4) makes it clear that if $r \geq 0$ is an integer such that $a_i \in \mathfrak{G}_c^{(r)}$ for $1 \leq i \leq s$, the map $y \mapsto E_y$ is analytic from U into $\mathfrak{G}_c^{(r)}$. Conversely, it is obvious that if $y \mapsto E_y$ is an analytic map of U into $\mathfrak{G}_c^{(r)}$, there is a unique analytic differential operator E on U with E_y as its local expression at $y \in U$; in fact, if $\{a_1, \ldots, a_s\}$ is a basis for $\mathfrak{G}_c^{(r)}$ over $\mathbf{C}$, there are analytic functions $f_1, \ldots, f_s$ on U such that $E_y = \sum_{1 \leq i \leq s} f_i(y) a_i$ for all $y \in U$; then one can define E as $\sum_{1 \leq i \leq s} f_i \circ a_i$, the uniqueness of E having been noted earlier.

Let us now take $U = G$ in the foregoing discussion. Suppose α $(x \mapsto x^\alpha)$ is an automorphism of the Lie group G. Then α induces an automorphism $X \mapsto X^\alpha$ of $\mathfrak{g}$, and the latter can be extended uniquely to an automorphism

$a \mapsto a^\alpha$ of $\mathfrak{G}_c$. More generally, α induces an automorphism $E \mapsto E^\alpha$ of the algebra of all analytic differential operators on G. Between the local expressions of E^α and E we have the relation

$$(E^\alpha)_{y^\alpha} = (E_y)^\alpha \quad (y \in G). \tag{3.4.6}$$

To see this, write E in the form (3.4.3). Then $E^\alpha = \sum_{1\le i\le s} f_i^\alpha \circ a_i^\alpha$, so $(E^\alpha)_{y^\alpha} = \sum_{1\le i\le s} f_i(y)\, a_i^\alpha = (E_y)^\alpha$. In particular, E is invariant under α if and only if

$$E_{y^\alpha} = (E_y)^\alpha \quad (y \in G). \tag{3.4.7}$$

The most important automorphisms of G are the inner ones. For $y \in G$, let $\mathrm{Ad}(y)$ denote the automorphism of $\mathfrak{g}_c$ that extends the automorphism $X \mapsto X^y$ of $\mathfrak{g}$; write $\mathrm{Ad}(y)$ for the extension of this to an automorphism of $\mathfrak{G}_c$. It is usual to write

$$\mathrm{Ad}(y)a = a^y \quad (a \in \mathfrak{G}_c, y \in G). \tag{3.4.8}$$

By Corollary 3.3.2, $\mathrm{Ad}(y)$ leaves $\mathfrak{G}^{(n)}$ invariant for each $n \geq 0$ and each $y \in G$. We now have the following equation valid on each $\mathfrak{G}_c^{(n)}$:

$$\mathrm{Ad}(\exp tX) = e^{t\,\mathrm{ad}\,X} \quad (X \in \mathfrak{g}, t \in \mathbf{R}). \tag{3.4.9}$$

To establish the relation (3.4.9), differentiate the relation $(X_1 \cdots X_n)^{y_t} = X_1^{y_t} \cdots X_n^{y_t}$ with respect to t at $t = 0$ ($X_i \in \mathfrak{g}, y_t = \exp tX$); we then obtain the relation $((d/dt)\,\mathrm{Ad}(\exp tX))_{t=0} = \mathrm{ad}\, X$, valid on $\mathfrak{G}_c^{(n)}$, leading at once to (3.4.9).

More generally, let π be a representation of G in a finite-dimensional vector space V. Replacing V by its complexification, we may assume that π is an analytic homomorphism of G into $GL(V)_{\mathbf{R}}$; here $GL(V)_{\mathbf{R}}$ (resp. $\mathfrak{gl}(V)_{\mathbf{R}}$) is the real Lie group (resp. Lie algebra) underlying $GL(V)$ (resp. $\mathfrak{gl}(V)$). The differential of π is then a homomorphism of $\mathfrak{g}$ into $\mathfrak{gl}(V)_{\mathbf{R}}$ and so can be extended to a homomorphism of $\mathfrak{g}_c$ into $\mathfrak{gl}(V)$; in turn, this can be extended to a representation of $\mathfrak{G}_c$ in V. We write π for this last representation. Then $\pi(\exp X) = \exp \pi(X)$ is $(X \in \mathfrak{g})$. The basic relation between these representations of G and $\mathfrak{G}_c$ is given by

$$\pi(a^y) = \pi(y)\pi(a)\pi(y)^{-1} \quad (a \in \mathfrak{G}_c, y \in G). \tag{3.4.10}$$

Clearly, it is sufficient to prove (3.4.10) with a replaced by an arbitrary element $X \in \mathfrak{g}$. Suppose then that $X \in \mathfrak{g}$. Then for any $y \in G$,

$$\begin{aligned}\pi(X^y) &= \left(\frac{d}{dt}\pi(y)\pi(\exp tX)\pi(y)^{-1}\right)_{t=0} \quad \text{(cf. (2.13.7))}\\ &= \pi(y)\pi(X)\pi(y)^{-1}.\end{aligned}$$

Let $\mathfrak{Z}_{0c}$ be the subalgebra of $\mathfrak{G}_c$ consisting of all $a \in \mathfrak{G}_c$ for which $a^y = a$ for all $y \in G$. If $\mathfrak{Z}_0 = \mathfrak{Z}_{0c} \cap \mathfrak{G}$, then $\mathfrak{Z}_{0c}$ is the complex span of $\mathfrak{Z}_0$ in $\mathfrak{G}$. It is clear from (3.4.7) that $\mathfrak{Z}_{0c}$ is the algebra of all analytic differential-operators which are invariant under all left translations and inner automorphisms. Since $r_a = i_{a^{-1}} l_a$ for any $a \in G$ (cf. (2.1.2), (2.1.3)), $\mathfrak{Z}_{0c}$ is precisely the algebra of all analytic differential operators on G invariant under all translations. Since G^0 is generated by exp[$\mathfrak{g}$], it follows from (3.4.9) that $\mathfrak{Z}$, the center of $\mathfrak{G}$, is the algebra of all $a \in \mathfrak{G}$ such that $a^y = a$ for all $y \in G^0$; thus $\mathfrak{Z}_0 \subseteq \mathfrak{Z}$, and if G is connected, $\mathfrak{Z} = \mathfrak{Z}_0$.

So far, we have allowed the elements of $\mathfrak{G}_c$ to act as differential operators only from the left. It is also possible to consider each element of $\mathfrak{G}_c$ as a differential operator acting on elements of $C^\infty(G)$ from the right. We now indicate how this is done. Let $a \in \mathfrak{G}_c$. If $r \geq 0$ is such that $a \in \mathfrak{G}^{(r)}$, the map $y \mapsto a^{y^{-1}}$ is analytic from G to $\mathfrak{G}^{(r)}$, so there is a unique analytic differential operator D_a on G such that $a^{y^{-1}}$ is the local expression of D_a at $y \in G$:

$$(D_a)_y = a^{y^{-1}} \quad (a \in \mathfrak{G}_c, y \in G). \tag{3.4.11}$$

It follows from (3.4.11) that if $a = X_1 \cdots X_s$, where the X_j are elements of $\mathfrak{g}$, then for all $f \in C^\infty(G)$ and $y \in G$,

$$f(y; D_a) = \left(\frac{\partial^s}{\partial t_1 \cdots \partial t_s} f(\exp t_1 X_1 \cdots \exp t_s X_s y)\right)_{t_1 = \cdots = t_s = 0} \tag{3.4.12}$$

The formula (3.4.12) makes it clear that each D_a is invariant under all right translations. If $b = Y_1 \cdots Y_r$, where the $Y_i \in \mathfrak{g}$, then a simple calculation based on (3.4.12) shows that $D_{ab}f = D_b D_a f$ for all $f \in C^\infty(G)$. Hence

$$D_{ab} = D_b D_a \quad (a, b \in \mathfrak{G}_c). \tag{3.4.13}$$

In view of (3.4.13), it is natural to write

$$D_a f = fa \quad (a \in \mathfrak{G}_c, f \in C^\infty(G)). \tag{3.4.14}$$

It is also convenient to write

$$f(y; D_a) = f(a; y) \quad (a \in \mathfrak{G}_c, y \in G, f \in C^\infty(G)). \tag{3.4.15}$$

Thus, if $a = X_1 \cdots X_s$, where the $X_i \in \mathfrak{g}$, $y \in G$, and $f \in C^\infty(G)$, then

$$f(a; y) = \left(\frac{\partial^s}{\partial t_1 \cdots \partial t_s} f(\exp t_1 X_1 \cdots \exp t_s X_s y)\right)_{t_1 = \cdots = t_s = 0} \tag{3.4.16}$$

We note that the operators $f \mapsto af$ and $f \mapsto fb$ ($a, b \in \mathfrak{G}_c$) commute. We also note the following easy consequence of (3.4.16):

$$f(a; 1) = f(1; a) \quad (f \in C^\infty(G), a \in \mathfrak{G}_c). \tag{3.4.17}$$

Let D be an analytic differential operator on G invariant under all right translations. Let $a \in \mathfrak{G}_c$ be the local expression of D at the identity 1 of G. Since a is also the local expression of D_a at 1 (cf. (3.4.17)), the operator $D - D_a$ has local expression 0 at 1. Since $D - D_a$ is also invariant under all right translations, the local expressions of $D - D_a$ at all $y \in G$ are zero. Hence $D = D_a$. Since $D_a = 0$ implies $a = 0$, we conclude that the map $a \mapsto D_a$ is an anti-isomorphism of $\mathfrak{G}$ onto the algebra of all analytic differential operators on G which are invariant under all right translations.

Let $\mathfrak{D}$ be the algebra of analytic differential operators on G generated by the left-invariant and right-invariant differential operators defined above. For $a, b \in \mathfrak{G}_c$, the endomorphism $f \mapsto afb^t$ ($f \in C^\infty(G)$) is an element of $\mathfrak{D}$, $c \mapsto c^t$ being the anti-automorphism of $\mathfrak{G}_c$ under which $X^t = -X$ for $X \in \mathfrak{g}_c$ (cf. Corollary 3.2.4); the map which assigns this differential operator to (a, b) extends to a homomorphism of the tensor product $\mathfrak{G}_c \otimes \mathfrak{G}_c$ onto $\mathfrak{D}$.

3.5. Nilpotent Lie Algebras

k is, as usual, a field of characteristic 0, $\mathfrak{g}$ a Lie algebra of finite dimension m over k. $\mathfrak{g}$ is said to be *nilpotent* if ad X is a nilpotent endomorphism of $\mathfrak{g}$ for all $X \in \mathfrak{g}$. A representation ρ of an arbitrary Lie algebra $\mathfrak{g}$ in a finite-dimensional vector space is called a *nil representation* if $\rho(X)$ is nilpotent for all $X \in \mathfrak{g}$.

Any abelian Lie algebra is nilpotent. More generally, let V be a finite-dimensional vector space over k and $\mathfrak{N}$ the set of all nilpotent endomorphisms of V. If $\mathfrak{g}$ is a subalgebra of $\mathfrak{gl}(V)$ such that $\mathfrak{g} \subseteq \mathfrak{N}$, then $\mathfrak{g}$ is nilpotent. For if $X \in \mathfrak{g}$, then ad X is a nilpotent endomorphism of $\mathfrak{gl}(V)$ (cf. §3.1), and therefore ad $X|\mathfrak{g}$ is also nilpotent. In particular, if $\{e_i\}_{1 \leq i \leq n}$ is a basis for V and $\mathfrak{g}$ is the Lie algebra of all endomorphisms of V whose matrices in this basis have zeros on and below the main diagonal, then $\mathfrak{g}$ is nilpotent.

Theorem 3.5.1. *Let $\mathfrak{g}$ be a Lie algebra over k, k' an extension field of k, and $\mathfrak{g}^{k'}$ the extension of $\mathfrak{g}$ to k'. Then $\mathfrak{g}$ is nilpotent if and only if $\mathfrak{g}^{k'}$ is. If $\mathfrak{g}$ is nilpotent, so are its subalgebras and quotient algebras.*

Proof. Since $\mathfrak{g} \subseteq \mathfrak{g}^{k'}$, $\mathfrak{g}^{k'}$ nilpotent $\Rightarrow$ $\mathfrak{g}$ nilpotent. Suppose that $\mathfrak{g}$ is nilpotent, and for any integer $r \geq 1$, let $g_r(X) = tr(\text{ad } X)^r$ ($X \in \mathfrak{g}^{k'}$). Then g_r are polynomials on $\mathfrak{g}^{k'}$ vanishing on $\mathfrak{g}$. Hence $g_r \equiv 0$ for $r \geq 1$. This proves that $\mathfrak{g}^{k'}$ is nilpotent.

Suppose $\mathfrak{g}$ is nilpotent and $\mathfrak{h}$ is a subalgebra of $\mathfrak{g}$. If $X \in \mathfrak{h}$, then $\text{ad}_{\mathfrak{h}} X = (\text{ad}_{\mathfrak{h}} X)|\mathfrak{h}$, showing that $\text{ad}_{\mathfrak{h}} X$ is nilpotent. Thus $\mathfrak{h}$ is nilpotent. Suppose $\mathfrak{g}'$ is a quotient of $\mathfrak{g}$ and $\lambda : \mathfrak{g} \longrightarrow \mathfrak{g}'$ the canonical homomorphism. If $X \in \mathfrak{g}$ and $X' \in \mathfrak{g}'$ are such that $X' = \lambda(X)$, then $\lambda \circ \text{ad } X = \text{ad } X' \circ \lambda$, from which we get $\lambda \circ (\text{ad } X)^r = (\text{ad } X')^r \circ \lambda$ for all integer $r \geq 1$. This shows that $\mathfrak{g}'$ is nilpotent.

The central result in the theory of nilpotent Lie algebras is the well-known Engel's theorem.

Theorem 3.5.2. *Let $\mathfrak{g}$ be a Lie algebra over k, ρ a nil representation of $\mathfrak{g}$ in a nonzero finite-dimensional vector space V over k. Then there is a nonzero vector $v \in V$ such that*

$$\rho(X)v = 0 \quad (X \in \mathfrak{g}). \tag{3.5.1}$$

Proof. Let $\mathfrak{a} = \text{kernel}(\rho)$. Then ρ induces a faithful representation of $\mathfrak{g}/\mathfrak{a}$. Since all elements of $\rho[\mathfrak{g}]$ are nilpotent, $\rho[\mathfrak{g}]$ is a nilpotent Lie algebra. Hence $\mathfrak{g}/\mathfrak{a}$ is nilpotent. Since we may replace $\mathfrak{g}$ by $\mathfrak{g}/\mathfrak{a}$ for the proof, we see that there is no loss of generality in assuming that $\mathfrak{g}$ is nilpotent. We shall do so and prove the theorem by induction on dim $\mathfrak{g}$. For dim $\mathfrak{g} = 1$ there is nothing to prove. So let dim $\mathfrak{g} \geq 2$ and assume the theorem for all nilpotent Lie algebras of lower dimension. Let $\mathfrak{S}$ be the set of all subalgebras $\mathfrak{h}$ of $\mathfrak{g}$ with $0 < \dim \mathfrak{h} < \dim \mathfrak{g}$. $\mathfrak{S}$ is nonempty, since $k \cdot X \in \mathfrak{S}$ for all $X \neq 0$ in $\mathfrak{g}$. Let $\mathfrak{h}$ be an element of $\mathfrak{S}$ of maximal dimension; $\mathfrak{h}$ is clearly nilpotent.

We claim that $\mathfrak{h}$ is an ideal and that $\dim(\mathfrak{g}/\mathfrak{h}) = 1$. To see this, let W be the vector space $\mathfrak{g}/\mathfrak{h}$. If $X \in \mathfrak{h}$, ad X leaves $\mathfrak{h}$ invariant, so it induces an endomorphism $\rho'(X)$ of W. ρ' $(X \mapsto \rho'(X))$ is obviously a nil representation of $\mathfrak{h}$ in W. So by the induction hypothesis, there is a nonzero $w \in W$ such that $\rho'(X)w = 0$ for $X \in \mathfrak{h}$. If $X_0 \in \mathfrak{g}$ lies above w, then $X_0 \notin \mathfrak{h}$ and $[X_0,\mathfrak{h}] \subseteq \mathfrak{h}$. This implies that $k \cdot X_0 + \mathfrak{h}$ is a subalgebra whose dimension is strictly larger than dim $\mathfrak{h}$. By the choice of $\mathfrak{h}$ we must have $\mathfrak{g} = k \cdot X_0 + \mathfrak{h}$, showing $\dim(\mathfrak{g}/\mathfrak{h}) = 1$. At the same time $[\mathfrak{g},\mathfrak{h}] \subseteq \mathfrak{h}$.

By the induction hypothesis applied to $\mathfrak{h}$, we conclude that the space

$$V' = \{u : u \in V, \rho(Y)\, u = 0 \text{ for all } Y \in \mathfrak{h}\}$$

is nonzero. If $u \in V'$ and $Y \in \mathfrak{h}$, then

$$\begin{aligned} \rho(Y)\rho(X_0)u &= \rho(X_0)\rho(Y)u + \rho([Y,X_0])u \\ &= 0, \end{aligned}$$

since $[Y,X_0] \in \mathfrak{h}$, showing that $\rho(X_0)u \in V'$. $\rho(X_0)$ thus leaves V' invariant. Since $\rho(X_0)$ is nilpotent, we can find a nonzero $v \in V'$ such that $\rho(X_0)v = 0$. Clearly such a v satisfies (3.5.1).

Theorem 3.5.2 leads at once to the following.

Theorem 3.5.3. *Let $\mathfrak{g}$ be a nilpotent Lie algebra over k, ρ a nil representation of $\mathfrak{g}$ in a finite-dimensional vector space V over k. Define $V_0 = 0$ and for*

$i \geq 1$, let

(3.5.2) $$V_i = \{v : v \in V, \rho(X)v \in V_{i-1} \text{ for all } X \in \mathfrak{g}\}.$$

Then $V_0 \subseteq V_1 \subseteq V_2 \subseteq \cdots, V_s = V$ for some integer s with $1 \leq s \leq \dim V$, $\dim V_i < \dim V_{i+1}, 0 \leq i \leq s-1$. Let $\mathfrak{A}_\rho$ be the associative algebra of all endomorphisms L of V such that $L[V_i] \subseteq V_i$ for $0 \leq i \leq s$, and $\mathfrak{M}_\rho$ the two-sided ideal in $\mathfrak{A}_\rho$ of all L with $L[V_i] \subseteq V_{i-1}$ for $1 \leq i \leq s$. Then $\rho[\mathfrak{g}] \subseteq \mathfrak{M}_\rho$. The product of any s elements of $\mathfrak{M}_\rho$ is 0; in particular, each element of $\mathfrak{M}_\rho$ is nilpotent. There is a basis for V with respect to which the matrix of each $\rho(X)$ $(X \in \mathfrak{g})$ has zeros on and below the main diagonal.

Proof. It is clear by induction on i that the V_i are well-defined subspaces of V invariant for the representation ρ, and $V_0 \subseteq V_1 \subseteq \cdots$. Suppose $i \geq 0$ and $V_i \neq V$. Then the quotient representation in V/V_i is also a nil representation of $\mathfrak{g}$. If we apply Theorem 3.5.2 to this, we find that there are $v \in V$, $v \notin V_i$ with $\rho(X)v \in V_i$ for all $X \in \mathfrak{g}$. Thus $V_i \subseteq V_{i+1}$, and $\dim V_i < \dim V_{i+1}$. This shows that $V_s = V$ for some s with $1 \leq s \leq \dim(V)$. Now it is obvious that $\mathfrak{M}_\rho$ is a two-sided ideal in $\mathfrak{A}_\rho$, and that $\rho[\mathfrak{g}] \subseteq \mathfrak{M}_\rho$. If $r \geq 1$, $L_1, \ldots, L_r, \in \mathfrak{M}_\rho$, and $L = L_1 \cdots L_r$, then L maps V_i into V_{i-r} for $r \leq i \leq s$; in particular, $L = 0$ for $r = s$. Let $n_i = \dim V_i$ and let $\{v_j\}_{1 \leq j \leq n_s}$ be a basis for V such that $\{v_j\}_{1 \leq j \leq n_i}$ is a basis for V_i $(1 \leq i \leq s)$. Then in this basis, the matrix of each $\rho(X)$ $(X \in \mathfrak{g})$ has zeros on and below the main diagonal.

The adjoint representation of a nilpotent Lie algebra is a nil representation, so Theorems 3.5.2 and 3.5.3 may be applied to it. This leads at once to the basic results on the structure of nilpotent Lie algebras.

Theorem 3.5.4. *Let $\mathfrak{g}$ be a nilpotent Lie algebra over k; then:*

(i) *Let $\mathfrak{g}_i$ $(i \geq 0)$ be defined inductively as follows. $\mathfrak{g}_0 = 0$, and for $i \geq 1$,*

(3.5.3) $$\mathfrak{g}_i = \{X : X \in \mathfrak{g}, [X, \mathfrak{g}] \subseteq \mathfrak{g}_{i-1}\}.$$

Then each $\mathfrak{g}_i$ is an ideal in $\mathfrak{g}$, $\mathfrak{g}_s = \mathfrak{g}$ for some integer s with $1 \leq s \leq m = \dim(\mathfrak{g})$, and $\mathfrak{g}_i \subseteq \mathfrak{g}_{i+1}$, $\dim(\mathfrak{g}_i) < \dim(\mathfrak{g}_{i+1})$ for $0 \leq i \leq s-1$. In particular, $\mathfrak{g}$ has nonzero center.

(ii) *Let s be as in* (i). *Then*

(3.5.4) $$\operatorname{ad} Z_1 \cdots \operatorname{ad} Z_s = 0 \quad (Z_1, \ldots, Z_s \in \mathfrak{g}).$$

(iii) *There is a basis $\{X_1, \ldots, X_m\}$ for $\mathfrak{g}$ such that for the structure constants c_{ijr} defined by $[X_i, X_j] = \sum_{1 \leq r \leq m} c_{ijr} X_r$ $(1 \leq i, j \leq m)$, one has*

(3.5.5) $$c_{ijr} = 0 \quad r \geq \min(i, j).$$

In particular, in this basis, the matrix of each ad X ($X \in \mathfrak{g}$) has zeros on and below the main diagonal.

Proof. (i) and (ii) are immediate consequence of Theorem 3.5.3 applied to the adjoint representation of $\mathfrak{g}$. Note that (3.5.3) displays the fact that the $\mathfrak{g}_i$ are ideals. To prove (iii), let $m_i = \dim(\mathfrak{g}_i)$ $(0 \leq i \leq s, m_s = m)$, and let $\{X_1, \ldots, X_m\}$ be a basis for $\mathfrak{g}$ such that $\{X_1, \ldots, X_{m_i}\}$ is a basis for $\mathfrak{g}_i (1 \leq i \leq s)$. Clearly, $1 \leq m_1 < m_2 < \cdots < m_s = m$; in particular, $m_i \geq i$, $1 \leq i \leq s$. If $m_i < j < m_{i+1}$, then $X_j \in \mathfrak{g}_{i+1}$, so $[X_j, X_p] \in \mathfrak{g}_i$ for $1 \leq p \leq m$. This shows that $c_{jpr} = 0$ for $r > m_i$; in particular, for $r \geq j$. As $c_{pjr} = -c_{jpr}$, $c_{jpr} = 0$ for $r \geq p$ also. Hence we get (3.5.5). This completes the proof of the theorem.

Corollary 3.5.5. *Let $\mathfrak{g}$ be a Lie algebra over k, and let $\mathfrak{C}^0\mathfrak{g}, \mathfrak{C}^1\mathfrak{g}, \ldots$ be defined inductively as follows: $\mathfrak{C}^0\mathfrak{g} = \mathfrak{g}$, and for $q \geq 1$, $\mathfrak{C}^q\mathfrak{g} = [\mathfrak{g}, \mathfrak{C}^{q-1}\mathfrak{g}]$. Then $\mathfrak{C}^0\mathfrak{g} \supseteq \mathfrak{C}^1\mathfrak{g} \supseteq \cdots$, the $\mathfrak{C}^q\mathfrak{g}$ are all ideals in $\mathfrak{g}$, and $[\mathfrak{C}^a\mathfrak{g}, \mathfrak{C}^q\mathfrak{g}] \subseteq \mathfrak{C}^{a+q+1}\mathfrak{g}$ for $a, q \geq 0$. $\mathfrak{g}$ is nilpotent if and only if $\mathfrak{C}^p\mathfrak{g} = 0$ for some $p \geq 1$. In this case, the $\mathfrak{C}^q\mathfrak{g}$ decrease strictly till they vanish; i.e., if $\mathfrak{C}^q\mathfrak{g} \neq 0$, then $\mathfrak{C}^{q+1}\mathfrak{g} \neq \mathfrak{C}^q\mathfrak{g}$.*

Proof. Write $\mathfrak{L}_q = \mathfrak{C}^q\mathfrak{g}$, $q \geq 0$. If $q \geq 0$ and $\mathfrak{L}_q$ is an ideal, then $\mathfrak{L}_{q+1} = [\mathfrak{g}, \mathfrak{L}_q] \subseteq \mathfrak{L}_q$ and $[\mathfrak{g}, \mathfrak{L}_{q+1}] \subseteq [\mathfrak{g}, \mathfrak{L}_q] \subseteq \mathfrak{L}_{q+1}$. So $\mathfrak{L}_{q+1}$ is an ideal and is contained in $\mathfrak{L}_q$. By induction on q we conclude that the $\mathfrak{L}_q$ are all ideals and that $\mathfrak{L}_0 \supseteq \mathfrak{L}_1 \supseteq \cdots$. By definition, $[\mathfrak{L}_0, \mathfrak{L}_q] \subseteq \mathfrak{L}_{q+1}$ for all $q \geq 0$. Suppose now that for some $a \geq 0$, $[\mathfrak{L}_a, \mathfrak{L}_q] \subseteq \mathfrak{L}_{a+q+1}$ for all $q \geq 0$. The identity $[[X, Y], Z] = [X, [Y, Z]] + [Y, [Z, X]]$, where $X \in \mathfrak{g}$, $Y \in \mathfrak{L}_a$, $Z \in \mathfrak{L}_q$, shows that $[\mathfrak{L}_{a+1}, \mathfrak{L}_q] \subseteq \mathfrak{L}_{a+q+2}$ for all $q \geq 0$. So, by induction on a, we see that $[\mathfrak{L}_a, \mathfrak{L}_q] \subseteq \mathfrak{L}_{a+q+1}$ for all $a \geq 0, q \geq 0$. For $X \in \mathfrak{g}$ and $p \geq 1$, $(\text{ad } X)^p$ maps $\mathfrak{g}$ into $\mathfrak{L}_p$. Hence if $\mathfrak{L}_p = 0$ for some $p \geq 1$, ad X is nilpotent for all $X \in \mathfrak{g}$, showing that $\mathfrak{g}$ is nilpotent. Conversely, suppose $\mathfrak{g}$ is nilpotent. Let $\mathfrak{g}_i$ $(1 \leq i \leq s)$ be the ideals defined by (3.5.3). Clearly, $\mathfrak{L}_1 \subseteq \mathfrak{g}_{s-1}$, $\mathfrak{L}_2 \subseteq \mathfrak{g}_{s-2}, \ldots$, so $\mathfrak{L}_s = 0$. In this case, suppose $q \geq 0$ is such that $\mathfrak{L}_q \neq 0$ but $\mathfrak{L}_{q+1} = \mathfrak{L}_q$. Then $\mathfrak{L}_p = \mathfrak{L}_q$ for $p \geq q$, contradicting the fact that $\mathfrak{L}_r = 0$ for all $r \geq s$. So $\mathfrak{L}_{q+1} \neq \mathfrak{L}_q$.

Corollary 3.5.6. *Let $\mathfrak{g}$ be nilpotent, $m = \dim \mathfrak{g}$. Then there are ideals $\mathfrak{h}_i$ of $\mathfrak{g}$ such that* (i) *$\dim \mathfrak{h}_i = m - i$ for $0 \leq i \leq m$,* (ii) *$\mathfrak{h}_0 = \mathfrak{g} \supseteq \mathfrak{h}_1 \supseteq \cdots \supseteq \mathfrak{h}_m = 0$, and* (iii) *$[\mathfrak{g}, \mathfrak{h}_i] \subseteq \mathfrak{h}_{i+1}$ for $0 \leq i \leq m - 1$.*

Proof. Let $\mathfrak{g}_0 = 0, \mathfrak{g}_1, \ldots, \mathfrak{g}_s = \mathfrak{g}$ be the ideals defined by the equation (3.5.3). If $\mathfrak{a}, \mathfrak{b}$ are any two linear subspaces of $\mathfrak{g}$ with $\mathfrak{g}_i \subseteq \mathfrak{b} \subseteq \mathfrak{a} \subseteq \mathfrak{g}_{i+1}$, then $[\mathfrak{g}, \mathfrak{a}] \subseteq [\mathfrak{g}, \mathfrak{g}_{i+1}] \subseteq \mathfrak{g}_i \subseteq \mathfrak{b} \subseteq \mathfrak{a}$, so $\mathfrak{a}$ is an ideal and $[\mathfrak{g}, \mathfrak{a}] \subseteq \mathfrak{b}$. By interpolating suitably many linear subspaces between the successive $\mathfrak{g}_i$ we obtain a sequence $\{\mathfrak{h}_i\}$ with the required properties.

It is clear from the definition of the ideals $\mathfrak{g}_i$ that $\mathfrak{g}_1$ is the center of $\mathfrak{g}$ and $\mathfrak{g}_i$ is the complete inverse image in $\mathfrak{g}$ of the center of $\mathfrak{g}/\mathfrak{g}_{i-1}$. For this reason, the increasing sequence $\mathfrak{z}_0 = 0, \mathfrak{z}_1, \mathfrak{z}_2, \ldots$ is called the *ascending central series* of $\mathfrak{g}$. Note that this can be defined for any Lie algebra, and the nilpotent Lie algebras are precisely those for which some member of the ascending central series coincides with $\mathfrak{g}$. The sequence $\{\mathfrak{C}^q\mathfrak{g}\}$ is called the *descending central series* for $\mathfrak{g}$.

We now use Theorem 3.5.3 to study finite-dimensional representations of nilpotent Lie algebras which are not necessarily nil representations.

Let $\mathfrak{g}$ be a nilpotent Lie algebra and ρ a representation of $\mathfrak{g}$ in a finite-dimensional vector space V over k. A linear function λ on $\mathfrak{g}$ with values in k is said to be a *weight* of ρ if there exist $v \neq 0$ in V and an integer $m = m(v) \geq 1$ such that

$$(\rho(X) - \lambda(X)1)^m v = 0 \quad (X \in \mathfrak{g}); \tag{3.5.6}$$

in this case, the set of all such v together with 0 forms a linear subspace of V, called the *weight subspace* of ρ corresponding to the weight λ, and is denoted by $V_{\rho,\lambda}$. It is obvious that ρ is a nil representation if and only if $V = V_{\rho,0}$. More generally, if $\lambda \in \mathfrak{g}^*$ (= dual of the vector space underlying $\mathfrak{g}$), ρ is called a *λ-representation* if $V = V_{\rho,\lambda}$.

Given a representation ρ of $\mathfrak{g}$ in V, the weight subspaces corresponding to distinct weights are linearly independent. For suppose $\lambda_1, \ldots, \lambda_r$ are distinct weights of ρ. Choose $X_0 \in \mathfrak{g}$ such that $\lambda_1(X_0), \ldots, \lambda_r(X_0)$ are distinct elements of k, and write $\mu_i = \lambda_i(X_0)$, $L = \rho(X_0)$; then $V_{\rho,\lambda_i} \subseteq V_{L,\mu_i}$(cf.(3.1.1)), so the linear independence of the V_{ρ,λ_i} follows from that of the V_{L,μ_i} (cf. Theorems 3.1.1 and 3.1.2).

Suppose ρ such that for each $X \in \mathfrak{g}$ there is a $\lambda(X) \in k$ such that $\rho(X) - \lambda(X)1$ is nilpotent. We then obtain the identity

$$\lambda(X)\cdot\dim(V) = tr\,\rho(X) \quad (X \in \mathfrak{g}). \tag{3.5.7}$$

This equation implies at once that λ is a linear function on $\mathfrak{g}$ vanishing on $[\mathfrak{g},\mathfrak{g}]$ and that ρ is a λ-representation. In particular, ρ gives rise to a nil representation of $[\mathfrak{g},\mathfrak{g}]$ by restriction. A simple calculation, based on the vanishing of λ on $[\mathfrak{g},\mathfrak{g}]$, shows that $\rho' : X \mapsto \rho(X) - \lambda(X)1$ is a nil representation of $\mathfrak{g}$. Conversely, if ρ' is a nil representation of $\mathfrak{g}$ in V, and λ is a linear function on $\mathfrak{g}$ with values in k and vanishing on $[\mathfrak{g},\mathfrak{g}]$, $\rho : X \mapsto \rho'(X) + \lambda(X)1$ is a λ-representation of $\mathfrak{g}$ in V. If ρ is a λ-representation of $\mathfrak{g}$ in V, there is a basis $\{v_1, \ldots, v_n\}$ for V in which the matrix of each $\rho(X)$ has the form

$$\begin{pmatrix} \lambda(X) & & & & \\ & \lambda(X) & & & \\ & & \cdot & & \\ & & & \cdot & \\ 0 & & & \cdot & \lambda(X) \end{pmatrix} \tag{3.5.8}$$

this follows at once on applying Theorem 3.5.3 to the nil representation $X \mapsto \rho(X) - \lambda(X)1$ of $\mathfrak{g}$.

Lemma 3.5.7. *Let $\mathfrak{g}$ be a nilpotent Lie algebra over k and let ρ_i be a λ_i-representation of $\mathfrak{g}$ in a finite-dimensional vector space V_i ($i = 1, 2$, $\lambda_i \in \mathfrak{g}^*$). Then $\rho_1 \otimes \rho_2$ is a $\lambda_1 + \lambda_2$-representation of $\mathfrak{g}$.*

Proof. Let 1_j be the identity endomorphism of V_j ($j = 1, 2$). Write $V = V_1 \otimes V_2$, $\rho = \rho_1 \otimes \rho_2$, $\lambda = \lambda_1 + \lambda_2$. Then $\rho(X) = \rho_1(X) \otimes 1_2 + 1_1 \otimes \rho_2(X)$ ($X \in \mathfrak{g}$). Fix $X \in \mathfrak{g}$, and let $L_j = \rho_j(X) - \lambda_j(X)1_j$. Then $L_1 \otimes 1_2$ and $1_1 \otimes L_2$ are commuting nilpotent endomorphisms of V, so $L_1 \otimes 1_2 + 1_1 \otimes L_2 = L$ is also nilpotent. But $L = \rho(X) - \lambda(X) \cdot 1$.

Theorem 3.5.8. *Let $\mathfrak{g}$ be a nilpotent Lie algebra, V a finite-dimensional vector space, both over k. Let ρ be a representation of $\mathfrak{g}$ in V. Then the weight subspaces of ρ corresponding to distinct weights are linearly independent. If k is algebraically closed, and $\lambda_1, \ldots, \lambda_r$ are all the distinct weights of ρ, then*

$$V = \sum_{1 \leq i \leq r} V_{\rho, \lambda_i}, \tag{3.5.9}$$

the sum being direct.

Proof. We have already proved the first assertion. Let k be algebraically closed. We prove the second assertion by induction on $\dim V$. Suppose that for each $X \in \mathfrak{g}$, $\rho(X)$ has exactly one eigenvalue, say $\lambda(X)$. Then, as we saw above, λ is a linear function on $\mathfrak{g}$ with values in k, zero on $[\mathfrak{g},\mathfrak{g}]$, and $V = V_{\rho,\lambda}$. This is the case, for example, if $\dim V = 1$. We may thus assume that for some $X_0 \in \mathfrak{g}$, $\rho(X_0)$ has at least two distinct eigenvalues. Let $\mu_1, \ldots, \mu_r$ ($r \geq 2$) be the distinct eigenvalues of $\rho(X_0)$. By Theorem 3.1.2,

$$V = \sum_{1 \leq i \leq r} V_{\rho(X_0), \mu_i},$$

the sum being direct. Further,

$$\dim(V_{\rho(X_0), \mu_i}) < \dim V \quad (1 \leq i \leq r). \tag{3.5.10}$$

Suppose $X \in \mathfrak{g}$. Since ad X_0 is nilpotent, we can find an integer $p \geq 1$ such that $(\text{ad } X_0)^p(X) = 0$, i.e.,

$$\underbrace{[X_0,[X_0,[\cdots[X_0}_{p \text{ times}},X]\cdots] = 0.$$

Since ρ is a representation, we then have

$$\underbrace{[\rho(X_0), [\rho(X_0), [\cdots[\rho(X_0)}_{p \text{ times}},\rho(X)]\cdots] = 0.$$

In other words,

$$(\mathrm{ad}\,\rho(X_0))^p(\rho(X)) = 0.$$

We may then conclude from Theorem 3.1.6 that $\rho(X)$ leaves each $V_{\rho(X_0),\mu_i}$, invariant. Let $V_i = V_{\rho(X_0),\mu_i}$, $\rho_i(X) = \rho(X)|V_i$ $(X \in \mathfrak{g}, 1 \leq i \leq r)$. Then ρ_i is a representation of $\mathfrak{g}$ in V_i; and, in view of (3.5.10), the induction hypothesis is applicable to it. The relation (3.5.9) then follows from the decompositions of the V_i relative to the ρ_i.

3.6. Nilpotent Analytic Groups

An analytic group (real or complex) is said to be *nilpotent* if its Lie algebra is nilpotent. We shall obtain in this section some basic results concerning the structure of nilpotent analytic groups. Our principal tools will be the results of the previous section and the Baker–Campbell–Hausdorff formula. Throughout this section, G will denote a nilpotent *analytic* group and $\mathfrak{g}$ its Lie algebra.

Theorem 3.6.1. (i) *There exists a polynomial mapping*[4] P *of* $\mathfrak{g} \times \mathfrak{g}$ *into* $\mathfrak{g}$ *such that*

(3.6.1) $$\exp X \exp Y = \exp P(X:Y) \quad (X, Y \in \mathfrak{g}).$$

(ii) *Let* $\mathfrak{z}$ *be the center of* $\mathfrak{g}$ *and*

(3.6.2) $$D = \{X : X \in \mathfrak{z}, \exp X = 1\}.$$

Then D *is a discrete additive subgroup of* $\mathfrak{g}$, *and the exponential map induces an analytic diffeomorphism of the manifold* $\mathfrak{g}/D$ *onto* G. *In particular,* D *is the fundamental group of* G, $\mathfrak{g}$ *is a covering manifold of* G *with exp as the covering map, and exp is surjective.*

Proof. We prove (i) using the results of §2.15. Let c_n $(n \geq 1)$ be the maps of $\mathfrak{g} \times \mathfrak{g}$ into $\mathfrak{g}$ such that $c_1(X:Y) = X + Y$ $(X, Y \in \mathfrak{g})$ and the recursion formulae (2.15.15) are satisfied. Each c_n is a polynomial map. Let $\mathfrak{L}_0 = \mathfrak{g}$, $\mathfrak{L}_1, \ldots$ be the descending central series of $\mathfrak{g}$, so that $\mathfrak{L}_{q+1} = [\mathfrak{g}, \mathfrak{L}_q]$ $(q \geq 0)$ and $\mathfrak{L}_s = 0$ for some $s \geq 1$. We now prove by induction on q that $c_q(X:Y) \in \mathfrak{L}_{q-1}$ for $q \geq 1$ and $X, Y \in \mathfrak{g}$. This is clear for $q = 1$. Suppose that $q > 1$ and that c_r maps $\mathfrak{g} \times \mathfrak{g}$ into $\mathfrak{L}_{r-1}$ for $1 \leq r < q$. By Corollary 3.5.5, if k_j are

[4]Let U, V be finite-dimensional vector spaces over a field k of characteristic 0 and φ a map of U into V. φ is said to be a *polynomial map* if there are bases $\{u_1, \ldots, u_m\}$ for U, $\{v_1, \ldots, v_n\}$ for V, and polynomials $p_1, \ldots, p_n$ such that $\varphi(\sum_{1 \leq i \leq m} x_i u_i) = \sum_{1 \leq j \leq n} p_j(x_1, \ldots, x_m) v_j$. It is easy to verify that this definition is independent of the choice of bases. If k' is an extension field of k, it is obvious that φ extends uniquely to a polynomial map of $U^{k'}$ into $V^{k'}$.

integers ≥ 1 and $X_j \in \mathfrak{L}_{k_j-1}$ $(1 \leq j \leq p)$, then for any $Z \in \mathfrak{g}$,

$$[X_1,[X_2,[\cdots[X_p,Z]\cdots]]] \in \mathfrak{L}_{k_1+\cdots+k_p}.$$

If we use this and the induction hypothesis, we may conclude at once from the recursion formulae (2.15.15) that c_q maps $\mathfrak{g} \times \mathfrak{g}$ into $\mathfrak{L}_{q-1}$. In particular, $c_{s+1} = 0$.

Let $P = \sum_{0 \leq q \leq s} c_q$. Then P is a polynomial map of $\mathfrak{g} \times \mathfrak{g}$ into $\mathfrak{g}$, and there is an open neighborhood $\mathfrak{u}$ of 0 in $\mathfrak{g}$ such that $\exp X \exp Y = \exp P(X : Y)$ for all $X, Y \in \mathfrak{u}$. By the analyticity of exp, this equation is valid for all $X, Y \in \mathfrak{g}$.

We now come to (ii). The formula (3.6.1) shows that $G' = \exp \mathfrak{g}$ is a subgroup of G. Since G' contains an open neighborhood of 1 in G, G' must be an open and hence also closed subgroup of G. Since G is connected, $G = \exp \mathfrak{g}$. Now use Theorem 2.14. For any $X \in \mathfrak{g}$, ad X is nilpotent and so has 0 as its sole eigenvalue. Hence the open set $\mathfrak{v}$ of Theorem 2.14.6 coincides with $\mathfrak{g}$ itself. The assertions of (ii) now follow immediately from that theorem.

These results lead at once to the following.

Theorem 3.6.2. *Let G be simply connected. Then exp is an analytic diffeomorphism of $\mathfrak{g}$ onto G. If H is an analytic subgroup of G and $\mathfrak{h}$ is the corresponding subalgebra of $\mathfrak{g}$, then H is closed in G, is simply connected, and is equal to exp* $[\mathfrak{h}]$.

Proof. Since $\mathfrak{g}$ is a covering manifold of G with exp as the covering map, D must be $\{0\}$ when G is simply connected. Hence exp is an analytic diffeomorphism of $\mathfrak{g}$ onto G. Let H be an analytic subgroup of G, $\mathfrak{h}$ the corresponding subalgebra of $\mathfrak{g}$. Since $\mathfrak{h}$ is nilpotent, $\exp \mathfrak{h} = H$ by (ii) of the previous theorem. Now exp is a homeomorphism of $\mathfrak{g}$ onto G, and $\mathfrak{h}$ is a closed simply connected subset of $\mathfrak{g}$. Hence $H = \exp \mathfrak{h}$ must be a closed simply connected subset of G.

Remarks 1. If G is not simply connected, its analytic subgroups need not always be closed. This is already the case, for instance, when G is a torus.

2. When G is simply connected it is customary to write "log" for the map of G onto $\mathfrak{g}$ that inverts the exponential map.

Our aim now is to prove that the simply connected nilpotent groups are precisely those which are isomorphic to unipotent subgroups of matrix groups. An endomorphism u of a finite-dimensional vector space V is said to be *unipotent* if $u - 1$ is nilpotent; it is then invertible, and u^{-1} is also unipotent. A subgroup of $GL(V)$ is said to be *unipotent* if it consists entirely of unipotent elements.

Theorem 3.6.3. *Let V be a finite-dimensional vector space (over* **R** *or* **C**), *and let $p = \dim V$. Let $\bar{\mathfrak{g}}$ be a subalgebra of $\mathfrak{gl}(V)$ consisting entirely of nilpotent endomorphisms of V, and $\bar{G}$ the analytic subgroup of $GL(V)$ defined by $\bar{\mathfrak{g}}$. Then $\bar{\mathfrak{g}}$ is nilpotent, and $\bar{G}$ is a simply connected, unipotent, algebraic subgroup of $GL(V)$. Moreover,*

$$(3.6.3)\qquad \begin{aligned} \exp X &= 1 + \sum_{1 \le s < p} \frac{X^s}{s!} \qquad (X \in \bar{\mathfrak{g}}) \\ \log x &= \sum_{1 \le s < p} (-1)^{s-1} \frac{(x-1)^s}{s} \qquad (x \in \bar{G}). \end{aligned}$$

Proof. We saw at the beginning of §3.5 that $\bar{\mathfrak{g}}$ is nilpotent. By Theorem 3.6.1, $\bar{G} = \exp[\bar{\mathfrak{g}}]$. Let $\{v_1, \ldots, v_p\}$ be a basis for V in which the matrices of $X \in \bar{\mathfrak{g}}$ have zeros on and below the main diagonal. It is then clear that the matrix of $x = \exp X$ $(X \in \bar{\mathfrak{g}})$ in this basis has the form

$$\begin{pmatrix} 1 & & & & \\ & 1 & & & \\ & & \cdot & & \\ & & & \cdot & \\ & 0 & & & 1 \end{pmatrix}$$

G is thus seen to be a unipotent subgroup of $GL(V)$.

Let $\mathcal{E}$ be the associative algebra of endomorphisms of V, and

$$(3.6.4)\qquad l(x) = \sum_{1 \le s < p} (-1)^{s-1} \frac{(x-1)^s}{s} \qquad (x \in \mathcal{E}).$$

Then l $(x \mapsto l(x))$ is a polynomial map of $\mathcal{E}$ into $\mathfrak{gl}(V)$. If x is unipotent, then

$$(3.6.5)\qquad l(x) = \sum_{s=1}^{\infty} (-1)^{s-1} \frac{(x-1)^s}{s}.$$

A straightforward verification then shows that if x is unipotent and X is nilpotent.

$$(3.6.6)\qquad \exp l(x) = x \qquad l(\exp X) = X$$

In other words, l is a continuous map of $\bar{G}$ onto $\bar{\mathfrak{g}}$ inverting exp. exp is thus a homeomorphism of $\bar{\mathfrak{g}}$ onto $\bar{G}$. $\bar{G}$ is thus simply connected, $l = \log$ on G, and we have (3.6.3).

It remains to prove that $\bar{G}$ is algebraic. Let $\{v_1, \ldots, v_p\}$ be the basis considered above, let $\mathfrak{g}'$ be the Lie algebra of all $X \in \mathfrak{gl}(V)$ whose matrices in this basis have zeros on and below the main diagonal, and let $G' = \exp[\mathfrak{g}']$. For any endomorphism x of V, let $u_{ij}(x)$ be the ijth entry of the matrix of x

in the basis $\{v_1, \ldots, v_p\}$. Then $x \in G'$ if and only if

$$u_{ij}(x) = \delta_{ij} \quad (i \geq j). \tag{3.6.7}$$

Let $\lambda_1, \ldots, \lambda_s$ be linear functions on $\mathfrak{gl}(V)$ such that $\bar{\mathfrak{g}}$ is precisely the set of all X for which $\lambda_1(X) = \cdots = \lambda_s(X) = 0$. Let

$$p_r(x) = \lambda_r(l(x)) \quad (x \in \mathcal{E}) \tag{3.6.8}$$

for $1 \leq r \leq s$ (cf. (3.6.4)). Then the p_r are polynomials on $\mathcal{E}$ and for any $x \in \mathcal{E}$, $x \in \bar{G}$ if and only if

$$u_{ij}(x) = \delta_{ij} \quad (i \geq j), \qquad p_r(x) = 0 \quad (1 \leq r \leq s). \tag{3.6.9}$$

The equations (3.6.9) show that $\bar{G}$ is algebraic.

Corollary 3.6.4. *The center of any nilpotent analytic group is connected.*

Proof. Let H be a nilpotent analytic group with Lie algebra $\mathfrak{h}$, Z the center of H. Applying Theorem 3.6.3 to the case when $V = \mathfrak{h}$ and $\bar{\mathfrak{g}} = \mathrm{ad}[\mathfrak{h}]$, we find that $H/Z \cong \mathrm{Ad}(H) = \exp[\mathrm{ad}[\mathfrak{h}]]$ is simply connected. Let Z^0 be the component of 1 in Z, and η the natural map of H/Z^0 onto H/Z. Then η is easily seen to be a covering map. So η is bijective, i.e., $Z^0 = Z$.

Our aim now is to obtain a converse to Theorem 3.6.3. We need a lemma.

Lemma 3.6.5. *Let A be an analytic manifold, H a Lie group acting analytically on A via the action $(h,a) \mapsto h \cdot a$ $(h \in H, a \in A)$. For any analytic function φ on A and $h \in H$, let $\varphi^h(a) = \varphi(h^{-1} \cdot a)$ $(a \in A)$, and let $\pi(h)$ be the map $\varphi \mapsto \varphi^h$. Suppose $\varphi_1, \ldots, \varphi_r$ are analytic functions[5] on A such that the linear span V of the functions φ_i^h $(1 \leq i \leq r, h \in H)$ is finite-dimensional. Then V is invariant under all $\pi(h)$ and π $(h \mapsto \pi(h))$ gives rise, by restriction to V, to an analytic homomorphism of H into $GL(V)$.*

Proof. Since $\varphi_i^{hh'} = (\varphi_i^{h'})^h$ $(h, h' \in H)$, it is clear that V is invariant under all the $\pi(h)$. Let $\{\psi_1, \ldots, \psi_r\}$ be a basis for V. Then there are functions d_{ij} on H such that, for $i \leq j \leq r$,

$$\psi_j(h^{-1} \cdot a) = \sum_{1 \leq i \leq r} d_{ij}(h)\psi_i(a) \quad (a \in A, h \in H).$$

Since the map $(h, a) \mapsto h^{-1} \cdot a$ is analytic, it is clear that the right side of this equation are analytic in h for each $a \in A$. On the other hand, since the ψ_i are linearly independent, it follows that $a \mapsto (\psi_1(a), \ldots, \psi_r(a))$ maps A onto $k^{(r)}(= \mathbf{R}^{(r)}$ or $\mathbf{C}^{(r)}$ according as we are in the real or complex analytic case). Consequently, if $\psi_{j,a}(h) = \psi_j(h^{-1} \cdot a)$, the functions d_{ij} $(1 \leq i \leq r)$ are in the

[5] With real or complex values according as we are in the real or complex analytic setup.

linear span of the $\psi_{j,a}$ $(a \in A)$, hence analytic. This leads easily to the statements of the lemma.

Theorem 3.6.6. *Any simply connected nilpotent analytic group G is isomorphic to a closed unipotent subgroup of GL(V) for some finite-dimensional vector space V.*

Proof. We write k for either **R** or **C**. For purposes of this proof, a function $\varphi : G \mapsto k$ is called linear (resp. polynomial) if $\varphi \circ \exp$ is a linear (resp. polynomial) function on $\mathfrak{g}$. Let $\mathfrak{L}$ (resp. $\mathcal{P}$) be the vector space over k of all linear (resp. polynomial) functions on G. For any integer $r \geq 0$, let $\mathcal{P}_r$ be the subspace of all $f \in \mathcal{P}$ such that $f \circ \exp$ is a polynomial of degree $\leq r$ on $\mathfrak{g}$. For $\varphi \in \mathcal{P}$, $h \in G$, let

$$\varphi^h(y) = \varphi(yh) \quad (y \in G). \tag{3.6.10}$$

If P is the polynomial map of $\mathfrak{g} \times \mathfrak{g}$ into $\mathfrak{g}$ satisfying (3.6.1), then

$$\varphi^h(\exp X) = \varphi(\exp P(X : \log h)) \quad (X \in \mathfrak{g})$$

for $\varphi \in \mathcal{P}$ and $h \in G$, so $\varphi^h \in \mathcal{P}$ for $\varphi \in \mathcal{P}$ and $h \in G$. Let V be the linear span of all φ^h with $h \in G, \varphi \in \mathfrak{L}$. We claim that $\dim V < \infty$. Let $d \geq 1$ be an integer with the property that for any $Y \in \mathfrak{g}$ and any linear function λ: $\mathfrak{g} \to k$, $X \mapsto \lambda(P(X : Y)))$ is a polynomial of degree $\leq d$ on $\mathfrak{g}$; it is clearly possible to choose such a d, since P is a polynomial map. But this implies that $\varphi^h \in \mathcal{P}_d$ for all $\varphi \in \mathfrak{L}$, $h \in G$. This proves that $\dim V < \infty$.

Lemma 3.6.4 now applies and gives rise to an analytic homomorphism π of G into $GL(V)$, where $\pi(h)\varphi = \varphi^h$ $(h \in G, \varphi \in V)$. The proof of the theorem will be complete if we show that π is injective and that $\pi[G]$ is a closed unipotent subgroup of $GL(V)$.

To prove that π is injective, let $h \in G$ be such that $\pi(h) = 1$, i.e., $\varphi^h = \varphi$ for all $\varphi \in V$. Then $\varphi(xh) = \varphi(x)$ for all linear φ and $x \in G$. Since the linear functions on G separate the points of G, $xh = x$ for $x \in G$, i.e., $h = 1$.

To prove the assertions about $\pi[G]$ it is enough, in view of Theorem 3.6.3, to show that $d\pi$ is a nil representation of $\mathfrak{g}$ in V. Suppose this is not true. We consider first the case $k = $ **C**. Then by Theorem 3.5.8 we can find a nonzero linear function $\lambda : \mathfrak{g} \to$ **C** and a nonzero $f \in V$ such that $d\pi(X)f = \lambda(X)f$ for all $X \in \mathfrak{g}$. We then have, for $X \in \mathfrak{g}$,

$$\pi(\exp X)f = (\exp d\pi(X))f = e^{\lambda(X)}f. \tag{3.6.11}$$

If we write F for the function $Z \mapsto f(\exp Z)$ on $\mathfrak{g}$, we get

$$F(P(Z : X)) = e^{\lambda(X)}F(Z) \quad (X, Z \in \mathfrak{g}).$$

Since P is a polynomial map, this equation implies that $(X,Z) \mapsto e^{\lambda(X)}F(Z)$ is a polynomial function. This is a contradiction, since both λ and F are nonzero.

Suppose now that $k = \mathbf{R}$ and assume as before that $d\pi$ is not a nil representation. We extend $d\pi$ to a representation, denoted by $d\pi$ again, of $\mathfrak{g}_c$ in the complex linear span V_c of V. Then we can find a nonzero complex linear function $\lambda : \mathfrak{g} \to \mathbf{C}$ and a nonzero $f \in V_c$ such that

$$f(x \exp X) = e^{\lambda(X)} f(x) \quad (x \in G, X \in \mathfrak{g}).$$

From this point on, the argument is the same as before.

3.7. Solvable Lie Algebras

As usual, k is a field of characteristic 0, $\mathfrak{g}$ a Lie algebra of finite dimension m over k. We write $\mathfrak{D}\mathfrak{g} = [\mathfrak{g},\mathfrak{g}]$ for the linear span of elements of the form $[X, Y]$, X, $Y \in \mathfrak{g}$. $\mathfrak{D}\mathfrak{g}$ is a subalgebra of $\mathfrak{g}$ and is called the *derived algebra* of $\mathfrak{g}$. We define $\mathfrak{D}^p\mathfrak{g}$ $(p \geq 0)$ inductively by

$$\begin{aligned} \mathfrak{D}^0\mathfrak{g} &= \mathfrak{g} \\ \mathfrak{D}^p\mathfrak{g} &= \mathfrak{D}(\mathfrak{D}^{p-1}\mathfrak{g}) \quad (p \geq 1). \end{aligned} \tag{3.7.1}$$

If $\mathfrak{a}$ is a subalgebra of $\mathfrak{g}$, $\mathfrak{D}\mathfrak{a} = [\mathfrak{a},\mathfrak{a}]$ is again a subalgebra, so (3.7.1) leads to a well-defined sequence $\mathfrak{D}^0\mathfrak{g} \supseteq \mathfrak{D}^1\mathfrak{g} \supseteq \cdots$ of subalgebras of $\mathfrak{g}$. $\mathfrak{D}^p\mathfrak{g}$ is called the *pth derived algebra* of $\mathfrak{g}$.

Theorem 3.7.1 (i) *If $\mathfrak{h}$ is an ideal in $\mathfrak{g}$, the $\mathfrak{D}^p\mathfrak{h}$ are ideals of $\mathfrak{g}$ for all $p \geq 0$. If D is a derivation of $\mathfrak{g}$ that leaves $\mathfrak{h}$ invariant, then D leaves each $\mathfrak{D}^p\mathfrak{h}$ invariant. In particular, the $\mathfrak{D}^p\mathfrak{g}$ are ideals of $\mathfrak{g}$ invariant under all derivations of $\mathfrak{g}$.*

(ii) *If k' is an extension field of k, then for all $p \geq 0$,*

$$\mathfrak{D}^p(\mathfrak{g}^{k'}) = (\mathfrak{D}^p\mathfrak{g})^{k'}, \qquad \mathfrak{D}^p\mathfrak{g} = \mathfrak{D}^p(\mathfrak{g}^{k'}) \cap \mathfrak{g}. \tag{3.7.2}$$

(iii) *If π is homomorphism of $\mathfrak{g}$ onto a Lie algebra $\mathfrak{h}$, then*

$$\pi[\mathfrak{D}^p\mathfrak{g}] = \mathfrak{D}^p\mathfrak{h} \quad (p \geq 0). \tag{3.7.3}$$

(iv) *The algebras $\mathfrak{D}^p\mathfrak{g}/\mathfrak{D}^{p+1}\mathfrak{g}$ are abelian for $p \geq 0$.*

Proof. (i) If $X, X' \in \mathfrak{h}$, $Y \in \mathfrak{g}$, then $[Y,[X,X']] = -[X,[X',Y]] - [X',[Y,X]]$. So $\mathfrak{D}\mathfrak{h}$ is an ideal in $\mathfrak{g}$. If D is a derivation of $\mathfrak{g}$ mapping $\mathfrak{h}$ into itself, $D[X,X'] = [DX,X'] + [X,DX']$, showing that D maps $\mathfrak{D}\mathfrak{h}$ into itself. By induction on p we now have (i).

(ii) is obvious.

For (iii), the surjectivity of π implies that $\pi[\mathfrak{D}\mathfrak{g}] = \mathfrak{D}\mathfrak{h}$; thus (3.7.3) follows by induction on p.

For (iv), let X, $X' \in \mathfrak{g}$. Then $[X,X'] \in \mathfrak{D}\mathfrak{g}$, showing that $\mathfrak{g}/\mathfrak{D}\mathfrak{g}$ is abelian. (iv) follows on using induction on p once again.

$\mathfrak{g}$ is said to be *solvable* if $\mathfrak{D}^p\mathfrak{g} = 0$ for some $p \geq 1$. In this case, if $\mathfrak{D}^q\mathfrak{g} \neq 0$, then $\mathfrak{D}^{q+1}\mathfrak{g} \neq \mathfrak{D}^q\mathfrak{g}$, so the $\mathfrak{D}^p\mathfrak{g}$ strictly decrease until they become 0. It is clear from (3.7.2) that if k' is an extension field of k, then $\mathfrak{g}$ is solvable if and only if $\mathfrak{g}^{k'}$ is. If $\mathfrak{g}$ is solvable and $p \geq 0$ is such that $\mathfrak{D}^p\mathfrak{g} \neq 0$ but $\mathfrak{D}^p\mathfrak{g} = 0$, $\mathfrak{D}^p\mathfrak{g}$ is a nonzero abelian ideal of $\mathfrak{g}$. If $\mathfrak{a}$ is any subspace of $\mathfrak{g}$ with $\mathfrak{D}\mathfrak{g} \subseteq \mathfrak{a} \subseteq \mathfrak{g}$, then $[\mathfrak{g},\mathfrak{a}] \subseteq \mathfrak{D}\mathfrak{g} \subseteq \mathfrak{a}$, so $\mathfrak{a}$ is an ideal. In particular, we can choose ideals $\mathfrak{a}$ with $\dim(\mathfrak{g}/\mathfrak{a}) = 1$ when $\mathfrak{g}$ is solvable.

Theorem 3.7.2. (i) *$\mathfrak{g}$ is solvable if and only if we can find ideals $\mathfrak{g}_0 = \mathfrak{g}$, $\mathfrak{g}_1, \ldots, \mathfrak{g}_{s+1} = 0$ such that $\mathfrak{g}_i \supseteq \mathfrak{g}_{i+1}$ and $\mathfrak{g}_i/\mathfrak{g}_{i+1}$ is abelian, for $0 \leq i \leq s$.*

(ii) *If $\mathfrak{g}$ is solvable, subalgebras and quotient algebras of $\mathfrak{g}$ are solvable.*

(iii) *If $\mathfrak{h}$ is an ideal in $\mathfrak{g}$ such that $\mathfrak{h}$ and $\mathfrak{g}/\mathfrak{h}$ are solvable, then $\mathfrak{g}$ is solvable.*

(iv) *Nilpotent Lie algebras are solvable.*

Proof. (i) If $\mathfrak{g}$ is solvable, $\mathfrak{g}_p = \mathfrak{D}^p\mathfrak{g}$ $(p \geq 0)$ have all the required properties. Conversely, let $\mathfrak{g}_0, \mathfrak{g}_1, \ldots$ be as in (i). Since $\mathfrak{g}_i/\mathfrak{g}_{i+1}$ is abelian, $\mathfrak{D}\mathfrak{g}_i \subseteq \mathfrak{g}_{i+1}$. Hence $\mathfrak{D}^p\mathfrak{g} \subseteq \mathfrak{g}_p$, $p = 0,1,\ldots$, showing that $\mathfrak{D}^{s+1}\mathfrak{g} = 0$. $\mathfrak{g}$ is thus solvable. If $\mathfrak{g}$ is solvable and $\mathfrak{h}$ is a subalgebra of $\mathfrak{g}$, then $\mathfrak{D}^p\mathfrak{h} \subseteq \mathfrak{h} \cap \mathfrak{D}^p\mathfrak{g}$; so $\mathfrak{D}^p\mathfrak{h} = 0$ for large p, showing that $\mathfrak{h}$ is solvable. From (3.7.3) we see that the quotient algebras of $\mathfrak{g}$ are all solvable. To prove (iii), let $\mathfrak{h}$ be a solvable ideal in $\mathfrak{g}$ such that $\mathfrak{g}/\mathfrak{h}$ is solvable. Let $r, s \geq 0$ be such that $\mathfrak{D}^r\mathfrak{h} = 0$, $\mathfrak{D}^s(\mathfrak{g}/\mathfrak{h}) = 0$. Then $\mathfrak{D}^s\mathfrak{g} = \mathfrak{h}$ by (3.7.3), and hence $\mathfrak{D}^{r+s}\mathfrak{g} = 0$. Suppose $\mathfrak{g}$ is nilpotent and $\mathfrak{g}_i$ are ideals of $\mathfrak{g}$ defined by (3.5.3). Since $\mathfrak{D}\mathfrak{g}_i \subseteq [\mathfrak{g},\mathfrak{g}_i] \subseteq \mathfrak{g}_{i+1}$, $\mathfrak{g}_i/\mathfrak{g}_{i+1}$ is abelian for $i \geq 0$. Since $\mathfrak{g}_p = 0$ for sufficiently large $p \geq 1$, $\mathfrak{g}$ is solvable by (i). This proves the theorem.

Let V be a vector space over k, $\{v_1, \ldots, v_n\}$ a basis for V. Let $\mathfrak{g}$ be the Lie algebra of all endomorphisms X of V whose matrices in the basis $\{v_1, \ldots, v_n\}$ have the form

$$\begin{pmatrix} \lambda_1 & & & \\ & \cdot & & \\ & & \cdot & \\ 0 & & & \lambda_n \end{pmatrix}.$$

Then $\mathfrak{g}$ is solvable. For $\mathfrak{D}\mathfrak{g}$ is contained in the nilpotent Lie algebra of all endomorphisms of V whose matrices have zeros on and below the main diagonal; hence $\mathfrak{D}\mathfrak{g}$ is solvable, and hence so is $\mathfrak{g}$. Any subalgebra of $\mathfrak{g}$ is thus solvable. The basic result in the theory of solvable Lie algebras is the theorem of Lie which asserts that if k is algebraically closed, then any solvable matrix Lie algebra can be obtained in the above manner.

Theorem 3.7.3. *Let k be algebraically closed, $\mathfrak{g}$ a solvable Lie algebra over k, and ρ a representation of $\mathfrak{g}$ in a vector space V of finite dimension n over k. Then there exist $\lambda_i \in \mathfrak{g}^*$ $(1 \leq i \leq n)$ and a basis $\{v_1, \ldots, v_n\}$ of V such that for each $X \in \mathfrak{g}$, the matrix of $\rho(X)$ in the basis has the form*

$$(3.7.4) \qquad \begin{pmatrix} \lambda_1(X) & & & * \\ & \cdot & & \\ & & \cdot & \\ 0 & & & \lambda_n(X) \end{pmatrix}.$$

In particular, for all $X \in \mathfrak{g}$,

$$(3.7.5) \qquad \rho(X)v_1 = \lambda_1(X)v_1.$$

Proof. By induction on $\dim \mathfrak{g}$. Since the case $\dim \mathfrak{g} = 1$ is trivial, assume that $\dim \mathfrak{g} \geq 2$.

First we prove the existence of a nonzero vector of V which is an eigenvector for all $\rho(X)$, $X \in \mathfrak{g}$. Let $\mathfrak{h}$ be an ideal in $\mathfrak{g}$ with $\dim(\mathfrak{g}/\mathfrak{h}) = 1$; let $X_0 \in \mathfrak{g}$, $X_0 \notin \mathfrak{h}$. By the induction hypothesis, we can select a nonzero $w_0 \in V$ and a $\lambda \in \mathfrak{h}^*$ such that $\rho(Y)w_0 = \lambda(Y)w_0$ for all $Y \in \mathfrak{h}$. Let $w_s = \rho(X_0)^s w_0$ $(s \geq 1)$. Let $p \geq 0$ be the largest of the integers s for which $w_0, \ldots, w_s$ are linearly independent. Let $W_{-1} = 0$, and let W_r be the linear span of $w_0, \ldots, w_r$ $(0 \leq r \leq p)$. Then $w_q \in W_p$ for $q \geq p$, so $\rho(X_0)$ leaves W_p invariant and maps W_r into W_{r+1} $(0 \leq r < p)$.

We claim that for $0 \leq r \leq p$ and $Y \in \mathfrak{h}$,

$$(3.7.6) \qquad \rho(Y)w_r \equiv \lambda(Y)w_r \pmod{W_{r-1}}.$$

For $r = 0$ this is obvious. Suppose (3.7.6) is true for some $r < p$. Then for $Y \in \mathfrak{h}$,

$$(3.7.7) \qquad \rho(Y)w_{r+1} = \rho(X_0)\rho(Y)w_r + \rho([Y,X_0])w_r;$$

and since $[X_0,\mathfrak{h}] \subseteq \mathfrak{h}$, we conclude easily from the assumption on r that (3.7.6) is true with r replaced by $r + 1$. In particular, W_p is invariant under ρ.

If $Y \in \mathfrak{h}$, both $\rho(Y)$ and $\rho(X_0)$ leave W_p invariant, and hence $tr(\rho([Y,X_0]) \mid W_p) = 0$. On the other hand, it is clear from (3.7.6) that $tr(\rho(Z) \mid W_p) = (p + 1)\lambda(Z)$ for all $Z \in \mathfrak{h}$. So, taking $Z = [Y,X_0]$, we have $\lambda([Y,X_0]) = 0$ for $Y \in \mathfrak{h}$. But then since $\rho(Y)w_0 = \lambda(Y)w_0$ for all $Y \in \mathfrak{h}$, an easy induction on r enables us to conclude from (3.7.7) that $\rho(Y)w_r = \lambda(Y)w_r$ for all $Y \in \mathfrak{h}$ and $0 \leq r \leq p$. Now choose a nonzero $v_1 \in W_p$ such that $\rho(X_0)v_1 = cv_1$ for some $c \in k$, and let λ_1 denote the extension of λ to the element of $\mathfrak{g}^*$ which takes the value c at X_0. Then λ_1 and v_1 satisfy (3.7.5) for all $X \in \mathfrak{g}$.

$k \cdot v_1$ is thus invariant under ρ. Considering the representation induced by

ρ in $V/k\cdot v_1$ and using induction on dim V, we obtain a basis $\{v_1,\ldots,v_n\}$ for V and elements $\lambda_1,\ldots,\lambda_n$ of $\mathfrak{g}^*$ such that

(3.7.8) $$\rho(X)v_r \equiv \lambda_r(X)v_r \quad (\text{mod} \sum_{1\leq j<r} k\cdot v_j).$$

But then the matrix of $\rho(X)$ has the form (3.7.4) for all $X \in \mathfrak{g}$. This proves the theorem.

Corollary 3.7.4. *Let assumptions be as in the above theorem. If ρ is irreducible, then dim $V = 1$.*

Proof. Obvious.

Corollary 3.7.5. *Let $\mathfrak{g}$ be a solvable Lie algebra over k. Then we can find subalgebras $\mathfrak{g}_1 = \mathfrak{g},\mathfrak{g}_2,\ldots,\mathfrak{g}_{m+1} = 0$ such that* (i) *$\mathfrak{g}_{i+1} \subseteq \mathfrak{g}_i$ and $\mathfrak{g}_{i+1}$ is an ideal of $\mathfrak{g}_i$ for $1 \leq i \leq m$, and* (ii) *dim $(\mathfrak{g}_i/\mathfrak{g}_{i+1}) = 1$ for $1 \leq i \leq m$. If k is algebraically closed we can choose the $\mathfrak{g}_i$ to be ideals in $\mathfrak{g}$ itself.*

Proof. We have seen that we can select an ideal $\mathfrak{g}_2$ of $\mathfrak{g}_1 = \mathfrak{g}$ such that $\dim(\mathfrak{g}_1/\mathfrak{g}_2) = 1$. The first assertion is now immediate by induction in $\dim(\mathfrak{g})$. Suppose now that k is algebraically closed. Applying the theorem above to the adjoint representation of $\mathfrak{g}$, we see that there are $\lambda_1,\ldots,\lambda_m \in \mathfrak{g}^*$ and a basis $\{X_1,\ldots,X_m\}$ for $\mathfrak{g}$ such that

(3.7.9) $$[X,X_r] \equiv \lambda_r(X)X_r \quad (\text{mod} \sum_{1\leq s<r} k\cdot X_s).$$

Clearly, it is then sufficient to take $\mathfrak{g}_r$ to be the linear span of $X_1,\ldots,X_{m-r+1}$.

As another consequence of the theorem of Lie we have

Theorem 3.7.6. *Let $\mathfrak{g}$ be a solvable Lie algebra over k. Let ρ be a representation of $\mathfrak{g}$ in a finite-dimensional vector space V. Then the set of all $X \in \mathfrak{g}$ with $\rho(X)$ nilpotent is an ideal in $\mathfrak{g}$ that contains $\mathfrak{D}\mathfrak{g}$. A Lie algebra over k is solvable if and only if its derived algebra is nilpotent.*

Proof. First, assume that k is algebraically closed. Let $\{v_1,\ldots,v_m\}$ be a basis for V and $\lambda_1,\ldots,\lambda_n$ elements of $\mathfrak{g}^*$ such that

(3.7.10) $$\rho(X)v_j \equiv \lambda_j(X)v_j \quad (\text{mod} \sum_{i<j} k\cdot v_i) \quad (X \in \mathfrak{g},\ 1 \leq j \leq n).$$

If $\mathfrak{a} = \{X : X \in \mathfrak{g}, \rho(X) \text{ is nilpotent}\}$, then $X \in \mathfrak{a}$ if and only if $\lambda_j(X) = 0$ for $1 \leq j \leq n$. It follows from this that $\mathfrak{a}$ is a linear subspace that contains $\mathfrak{D}\mathfrak{g}$. In particular, $\mathfrak{a}$ is an ideal. If k is not algebraically closed, let k' be an algebraic closure of k and denote by $\mathfrak{g}'$, V', and ρ' the respective k'-extensions of $\mathfrak{g}$, V, and ρ. If $\mathfrak{a}$ (resp. $\mathfrak{a}'$) is the set of all $X \in \mathfrak{g}$ (resp. $X \in \mathfrak{g}'$) such that

$\rho(X)$ (resp. $\rho'(X)$) is nilpotent, then $\mathfrak{a}'$ is an ideal of $\mathfrak{g}'$ containing $\mathfrak{D}\mathfrak{g}'$ and $\mathfrak{a} = \mathfrak{g} \cap \mathfrak{a}'$. So $\mathfrak{a}$ is an ideal of $\mathfrak{g}$ containing $\mathfrak{D}\mathfrak{g}$. This proves the first assertion.

To prove the second, let $\mathfrak{D}\mathfrak{g}$ be nilpotent. Since $\mathfrak{g}/\mathfrak{D}\mathfrak{g}$ is abelian, $\mathfrak{g}$ is solvable by (iii) of Theorem 3.7.2. Conversely, let $\mathfrak{g}$ be solvable. By the first result applied to the adjoint representation, we see that ad X is nilpotent for any $X \in \mathfrak{D}\mathfrak{g}$. This implies at once that $\mathfrak{D}\mathfrak{g}$ is nilpotent.

An analytic group is called *solvable* if its Lie algebra is solvable. Despite the similarity of the concepts of solvability and nilpotency for Lie algebras, there are many differences in the structure of solvable and nilpotent groups. Examples of some of these may be found in the exercises at the end of this chapter.

3.8. The Radical and the Nil Radical

Let k be a field of characteristic zero. Let $\mathfrak{g}$ be a finite-dimensional Lie algebra over k. Suppose $\mathfrak{a}$ and $\mathfrak{b}$ are two solvable ideals of $\mathfrak{g}$. Then $\mathfrak{a} + \mathfrak{b}$ is an ideal, and since $(\mathfrak{a} + \mathfrak{b})/\mathfrak{a}$ is isomorphic to $\mathfrak{b}/(\mathfrak{a} \cap \mathfrak{b})$, $(\mathfrak{a} + \mathfrak{b})/\mathfrak{a}$ is solvable, so $\mathfrak{a} + \mathfrak{b}$ is solvable. This shows that there is a unique solvable ideal $\mathfrak{q}$ of $\mathfrak{g}$ containing all solvable ideals of $\mathfrak{g}$. $\mathfrak{q}$ is called the *radical* of $\mathfrak{g}$(rad $\mathfrak{g}$). rad $\mathfrak{g} = \mathfrak{g}$ if and only if $\mathfrak{g}$ is solvable. $\mathfrak{g}$ is said to be *semisimple* if rad $\mathfrak{g} = 0$. The radical of a Lie algebra is obviously invariant under all automorphisms of the Lie algebra.

Theorem 3.8.1. (i) *If k' is an extension field of k, then $(\text{rad}\, \mathfrak{g})^{k'} = \text{rad}\, \mathfrak{g}^{k'}$.*
(ii) *rad $\mathfrak{g}$ is invariant under all derivations of $\mathfrak{g}$.*
(iii) *If $\mathfrak{h}$ is an ideal of $\mathfrak{g}$, so is rad $\mathfrak{h}$, and rad $\mathfrak{h} = (\text{rad}\, \mathfrak{g}) \cap \mathfrak{h}$.*
(iv) *$\mathfrak{g}/\text{rad}\, \mathfrak{g}$ is semisimple.*

Proof. To prove (i), let $\bar{k}$ be an algebraic closure of k'. Let $\mathfrak{q}$, $\mathfrak{q}'$, $\bar{\mathfrak{q}}$ be the respective radicals of $\mathfrak{g}$, $\mathfrak{g}^{k'}$, $\mathfrak{g}^{\bar{k}}$. Since $\mathfrak{q}^{\bar{k}}$ is a solvable ideal of $\mathfrak{g}^{\bar{k}}$, $\mathfrak{q}^{\bar{k}} \subseteq \bar{\mathfrak{q}}$. On the other hand, let s be a k-automorphism of $\bar{k}$. It is then easily seen that the corresponding s-linear automorphism of $\mathfrak{g}^{\bar{k}}$ maps solvable ideals into solvable ideals, and consequently leaves $\bar{\mathfrak{q}}$ invariant. So $\bar{\mathfrak{q}} = (\bar{\mathfrak{q}} \cap \mathfrak{g})^{\bar{k}}$. Since $\bar{\mathfrak{q}} \cap \mathfrak{g}$ is a solvable ideal of $\mathfrak{g}$, $\bar{\mathfrak{q}} \cap \mathfrak{g} \subseteq \mathfrak{q}$, showing that $\bar{\mathfrak{q}} \subseteq \mathfrak{q}^{\bar{k}}$. Thus $\bar{\mathfrak{q}} = \mathfrak{q}^{\bar{k}}$; similarly, $\bar{\mathfrak{q}} = \mathfrak{q}'^{\bar{k}}$. But then $\mathfrak{q}' = \mathfrak{q}^{k'}$.

Let D be a derivation of $\mathfrak{g}$. If we choose a basis for $\mathfrak{g}$, the structure constants of $\mathfrak{g}$, as well as the entries of the matrix of D in this basis, all belong to a subfield of k which is finitely generated over the prime field $\mathbf{Q}$. So, to prove (ii), in view of (i), we may assume that k itself is finitely generated over $\mathbf{Q}$. But then k may be regarded as a subfield of $\mathbf{C}$. So using (i) again, we see that k may be assumed to be $\mathbf{C}$ without any loss of generality. In this case, the endomorphisms exp tD ($t \in \mathbf{C}$) are automorphisms of $\mathfrak{g}$, and therefore leave

$\mathfrak{q}$ invariant, $\mathfrak{q}$ being the radical of $\mathfrak{g}$. Since $D = (d/dt)(\exp tD)_{t=0}$, it is clear that D leaves $\mathfrak{q}$ invariant.

Now let $\mathfrak{h}$ be an ideal in $\mathfrak{g}$. If $X \in \mathfrak{g}$, ad $X|\mathfrak{h}$ is a derivation of $\mathfrak{h}$, so by (ii) it must leave rad $\mathfrak{h}$ invariant. This shows that rad $\mathfrak{h}$ is an ideal of $\mathfrak{g}$. Since it is solvable, rad $\mathfrak{h} \subseteq$ (rad $\mathfrak{g}$) $\cap\, \mathfrak{h}$. On the other hand, (rad $\mathfrak{g}$) $\cap\, \mathfrak{h}$ is a solvable ideal of $\mathfrak{h}$, showing that (rad $\mathfrak{g}$) $\cap\, \mathfrak{h} \subseteq$ rad $\mathfrak{h}$. Thus rad $\mathfrak{h} = \mathfrak{h} \cap$ rad $\mathfrak{g}$.

To prove (iv), let π be the natural map of $\mathfrak{g}$ onto $\mathfrak{h} = \mathfrak{g}/\text{rad}\,\mathfrak{g}$. If $\mathfrak{a}$ is a solvable ideal of $\mathfrak{h}$, $\pi^{-1}(\mathfrak{a})$ is a solvable ideal of $\mathfrak{g}$ by (iii) of Theorem 3.7.2. Since rad $\mathfrak{g} \subseteq \pi^{-1}(\mathfrak{a})$, we must have $\pi^{-1}(\mathfrak{a}) =$ rad $\mathfrak{g}$, showing that $\mathfrak{a} = 0$. So $\mathfrak{h}$ is semisimple.

This proves the theorem.

Lemma 3.8.2. *Let $\mathfrak{g}$ be a Lie algebra over k and ρ a representation of $\mathfrak{g}$ in a finite-dimensional vector space V over k. Let $\mathfrak{G}$ be the universal enveloping algebra of $\mathfrak{g}$ and let σ be the extension of ρ to a representation of $\mathfrak{G}$ in V. Let $\mathfrak{K}$ be the kernel of σ. Then:*

(i) *if $\mathfrak{S}$ is the set of all ideals $\mathfrak{n} \subseteq \mathfrak{g}$ with the property that $\rho(X)$ is nilpotent for all $X \in \mathfrak{n}$, then there is a unique element $\mathfrak{n}_\rho \in \mathfrak{S}$ such that $\mathfrak{n} \subseteq \mathfrak{n}_\rho$ for all $\mathfrak{n} \in \mathfrak{S}$.*

(ii) *if V_i $(0 \leq i \leq r)$ are invariant subspaces for ρ with $V_0 = V \supseteq V_1 \supseteq \cdots \supseteq V_r = 0$ such that the representations ρ_i of $\mathfrak{g}$ in V_{i-1}/V_i are irreducible $(1 \leq i \leq r)$, then $\mathfrak{n}_\rho$ is the intersection of the kernels of the ρ_i.*

(iii) *if $\mathfrak{N}_\rho = \mathfrak{G}\mathfrak{n}_\rho\mathfrak{G}$, then $\mathfrak{N}_\rho$ is a proper two-sided ideal of $\mathfrak{G}$, and if $a_i \in \mathfrak{N}_\rho$ for $1 \leq i \leq r$, then $a_1 \cdots a_r \in \mathfrak{K}$; in particular, $\sigma(a)$ is nilpotent for each $a \in \mathfrak{N}_\rho$.*

Proof. We begin with the following simple result. Let τ be an irreducible representation of $\mathfrak{g}$ in a finite-dimensional vector space W, $\mathfrak{m}$ an ideal of $\mathfrak{g}$ such that $\tau(X)$ is nilpotent for all $X \in \mathfrak{m}$; then $\tau[\mathfrak{m}] = 0$. For if we write W' for the subspace $\{v : v \in W, \tau(X)v = 0 \text{ for all } X \in \mathfrak{m}\}$, then $W' \neq 0$ by Theorem 3.5.2. On the other hand, if $w \in W'$, $X \in \mathfrak{g}$, $Y \in \mathfrak{m}$, then $\tau(Y)\tau(X)w = \tau(X)\tau(Y)w + \tau([Y,X])w = 0$, so that W' is τ-invariant. So $W' = W$; i.e., $\tau[\mathfrak{m}] = 0$.

Suppose now that $\mathfrak{m}$ is any ideal in $\mathfrak{g}$ such that $\rho(X)$ is nilpotent for all $X \in \mathfrak{m}$; the observation made above shows that $\rho_i(X) = 0$ for $1 \leq i \leq r$, $X \in \mathfrak{m}$. So writing $\mathfrak{n}_\rho = \bigcap_{1 \leq i \leq r} \text{kernel}(\rho_i)$, we have $\mathfrak{m} \subseteq \mathfrak{n}_\rho$. On the other hand, $\mathfrak{n}_\rho$ is an ideal of $\mathfrak{g}$, and it is clear that

$$\mathfrak{n}_\rho = \{X : X \in \mathfrak{g}, \rho(X)[V_{i-1}] \subseteq V_i, 1 \leq i \leq r\}. \tag{3.8.1}$$

It follows from (3.8.1) that $\rho(X)^r = 0$ for all $X \in \mathfrak{n}_\rho$. Thus $\mathfrak{n}_\rho$ has the properties (i) and (ii). Moreover, for any $a \in \mathfrak{G}$, $\sigma(a)$ maps V_i into itself for $0 \leq i \leq r$, and hence we deduce from (3.8.1) that

$$\mathfrak{N}_p \subseteq \{a : a \in \mathfrak{G}, \sigma(a)[V_{i-1}] \subseteq V_i, 1 \leq i \leq r\}. \tag{3.8.2}$$

The assertions in (iii) follow immediately from (3.8.2).

A *nil ideal* of $\mathfrak{g}$ is an ideal $\mathfrak{m}$ of $\mathfrak{g}$ such that ad X is nilpotent for $X \in \mathfrak{m}$. An ideal $\mathfrak{m}$ of $\mathfrak{g}$ is a nil ideal if and only if $\mathfrak{m}$, as an algebra, is nilpotent. Applying the above lemma to the adjoint representation, we see that any Lie algebra $\mathfrak{g}$ has a unique maximal nil ideal that contains every nil ideal. We call it the *nil radical* of $\mathfrak{g}$ (*nil rad* $\mathfrak{g}$). It is clear that nil rad $\mathfrak{g} \subseteq$ rad $\mathfrak{g}$ and that nil rad $\mathfrak{g}$ is invariant under all automorphisms of $\mathfrak{g}$.

Theorem 3.8.3. (i) *If k' is an extension field of k, then (nil rad $\mathfrak{g})^{k'}$ = nil rad $\mathfrak{g}^{k'}$.*

(ii) *If $\mathfrak{h}$ is an ideal of $\mathfrak{g}$, so is nil rad $\mathfrak{h}$, and nil rad $\mathfrak{h} = \mathfrak{h} \cap$ nil rad $\mathfrak{g}$.*

(iii) *If $\mathfrak{q}$ = rad $\mathfrak{g}$ and $\mathfrak{n}$ = nil rad $\mathfrak{g}$, then $\mathfrak{n}$ = nil rad $\mathfrak{q}$ = the set of all $X \in \mathfrak{q}$ such that ad X (or equivalently $\mathrm{ad}_{\mathfrak{q}}\, X$) is nilpotent, and any derivation of $\mathfrak{g}$ or $\mathfrak{q}$ maps $\mathfrak{q}$ into $\mathfrak{n}$. In particular, $[\mathfrak{q},\mathfrak{g}] \subseteq \mathfrak{n}$.*

Proof. (i) is proved exactly as the assertion (i) of Theorem 3.8.1. Moreover, as in that theorem, we can prove that $\mathfrak{n}$ is invariant under all derivations of $\mathfrak{g}$. The proof of (ii) can now be carried out exactly as the proof of (iii) of Theorem 3.8.1. In particular, since $\mathfrak{q}$ is an ideal of $\mathfrak{g}$ and $\mathfrak{n} \subseteq \mathfrak{q}$, we see by (ii) that $\mathfrak{n}$ = nil rad $\mathfrak{q}$. Let $\mathfrak{n}' = \{X : X \in \mathfrak{q}$, ad X is nilpotent$\}$. By Theorem 3.7.6, $\mathfrak{n}'$ is an ideal of $\mathfrak{q}$. Since $\mathfrak{n}'$ is nilpotent, $\mathfrak{n}' \subseteq$ nil rad $\mathfrak{q} = \mathfrak{n}$. Since it is obvious that $\mathfrak{n} \subseteq \mathfrak{n}'$, $\mathfrak{n} = \mathfrak{n}'$.

Let D be a derivation of $\mathfrak{q}$. We regard k, trivially, as a Lie algebra of dimension 1 over itself. Let $\mathfrak{q}' = \mathfrak{q} \times k$ and let us define, for X, $X' \in \mathfrak{q}$ and c, $c' \in k$,

$$[(X,c),(X',c')] = ([X,X'] + cDX' - c'DX,0).$$

It is then easily verified that $\mathfrak{q}'$ becomes a Lie algebra with this definition of the bracket, that $\mathfrak{q} \times \{0\}$ is an ideal of $\mathfrak{q}'$, and that $X \mapsto (X,0)$ is an injection of $\mathfrak{q}$ into $\mathfrak{q}'$. $\mathfrak{q}'$ is thus solvable. Let $\mathfrak{n}'$ = nil rad $\mathfrak{q}'$. Since by Theorem 3.7.6, $\mathfrak{D}\mathfrak{q}'$ is a nilpotent ideal of $\mathfrak{q}'$, $\mathfrak{D}\mathfrak{q}' \subseteq \mathfrak{n}'$. Hence $\mathfrak{D}\mathfrak{q}' \cap (\mathfrak{q} \times \{0\}) \subseteq \mathfrak{n}' \cap (\mathfrak{q} \times \{0\}) = \mathfrak{n} \times \{0\}$ by (ii). So for $X \in \mathfrak{q}$, $[(X,0),(0,1)] = (-DX,0) \in \mathfrak{n} \times \{0\}$, from which we conclude that $DX \in \mathfrak{n}$ for $X \in \mathfrak{q}$. In particular, taking D = ad X ($X \in \mathfrak{g}$), we find that $[X,\mathfrak{q}] \subseteq \mathfrak{n}$. Since any derivation of $\mathfrak{g}$ leaves $\mathfrak{q}$ invariant and induces a derivation of $\mathfrak{q}$, we also have $D[\mathfrak{q}] \subseteq \mathfrak{n}$ for all derivations D of $\mathfrak{g}$.

Corollary 3.8.4. *If $\mathfrak{g}$ is solvable, nil rad $\mathfrak{g}$ is the set of all $X \in \mathfrak{g}$ for which ad X is nilpotent, and $\mathfrak{D}\mathfrak{g} \subseteq$ nil rad $\mathfrak{g}$.*

Proof. Follows from (iii) above.

3.9. Cartan's Criteria for Solvability and Semisimplicity

The aim of this section is to derive the well-known criteria of Cartan for a Lie algebra to be solvable or semisimple. These criteria are formulated in terms of the *Cartan–Killing form* of the Lie algebra. The Cartan–Killing form is the bilinear form associated with a canonically defined quadratic form on the Lie algebra which is invariant under all its automorphisms. We shall therefore begin with a discussion of the ring of invariants attached to a representation of a Lie algebra. The field k is of characteristic 0.

Let $\mathfrak{g}$ be a Lie algebra of dimension m over k, $1 \leq m < \infty$. For any indeterminate T and any $X \in \mathfrak{g}$, let

$$F(T:X) \equiv \det(T \cdot 1 - \text{ad}\, X) \equiv \sum_{0 \leq i \leq m} (-1)^{m-i} p_i(X) T^i. \tag{3.9.1}$$

It is obvious that $p_m = 1$ and that the p_i are polynomial functions on $\mathfrak{g}$ with values in k. From elementary linear algebra we find that for $X \in \mathfrak{g}$

$$\begin{aligned} p_{m-1}(X) &= tr(\text{ad}\, X) \\ p_{m-2}(X) &= \tfrac{1}{2}\{[tr(\text{ad}\, X)]^2 - [tr(\text{ad}\, X)^2]\}. \end{aligned} \tag{3.9.2}$$

Changing X to cX in (3.9.1) ($c \in k$), we deduce that the p_i are homogeneous polynomials, with $\deg(p_i) = m - i$, $0 \leq i \leq m$.

If α ($X \mapsto X^\alpha$) is an automorphism of $\mathfrak{g}$, ad $X^\alpha = \alpha \circ \text{ad}\, X \circ \alpha^{-1}$ for all $X \in \mathfrak{g}$, and hence

$$F(T:X^\alpha) = F(T:X) \quad (X \in \mathfrak{g}). \tag{3.9.3}$$

It follows from this that the p_i are invariant under all the automorphisms of $\mathfrak{g}$. If k' is an extension field of k and p_i' ($0 \leq i \leq m$) are the polynomials on $\mathfrak{g}^{k'}$ defined by (3.9.1), it is obvious that

$$p_i = p_i' | \mathfrak{g} \quad (0 \leq i \leq m). \tag{3.9.4}$$

Let D be an endomorphism of the vector space underlying $\mathfrak{g}$. We write $\bar{D}$ for the derivation of the algebra of polynomial functions on $\mathfrak{g}$ (with values in k) such that if $\lambda : \mathfrak{g} \mapsto k$ is a linear function, $(\bar{D}\lambda)(X) = -\lambda(DX)$ ($X \in \mathfrak{g}$). If $k = \mathbf{C}$ and D is a derivation of $\mathfrak{g}$, the invariance of the p_i with respect to the automorphisms exp tD ($t \in \mathbf{C}$) leads at once, through differentiation, to the result

$$\bar{D}p_i = 0 \quad (0 \leq i \leq m). \tag{3.9.5}$$

We may now argue as in the proof of (ii) of Theorem 3.8.1 to conclude that (3.9.5) is valid with k instead of $\mathbf{C}$.

Write now, for $X, Y \in \mathfrak{g}$,

$$\begin{aligned} \xi(X) &= tr(\text{ad}\,X)^2 \\ \langle X,Y \rangle &= tr(\text{ad}\,X\,\text{ad}\,Y). \end{aligned} \tag{3.9.6}$$

ξ is a quadratic form. Since by (3.9.2) $\xi = p_{m-1}^2 - 2p_{m-2}$, it follows that ξ is invariant under all automorphisms of $\mathfrak{g}$. ξ is known as the *Casimir polynomial* of $\mathfrak{g}$. $\langle \cdot , \cdot \rangle$ is obviously the symmetric bilinear form on $\mathfrak{g} \times \mathfrak{g}$ that is associated with ξ. It is called the *Cartan–Killing form* of $\mathfrak{g}$, and it is also invariant under all automorphisms of $\mathfrak{g}$. It follows from (3.9.5) that if D is any derivation of $\mathfrak{g}$, then

$$\langle DX,Y \rangle + \langle X,DY \rangle = 0 \quad (X, Y \in \mathfrak{g}). \tag{3.9.7}$$

Since $[DX, Y] = D([X, Y]) - [X,DY]$ for all $X, Y \in \mathfrak{g}$, we have

$$\text{ad}\,DX = [D, \text{ad}\,X] \quad (X \in \mathfrak{g}); \tag{3.9.8}$$

using this, (3.9.7) follows from (3.9.6) by direct calculation. In particular,

$$\langle [X,Y],Z \rangle + \langle Y,[X,Z] \rangle = 0 \quad (X, Y, Z \in \mathfrak{g}). \tag{3.9.9}$$

Note that if $\xi = 0$, the Cartan–Killing form is also 0. This is the case, for example, if $\mathfrak{g}$ is nilpotent.

The above construction which leads to the polynomials p_i is a special case of a more general one. Let ρ be a representation of $\mathfrak{g}$ in a finite-dimensional vector space V over k. Let $d = \dim V$, and for any $X \in \mathfrak{g}$, let

$$F^\rho(T : X) \equiv \det(T \cdot 1 - \rho(X)) \equiv \sum_{0 \leq i \leq d} (-1)^{d-i} p_i^\rho(X) T^i, \tag{3.9.10}$$

T being an indeterminate as before. The p_i^ρ are polynomials on $\mathfrak{g}$, and we have the obvious analogue of (3.9.4). Let

$$B^\rho(X,Y) = tr\,\rho(X)\rho(Y) \quad (X,Y \in \mathfrak{g}); \tag{3.9.11}$$

B^ρ is a symmetric bilinear form on $\mathfrak{g} \times \mathfrak{g}$. It is said to be *defined by* ρ.

Unlike the p_i, the p_i^ρ are not in general invariant under all automorphisms of $\mathfrak{g}$. (For example, let $\mathfrak{g} = \mathfrak{gl}(n,k)$, $V = k^n$, and $\rho(X) = X$ for $X \in \mathfrak{g}$; then

$p_0(X) \equiv \det X$, and if n is odd, p_0 is not invariant under the automorphism $X \mapsto -X^t$ of $\mathfrak{g}$.) However, if α ($X \mapsto X^\alpha$) is an automorphism of $\mathfrak{g}$ with the property that the representations ρ and ρ^α ($X \mapsto \rho(X^{\alpha^{-1}})$) are equivalent, then $F(T: X^{\alpha^{-1}}) \equiv F(T: X)$, so the p_i^ρ are invariant under α.

Let $k = \mathbf{R}$ or $\mathbf{C}$, and let G be a simply connected analytic group with Lie algebra $\mathfrak{g}$. If ρ and V are as above, there is a representation of G in V whose differential is ρ. We also denote this representation by ρ. Then $\rho(X^y) = \rho(y)\rho(X)\rho(y)^{-1}$ for all $X \in \mathfrak{g}$ and $y \in G$, by (3.4.10). So it follows from what we said above that the p_i^ρ are invariant under the adjoint group:

$$p_i^\rho(X^y) = p_i^\rho(X) \quad (0 \le i \le d, X \in \mathfrak{g}, y \in G). \tag{3.9.12}$$

For arbitrary k, it follows from (3.9.12) in the usual way (cf. proof of Theorems 3.8.1 and 3.8.3) that

$$(\operatorname{ad} X)^{-}(p_i^\rho) = 0 \quad (0 \le i \le d, X \in \mathfrak{g}), \tag{3.9.13}$$

$(\operatorname{ad} X)^-$ being the derivation $\bar{D}$ of the algebra of polynomials on $\mathfrak{g}$ defined above when $D = \operatorname{ad} X$. In particular,

$$B^\rho([X,Y],Z) + B^\rho(Y,[X,Z]) = 0 \tag{3.9.14}$$

for all $X, Y, Z \in \mathfrak{g}$. This can also be established by a simple direct calculation based on (3.9.11). In fact,

$$\begin{aligned} tr(\rho([X,Y])\rho(Z)) &= -tr(\rho(Y)\rho(X)\rho(Z)) + tr(\rho(X)\rho(Y)\rho(Z)) \\ &= -tr(\rho(Y)\rho(X)\rho(Z)) + tr(\rho(Y)\rho(Z)\rho(X)) \\ &= -tr(\rho(Y)\rho([X,Z])), \end{aligned}$$

for $X, Y, Z \in \mathfrak{g}$.

We are now in a position to formulate and prove Cartan's criteria. Our proof is essentially Chevalley's and relies on the theory of replicas.

Theorem 3.9.1. *Let $\mathfrak{g}$ be a Lie algebra over k. Then $\mathfrak{g}$ is solvable if and only if*

$$\langle X,[Y,Z]\rangle = 0 \quad (X,Y,Z \in \mathfrak{g}). \tag{3.9.15}$$

In particular, if the Cartan-Killing form of $\mathfrak{g}$ is identically zero, then $\mathfrak{g}$ is solvable.

Proof. Let $\mathfrak{g}$ be solvable, $X \in \mathfrak{g}$, $X' \in \mathfrak{D}\mathfrak{g}$. By Theorem 3.7.6, $\mathfrak{D}\mathfrak{g}$ is a nilpotent ideal of $\mathfrak{g}$, and hence $\mathfrak{D}\mathfrak{g} \subseteq \text{nil rad } \mathfrak{g}$. Consequently, by Lemma 3.8.2, $\operatorname{ad} X \operatorname{ad} X'$ is nilpotent; in particular, $\langle X,X'\rangle = tr \operatorname{ad} X \operatorname{ad} X' = 0$. This proves (3.9.15). Conversely, let (3.9.15) hold for all $X,Y,Z \in \mathfrak{g}$. To prove that $\mathfrak{g}$ is solvable it is sufficient to prove that $\mathfrak{D}\mathfrak{g}$ is solvable. On the other hand, if $X, X' \in \mathfrak{D}\mathfrak{g}$,

$$\begin{aligned} tr(\mathrm{ad}_{\mathfrak{D}\mathfrak{g}}X\, \mathrm{ad}_{\mathfrak{D}\mathfrak{g}}X') &= tr\, \mathrm{ad}\, X\, \mathrm{ad}\, X' \\ &= 0 \end{aligned}$$

since ad X ad X' maps $\mathfrak{g}$ into $\mathfrak{D}\mathfrak{g}$. So $\mathfrak{D}\mathfrak{g}$ is a Lie algebra with identically vanishing Cartan-Killing form. In other words, we may assume without any loss of generality that $\mathfrak{g}$ itself has identically vanishing Cartan-Killing from. Assuming this we now prove that $\mathfrak{D}\mathfrak{g}$ is nilpotent. To prove this it is clearly sufficient to prove that for any $X \in \mathfrak{D}\mathfrak{g}$, ad X is nilpotent. Write $X = \sum_{1 \leq i \leq r} [Y_i, Z_i]$ $(Y_i, Z_i \in \mathfrak{g})$. Since ad X is a derivation of $\mathfrak{g}$, we may use Corollary 3.1.17 to reduce the proof of the nilpotency of ad X to showing that $tr(\mathrm{ad}\, X\, M) = 0$ for every derivation M of $\mathfrak{g}$. But

$$\begin{aligned} tr(\mathrm{ad}\, X\, M) &= \sum_{1 \leq i \leq r} tr([\mathrm{ad}\, Y_i, \mathrm{ad}\, Z_i] M) \\ &= \sum_{1 \leq i \leq r} tr(\mathrm{ad}\, Y_i\, \mathrm{ad}\, Z_i\, M - \mathrm{ad}\, Z_i\, \mathrm{ad}\, Y_i\, M) \\ &= \sum_{1 \leq i \leq r} tr(\mathrm{ad}\, Z_i\, M\, \mathrm{ad}\, Y_i - \mathrm{ad}\, Z_i\, \mathrm{ad}\, Y_i\, M) \\ &= \sum_{1 \leq i \leq r} tr(\mathrm{ad}\, Z_i [M, \mathrm{ad}\, Y_i]) \\ &= \sum_{1 \leq i \leq r} tr(\mathrm{ad}\, Z_i\, \mathrm{ad}\, M Y_i) \quad \text{(by (3.9.8))} \\ &= \sum_{1 \leq i \leq r} \langle Z_i, M Y_i \rangle \\ &= 0 \end{aligned}$$

This proves that ad X is nilpotent.

Theorem 3.9.2. *Let $\mathfrak{g}$ be a Lie algebra over k. Then $\mathfrak{g}$ is semisimple if and only if the Cartan–Killing form of $\mathfrak{g}$ is nonsingular.*

Proof. Suppose $\mathfrak{g} = \mathrm{rad}\, \mathfrak{g} \neq 0$. Let $p \geq 0$ be such that $\mathfrak{a} = \mathfrak{D}^p\mathfrak{g} \neq 0$ but $\mathfrak{D}\mathfrak{a} = 0$. Then $\mathfrak{a}$ is abelian and is an ideal of $\mathfrak{g}$ by (i) of Theorem 3.7.1. Suppose $X \in \mathfrak{a}$, $Y \in \mathfrak{g}$. Then

$$\begin{aligned} \langle X, Y \rangle &= tr\, \mathrm{ad}\, X\, \mathrm{ad}\, Y \\ &= tr\, ((\mathrm{ad}\, X\, \mathrm{ad}\, Y) | \mathfrak{a}) \end{aligned}$$

since ad X ad Y maps $\mathfrak{g}$ into $\mathfrak{a}$. On the other hand, as $\mathfrak{a}$ is abelian,

$$[X,[Y,Z]] = 0 \quad (X, Z \in \mathfrak{a}, Y \in \mathfrak{g})$$

so that

$$\mathrm{ad}\, X\, \mathrm{ad}\, Y | \mathfrak{a} = 0.$$

This shows that

$$\langle X, Y \rangle = 0 \qquad (X \in \mathfrak{a},\ Y \in \mathfrak{g}),$$

i.e., that $\langle \cdot, \cdot \rangle$ is singular.

Suppose conversely that $\langle \cdot, \cdot \rangle$ is singular. Let

$$\mathfrak{m} = \{X : X \in \mathfrak{g}, \langle X, Y \rangle = 0 \text{ for all } Y \in \mathfrak{g}\} \tag{3.9.16}$$

Then $\mathfrak{m} \neq 0$ and it follows from (3.9.9) that $\mathfrak{m}$ is an ideal of $\mathfrak{g}$. If $X, X' \in \mathfrak{m}$, then ad X ad X' maps $\mathfrak{g}$ into $\mathfrak{m}$, and so

$$tr(\text{ad}\, X \,\text{ad}\, X') = tr(\text{ad}_{\mathfrak{m}} X \,\text{ad}_{\mathfrak{m}} X')$$

Consequently

$$tr(\text{ad}_{\mathfrak{m}} X \,\text{ad}_{\mathfrak{m}} X') = 0 \quad (X, X' \in \mathfrak{m}),$$

i.e., $\mathfrak{m}$ is a Lie algebra with identically vanishing Cartan–Killing form. By the previous theorem, $\mathfrak{m}$ is solvable. So rad $\mathfrak{g} \neq 0$, proving that $\mathfrak{g}$ is not semisimple.

Corollary 3.9.3. *Let $\mathfrak{g}$ admit no ideals other than 0 and $\mathfrak{g}$. Then $\mathfrak{g}$ is either of dimension 1 or semisimple.*

Proof. Let $\mathfrak{m}$ be as in (3.9.16). Then $\mathfrak{m}$ is an ideal. If $\mathfrak{m} = 0$, $\langle \cdot, \cdot \rangle$ is nonsingular, so $\mathfrak{g}$ is semisimple. If $\mathfrak{m} = \mathfrak{g}$, $\mathfrak{g}$ is solvable. In this case, $\mathfrak{D}\mathfrak{g} \neq \mathfrak{g}$, and if $\mathfrak{a}$ is any subspace such that $\mathfrak{D}\mathfrak{g} \subseteq \mathfrak{a} \subseteq \mathfrak{g}$, $\mathfrak{a}$ is an ideal. So $\mathfrak{g}$ must have dimension 1.

A Lie algebra $\mathfrak{g}$ over k is said to be *simple* if it is not abelian and if 0 and $\mathfrak{g}$ are its only ideals. $\mathfrak{g}$ is simple if and only if it is semisimple and has no proper ideals.

Corollary 3.9.4. *Let $\mathfrak{g}$ be a Lie algebra over k, $\{X_1, \ldots, X_m\}$ a basis for $\mathfrak{g}$. Then $\mathfrak{g}$ is semisimple if and only if*

$$\det(\langle X_i, X_j \rangle)_{1 \le i,j \le m} \neq 0. \tag{3.9.17}$$

In particular, if k' is an extension field of k, then $\mathfrak{g}$ is semisimple if and only if $\mathfrak{g}^{k'}$ is.

Proof. The relation (3.9.17) is the criterion for $\langle \cdot, \cdot \rangle$ to be nonsingular. The second statement follows trivially from the first.

We remark that if $\mathfrak{g}$ is a Lie algebra over k, if $\mathfrak{a}$ is an ideal of $\mathfrak{g}$, and if $\langle \cdot, \cdot \rangle_{\mathfrak{a}}$ is the Cartan–Killing form of $\mathfrak{a}$, then

$$\langle X, Y \rangle_{\mathfrak{a}} = \langle X, Y \rangle \quad (X, Y \in \mathfrak{a}). \tag{3.9.18}$$

We have already used this result implicitly in some of the preceding proofs. Note also the following fact established in the course of proving Theorem 3.9.2: $\mathfrak{g}$ is semisimple if and only if $\mathfrak{g}$ has no nonzero abelian ideals.

Finally the argument used in the proof of Theorems 3.9.1 and 3.9.2 can be isolated and formulated in the following manner, in order to facilitate subsequent applications.

Lemma 3.9.5. *Let $\mathfrak{g}$ be a Lie algebra over k, and ρ a representation of $\mathfrak{g}$ in a finite dimensional vector space V over k. Let B^ρ be as in (3.9.11) and let $\mathfrak{p}$ be defined by*

$$\mathfrak{p} = \{X : X \in \mathfrak{g}, B^\rho(X,Y) = 0 \text{ for all } Y \in \mathfrak{g}\} \tag{3.9.19}$$

Then $\mathfrak{p}$ is an ideal of $\mathfrak{g}$, and $\rho(X)$ is nilpotent for all $X \in [\mathfrak{p},\mathfrak{g}]$. In particular, $\mathfrak{m}$ is as in (3.9.16), $\mathfrak{m}$ is an ideal of $\mathfrak{g}$ and $[\mathfrak{m},\mathfrak{g}] \subseteq$ nil rad $\mathfrak{g}$.

Proof. The relation (3.9.14) implies at once that $\mathfrak{p}$ is an ideal of $\mathfrak{g}$. Let $X \in [\mathfrak{p},\mathfrak{g}]$ and write

$$X = \sum_{1\leq i\leq r} [Y_i,Z_i] \quad (Y_i \in \mathfrak{p}, Z_i \in \mathfrak{g})$$

In order to prove that $\rho(X)$ is nilpotent, it is sufficient to prove, in view of Theorem 3.1.16, that $tr(\rho(X)R) = 0$ for each replica R of $\rho(X)$. Let R be a replica of $\rho(X)$. Since ad $R : \mathfrak{gl}(V) \longrightarrow \mathfrak{gl}(V)$ is a replica of ad $\rho(X)$ by (v) of Theorem 3.1.13, ad R is a polynomial in ad $\rho(X)$, and so ad R leaves $\rho[\mathfrak{g}]$ invariant. Therefore we can find $U_i \in \mathfrak{g}$ such that

$$[\rho(Z_i),R] = \rho(U_i) \quad (1 \leq i \leq r).$$

But then,

$$\begin{aligned}
tr(\rho(X)R) &= \sum_{1\leq i\leq r} tr([\rho(Y_i), \rho(Z_i)]R) \\
&= \sum_{1\leq i\leq r} tr(\rho(Y_i)\rho(Z_i)R - \rho(Z_i)\rho(Y_i)R) \\
&= \sum_{1\leq i\leq r} tr(\rho(Y_i)\rho(Z_i)R - \rho(Y_i)R\rho(Z_i)) \\
&= \sum_{1\leq i\leq r} tr(\rho(Y_i)[\rho(Z_i),R]) \\
&= \sum_{1\leq i\leq r} tr(\rho(Y_i)\rho(U_i)) \\
&= \sum_{1\leq i\leq r} B^\rho(Y_i,U_i) \\
&= 0.
\end{aligned}$$

This proves that $\rho(X)$ is nilpotent.

3.10. Semisimple Lie Algebras

We devote this section to a discussion of some of the elementary properties of semisimple Lie algebras over k. Let $\mathfrak{g}$ be a Lie algebra over k, $\langle \cdot,\cdot \rangle$ its Cartan–Killing form; for any linear subspace $\mathfrak{a}$ of $\mathfrak{g}$, write

$$\mathfrak{a}^{\perp} = \{X : X \in \mathfrak{g}, \langle X,Y \rangle = 0 \text{ for all } Y \in \mathfrak{a}\}. \tag{3.10.1}$$

$\mathfrak{a}^{\perp}$ is called the *orthocomplement* of $\mathfrak{a}$. In case $\mathfrak{g}$ is semisimple, the nonsingularity of $\langle \cdot,\cdot \rangle$ implies that $\dim \mathfrak{a} + \dim \mathfrak{a}^{\perp} = \dim \mathfrak{g}$; however, it is not in general true that $\mathfrak{a}^{\perp}$ is complementary to $\mathfrak{a}$.

Theorem 3.10.1 *Let $\mathfrak{g}$ be semisimple. If $\mathfrak{h}$ is an ideal of $\mathfrak{g}$, then $\mathfrak{h}^{\perp}$ is also an ideal, $[\mathfrak{h},\mathfrak{h}^{\perp}] = 0$, and $\mathfrak{g}$ is the direct sum of $\mathfrak{h}$ and $\mathfrak{h}^{\perp}$. Moreover, both $\mathfrak{h}$ and $\mathfrak{g}/\mathfrak{h}$ are semisimple.*

Proof. That $\mathfrak{h}^{\perp}$ is an ideal follows from (3.9.9). Suppose $\mathfrak{h} \cap \mathfrak{h}^{\perp} \neq 0$. Then $\langle X,X' \rangle = 0$ for $X, X' \in \mathfrak{h} \cap \mathfrak{h}^{\perp}$. So $\langle \cdot,\cdot \rangle_{\mathfrak{h} \cap \mathfrak{h}^{\perp}} = 0$ by (3.9.18). Theorem 3.9.1 now implies that $\mathfrak{h} \cap \mathfrak{h}^{\perp}$ is solvable, contradicting the semisimplicity of $\mathfrak{g}$. The relation $\dim \mathfrak{g} = \dim \mathfrak{h} + \dim \mathfrak{h}^{\perp}$ now shows that $\mathfrak{g}$ is the direct sum of $\mathfrak{h}$ and $\mathfrak{h}^{\perp}$. Since $[\mathfrak{h},\mathfrak{h}^{\perp}] \subseteq \mathfrak{h} \cap \mathfrak{h}^{\perp}$, it follows that $[\mathfrak{h},\mathfrak{h}^{\perp}] = 0$, and hence that the $(X,X') \mapsto X + X'$ is a Lie algebra isomorphism of $\mathfrak{h} \times \mathfrak{h}^{\perp}$ onto $\mathfrak{g}$. If $\mathfrak{a}$ is an ideal of $\mathfrak{g}$ and $\operatorname{rad} \mathfrak{a} \neq 0$, $\operatorname{rad} \mathfrak{a}$ will be a nonzero solvable ideal of $\mathfrak{g}$ by Theorem 3.8.1; consequently, either $\operatorname{rad} \mathfrak{a} = 0$ or $\mathfrak{a}$ is semisimple. So both $\mathfrak{h}$ and $\mathfrak{g}/\mathfrak{h}$, which is isomorphic to $\mathfrak{h}^{\perp}$, are semisimple.

Corollary 3.10.2. *If $\mathfrak{g}$ is semisimple, then*

$$\mathfrak{g} = \mathfrak{D}\mathfrak{g}. \tag{3.10.2}$$

Proof. If $\mathfrak{D}\mathfrak{g} \neq \mathfrak{g}$, $\mathfrak{g}/\mathfrak{D}\mathfrak{g}$ will be nonzero, abelian, and semisimple all at once, which is impossible.

Corollary 3.10.3. *Let $\mathfrak{g}$ be semisimple, $\mathfrak{h}$ an ideal of $\mathfrak{g}$, and $\mathfrak{a}$ an ideal of $\mathfrak{h}$. Then $\mathfrak{a}$ is an ideal of $\mathfrak{g}$. In particular, if $\mathfrak{h}$ is a minimal element of the set of all ideals of $\mathfrak{g}$ partially ordered by inclusion, $\mathfrak{h}$ is a simple Lie algebra.*

Proof. Since $\mathfrak{g} = \mathfrak{h} + \mathfrak{h}^{\perp}$ and $[\mathfrak{h},\mathfrak{h}^{\perp}] = 0$, we have $[\mathfrak{a},\mathfrak{g}] = [\mathfrak{a},\mathfrak{h}] \subseteq \mathfrak{a}$.

Theorem 3.10.4. *Any semisimple Lie algebra over k is isomorphic to a direct sum of simple Lie algebras. More precisely, let $\mathfrak{g}$ be semisimple. Let $\mathfrak{S}$ be the set of minimal elements in the set of all ideals of $\mathfrak{g}$ partially ordered by inclusion. Then $\mathfrak{S}$ is finite; if $\mathfrak{S} = \{\mathfrak{g}_1, \ldots, \mathfrak{g}_r\}$, the $\mathfrak{g}_i$ are mutually orthogonal simple algebras, and $(X_1, \ldots, X_r) \mapsto X_1 + \cdots + X_r$ is a Lie algebra isomor-*

phism of $\mathfrak{g}_1 \times \cdots \times \mathfrak{g}_r$ *onto* $\mathfrak{g}$. *The only ideals of* $\mathfrak{g}$ *are the direct sums of the members of subfamilies of* $\mathfrak{S}$.

Proof. If $\mathfrak{a} \in \mathfrak{S}$ and $\mathfrak{b}$ is any ideal, the minimal property of $\mathfrak{a}$ implies that either $\mathfrak{a} \subseteq \mathfrak{b}$ or $\mathfrak{a} \cap \mathfrak{b} = 0$. If $\mathfrak{a} \cap \mathfrak{b} = 0$, then $[\mathfrak{a},\mathfrak{b}] \subseteq \mathfrak{a} \cap \mathfrak{b} = 0$; in this case $\mathfrak{a} \subseteq \mathfrak{b}^{\perp}$ because for $X \in \mathfrak{a}$ and $Y \in \mathfrak{b}$, ad X ad $Y = 0$. The elements of $\mathfrak{S}$ are thus mutually orthogonal under $\langle \cdot,\cdot \rangle$. We claim that they are also linearly independent. To see this, let $\{\mathfrak{g}_1, \ldots ,\mathfrak{g}_r\}$ be a maximal family of linearly independent members of $\mathfrak{S}$. Write $\mathfrak{h} = \mathfrak{g}_1 + \cdots + \mathfrak{g}_r$. If $\mathfrak{S} \neq \{\mathfrak{g}_1, \ldots ,\mathfrak{g}_r\}$ and $\mathfrak{g}_0$ is a member of $\mathfrak{S}$ distinct from all the $\mathfrak{g}_i$ $(1 \leq i \leq r)$, then $\mathfrak{g}_0^{\perp}$ constains all the $\mathfrak{g}_i$, and hence $\mathfrak{h} \subseteq \mathfrak{g}_0^{\perp}$. Then by Theorem 3.10.1, we may conclude that $\mathfrak{g}_0 \cap \mathfrak{h} = 0$, contradicting the maximality of $\{\mathfrak{g}_1, \ldots ,\mathfrak{g}_r\}$ Thus $\mathfrak{S} = \{\mathfrak{g}_1, \ldots ,\mathfrak{g}_r\}$. Suppose $\mathfrak{h}$ is an ideal of $\mathfrak{g}$ and $\mathfrak{h}'$ is the sum of all members of $\mathfrak{S}$ contained in $\mathfrak{h}$; put $\mathfrak{h}' = 0$ if there is no such member. If $\mathfrak{h}' \neq \mathfrak{h}$, then the fact that $\mathfrak{g}$ is the direct sum of $\mathfrak{h}'$ and $\mathfrak{h}'^{\perp}$ implies that $\mathfrak{a} = \mathfrak{h}'^{\perp} \cap \mathfrak{h}$ is a nonzero ideal of $\mathfrak{g}$. Clearly, there would be a member, say $\mathfrak{g}_j$, of $\mathfrak{S}$ such that $\mathfrak{g}_j \subseteq \mathfrak{a}$. But since $\mathfrak{g}_j \subseteq \mathfrak{h}$, we also have $\mathfrak{g}_j \subseteq \mathfrak{h}'$. So $\mathfrak{g}_j \subseteq \mathfrak{h}' \cap \mathfrak{h}'^{\perp} = 0$, a contradiction. In particular, $\mathfrak{g} = \sum_{1 \leq j \leq r} \mathfrak{g}_j$. Since $[\mathfrak{g}_i,\mathfrak{g}_j] = 0$ for $1 \leq i \neq j \leq r$, it is obvious that the map $(X_1, \ldots ,X_r) \mapsto X_1 + \cdots + X_r$ is a Lie algebra isomorphism of $\mathfrak{g}_1 \times \cdots \times \mathfrak{g}_r$ onto $\mathfrak{g}$. In particular, any ideal of $\mathfrak{g}_i$ is an ideal of $\mathfrak{g}$, showing that each $\mathfrak{g}_i$ is simple. This completes the proof of the theorem.

In essence, this theorem reduces the study of semisimple algebras to that of simple algebras. It turns out that when k is algebraically closed, the simple Lie algebras over k can be completely classified, thereby opening the way for a very intensive study of the semisimple Lie algebras. The classification of simple Lie algebras over an algebraically closed k, which is the great achievement of the classical work of Cartan and Killing, will be taken up in the next chapter.

Theorem 3.10.5. *Let* $\mathfrak{h}$ *be a Lie algebra over* k, $\mathfrak{q}$ *the radical of* $\mathfrak{h}$. *If* $\mathfrak{a}$ *is an ideal such that* $\mathfrak{h}/\mathfrak{a}$ *is semisimple, then* $\mathfrak{q} \subseteq \mathfrak{a}$. *If* π *is a homomorphism of* $\mathfrak{h}$ *onto a Lie algebra* $\mathfrak{h}'$, *then* $\pi[\mathfrak{q}]$ *is the radical of* $\mathfrak{h}'$.

Proof. Let τ be the natural map of $\mathfrak{h}$ onto $\mathfrak{h}/\mathfrak{a}$. If $\mathfrak{q} \not\subseteq \mathfrak{a}$, $\tau[\mathfrak{q}]$ will be a nonzero solvable ideal of $\mathfrak{h}/\mathfrak{a}$. So we must have $\mathfrak{q} \subseteq \mathfrak{a}$. Let $\mathfrak{q}' =$ rad $\mathfrak{h}'$. Then π induces in a natural fashion a homomorphism of $\mathfrak{h}/\mathfrak{q}$ into $\mathfrak{h}'/\pi[\mathfrak{q}]$. So since $\mathfrak{h}/\mathfrak{q}$ is semisimple, Theorem 3.10.1 implies that $\mathfrak{h}'/\pi[\mathfrak{q}]$ is semisimple. By the previous result, $\mathfrak{q}' \subseteq \pi[\mathfrak{q}]$. On the other hand, $\pi[\mathfrak{q}]$ is a solvable ideal of $\mathfrak{h}'$, so $\pi[\mathfrak{q}] \subseteq \mathfrak{q}'$. Hence $\pi[\mathfrak{q}] = \mathfrak{q}'$.

If $\mathfrak{g}$ is semisimple, the center of $\mathfrak{g}$ has to be zero, as otherwise it would be a nonzero abelian ideal of $\mathfrak{g}$. Thus the adjoint representation of $\mathfrak{g}$ is faithful.

Further, by (3.10.2),

$$tr((\text{ad } X) = 0 \quad (X \in \mathfrak{g}). \tag{3.10.3}$$

Theorem 3.10.6. *Let $\mathfrak{g}$ be semisimple. If D is a derivation of $\mathfrak{g}$, there is a unique $X \in \mathfrak{g}$ such that $D = ad\, X$. Any $X \in \mathfrak{g}$ can be written as $S + N$, where $[S,N] = 0$, $ad\, S$ is semisimple, and $ad\, N$ is nilpotent; S and N are, moreover, uniquely determined by these requirements.*

Proof. Since $\langle \cdot,\cdot \rangle$ is nonsingular, we can find $X \in \mathfrak{g}$ such that $\langle X,Y \rangle = tr(D \text{ ad } Y)$ for all $Y \in \mathfrak{g}$. Let $D' = D - \text{ad } X$. Then D' is a derivation of $\mathfrak{g}$, and $tr(D' \, ad\, Y) = 0$ for all $Y \in \mathfrak{g}$. We prove that $D' = 0$. Fix $Z \in \mathfrak{g}$. Then $tr(\text{ad } Y \cdot [D',\text{ad } Z]) = -tr(D' \text{ ad } [Y,Z]) = 0$ for all $Y \in \mathfrak{g}$. But since $[D', \text{ad } Z] = \text{ad } (D'Z)$ by (3.9.8), we may conclude that $\langle Y,D'Z \rangle = 0$ for all $Y \in \mathfrak{g}$. Hence $D'Z = 0$.

We now come to the second assertion. Let $X \in \mathfrak{g}$. Let ad $X = Y + Z$ be the Jordan decomposition of ad X, with Y semisimple, Z nilpotent, and $[Y,Z] = 0$. By Corollary 3.1.14, Y and Z are both derivations of $\mathfrak{g}$. Consequently, by the previous result, we can find $S, N \in \mathfrak{g}$ such that $Y = \text{ad } S$ and $Z = \text{ad } N$. Since the adjoint representation is faithful, we must have $X = S + N$ and $[S,N] = 0$. The uniqueness of S and N follows from the uniqueness of Y and Z, and the fact that the adjoint representation is faithful. This proves the theorem.

The decomposition $X = S + N$ is known as the *Jordan decomposition* of X; S (resp. N) is known as the *semisimple* (resp. *nilpotent*) *component* of X; X is called *semisimple* (resp. *nilpotent*) if ad X is.

Lemma 3.10.7. *Let $\mathfrak{g}$ be semisimple, ρ a finite-dimensional representation of $\mathfrak{g}$ in a vector space V, and B^ρ the bilinear form on $\mathfrak{g} \times \mathfrak{g}$ defined by ρ. If $\mathfrak{h}$ is the kernel of ρ and $\mathfrak{p} = \mathfrak{h}^\perp$, then B^ρ is nonsingular on $\mathfrak{p} \times \mathfrak{p}$.*

Proof. $\mathfrak{h}$ is an ideal and $\mathfrak{g}$ is the direct sum of $\mathfrak{h}$ and $\mathfrak{p}$ by Theorem 3.10.1. Let

$$\mathfrak{m} = \{X : X \in \mathfrak{p}, B^\rho(X,Y) = 0 \text{ for all } Y \in \mathfrak{p}\}.$$

Then $\mathfrak{m}$ is an ideal in $\mathfrak{p}$. Applying Lemma 3.9.5, we conclude that $\rho(X)$ is nilpotent for all $X \in [\mathfrak{m},\mathfrak{p}]$. So $\rho([\mathfrak{m},\mathfrak{p}])$ is a nil ideal of $\rho[\mathfrak{p}]$. But by Theorem 3.10.1, $\mathfrak{p}$ and $\rho[\mathfrak{p}]$ are both semisimple. Hence $\rho([\mathfrak{m},\mathfrak{p}]) = 0$. Since ρ is obviously faithful on $\mathfrak{p}$, $[\mathfrak{m},\mathfrak{p}] = 0$. So $\mathfrak{m} \subseteq \text{center}(\mathfrak{p})$. Since $\text{center}(\mathfrak{p}) = 0$, we must have $\mathfrak{m} = 0$.

A Lie group G (real or complex) is said to be *semisimple* if its Lie aglebra is semisimple.

Now assume that $k = \mathbf{R}$ or $\mathbf{C}$ and that $\mathfrak{g}$ is a semisimple Lie algebra over k. Let $Aut(\mathfrak{g})$ be the group of all automorphisms of $\mathfrak{g}$. Then $Aut(\mathfrak{g})$ is an algebraic subgroup of $GL(\mathfrak{g})$, and its Lie algebra consists of all the elements in $\mathfrak{gl}(\mathfrak{g})$ that are derivations of $\mathfrak{g}$ (cf. Exercise 21, Chapter 2).

Theorem 3.10.8. *Let G be a semisimple analytic group over k with Lie algebra $\mathfrak{g}$. Then $Ad[G] = Aut(\mathfrak{g})^0$, the component of* 1 *in $Aut(\mathfrak{g})$, and $Aut(\mathfrak{g})^0 \subseteq SL(\mathfrak{g})$.*

Proof. It is obvious that $\mathrm{Ad}[G] \subseteq Aut(\mathfrak{g})^0$. To prove the first assertion it is therefore enough to prove that the Lie algebra of $Aut(\mathfrak{g})^0$ is contained in $\mathrm{ad}[\mathfrak{g}]$. But this is immediate since, by Theorem 3.10.6, $\mathrm{ad}[\mathfrak{g}]$ is precisely the set of all derivations of $\mathfrak{g}$. By (3.10.3), $\mathrm{ad}[\mathfrak{g}] \subseteq \mathfrak{sl}(\mathfrak{g})$, and hence $\mathrm{Ad}[G] \subseteq SL(\mathfrak{g})$. This proves the second assertion.

3.11. The Casimir Element

Our aim in Sections 3.11–3.14 is to prove the famous theorems of Weyl and Levi-Mal'čev. The proofs of both these theorems use the cohomological result known as *Whitehead's Lemma.* The entire argument hinges on a consideration of a remarkable element of the center of the universal enveloping algebra of a semisimple Lie algebra, known as the *Casimir element.* We now define this element and obtain some of its fundamental properties.

Throughout this section, $\mathfrak{g}$ will denote a fixed semisimple Lie algebra over k. Let $\mathcal{P}$ (resp. $\mathfrak{S}$) be the polynomial (resp. symmetric) algebra over $\mathfrak{g}$. Denote by $\mathfrak{G}$ the universal enveloping algebra of $\mathfrak{g}$. We use the same notation for products, in both $\mathfrak{G}$ and $\mathfrak{S}$, of elements from $\mathfrak{g}$; it will usually be clear from the context whether we are operating in $\mathfrak{S}$ or $\mathfrak{G}$. Since the Cartan–Killing form is nonsingular, we have a canonical linear isomorphism of $\mathfrak{g}^*$ onto $\mathfrak{g}$. We extend this to an algebra isomorphism $p \mapsto \tilde{p}$ of $\mathcal{P}$ onto $\mathfrak{S}$. Thus, for $f \in \mathfrak{g}^*$, the element $\tilde{f}$ of $\mathfrak{g}$ is defined by

$$\langle \tilde{f}, X \rangle = f(X) \quad (X \in \mathfrak{g}). \tag{3.11.1}$$

Given any endomorphism L of $\mathfrak{g}$, the endomorphism $\tilde{L}$ of $\mathfrak{g}^*$ is defined by

$$(\tilde{L}f)(X) = -f(LX) \quad (X \in \mathfrak{g}, f \in \mathfrak{g}^*). \tag{3.11.2}$$

For any $X \in \mathfrak{g}$, let D_X (resp. $\tilde{D}_X$) be the derivation of $\mathfrak{S}$ (resp. $\mathcal{P}$) that extends $\mathrm{ad}\, X$ (resp. $(\mathrm{ad}\, X)^\sim$). It follows easily from (3.9.9) and (3.11.1) that the isomorphism $p \mapsto \tilde{p}$ intertwines D_X and $\tilde{D}_X$:

$$D_X\tilde{p} = (\tilde{D}_X p)^\sim \quad (X \in \mathfrak{g}, p \in \mathcal{P}). \tag{3.11.3}$$

If ξ is the Casimir polynomial (3.9.6) of $\mathfrak{g}$, it then follows from (3.11.3) that $D_X\xi = 0$ for all $X \in \mathfrak{g}$. Let λ be the symmetrizer map of $\mathcal{S}$ onto $\mathfrak{G}$ (cf. §3.3), and let

$$\omega = \lambda(\xi); \tag{3.11.4}$$

ω is called the *Casimir element* of $\mathfrak{G}$.

Theorem 3.11.1. *The Casimir element ω belongs to the center of $\mathfrak{G}$. Let $\{X_1, \ldots, X_m\}$ be a basis for $\mathfrak{g}$ and let $\{X^1, \ldots, X^m\}$ be the dual basis defined by*

$$\langle X_i, X^j \rangle = \delta_{ij} \quad \text{(the Kronecker delta).} \tag{3.11.5}$$

Then

$$\omega = \sum_{1 \le i \le m} X_i X^i. \tag{3.11.6}$$

Proof. Since $D_X\xi = 0$ for all $X \in \mathfrak{g}$, ω lies in the center of $\mathfrak{G}$ by Theorem 3.3.8. We now prove (3.11.6). Fix the basis $\{X_1, \ldots, X_m\}$ of $\mathfrak{g}$. The existence of the dual basis $\{X^1, \ldots, X^m\}$ follows, of course, from the nonsingularity of $\langle \cdot, \cdot \rangle$. Now, it is obvious that $X = \sum_{1 \le j \le m} \langle X, X^j \rangle X_j$ for any $X \in \mathfrak{g}$. So $(\operatorname{ad} X)^2 = \sum_{r,s} \langle X, X^r \rangle \langle X, X^s \rangle \operatorname{ad} X_r \operatorname{ad} X_s$, and from (3.9.6) we get

$$\xi(X) = \sum_{1 \le r,s \le m} \langle X, X^r \rangle \langle X, X^s \rangle \langle X_r, X_s \rangle \quad (X \in \mathfrak{g}).$$

Consequently,

$$\xi = \sum_{1 \le r,s \le m} \langle X_r, X_s \rangle X^r X^s.$$

Now apply the symmetrizer map λ to this, remembering that $\langle X_r, X_s \rangle = \langle X_s, X_r \rangle$; we obtain, in $\mathfrak{G}$,

$$\begin{aligned} \omega &= \sum_{1 \le r,s \le m} \langle X_r, X_s \rangle X^r X^s \\ &= \sum_{1 \le s \le m} \Big(\sum_{1 \le r \le m} \langle X_r, X_s \rangle X^r \Big) X^s \\ &= \sum_{1 \le s \le m} X_s X^s. \end{aligned}$$

The construction above can be generalized substantially. Let ρ be a representation of $\mathfrak{g}$ in a finite-dimensional vector space V. Let $\mathfrak{h}$ be the kernel of ρ and $\mathfrak{p} = \mathfrak{h}^\perp$. Write $B^\rho(X,Y) = \operatorname{tr}(\rho(X)\rho(Y))$ for all $X, Y \in \mathfrak{g}$. Then $\mathfrak{p}$ is an ideal of $\mathfrak{g}$ and B^ρ is nonsingular on $\mathfrak{p} \times \mathfrak{p}$ (Lemma 3.10.7). If $\mathcal{P}(\mathfrak{p})$ and $\mathcal{S}(\mathfrak{p})$ are respectively the polynomial and symmetric algebras over $\mathfrak{p}$, there is a unique isomorphism $f \mapsto f'$ of $\mathcal{P}(\mathfrak{p})$ onto $\mathcal{S}(\mathfrak{p})$ such that

$$B^\rho(f', X) = f(X) \quad (X \in \mathfrak{p}, f \in \mathfrak{p}^*). \tag{3.11.7}$$

Let ξ^ρ be the function $X \mapsto B^\rho(X,X)$ ($X \in \mathfrak{p}$) in $\mathcal{P}(\mathfrak{p})$. It is then clear from (3.9.14) that $D_X(\xi^\rho)' = 0$ for all $X \in \mathfrak{p}$. On the other hand, since $[\mathfrak{h},\mathfrak{p}] = 0$, $D_X(\xi^\rho)' = 0$ for all $X \in \mathfrak{h}$ (here we are regarding $\mathcal{S}(\mathfrak{p})$ as imbedded in $\mathcal{S}(\mathfrak{g})$). Let

$$\omega^\rho = \lambda((\xi^\rho)). \tag{3.11.8}$$

Theorem 3.11.2. *Let ρ be a finite-dimensional representation of $\mathfrak{g}$ and $\mathfrak{p} = (\text{kernel } \rho)^\perp$. Let ω^ρ be defined as above. Then ω^ρ lies in the center of $\mathfrak{G}$. Let $\{X_1, \ldots, X_p\}$ be a basis for $\mathfrak{p}$ and let $\{X^1, \ldots, X^p\}$ be the basis of $\mathfrak{p}$ such that $B^\rho(X_i,X^j) = \delta_{ij}$ $(1 \leq i,j \leq p)$. Then*

$$\omega^\rho = \sum_{1 \leq i \leq p} X_i X^i. \tag{3.11.9}$$

In particular, denoting by ρ the representation of $\mathfrak{G}$ that extends the given representation of $\mathfrak{g}$,

$$tr\rho(\omega^\rho) = \dim \mathfrak{p}. \tag{3.11.10}$$

Proof. Proceed as in the preceding theorem. Since $tr\rho(X_iX^i) = B^\rho(X_i,X^i) = 1$, we have (3.11.10).

ω^ρ is called the *Casimir element associated with ρ*. Its consideration has turned out to be one of the most fruitful ideas in harmonic analysis on semisimple Lie groups and Lie algebras.

Corollary 3.11.3. *Let ρ be as above and let*

$$\begin{aligned} V_n &= \{v : v \in V, \rho(X)v = 0 \text{ for all } X \in \mathfrak{g}\} \\ V_r &= \sum_{X \in \mathfrak{g}} \rho(X)[V]. \end{aligned} \tag{3.11.11}$$

Then V_n and V_r are invariant subspaces for ρ, and V is their direct sum.

Proof. The ρ-invariance of V_n and V_r is trivial. It remains to prove that V is the direct sum of V_n and V_r. We use induction on dim ρ. If V is the direct sum of two nonzero invariant subspaces W_1 and W_2, then the relations $V_n = W_{1,n} + W_{2,n}$, $V_r = W_{1,r} + W_{2,r}$, together with the induction hypothesis, imply the result for ρ. In what follows we use this observation without comment.

If $\rho = 0$, then $V_n = V$, $V_r = 0$, and there is nothing to prove. Let $\rho \neq 0$. By (3.11.10), $tr\rho(\omega^\rho) \neq 0$, and hence if $C = \rho(\omega^\rho)$, then C cannot be nilpotent. Let $N(C^s)$ and $R(C^s)$ be the null space and range of C^s for $s \geq 1$, and let

$$W_1 = \bigcup_{s \geq 1} N(C^s), \qquad W_2 = \bigcap_{s \geq 1} R(C^s).$$

Then by Theorem 3.1.5, V is the direct sum of W_1 and W_2. On the other hand, ω^p lies in the center of $\mathfrak{G}$, so the endomorphisms C^s $(s \geq 1)$ commute with $\rho(X)$ for all $X \in \mathfrak{g}$. W_1 and W_2 are therefore invariant subspaces for ρ. If both of them are nonzero, we have the result by the induction hypothesis. Since C is not nilpotent, W_2 cannot be 0. So we are left with the case when $W_1 = 0$ and $W_2 = V$. Then C must be invertible. But then, using the notation of the theorem, we see that for any $v \in V$,

$$\begin{aligned} v &= CC^{-1}v \\ &= \sum_{1 \leq i \leq p} \rho(X_i)\rho(X^i)C^{-1}v. \end{aligned}$$

Furthermore, if $v \in V_n$, then $\rho(X^i)v = 0$ for $1 \leq i \leq p$, so

$$\begin{aligned} Cv &= \sum_{1 \leq i \leq p} \rho(X_i)\rho(X^i)v \\ &= 0. \end{aligned}$$

In other words, $V = V_r$ and $V_n = 0$ in this case. The induction thus goes forward. The corollary is proved.

3.12. Some Cohomology

The aim of this section is to prove the cohomological lemmas of Whitehead from which the theorems of Weyl and Levi-Mal'čev follow quickly. It turns out not to be difficult to develop a general cohomology theory of semisimple Lie algebras. We shall not take this up here, but prove the Whitehead lemmas only in the cases of immediate interest to us.

Throughout this section, fix a semisimple Lie algebra $\mathfrak{g}$ over k and a representation ρ of $\mathfrak{g}$ in a vector space F of finite dimension over k. Let $V^1(\mathfrak{g},\rho)$ be the space of linear maps of $\mathfrak{g}$ into F, and for $s > 1$, let $V^s(\mathfrak{g},\rho)$ be the space of s-linear skew symmetric maps of $\mathfrak{g} \times \cdots \times \mathfrak{g}$ (s factors) into F. Given $\theta \in V^1(\mathfrak{g},\rho)$, define $d\theta \in V^2(\mathfrak{g},\rho)$ by

$$d\theta(X,Y) = \rho(X)\theta(Y) - \rho(Y)\theta(X) - \theta([X,Y]) \quad (X,Y \in \mathfrak{g}). \tag{3.12.1}$$

The map $\theta \mapsto d\theta$ is linear, and the set of all $\theta \in V^1(\mathfrak{g},\rho)$ such that $d\theta = 0$ is a linear subspace of $V^1(\mathfrak{g},\rho)$. Denote this by $C^1(\mathfrak{g},\rho)$. Suppose that $v \in F$ and that

$$\theta_v(X) = \rho(X)v \quad (X \in \mathfrak{g}). \tag{3.12.2}$$

It then follows from a trivial calculation that

$$d\theta_v = 0 \quad (v \in F). \tag{3.12.3}$$

The set of all θ_v ($v \in F$) is thus a subspace of $C^1(\mathfrak{g},\rho)$, denoted by $B^1(\mathfrak{g},\rho)$. Put

$$H^1(\mathfrak{g},\rho) = C^1(\mathfrak{g},\rho)/B^1(\mathfrak{g},\rho). \tag{3.12.4}$$

Now consider an element $\theta \in V^2(\mathfrak{g},\rho)$. Define the trilinear map $d\theta$ of $\mathfrak{g} \times \mathfrak{g} \times \mathfrak{g}$ into F by

$$d\theta(X,Y,Z) = -\sum \theta(X,[Y,Z]) - \sum \rho(X)\theta(Y,Z) \quad (X, Y, Z \in \mathfrak{g}), \tag{3.12.5}$$

where $\sum$ denotes summation over the set of cyclic permutations of X, Y, Z. It is easily verified that $d\theta \in V^3(\mathfrak{g},\rho)$. The map $\theta \mapsto d\theta$ is linear, so the set $C^2(\mathfrak{g},\rho)$ of all $\theta \in V^2(\mathfrak{g},\rho)$ such that $d\theta = 0$ is a linear subspace of $V^2(\mathfrak{g},\rho)$. On the other hand, if $\varphi \in V^1(\mathfrak{g},\rho)$ and if we define $d\varphi$ by (3.12.1), a direct calculation shows that

$$d(d\varphi) = 0 \quad (\varphi \in V^1(\mathfrak{g},\rho)). \tag{3.12.6}$$

So d ($\varphi \mapsto d\varphi$) maps $V^1(\mathfrak{g},\rho)$ onto a subspace of $C^2(\mathfrak{g},\rho)$. Denote this subspace by $B^2(\mathfrak{g},\rho)$. Let

$$H^2(\mathfrak{g},\rho) = C^2(\mathfrak{g},\rho)/B^2(\mathfrak{g},\rho). \tag{3.12.7}$$

The Whitehead lemmas may now be formulated as follows.

Theorem 3.12.1. *Let $\mathfrak{g}$ be a semisimple Lie algebra over k, ρ a representation of $\mathfrak{g}$ in a vector space F of finite dimension over k. Then*

$$H^1(\mathfrak{g},\rho) = 0, \qquad H^2(\mathfrak{g},\rho) = 0. \tag{3.12.8}$$

Proof. To start with, we take up the proof that $H^1(\mathfrak{g},\rho) = 0$. This is equivalent to proving that

$$C^1(\mathfrak{g},\rho) = B^1(\mathfrak{g},\rho). \tag{3.12.9}$$

In what follows, we write C^i and B^i for $C^i(\mathfrak{g},\rho)$ and $B^i(\mathfrak{g},\rho)$ respectively.

For any $X \in \mathfrak{g}$, let $\pi(X)$ be the endomorphism of $V^1(\mathfrak{g},\rho)$ defined by setting, for each $\varphi \in V^1(\mathfrak{g},\rho)$,

$$(\pi(X)\varphi)(Y) = -\varphi([X,Y]) + \rho(X)\varphi(Y) \quad (Y \in \mathfrak{g}). \tag{3.12.10}$$

A straightforward calculation shows that π ($X \mapsto \pi(X)$) is a representation of $\mathfrak{g}$. If $\varphi \in C^1$, we see from (3.12.1) and (3.12.2) that

$$\pi(X)\varphi = \theta_{\varphi(X)} \quad (X \in \mathfrak{g}). \tag{3.12.11}$$

Thus $\pi(X)$ maps C^1 into B^1 for all $X \in \mathfrak{g}$. In particular, C^1 is invariant under

π, and we write $\pi^1(X) = \pi(X)|C^1$. Now apply Corollary 3.11.3 to the representation π^1. We may then conclude that C^1 is the direct sum of C_n^1 and C_r^1 (cf. (3.11.11)). But (3.12.11) shows that $C_r^1 \subseteq B^1$. So (3.12.9) will be proved if we show that $C_n^1 = 0$. Suppose $\beta \in C_n^1$. Then

(3.12.12) $$\rho(X)\beta(Y) - \beta([X,Y]) = 0$$

for all $X, Y \in \mathfrak{g}$. On the other hand, as $d\beta = 0$, we also have $\rho(X)\beta(Y) - \rho(Y)\beta(X) - \beta([X,Y]) = 0$ for all $X, Y \in \mathfrak{g}$. So $\rho(Y)\beta(X) = 0$ for all $X, Y \in \mathfrak{g}$. Using this in (3.12.12) we get $\beta([X,Y]) = 0$ for all $X, Y \in \mathfrak{g}$. It now follows from (3.10.2) that $\beta = 0$.

We now prove that $H^2(\mathfrak{g},\rho) = 0$. This is the same as proving

(3.12.13) $$C^2 = B^2.$$

As before, for any $X \in \mathfrak{g}$ we define the endomorphism $\pi(X)$ of $V^2(\mathfrak{g},\rho)$ by

(3.12.14) $$(\pi(X)\varphi(Y,Z)) = \rho(X)\varphi(Y,Z) + \varphi(Y,[Z,X]) + \varphi(Z,[X,Y])$$

for $Y, Z \in \mathfrak{g}$ and $\varphi \in V^2(\mathfrak{g},\rho)$. A straightforward calculation shows that π $(X \mapsto \pi(X))$ is a representation of $\mathfrak{g}$ in $V^2(\mathfrak{g},\rho)$. Suppose $\varphi \in C^2$ and that

(3.12.15) $$\varphi_X(Y) = \varphi(X,Y) \quad (X, Y \in \mathfrak{g}).$$

Then since $d\varphi = 0$, we find from (3.12.1), (3.12.5), and (3.12.15) that

(3.12.16) $$\pi(X)\varphi = d\varphi_X \quad (X \in \mathfrak{g}).$$

This equation shows that $\pi(X)$ maps C^2 into B^2 for all $X \in \mathfrak{g}$. In particular, C^2 is invariant under π, and we write $\pi^2(X) = \pi(X)|C^2$ $(X \in \mathfrak{g})$. By Corollary 3.11.3, C^2 is the direct sum of C_n^2 and C_r^2. On the other hand, (3.12.16) shows that $C_r^2 \subseteq B^2$. So in order to prove (3.12.13), it is enough to prove that $C_n^2 \subseteq B^2$.

Suppose $\beta \in C_n^2$, so that $\pi(X)\beta = 0$ for all $X \in \mathfrak{g}$. Since $d\beta = 0$, we see that, on summing (3.12.14) over all cyclic permutations of X, Y, Z (and denoting such sums by $\sum$),

$$\sum \beta(X,[Y,Z]) = 0 \quad (X, Y, Z \in \mathfrak{g}).$$

Substituting this in the relation $\pi(X)\beta = 0$, we get

(3.12.17) $$\rho(X)\beta(Y,Z) = \beta(X,[Y,Z]) \quad (X, Y, Z \in \mathfrak{g}).$$

Now $\mathfrak{D}\mathfrak{g} = \mathfrak{g}$, so we conclude from this relation that β maps $\mathfrak{g} \times \mathfrak{g}$ into the linear span F_r of ranges of $\rho(X)$, $X \in \mathfrak{g}$ (cf. (3.11.11)). On the other hand,

(3.12.16) implies that $d\beta_X = 0$, $X \in \mathfrak{g}$. If we now write ρ_r for the representation of $\mathfrak{g}$ obtained by restriction to F_r, we can use the previous result that $H^1(\mathfrak{g},\rho_r) = 0$ to obtain, for each $X \in \mathfrak{g}$, an element $v(X) \in F_r$ such that $\beta_X = \theta_{v(X)}$; i.e.,

$$\beta(X,Y) = \rho(Y)v(X) \quad (X, Y \in \mathfrak{g}). \tag{3.12.18}$$

Since $F_r \cap F_n = 0$, it is obvious that $v(X)$ is uniquely determined. Consequently, the map v ($X \mapsto v(X)$) is linear. If we substitute (3.12.18) in (3.12.17), we get

$$\rho(X)\{\rho(Y)v(Z) - v([Y,Z]\} = 0 \quad (X, Y, Z \in \mathfrak{g}).$$

Thus $\rho(Y)v(Z) - v([Y,Z]) \in F_n$. Since $\rho(Y)v(Z) - v([Y,Z]) \in F_r$, we have

$$\rho(Y)v(Z) - v([Y,Z]) = 0 \quad (Y, Z \in \mathfrak{g}).$$

Using this in (3.12.18), we obtain

$$\beta(X,Y) = \rho(Y)v(X) = v([Y,X]) \quad (X, Y \in \mathfrak{g}). \tag{3.12.19}$$

It follows easily from (3.12.19) that $\beta = -dv$, so $\beta \in B^2$. Thus $C_n^2 \subseteq B^2$. As observed earlier, this completes the proof of the theorem.

3.13. The Theorem of Weyl

We now use the relation $H^1(\mathfrak{g},\rho) = 0$ to prove one of the most fundamental theorems in the theory of semisimple Lie algebras, namely, the theorem of H. Weyl, which asserts that every finite-dimensional representation of a semisimple Lie algebra is semisimple. Weyl proved this by transcendental arguments based on his theory of compact semisimple Lie groups. Our present method is algebraic.

Theorem 3.13.1. *Let $\mathfrak{g}$ be a semisimple Lie algebra over k. Then all finite-dimensional representations of $\mathfrak{g}$ are semisimple.*

Proof. Let V be a finite-dimensional vector space over k and σ a representation of $\mathfrak{g}$ in V. Let W be a subspace of V that is invariant under σ. We show that there is a subspace W' of V that is complementary to W and invariant under σ. Assume that $W \neq 0$, $W \neq V$, and select some subspace $\bar{W}$ of V complementary to W. Let B_0 be the projection of V onto W parallel to $\bar{W}$. If A is any projection of V onto W, i.e., $A^2 = A$ and $A[V] = W$, then the null space N_A of A is complementary to W; N_A is invariant under σ if and only if A commutes with σ, i.e.,

$$[\sigma(X),A] = 0 \quad (X \in \mathfrak{g}). \tag{3.13.1}$$

Now B_0 may not satisfy (3.13.1), so we have to modify it in order to be able to construct a projection A of V onto W satisfying (3.12.1). To this end, therefore, we introduce the vector space F of all endomorphisms C of V such that $C[V] \subseteq W$ and $C[W] = 0$. Since $W \neq 0$ and $W \neq V$, it follows that $F \neq 0$. It is easy to see that an endomorphism A of V is a projection of V onto W if and only if it is of the form $B_0 - C$ for a suitable $C \in F$. Consequently, in order to construct a projection A of V onto W satisfying (3.13.1), it is sufficient to construct an element $C \in F$ such that

$$[\sigma(X),B_0] = [\sigma(X),C] \quad (X \in \mathfrak{g}). \tag{3.13.2}$$

It follows from the definition of F that if $D \in F$, then for any endomorphism L of V that leaves W invariant, both LD and DL belong to F. In particular, if we set

$$\rho(X)\cdot D = [\sigma(X),D] \quad (X \in \mathfrak{g}, D \in F), \tag{3.13.3}$$

then, $\rho(X)$: $D \mapsto \rho(X)\cdot D$ is an endomorphism of F for any $X \in \mathfrak{g}$. It is elementary to verify that ρ is a representation. On the other hand, it follows from the relations $B_0 v \in W$ $(v \in V)$ and $B_0 w = w$ $(w \in W)$ that for any $X \in \mathfrak{g}$, $[\sigma(X),B_0]$ is an element of F. Let

$$\theta(X) = [\sigma(X),B_0] \quad (X \in \mathfrak{g}). \tag{3.13.4}$$

Then θ is a linear map of $\mathfrak{g}$ into F.

We now calculate $d\theta$. From (3.12.1) we have, for $X, Y \in \mathfrak{g}$,

$$\begin{aligned}(d\theta)(X,Y) &= [\sigma(X),[\sigma(Y),B_0]] - [\sigma(Y),[\sigma(X),B_0]] - [\sigma([X,Y]),B_0] \\ &= 0.\end{aligned}$$

So $\theta \in C^1(\mathfrak{g},\rho)$. Since $H^1(\mathfrak{g},\rho) = 0$ by Theorem 3.12.1, it follows that there is an element $C \in F$ such that $\theta(X) = \rho(X)\cdot C$ for all $X \in \mathfrak{g}$. Thus $[\sigma(X),B_0] = [\sigma(X),C]$ for all $X \in \mathfrak{g}$, which is just (3.13.2). This completes the proof of the theorem.

Weyl's theorem reduces the study of arbitrary representations of a semisimple Lie algebra to the study of its irreducible representations. It was Cartan who first obtained a description of all the irreducible representations of a complex semisimple Lie algebra. His method involved an extremely detailed consideration of the classification of simple Lie algebras; a general algebraic method for this problem was devised only recently by Harish-Chandra. The same question had been solved earlier by H. Weyl using transcendental methods. We treat these developments in the next chapter.

3.14. The Levi Decomposition

We now prove that any Lie algebra is a *semidirect product* of its radical and a semisimple subalgebra. Such decompositions are known as *Levi decompositions* and are very useful in reducing problems about general Lie algebras to problems of solvable and semisimple Lie algebras. We begin with the definition of semidirect products.

Let $\mathfrak{q}$, $\mathfrak{m}$ be Lie algebras over k, σ a representation of $\mathfrak{m}$ in $\mathfrak{q}$ such that $\sigma(Y)$ is a derivation of $\mathfrak{q}$ for all $Y \in \mathfrak{m}$. For $X,X' \in \mathfrak{q}$ and $Y,Y' \in \mathfrak{m}$, let

$$[(X,Y),(X',Y')] = ([X,X'] + \sigma(Y)X' - \sigma(Y')X,[Y,Y']). \tag{3.14.1}$$

It is then easily verified that this converts the vector space $\mathfrak{q} \times \mathfrak{m}$ into a Lie algebra. We denote it by $\mathfrak{q} \times_\sigma \mathfrak{m}$ and call it the *semidirect product of* $\mathfrak{q}$ *with* $\mathfrak{m}$ *relative to* σ. If $\sigma = 0$, we obtain the direct product, denoted by $\mathfrak{q} \times \mathfrak{m}$. If $\mathfrak{q}' = \mathfrak{q} \times \{0\}$ and $\mathfrak{m}' = \{0\} \times \mathfrak{m}$, it is obvious from (3.14.1) that $\mathfrak{q}'$ is an ideal and $\mathfrak{m}'$ a subalgebra of $\mathfrak{q} \times_\sigma \mathfrak{m}$, and that

$$\mathfrak{q}' + \mathfrak{m}' = \mathfrak{q} \times_\sigma \mathfrak{m}, \qquad \mathfrak{q}' \cap \mathfrak{m}' = 0. \tag{3.14.2}$$

Conversely, let $\mathfrak{g}$ be any Lie algebra over k, $\mathfrak{q}$ an ideal and $\mathfrak{m}$ a subalgebra of $\mathfrak{g}$ such that

$$\mathfrak{q} + \mathfrak{m} = \mathfrak{g}, \qquad \mathfrak{q} \cap \mathfrak{m} = 0. \tag{3.14.3}$$

For $Y \in \mathfrak{m}$ and $X \in \mathfrak{q}$, let $\sigma(Y)X = -[X,Y]$. Then σ $(Y \mapsto \sigma(Y))$ is a representation of $\mathfrak{m}$ in $\mathfrak{q}$ and $\sigma(Y)$ is a derivation of $\mathfrak{g}$ for all $Y \in \mathfrak{m}$. It is then easy to verify that

$$\tau:(X,Y) \mapsto X + Y \quad (X \in \mathfrak{q}, Y \in \mathfrak{m}) \tag{3.14.4}$$

is a Lie algebra isomorphism of $\mathfrak{q} \times_\sigma \mathfrak{m}$ with $\mathfrak{g}$ and that

$$\tau[\mathfrak{q} \times \{0\}] = \mathfrak{q}, \qquad \tau[\{0\} \times \mathfrak{m}] = \mathfrak{m}. \tag{3.14.5}$$

Let $\mathfrak{g}$ be any Lie algebra over k and $\mathfrak{q}$ its radical. By a *Levi subalgebra* of $\mathfrak{g}$ we mean a subalgebra $\mathfrak{m}$ such that (3.14.3) is satisfied. Since $\mathfrak{g}/\mathfrak{q}$ is semisimple and $\mathfrak{m}$ is isomorphic to it, a Levi subalgebra is necessarily semisimple. The relation (3.14.3) is then called a *Levi decomposition* of $\mathfrak{g}$. The main theorem of this section is that of Levi–Mal'čev, which asserts that any Lie algebra admits Levi subalgebras. We shall also prove the result of Mal'čev–Harish-Chandra, which asserts that any two Levi subalgebras are conjugate under a naturally defined subgroup of the group of automorphisms of the Lie algebra.

Theorem 3.14.1. *Let* $\mathfrak{g}$ *be a Lie algebra over* k, $\mathfrak{q}$ *its radical. Then* $\mathfrak{g}$ *admits Levi subalgebras. If* $\mathfrak{m}$ *is a Levi subalgebra of* $\mathfrak{g}$, *then it is also a Levi subalgebra of* $\mathfrak{D}\mathfrak{g}$, *and* $\mathfrak{D}\mathfrak{g} = [\mathfrak{q},\mathfrak{g}] + \mathfrak{m}$ *is a Levi decomposition of* $\mathfrak{D}\mathfrak{g}$.

Proof. We prove the existence of a Levi subalgebra of $\mathfrak{g}$ by induction on $\dim \mathfrak{q}$. If $\dim \mathfrak{q} = 0$, $\mathfrak{g}$ itself is a Levi subalgebra. So let $\dim \mathfrak{q} \geq 1$, and assume the existence of Levi subalgebras for any Lie algebra whose radical has dimension $< \dim \mathfrak{q}$. We consider two cases.

Case 1: $\mathfrak{D}\mathfrak{q} \neq 0$. Let $\mathfrak{g}' = \mathfrak{g}/\mathfrak{D}\mathfrak{q}$ and let π be the natural map of $\mathfrak{g}$ onto $\mathfrak{g}'$. Then $\pi[\mathfrak{q}] = \mathfrak{q}'$ is the radical of $\mathfrak{g}'$ by Theorem 3.10.5. By the induction hypothesis, $\mathfrak{g}'$ admits Levi subalgebras. Let $\mathfrak{m}'$ be one of them, and let $\mathfrak{m}_0 = \pi^{-1}(\mathfrak{m}')$. Then $\mathfrak{g} = \mathfrak{q} + \mathfrak{m}_0$ and $\mathfrak{D}\mathfrak{q} = \mathfrak{q} \cap \mathfrak{m}_0$. Now, $\mathfrak{D}\mathfrak{q}$ is a solvable ideal of $\mathfrak{m}_0$, and $\mathfrak{m}_0/\mathfrak{D}\mathfrak{q}$ is isomorphic to $\mathfrak{m}'$, which is semisimple. So $\mathfrak{D}\mathfrak{q} = \operatorname{rad} \mathfrak{m}_0$ by Theorem 3.10.5 again. Further, since $\mathfrak{q}$ is solvable, $\dim \mathfrak{D}\mathfrak{q} < \dim \mathfrak{q}$. So by the induction hypothesis we can find a Levi subalgebra $\mathfrak{m}$ of $\mathfrak{m}_0$. It is now obvious that $\mathfrak{m}$ satisfies (3.14.3) and is thus a Levi subalgebra of $\mathfrak{g}$.

Case 2: $\mathfrak{D}\mathfrak{q} = 0$. $\mathfrak{q}$ is thus abelian. Let $\mathfrak{g}_1 = \mathfrak{g}/\mathfrak{q}$, and let π be the natural map of $\mathfrak{g}$ onto $\mathfrak{g}_1$. Select a linear map μ of $\mathfrak{g}_1$ into $\mathfrak{g}$ such that $\pi \circ \mu$ is the identity. For any $X_1 \in \mathfrak{g}_1$, write $\rho(X_1)$ for the endomorphism $\operatorname{ad} X | \mathfrak{q}$ where $X \in \mathfrak{g}$ is such that $\pi(X) = X_1$; the fact that $\mathfrak{q}$ is abelian implies easily that this is a valid definition. Obviously, ρ $(X_1 \mapsto \rho(X_1))$ is a representation of the semisimple Lie algebra $\mathfrak{g}_1$ in $\mathfrak{q}$, and

$$\rho(X_1) = \operatorname{ad} \mu(X_1) | \mathfrak{g} \quad (X_1 \in \mathfrak{g}). \tag{3.14.6}$$

For $X, Y \in \mathfrak{g}_1$, let

$$\theta(X,Y) = [\mu(X),\mu(Y)] - \mu([X,Y]). \tag{3.14.7}$$

Since π is a homomorphism and $\pi \circ \mu$ is the identity, it is clear that $\mu([X,Y])$ and $[\mu(X),\mu(Y)]$ both lie above the same element, namely $[X,Y]$, of $\mathfrak{g}_1$. So $\theta(X,Y) \in \mathfrak{q}$. A trivial verification shows that θ is a skew-symmetric bilinear map of $\mathfrak{g}_1 \times \mathfrak{g}_1$ into $\mathfrak{q}$. We claim that $d\theta = 0$, $d\theta$ being defined by (3.12.5). In fact, if $X, Y, Z \in \mathfrak{g}_1$ and Σ denotes summation over the cyclic permutations of X, Y, Z, we see from (3.14.6) and (3.14.7) that

$$\begin{aligned} -d\theta(X,Y,Z) &= \Sigma\{\theta(X,[Y,Z]) + \rho(X)\theta(Y,Z)\} \\ &= \Sigma\{-\mu([X,[Y,Z]]) + [\mu(X),[\mu(Y),\mu(Z)]]\} \\ &= -\mu(\Sigma[X,[Y,Z]]) + \Sigma[\mu(X),[\mu(Y),\mu(Z)]] \\ &= 0, \end{aligned}$$

by the Jacobi identity. Now, $H^2(\mathfrak{g}_1,\rho) = 0$, since $\mathfrak{g}_1$ is semisimple. Hence there is a linear map ν of $\mathfrak{g}_1$ into $\mathfrak{q}$ such that $\theta = d\nu$, i.e.,

$$\theta(X,Y) = \rho(X)\nu(Y) - \rho(Y)\nu(X) - \nu([X,Y]) \quad (X,Y \in \mathfrak{g}_1).$$

In other words, using (3.14.6) and (3.14.7), we have

$$[\mu(X),\mu(Y)] - \mu([X,Y]) = [\mu(X),\nu(Y)] - [\mu(Y),\nu(X)] - \nu([X,Y]) \tag{3.14.8}$$

for all X, $Y \in \mathfrak{g}_1$. Let us now write

$$\lambda(X) = \mu(X) - \nu(X) \quad (X \in \mathfrak{g}_1).$$

Since the values of ν are in $\mathfrak{q}$, $\pi \circ \lambda$ is also the identity, and hence λ is also a linear injection of $\mathfrak{g}_1$ into $\mathfrak{g}$. If we now remember that $[\nu(X),\nu(Y)] = 0$ for all X, $Y \in \mathfrak{g}_1$, we can conclude at once from (3.14.8) that λ is a *homomorphism* of $\mathfrak{g}_1$ into $\mathfrak{g}$. Consequently, if $\mathfrak{m} = \lambda[\mathfrak{g}_1]$, then $\mathfrak{m}$ is a subalgebra of $\mathfrak{g}$, and the relations (3.14.3) are satisfied. $\mathfrak{m}$ is thus a Levi subalgebra.

The induction argument is completed, so we have proved the existence of a Levi subalgebra for any Lie algebra.

Now let $\mathfrak{p} = [\mathfrak{q},\mathfrak{g}]$, and let $\mathfrak{m}$ be a Levi subalgebra of $\mathfrak{g}$. Since $\mathfrak{g} = \mathfrak{q} + \mathfrak{m}$, $\mathfrak{D}\mathfrak{g} = [\mathfrak{g},\mathfrak{g}] = [\mathfrak{q},\mathfrak{g}] + [\mathfrak{m},\mathfrak{m}]$. But $\mathfrak{D}\mathfrak{m} = \mathfrak{m}$, because $\mathfrak{m}$ is semisimple. Hence $\mathfrak{D}\mathfrak{g} = \mathfrak{p} + \mathfrak{m}$, while $\mathfrak{p} \cap \mathfrak{m} \subseteq \mathfrak{g} \cap \mathfrak{m} = 0$. Thus $\mathfrak{p}$ is the radical of $\mathfrak{D}\mathfrak{g}$, and $\mathfrak{m}$ is a Levi subalgebra of $\mathfrak{D}\mathfrak{g}$. Note that

$$\mathfrak{p} = \mathfrak{q} \cap \mathfrak{D}\mathfrak{g}. \tag{3.14.9}$$

The theorem is completely proved.

Let notation be as above. Then $\mathfrak{p} = [\mathfrak{q},\mathfrak{g}]$ is contained in nil rad $\mathfrak{g}$, by (iii) of Theorem 3.8.3. In other words, $\mathfrak{p}$ is an ideal in $\mathfrak{g}$ and ad X is nilpotent for any $X \in \mathfrak{p}$. Now, if A is any finite-dimensional (not necessarily associative) algebra over k and D is a nilpotent derivation of A, a direct calculation shows that $\exp D = 1 + \sum_{s \geq 1} (1/s!)D^s$ is a well-defined automorphism of A. Applying this to the present situation, we see that exp ad Z is an automorphism of the Lie algebra $\mathfrak{g}$ for any $Z \in \mathfrak{p}$. Let $G_\mathfrak{p}$ denote the group of automorphisms of $\mathfrak{g}$ generated by the exp ad Z, $Z \in \mathfrak{p}$. If $y \in G_\mathfrak{p}$, it is obvious that $\mathfrak{q}^y = \mathfrak{q}$. So for any Levi subalgebra $\mathfrak{m}$ of $\mathfrak{g}$ and any $y \in G_\mathfrak{p}$, $\mathfrak{m}^y$ is also a Levi subalgebra. We now prove the conjugacy theorem of Mal'čev–Harish-Chandra.

Theorem 3.14.2. *Let $\mathfrak{g}$ be a Lie algebra over k, $\mathfrak{q}$ its radical, $\mathfrak{p} = [\mathfrak{q},\mathfrak{g}]$, and $G_\mathfrak{p}$ the group generated by exp ad Z, $Z \in \mathfrak{p}$. If $\mathfrak{m}_1,\mathfrak{m}_2$ are two Levi subalgebras of $\mathfrak{g}$, there is $y \in G_\mathfrak{p}$ such that $\mathfrak{m}_1^y = \mathfrak{m}_2$.*

Proof. We prove this by induction on dim $\mathfrak{g}$. Assume the theorem to be true for all Lie algebras of dimension $<$dim $\mathfrak{g}$. If dim $\mathfrak{D}\mathfrak{g} <$ dim $\mathfrak{g}$, and $\mathfrak{m}_i$

($i = 1, 2$) are Levi subalgebras of $\mathfrak{g}$, they are also Levi subalgebras of $\mathfrak{D}\mathfrak{g}$, and the result follows from the induction hypothesis. We may thus assume that $\mathfrak{D}\mathfrak{g} = \mathfrak{g}$. Then by 3.14.9, $\mathfrak{p} = \mathfrak{q}$; in particular, $\mathfrak{q} = \text{nil rad } \mathfrak{g}$. Let $\mathfrak{c}$ be the center of $\mathfrak{q}$. $\mathfrak{c}$ is easily seen to be a nonzero ideal of $\mathfrak{g}$. Two cases arise.

Case 1: $\mathfrak{c} \neq \mathfrak{q}$. Let $\mathfrak{m}_i$ ($i = 1, 2$) two Levi subalgebras of $\mathfrak{g}$. Let $\mathfrak{g}' = \mathfrak{g}/\mathfrak{c}$, and $\mathfrak{m}'_i$ (resp. $\mathfrak{q}'$) ($i = 1,2$) be the image of $\mathfrak{m}_i$ (resp. $\mathfrak{q}$) in $\mathfrak{g}'$. Then $\mathfrak{m}'_i$ is a Levi subalgebra of $\mathfrak{g}'$ for $i = 1, 2$, and $\mathfrak{q}' = \text{nil rad } \mathfrak{g}' = \text{rad } \mathfrak{g}'$. So by the induction hypothesis, we can find $Z'_i, \ldots, Z'_s \in \mathfrak{q}'$ such that $(\mathfrak{m}'_1)^{x'} = \mathfrak{m}'_2$, where $x' = \exp \text{ad } Z'_1 \cdots \exp \text{ad } Z'_s$. Let Z_j be an element of $\mathfrak{q}$ such that its image in $\mathfrak{g}'$ is Z'_j ($1 \leq j \leq s$). If we write $x = \exp \text{ad } Z_1 \cdots \exp \text{ad } Z_s$, then $\mathfrak{m}_1^x \subseteq \mathfrak{m}_2 + \mathfrak{c}$. Now, $\mathfrak{h} = \mathfrak{m}_2 + \mathfrak{c}$ is a Lie algebra with $\dim \mathfrak{h} < \dim \mathfrak{g}$, $\mathfrak{c} = \text{rad } \mathfrak{h}$, and $\mathfrak{m}_1^x, \mathfrak{m}_2$ are two Levi subalgebras of $\mathfrak{h}$. So by the induction hypothesis, we can find elements $Y_1, \ldots, Y_t \in \mathfrak{c}$ such that $(\mathfrak{m}_1^x)^z = \mathfrak{m}_2$, where $z = \exp \text{ad } Y_1 \cdots \exp \text{ad } Y_t$. If $y = zx$, then $y \in G_\mathfrak{p}$ and $\mathfrak{m}_1^y = \mathfrak{m}_2$.

Case 2: $\mathfrak{c} = \mathfrak{q}$. In this case $\mathfrak{q}$ is abelian. Also, as before, $[\mathfrak{q},\mathfrak{g}] = \mathfrak{q}$. Let $\mathfrak{m}_i$ ($i = 1, 2$) be Levi subalgebras of $\mathfrak{g}$. If $Y \in \mathfrak{q}$, $(\text{ad } Y)^2 = 0$, so

$$\exp \text{ad } Y = 1 + \text{ad } Y \quad (Y \in \mathfrak{q}). \tag{3.14.10}$$

Let $E_\mathfrak{q}$ and $E_{\mathfrak{m}_1}$ be the projections of $\mathfrak{g}$ onto $\mathfrak{q}$ and $\mathfrak{m}_1$ respectively, corresponding to the direct sum decomposition $\mathfrak{g} = \mathfrak{q} + \mathfrak{m}_1$. $E_{\mathfrak{m}_1}$ is obviously a homomorphism. Moreover, $E_{\mathfrak{m}_1}$ is zero precisely on $\mathfrak{q}$, so $E_{\mathfrak{m}_1}$ is injective on $\mathfrak{m}_2$. So the restriction E of $E_{\mathfrak{m}_1}$ to $\mathfrak{m}_2$ is a Lie algebra isomorphism of $\mathfrak{m}_2$ onto $\mathfrak{m}_1$, and we have

$$Z = E(Z) + E_\mathfrak{q}(Z) \quad (Z \in \mathfrak{m}_2). \tag{3.14.11}$$

If we now write the condition that E is a homomorphism, and remember that $\mathfrak{q}$ is abelian, we can conclude from (3.14.11) that

$$E_\mathfrak{q}([Z,Z']) = [Z,E_\mathfrak{q}(Z')] + [E_\mathfrak{q}(Z),Z'] \tag{3.14.12}$$

for all $Z, Z' \in \mathfrak{m}_2$. Let $\rho(Z) = \text{ad } Z|\mathfrak{q}$ for $Z \in \mathfrak{m}_2$. Then (3.14.12) becomes

$$\rho(Z)E_\mathfrak{q}(Z') - \rho(Z')E_\mathfrak{q}(Z) - E_\mathfrak{q}([Z,Z']) = 0 \tag{3.14.13}$$

for $Z, Z' \in \mathfrak{m}_2$. In other words, $dE_\mathfrak{q} = 0$. Since $H^1(\mathfrak{m}_2,\rho) = 0$, it follows that there is an element $Y \in \mathfrak{q}$ such that $E_\mathfrak{q}(Z) = [Z,Y]$ for all $Z \in \mathfrak{m}_2$. But then, by (3.14.10),

$$\begin{aligned} E(Z) &= Z + (\text{ad } Y)(Z) \\ &= (\exp \text{ad } Y)(Z). \end{aligned}$$

If $y = (\exp \text{ad } Y)^{-1}$, then $y \in G_\mathfrak{p}$ and $\mathfrak{m}_1^y = \mathfrak{m}_2$. This completes the proof of the theorem.

Corollary 3.14.3. *Let $\mathfrak{m}$ be a Levi subalgebra of $\mathfrak{g}$, $\mathfrak{a}$ a semisimple subalgebra. Then there is $y \in G_{\mathfrak{p}}$ such that $\mathfrak{a}^y \subseteq \mathfrak{m}$. In particular, a maximal semisimple subalgebra is a Levi subalgebra.*

Proof. $\mathfrak{g}_0 = \mathfrak{q} + \mathfrak{a}$ is a subalgebra, and $\mathfrak{a}$ is a Levi subalgebra of $\mathfrak{g}_0$, while $\mathfrak{q} = \mathrm{rad}(\mathfrak{g}_0)$. Let $\mathfrak{a}' = \mathfrak{m} \cap \mathfrak{g}_0$. Then $\mathfrak{g}_0 = \mathfrak{q} + \mathfrak{a}'$, and $\mathfrak{q} \cap \mathfrak{a}' = 0$, so $\mathfrak{a}'$ is also a Levi subalgebra of $\mathfrak{g}_0$. Since $[\mathfrak{q},\mathfrak{g}_0] \subseteq [\mathfrak{q},\mathfrak{g}]$, we can find $y \in G_{\mathfrak{p}}$ such that $\mathfrak{a}^y = \mathfrak{a}' \subseteq \mathfrak{m}$.

3.15. The Analytic Group of a Lie Algebra

We shall now use the Levi–Mal'čev theorem to prove the global version of the third fundamental theorem of Lie, namely that given a real or complex Lie algebra, there is an analytic group whose Lie algebra is isomorphic to the given one. Throughout this section, $k = \mathbf{R}$ or $\mathbf{C}$.

We begin with the concept of semidirect products for Lie groups. Let A and B be analytic groups and let t $(b \mapsto t_b)$ be a homomorphism of B into the group of automorphism of the analytic group A. For $b \in B$, $a \in A$, write $t_b[a]$ for the image of a under t_b. We assume that the map $(a,b) \mapsto t_b[a]$ is analytic from $A \times B$ into A. For $a_1,a_2 \in A$ and $b_1,b_2 \in B$, let

$$(3.15.1) \qquad (a_1,b_1)(a_2,b_2) = (a_1 t_{b_1}[a_2], b_1 b_2).$$

Let 1_A and 1_B be the respective identities of A and B. It is then easily verified that (3.15.1) converts the set $A \times B$ into a group and that

$$(3.15.2) \qquad (a,b)^{-1} = (t_{b^{-1}}[a^{-1}], b^{-1}) \quad (a \in A, b \in B).$$

It is clear from the analyticity of the map $(a,b) \mapsto t_b[a]$ that on equipping $A \times B$ with the product analytic structure, we obtain an analytic group. We denote this analytic group by $A \times_t B$ and call it the *semidirect product of A with B relative to t.* If t_b is the identity for all $b \in B$, this reduces to the usual direct product.

For $a \in A$ and $b \in B$, let

$$(3.15.3) \qquad a' = (a, 1_B), \qquad b' = (1_A, b)$$

and let

$$(3.15.4) \qquad A' = A \times \{1_B\}, \qquad B' = \{1_A\} \times B.$$

Since for $a_1,a_2 \in A$ and $b_1,b_2 \in B$ one has

$$(3.15.5) \qquad (a_1,b_1)(a_2,b_2)(a_1,b_1)^{-1} = (a_1 t_{b_1}[a_2] t_{b_1 b_2 b_1^{-1}}[a_1^{-1}], b_1 b_2 b_1^{-1}),$$

it follows that A' is a closed normal subgroup of $A \times_t B$, that B' is a closed subgroup, and that

$$b'a'b'^{-1} = t_b[a]' \quad (a \in A, b \in B). \tag{3.15.6}$$

Let $\mathfrak{a}$ and $\mathfrak{b}$ be the respective Lie algebras of A and B. Let $G = A \times_t B$. For any $b \in B$, t_b is an automorphism of A and its differential τ_b is an automorphism of $\mathfrak{a}$; moreover

$$t_b[\exp X] = \exp \tau_b(X) \quad (X \in \mathfrak{a}). \tag{3.15.7}$$

Since $b \mapsto t_b$ is a homomorphism, the map τ $(b \mapsto \tau_b)$ is a homomorphism of B into the group of automorphism of $\mathfrak{a}$. It follows easily from (3.15.7) that τ is an analytic map of B into $GL(\mathfrak{a})$. Let σ be its differential. Since each τ_b is an automorphism of $\mathfrak{a}$, it follows that $\sigma(Y)$ is a derivation of $\mathfrak{a}$ for each $Y \in \mathfrak{b}$. We may therefore form the semidirect product $\mathfrak{g} = \mathfrak{a} \times_\sigma \mathfrak{b}$. $\mathfrak{g}$ is said to be *associated with G*.

We now show that there is a natural isomorphism of the Lie algebra of G with $\mathfrak{a} \times_\sigma \mathfrak{b}$. Let $\mathfrak{g}'$ be the Lie algebra of G, and let $\mathfrak{a}'$ and $\mathfrak{b}'$ be the respective subalgebras of $\mathfrak{g}'$ defined by A' and B'. $\mathfrak{a}'$ is an ideal of $\mathfrak{g}'$, and

$$\mathfrak{a}' + \mathfrak{b}' = \mathfrak{g}', \qquad \mathfrak{a}' \cap \mathfrak{b}' = 0. \tag{3.15.8}$$

Let $X \mapsto X'$ (resp. $Y \mapsto Y'$) denote the isomorphism of $\mathfrak{a}$ onto $\mathfrak{a}'$ (resp. $\mathfrak{b}$ onto $\mathfrak{b}'$) corresponding to the isomorphism $a \mapsto a'$ (resp. $b \mapsto b'$) of A onto A' (resp. B onto B'). It is clear from (3.15.6) that for any $b \in B$ and $X \in \mathfrak{a}$

$$(\exp \tau_b(X))' = b' \exp X' b'^{-1},$$

from which we conclude that

$$\tau_b(X)' = \mathrm{Ad}_G(b')(X'). \tag{3.15.9}$$

If we differentiate this relation, we get

$$(\sigma(Y)X)' = [Y',X'] \quad (Y \in \mathfrak{b}, X \in \mathfrak{a}). \tag{3.15.10}$$

The last equation makes it clear that the map $(X,Y) \mapsto X' + Y'$ is a Lie algebra isomorphism of $\mathfrak{a} \times_\sigma \mathfrak{b}$ onto $\mathfrak{g}'$.

Now consider the converse problem of associating a semidirect product of groups with a given semidirect product of Lie algebras. Let A and B be simply connected analytic groups with Lie algebras $\mathfrak{a}$ and $\mathfrak{b}$ respectively, and let σ be a representation of $\mathfrak{b}$ in $\mathfrak{a}$ such that $\sigma(Y)$ is a derivation of $\mathfrak{a}$ for all $Y \in \mathfrak{b}$. Since B is simply connected, there is a representation τ $(b \mapsto \tau_b)$ of B in $\mathfrak{a}$ whose differential is σ. Clearly, each τ_b is an automorphism of $\mathfrak{a}$ and thus is the differential of an automorphism τ_b of A, the existence and uniqueness of t_b being an immediate consequence of the simple connectedness of A.

We leave the easy verification of the analyticity of the map $(a,b) \mapsto t_b[a]$ to the reader. We may thus form the semidirect product $G = A \times_t B$. It is said to be *associated with* $\mathfrak{g} = \mathfrak{a} \times_\sigma \mathfrak{b}$. Clearly, $\mathfrak{a} \times_\sigma \mathfrak{b}$ is associated with $A \times_t B$.

Semidirect products occur naturally in many problems. As an interesting class of examples, we mention the case when A is a vector space and $b \mapsto t_b$ is a representation of B in A.

We now state and prove the global version of the third fundamental theorem of Lie.

Theorem 3.15.1. *Let* $\mathfrak{g}$ *be a Lie algebra over* k (= **R** *or* **C**). *Then there is a simply connected analytic group whose Lie algebra is isomorphic to* $\mathfrak{g}$.

Proof. Note that if the theorem is true for two Lie algebras $\mathfrak{a}$ and $\mathfrak{b}$, and if σ is a representation of $\mathfrak{a}$ in $\mathfrak{b}$ such that $\sigma(Y)$ is a derivation of $\mathfrak{a}$ for all $Y \in \mathfrak{b}$, then it is also true for $\mathfrak{a} \times_\sigma \mathfrak{b}$. For let A and B be simply connected analytic groups whose Lie algebras are respectively isomorphic to $\mathfrak{a}$ and $\mathfrak{b}$. Let us assume, by appropriate identification, that $\mathfrak{a}$ and $\mathfrak{b}$ are actually the respective Lie algebras of A and B. Then the semidirect product of A with B that is associated to $\mathfrak{a} \times_\sigma \mathfrak{b}$ is a simply connected analytic group whose Lie algebra is isomorphic to $\mathfrak{a} \times_\sigma \mathfrak{b}$.

This said, we come to the proof of the theorem. Consider first two special cases. *Case 1:* $\mathfrak{g}$ solvable. We prove the theorem in this case by induction on $\dim \mathfrak{g}$. For $\dim \mathfrak{g} = 1$ this is trivial. Since $\mathfrak{D}\mathfrak{g} \neq \mathfrak{g}$, we can select a subspace $\mathfrak{a} \subseteq \mathfrak{g}$ such that $\mathfrak{D}\mathfrak{g} \subseteq \mathfrak{a}$ and $\dim(\mathfrak{g}/\mathfrak{a}) = 1$. Let $\mathfrak{b}$ be a subspace of dimension 1 complementary to $\mathfrak{a}$, and for $X \in \mathfrak{a}$, $Y \in \mathfrak{b}$, let $\sigma(Y)X = [Y,X]$. Since $[\mathfrak{a},\mathfrak{g}] \subseteq \mathfrak{D}\mathfrak{g} \subseteq \mathfrak{a}$, $\mathfrak{a}$ is an ideal of $\mathfrak{g}$, $\sigma(Y)$ is a derivation of $\mathfrak{a}$ for $Y \in \sigma$, and σ is a representation of $\mathfrak{b}$ in $\mathfrak{a}$. The theorem is true for $\mathfrak{a}$ and $\mathfrak{b}$ by the induction hypothesis. So since $\mathfrak{g}$ is isomorphic to $\mathfrak{a} \times_\sigma \mathfrak{b}$, the theorem is true for $\mathfrak{g}$ too. *Case 2:* $\mathfrak{g}$ semisimple. Then the adjoint representation of $\mathfrak{g}$ is faithful and so gives rise to an isomorphism of $\mathfrak{g}$ with a subalgebra $\mathfrak{g}'$ of $\mathfrak{gl}(\mathfrak{g})$. Let G' be an analytic subgroup of $GL(\mathfrak{g})$ defined by $\mathfrak{g}'$, G a universal covering group of G'. Then G is a simply connected analytic group whose Lie algebra is isomorphic to $\mathfrak{g}$.

We now come to the general case. Let $\mathfrak{a} = \operatorname{rad} \mathfrak{g}$ and $\mathfrak{b}$ a Levi subalgebra of $\mathfrak{g}$. For $X \in \mathfrak{a}$, $Y \in \mathfrak{b}$, let $\sigma(Y)X = [Y,X]$. Then $\mathfrak{g}$ is obviously isomorphic to $\mathfrak{a} \times_\sigma \mathfrak{b}$. Since $\mathfrak{a}$ is solvable and $\mathfrak{b}$ is semisimple, the theorem is true for them. So the theorem is true for $\mathfrak{g}$. This completes the proof.

3.16. Reductive Lie Algebras

It is possible to use the foregoing results to obtain some general results concerning semisimple representations of arbitrary Lie algebras and the structure of reductive Lie algebras.

Lemma 3.16.1. *Let A be an associative algebra with unit 1 over k, S a subset of A such that S and 1 generate A. Let D be a derivation of A with the property that for each $u \in S$, there is an integer $n(u) > 0$ such that $D^{n(u)}u = 0$. Then for each $v \in A$ we can find an integer $n(v) > 0$ such that $D^{n(v)}v = 0$. In particular, if A is finite-dimensional, D is a nilpotent derivation of A.*

Proof. For $u, v \in A$ and any integer $m > 0$, we have the Leibniz formula

$$D^m(uv) = \sum_{0 \leq r \leq m} \binom{m}{r} D^r u D^{m-r} v.$$

Let A' be the set of all $u \in A$ with the property that for some integer $m(u) > 0$, $D^{m(u)}u = 0$. Then the above formula shows that A' is a subalgebra of A containing S. Hence $A' = A$. The second assertion follows trivially from the first.

Theorem 3.16.2. *Let $\mathfrak{g}$ be a Lie algebra over k and let $\mathfrak{q} = rad\ \mathfrak{g}$. Then $[\mathfrak{q},\mathfrak{g}]$ is a ρ-nil ideal for any representation ρ of $\mathfrak{g}$ and is the intersection of the kernels of the semisimple representations of $\mathfrak{g}$. Moreover, there is a semisimple representation whose kernel is precisely $[\mathfrak{q},\mathfrak{g}]$. In particular, $\mathfrak{g}$ has a faithful semisimple representation if and only if $\mathfrak{q}$ is the center of $\mathfrak{g}$. In this case, $\mathfrak{g}$ is the direct sum of $\mathfrak{g}$ and $\mathfrak{D}\mathfrak{g}$, and $\mathfrak{D}\mathfrak{g}$ is semisimple.*

Proof. Let $\mathfrak{p} = [\mathfrak{q},\mathfrak{g}]$. We begin by proving that if ρ is any representation of $\mathfrak{g}$ in a finite-dimensional vector space V over k, then $\rho(X)$ is nilpotent for $X \in \mathfrak{p}$. Clearly, we may assume that k is algebraically closed. Suppose first that ρ is irreducible. $\mathfrak{p}$ being a nil ideal of $\mathfrak{g}$, $\rho[\mathfrak{p}]$ is a nil ideal of $\rho[\mathfrak{g}]$. Let E be the associative algebra of all endomorphism of V, and for $X \in \mathfrak{p}$, let A_X be the endomorphism $M \mapsto \rho(X)M - M\rho(X)$ of E. Fix $X \in \mathfrak{p}$. Then A_X is a derivation of E and A_X induces a nilpotent endomorphism of $\rho[\mathfrak{g}]$. On the other hand, ρ being irreducible, $\rho[\mathfrak{g}]$ generates E. So by Lemma 3.16.1, A_X is nilpotent. Using the Jordan decomposition of $\rho(X)$, we conclude easily that $\rho(X)$ is of the form $c \cdot 1 + N$, where $c \in k$ and N is nilpotent. But $tr\ \rho(X) = 0$ because $X \in \mathfrak{D}\mathfrak{g}$. So $c = 0$, and $\rho(X)$ is nilpotent. It now follows easily from Lemma 3.8.2 that $\rho[\mathfrak{p}] = 0$ if ρ is semisimple and that $\mathfrak{p}$ is a ρ-nil ideal, for arbitrary ρ.

Let $\mathfrak{g}' = \mathfrak{g}/[\mathfrak{q},\mathfrak{g}]$, and let π be the natural map of $\mathfrak{g}$ onto $\mathfrak{g}'$. Write $\mathfrak{q}' = \pi[\mathfrak{q}]$. By Theorem 3.10.5, $\mathfrak{q}'$ is the radical of $\mathfrak{g}'$. Since $[\mathfrak{q}',\mathfrak{g}'] = 0$, $\mathfrak{q}'$ is the center of $\mathfrak{g}'$. Let $\mathfrak{m}'$ be a Levi subalgebra of $\mathfrak{g}'$. Then

$$\begin{aligned}\mathfrak{D}\mathfrak{g}' &= [\mathfrak{g}',\mathfrak{q}'] + [\mathfrak{g}',\mathfrak{m}'] \\ &= [\mathfrak{q}',\mathfrak{m}'] + [\mathfrak{m}',\mathfrak{m}'] \\ &= \mathfrak{m}',\end{aligned}$$

showing that $\mathfrak{m}' = \mathfrak{D}\mathfrak{g}'$ is an ideal of $\mathfrak{g}' \cdot \mathfrak{g}'$ is thus the direct sum of the ideals $\mathfrak{q}'$ and $\mathfrak{m}'$. $\mathfrak{m}'$, being semisimple, has a faithful semisimple representation.

$\mathfrak{q}'$, being abelian, is trivially seen to possess a faithful semisimple representation too. So $\mathfrak{g}'$ has a faithful semisimple representation. This gives rise to a semisimple representation of $\mathfrak{g}$ whose kernel is precisely $[\mathfrak{q},\mathfrak{g}]$. The remaining assertions follow trivially from this.

A Lie algebra $\mathfrak{g}$ is called *reductive* if its radical coincides with its center (cf. Koszul [1]).

Theorem 3.16.3. *Let $\mathfrak{g}$ be a Lie algebra over k. Then the following statements are equivalent.*

(i) *$\mathfrak{g}$ is reductive*
(ii) *$\mathfrak{g}$ has a faithful semisimple representation*
(iii) *The adjoint representation of $\mathfrak{g}$ is semisimple.*
(iv) *$\mathfrak{D}\mathfrak{g}$ is semisimple.*

In this case, $\mathfrak{g}$ is the direct sum of its center and $\mathfrak{D}\mathfrak{g}$.

Proof. (i) $\Leftrightarrow$ (ii) and (i) $\Rightarrow$ (iv) by the previous theorem. If $\mathfrak{D}\mathfrak{g}$ is semisimple, it must be a Levi subalgebra of $\mathfrak{g}$ by Theorem 3.14.1. So $\mathfrak{g} = \mathfrak{q} + \mathfrak{D}\mathfrak{g}$ is a direct sum. But then $[\mathfrak{q},\mathfrak{g}] \subseteq \mathfrak{q} \cap \mathfrak{D}\mathfrak{g} = 0$, so $\mathfrak{q} = \text{center}(\mathfrak{g})$. Thus (iv) $\Leftrightarrow$ (i). If $\mathfrak{g}$ is reductive, then $\text{ad}[\mathfrak{g}] = \text{ad}[\mathfrak{D}\mathfrak{g}]$ and $\mathfrak{D}\mathfrak{g}$ is semisimple, so by Weyl's theorem, the adjoint representation of $\mathfrak{g}$ is semisimple. Thus (i) $\Rightarrow$ (iii). Conversely, let (iii) be true. Let $\mathfrak{c}$ be the center of $\mathfrak{g}$, and $\mathfrak{v}$ a subspace of $\mathfrak{g}$ complementary to $\mathfrak{c}$ and invariant under $\text{ad}[\mathfrak{g}]$. Then $\mathfrak{v}$ is an ideal. Obviously, as $[\mathfrak{c},\mathfrak{v}] = 0$, $\text{center}(\mathfrak{v}) \subseteq \text{center}(\mathfrak{g})$, so $\text{center}(\mathfrak{v}) = 0$. Thus the adjoint representation of $\mathfrak{v}$ is faithful; since $\text{ad}[\mathfrak{v}] = \text{ad}[\mathfrak{g}]$, it is even semisimple. So by the previous theorem $\mathfrak{v}$ is reductive. Since $\text{center}(\mathfrak{v}) = 0$, $\mathfrak{v}$ is semisimple, and since $\mathfrak{v} = \mathfrak{D}\mathfrak{g}$, we have (iv). This completes the proof.

Corollary 3.16.4. *Let $\mathfrak{n}$ be the nil radical of $\mathfrak{g}$, $\mathfrak{h} = \mathfrak{g}/\mathfrak{n}$, and let γ be the natural map of $\mathfrak{g}$ onto $\mathfrak{h}$. Then $\mathfrak{h}$ is reductive, and $\gamma[\mathfrak{q}]$ is the center of $\mathfrak{h}$. If $\mathfrak{g} = \mathfrak{q} + \mathfrak{m}$ is a Levi decomposition of $\mathfrak{g}$, then $\mathfrak{h} = \gamma[\mathfrak{q}] + \gamma[\mathfrak{m}]$ is a Levi decomposition of $\mathfrak{h}$, and γ is a bijection of $\mathfrak{m}$ onto $\gamma[\mathfrak{m}]$.*

Proof. By Theorem 3.10.5, $\gamma[\mathfrak{q}] = \text{rad}\,\mathfrak{h}$. Since $[\mathfrak{q},\mathfrak{g}] \subseteq \mathfrak{n}$, we have $[\gamma[\mathfrak{q}],\mathfrak{h}] = 0$. This implies that $\mathfrak{h}$ is reductive and that $\gamma[\mathfrak{q}]$ is the center of $\mathfrak{h}$. Since $\mathfrak{m} \cap \mathfrak{n} = 0$, γ is a bijection on $\mathfrak{m}$. The rest is trivial.

From these two theorems we can obtain the following decisive criterion for the semisimplicity of a representation of an arbitrary Lie algebra.

Theorem 3.16.5. *Let $\mathfrak{g}$ be a Lie algebra over k, ρ a finite-dimensional representation of $\mathfrak{g}$. Then ρ is semisimple if and only if $\rho(X)$ is a semisimple endomorphism for every element X in the radical of $\mathfrak{g}$.*

Proof. Let $\mathfrak{q} = \text{rad}\,\mathfrak{g}$. To prove that $\rho(X)$ is semisimple for all $X \in \mathfrak{q}$ when ρ is semisimple, we may assume that k is algebraically closed. Let ρ be semisimple. Then $\rho\,|\,[\mathfrak{q},\mathfrak{g}] = 0$, by Theorem 3.16.2, so ρ induces a semisimple representation of $\mathfrak{g}/[\mathfrak{q},\mathfrak{g}]$. We may therefore assume that $\mathfrak{q}$ is the center of $\mathfrak{g}$ (and hence that $\mathfrak{g}$ is reductive) without loss of generality. Let $X \in \mathfrak{q}$. To prove that $\rho(X)$ is semisimple it is enough to prove that each eigen subspace of $\pi(X)$ admits a complementary subspace invariant under $\rho(X)$. Let V be the space on which ρ acts and let λ be an eigenvalue of $\rho(X)$. Then $V_\lambda = \{v: v \in V, \rho(X)v = \lambda v\}$ is invariant under ρ, since $[X,\mathfrak{g}] = 0$. So there is a subspace of V complementary to V_λ and invariant under $\rho[\mathfrak{g}]$, in particular under $\rho(X)$. We now turn to the converse. Suppose ρ is a representation of $\mathfrak{g}$ in a finite-dimensional vector space V over k such that $\rho(X)$ is semisimple for each $X \in \mathfrak{q}$. Since $\rho[\mathfrak{q}]$ is the radical of $\rho[\mathfrak{g}]$ by Theorem 3.10.5, we may, by replacing $\mathfrak{g}$ with $\rho[\mathfrak{g}]$, assume that $\mathfrak{g} \subseteq \mathfrak{gl}(V)$ and $\rho(X) = X\,(X \in \mathfrak{g})$. Let $X \in [\mathfrak{q},\mathfrak{g}]$. Then $\text{ad}_\mathfrak{g}\, X$ is nilpotent. On the other hand, X is semisimple, so by Lemma 3.1.11 $\text{ad}\, X$ is a semisimple endomorphism of $\mathfrak{gl}(V)$; in particular, $\text{ad}_\mathfrak{g}\, X$ is semisimple. This shows that $\text{ad}_\mathfrak{g}\, X = 0$, proving that $[\mathfrak{q},\mathfrak{g}] = 0$. $\mathfrak{g}$ is thus reductive, and $\mathfrak{q} = \text{center}(\mathfrak{g})$. By passing if necessary to an algebraic closure of k, we come down to the case when k is algebraically closed. We can then find distinct elements $\lambda_1, \ldots, \lambda_p$ of $\mathfrak{q}^*$ and subspaces $V_1, \ldots, V_p$ of V such that V is the direct sum of the V_i and $V_i = \{v: v \in V, Xv = \lambda_i(X)v \text{ for all } X \in \mathfrak{q}\}$. Since $\mathfrak{q} = \text{center}(\mathfrak{g})$, each V_i is invariant under $\mathfrak{g}$. For any i, $\mathfrak{g}\,|\,V_i = \mathfrak{D}\mathfrak{g}\,|\,V_i$, so by Weyl's theorem applied to the semisimple Lie algebra $\mathfrak{D}\mathfrak{g}$, $X \mapsto X\,|\,V_i$ is a semisimple representation of $\mathfrak{g}$. Hence $X \mapsto X$ is a semisimple representation of $\mathfrak{g}$.

Theorem 3.16.6. *Let $\mathfrak{g}$ be a Lie algebra over k and let $\rho, \rho_1, \ldots, \rho_s$ be finite-dimensional semisimple representations of $\mathfrak{g}$. Then the representations ρ^* and $\rho_1 \otimes \cdots \otimes \rho_s$ are semisimple.*

Proof. Let $\mathfrak{q} = \text{rad}\,\mathfrak{g}$. For $X \in \mathfrak{q}$, $\rho(X)$ is semisimple. Hence $\rho^*(X)$ is semisimple by Lemma 3.1.11. This proves that ρ^* is semisimple. For the next assertion it is enough to consider the case $s = 2$; the general case follows by induction on s. Let ρ_i act in the space V_i and let 1_i be the identity of V_i. Then if $\tau = \rho_1 \otimes \rho_2$, we have $\tau(X) = \rho_1(X) \otimes 1_2 + 1_1 \otimes \rho_2(X)\,(X \in \mathfrak{g})$. For $X \in \mathfrak{q}$, $\rho_1(X)$ and $\rho_2(X)$ are semisimple. Then $\tau(X)$ is semisimple, and hence τ is a semisimple representation.

3.17. The Theorem of Ado

The aim of this section is to prove the theorem of Ado, which asserts that a finite-dimensional Lie algebra over a field k of characteristic zero always possesses at least one faithful finite-dimensional representation. This was prove by Ado in 1935 [1,2], but his proof appears to be incomplete at some places. A complete proof was given by Cartan in 1938 [5] using global transcendental methods; more recently, proofs using algebraic methods were given by Harish-Chandra [2] (cf. also Hochschild [1]. Essentially, we follow Harish-Chandra's method in this section. All Lie algebras considered in this section are finite-dimensional.

Lemma 3.17.1. *Let $\mathfrak{a}$ be a Lie algebra over k, and let $\mathfrak{A}$ be its universal enveloping algebra.*

(i) *Suppose $\mathfrak{M}$ is a proper two-sided ideal of $\mathfrak{A}$. Then in order that dim $(\mathfrak{A}/\mathfrak{M}) < \infty$, it is necessary and sufficient that for any $a \in \mathfrak{A}$ there exist an element p, in the algebra $k[T]$ of polynomials in an indeterminate T with coefficients from k, such that $p(a) \in \mathfrak{M}$*

(ii) *Suppose $\mathfrak{M}_i$ are proper two-sided ideal of $\mathfrak{A}$ such that $\dim(\mathfrak{A}/\mathfrak{M}_i) < \infty$ $(1 \leq i \leq r)$. Let $\mathfrak{M} = \mathfrak{M}_1 \cdots \mathfrak{M}_r$. Then $\mathfrak{M}$ is a proper two-sided ideal of $\mathfrak{A}$, and $\dim(\mathfrak{A}/\mathfrak{M}) < \infty$.*

Proof. (i) Suppose $\dim(\mathfrak{A}/\mathfrak{M}) < \infty$ and $a \in \mathfrak{A}$. Then for some integer $N > 0$, $1, a, a^2, \ldots, a^N$ are linearly dependent modulo $\mathfrak{M}$. So $p(a) \in \mathfrak{M}$ for some $p \in k[T]$. For the converse, let $\mathfrak{M}$ be a proper two-sided ideal of $\mathfrak{A}$, and let $\{X_1, \ldots, X_n\}$ be a basis for $\mathfrak{a}$. For each i we can choose an integer $N_i > 0$ and an element $p_i \in k[T]$, of the form

$$p_i = T^{N_i} + \sum_{0 < j \leq N_i} c_{ij} T^{N_i - j} \quad (c_{ij} \in k),$$

such that $p_i(X_i) \in \mathfrak{M}$. A simple induction on r shows that

$$X_i^r \in \mathfrak{M} + \sum_{0 \leq s < N_i} k \cdot X_i^s \quad (r \geq 0). \tag{3.17.1}$$

Now use induction on m $(1 \leq m \leq n)$ to conclude that for all integers $r_1, \ldots, r_m \geq 0$,

$$X_1^{r_1} \cdots X_m^{r_m} \in \mathfrak{M} + \sum_{\substack{0 \leq s_i < N_i \text{ for} \\ 1 \leq i \leq m}} k \cdot X_1^{s_1} \cdots X_m^{s_m}. \tag{3.17.2}$$

Taking $m = n$ in (3.17.2) and using the Poincaré–Birkhoff–Witt theorem, we find that the monomials $X_1^{s_1} \cdots X_n^{s_n}$ $(0 \leq s_i < N_i$ for $1 \leq i \leq n)$ span $\mathfrak{A}$ modulo $\mathfrak{M}$. Thus $\dim(\mathfrak{A}/\mathfrak{M}) \leq N_1 \cdots N_n < \infty$.

(ii) Clearly, $\mathfrak{M} \subseteq \bigcap_{1 \leq i \leq r} \mathfrak{M}_i$, so $\mathfrak{M}$ is proper. Suppose $a \in \mathfrak{A}$. We can choose $p_i \in k[T]$ such that $p_i(a) \in \mathfrak{M}_i$. Write $p = p_1 \cdots p_r$. Then $p \in k[T]$ and $p(a) \in \mathfrak{M}$. So by (i), $\dim(\mathfrak{A}/\mathfrak{M}) < \infty$.

Lemma 3.17.2. *Let $\mathfrak{a}$, $\mathfrak{A}$ be as above. Suppose that $\mathfrak{n}$ is an ideal of $\mathfrak{a}$ and ρ is a finite-dimensional representation of $\mathfrak{a}$ such that $\rho(X)$ is nilpotent for all $X \in$*

$\mathfrak{n}$. *Let σ be the representation of $\mathfrak{A}$ that extends ρ and let $\mathfrak{K}$ denote the kernel of ρ. Let*

(3.17.3) $$\mathfrak{M}_p = (\mathfrak{K} + \mathfrak{A}\mathfrak{n}\mathfrak{A})^p \quad (p = 1,2, \ldots).$$

Then we have the following properties.

(i) *$\mathfrak{M}_p$ is a proper two-sided ideal of $\mathfrak{A}$, and $\dim(\mathfrak{A}/\mathfrak{M}_p) < \infty$ for all $p \geq 1$*

(ii) *If D is a derivation of $\mathfrak{a}$ that maps $\mathfrak{a}$ into $\mathfrak{n}$, and if $\tilde{D}$ is the derivation of $\mathfrak{A}$ that extends D, then $\tilde{D}[\mathfrak{M}_p] \subseteq \mathfrak{M}_p$ for $p \geq 1$*

(iii) *There exists $r \geq 1$ such that $\mathfrak{M}_p \subseteq \mathfrak{K}$*

Proof. Let V denote the vector space on which σ acts. Let V_i $(0 \leq i \leq r)$ be invariant subspaces for σ such that $V_0 = V \supseteq V_1 \supseteq \cdots \supseteq V_r = 0$ and such that the representation σ_i of $\mathfrak{A}$ in V_{i-1}/V_i is irreducible $(1 \leq i \leq r)$. Write $\mathfrak{M} = \mathfrak{K} + \mathfrak{A}\mathfrak{n}\mathfrak{A}$. Then by (3.8.2),

(3.17.4) $$\mathfrak{M} \subseteq \{a : a \in \mathfrak{A}, \sigma(a)[V_{i-1}] \subseteq V_i \text{ for } 1 \leq i \leq r\}.$$

From (3.17.4) we find that $1 \notin \mathfrak{M}$. Hence $\mathfrak{M}$ is a proper two-sided ideal of $\mathfrak{A}$. As $\mathfrak{K} \subseteq \mathfrak{M}$, we must have $\dim(\mathfrak{A}/\mathfrak{M}) < \infty$. Hence by Lemma 3.17.1, $\dim(\mathfrak{A}/\mathfrak{M}_p) < \infty$ for all $p \geq 1$. Further, it follows from (3.17.4) that if $a_i \in \mathfrak{M}$ $(1 \leq i \leq r)$, then $\sigma(a_1 \cdots a_r) = 0$. This proves (iii). Suppose finally that D is a derivation of $\mathfrak{a}$ that maps $\mathfrak{a}$ into $\mathfrak{n}$ and that $\tilde{D}$ is its extension to a derivation of $\mathfrak{M}$. Since $\tilde{D}1 = 0$, $D[\mathfrak{a}] \subseteq \mathfrak{n}$, and $\mathfrak{a}$ generates $\mathfrak{A}$, we see that $\tilde{D}$ maps $\mathfrak{A}$ into $\mathfrak{A}\mathfrak{n}\mathfrak{A}$, so $\tilde{D}[\mathfrak{M}] \subseteq \mathfrak{M}$. Consequently $\tilde{D}$ leaves $\mathfrak{M}_p$ invariant for all $p \geq 1$.

Lemma 3.17.3. *Let notation be as above and let $r \geq 1$ be such that $\mathfrak{M}_r \subseteq \mathfrak{K}$. Let $U = \mathfrak{A}/\mathfrak{M}_r$, let γ be the natural map of $\mathfrak{A}$ onto U, and let ξ be the natural representation of $\mathfrak{A}$ in U. Let $\mathfrak{d}$ be the Lie algebra of all derivations of $\mathfrak{a}$ that map $\mathfrak{a}$ into $\mathfrak{n}$. For $D \in \mathfrak{d}$, let $\hat{D}$ be the endomorphism of U induced by $\tilde{D}$. Then we have the following:*

(i) *$D \mapsto \hat{D}$ is a representation of $\mathfrak{d}$ in U.*

(ii) *For $a \in \mathfrak{A}$ and $D \in \mathfrak{d}$, $\xi(\tilde{D}a) = [\hat{D},\xi(a)]$*

(iii) *If $D \in \mathfrak{d}$ is nilpotent, $\hat{D}$ is nilpotent.*

Proof. Since $D \mapsto \tilde{D}$ is a representation of $\mathfrak{d}$ in $\mathfrak{A}$, (i) is immediate. Suppose that $a, b \in \mathfrak{A}$. Then

$$\begin{aligned} \xi(\tilde{D}a)(\gamma(b)) &= \gamma((\tilde{D}a)b) \\ &= \gamma(\tilde{D}(ab) - \gamma(a\tilde{D}b) \\ &= \hat{D}\gamma(ab) - \xi(a)\gamma(\tilde{D}b) \\ &= \hat{D}\xi(a)\gamma(b) - \xi(a)\hat{D}\gamma(b) \end{aligned}$$

This proves (ii). To prove (iii), let $l \geq 1$ be such that $D^l X = 0$ for all $X \in \mathfrak{a}$. If $X_1, \dots, X_q \in \mathfrak{a}$, then $D^p(X_1 \cdots X_q)$ is a linear combination of terms of the form $(D^{s_1}X_1)\cdots(D^{s_q}X_q) = a_{s_1,\dots,s_q}$ with $s_1 \geq 0$ for all i and $s_1 + \cdots + s_q = p$. Now, $a_{s_1,\dots,s_q} = 0$ if some $s_i \geq l$. Suppose $s_i < l$ for all i. If $p \geq rl$, then the number of indices i with $s_i > 0$ is at least r, so $a_{s_1,\dots,s_q} \in (\mathfrak{A}\mathfrak{n}\mathfrak{A})^r \subseteq \mathfrak{M}_r$. So if $p \geq rl$, D^p maps $\mathfrak{A}$ into $\mathfrak{M}_r$, showing that $(\hat{D})^p = 0$. This proves (iii).

Lemma 3.17.4. *Let $\mathfrak{g}$ be a Lie algebra over k, and let $\mathfrak{g} = \mathfrak{a} + \mathfrak{b}$, where $\mathfrak{a}$ is an ideal and $\mathfrak{b}$ is a subalgebra of $\mathfrak{g}$, the sum being direct. Suppose $\mathfrak{n}$ is an ideal of $\mathfrak{a}$ such that $[\mathfrak{b},\mathfrak{a}] \subseteq \mathfrak{n}$. Let $\mathfrak{G}$ be the universal enveloping algebra of $\mathfrak{g}$, $\mathfrak{A}$ the subalgebra of $\mathfrak{G}$ generated by $\mathfrak{a}$. Suppose σ is a finite-dimensional representation of $\mathfrak{A}$ such that $\sigma(X)$ is nilpotent for all $X \in \mathfrak{n}$. Then there exists a finite-dimensional representation σ' of $\mathfrak{G}$ with the following properties.*

(i) *The kernel of $\sigma' | \mathfrak{A}$ is contained in the kernel of σ*

(ii) *If $X \in \mathfrak{g}$, then $\sigma'(X)$ is nilpotent provided either that $X \in \mathfrak{a}$ and $\sigma(X)$ is nilpotent, or that $X \in \mathfrak{b}$ and $(ad\, X)|\mathfrak{a}$ is nilpotent.*

(iii) *Furthermore if $\mathfrak{n}$ is nilpotent, then for any $X \in \mathfrak{n}$ and $Y \in \mathfrak{b}$ such that $(ad\, Y)|\mathfrak{a}$ is nilpotent, $\sigma'(X + Y)$ is nilpotent.*

Proof. We use the notation of the previous lemma. Let $\mathfrak{K} = \text{kernel}(\sigma)$. For $Y \in \mathfrak{b}$, let D_Y be the derivation $X \mapsto [Y,X]$ of $\mathfrak{a}$. Then $D_Y \in \mathfrak{d}$ for all $Y \in \mathfrak{b}$. We now define the map σ' $(\mathfrak{g} \longrightarrow \mathfrak{gl}(U))$ by

$$\sigma'(X + Y) = \xi(X) + \hat{D}_Y \quad (X \in \mathfrak{a}, Y \in \mathfrak{b}). \tag{3.17.5}$$

It follows from (i) and (ii) of Lemma 3.17.3 that σ' is a representation of $\mathfrak{g}$ in U. We extend σ' to a representation of $\mathfrak{G}$ in U and denote the extension by σ' also. If $a \in \mathfrak{A}$, then $\sigma'(a) = \xi(a)$, so the kernel of $\sigma' | \mathfrak{A}$ is precisely $\mathfrak{M}_r$. Since $\mathfrak{M}_r \subseteq \mathfrak{K}$, we have (i). If $X \in \mathfrak{a}$ and $\sigma(X)$ is nilpotent, $X^l \in \mathfrak{K}$ for some $l > 0$. So $X^{rl} \in \mathfrak{M}_r$, proving that $\sigma'(X)^{rl} = 0$. If $X \in \mathfrak{b}$ is such that D_X is nilpotent, then $\sigma'(X) = \hat{D}_X$ is nilpotent by (iii) of previous lemma. We thus have (ii). Suppose now that $\mathfrak{n}$ is nilpotent and $Y \in \mathfrak{b}$ is such that D_Y is nilpotent. Let $\mathfrak{h} = \mathfrak{n} + k \cdot Y$. Then $\mathfrak{h}$ is a subalgebra and $\mathfrak{n}$ is an ideal in $\mathfrak{h}$. Since $\mathfrak{n}$ and $\mathfrak{h}/\mathfrak{n}$ are nilpotent, $\mathfrak{h}$ is solvable. Therefore, if

$$\mathfrak{c} = \{Z : Z \in \mathfrak{h}, \sigma'(Z) \text{ is nilpotent}\},$$

then $\mathfrak{c}$ is an ideal of $\mathfrak{h}$ (Theorem 3.7.6). Since $\mathfrak{n} \subseteq \mathfrak{c}$ and $Y \in \mathfrak{c}$, $\mathfrak{c} = \mathfrak{h}$. In particular, $\sigma'(X + Y)$ is nilpotent for all $X \in \mathfrak{n}$.

This proves the lemma.

Corollary 3.17.5. *Suppose σ is faithful on $\mathfrak{a}$. Then σ' is faithful on $\mathfrak{a}$.*

Proof. This follows at once from (i).

Corollary 3.17.6. *Suppose $\mathfrak{g}$ is a nilpotent Lie algebra over k. Then $\mathfrak{g}$ has a faithful finite-dimensional nil representation.*

Proof. We use induction on $\dim(\mathfrak{g})$. If $\dim(\mathfrak{g}) = 1$, this is obvious. For $\dim(\mathfrak{g}) > 1$, we select an ideal $\mathfrak{a}$ of $\mathfrak{g}$ such that $\dim(\mathfrak{a}) = \dim(\mathfrak{g}) - 1$. Let X be an element of $\mathfrak{g}$ not in $\mathfrak{a}$, and let $\mathfrak{b} = k \cdot X$. We use the above lemma with $\mathfrak{a} = \mathfrak{n}$ and σ as a finite-dimensional representation of $\mathfrak{A}$ such that $\sigma|\mathfrak{a}$ is a faithful nil representation. Then $\rho' = \sigma'|\mathfrak{g}$ is a nil representation that is faithful on $\mathfrak{a}$. On the other hand, since $\dim(\mathfrak{g}/\mathfrak{a}) = 1$, we can find a nil-representation ρ'' of $\mathfrak{g}$ such that $\mathfrak{a}$ is the kernel of ρ''. The direct sum of ρ' and ρ'' is a faithful nil representation of $\mathfrak{g}$.

We now state and prove the following version of Ado's theorem.

Theorem 3.17.7. *Let $\mathfrak{g}$ be a Lie algebra over k and $\mathfrak{n}$ its nil radical. Then there exists a faithful finite-dimensional representation ρ of $\mathfrak{g}$ such that $\rho(X)$ is nilpotent for all $X \in \mathfrak{n}$.*

Proof. We use induction on $\dim(\mathfrak{g})$. Suppose first that $\mathfrak{g}$ is solvable. In view of the preceding corollary, we may assume that $\mathfrak{n} \neq \mathfrak{g}$. Now $\mathfrak{D}\mathfrak{g} \subseteq \mathfrak{n}$; hence if $\mathfrak{a}$ is any linear subspace of $\mathfrak{g}$ such that $\mathfrak{a} \subseteq \mathfrak{n}$ and $\dim(\mathfrak{a}) = \dim(\mathfrak{g}) - 1$, then $\mathfrak{a}$ is an ideal of $\mathfrak{g}$ and $[\mathfrak{g},\mathfrak{a}] \subseteq \mathfrak{D}\mathfrak{g} \subseteq \mathfrak{n}$. Choose any such $\mathfrak{a}$ and an element $X \in \mathfrak{g}$ not in $\mathfrak{a}$. We write $\mathfrak{b} = k \cdot X$. By the induction hypothesis, there is a finite-dimensional representation σ of $\mathfrak{A}$ such that $\sigma|\mathfrak{a}$ is faithful and $\sigma|\mathfrak{n}$ is a nil representation. The direct sum ρ of ρ' and ρ'', where ρ'' is a nil representation of $\mathfrak{g}$ with kernel $\mathfrak{a}$, has the required properties to carry the induction forward.

Suppose that $\mathfrak{g}$ is not solvable. Let $\mathfrak{q} = \operatorname{rad} \mathfrak{g}$, and let $\mathfrak{g} = \mathfrak{q} + \mathfrak{m}$ be a Levi decomposition of $\mathfrak{g}$. Then $\dim(\mathfrak{q}) < \dim(\mathfrak{g})$, so by the induction hypothesis there is a faithful finite-dimensional representation σ of $\mathfrak{q}$ such that $\sigma|\mathfrak{n}$ is a nil representation. By Theorem 3.8.3, $[\mathfrak{m},\mathfrak{q}] \subseteq \mathfrak{n}$. We apply Lemma 3.17.4 and its corollaries to derive the existence of a finite-dimensional representation ρ' of $\mathfrak{g}$ such that ρ' is faithful on $\mathfrak{q}$ and $\rho'|\mathfrak{n}$ is a nil representation. On the other hand, $\mathfrak{g}/\mathfrak{q}$ is semisimple, so its adjoint representation is already faithful. So we can find a finite-dimensional representation ρ'' of $\mathfrak{g}$ such that $\mathfrak{q} = \operatorname{kernel}(\rho'')$. If ρ is the direct sum of ρ' and ρ'', then ρ is faithful and $\rho|\mathfrak{n}$ is a nil representation. This completes the induction argument.

The theorem is completely proved.

Let us now assume that $k = \mathbf{R}$ or $\mathbf{C}$ and that $\mathfrak{g}$ is a Lie algebra over k. From Ado's theorem we immediately have the following.

Theorem 3.17.8. *Let $\mathfrak{g}$ be a Lie algebra over $k = \mathbf{R}$ or $\mathbf{C}$. Then there exists an integer $n \geq 1$ and an analytic subgroup G of $GL(n,k)$ such that the Lie algebra of G is isomorphic to $\mathfrak{g}$.*

We thus have the remarkable result that any analytic group is locally isomorphic to a matrix group. At the same time the above theorem yields another proof of the global version of the third fundamental theorem of Lie.

It must be remarked, however, that an analytic group may not always be *globally* isomorphic to some matrix group. More precisely, let G be an analytic group with Lie algebra $\mathfrak{g}$, and let π be an analytic homomorphism of G into $GL(n,k)$; write $\rho = d\pi$. If π is faithful, then ρ is a faithful representation of $\mathfrak{g}$, but the converse is not true in general. From the fact that ρ is faithful we can conclude only that π has discrete kernel.

There are analytic groups which do not possess faithful finite-dimensional representations (cf. Exercise 15, Chapter 2). In the next section we prove that simply connected solvable groups have faithful finite-dimensional representations. We have already proved this for nilpotent groups (Theorem 3.6.6). For some of the subtler aspects of globally faithful representations we refer the reader to the exercises at the end of this chapter.

3.18. Some Global Results

Our concern so far has been almost exclusively with the structure of Lie algebras. In this section we discuss some of the group-theoretic implications of the preceding theory. Throughout this section k will be **R** or **C**; 'analytic' means k-analytic unless we state otherwise.

Theorem 3.18.1. *Let G be a simply connected analytic group with Lie algebra $\mathfrak{g}$. Let $\mathfrak{a}$ be an ideal in $\mathfrak{g}$, A the analytic subgroup of G defined by $\mathfrak{a}$. Then A is a closed normal subgroup of G.*

Proof. We need prove only that A is closed. By the global form of the third fundamental theorem of Lie, there exists an analytic group H whose Lie algebra $\mathfrak{h}$ is isomorphic to $\mathfrak{g}/\mathfrak{a}$. Then there exists a homomorphism λ of $\mathfrak{g}$ onto $\mathfrak{h}$ such that $\mathfrak{a} = \text{kernel}(\lambda)$. Since G is simply connected, we can find an analytic homomorphism π of G onto H such that $d\pi = \lambda$. Then A is the component of the identity of the kernel of π, which is closed in G. So A is closed.

Actually, A is simply connected, as was proved by Mal'čev [1]. We have, in fact, the following theorem.

Theorem 3.18.2. *Let G be a simply connected analytic group and A a normal analytic subgroup. Then A is closed, A and G/A are both simply connected, and the coset space G/A admits a global analytic section.*

Essentially, we follow Hochschild's method of proof [1]. We need some lemmas. Note that, by the preceding theorem, A is closed.

Lemma 3.18.3. *Let $\mathfrak{h}$ be a Lie algebra over a field k' of characteristic zero and $\mathfrak{a} \subseteq \mathfrak{h}$ an ideal which is maximal among the ideals of $\mathfrak{h}$ that are properly contained in $\mathfrak{h}$. Then there exists a subalgebra $\mathfrak{b}$ of $\mathfrak{h}$ such that*

$$\mathfrak{a} + \mathfrak{b} = \mathfrak{h}, \qquad \mathfrak{a} \cap \mathfrak{b} = 0. \tag{3.18.1}$$

Proof. As $\mathfrak{h}/\mathfrak{a}$ has no ideals other than 0 and $\mathfrak{h}/\mathfrak{a}$, either $\dim(\mathfrak{h}/\mathfrak{a}) = 1$ or $\mathfrak{h}/\mathfrak{a}$ is semisimple. If $\dim(\mathfrak{h}/\mathfrak{a}) = 1$, we can take $\mathfrak{b} = k' \cdot X$, where X is any element of $\mathfrak{h}$ not in $\mathfrak{a}$. Suppose now that $\mathfrak{h}/\mathfrak{a}$ is semisimple. Then $\operatorname{rad}(\mathfrak{h}/\mathfrak{a}) = 0$, so by Theorem 3.10.5, $\mathfrak{q} = \operatorname{rad} \mathfrak{h} \subseteq \mathfrak{a}$. Let $\mathfrak{h} = \mathfrak{q} + \mathfrak{m}$ be a Levi decomposition of $\mathfrak{h}$. Clearly, $\mathfrak{m} \cap \mathfrak{a}$ is an ideal of $\mathfrak{m}$, so by Theorem 3.10.1 we can find an ideal $\mathfrak{b}$ of $\mathfrak{m}$ such that $\mathfrak{m}$ is the direct sum of $\mathfrak{m} \cap \mathfrak{a}$ and $\mathfrak{b}$. It is obvious that $\mathfrak{b}$ satisfies (3.18.1).

Lemma 3.18.4. *Let H be a simply connected analytic group with Lie algebra $\mathfrak{h}$. Suppose that $\mathfrak{a}$ is an ideal of $\mathfrak{h}$ and that $\mathfrak{b}$ is a subalgebra of $\mathfrak{h}$ such that (3.18.1) is satisfied. Let A and B be the respective analytic subgroups of H defined by $\mathfrak{a}$ and $\mathfrak{b}$. Write $t_b[a] = bab^{-1}$ $(a \in A, b \in B)$. Then*

$$(a,b) \mapsto ab \quad (a \in A, b \in B)$$

is an analytic group isomorphism of $A \times_t B$ onto H. In particular, A and B are both closed and simply connected, and we have

$$AB = H, \qquad A \cap B = \{1\}. \tag{3.18.2}$$

Proof. For $Y \in \mathfrak{b}$ let $\sigma(Y) = (\operatorname{ad} Y)|\mathfrak{a}$. Then the map ξ $((X,Y) \mapsto X + Y)$ is an isomorphism of $\mathfrak{a} \times_\sigma \mathfrak{b}$ onto $\mathfrak{h}$. Let $H' = A \times_t B$ be the semidirect product associated with $\mathfrak{a} \times_\sigma \mathfrak{b}$ (cf. §3.15). If $\mathfrak{h}'$ is the Lie algebra of H', we have an isomorphism ξ' of $\mathfrak{h}'$ onto $\mathfrak{a} \times_\sigma \mathfrak{b}$ such that $\mathfrak{a}' = \xi'^{-1}(\mathfrak{a} \times \{0\})$ and $\mathfrak{b}' = \xi'^{-1}(\{0\} \times \mathfrak{b})$ are the subalgebras of $\mathfrak{h}'$ defined by $A \times \{1\}$ and $\{1\} \times B$ (cf. §3.15). Let $\eta = \xi'^{-1}\xi^{-1}$. Then η is an isomorphism of $\mathfrak{h}$ onto $\mathfrak{h}'$ that maps $\mathfrak{a}$ onto $\mathfrak{a}'$ and $\mathfrak{b}$ onto $\mathfrak{b}'$. Since H and H' are both simply connected, we have an analytic isomorphism π of H onto H' such that $d\pi = \eta$. Then $A = \pi^{-1}(A \times \{1\})$ and $B = \pi^{-1}(\{1\} \times B)$, so all assertions of the lemma follow at once.

Lemma 3.18.5. *Let G be a simply connected analytic group with Lie algebra $\mathfrak{g}$. Suppose $\mathfrak{a},\mathfrak{b}_1,\ldots,\mathfrak{b}_r$ are subalgebras of $\mathfrak{g}$ such that* (i) *$\mathfrak{g} = \mathfrak{a} + \mathfrak{b}_1 + \cdots + \mathfrak{b}_r$ is a direct sum (of vector spaces), and* (ii) *if $\mathfrak{h}_0 = \mathfrak{a}$ and $\mathfrak{h}_i = \mathfrak{a} + \mathfrak{b}_1 + \cdots + \mathfrak{b}_i$, then the $\mathfrak{h}_i$ are all subalgebras of $\mathfrak{g}$ and $\mathfrak{h}_i$ is an ideal of $\mathfrak{h}_{i+1}$* $(0 \leq i \leq r-1)$. Let *$A,B_1,\ldots,B_r$ be the respective analytic subgroups of G defined by $\mathfrak{a},\mathfrak{b}_1,\ldots,\mathfrak{b}_r$. Then these are all closed and simply connected, and the* *map*

$$(a,b_1,\ldots,b_r)\mapsto ab_1\cdots,b_r \quad (a\in A, b_j\in B_j)$$

is an analytic diffeomorphism of $A\times B_1\times\cdots\times B_r$ *onto* G.

Proof. For $r=1$, this follows from the preceding lemma. We use induction on r. Assume that $r\geq 2$. Let H_{r-1} be the analytic subgroup of G defined by $\mathfrak{h}_{r-1}$. By the previous lemma, H_{r-1} and B_r are closed and simply connected in G, and the map $(h,b)\mapsto hb$ is an analytic diffeomorphism of $H_{r-1}\times B_r$ onto G. On the other hand, by the induction hypothesis, $A,B_1,\ldots,$ B_{r-1} are all closed in H_{r-1} and simply connected, and the map

$$(a,b_1,\ldots,b_{r-1})\mapsto ab_1b_2\cdots b_{r-1}$$

is an analytic diffeomorphism of $A\times B_1\times\cdots\times B_{r-1}$ onto H_{r-1}. Combining these two facts, we get the result at once.

We are now in a position to prove Theorem 3.18.2. It is obvious that we can choose subalgebras $\mathfrak{g}=\mathfrak{g}_0\supseteq\mathfrak{g}_1\supseteq\cdots\supseteq\mathfrak{g}_r=\mathfrak{a}$ such that for each $i=1,2,\ldots,r$, $\mathfrak{g}_i$ is an ideal of $\mathfrak{g}_{i-1}$ containing $\mathfrak{a}$ and $\mathfrak{g}_i$ is a maximal element of the set of ideals of $\mathfrak{g}_{i-1}$ that are properly contained in $\mathfrak{g}_{i-1}$. Write $\mathfrak{h}_i=\mathfrak{g}_{r-i}$. By Lemma 3.18.3, we can find subalgebras $\mathfrak{b}_i$ such that $\mathfrak{h}_i=\mathfrak{h}_{i-1}+\mathfrak{b}_i,\mathfrak{h}_{i-1}\cap\mathfrak{b}_i=0$ $(1\leq i\leq r)$. From Lemma 3.18.5 we conclude that (i) A is closed and simply connected in G, (ii) the map $(b_1,\ldots,b_r)\mapsto Ab_1\cdots b_r$ is an analytic diffeomorphism of $B_1\times\cdots\times B_r$ onto G/A, and (iii) the map $Ab_1\cdots b_r\mapsto b_1\cdots b_r$ is a global analytic section for G/A. This proves everything stated in Theorem 3.18.2.

Corollary 3.18.6. *Assume that G is simply connected. Let $\mathfrak{z}$ be the center of $\mathfrak{g}$ and Z the analytic subgroup of G defined by $\mathfrak{z}$. Then Z is simply connected.*

Next we examine some global properties of commutator subgroups. This will lead naturally to alternative definitions of solvable and nilpotent groups.

If a, b are two elements of a group, define $[a,b]$ to be the commutator $aba^{-1}b^{-1}$. If A and B are two subgroups of a group, define $[A,B]$ to be the group generated by $[a,b]$, with $a\in A$, $b\in B$. Note that $[b,a]=[a,b]^{-1}$ and that $[A,B]$ is the set consisting of 1 and all elements of the form $[a_1,b_1]\cdots[a_r,b_r]$ $(r\geq 1, a_i\in A, b_i\in B$ for all $i)$. We recall also that if $\mathfrak{a}$ and $\mathfrak{b}$ are two subalgebras of a Lie algebra, $[\mathfrak{a},\mathfrak{b}]$ is the linear span of all elements of the form $[X,Y]$ with $X\in\mathfrak{a}$ and $Y\in\mathfrak{b}$.

If A and B are connected subgroups of a connected topological group G, then $[A,B]$ is connected. For $[A,B]=\bigcup_{r\geq 0}[A,B]_r$, where $[A,B]_0=\{1\}$ and $[A,B]_r=\{[a_1,b_1]\cdots[a_r,b_r]: a_i\in A, b_i\in B \text{ for } 1\leq i\leq r\}$ for $r\geq 1$. Since $[A,B]_r$ is the image of $(A\times B)\times(A\times B)\times\cdots\times(A\times B)$ (r factors) under the continuous map

$$((a_1,b_1),\ldots,(a_r,b_r)) \mapsto [a_1,b_1]\cdots[a_r,b_r],$$

$[A,B]_r$ is connected; since $1 \in [A,B]_r$ for all $r \geq 0$, $[A,B]$ is connected.

Theorem 3.18.7. *Let G be an analytic group with Lie algebra $\mathfrak{g}$. Suppose $\mathfrak{a},\mathfrak{b},\mathfrak{h}$ are subalgebras of $\mathfrak{g}$ with the following properties:* (i) $[\mathfrak{a},\mathfrak{h}] \subseteq \mathfrak{h},[\mathfrak{b},\mathfrak{h}] \subseteq \mathfrak{h}$, *and* (ii) $[\mathfrak{a},\mathfrak{b}] = \mathfrak{h}$. *Let A,B,H be the respective analytic subgroups of G defined by $\mathfrak{a},\mathfrak{b},\mathfrak{h}$. Then $H = [A,B]$.*

Proof. Note that if $a \in A$, $b \in B$, then $aHa^{-1} = H$ and $bHb^{-1} = H$, by (i). The proof consists of three steps.

First we show that for any $Y \in \mathfrak{b}$ and $a \in A$, $Y^a - Y$ belongs to $\mathfrak{h}$, and $\mathfrak{h}$ is spanned by such elements. If $X \in \mathfrak{a}$ and $t \in k$, then for any $Y \in \mathfrak{b}$,

$$Y^{\exp tX} - Y = \sum_{n \geq 1} \frac{t^n}{n!}(\operatorname{ad} X)^n(Y), \tag{3.18.3}$$

so since $(\operatorname{ad} X)^n (Y) \in \mathfrak{h}$ for all $n \geq 1$, we find that $Y^{\exp tX} - Y \in \mathfrak{h}$. On the other hand, $\{a: a \in A, Y^a - Y \in \mathfrak{h}\}$ is a subgroup of A. Hence $Y^a - Y \in \mathfrak{h}$ for all $a \in A$. To prove that these elements span $\mathfrak{h}$, it is enough to prove that if $\lambda: \mathfrak{h} \longrightarrow k$ is a linear function such that $\lambda(Y^a - Y) = 0$ for all $Y \in \mathfrak{b}$, $a \in A$, then $\lambda = 0$. Now, (3.18.3) gives the result that

$$\left(\frac{d}{dt}(\lambda(Y^{\exp tX} - Y))\right)_{t=0} = \lambda([X,Y]),$$

from which we get that $\lambda([X,Y]) = 0$ for all $X \in \mathfrak{a}$, $Y \in \mathfrak{b}$. This shows that $\lambda = 0$.

Next we show that $[A,B] \subseteq H$. To this end we begin by exhibiting a neighborhood $\mathfrak{m}$ of 0 in $\mathfrak{g}$ such that if $X \in \mathfrak{a} \cap \mathfrak{m}$ and $Y \in \mathfrak{b} \cap \mathfrak{m}$, then $[\exp X, \exp Y] \in H$. We use the Baker–Campbell–Hausdorff formula (cf. §2.15). Let $c_n (n \geq 1)$ be the maps of $\mathfrak{g} \times \mathfrak{g}$ into $\mathfrak{g}$ defined by (2.15.15), and let $\mathfrak{m}_1 = -\mathfrak{m}_1$ be an open neighborhood of 0 in $\mathfrak{g}$ with the following property: if $X,Y \in \mathfrak{m}_1$, the series $\sum_{n\geq 1} c_n(X: Y)$ converges absolutely (with respect to some norm on $\mathfrak{g}$) to the sum $C(X: Y)$, and

$$\exp X \exp Y = \exp C(X: Y).$$

Now select a neighborhood $\mathfrak{m} = -\mathfrak{m}$ of 0 in $\mathfrak{g}$ such that $\mathfrak{m} \subseteq \mathfrak{m}_1$ and if $X,Y \in \mathfrak{m}$, then $Y^{\exp X} \in \mathfrak{m}_1$. In particular, for $Y \in \mathfrak{m} \cap \mathfrak{b}$ and $X \in \mathfrak{m} \cap \mathfrak{a}$, we have $Y^{\exp X} \in \mathfrak{m}_1$, so

$$[\exp X, \exp Y] = \exp C(Y^{\exp X}: -Y).$$

Let us write $X' = Y^{\exp X}$, $Y' = -Y$. Then from (2.15.15) we have

$$c_1(X': Y') = X' + Y' = Y^{\exp X} - Y \in \mathfrak{h}.$$

Suppose that for some $n \geq 1$, $c_m(X' : Y') \in \mathfrak{h}$ for all $m = 1,2,\ldots,n$. Then $[X' - Y', c_n(X' : Y')] = [X' + Y', c_n(X' : Y')] - 2[Y', c_n(X' : Y')]$ belongs to $\mathfrak{h}$ because $[\mathfrak{b},\mathfrak{h}] \subseteq \mathfrak{h}$, and we also conclude from (2.15.15) that $c_{n+1}(X' : Y') \in \mathfrak{h}$. By induction, $c_n(X' : Y') \in \mathfrak{h}$ for all $n \geq 1$. Thus $C(X' : Y') \in \mathfrak{h}$, proving that $[\exp X, \exp Y] \in H$. This proves that for suitable neighborhoods $A_1 = A_1^{-1}$ and $B_1 = B_1^{-1}$ of the respective identities in A and B, $[a,b] \in H$ if $a \in A_1$, $b \in B_1$. Now if $a \in A, b', b'' \in B$, then

$$[a,b'b''] = [a,b'] \cdot b'[a,b'']b'^{-1}.$$

Consequently, since $b'Hb'^{-1} = H$ for all $b' \in B$ and B_1 generates B, we conclude from the above relation that $[a,b] \in H$ for $b \in B$, $a \in A_1$. Interchanging the roles of A and B in the above argument, we see now that $[b,a] \in H$ for all $b \in B$, $a \in A$. This proves that $[A,B] \subseteq H$.

The third and final step consists in proving that $[A,B]$ contains a neighborhood of 1 in H. To do this, select elements $Y_i \in \mathfrak{b}$ and $a_i \in A$ $(1 \leq i \leq r)$ such that if $Z_i = Y_i^{a_i} - Y_i$, $Z_1, \ldots, Z_r$ span $\mathfrak{h}$. Let us consider the analytic map

$$\psi: ((x_1,y_1),\ldots,(x_r,y_r)) \mapsto [x_1,y_1]\cdots[x_1,y_r]$$

of the manifold $M = (A \times B) \times (A \times B) \times \cdots \times (A \times B)$ (r factors) into G. Then $\psi[M] \subseteq H$ by the second step, so ψ is an analytic map of M into H with $\psi[M] \subseteq [A,B]$. Now

$$\psi((a_1,1),(a_2,1),\ldots,(a_r,1)) = 1,$$

so in order to prove that $[A,B]$ contains an open neighborhood of 1 in H, it suffices to show that the differential of ψ is surjective at $m = ((a_1,1),(a_2,1),\ldots (a_r,1))$. If $m_i^{(i)}$ is the point obtained from m by replacing $(a_i,1)$ with $(a_i, \exp tY_i)$ $(1 \leq i \leq r)$, then

$$\begin{aligned}\left(\frac{d}{dt}\psi(m_i^{(i)})\right)_{t=0} &= \left(\frac{d}{dt}\exp(tY_i^{a_i})\exp(-tY_i)\right)_{t=0}\\ &= Y_i^{a_i} - Y_i,\end{aligned}$$

by (2.12.10). So the range of $(d\psi)_m$ contains Z_i for all i, proving that $(d\psi)_m$ is subjective. The theorem is completely proved.

Let G be an analytic group and $\mathfrak{g}$ its Lie algebra. Define

$$\begin{aligned} &DG = [G,G], \qquad D^rG = D(D^{r-1}G) \quad (r > 1),\\ &CG = [G,G], \qquad C^rG = [G,C^{r-1}G] \quad (r > 1). \end{aligned} \tag{3.18.4}$$

Theorem 3.18.8. *For any $r \geq 1$, D^rG (resp. C^rG) is the analytic subgroup of G defined by $\mathfrak{D}^r\mathfrak{g}$ (resp. $\mathfrak{C}^r\mathfrak{g}$). These are all normal, and if G is simply connected, they are all closed and simply connected.*

Proof. This follows at once from Theorems 3.18.7 and 3.18.2.

Corollary 3.18.9. *G is nilpotent if and only if $C^rG = \{1\}$ for some $r \geq 1$. G is solvable if and only if $D^rG = \{1\}$ for some $r \geq 1$.*

Corollary 3.18.10. *Suppose G is semisimple. Then $G = [G,G]$.*

We shall now obtain some results on the structure of solvable groups analogous to the results of §3.6 concerning nilpotent groups. Unlike the nilpotent case, the exponential map is no longer an analytic diffeomorphism. It turns out, however, that the canonical coordinates of the second kind serve the same purpose.

Theorem 3.18.11. *Let G be a simply connected solvable analytic group with Lie algebra $\mathfrak{g}$. Suppose*[6] *$\{X_1, \ldots, X_m\}$ is a basis of $\mathfrak{g}$ with the following property: $\mathfrak{h}_i = \sum_{1 \leq j \leq i} k \cdot X_j$ is a subalgebra of $\mathfrak{g}$ and $\mathfrak{h}_i$ is an ideal of $\mathfrak{h}_{i+1}$ for $1 \leq i \leq m - 1$. Then the map*

$$\psi: (t_1, \ldots, t_m) \mapsto \exp(t_1X_1) \cdots \exp(t_mX_m) \tag{3.18.5}$$

is an analytic diffeomorphism of k^m onto G.

Proof. Write $\mathfrak{b}_i = k \cdot X_i$. Let B_i be the analytic subgroup of G defined by $\mathfrak{b}_i$. Then Lemma 3.18.5 is applicable (with $\mathfrak{a} = 0$, $r = m$), and we conclude that (i) the B_i are all closed and simply connected in G, and (ii) the map $(b_1, \ldots, b_m) \mapsto b_1 \cdots b_m$ is an analytic diffeomorphism of $B_1 \times \cdots \times B_m$ onto G. On the other hand, since B_i is one-dimensional and simply connected, $t \mapsto \exp tX_i$ is an analytic isomorphism of k onto B_i. The theorem follows at once from this.

From this theorem we obtain the following result, which generalizes Theorem 3.6.2 to solvable groups.

Theorem 3.18.12. *Let G be a solvable analytic group. If G is simply connected, then every analytic subgroup of G is closed and simply connected.*

Proof. Let $\mathfrak{g}$ be the Lie algebra of G. Let A be an analytic subgroup of G, $\mathfrak{a}$ the corresponding subalgebra of $\mathfrak{g}$. Choose a basis $\{X_1, \ldots, X_m\}$ for $\mathfrak{g}$ satisfying the conditions of the previous theorem. If $\dim(\mathfrak{a}) = r$ (we may assume $r \geq 1$), we can find integers $i_1, \ldots, i_r$ (with $1 \leq i_1 < i_2 < \cdots < i_r \leq m$) with the following property: if $d_i = \dim(\mathfrak{h}_i \cap \mathfrak{a})$, then $d_i = 0$ for $i < i_1$, $d_i = \nu$ for $i_\nu \leq i < i_{\nu+1}$ $(1 \leq \nu \leq r - 1)$, $d_i = r$ for $i \geq i_r$. Let $\{Y_1, \ldots, Y_r\}$ be a basis for $\mathfrak{a}$ such that $\{Y_1, \ldots, Y_\nu\}$ span $\mathfrak{h}_{i_\nu} \cap \mathfrak{a}$ $(1 \leq \nu \leq r)$. Replacing

[6] Such bases exist by Corollary 3.7.5.

X_{i_ν} with Y_ν for $1 \leq \nu \leq r$ does not change the $\mathfrak{h}_i$. We may therefore assume without losing generality that $X_{i_\nu} = Y_\nu$ $(1 \leq \nu \leq r)$. If $\mathfrak{a}_\nu = \sum_{1 \leq \mu \leq \nu} k \cdot Y_\mu$, then $\mathfrak{a}_\nu = \mathfrak{h}_{i_\nu} \cap \mathfrak{a}$, so the $\mathfrak{a}_\nu$ are subalgebras of $\mathfrak{a}$ for all ν. If $1 \leq \nu \leq r - 1$, then $\mathfrak{h}_i \cap \mathfrak{a} = \mathfrak{h}_{i_\nu} \cap \mathfrak{a}$ for $i_\nu \leq i < i_{\nu+1}$, so

$$\begin{aligned}[\mathfrak{a}_{\nu+1}, \mathfrak{a}_\nu] &= [\mathfrak{h}_{i_{\nu+1}} \cap \mathfrak{a}, \mathfrak{h}_{i_{\nu+1}-1} \cap \mathfrak{a}] \\ &\subseteq [\mathfrak{h}_{i_{\nu+1}}, \mathfrak{h}_{i_{\nu+1}-1}] \cap \mathfrak{a} \\ &\subseteq \mathfrak{h}_{i_{\nu+1}-1} \cap \mathfrak{a} \\ &= \mathfrak{a}_\nu.\end{aligned}$$

This proves that $\mathfrak{a}_\nu$ is an ideal of $\mathfrak{a}_{\nu+1}$.

Let A^* be a simply connected analytic group that is a covering group of A, and let π be the covering homomorphism. The properties of the Y_ν established above show that Theorem 3.18.11 is applicable to A^*. Consequently, the map

$$f: (u_1, \ldots, u_r) \mapsto \exp_{A^*}(u_1 Y_1) \cdots \exp_{A^*}(u_r Y_r)$$

is an analytic diffeomorphism of k^r onto A^*; here we write $\exp_{A^*}$ for the exponential map into A^*. Since $\pi[A^*] = A$, we conclude from this that the map

$$g: (u_1, \ldots, u_r) \mapsto \exp(u_1 Y_1) \cdots \exp(u_r Y_r)$$

maps k^r onto A. But then we find from (3.18.5) that A is the image under ψ of the subset of all $(t_1, \ldots, t_m) \in k^m$ such that $t_i = 0$ for $i \notin \{i_1, \ldots, i_r\}$. So A is closed and simply connected. This proves the theorem.

Let G be an analytic group with Lie algebra $\mathfrak{g}, \mathfrak{q} = \text{rad } \mathfrak{g}, \mathfrak{n} = \text{nil rad } \mathfrak{g}$, and Q, and N the respective analytic subgroups of G defined by $\mathfrak{q}$ and $\mathfrak{n}$. Clearly, Q is the largest solvable normal analytic subgroup of G, and N is the normal analytic subgroup of G. Q is called the *radical* of G and N, the *nil radical* of G. Obviously, G is semisimple if and only if $Q = \{1\}$.

Theorem 3.18.13. *Let G be an analytic group with Lie algebra $\mathfrak{g}$, and Q (resp. N) the radical (resp. nil radical) of G. Then Q and N are closed. Suppose that $\mathfrak{g} = \mathfrak{q} + \mathfrak{m}$ is a Levi decomposition of $\mathfrak{g}$ and that M is the analytic subgroup of G defined by $\mathfrak{m}$. Then $G = QM$, and M is a maximal semisimple analytic subgroup of G. If M' is another maximal semisimple analytic subgroup of G, then $G = QM'$, and there is $y \in [Q,N]$ such that $yMy^{-1} = M'$. If G is simply connected, then M is closed in G and simply connected, and $(q,m) \mapsto qm$ is an analytic diffeomorphism of $Q \times M$ onto G. In particular,*

$$QM = G, \qquad Q \cap M = \{1\}. \tag{3.18.6}$$

We require a lemma.

Lemma 3.18.14. *Let H be an analytic group and $A \subseteq H$ an analytic subgroup. Suppose A is solvable (resp. nilpotent). Then $Cl(A)$ is a solvable (resp. nilpotent) analytic group.*

Proof. The arguments for the two cases are quite similar, so we treat only the solvable case. Let $A_0 = A$ and $A_p = D^pA (p \geq 1)$. Since A is solvable, we can find $a \geq 0$ such that $A_{s+1} = \{1\}$. Let $B_p = Cl(A_p)$. Then B_0 is an analytic group and $[B_p, B_p] \subseteq B_{p+1}$ $(0 \leq p \leq s)$. This implies that $D^pB_0 \subseteq B_p$ for $0 \leq p \leq s + 1$. In particular, $D^{s+1}B_0 = \{1\}$, proving that B_0 is solvable.

We can now prove Theorem 3.18.13. By the above lemma, $Cl(Q)$ is a solvable analytic group. Since it is obviously normal, the maximality of Q implies that $Q = Cl(Q)$. Similarly, $N = Cl(N)$. The assertions concerning M follow from Theorem 3.14.2 and Corollary 3.14.3, except for the proof that $G = QM$. To prove this, observe that since Q is normal, QM is a subgroup of G. Now consider the map $\mathfrak{g}: (q,m) \mapsto qm$ of $Q \times M$ into G. With the usual identification of tangent spaces, we have

$$(dg)_{(1,1)}(X, Y) = X + Y \quad (X \in \mathfrak{g},\ Y \in \mathfrak{m}),$$

so that $(dg)_{(1,1)}$ is surjective. Thus QM contains a neighborhood of 1 in G, proving that $G = QM$. If G is simply connected, the assertions concening M and Q follow from Lemma 3.18.4.

We refer to any maximal semisimple analytic subgroup of G as a *Levi subgroup*. If G has a decomposition of the form (3.18.6), we call it a *Levi decomposition*.

We conclude this section with some remarks concerning faithful representations. In what follows, by a representation of an analytic group G we mean an analytic homomorphism π of G into $GL(V)$, where V is a finite-dimensional vector space over k. π is called *unipotent* if $\pi(x)$ is unipotent for all $x \in G$. Clearly, π is unipotent if and only if $d\pi$ is a nil representation of the Lie algebra of G.

Lemma 3.18.15. *Let G be a nilpotent analytic group with Lie algebra $\mathfrak{g}$. Suppose π is a unipotent representation of G. Then*

$$\text{kernel}(\pi) = \exp[\text{kernel}(d\pi)]. \tag{3.18.6}$$

In particular, π is faithful if $d\pi$ is faithful.

Proof. Let $Z = \text{kernel}(\pi)$, $\mathfrak{z} = \text{kernel}(d\pi)$. Write $N = \pi[G]$, $\mathfrak{n} = (d\pi)[\mathfrak{g}]$. Let V be the vector space on which π acts. Then N is the analytic subgroup of $GL(V)$ defined by $\mathfrak{n}$ and all elements of $\mathfrak{n}$ are nilpotent. So by Theorem 3.6.3, N is closed and simply connected. On the other hand, $Z^0 = \exp[\mathfrak{z}]$, while

$G/Z^0 \longrightarrow G/Z \approx N$ is a covering map. So $Z = Z^0$. In particular, if $\mathfrak{z} = 0$, $Z = \{1\}$.

Theorem 3.18.16. *Let G be an analytic group. Assume that there are closed analytic subgroups A and B such that (i) A is normal, simply connected, and solvable, and (ii) $G = AB$ and $A \cap B = \{1\}$. Then there exists a representation of G that is faithful on A and unipotent on the nil radical N of A. If B has a faithful representation, then G has a faithful representation that is unipotent on N.*

Proof. Let $\mathfrak{g}$ be the Lie algebra of G, and let $\mathfrak{a},\mathfrak{b},\mathfrak{n}$ be the respective subalgebras defined by A, B, N. First, consider the case $A = G$. By Theorem 3.17.7, there is a faithful representation ρ of $\mathfrak{g}$ that is a nil representation on $\mathfrak{n}$. Let π_1 be the representation of G such that $d\pi_1 = \rho$. Then by the previous lemma, $\pi_1 | N$ is a faithful unipotent representation. On the other hand, N is closed and G/N is an abelian group, so there is a representation π_2 of G such that N is the kernel of π_2. If π is the direct sum of π_1 and π_2, then π is faithful and $\pi | N$ is unipotent.

We now take up the general case. Write $\mathfrak{G}$ for the universal enveloping algebra of $\mathfrak{g}$, and let $\mathfrak{A}$ be the subalgebra of $\mathfrak{G}$ generated by $\mathfrak{a}$. By the preceding argument, there is a faithful representation ζ of A that is unipotent on N. Extend $d\zeta$ to a representation σ of $\mathfrak{A}$, and let $\mathfrak{K}$ be the kernel of σ. By Theorem 3.8.3, $[\mathfrak{b},\mathfrak{a}] \subseteq \mathfrak{n}$. Now use the results and notation of Lemma 3.17.2–3.17.4.

We assert the existence of a representation π_B of B in U such that $d\pi_B(Y) = \hat{D}_Y$ for all $Y \in \mathfrak{b}$. To prove this, observe that since the adjoint representation of $\mathfrak{g}$ in $\mathfrak{G}$ leaves both $\mathfrak{A}$ and $\mathfrak{M}_r$ invariant, the same is true for the adjoint representation of G in $\mathfrak{G}$. So we have a representation λ of G acting on $\mathfrak{A}/\mathfrak{M}_r$ such that $d\lambda$ is the representation of $\mathfrak{g}$ in $\mathfrak{A}/\mathfrak{M}_r$ induced by the adjoint representation of $\mathfrak{g}$. If $\pi_B = \lambda | B$, the $d\pi_B(Y) = \hat{D}_Y$ ($Y \in \mathfrak{b}$).

A being simply connected, there is a representation π_A of A in U such that $d\pi_A = \xi | \mathfrak{a}$. From (3.17.5) and (ii) of Lemma 3.17.3 we have

$$d\pi_A([Y,X]) = [d\pi_B(Y), d\pi_A(X)] \quad (X \in \mathfrak{a},\ Y \in b). \tag{3.18.7}$$

Replacing X by $(\text{ad } Y)^n(X)$ ($n = 0,1, \ldots$) in succession in this relation, we get

$$\begin{aligned} d\pi_A(X^{\exp Y}) &= \exp(\text{ad } d\pi_B(Y))(d\pi_A(X)) \\ &= \exp(d\pi_B(Y))\, d\pi_A(X) \exp(-d\pi_B(Y)) \\ &= \pi_B(\exp Y)\, d\pi_A(X)\, \pi_B(\exp Y)^{-1} \end{aligned}$$

for $X \in \mathfrak{a}$, $Y \in \mathfrak{b}$. This implies that

$$d\pi_A(X^b) = \pi_B(b)\, d\pi_A(X)\, \pi_B(b)^{-1} \quad (X \in \mathfrak{a},\ b \in \mathfrak{B}), \tag{3.18.8}$$

from which we finally get

(3.18.9) $$\pi_A(bab^{-1}) = \pi_B(b)\pi_A(a)\pi_B(b)^{-1} \quad (a \in A, b \in B).$$

As an immediate consequence of (3.18.9) we find that

(3.18.10) $$\pi: ab \mapsto \pi_A(a)\pi_B(b) \quad (a \in A, b \in B)$$

is a representation of G in U. Clearly, $d\pi|\mathfrak{a} = \xi$, so it follows from Lemma 3.17.4 that $d\pi|\mathfrak{n}$ is a nil representation. π is thus unipotent on N. Now, if $v \neq 0$ is any vector in the vector space V on which σ acts, the map $a \mapsto \sigma(a)v$ induces a map of U onto V that intertwines ξ and σ. So the same map also intertwines the representations π_A and ζ. ζ is thus a quotient of π_A. This proves that π_A is faithful. If B has a faithful representation, G has a representation π' whose kernel is A, and the direct sum of π and π' is a faithful representation of G that is unipotent on N. This proves all statements of the theorem.

EXERCISES

Unless otherwise stated, all vector spaces and Lie algebras considered are finite-dimensional; k denotes a field of characteristic zero.

1. Let V be a vector space of dimension m over k, L an endomorphism of V, and $\mathfrak{D}$ the algebra of all endomorphisms of V that commute with L. The following observations lead to a proof of Theorem 3.1.7.
   (a) Let v_i, J_i be as in Theorem 3.1.3, $1 \leq i \leq n$. Suppose $B \in \mathfrak{D}'$. Show that $Bv_1 \in [v_1]$, and hence deduce the existence of a $p \in k[T]$ such that $Bv_1 = p(L)v_1$.
   (b) Let $v \in V$. Prove the existence of a $D \in \mathfrak{D}$ such that $Dv_1 = v$. (Hint: Observe that $J_1 \subseteq J_v$).
   (c) Let p be as in (a). Prove that $B = p(L)$.

2. Let V be a vector space of dimension m over **C**. For $x \in SL(V)$ and $X \in \mathfrak{sl}(V)$, write $X^x = xXx^{-1}$. We thus have an action of $SL(V)$ on $\mathfrak{sl}(V)$.
   (a) Let L be a nilpotent endomorphism of V. Prove that the minimal polynomial of L is T^m if and only if there is a basis $\{v_1, \ldots, v_m\}$ for V such that $Lv_1 = 0$ and $Lv_s = v_{s-1}$ $(1 < s \leq m)$. Deduce that $SL(V)$ acts transitively on the set of all such nilpotent endomorphisms. We call them *principal nilpotent endomorphisms*.
   (b) Prove that the set of principal nilpotent endomorphisms is a regular submanifold of dimension $m^2 - m$ in $\mathfrak{sl}(V)$ and is a dense open subset of the set of all nilpotent endomorphisms.
   (c) Let L be an arbitrary endomorphism. Prove that its minimal and characteristic polynomials coincide if and only if $(L - \lambda \cdot 1)|V_{L,\lambda}$ is a principal nilpotent endomorphism of $V_{L,\lambda}$ for each $\lambda \in \sigma(L)$.

(d) Deduce from (b) and (c) that the orbit of an element $L \in \mathfrak{sl}(V)$ under $SL(V)$ is closed if and only if L is a semisimple endomorphism of V.

(e) Extend the result of (d) to the real case.

3. (a) Let V be a vector space of dimension m over k, and let L be a nilpotent endomorphism of V. Let $\mathfrak{n}$ be the centralizer of L in $\mathfrak{gl}(V)$, i.e., the set of all elements of $\mathfrak{gl}(V)$ that commute with L. Prove that $\dim(\mathfrak{n}) \geq m$ and that L is principal (cf. Exercise 2) if and only if $\dim(\mathfrak{n}) = m$.

(b) Let L be an arbitrary element of $\mathfrak{gl}(V)$ and let $\mathfrak{n}$ be the centralizer of L in $\mathfrak{gl}(V)$. Prove that $\dim(\mathfrak{n}) \geq m$.

4. Let V be a vector space of finite dimension over k, and let $x \in GL(V)$. Prove that one can write $x = su$ where (i) s is semisimple, u is unipotent, and both s and u are in $GL(V)$, and (ii) s and u commute. Prove also that s and u are uniquely determined by these conditions and that $s = p(x)u = q(x)$ for suitable $p,q \in k[T]$.

In Exercises 5–11, $\mathfrak{A}$ is a fixed associative algebra over k. All representations of $\mathfrak{A}$ to be considered are finite-dimensional unless otherwise stated.

5. Let ρ be a representation of $\mathfrak{A}$ acting on W. Prove that the following statements on ρ are equivalent: (i) ρ is semisimple, (ii) given any invariant subspace, $W' \neq W$, of W, one can find an invariant subspace $W'' \neq 0$ such that $W' \cap W'' = 0$, (iii) W is the linear span of invariant subspaces W' such that $\rho_{W'}$ is irreducible, and (iv) W is the direct sum of invariant subspaces $W_1, \ldots, W_s$ such that ρ_{W_i} is irreducible for $i = 1, \ldots, s$.

6. The following observations are designed to lead to a proof of Theorem 3.1.9. We use the notation therein. ρ is semisimple.

(a) Let $v \in W$, and let $[v] = \rho[\mathfrak{A}]v$. Prove the existence of a projection belonging to $\rho[\mathfrak{A}]'$ that maps W onto $[v]$. Deduce that if $B \in \rho[\mathfrak{A}]''$, one can find $a \in \mathfrak{A}$ such that $\rho(a)v = Bv$.

(b) Let r be an integer ≥ 1 and let $\bar{W} = W \times \cdots \times W$ (r factors). For any endomorphism A of W let $\bar{A}$ be the endomorphism $(v_1, \ldots, v_r) \mapsto (Av_1, \ldots, Av_r)$ of $\bar{W}$. Let $\bar{\rho}$ be the representation $a \mapsto \overline{\rho(a)}$ of $\mathfrak{A}$ acting on $\bar{W}$. Prove that $\bar{\rho}$ is semisimple. Apply the result of (a) to $\bar{\rho}$ and deduce that give $v_1, \ldots, v_r \in W$ and a $B \in \rho[\mathfrak{A}]''$, there exists an $a \in \mathfrak{A}$ such that $Bv_s = \rho(a)v_s$ for $1 \leq s \leq r$.

7. Let ρ be a representation of $\mathfrak{A}$ acting on W, k' a Galois extension of k, and Γ the Galois group of k' over k. We write $W' = W^{k'}$, $\rho' = \rho^{k'}$, $\mathfrak{A}' = \mathfrak{A}^{k'}$.

(a) Assume that ρ is irreducible, and let W_1 be a subspace of W' invariant under ρ'. Prove that $W' = \sum_{s \in \Gamma} s \cdot W_1$, and hence deduce from Exercise 5 that ρ' is semisimple. Prove that the number of irreducible constituents of ρ' cannot exceed the order of Γ.

(b) Assume that ρ' is semisimple, and let W_1 be a subspace of W invariant under ρ, with $W_1 \neq 0$, $W_1 \neq W$. Let $\mathfrak{C}$ (resp. $\mathfrak{C}'$) be the algebra of all endomorphisms A of W lying in $\rho[\mathfrak{A}]'$ and vanishing on W_1 (resp. A' of W' lying in $\rho'[\mathfrak{A}']'$ and vanishing on $W_1^{k'}$). Prove that $\mathfrak{C}'$ contains nonzero semisimple elements, and hence deduce that $\mathfrak{C}$ contains nonzero elements which are not nilpotent. (Hint: Observe that $\mathfrak{C}'$, considered as a subspace of the vector space of endomorphisms of W', is defined over k).

(c) Let A be a nonzero element of $\mathfrak{C}$ which is not nilpotent. Prove that the semisimple part of A belongs to $\mathfrak{C}$.

(d) Deduce that ρ is semisimple.

8. Let $k = \mathbf{R}$, and let ρ be a representation of $\mathfrak{A}$ in W. Prove that if ρ is irreducible, the division algebra $\rho[\mathfrak{A}]'$ is isomorphic to $\mathbf{R}$, $\mathbf{C}$, or the algebra of quaternions. Give examples to show that all the three possibilities can arise.

9. Let $k = \mathbf{R}$ or $\mathbf{C}$ and let ρ be a representation of $\mathfrak{A}$ in a Hilbert space W of finite dimension over k. Let $(\cdot,\cdot)$ denote the scalar product in W. Suppose there is a subset E of $\mathfrak{A}$ such that (i) E generates $\mathfrak{A}$, and (ii) $\rho(a)$ is symmetric (or skew-symmetric) with respect to $(\cdot,\cdot)$ for all $a \in E$. Then prove that ρ is semisimple.

10. Let ρ be an irreducible representation of $\mathfrak{A}$. Prove that the equivalence class of ρ is uniquely determined by the character of ρ. Hence (or otherwise) show that if $\rho_1, \ldots, \rho_s$ are mutually inequivalent irreducible representations of $\mathfrak{A}$ with respective characters $\chi_1, \ldots, \chi_s$, the linear functions χ_j on $\mathfrak{A}$ are linearly independent.

11. Let $\mathfrak{A}$ be commutative, ρ a representation of $\mathfrak{A}$ in W.

(a) Assume that k is algebraically closed and that $\rho(a)$ is semisimple for each $a \in \mathfrak{A}$. Prove that there exist a basis $\{v_1, \ldots, v_m\}$ of W and homomorphisms $\chi_1, \ldots, \chi_m$ of $\mathfrak{A}$ into k such that $\rho(a)v_s = \chi_s(a)v_s$ for all $a \in \mathfrak{A}$ and $1 \leq s \leq m$.

(b) Let k and ρ be arbitrary. For each $a \in \mathfrak{A}$, let $\rho_s(a)$ (resp. $\rho_n(a)$) be the semisimple (resp. nilpotent) part of $\rho(a)$. Prove that ρ_s and ρ_n are representations of $\mathfrak{A}$ in W and that ρ_s is a semisimple representation. Deduce that ρ is a semisimple representation if and only if $\rho(a)$ is semisimple for each $a \in \mathfrak{A}$.

12. Let ρ be a representation of a Lie algebra $\mathfrak{g}$ in a vector space W of possibly infinite dimension. Assume that $W = \sum_i W_i$, where the W_i are finite-dimensional subspaces invariant under ρ such that the representations ρ_{W_i} are all semisimple. Prove that W is the direct sum of the subspaces W^0 and W^1, where $W^0 = \{v : v \in W, \rho(X)v = 0 \text{ for all } X \in \mathfrak{g}\}$, and W^1 is the linear span of the ranges of all the $\rho(X)$, $X \in \mathfrak{g}$. Prove further that W^0 and W^1 are invariant and that each is the unique complementary invariant subspace in W of the other.

13. Let V be a finite-dimensional vector space over k, E the exterior algebra over V. We denote by $\wedge$ the operation of multiplication in E and by E_p the homogeneous subspace of E of degree p.

(a) Let $x_1, \ldots, x_s \in V$; then they are linearly independent if and only if $x_1 \wedge x_2 \wedge \cdots \wedge x_s \neq 0$. In this case, if U is the linear space spanned by the $x_j (1 \leq j \leq s)$, then for any vector y to belong to U, it is necessary and sufficient that $y \wedge x_1 \wedge \cdots \wedge x_s = 0$. Assume from here on that the x_i form a basis for U.

(b) Write $u = x_1 \wedge \cdots \wedge x_s$. Let L be an endomorphism of V, and let $\tilde{L}$ be the derivation of E which extends L. Prove that U is invariant under L if and only if u is an eigenvector for $\tilde{L}$.

(c) Suppose F is a subspace of E_s that is complementary to $k \cdot u$. Denote by U' the set of all vectors $v \in V$ having the following property: if $y_1, \ldots, y_{s-1} \in U$, then $v \wedge y_1 \wedge \cdots \wedge y_{s-1} \in F$. Show that U' is a subspace of V that is complementary to U.

(d) Suppose F is invariant under $\tilde{L}$; show that U' is invariant under L.

14. (a) Let $\mathfrak{g}$ be a Lie algebra of arbitrary dimension over k, $\mathfrak{G}$ its universal enveloping algebra. Prove that the map $X \mapsto X \otimes 1 + 1 \otimes X$ of $\mathfrak{g}$ into $\mathfrak{G} \otimes \mathfrak{G}$ extends to a homomorphism δ of $\mathfrak{G}$ into $\mathfrak{G} \otimes \mathfrak{G}$. Prove also that δ is an injection.

(b) Let $a \in \mathfrak{G}$. Prove that $a \in \mathfrak{g}$ if and only if $\delta(a) = a \otimes 1 + 1 \otimes a$. (Hint: Let $X_1, X_2, \ldots$ be linearly independent in $\mathfrak{g}$. For $M = (m_1, m_2, \ldots)$, where the m_i are integers ≥ 0 with $\sum_i m_i < \infty$, let $x^M = X_1^{m_1} X_2^{m_2} \cdots$. Calculate $\delta(x^M)$).

15. Let $\mathfrak{g}$ (resp. $\mathfrak{G}$) be the free Lie (resp. associative) algebra generated by $\{X_1, \ldots, X_m\}$; let $\mathfrak{G}_n$ be the subspace of $\mathfrak{G}$ spanned by elements of degree n; and let $\mathfrak{g}_n = \mathfrak{G}_n \cap \mathfrak{g}$. Prove that $\dim(\mathfrak{g}_n) = (1/n) \sum_{d|n} \mu(d) m^{n/d}$ for all n, where μ is the usual Möbius function defined on the set of positive integers as follows: $\mu(1) = 1$, $\mu(n) = 0$ if n is divisible by the square of a prime, and $\mu(n) = (-1)^k$ if $n = p_1 \cdots p_k$, where $p_1, \ldots, p_k$ are distinct primes. (Hint: Let $\nu_n = \dim(\mathfrak{g}_n)$, and let T be an indeterminate. Calculate $\dim(\mathfrak{G}_n)$ using Theorem 3.2.8 to get $(1 - mT)^{-1} = \prod_{n \geq 1} (1 - T^n)^{-\nu_n}$. Take logarithms to deduce that $m^n = \sum_{d|n} d\nu_d$. Now use the Möbius inversion formula.)

16. Let $\mathfrak{g}$ be a Lie algebra over k, $\mathfrak{a}$ a subalgebra. Let $\mathfrak{G}$ be the universal enveloping algebra of $\mathfrak{g}$, $\mathfrak{A}$ the subalgebra of $\mathfrak{G}$ generated by $\mathfrak{a}$. Suppose that $\mathfrak{M}$ is a left ideal of $\mathfrak{A}$. Prove that $\mathfrak{G}\mathfrak{M} \cap \mathfrak{A} = \mathfrak{M}$.

17. Let $\mathfrak{g} = \mathfrak{gl}(n,k)$, $G = GL(n,k)$, and for $X \in \mathfrak{g}$, $x \in G$, write $X^x = xXx^{-1}$. Let Z and $\mathfrak{Z}$ be as in Theorem 3.3.8. Denote by $P(\mathfrak{g})$ the algebra of polynomial functions on $\mathfrak{g}$ and by J the subalgebra of all $p \in P(\mathfrak{g})$ such that $p(X^x) = p(X)$ for all $X \in \mathfrak{g}$, $x \in G$. Let $S(\mathfrak{g})$ be the symmetric algebra over $\mathfrak{g}$.

(a) Prove the existence of a unique algebra isomorphism ξ of $S(\mathfrak{g})$ onto $P(\mathfrak{g})$ such that for each $X \in \mathfrak{g}$, $\xi(X)$ is the element of $\mathfrak{g}^*$ (the dual of $\mathfrak{g}$) given by $\xi(X)(X') = tr(XX')$ $(X' \in \mathfrak{g})$.

(b) Prove that $\xi[\mathrm{Z}] = J$.

(c) Deduce from (b) that Z and $\mathfrak{Z}$ are isomorphic to $k[T_1, \ldots, T_n]$.

18. Let $\mathfrak{g}$ be a Lie algebra over k, $\mathfrak{G}$ its universal enveloping algebra.

(a) Prove that $\mathfrak{g}$ has a faithful representation ρ such that $tr\, \rho(X) = 0$ for all $X \in \mathfrak{g}$.

(b) Let ρ be as in (a). Denote by $\varphi^{(r)}$ the tensor product $\rho \otimes \rho \otimes \cdots \otimes \rho$ (r factors), and let σ_r be its extension to a representation of $\mathfrak{G}$. Prove that $\bigcap_{r \geq 1}$ kernel $(\sigma_r) = 0$. (cf. Harish-Chandra [1]).

19. Let $\mathfrak{g}$ be the Lie algebra of dimension 3 with basis $\{X, Y, Z\}$ such that $[X,Z] = [Y,Z] = 0$, $[X,Y] = Z$. Prove that $\mathfrak{Z} = K[Z]$.

20. Prove that the members of the ascending and descending central series of a Lie algebra $\mathfrak{g}$ are invariant under each derivation of $\mathfrak{g}$.

21. Let $\mathfrak{g}$ be a nilpotent Lie algebra over k and ρ a representation of $\mathfrak{g}$ in a finite-dimensional nonzero vector space V over k. Suppose that $\det \rho(X) = 0$ for all $X \in \mathfrak{g}$. Show that there is a nonzero vector $v \in V$ such that $\rho(X)v = 0$ for all $X \in \mathfrak{g}$.

22. Let G be a nilpotent analytic group, $\mathfrak{g}$ its Lie algebra. Let dx be a Haar measure on G, and dX a Lebesgue measure on $\mathfrak{g}$. Prove that dx is both right- and left-invariant and that there is a constant $c > 0$ such that cdx is the measure on G that corresponds to dX under exp.

23. Prove that compact subgroups of a nilpotent analytic group G are necessarily central, and that they are equal to $\{1\}$ if G is simply connected.

24. Let G be a simply connected nilpotent analytic group, $\mathfrak{g}$ its Lie algebra. $\{X_1, \ldots, X_m\}$ is a basis for $\mathfrak{g}$ such that if $\mathfrak{h}_i$ is the linear span of $\{X_{i+1}, \ldots, X_m\}$ $(0 \leq i \leq m)$, then $\mathfrak{h}_i$ is an ideal in $\mathfrak{g}$ with $[\mathfrak{g},\mathfrak{h}_i] \subseteq \mathfrak{h}_{i+1}$ $(0 \leq i \leq m-1)$. Let $\omega_1, \ldots, \omega_m$ be the left-invariant 1-forms on G such that $\omega_i(X_j) = \delta_{ij}$ $(1 \leq i, j \leq m)$. Let $\hat{\omega}_i$ be the form on $\mathfrak{g}$ that corresponds to ω_i under exp.

(a) If $x_1, \ldots, x_m$ are the linear coordinates on $\mathfrak{g}$ with respect to the basis $\{X_1, \ldots, X_m\}$, prove that $\hat{\omega}_1 = dx_1$ and that for $1 < i \leq m$, $\hat{\omega}_i = dx_i + \sum_{1 \leq j \leq i-1} Q_{ij}(x_1, \ldots, x_{i-1})\, dx_j$, Q_{ij} being a polynomial in $i-1$ variables of degree $\leq m-1$.

(b) For $x \in G$, let β_x be the analytic diffeomorphism $X \mapsto \log(\exp X \cdot x)$ of $\mathfrak{g}$. Prove that the β_x $(x \in G)$ are precisely the analytic diffeomorphisms of $\mathfrak{g}$ that leave each $\hat{\omega}_i$ invariant.

(c) Let P be the polynomial map of $\mathfrak{g} \times \mathfrak{g}$ into $\mathfrak{g}$ satisfying (3.6.1). Let B_i be the polynomials of $x_1, \ldots, x_m$ such that $P(\sum_{1 \leq i \leq m} x_i X_i, \sum_{1 \leq i \leq m} y_i X_i) = \sum_{1 \leq i \leq m} B_i(x_1, \ldots, x_m : y_1, \ldots, y_m) X_i$ for all $x_1, \ldots, x_m, y_1, \ldots, y_m$. Prove that $B_1(x_1, \ldots, x_m : y_1, \ldots, y_m) \equiv x_1 + y_1$ and that for $1 < i \leq m$, $B_i(x_1, \ldots, x_m : y_1, \ldots, y_m) \equiv x_i + y_i + Q_i(x_1, \ldots, x_{i-1} : y_1, \ldots, y_{i-1})$, where Q_i is a polynomial in $2i-2$ variables.

(d) Prove that the differential operators on $\mathfrak{g}$ that correspond to $\mathfrak{G}$ under exp have polynomial coefficients.

25. Let $\mathfrak{a}$ and $\mathfrak{g}$ be Lie algebras over k, $\mathfrak{a}$ being abelian. Let ρ be a representation of $\mathfrak{g}$ in $\mathfrak{a}$. A triple $(\mathfrak{b},\lambda,\mu)$ is called an *extension of* $\mathfrak{g}$ *by* $\mathfrak{a}$ *associated with* ρ if (a) $\mathfrak{b}$ is a Lie algebra, λ is an injection of $\mathfrak{a}$ onto an ideal of $\mathfrak{b}$, and μ is a homomorphism of $\mathfrak{b}$ onto $\mathfrak{g}$; (b) $\lambda[\mathfrak{a}] = \text{kernel}(\mu)$; and (c) if $Z \in \mathfrak{b}$ and $X \in \mathfrak{a}$, $[Z,\lambda(X)] = \rho(\mu(Z))X$. Extensions $(\mathfrak{b},\lambda,\mu)$ and $(\mathfrak{b}',\lambda',\mu')$ are equivalent if there is an isomorphism ζ of $\mathfrak{b}$ onto $\mathfrak{b}'$ such that $\lambda' = \zeta \circ \lambda$, $\mu = \mu' \circ \zeta$. The extension $(\mathfrak{b},\lambda,\mu)$ is called *inessential* if there is a subalgebra $\mathfrak{h}$ of $\mathfrak{b}$ such that $\lambda[\mathfrak{a}] + \mathfrak{h} = \mathfrak{b}$, $\lambda[\mathfrak{a}] \cap \mathfrak{h} = 0$.

(a) Let $(\mathfrak{b},\lambda,\mu)$ be an extension and ν a linear map of $\mathfrak{g}$ into $\mathfrak{b}$ such that $\mu \circ \nu = \text{identity}$. Let $\varphi_\nu(X,Y) = [\nu(X),\nu(Y)] - \nu([X,Y])$, $X, Y \in \mathfrak{g}$. Prove that $\varphi_\nu \in C^2(\mathfrak{g},\rho)$. If ν' is another linear map of $\mathfrak{g}$ into $\mathfrak{b}$ such that $\mu \circ \nu' = \text{identity}$, prove that $\varphi_\nu - \varphi_{\nu'} \in B^2(\mathfrak{g}, \rho)$.

(b) Let $\bar{\mathfrak{b}} = \mathfrak{a} \times \mathfrak{g}$, $\bar{\lambda}(X) = (X, 0)$ $(X \in \mathfrak{a})$, $\bar{\mu}((X,Y)) = Y$ $(X \in \mathfrak{a}, Y \in \mathfrak{g})$. Let $\varphi \in C^2(\mathfrak{g},\rho)$. For (X,Y), $(X',Y') \in \bar{\mathfrak{b}}$, let $[(X,Y),(X',Y')] = (\varphi(Y,Y') + \rho(Y)X' - \rho(Y')X, [Y,Y'])$. Prove that $[\cdot, \cdot]$ converts $\bar{\mathfrak{b}}$ into a Lie algebra. Denote this Lie algebra by $\mathfrak{b}_\varphi$, and prove that $(\mathfrak{b}_\varphi,\bar{\lambda},\bar{\mu})$ is an extension of $\mathfrak{g}$ by $\mathfrak{a}$ associated with ρ. Prove also that the above extension corresponding to φ in $C^2(\mathfrak{g},\rho)$ is inessential if and only if $\varphi \in B^2(\mathfrak{g},\rho)$.

(c) Deduce that the elements of $H^2(\mathfrak{g},\rho)$ are in natural one-to-one correspondence with the equivalence classes of extensions of $\mathfrak{g}$ by $\mathfrak{a}$ associated with ρ.

26. Let $\mathfrak{g}$ be a Lie algebra over k, $\mathfrak{q}$ its radical. Prove that $[\mathfrak{q},\mathfrak{g}]$ is the smallest of the ideals $\mathfrak{a}$ such that $\mathfrak{g}/\mathfrak{a}$ is reductive.

27. Let $\mathfrak{g}$ be a semisimple Lie algebra over k, $\mathfrak{g}_i$ $(1 \leq i \leq r)$ the simple ideals whose direct sum is $\mathfrak{g}$, π_i the projections $\mathfrak{g} \longrightarrow \mathfrak{g}_i$ corresponding to this direct sum, and $\langle\cdot,\cdot\rangle$ the Cartan–Killing form of $\mathfrak{g}$. Let $\mathcal{B}$ be the vector space of all bilinear forms B on $\mathfrak{g} \times \mathfrak{g}$ such that $B([X,Y],Z) + B(Y,[X,Z]) = 0$ for all $X, Y, Z \in \mathfrak{g}$. Let $B_i(X,Y) = \langle \pi_i X, \pi_i Y\rangle$ $(X, Y \in \mathfrak{g}, 1 \leq i \leq r)$.

(a) Prove that the restriction of $\langle\cdot,\cdot\rangle$ to $\mathfrak{g}_i \times \mathfrak{g}_i$ is the Cartan–Killing form of $\mathfrak{g}_i$.

(b) Prove that $\{B_1, \ldots, B_r\}$ is a basis for $\mathcal{B}$ if k is algebraically closed.

28. Give an example of a solvable Lie algebra whose Cartan–Killing form is not identically zero, and of a solvable but non-nilpotent Lie algebra whose Cartan–Killing form is identically zero.

29. Let $\mathfrak{g}$ be a solvable Lie algebra over $\mathbf{R}$, V a vector space over $\mathbf{R}$, and ρ a representation of $\mathfrak{g}$ in V. Prove that there are invariant subspaces V_i $(0 \leq i \leq s)$ of V such that (i) $V_0 = 0 \subseteq V_1 \subseteq \cdots \subseteq V_s = V$, (ii) the representations induced on V_i/V_{i-1} are irreducible for $1 \leq i \leq s$, and (iii) $\dim(V_i/V_{i-1}) \leq 2$, $1 \leq i \leq s$. Deduce the existence of ideals $\mathfrak{g}_i$ $(0 \leq i \leq r)$ of $\mathfrak{g}$ such that $\mathfrak{g}_0 = 0 \subseteq \mathfrak{g}_1 \subseteq \cdots \subseteq \mathfrak{g}_r = \mathfrak{g}$ with $\dim(\mathfrak{g}_i/\mathfrak{g}_{i-1}) \leq 2$, $1 \leq i \leq r$.

30. Let $\mathfrak{g}$ be a Lie algebra over k, $\mathfrak{q} = \operatorname{rad}\mathfrak{g}$, $\mathfrak{p} = [\mathfrak{q},\mathfrak{g}]$. Let $G_\mathfrak{p}$ be the group considered in Theorem 3.14.2. Prove that $\mathbf{G}_\mathfrak{p} = \{\exp \operatorname{ad} Z : Z \in \mathfrak{p}.\}$

The next exercise leads to an alternative proof of Weyl's theorem that does not use cohomology (cf. Chevalley [3], pp. 70–73).

31. Let $\mathfrak{g}$ be a semisimple Lie algebra over k, $\mathfrak{G}$ its universal enveloping algebra.

(a) Suppose that ρ is an arbitrary representation of $\mathfrak{G}$. Prove that $tr(\rho(X)) = 0$ for all $X \in \mathfrak{g}$. Deduce that if ρ is one-dimensional, $\rho(X) = 0$ for all $X \in \mathfrak{g}$.

(b) Let ρ be a representation of $\mathfrak{G}$ in W and let W_1 be a subspace of dimension 1 invariant under ρ. Prove the existence of an invariant subspace that is complementary to W_1. (Hint: Use induction on $\dim(W)$. Let $\rho|\mathfrak{g} \neq 0$ and let $R^- = \cap_{n\geq 1}$ range $(\rho(\omega^\rho)^n)$, $N^+ = \cup_{n\geq 1}$ (null space of $\rho(\omega^\rho)^n$). Then R^- and N^+ are invariant, $W = R^- + N^+$ is a direct sum, $R^- \neq 0$, and $W_1 \subseteq N^+$.

(c) Let σ be a representation of $\mathfrak{G}$ in V. Let E be the exterior algebra over V and E_k $(k \geq 0)$ the homogeneous subspaces of E. For $X \in \mathfrak{g}$ let $\tilde{\sigma}(X)$ be the derivation of E that extends $\sigma(X)$. Prove that $\tilde{\sigma}$ is a representation of $\mathfrak{g}$ in E leaving each of the E_k invariant.

(d) Let U be a subspace of V invariant under σ. Let $u = x_1 \wedge \cdots \wedge x_s$ where $\{x_1, \ldots, x_s\}$ is a basis for U. Prove that $\tilde{\sigma}(X)u = 0$ for all $X \in \mathfrak{g}$.

(e) Show that there is a subspace of V complementary to U and invariant under σ. (Hint: Use Exercise 13).

32. Let A be a finitely generated associate algebra over k. Suppose M is a two-sided ideal in A such that $\dim(A/M) < \infty$. Prove that M has a finite ideal basis, i.e., that there are $a_1, \ldots, a_r \in M$ such that $M = \sum_{1 \leq i \leq r} Aa_iA$. Prove also that $\dim(A/M^s) < \infty$ for all $s \geq 1$.

Exercises 33 and 34 lead to Hilbert's theorem on invariants (cf. Cartier, Exposé *nº* 7, Séminaire "Sophus Lie" [1]; also Harish-Chandra [5]).

33. Let $\mathfrak{g}$ be a Lie algebra over k, ρ a semisimple representation of $\mathfrak{g}$ in a vector space V over k. $\mathfrak{S}$ is the symmetric algebra over V and $\mathfrak{S}_n$ $(n \geq 0)$ is the homogeneous subspace of $\mathfrak{S}$ of degree n. For each $X \in \mathfrak{g}$, $\tilde{\rho}(X)$ is the derivation of $\mathfrak{S}$ that extends $\rho(X)$. J is the algebra of all $u \in \mathfrak{S}$ such that $\tilde{\rho}(X)u = 0$ for all $X \in \mathfrak{g}$. $\mathcal{E}$ (resp. $\mathcal{E}^\times$) is the set of all equivalence classes of irreducible (resp. nontrivial irreducible) representations of $\mathfrak{g}$ over k. For each $\mathfrak{d} \in \mathcal{E}$, $\mathfrak{S}_\mathfrak{d}$ is the linear span of all subspaces $U \subseteq \mathfrak{S}$ with the following property: U is invariant under $\tilde{\rho}$ and the corresponding subrepresentation is irreducible and belongs to $\mathfrak{d}$. If $u \in \mathfrak{S}$, we say that u *transforms according to* $\mathfrak{d}$ if $u \in \mathfrak{S}_\mathfrak{d}$.

(a) Prove that for each $n \geq 0$, $\mathfrak{S}_n$ is invariant under $\tilde{\rho}$ and the corresponding subrepresentation of $\mathfrak{g}$ is semisimple.

(b) Prove that $\mathfrak{S} = \sum_{\mathfrak{d} \in \mathcal{E}} \mathfrak{S}_\mathfrak{d}$, the sum being direct. Writing $\mathfrak{S}^\times = \sum_{\mathfrak{d} \in \mathcal{E}^\times} \mathfrak{S}_\mathfrak{d}$, deduce that $\mathfrak{S}^\times$ is the linear span of all elements of the form $\tilde{\rho}(X)u$ $(X \in \mathfrak{g}, u \in \mathfrak{S})$, that $\mathfrak{S}$ is the direct sum of J and $\mathfrak{S}^\times$, and that $\mathfrak{S}^\times$ is the unique $\tilde{\rho}$-invariant subspace of $\mathfrak{S}$ complementary to J.

(c) Show that each $\mathfrak{S}_\mathfrak{d}$ is a J-module.

(d) Let $J^+ = J \cap (\sum_{n>0} \mathfrak{S}_n)$. Prove that there are families $\{u_1, \ldots, u_r\}$ of homogeneous elements in J^+ such that $\mathfrak{S}J^+ = \sum_{1 \leq i \leq r} \mathfrak{S}u_i$. For any such $\{u_1, \ldots, u_r\}$, prove that $J = k[u_1, \ldots, u_r]$. (Hint: Let $u \mapsto \bar{u}$ be the projection $\mathfrak{S} \longrightarrow J$ modulo $\mathfrak{S}^\times$. Show that $\overline{vu} = v\bar{u}$ for $v \in J$, $u \in \mathfrak{S}$. Hence, if $d_i = \deg(u_i)$, $u \in \mathfrak{S}_n$, then $u = \sum_{1 \leq i \leq r} f_i u_i$ with $f_i \in \mathfrak{S}_{n-d_i}$, and $\bar{u} = \sum_{1 \leq i \leq r} \bar{f}_i u_i$. Now use induction on n.)

(e) Fix $\mathfrak{d} \in \mathcal{E}$, and let λ be an irreducible representation of $\mathfrak{g}$ in a vector space U over k such that the class of λ is contragredient to $\mathfrak{d}$. Let $\sigma = \tilde{\rho} \otimes \lambda$. Put

$$W = \{w : w \in \mathfrak{S} \otimes U, \sigma(X)w = 0 \text{ for all } X \in \mathfrak{g}\}.$$

Let $\{e_1, \ldots, e_m\}$ be a basis for U. For any $w = \sum_{1 \leq i \leq m} f_i \otimes e_i$ in $\mathfrak{S} \otimes U$, let $L(w)$ be the linear span of the f_i. Prove that $L(w)$ does not depend on the choice of the basis of U. Prove further that if $w \neq 0$ and w lies in W, then $L(w)$ is irreducibly invariant under $\tilde{\rho}$ and $L(w) \subseteq \mathfrak{S}_\mathfrak{d}$. Prove, finally, that if T is a subspace of $\mathfrak{S}_\mathfrak{d}$ that is irreducibly invariant under $\tilde{\rho}$, there is a $w \in W$ such that $T = L(w)$.

(f) Regard $\mathfrak{S} \otimes U$ as an $\mathfrak{S}$-module by the rule $f \cdot (g \otimes u) = fg \otimes u$ $(f, g \in \mathfrak{S}, u \in U)$. Let $\bar{W}$ be the smallest sub $\mathfrak{S}$-module of $\mathfrak{S} \otimes U$ containing W. Prove that W is a J-module and that there are $w_1, \ldots, w_p \in W$ such that (i) $\bar{W} = \sum_{1 \leq i \leq p} \mathfrak{S} \cdot w_i$, and (ii) for each i, $L(w_i)$ consists entirely of homogeneous elements. (Hint: $\mathfrak{S}$ is Noetherian and $\mathfrak{S} \otimes U$ is a finite $\mathfrak{S}$-module)

(g) Let $\bar{w}_1, \ldots, \bar{w}_q \in W$ be such that $\bar{W} = \sum_{1 \le i \le q} \mathfrak{S} \cdot \bar{w}_i$. Then prove that $W = \sum_{1 \le i \le q} J \cdot \bar{w}_i$. (Hint: Let T be the linear span of all $\sigma(X)w$, with $X \in \mathfrak{g}$, $w \in \mathfrak{S} \otimes U$. Use Exercise 12 to prove that $\mathfrak{S} \otimes U$ is the direct sum of W and T. Now let $w \in W$, and write $w = \sum_{1 \le i \le q} f_i \cdot \bar{w}_i$. Observe that $(f_i - \bar{f}_i) \cdot \bar{w}_i \in T$ for all i, and deduce that $w = \sum_{1 \le i \le q} \bar{f}_i \cdot \bar{w}_i$.)

(h) Prove that there exist homogeneous elements $f_1, \ldots, f_p$ in $\mathfrak{S}_\mathfrak{d}$ such that $\mathfrak{S}_\mathfrak{d} = \sum_{1 \le i \le p} Jf_i$.

34. Let $\mathfrak{g}, \mathfrak{h}$ be Lie algebras over k, ρ a semisimple representation of $\mathfrak{g}$ in $\mathfrak{h}$ such that for each $X \in \mathfrak{g}$, $\rho(X)$ is a derivation of $\mathfrak{h}$. Let $\mathfrak{H}$ be the universal enveloping algebra of $\mathfrak{h}$, and for each $X \in \mathfrak{g}$, let $\tilde{\rho}(X)$ be the derivation of $\mathfrak{H}$ that extends $\rho(X)$. Let $\mathfrak{A} = \{u : u \in \mathfrak{H}, \tilde{\rho}(X)u = 0 \text{ for all } X \in \mathfrak{g}\}$. For each $\mathfrak{d} \in \mathcal{E}$ let $\mathfrak{H}_\mathfrak{d}$ be the linear span of all subspaces U of $\mathfrak{H}$ with the following property: U is invariant under $\tilde{\rho}$, and the corresponding subrepresentation is irreducible and belongs to $\mathfrak{d}$.

(a) Prove that $\mathfrak{A}$ is an algebra and that $\mathfrak{H}$ is the direct sum of the $\mathfrak{H}_\mathfrak{d}$.

(b) Let $\mathfrak{S} = \mathfrak{S}(\mathfrak{h})$, and let J and $\mathfrak{S}_\mathfrak{d}$ be defined as in the previous exercise. Prove that $\boldsymbol{\lambda}$ maps J onto $\mathfrak{A}$ and $\mathfrak{S}_\mathfrak{d}$ onto $\mathfrak{H}_\mathfrak{d}$ for all $\mathfrak{d}$.

(c) Prove that $\mathfrak{A}$ is finitely generated.

(d) Prove that each $\mathfrak{H}_\mathfrak{d}$ is a finite $\mathfrak{A}$-module. (Hint: Choose homogeneous $f_1, \ldots, f_q$ such that $\mathfrak{S}_\mathfrak{d} = \sum_{1 \le i \le q} J \cdot f_i$. Then prove that $\mathfrak{H}_\mathfrak{d} = \sum_{1 \le i \le q} \mathfrak{A} \cdot \boldsymbol{\lambda}(f_i)$.)

35. Let G be an analytic group and A an analytic subgroup. Let $B = Cl(A)$. Prove that $DB = DA$. (Hint: We may assume $B = G$. Let $\tilde{G}$ be the universal covering group of G, $\tilde{A}$ the analytic subgroup of $\tilde{G}$ lying above A. Observe that $\tilde{A}$ is normal and that $D\tilde{G} \subset Cl(D\tilde{A}) = D\tilde{A}$.)

36. Prove that if G is a simply connected solvable analytic group, then G does not have nontrivial compact subgroups.

37. G is a real analytic group and H is a closed simply connected normal analytic subgroup of G. If H is solvable and G/H is compact, prove that there exist a compact analytic subgroup B of G such that $HB = G$, $H \cap B = \{1\}$. (Hint: Use induction on dim (H). First prove the result when H is a vector group, i.e., when $DH = \{1\}$. If $\dim(DH) > 0$, choose a closed analytic subgroup T such that $DH \subseteq T$, $HT = G$, $H \cap T = DH$, and T/DH is compact.)

38. G is a real analytic group and $H \subseteq G$ a closed normal subgroup such that G/H is compact. Suppose ρ is a representation of G such that $\rho|H$ is semisimple. Prove that ρ is semisimple. (Hint: Let V be the space of ρ and $W \neq V$ a subspace invariant under $\rho[G]$. Let L be a projection operator in V vanishing on W and commuting with $\rho[H]$. Let $M = \int_{G/H} (xLx^{-1})\, d\bar{x}$ where $x \mapsto \bar{x}$ is the canonical map of G onto G/H; consider the subspaces $\cup_{n \ge 1}$ (null space of M^n) and $\cap_{n \ge 1}$ range (M^n), and use induction on dim (V).)

39. Let G be a real analytic group with Lie algebra $\mathfrak{g}$.

(a) Prove that $\mathfrak{g}$ is reductive if and only if the adjoint representation of G in $\mathfrak{g}$ is semisimple.

(n) Let $\mathfrak{g}$ be reductive. Prove that a necessary and sufficient condition for every representation of G to be semisimple is that $G/Cl(DG)$ be compact.

40. Let G be a semisimple real or complex analytic group having a faithful representation. Prove that G has finite center. Give an example to show that the converse is not necessarily true.

Exercises 41–45 deal with the structure of real analytic groups possessing faithful linear representations. For a systematic study of these and other structural questions we refer the reader to the book of Hochschild [1] (cf. also Harish-Chandra [3]).

41. Let G be a real analytic group, Q the radical of G, and M a Levi subgroup of G.
(a) Prove that $Q \cap M \subseteq$ center (M).
(b) Write $\iota[m](q) = mqm^{-1}$. Then show that $\gamma((q,m) \mapsto qm)$ is an analytic homomorphism of $Q \times_\iota M$ onto G and that γ is a covering map.
(c) Suppose center (M) is finite. Prove that γ is a closed map. Deduce that M is closed.
(d) Let ζ be a faithful representation of G. If N is any analytic subgroup of G such that $\zeta \mid N$ is a unipotent representation, prove that N is closed and simply connected.
(e) Let G have a faithful representation. Prove that M, $[Q,G]$, and DG are all closed, that $[Q,G]$ is simply connected, and that DG is isomorphic to $[Q,G] \times_\iota M$. (Hint: Observe that $[Q,G]$ is closed and simply connected by (d), and that $[Q,G] \cap M = \{1\}$ by Exercise 36.)
(f) Deduce from (e) that $\pi_1(DG)$ and $\pi_1(M)$ are isomorphic.

42. Let G be a solvable real analytic group. Prove the equivalence of the following statements:
(i) G has a faithful representation.
(ii) DG is closed and simply connected.
(iii) G is a semidirect product of a simply connected solvable group and a compact abelian group.
(Hint: For (ii) $\Longrightarrow$ (iii) write G/DG as $V \times T$ where V is a vector group and T is a torus. Let M be the preimage of $V \times \{1_T\}$ in G. Prove, by considering covering groups, that M is simply connected. Now use Exercise 37.)

43. A real analytic group is said to be *reductive* if it has a faithful representation and all its representations are semisimple. Prove the equivalence of the following statements concerning a real analytic group G:
(i) G is reductive.
(ii) DG is closed in G, center (G) is compact, and DG is a semisimple group with a faithful representation.
(iii) $G = MA$, where M and A are closed subgroups of G, M is a semisimple analytic group with a faithful representation, and A is a compact abelian group centralizing M.

44. Let G be a real analytic group with a faithful representation. Let Q be the radical of G. Denote by $\mathfrak{g}$ the Lie algebra of G and by $\mathfrak{q}$ the subalgebra of $\mathfrak{g}$ defined by Q.
(a) Prove that $Q = PA$, where P is a simply connected closed normal subgroup of G that contains $[Q,G]$, A is a compact abelian subgroup of G, and $P \cap A = \{1\}$. (Hint: $[Q,G]$ is closed and simply connected by Exercise 41. Now argue as in Exercise 42.)

(b) Prove the existence of a Levi subgroup M centralizing A. (Hint: Let $\mathfrak{a}$ be the subalgebra of $\mathfrak{g}$ defined by A. If $\mathfrak{z}$ is the centralizer of $\mathfrak{a}$ in $\mathfrak{g}$, then $\mathfrak{g} = \mathfrak{z} + [\mathfrak{a},\mathfrak{g}]$, the sum being direct. Show that $\mathfrak{g} = \mathfrak{z} + \mathfrak{q}$.)

(c) Let $H = MA$. Prove that H is a closed analytic subgroup of G with $G = PH$ and $P \cap H = \{1\}$.

(d) Prove that H is reductive.

(e) Deduce that an analytic group has a faithful representation if and only if it is a semidirect product of a simply connected solvable group and a reductive group.

45. Let G be a real analytic group and let π be a representation of G in a vector space V. Prove that $[\pi[G],\pi[G]]$ is a closed subgroup of $GL(V)$. Deduce that a reductive analytic subgroup of $GL(V)$ is necessarily closed in $GL(V)$. (Hint: Assume that $G \subseteq GL(V)$ and that π is the identity. Write $\bar{G} = Cl(G)$ and use Exercises 35 and 41. If G is reductive, $G = (DG)\cdot(\text{center}\,(G))$ and center (G) is compact.)

46. Let G be a simply connected solvable analytic group with Lie algebra $\mathfrak{g}$. Let $\mathfrak{z}$ be the center of $\mathfrak{g}$, Z the center of G.

(a) Suppose that $\mathfrak{z} \cap \mathfrak{D}\mathfrak{g} \neq 0$. Then show that there are discrete nontrivial subgroups of $Z \cap DG$ and that if D is any such subgroup, G/D does not have a faithful representation.

(b) Prove that if $\mathfrak{g}$ is nilpotent and nonabelian, $\mathfrak{z} \cap \mathfrak{D}\mathfrak{g} \neq 0$.

(c) Assume that $\mathfrak{z} = 0$. Then prove that any analytic group locally isomorphic to G has a faithful representation. (Hint: Consider $G/Z \approx \text{Ad}\,[G]$ and its covering groups, and use Exercise 42.)

(d) Let $n \geq 2$, $k = \mathbf{R}$ or $\mathbf{C}$, and P the subgroup of $SL(n,k)$ consisting of all matrices $(a_{ij}) \in SL(n,k)$ with $a_{ij} = 0$ for $i > j$. Let G be the universal covering group of P. Prove that Z is discrete and that P is isomorphic to G/Z.

47. The following statements are designed to lead to an example of an analytic group G with a reductive Lie algebra for which $[G,G]$ is not closed in G.

(a) Let G_0 be the universal covering group of $SL(2,\mathbf{R})$, and Z the center of G_0. Prove that Z is isomorphic to the additive group $\mathbf{Z}$ of integers.

(b) Show that there is a discrete subgroup D of $\mathbf{R} \times Z$ such that $(\mathbf{R} \times Z)/D$ is isomorphic to the circle group $\mathbf{T}$ and the image of $\{0\} \times Z$ is dense in $\mathbf{T}$. (Hint: Let c be an irrational number, and let D be the kernel of the homomorphism $(t,v) \mapsto \exp 2i\pi(t + c\bar{v})$ of $\mathbf{R} \times Z$ onto $\mathbf{T}$; here $v \mapsto \bar{v}$ is an isomorphism of Z with $\mathbf{Z}$.)

(c) Let $G = (\mathbf{R} \times G_0)/D$, $C = \text{center}\,(G)$, and let π be the canonical map of $\mathbf{R} \times G_0$ onto G. Prove that $[G,G] = \pi[\{0\} \times G_0]$ and that $C = \pi[\mathbf{R} \times Z]$.

Exercises 48–52 discuss some aspects of complexifications of real analytic groups. The development begun here will be completed in the exercises for Chapter 4. Let G be a real analytic group. A pair (G_c,γ) consisting of a complex analytic group G_c and an $\mathbf{R}$-analytic homomorphism γ of G into G_c will be called a *universal complexification* of G if the following is satisfied: given any complex analytic group H_c and an **R**-analytic homomorphism ν of G into H_c, there is a unique **C**-analytic homomorphism ν_c of G_c into H_c such that $\nu_c \circ \gamma = \nu$.

48. (a) Suppose that G is a real analytic group and that (G_c,γ), (G_c',γ') are two universal complexifications of G. Prove the existence of a unique **C**-analytic isomorphism θ of G_c onto G_c' such that $\theta \circ \gamma = \gamma'$.

(b) Suppose $\mathfrak{g}$ is a Lie algebra over **R** and $\mathfrak{g}_c$ is its complexification. Let G_c be a simply connected complex analytic group with Lie algebra $\mathfrak{g}_c$. Prove that there is an **R**-analytic automorphism ζ of the **R**-analytic group underlying G_c such that $d\zeta$ is the conjugation of $\mathfrak{g}_c$ induced by the real form $\mathfrak{g}$. If $\bar{G}$ is the component of 1 in the subgroup of G_c consisting of all points left fixed by ζ, prove that $\bar{G}$ is closed in G_c and is the **R**-analytic subgroup defined by $\mathfrak{g}$.

(c) Let G be a covering group of $\bar{G}$ and let π be the covering homomorphism. Prove that (G_c,π) is a universal complexification of G.

49. Let G be a real analytic group with Lie algebra $\mathfrak{g}$, $\tilde{G}$ the universal covering group of G, and π $(\tilde{G} \longrightarrow G)$ the covering map. We assume that the Lie algebra of $\tilde{G}$ is identified with $\mathfrak{g}$ in such a way that $d\pi$ is the identity. Let F be the kernel of π. Let $\mathfrak{g}_c$ be the complexification of $\mathfrak{g}$, and $\tilde{G}_c$ a simply connected complex analytic group with Lie algebra $\mathfrak{g}_c$.

(a) Prove that there is a unique **R**-analytic homomorphism σ of $\tilde{G}$ into $\tilde{G}_c$ such that $d\sigma$ is the natural inclusion map of $\mathfrak{g}$ into $\mathfrak{g}_c$.

(b) Suppose H_c is a complex analytic group and ν is an **R**-analytic homomorphism of G into H_c. Prove that there is a unique **C**-analytic homomorphism $\tilde{\nu}_c$ of $\tilde{G}_c$ into H_c such that $\nu \circ \pi = \tilde{\nu}_c \circ \sigma$. Deduce that $\sigma[F] \subseteq$ kernel $(\tilde{\nu}_c)$.

(c) Let P be the intersection of the kernels of $\tilde{\nu}_c$ as H_c and ν vary. Prove that P is a closed normal complex Lie subgroup of $\tilde{G}_c$ containing $\sigma[F]$.

(d) Let ζ be the **R**-analytic automorphism of $\tilde{G}_c$ such that $d\zeta$ is the conjugation of $\mathfrak{g}_c$ induced by $\mathfrak{g}$ (cf. Exercise 48). Prove that $\zeta[P] = P$. (Hint: With notation as in (b), let $\bar{H}_c$ be the complex analytic group opposite to H_c. Prove that $\tilde{\nu}_c \circ \zeta$ is the **C**-analytic homomorphism of $\tilde{G}_c$ into $\bar{H}_c$ such that $\tilde{\nu} \circ \zeta = (\tilde{\nu}_c \circ \pi) \circ \sigma$. Observe now that kernel $(\tilde{\nu}_c \circ \zeta) = \zeta[\text{kernel }(\tilde{\nu}_c)]$.)

(e) Let $G_c = \tilde{G}_c/P$ and let η be the natural map $\tilde{G}_c \longrightarrow G_c$. Define γ $(G \longrightarrow G_c)$ by $\gamma(\pi(x)) = \eta(\sigma(x))$ $(x \in \tilde{G})$. Prove that (G_c,γ) is a universal complexification of G.

(f) Let $\hat{\mathfrak{g}}_c$ be the Lie algebra of G_c and let $\hat{\mathfrak{g}} = d\eta[\mathfrak{g}]$. Prove that $\hat{\mathfrak{g}}$ is a real form of $\hat{\mathfrak{g}}_c$ and that $d\eta$ intertwines the conjugations of $\mathfrak{g}_c$ and $\hat{\mathfrak{g}}_c$. Prove further that there is a unique **R**-analytic automorphism $\hat{\zeta}$ of the underlying **R**-analytic group of G_c such that $d\hat{\zeta}$ is the conjugation of $\hat{\mathfrak{g}}_c$ induced by $\hat{\mathfrak{g}}$, and that η intertwines ζ and $\hat{\zeta}$. Prove, finally, that $\gamma[G]$ is the **R**-analytic subgroup of G_c defined by $\hat{\mathfrak{g}}$ and coincides with the component of the identity of the set of fixed points of $\hat{\zeta}$.

50. Let notation be as in Exercise 49. Prove that the following conditions are equivalent: (i) $\sigma[F]$ is a discrete subgroup of $\tilde{G}_c$, (ii) $P = \sigma[F]$, and (iii) $\dim_{\mathbf{C}}(G_c) = \dim_{\mathbf{R}}(G)$. Prove also that this happens when G has a faithful representation and that in this case γ is an **R**-analytic isomorphism of G onto $\gamma[G]$. If conditions (i)–(iii) are satisfied, (G_c,γ) is called a *regular* universal complexification.

51. Let G be a solvable real analytic group, and let (G_c,γ) be its universal complexification. Prove that (G_c,γ) is regular, that $\gamma[G]$ is closed in G_c, and that γ is an **R**-analytic isomorphism of G onto $\gamma[G]$. Prove also that G_c is simply connected if and only if G is. (Hint: Use Exercise 49 and observe that $\sigma[\tilde{G}]$ is closed and simply connected by the results of §3.18; σ is thus injective. Observe now that $G_c = \tilde{G}_c/\sigma[F]$.)

52. Let G be a solvable real analytic group and (G_c,γ) its universal complexification. Prove that if G has a faithful representation, then G_c has a faithful complex analytic representation. (Hint: Use Exercise 42 to write G as a semidirect product $A \times_\eta B$, where A is simply connected and solvable and B is compact abelian. Let (A_c,α) and (B_c,β) be the universal complexifications of A and B. Then prove that $G_c = A_c \times_{\eta_c} B_c$ for a suitably chosen η_c. Prove, finally, that B_c is isomorphic to $\mathbf{C}^{*n}$ for some n and hence that B_c has a faithful complex analytic representation. Now use the results of §3.18.)

53. G is a real analytic group with Lie algebra $\mathfrak{g}$. Let $\mathfrak{G}$ be the universal enveloping algebra of the complexification $\mathfrak{g}_c$ of $\mathfrak{g}$. $\mathfrak{Z}$ is the center of $\mathfrak{G}$.

(a) Let π $(x \mapsto \pi(x))$ be an irreducible representation of G in a complex vector space V of finite dimension. Let $\mathfrak{M}_\pi$ be the linear space spanned by the functions on G of the form $x \mapsto v^*(\pi(x)v)$ $(v \in V, v^* \in V^*)$. Prove that the elements of $\mathfrak{M}_\pi$ are analytic and that if π is irreducible, $\dim \mathfrak{M}_\pi = \dim(V)^2$.

(b) Let π be as above, $[v_1, \ldots, v_d]$ be a basis for V, and the functions a_{ij} be defined on G by $\pi(x)v_j = \sum_{1 \le i \le d} a_{ij}(x)v_i$ $(x \in G)$. Let r $(x \mapsto r(x))$ denote the representation of G in the space of all analytic functions on G given by $(r(x)f)(y) = f(yx)$ (f analytic, $x, y \in G$). Prove that $\mathfrak{M}_\pi$ is invariant under r, the subspaces $\mathfrak{R}_i$ spanned by a_{ij} $(1 \le j \le d)$ are invariant under r for $1 \le i \le d$, and the representation induced on $\mathfrak{R}_i$ by r is equivalent to π.

(c) Let s be an integer ≥ 1, V^s the Cartesian product $V \times \cdots \times V$ (s times), and π^s, the representation $x \mapsto \pi^s(x)$ of G in V^s where $\pi^s(x)(v_1, \ldots, v_s) = (\pi(x)v_1, \ldots, \pi(x)v_s)$ $((v_1, \ldots, v_s) \in V^s)$. If $(u_1, \ldots, u_s) \in V^s$, then it is cyclic for π^s (i.e., $\pi^s(x)(u_1, \ldots, u_s)$ span V as x runs through G) if and only if $u_1, \ldots, u_s$ are linearly independent in V.

(d) Let $\psi_\pi(x) = tr\, \pi(x)$ $(x \in G)$. Prove that $\psi_\pi \in \mathfrak{M}_\pi$ and that it is cyclic for the representation r on $\mathfrak{M}_\pi$.

(e) Let $1 \le i \le d$ and let ζ_i be an isomorphism of $\mathfrak{R}_i$ with V that intertwines the restriction of r to $\mathfrak{R}_i$ and π. Let r and π denote the respective associated representations of $\mathfrak{G}$ in $\mathfrak{M}_\pi$ and V. Prove that $r(a)f = af$ $(f \in \mathfrak{M}_\pi, a \in \mathfrak{G})$ and that $\zeta_i(r(a)f) = \pi(a)\zeta_i(f)$ $(a \in \mathfrak{G}, f \in \mathfrak{R}_i)$. Hence deduce that $zf = \chi_\pi(z)f$ for $f \in \mathfrak{M}_\pi$, $z \in \mathfrak{Z}$, χ_π $(\mathfrak{Z} \to \mathbf{C})$ being the infinitesimal character of π.

(f) Prove that ψ_π is the unique element of $\mathfrak{M}_\pi$ that is invariant under all the inner automorphisms of G and takes the value $\dim(V)$ at 1.

54. Let the general notation be as in the previous exercise. Let $G = SU(2,\mathbf{C})$, and identify $\mathfrak{g}$ canonically with the Lie algebra of all 2×2 skew Hermitian matrices of trace 0. A function on G is said to be invariant if it is invariant under all the inner automorphisms of G.

(a) Prove that $\mathfrak{g}$ is semisimple. Let $A = \begin{pmatrix} 0 & \frac{1}{2} \\ -\frac{1}{2} & 0 \end{pmatrix}$, $B = \begin{pmatrix} 0 & i/2 \\ i/2 & 0 \end{pmatrix}$, and $C = \begin{pmatrix} i/2 & 0 \\ 0 & -i/2 \end{pmatrix}$. Prove that $\omega = A^2 + B^2 + C^2$ generates $\mathfrak{Z}$.

(b) Write any element $x \in G$ in the form $x = \begin{pmatrix} a + ib & c + id \\ -c + id & a - ib \end{pmatrix}$, where $a, b, c, d \in \mathbf{R}$ and $a^2 + b^2 + c^2 + d^2 = 1$. Let G^+ be the set of all x where $b \neq 0$. Prove that a, c, and d are coordinates on G^+. Determine the expression for ω in these coordinates. Prove also that if φ is any invariant analytic function on G, $\partial/\partial c\ \varphi = \partial/\partial d\ \varphi = 0$ on G^+.

(c) Let T be the diagonal subgroup of G, $T^+ = T \cap G^+$. For any analytic function φ on G let $\tilde{\varphi}$ be $\varphi \mid T^+$. Prove that for any invariant analytic function φ on G, $(\omega\varphi)^{\sim} = \tilde{\omega}\tilde{\varphi}$, where $\tilde{\omega}$ is the differential operator on T^+ whose expression in the coordinate a is $\frac{1}{4}(1 - a^2)(\partial^2/\partial a^2) - \frac{3}{4}a(\partial/\partial a)$.

(d) For $\theta \in \mathbf{R}$ let $u_\theta = \begin{pmatrix} e^{i\theta} & 0 \\ 0 & e^{-i\theta} \end{pmatrix}$, and let $(d/d\theta)$ be the usual differential operator on T. Let π be an irreducible representation of G with infinitesimal character χ_π and global character χ_π. Let $f_\pi = \psi_\pi \mid T$, $\lambda = \chi_\pi(\omega)$. Prove that f_π is a finite Fourier series with integral coefficients, invariant under the involution $u_\theta \mapsto u_{-\theta}$ of T. Prove further, using (e) of the previous exercise and (c) above, that the function $g_\pi \colon u_\theta \mapsto f_\pi(u_\theta) \sin \theta$ satisfies the differential equation $(d^2/d\theta^2)g_\pi = (4\lambda - 1)g_\pi$.

(e) Deduce from (d) that, for some integer $m \geq 1$, $4\lambda - 1 = -m^2$ and $f_\pi(u_\theta) = c_\pi(e^{im\theta} - e^{-im\theta})/(e^{i\theta} - e^{-i\theta})$ $(u_\theta \in T)$, c_π being a nonzero constant.

(f) Let $V = \mathbf{C}^2$, $\pi = \pi^{(1)}$ the representation $x \mapsto x$ of G in V, $V^{(r)} = V \otimes \cdots \otimes V$ (r factors), and $\pi^{(r)}$ the representation $x \mapsto x \otimes \cdots \otimes x$ of G in $V^{(r)}$. Let W_r denote the subspace of all symmetric tensors of $V^{(r)}$. Prove that W_r is invariant under $\pi^{(r)}$. Let σ_r denote the representation of G induced on W_r, ψ_r the global character of σ_r. Prove that $\psi_r(u_\theta) = (e^{i(r+1)\theta} - e^{-i(r+1)\theta})/(e^{i\theta} - e^{-i\theta})$ $(u_\theta \in T)$. Deduce from (e) that the σ_r $(r \geq 1)$, together with the trivial representation of G, exhaust the irreducible representations of G up to equivalence, and that the constant c_π of (e) is 1.

55. Let G be an analytic group, $\mathfrak{g}$ its Lie algebra, and $\{X_1, \ldots, X_m\}$ a basis for $\mathfrak{g}$. Assume that $\mathfrak{g}$ is solvable and that for any i, $1 \leq i \leq m$, the linear span of $X_i, \ldots, X_m$ is an ideal in $\mathfrak{g}$. Let $k = \mathbf{R}$ or $\mathbf{C}$ according as G is real or complex. Let $\omega_1, \ldots, \omega_m$ be the left invariant 1-forms on G such that $\omega_i(X_j) = \delta_{ij}$, $1 \leq i, j \leq m$, and let $\bar{\omega}_i$ be the 1-form on k^m that corresponds to ω_i under the map $(t_1, \ldots, t_m) \mapsto \exp t_1X_1 \cdots \exp t_mX_m$. Prove that $\bar{\omega}_1 = dt_1$ and that for $r > 1$, $\bar{\omega}_r = dt_r + \sum_{1 \leq s \leq r} A_{sr}\, dt_s$, where $A_{sr}(t_1, \ldots, t_m) = B_{sr}(t_{s+1}, \ldots, t_{r-1}) + t_r C_{sr}(t_{s+1}, \ldots, t_{r-1})$, B_{sr} and C_{sr} being analytic on k^m.

CHAPTER 4

COMPLEX SEMISIMPLE LIE ALGEBRAS AND LIE GROUPS: STRUCTURE AND REPRESENTATIONS

From now on we shall be concerned almost exclusively with semisimple Lie algebras and their representations. In this chapter we develop the structure theory of semisimple Lie algebras in full detail. This will then be followed by a treatment of the finite-dimensional representations of semisimple groups, both from the infinitesimal and global points of view.

Much of the algebraic theory is valid in an arbitrary field of characteristic 0. However, we work exclusively in the real or complex case. In view of the interplay between the global and infinitesimal aspects of the theory, this restriction is a natural one, if one does not want to get involved with the theory of algebraic groups.

In order that the exposition not be interrupted, we have discussed in an appendix to this chapter certain results in the theory of finite linear groups generated by reflections that are very useful in the theory of semisimple Lie groups and Lie algebras.

Throughout this chapter, k is either **R** *or* **C**. If V is a vector space over **R**, we write V_c for its complexification and regard V as canonically imbedded in V_c. As usual, all Lie algebras are of finite dimension, unless we state otherwise.

4.1. Cartan Subalgebras

Let $\mathfrak{g}$ be a Lie algebra over k of dimension m. Let T be an indeterminate, and for $X \in \mathfrak{g}$ let (cf. (3.9.1))

$$\det(T\cdot 1 - \operatorname{ad} X) = \sum_{0\leq i\leq m} (-1)^{m-i} p_i(X) T^i. \tag{4.1.1}$$

$p_m \equiv 1$, and the p_i are polynomial functions on $\mathfrak{g}$; since $\det \operatorname{ad} X \equiv 0$, $p_0 \equiv 0$. We denote by l the smallest integer $r \geq 0$ such that $p_r \not\equiv 0$, and call it the *rank of* $\mathfrak{g}$ ($\operatorname{rk}(\mathfrak{g})$). Clearly, $1 \leq l \leq m$, and $l = m$ if and only if $\mathfrak{g}$ is nilpotent. It is obvious that when $k = \mathbf{R}$, the rank of $\mathfrak{g}$ is the same as the rank of $\mathfrak{g}_c$.

We write

$$\eta = p_l. \tag{4.1.2}$$

An element $X \in \mathfrak{g}$ is said to be *regular* or *singular* according as $\eta(X) \neq 0$ or $\eta(X) = 0$. We put

$$\mathfrak{g}' = \{X : X \in \mathfrak{g},\ X \text{ regular}\}. \tag{4.1.3}$$

In view of (3.9.4), it is clear that if $k = \mathbf{R}$,

$$\mathfrak{g}' = (\mathfrak{g}_c)' \cap \mathfrak{g}. \tag{4.1.4}$$

Since the p_i are invariant under the group of all automorphisms of $\mathfrak{g}$ (cf. (3.9.3)), it follows that $\mathfrak{g}'$ is an open set that is invariant under the adjoint group of $\mathfrak{g}$ (i.e., the analytic subgroup of $GL(\mathfrak{g})$ defined by ad $\mathfrak{g} \subseteq \mathfrak{gl}(\mathfrak{g})$).

For any $X \in \mathfrak{g}$, let

(4.1.5) $\nu(X)$ = multiplicity of the root 0 of the characteristic equation of ad X.

Then $\nu(X)$ is the smallest integer $r \geq 0$ such that $p_r(X) \neq 0$. So $\nu(X) \geq \text{rk}(\mathfrak{g})$ for all $X \in \mathfrak{g}$, and $\nu(X) = \text{rk}(\mathfrak{g})$ is and only if X is regular.

To get an idea of what is meant by regularity, we consider an example. Let l be an integer ≥ 1. V a complex vector space of dimension $l + 1$, and $\mathfrak{g} = \mathfrak{sl}(V)$. For any $X \in \mathfrak{g}$, let $\tilde{X}$ be the endomorphism of the dual V^* of V given by $(\tilde{X}v^*)(v) = -v^*(Xv)$ $(v \in V, v^* \in V^*)$. If $\lambda_1, \ldots, \lambda_r$ are the distinct eigenvalues of X and ν_j is the multiplicity of λ_j as a root of the characteristic equation of $X (1 \leq j \leq r)$, it is easily seen that 0 is a root of the characteristic equation of $X_{1,1} = X \otimes 1 + 1 \otimes \tilde{X}$, of multiplicity $\nu_1^2 + \cdots + \nu_r^2$. Now $\mathfrak{gl}(V)$ is the direct sum of $\mathfrak{g}$ and $\mathbf{C} \cdot 1$, and $[X, 1] = 0$; conseqnently, we may conclude from Lemma 3.1.10 that $\nu(X) = \nu_1^2 + \cdots + \nu_r^2 - 1$. Since $\sum_{1 \leq i \leq r} \nu_i = l + 1$, it follows that $\nu(X) \geq l$, and that $\nu(X) = l$ if and only if $\nu_1 = \cdots = \nu_r = 1$, i.e., X has $l + 1$ distinct eigenvalues. If $\lambda_1, \ldots, \lambda_{l+1}$ are these eigenvalues, an easy calculation shows that

$$\eta(X) = \prod_{i \neq j} (\lambda_i - \lambda_j). \tag{4.1.6}$$

For arbitrary $\mathfrak{g}$, Whitney's theorem [1] implies that $\mathfrak{g}'$ has finitely many connected components if $k = \mathbf{R}$; if $k = \mathbf{C}$, $\mathfrak{g}'$ is actually connected.[1]

[1]The fact that $\mathfrak{g}'$ is connected if $k = \mathbf{C}$ is immediate from the following result: if V is a vector space of finite dimension over $\mathbf{C}$, and if f is a polynomial on V, then $V' = \{v : v \in V, f(v) \neq 0\}$ is connected. To prove this it is enough to assume $f \not\equiv 0$ and show that given $v, v' \in V'$, there is a connected set $W_{v,v'} \subseteq V'$ containing both v and v'. Let $g(t) = f(tv + (1 - t)v')$ $(t \in \mathbf{C};\ v, v' \in V', \text{fixed})$. Then g is a polynomial and $g \not\equiv 0$. Let Z be the set of zeros of g and $W_{v,v'} = \{tv + (1 - t)v' : t \in \mathbf{C} \backslash Z\}$. Then $W_{v,v'}$, being the continuous image of the complement in $\mathbf{C}$ of a finite set, is connected; and $v, v' \in W_{v,v'}$.

Let $\mathfrak{h}$ be a subalgebra of $\mathfrak{g}$. By the *normalizer* of $\mathfrak{h}$ in $\mathfrak{g}$ we mean the set of all $X \in \mathfrak{g}$ such that $[X,\mathfrak{h}] \subseteq \mathfrak{h}$; it is a subalgebra of $\mathfrak{g}$ that contains $\mathfrak{h}$ as an ideal in it. $\mathfrak{h}$ is said to be a *Cartan subalgebra* (*CSA*) of $\mathfrak{g}$ if (i) $\mathfrak{h}$ is nilpotent, and (ii) $\mathfrak{h}$ is its own normalizer in $\mathfrak{g}$. These subalgebras play a fundamental role in the theory of semisimple Lie algebras.

Lemma 4.1.1. *Let $\mathfrak{g}$ be a Lie algebra over k, $\mathfrak{h}$ a CSA. Then: (i) $\mathfrak{h}$ is maximal nilpotent. (ii) If $\mathfrak{z}$ is any subalgebra of $\mathfrak{g}$ such that $\mathfrak{h} \subseteq \mathfrak{z} \subseteq \mathfrak{g}$, then $\mathfrak{h}$ is a CSA of $\mathfrak{z}$. (iii) If $k = \mathbf{R}$, $\mathfrak{h}_c$ is a CSA of $\mathfrak{g}_c$.* (iv) *Let*

$$\zeta_{\mathfrak{h}}(X) = \det(\operatorname{ad} X)_{\mathfrak{g}/\mathfrak{h}} \quad (X \in \mathfrak{h}); \tag{4.1.7}$$

then $\zeta_{\mathfrak{h}}$ is a polynomial function $\not\equiv 0$ on $\mathfrak{h}$ that does not vanish identically.

Proof. (i) Let $\mathfrak{h}$ be a CSA of $\mathfrak{g}$, $\mathfrak{n} \neq \mathfrak{h}$ a nilpotent subalgebra containing $\mathfrak{h}$. By Engel's theorem (cf. §3.5) applied to the nil representation $H \mapsto (\operatorname{ad} H)_{\mathfrak{n}/\mathfrak{h}}$ $(H \in \mathfrak{n})$ of $\mathfrak{h}$ in $\mathfrak{n}/\mathfrak{h}$, we see that there is an $X \in \mathfrak{n}$, $X \notin \mathfrak{h}$, such that $[X,\mathfrak{h}] \subseteq \mathfrak{h}$; this is a contradiction. (ii) is obvious. If $\mathfrak{b}$ is any subalgebra of $\mathfrak{g}$ and $\mathfrak{m}$ is its normalizer in $\mathfrak{g}$, then in the case $k = \mathbf{R}$, it is obvious that $\mathfrak{m}_c$ is the normalizer of $\mathfrak{b}_c$ in $\mathfrak{g}_c$. This proves (iii). We come to (iv). In view of (iii) we may assume that $k = \mathbf{C}$. Then ρ $(Y \mapsto (\operatorname{ad} Y)_{\mathfrak{g}/\mathfrak{h}})$ is a representation of $\mathfrak{h}$ in $\mathfrak{g}/\mathfrak{h}$. If $\zeta_{\mathfrak{h}} \equiv 0$, $\det \rho(Y) = 0$ for all $Y \in \mathfrak{h}$. From Theorem 3.5.8 we can conclude that 0 must be a weight of ρ. So the subrepresentation of ρ defined by the weight subspace corresponding to the weight 0 is a nil-representation. So by Engel's theorem we conclude that there is an $X \notin \mathfrak{h}$ such that $[X,\mathfrak{h}] \subseteq \mathfrak{h}$, a contradiction.

We shall now describe a method of constructing CSA's.

Theorem 4.1.2. *Let $\mathfrak{g}$ be a Lie algebra over k. For $X \in \mathfrak{g}$, let*

$$\mathfrak{h}_X = \{Y : Y \in \mathfrak{g}, (\operatorname{ad} X)^s(Y) = 0 \text{ for some integer } s \geq 1\}. \tag{4.1.8}$$

Then for any regular X, $\mathfrak{h}_X$ is a CSA, and $\dim \mathfrak{h}_X = rk(\mathfrak{g})$. In this case, if $\zeta_{\mathfrak{h}_X}$ is defined by (4.1.7) *and η is given by* (4.1.2),

$$\eta \mid \mathfrak{h}_X = \zeta_{\mathfrak{h}_X}. \tag{4.1.9}$$

In particular, $Y \in \mathfrak{h}_X$ is regular if and only if $\zeta_{\mathfrak{h}_X}(Y) \neq 0$.

Proof. Let $X \in \mathfrak{g}$ be regular and let us write $\mathfrak{h}$ for $\mathfrak{h}_X$. If $Y \in \mathfrak{h}$, $(\operatorname{ad} X)^s$ $(Y) = \underbrace{[X,[X,\ldots,[X,}_{s \text{ terms}} Y]] \cdots] = 0$ for some $s \geq 1$. So

$$\underbrace{[\operatorname{ad} X,[\operatorname{ad} X,[\cdots[\operatorname{ad}}_{s \text{ terms}} X, \operatorname{ad} Y]\cdots] = 0$$

Theorem 3.1.5 then shows that ad Y leaves $\mathfrak{h}$ invariant. $\mathfrak{h}$ is thus a subalgebra.

We assert that $\mathfrak{h}$ is nilpotent. Let $\zeta = \zeta_{\mathfrak{h}_X}$ be defined by (4.1.7), so that

$\zeta(Y) = \det(\text{ad } Y)_{\mathfrak{g}/\mathfrak{h}}$, and let $\mathfrak{h}' = \{Y: Y \in \mathfrak{h}, \zeta(Y) \neq 0\}$. By Theorem 3.1.5, $\mathfrak{g}$ is the direct sum of $\mathfrak{h}$ and $\mathfrak{m}$, where $\mathfrak{m} = \bigcap_{s \geq 1} \text{range}((\text{ad } X)^s)$, and for any $Y \in \mathfrak{h}$, ad Y leaves both $\mathfrak{h}$ and $\mathfrak{m}$ invariant. If $Y \in \mathfrak{h}'$, $(\text{ad } Y)|\mathfrak{m}$ is invertible, so $\mathfrak{h}_Y \subseteq \mathfrak{h}$. On the other hand, since X is regular, $\dim(\mathfrak{h}) = \text{rk}(\mathfrak{g}) \leq \nu(Y) = \dim(\mathfrak{h}_Y)$ (cf. (4.1.5)), so $\mathfrak{h}_Y = \mathfrak{h}$. In other words, $(\text{ad } Y)|\mathfrak{h}$ is nilpotent for $Y \in \mathfrak{h}'$. Now $\zeta(X) \neq 0$, so $\zeta \not\equiv 0$. So $\mathfrak{h}'$ is dense in $\mathfrak{h}$. This implies that $(\text{ad } Y)|\mathfrak{h}$ is nilpotent for all $Y \in \mathfrak{h}$.

Let $\mathfrak{n}$ be the normalizer of $\mathfrak{h}$ in $\mathfrak{g}$. If $X' \in \mathfrak{n}$, then $[X,X'] \in \mathfrak{h}$, so $X' \in \mathfrak{h}$ too. So $\mathfrak{n} \subseteq \mathfrak{h}$. This completes the proof that $\mathfrak{h}$ is a CSA. Note that $\dim(\mathfrak{h}) = \text{rk}(\mathfrak{g})(= l$, say).

It remains to prove (4.1.9). For $Y \in \mathfrak{h}$, $(\text{ad } Y)|\mathfrak{h}$ is nilpotent, so there is a basis for $\mathfrak{h}$ in which the matrix of $(\text{ad } Y)|\mathfrak{h}$ has zeros on and below the main diagonal. Combining this with a basis for $\mathfrak{m}$, we obtain a basis for $\mathfrak{g}$. If we calculate the determinants in this basis, we easily obtain (4.1.9) from (4.1.1).

We now prove Chevalley's theorem on the conjugacy of the Cartan subalgebras of a complex Lie algebra under the adjoint group (Chevalley [3]).

Theorem 4.1.3. *Let $\mathfrak{g}$ be a Lie algebra over k. Then any CSA of $\mathfrak{g}$ is of the form $\mathfrak{h}_X$ for some regular element X. There are finitely many CSA's $\mathfrak{h}_1, \ldots, \mathfrak{h}_r$ such that any CSA of $\mathfrak{g}$ is conjugate to one of the $\mathfrak{h}_i$ through an element of the adjoint group G of $\mathfrak{g}$. If $k = \mathbf{C}$, all CSA's are mutually conjugate under G.*

Proof. We note that if $\mathfrak{h}$ is a CSA and $X \in \mathfrak{h}$ is a regular element, then $\mathfrak{h} = \mathfrak{h}_X$. In fact, since $\mathfrak{h}$ is nilpotent, $\mathfrak{h} \subseteq \mathfrak{h}_X$; since $\mathfrak{h}_X$ is nilpotent and $\mathfrak{h}$ is maximal nilpotent, $\mathfrak{h} = \mathfrak{h}_X$. In particular, if two CSA's contain a regular element in common, they must coincide.

This said, we take up the proof of the theorem. Let $\mathfrak{h}$ be a CSA of $\mathfrak{g}$ and $\zeta = \zeta_{\mathfrak{h}}$ be defined by (4.1.7). Let

$$\mathfrak{h}' = \{Y : Y \in \mathfrak{h},\ \zeta(Y) \neq 0\}. \tag{4.1.10}$$

Then, by Lemma 4.1.1, $\mathfrak{h}'$ is a dense open subset of $\mathfrak{h}$. We now introduce the analytic map

$$\psi : (x,X) \mapsto X^x \tag{4.1.11}$$

of $G \times \mathfrak{h}$ into $\mathfrak{g}$, and calculate its differential. We canonically identify the tangent spaces to G, $\mathfrak{g}$, and $\mathfrak{h}$ at each of their points with $\mathfrak{g}$, $\mathfrak{g}$, and $\mathfrak{h}$ respectively. Then, for $Y \in \mathfrak{g}$, $H \in \mathfrak{h}$,

$$\begin{aligned}(d\psi)_{(x,X)}(Y,H) &= (d\psi)_{(x,X)}(Y,0) + (d\psi)_{(x,X)}(0,H)\\ &= \left(\frac{d}{dt}X^{x \exp tY}\right)_{t=0} + \left(\frac{d}{dt}(X + tH)^x\right)_{t=0}\\ &= [Y,X]^x + H^x.\end{aligned}$$

Let

(4.1.12) $$L_X(Y,H) = [Y,X] + H \quad (Y \in \mathfrak{g},\ H \in \mathfrak{h}).$$

Then

(4.1.13) $$(d\psi)_{(x,X)} = \mathrm{Ad}(x) \circ L_X \quad (x \in G,\ X \in \mathfrak{h}).$$

In particular, for $x \in G$ and $X \in \mathfrak{h}$,

(4.1.14) $$\text{range}\,((d\psi)_{(x,X)}) = (\mathfrak{h} + \text{range}\,(\mathrm{ad}\,X))^x.$$

Now, for any $X \in \mathfrak{h}$, $\mathfrak{h} + \text{range}\,(\mathrm{ad}\,X) = \mathfrak{g}$ if and only if ad X induces an invertible endomorphism of $\mathfrak{g}/\mathfrak{h}$, i.e., $\zeta(X) \neq 0$. So the relation (4.1.14) shows that $(d\psi)_{(x,X)}$ is surjective if and only if $X \in \mathfrak{h}'$.

It follows from this last assertion that $(\mathfrak{h}')^G = \psi[G \times \mathfrak{h}']$ is an open subset of $\mathfrak{g}$ and that ψ is a submersion of $\mathfrak{g} \times \mathfrak{h}'$ onto $(\mathfrak{h}')^G$. Since the set $\mathfrak{g}'$ is dense in $\mathfrak{g}$, it follows that $(\mathfrak{h}')^G \cap \mathfrak{g}' \neq \varnothing$. Since $\mathfrak{g}'$ is invariant, we must have $\mathfrak{h}' \cap \mathfrak{g}' \neq \varnothing$. So $\mathfrak{h}$ contains a regular element, say X. Then $\mathfrak{h} = \mathfrak{h}_X$, as we had observed at the outset.

It remains to examine the conjugacy question. Let $\mathfrak{g}_1, \ldots, \mathfrak{g}_r$ be the connected components of $\mathfrak{g}'$, and let $X_i \in \mathfrak{g}_i$ be arbitrary, $1 \leq i \leq r$. Write $\mathfrak{h}_i = \mathfrak{h}_{X_i}$ $(1 \leq i \leq r)$. Since G is connected, it is clear that the $\mathfrak{g}_i$ are invariant under G. Fix i, $1 \leq i \leq r$. For any $X \in \mathfrak{g}_i$, write $\mathfrak{h}'_X = \mathfrak{h}_X \cap \mathfrak{g}'$, and let $\mathfrak{h}^+_X$ be the connected component of $\mathfrak{h}'_X$ that contains X; let $\mathfrak{v}_X = (\mathfrak{h}^+_X)^G$. Then $\mathfrak{h}^+_X$ is open in $\mathfrak{h}'_X$, and it follows from the submersive nature of the map $(y,Y) \mapsto Y^y$ of $G \times \mathfrak{h}'_X$ onto $(\mathfrak{h}'_X)^G$ that $\mathfrak{v}_X$ is a connected open subset of $\mathfrak{g}'$. Thus $\mathfrak{v}_X \subseteq \mathfrak{g}_i$. If $Z \in \mathfrak{h}^+_X$, it is clear that $\mathfrak{h}^+_Z = \mathfrak{h}^+_X$, so $\mathfrak{v}_Z = \mathfrak{v}_X$. If X, $Y \in \mathfrak{g}_i$ and $\mathfrak{v}_X \cap \mathfrak{v}_Y \neq \varnothing$, it is obvious that Y is conjugate to some element Z in $\mathfrak{h}^+_X$, so $\mathfrak{v}_Y = \mathfrak{v}_Z = \mathfrak{v}_X$. In other words, two members of the family $\{\mathfrak{v}_X \colon X \in \mathfrak{g}_i\}$ are either identical or disjoint. Since they are all open and $\mathfrak{g}_i$ is connected, all of them must coincide with $\mathfrak{g}_i$. Thus $\mathfrak{g}_i = \mathfrak{v}_{X_i}$, $1 \leq i \leq r$. Suppose $\mathfrak{h}$ is a CSA, X a regular point such that $\mathfrak{h} = \mathfrak{h}_X$. Then $X \in \mathfrak{g}_i$ for some i, so $X = Z^y$ for some $Z \in \mathfrak{h}^+_{X_i}$, $y \in G$ $(1 \leq i \leq r)$. This implies that $\mathfrak{h}_i^y = \mathfrak{h}$. Finally, if $k = \mathbf{C}$, $\mathfrak{g}'$ is connected. So $r = 1$ in the above discussion, and any CSA of $\mathfrak{g}$ is conjugate, via G, to $\mathfrak{h}_1$. This proves the theorem.

Corollary 4.1.4. *The dimension of any CSA is the rank of* $\mathfrak{g}$. *If* $\mathfrak{h}$ *is any CSA*

$$\eta(Y) = \det\,(\mathrm{ad}\,Y)_{\mathfrak{g}/\mathfrak{h}} \quad (Y \in \mathfrak{h}).$$

We give two examples.

(1) Let l be an integer ≥ 1, V a vector space of dimension $l + 1$ over $\mathbf{C}$, and $\mathfrak{g} = \mathfrak{sl}(V)$. Let $X \in \mathfrak{g}$ be regular. We have then seen that X has $l + 1$ distinct eigenvalues. So there is a basis $\{v_i \colon 1 \leq i \leq l + 1\}$ of V in which the matrix of X is diagonal. X is clearly semisimple, so ad X is semisimple. Thus

$\mathfrak{h}_X$ becomes the centralizer of X in $\mathfrak{g}$. $\mathfrak{h}_X$ is thus the space of all endomorphisms of trace 0 of V whose matrices in the above basis are diagonal; $\mathrm{rk}(\mathfrak{g}) = l$.

(2) Let $\mathfrak{g} = \mathfrak{sl}(2,\mathbf{R})$; $k = \mathbf{R}$. Let $H = \begin{pmatrix} 1 & 0 \\ 0 & -1 \end{pmatrix}$, $X = \begin{pmatrix} 0 & 1 \\ 0 & 0 \end{pmatrix}$, $Y = \begin{pmatrix} 0 & 0 \\ 1 & 0 \end{pmatrix}$. It is then easy to verify that H and $X - Y$ are regular, and that $\mathfrak{a} = \mathbf{R}\cdot H$ and $\mathfrak{b} = \mathbf{R}\cdot(X - Y)$ are the CSA's containing H and $X - Y$ respectively. But $\mathfrak{a}$ and $\mathfrak{b}$ are not conjugate because, for any $t \in \mathbf{R}$, the eigenvalues of $\mathrm{ad}(tH)$ are 0, $2t$, and $-2t$ while those of $\mathrm{ad}(t(X - Y))$ are 0, $2it$, and $-2it (i^2 = -1)$. However, $\mathfrak{a}_c$ and $\mathfrak{b}_c$ are conjugate in $\mathfrak{g}_c$ under the adjoint group of $\mathfrak{g}_c$.

From now on, $\mathfrak{g}$ is semisimple, and $\langle\cdot,\cdot\rangle$ is its Cartan–Killing form.

Theorem 4.1.5. *Let $\mathfrak{g}$ be a semisimple Lie algebra over k and $\mathfrak{h} \subseteq \mathfrak{g}$ a subalgebra. Then $\mathfrak{h}$ is a CSA if and only if (a) $\mathfrak{h}$ is maximal abelian, and (b) ad H is semisimple for any $H \in \mathfrak{h}$. Moreover, in this case, the restriction of $\langle\cdot,\cdot\rangle$ to $\mathfrak{h} \times \mathfrak{h}$ is also non-singular.*

Proof. We prove first that if $\mathfrak{h}$ is any CSA of $\mathfrak{g}$, the restriction of the Cartan–Killing form to $\mathfrak{h} \times \mathfrak{h}$ is non-singular. Let X be a regular element such that $\mathfrak{h} = \mathfrak{h}_X$. If S is the semisimple component of X (cf. Theorem 3.10.6), then $[X,S] = 0$, so $S \in \mathfrak{h}$; moreover, since ad S is the semisimple component of ad X, they both have the same characteristic polynomials, from which we conclude that S is regular. So $\mathfrak{h} = \mathfrak{h}_S$, and the semisimplicity of S shows that $\mathfrak{h}$ is the centralizer of S in $\mathfrak{g}$. Let $\mathfrak{q} = [S,\mathfrak{g}]$. Then Theorem 3.1.5 implies that $\mathfrak{g}$ is the direct sum of $\mathfrak{h}$ and $\mathfrak{q}$. If $H \in \mathfrak{h}$ and $Y \in \mathfrak{g}$, we see from (3.9.9) that

$$\langle H,[S,Y]\rangle = -\langle [S,H],Y\rangle$$

Therefore, $\mathfrak{h}$ and $\mathfrak{q}$ are mutually orthogonal with respect to $\langle\cdot,\cdot\rangle$. We may therefore conclude that $\mathfrak{q}$ is the orthogonal complement of $\mathfrak{h}$ with respect to $\langle\cdot,\cdot\rangle$, and therefore that $\langle\cdot,\cdot\rangle$ is nonsingular when restricted to $\mathfrak{h} \times \mathfrak{h}$.

Let N be an element of $\mathfrak{h}$ such that ad N is nilpotent. By Theorem 3.7.3, there is a basis—for $\mathfrak{g}_c$ when $k = \mathbf{R}$, and for $\mathfrak{g}$ itself when $k = \mathbf{C}$—with respect to which the matrices of ad H, $H \in \mathfrak{h}$, have zeros below the main diagonal; since ad N is nilpotent, its matrix will then have zeros even on the main diagonal. Consequently, $tr(\mathrm{ad}\, N\ \mathrm{ad}\, H) = 0$ for all $H \in \mathfrak{h}$, i.e., $\langle N,H\rangle = 0$, $H \in \mathfrak{h}$. From the previous result we see that N must be 0. On the other hand, let $\mathfrak{h}'$ be the set of all elements of $\mathfrak{h}$ which are regular in $\mathfrak{g}$. If $X \in \mathfrak{h}'$ and N is its nilpotent component, $N \in \mathfrak{h}$ as $[X,N] = 0$, so by the result proved just now, $N = 0$. So $\mathfrak{h}'$ consists of semisimple elements. If $X \in \mathfrak{h}'$, $\mathfrak{h}$ is then the centralizer of X in $\mathfrak{g}$. So $[X, Y] = 0$ for X, $Y \in \mathfrak{h}'$. Now since $\mathfrak{h}'$ is open in $\mathfrak{h}$, we can find a basis for $\mathfrak{h}$ whose members belong to $\mathfrak{h}'$. Since these elements commute and are semisimple, it follows that $\mathfrak{h}$ is abelian and that ad H is semisimple for all $H \in \mathfrak{h}$. Since $\mathfrak{h}$ is maximal nilpotent, it must a fortiori be maximal abelian.

Conversely, let $\mathfrak{h}$ be a maximal abelian subalgebra such that ad H is semisimple for all $H \in \mathfrak{h}$. Then $H \mapsto \text{ad } H$ is a semisimple representation of $\mathfrak{h}$ in $\mathfrak{g}$. So we can find a subspace $\mathfrak{q}$ of $\mathfrak{g}$ that is invariant under ad $\mathfrak{h}$ and complementary to $\mathfrak{h}$. Suppose now that $\mathfrak{h}$ is not a CSA. Then there is a $Y \in \mathfrak{g}$ such that $Y \notin \mathfrak{h}$ and $[Y,\mathfrak{h}] \subseteq \mathfrak{h}$. If we write $Y = H + Y'$, $H \in \mathfrak{h}$, $Y' \in \mathfrak{q}$, we see that $Y' \neq 0$ and $[Y',\mathfrak{h}] \subseteq \mathfrak{h}$. On the other hand, $[Y',\mathfrak{h}] \subseteq \mathfrak{q}$, so $[Y',\mathfrak{h}] = 0$. This contradicts the fact that $\mathfrak{h}$ is maximal abelian. The theorem is completely proved.

Note that not every maximal abelian subalgebra of $\mathfrak{g}$ is a CSA. For example, let $\mathfrak{g} = \mathfrak{sl}(2,\mathbf{R})$ and $\mathfrak{n} = \mathbf{R}\cdot X$, where $X = \begin{pmatrix} 0 & 1 \\ 0 & 0 \end{pmatrix}$. Then $\mathfrak{n}$ is maximal abelian but not a CSA, because ad X is not semisimple.

Roughly the same arguments as above lead to the following theorem.

Theorem 4.1.6. *Let $\mathfrak{g}$ be a semisimple Lie algebra over k. Then an element of $\mathfrak{g}$ is semisimple if and only if it belongs to some CSA of $\mathfrak{g}$. Let X be a semisimple element of $\mathfrak{g}$, $\mathfrak{z}$ the centralizer of X, and $\mathfrak{q} = [X,\mathfrak{g}]$. Then $\mathfrak{z}$ and $\mathfrak{q}$ are orthogonal complements of each other with respect to $\langle\cdot,\cdot\rangle$, $\mathfrak{z}$ is a subalgebra of the same rank as $\mathfrak{g}$, and $\mathfrak{g}$ is the direct sum of $\mathfrak{z}$ and $\mathfrak{q}$. Moreover, $\langle\cdot,\cdot\rangle$ is non-singular on $\mathfrak{z} \times \mathfrak{z}$.*

Proof. Let X be semisimple. $\mathfrak{z}$ is obviously a subalgebra. As in the previous theorem, we prove that $\mathfrak{g} = \mathfrak{z} + \mathfrak{q}$ is a direct sum, and that $\mathfrak{z}$ and $\mathfrak{q}$ are orthogonal complements of each other with respect to $\langle\cdot,\cdot\rangle$. This already implies that $\langle\cdot,\cdot\rangle$ is non-singular on $\mathfrak{z} \times \mathfrak{z}$. Note also that if there is a CSA of $\mathfrak{g}$, say $\mathfrak{h}$, containing X, then $X \in \mathfrak{h} \subseteq \mathfrak{z}$, so $\mathfrak{h}$ will also be a CSA of $\mathfrak{z}$; this will imply that $\text{rk}(\mathfrak{z}) = \text{rk}(\mathfrak{g})$. So it remains to prove that there is a CSA of $\mathfrak{g}$ containing X. It is sufficient to prove that $\mathfrak{z} \cap \mathfrak{g}' \neq \varnothing$. For if $Y \in \mathfrak{g}'$ and lies in $\mathfrak{z}$, then $\mathfrak{h} = \mathfrak{h}_Y$ is a CSA containing X.

For $Z \in \mathfrak{z}$, ad Z leaves $\mathfrak{z}$ and $\mathfrak{q}$ invariant. Let

$$\zeta(Z) = \det(\text{ad } Z)_{\mathfrak{q}}.$$

Then ζ is a polynomial function on $\mathfrak{z}$ and $\zeta(X) \neq 0$. Put

$$'\mathfrak{z} = \{Z\colon Z \in \mathfrak{z}, \zeta(Z) \neq 0\}. \tag{4.1.15}$$

Then $'\mathfrak{z}$ is a nonnull open subset of $\mathfrak{z}$ containing X. Let ψ be the map $(z,Z) \mapsto Z^z$ of $G \times \mathfrak{z}$ into $\mathfrak{g}$. As in Theorem 4.1.2, we find that

$$(d\psi)_{(z,Z)}(X',Z') = \{[X',Z] + Z'\}^z \quad (X' \in \mathfrak{g}, Z' \in \mathfrak{z}). \tag{4.1.16}$$

This relation shows that $d\psi$ is surjective at (z,Z) if and only if $\mathfrak{z} + \text{range}\,(\text{ad}\, Z) = \mathfrak{g}$, i.e., if and only if ad Z maps $\mathfrak{q}$ onto itself, i.e., if and only if $Z \in {}'\mathfrak{z}$. But then $({}'\mathfrak{z})^G = \psi[G \times {}'\mathfrak{z}]$ is an open subset of $\mathfrak{g}$. In particular, since $\mathfrak{g}'$ is dense in $\mathfrak{g}$, $({}'\mathfrak{z})^G \cap \mathfrak{g}' \neq \varnothing$, so that ${}'\mathfrak{z} \cap \mathfrak{g}' \neq \varnothing$.

It is quite useful to isolate the submersive nature of ψ as a corollary.

Corollary 4.1.7. *Let $X \in \mathfrak{g}$ be semisimple, $\mathfrak{z}$ the centralizer of X in $\mathfrak{g}$, and ${}'\mathfrak{z}$ as in (4.1.15). Then $\psi: G \times {}'\mathfrak{z} \to \mathfrak{g}$ is everywhere submersive. In particular, $({}'\mathfrak{z})^G$ is an open subset of $\mathfrak{g}$ invariant under G.*

4.2. The Representations of $\mathfrak{sl}(2,\mathbf{C})$

The Lie algebra $\mathfrak{sl}(2,\mathbf{C})$ is simple, and its properties are of fundamental importance in the theory of semisimple Lie algebras. The object of this section is to present the basic results concerning the (finite-dimensional) representations of this Lie algebra.

Let $\mathfrak{g} = \mathfrak{sl}(2,\mathbf{C})$, $G = SL(2,\mathbf{C})$, and let $\mathfrak{G}$ denote the universal enveloping algebra of $\mathfrak{g}$. Write

$$H = \begin{pmatrix} 1 & 0 \\ 0 & -1 \end{pmatrix}, \qquad X = \begin{pmatrix} 0 & 1 \\ 0 & 0 \end{pmatrix}, \qquad Y = \begin{pmatrix} 0 & 0 \\ 1 & 0 \end{pmatrix}; \tag{4.2.1}$$

then

$$[H,X] = 2X, \qquad [H,Y] = -2Y, \qquad [X,Y] = H. \tag{4.2.2}$$

Also

$$\langle H,H \rangle = 8 \qquad \langle X,Y \rangle = 4. \tag{4.2.3}$$

The representations π_j. Since $\mathfrak{g}$ is semisimple, all of its representations are semisimple by Weyl's theorem. We shall now determine the irreducible representations of $\mathfrak{g}$. We need a Lemma.

Lemma 4.2.1. *Let p be an integer ≥ 0. Then we have the following identities in $\mathfrak{G}$:*

$$\begin{aligned} XY^{p+1} &= Y^{p+1}X + (p+1)Y^p(H-p) \\ YX^{p+1} &= X^{p+1}Y - (p+1)X^p(H+p). \end{aligned} \tag{4.2.4}$$

Proof. We use induction on p. For $p = 0$, this is clear, since $XY - YX = H$. Assume (4.2.4) for some $p \geq 0$. Since $HY = Y(H-2)$,

$$\begin{aligned}XY^{p+2} &= (Y^{p+1}X + (p+1)Y^p(H-p))Y \\ &= Y^{p+2}X + Y^{p+1}H + (p+1)Y^{p+1}(H-p-2) \\ &= Y^{p+2}X + (p+2)Y^{p+1}(H-p-1).\end{aligned}$$

The second relation in (4.2.4) is established similarly.

Theorem 4.2.2. *Let ρ be a finite-dimensional irreducible representation of $\mathfrak{g}$ in a complex vector space V. Then $\rho(H)$ is semisimple, its eigenvalues being all simple and integral. Moreover, there is an integer $j \geq 0$ and a basis $\{v_0,v_1, \ldots ,v_j\}$ for V such that*

$$\begin{aligned}&\rho(H)v_p = (j-2p)v_p && (p = 0,1,\ldots,j) \\ (4.2.5)\quad &\rho(X)v_0 = 0, \quad \rho(X)v_p = p(j-p+1)v_{p-1} && (p = 1,\ldots,j) \\ &\rho(Y)v_j = 0, \quad \rho(Y)v_p = v_{p+1} && (p = 0,1,\ldots,j-1)\end{aligned}$$

Conversely let $j \geq 0$ be any integer. Then there is exactly one equivalence class of irreducible representations of $\mathfrak{g}$ of dimension $j+1$, and (4.2.5) defines a member of this class. Finally, each of these representations is equivalent to its contragredient.

Proof. Let ρ be an irreducible representation of $\mathfrak{g}$ in a finite-dimensional vector space V. For any $\lambda \in \mathbf{C}$, write V_λ for the eigenspace of $\rho(H)$ corresponding to the eigenvalue λ, if λ is an eigenvalue of $\rho(H)$; otherwise, put $V_\lambda = 0$. Since $[\rho(H),\rho(X)] = 2\rho(X)$ and $[\rho(H),\rho(Y)] = -2\rho(Y)$, we see easily that for any $\lambda \in \mathbf{C}$

$$(4.2.6)\qquad \rho(X)[V_\lambda] \subseteq \lambda_{\lambda+2} \qquad \rho(Y)[V_\lambda] \subseteq V_{\lambda-2},$$

Now, we can find an eigenvalue j of $\rho(H)$ such that $j+2$ is not an eigenvalue. If $v_0 \neq 0$ is an eigenvector corresponding to the eigenvalue j, we get from (4.2.6)

$$(4.2.7)\qquad \rho(H)v_0 = jv_0, \qquad \rho(X)v_0 = 0.$$

Let

$$(4.2.8)\qquad v_s = \rho(Y)^s v_0 \quad (s = 0,1,\ldots).$$

By (4.2.6), $\rho(H)v_s = (j-2s)v_s$ for $s \geq 0$. As the numbers $j - 2s$ cannot all be eigenvalues of $\rho(H)$, we can find $s \geq 1$ such that $v_s = 0$. Let $m \geq 0$ be chosen so that

$$v_p \neq 0 \quad (0 \leq p \leq m), \qquad v_{m+1} = 0. \tag{4.2.9}$$

The $v_p (0 \leq p \leq m)$ are linearly independent, since they are eigenvectors of $\rho(H)$ for distinct eigenvalues. Moreover, by the first formula of (4.2.4) we have, for $0 \leq p \leq m$,

$$\begin{aligned}\rho(X)v_p &= \rho(XY^p)v_0 \\ &= p(j - p + 1)v_{p-1}\end{aligned}$$

Consequently, the subspace $\sum_{0 \leq p \leq m} \mathbf{C} \cdot v_p$ is invariant under $\rho(H)$, $\rho(X)$, and $\rho(Y)$. Since ρ is irreducible, we must have

$$V = \sum_{0 \leq p \leq m.} \mathbf{C} \cdot v_p.$$

At the same time, since $v_{m+1} = \rho(Y)^{m+1}v_0 = 0$, we have

$$\begin{aligned}0 &= \rho(XY^{m+1})v_0 \\ &= (m + 1)(j - m)v_m.\end{aligned}$$

This implies that $j = m$. Therefore j is an integer ≥ 0, the dimension of V is $j + 1$, and one has the relations (4.2.5). The semisimplicity of $\rho(H)$ is obvious.

Conversely, let j be an integer ≥ 0. Let V be a vector space over $\mathbf{C}$ of dimension $j + 1$. We choose a basis $\{v_0, \ldots, v_j\}$ for V and define the endomorphisms $\rho(H)$, $\rho(Y)$, and $\rho(Y)$ by (4.2.5). A straightforward calculation shows that the map $hH + xX + yY \mapsto h\rho(H) + x\rho(X) + y\rho(Y)$ $(h, x, y \in \mathbf{C})$ is a representation of $\mathfrak{g}$ in V. This representation must be irreducible. For, if $W \neq 0$ is an invariant subspace of V, the invariance of W under $\rho(H)$ implies that W is spanned by the v_p it contains. If s is the smallest of the integers $q \geq 0$ for which $v_q \in W$, then $s = 0$; for if $s > 0$, $\rho(X)v_s$ is a nonzero multiple of v_{s-1} and lies in W, showing that $v_{s-1} \in W$. So $v_0 \in W$. But then $v_p = \rho(Y)^p v_0 \in W$ for $0 \leq p \leq j$. Thus $W = V$.

For each integer $j \geq 0$, we thus have a unique equivalence class of irreducible representations of $\mathfrak{g}$, of dimension $j + 1$. Since the contragredient of an irreducible representation is an irreducible representation of the same dimension, the last assertion follows immediately. This proves the theorem.

Corollary 4.2.3. *Let ρ be a representation of $\mathfrak{g}$. Then $\rho(H)$ is semisimple, while $\rho(X)$ and $\rho(Y)$ are nilpotent. The eigenvalues of $\rho(H)$ are all integers; and the set of eigenvalues of $\rho(H)$ is invariant under the symmetry $s \mapsto -s$ of the additive group of integers.*

It is enough to prove these assertions for irreducible ρ; for such ρ, they are immediate from (4.2.5).

Corollary 4.2.4. *Let π be a representation of $\mathfrak{g}$ in a finite-dimensional vector space V. Assume that (i) all eigenvalues of $\pi(H)$ are of multiplicity 1, and (ii) the difference between any two eigenvalues of $\pi(H)$ is even. Then π is irreducible.*

Proof. π is seen to be either irreducible or equivalent to the direct sum of π_j and $\pi_{j'}$ where $|j - j'|$ is odd, on noting that in the irreducible representation ρ corresponding to $j \geq 0$ the eigenvalues of $\rho(H)$ are $j, j-2, \ldots, -j$; the second alternative is impossible in view of (ii).

The following corollary is immediate from the proof of Theorem 4.2.2.

Corollary 4.2.5. *Let ρ be a representation of $\mathfrak{g}$ in a (not necessarily finite-dimensional) vector space V, and v a nonzero vector in V such that:* (i) $\rho(X)v = 0$ *and* $\rho(Y)^s v = 0$ *for some integer* $s \geq 1$, *and* (ii) $\rho(H)v = \lambda v$ *for some* $\lambda \in \mathbf{C}$. *Then λ is an integer ≥ 0; moreover, the subspace $\sum_{0 \leq k \leq \lambda} \mathbf{C} \cdot \rho(Y)^k v$ is ρ-invariant, and it defines the irreducible representation of dimension $\lambda + 1$. In particular, $\rho(Y)^{\lambda+1} v = 0$.*

The eigenvalues of $\rho(H)$ are called *weights* of ρ; the eigenspaces are called the *weight spaces*. If ρ is irreducible and of dimension $j + 1$, j is the largest of the weights of ρ; j is called the *highest weight* of ρ, and we denote by π_j the irreducible representation of $\mathfrak{g}$ with highest weight j.

The left ideals $\mathfrak{M}_j$. The representations π_j are so important that it may be worthwhile looking at them from other angles. From the point of view of the universal enveloping algebra, we have

Theorem 4.2.6. *Let v_0 be a nonzero vector of weight j for π_j. Then $\mathbf{C} \cdot v_0$ is the null space of $\pi_j(X)$. If*

$$\mathfrak{M}_j = \{a : a \in \mathfrak{G}, \pi_j(a)v_0 = 0\}, \tag{4.2.10}$$

then $\mathfrak{M}_j$ is a maximal left ideal of $\mathfrak{G}$, and

$$\mathfrak{M}_j = \mathfrak{G}X + \mathfrak{G}(H - j) + \mathfrak{G}Y^{j+1}. \tag{4.2.11}$$

Proof. It is clear from (4.2.5) that $\mathbf{C} \cdot v_0$ is the null space of $\pi_j(X)$. Since π_j is irreducible, $\mathfrak{M}_j$ is a maximal left ideal of $\mathfrak{G}$. Let $\mathfrak{M}$ be the subspace of $\mathfrak{G}$ given by the right side of (4.2.11). Then, since X, $H - j$, and $Y^{j+1} \in \mathfrak{M}_j$, (cf. (4.2.5)), $\mathfrak{M} \subseteq \mathfrak{M}_j$. On the other hand, if r, s, and t are integers ≥ 0, we have, modulo $\mathfrak{M}$,

$$Y^r H^s X^t \equiv \begin{cases} 0 & \text{if } t > 0 \\ j^s Y^r & \text{if } s \geq 0, r \leq j, t = 0 \\ 0 & \text{if } s \geq 0, r \geq j+1, t = 0. \end{cases}$$

So, since $\mathfrak{G}$ is spanned by the monomials $Y^r H^s X^t$, we have

$$\mathfrak{G} = \mathfrak{M} + \sum_{0 \leq r \leq j} \mathbf{C} \cdot Y^r.$$

In other words, $\dim(\mathfrak{G}/\mathfrak{M}) \leq j + 1$. But $\dim(\mathfrak{G}/\mathfrak{M}_j) = \dim(\pi_j) = j + 1$, so we must have $\mathfrak{M} = \mathfrak{M}_j$.

Remark. Given any integer $j \geq 0$, it is possible to prove directly that $\mathfrak{M}_j$, as defined by (4.2.11), is a maximal left ideal of $\mathfrak{G}$, thus providing one with an alternative method of constructing the π_j. We shall take this method up in the case of arbitrary $\mathfrak{g}$ later on.

Corollary 4.2.7. *Let* $\omega = H^2 + 2H + 4YX + 1$. *Then* ω *lies in the center of* $\mathfrak{G}$, *and*

$$\pi_j(\omega) = (j + 1)^2 \cdot 1. \tag{4.2.12}$$

Proof. It is an easy verification that ω lies in the center of $\mathfrak{G}$. So, by Schur's lemma, $\pi_j(\omega) = c_j \cdot 1$ for some constant c_j. In particular, $\pi_j(\omega) \cdot v_0 = c_j v_0$. But

$$\begin{aligned} \pi_j(\omega)v_0 &= \pi_j(H)^2 v_0 + 2\pi_j(H)v_0 + v_0 + 4\pi_j(Y)\pi_j(X)v_0 \\ &= (j + 1)^2 v_0. \end{aligned}$$

Concrete realizations. We shall proceed to the concrete realizations of the π_j. Since $G = SL(2,\mathbf{C})$ is simply connected, the π_j can be lifted to complex analytic representations of G; these will also be denoted by π_j.

Let $V_1 = \mathbf{C}^2$, and let us write elements of V_1 as 2×1 column vectors; the elements of $\mathfrak{g}$ and G act as endomorphisms of V_1 through matrix multiplication. Put

$$v_{1,0} = \begin{pmatrix} 1 \\ 0 \end{pmatrix} \qquad v_{1,1} = \begin{pmatrix} 0 \\ 1 \end{pmatrix} \tag{4.2.13}$$

Then

$$Hv_{1,0} = v_{1,0}, \qquad Xv_{1,0} = 0, \qquad Yv_{1,0} = v_{1,1}. \tag{4.2.14}$$

So the representation $Z \mapsto Z$ is equivalent to π_1. Let $\mathcal{S}$ be the symmetric algebra over V_1, and for any integer $r \geq 0$, let $\mathcal{S}_r$ be the homogeneous subspace of degree r of $\mathcal{S}$. For $Z \in \mathfrak{g}$, let d_Z be the derivation of $\mathcal{S}$ that extends the endomorphism $v \mapsto Zv$ of V_1. The d_Z leave $\mathcal{S}_r$ invariant and so induce a representation $Z \mapsto d_Z | \mathcal{S}_r$ on $\mathcal{S}_r$. For $r = 0$, this is the trivial representation of $\mathfrak{g}$; for $r = 1$, this is the representation $Z \mapsto Z$ considered above. Let $r \geq 2$, and let

$$v_{r,p} = v_{1,0}^{r-p} v_{1,1}^{p} \quad (0 \leq p \leq r). \tag{4.2.15}$$

The $v_{r,p}$ $(0 \leq p \leq r)$ form a basis for $\mathcal{S}_r$, so $\dim(\mathcal{S}_r) = r + 1$. A straightforward calculation shows that

$$\begin{aligned} d_H v_{r,p} &= (r - 2p) v_{r,p} && (0 \leq p \leq r) \\ d_X v_{r,0} &= 0, \qquad d_Y v_{r,r} = 0 && \\ d_X v_{r,p} &= p v_{r,p-1} && (1 \leq p \leq r), \\ d_Y v_{r,p} &= (r - p) v_{r,p+1} && (0 \leq p \leq r - 1). \end{aligned} \tag{4.2.16}$$

(4.2.16) makes it clear that the representation defined by $\mathcal{S}_r$ is equivalent to π_r.

$\mathcal{S}$ is canonically isomorphic to the polynomial algebra on V_1^*; identifying V_1^* naturally with V_1, we obtain an isomorphism of $\mathcal{S}$ with the algebra P of polynomials on V_1. A straightforward verification shows that G acts on P in the following way:

$$(x \cdot p)(\mathbf{u}) = p(x^t \cdot \mathbf{u}) \quad (\mathbf{u} \in V_1, p \in P) \tag{4.2.17}$$

(t denotes matrix transposition). The space P_r of homogeneous polynomials of degree r is then invariant under this action and defines a representation equivalent to π_r.

Characters. Let φ_1 and φ_2 be respectively the linear functions $\begin{pmatrix} u_1 \\ u_2 \end{pmatrix} \mapsto u_1$ and $\begin{pmatrix} u_1 \\ u_2 \end{pmatrix} \mapsto u_2$ on V_1. For any $\lambda \in \mathbf{C}^*$ let

$$x_\lambda = \begin{pmatrix} \lambda & 0 \\ 0 & \lambda^{-1} \end{pmatrix}. \tag{4.2.18}$$

It is then clear from (4.2.17) that

$$x_\lambda \cdot \varphi_1^{j-p} \varphi_2^{p} = \lambda^{j-2p} \varphi_1^{j-p} \varphi_2^{p} \quad (0 \leq p \leq j).$$

So the trace of the endomorphism $q \mapsto x_\lambda \cdot q$ $(q \in P_j)$ is found to be $\lambda^j + \lambda^{j-2}$

$+ \cdots + \lambda^{-j}$. Let

$$\psi_j(x) = \operatorname{tr} \pi_j(x) \quad (x \in G). \tag{4.2.19}$$

Then

$$\psi_j(x_\lambda) = (\lambda^{j+1} - \lambda^{-(j+1)})/(\lambda - \lambda^{-1}) \quad (\lambda \in \mathbf{C}^*, \lambda \neq \pm 1). \tag{4.2.20}$$

ψ_j is a continuous function invariant under all the inner automorphisms of G. Further, any $x \in G$ with distinct eigenvalues is conjugate, via some inner automorphism of G, to an x_λ with $\lambda \neq \pm 1$. As the set of such $x \in G$ is a dense open subset of G, (4.2.20) determines ψ_j completely. From (4.2.12) we see (cf. Exercise 54, Chapter 3) that ψ_j satisfies the following differential equation:

$$\omega\psi_j = (j + 1)^2\psi_j. \tag{4.2.21}$$

4.3. Structure Theory

Let $\mathfrak{g}$ be a complex semisimple Lie algebra, $\mathfrak{h}$ a CSA. We fix $\mathfrak{g}$ and $\mathfrak{h}$ throughout this section. The aim of this section is to make a detailed analysis of the structure of $\mathfrak{g}$ through an examination of the spectral theory of ad $\mathfrak{h}$. This is not difficult, since $\mathfrak{h}$ is abelian, and ad H is semisimple for all $H \in \mathfrak{h}$.

We write $\langle \cdot, \cdot \rangle$ for the Cartan–Killing form of $\mathfrak{g}$; if $\mathfrak{a}$ and $\mathfrak{b}$ are two subspaces of $\mathfrak{g}$, we write $\mathfrak{a} \perp \mathfrak{b}$ when $\langle X, Y \rangle = 0$ for $X \in \mathfrak{a}$, $Y \in \mathfrak{b}$; $\mathfrak{a}^\perp = \{Y : \langle X, Y \rangle = 0 \text{ for all } X \in \mathfrak{a}\}$. $\mathfrak{h}'$ is the set of regular points of $\mathfrak{h}$. As usual, $\dim(\mathfrak{h}) = l$.

1. Roots and root subspaces. Let $\mathfrak{h}^*$ be the dual of $\mathfrak{h}$. For $\lambda \in \mathfrak{h}^*$ let

$$\mathfrak{g}_\lambda = \{X : X \in \mathfrak{g}, [H, X] = \lambda(H)X \text{ for all } H \in \mathfrak{h}\}. \tag{4.3.1}$$

Since $\mathfrak{h}$ is the centralizer of any element of $\mathfrak{h}'$, it is obvious that $\mathfrak{g}_0 = \mathfrak{h}$. $\lambda \in \mathfrak{h}^*$ is called a *root of* $(\mathfrak{g}, \mathfrak{h})$, or simply a *root*, if $\lambda \neq 0$ and $\mathfrak{g}_\lambda \neq 0$. We write Δ for the set of roots.

$$\Delta = \{\lambda : \lambda \in \mathfrak{h}^*, \lambda \text{ a root}\}. \tag{4.3.2}$$

For any $\lambda \in \Delta$, $\mathfrak{g}_\lambda$ is called the *root subspace corresponding to* λ. To the representation $H \mapsto \operatorname{ad} H$ we apply Theorem 3.5.8. Since ad H is semisimple for $H \in \mathfrak{h}$, it is clear that the weight subspaces are the root subspaces. Hence we have the decomposition

$$\mathfrak{g} = \mathfrak{h} + \sum_{\alpha \in \Delta} \mathfrak{g}_\alpha, \tag{4.3.3}$$

the sum being direct. We call (4.3.3) the *root space decomposition* of $\mathfrak{g}$ with respect to $\mathfrak{h}$.

Lemma 4.3.1. *Let* $\lambda, \mu \in \mathfrak{h}^*$. *Then* $[\mathfrak{g}_\lambda,\mathfrak{g}_\mu] \subseteq \mathfrak{g}_{\lambda+\mu}$; *if* $\lambda + \mu \neq 0$, $\mathfrak{g}_\lambda \perp \mathfrak{g}_\mu$

Proof. Since ad H is a derivation of $\mathfrak{g}$, we have, for $X \in \mathfrak{g}_\lambda$ and $Y \in \mathfrak{g}_\mu$, $[H,[X,Y]] = [[H,X],Y] + [X,[H,Y]] = (\lambda + \mu)(H)[X,Y]$, showing that $[X,Y] \in \mathfrak{g}_{\lambda+\mu}$. If $\lambda + \mu \neq 0$ and X and Y are as above, then for any $\nu \in \mathfrak{h}^*$, ad X ad Y maps $\mathfrak{g}_\nu$ into $\mathfrak{g}_{\lambda+\mu+\nu}$, which is different from $\mathfrak{g}_\nu$. Hence $\langle X, Y\rangle = tr(\text{ad } X \text{ ad } Y) = 0$.

Corollary 4.3.2. $\mathfrak{g}_\lambda \perp \mathfrak{h}$ $(\lambda \in \Delta)$. *If* $\alpha \in \Delta$, *then* $-\alpha \in \Delta$. *Moreover,* $\langle\cdot,\cdot\rangle$ *is non-singular on* $\mathfrak{g}_\alpha \times \mathfrak{g}_{-\alpha}$ *for any* $\alpha \in \Delta$.

Proof. The first relation is clear, since $\lambda \neq 0$. If $\alpha \in \Delta$ but $-\alpha \notin \Delta$, then $\mathfrak{g}_\alpha \perp \mathfrak{g}_\beta$ for all $\beta \in \Delta$. So, since $\mathfrak{g}_\alpha \perp \mathfrak{h}$ too, $\mathfrak{g}_\alpha \perp \mathfrak{g}$, contradicting the non-singularity of $\langle\cdot,\cdot\rangle$. If $X \in \mathfrak{g}_\alpha$ and $X \perp \mathfrak{g}_{-\alpha}$, then the same argument shows that $X \perp \mathfrak{g}$, so $X = 0$. This gives the last result.

Lemma 4.3.3. $\mathbf{C}\cdot\Delta = \mathfrak{h}^*$

Proof. For $H, H' \in \mathfrak{h}$ we have

$$\langle H,H'\rangle = \sum_{\alpha\in\Delta} \dim(\mathfrak{g}_\alpha)\alpha(H)\alpha(H'). \tag{4.3.4}$$

If $\mathbf{C}\cdot\Delta \neq \mathfrak{h}^*$, there will be an $H \neq 0$ in $\mathfrak{h}$ such that $\alpha(H) = 0$ for all $\alpha \in \Delta$. Then $H \perp \mathfrak{h}$ by (4.3.4), contradicting the non-singularity of $\langle\cdot,\cdot\rangle$ when restricted to $\mathfrak{h} \times \mathfrak{h}$.

Since $\langle\cdot,\cdot\rangle|\mathfrak{h} \times \mathfrak{h}$ is non-singular, there is a natural isomorphism of $\mathfrak{h}^*$ onto $\mathfrak{h}$. For any $\lambda \in \mathfrak{h}^*$ its image in $\mathfrak{h}$ will be denoted by H_λ. Thus

$$\lambda(H) = \langle H,H_\lambda\rangle \quad (\lambda \in \mathfrak{h}^*, H \in \mathfrak{h}). \tag{4.3.5}$$

We also transfer $\langle\cdot,\cdot\rangle$ to a symmetric non-singular bilinear form $\langle\cdot,\cdot\rangle$ on $\mathfrak{h}^* \times \mathfrak{h}^*$ via this isomorphism. Thus

$$\langle\lambda,\mu\rangle = \langle H_\lambda,H_\mu\rangle \quad (\lambda, \mu \in \mathfrak{h}^*). \tag{4.3.6}$$

Note that $H_\alpha \neq 0$ for any $\alpha \in \Delta$.

Lemma 4.3.4. *Let* $\alpha \in \Delta$, $X \in \mathfrak{g}_\alpha$, $Y \in \mathfrak{g}_{-\alpha}$. *Then*

$$[X,Y] = \langle X,Y\rangle H_\alpha. \tag{4.3.7}$$

In particular,

$$[\mathfrak{g}_\alpha, \mathfrak{g}_{-\alpha}] = \mathbf{C}\cdot H_\alpha. \tag{4.3.8}$$

Proof. By Lemma 4.3.1, $[\mathfrak{g}_\alpha, \mathfrak{g}_{-\alpha}] \subseteq \mathfrak{h}$. If X, Y are as above, then $\langle H,[X,Y]\rangle = \alpha(H)\langle X,Y\rangle = \langle H,\langle X,Y\rangle H_\alpha\rangle$ for any $H \in \mathfrak{h}$. This proves (4.3.7) and shows that $[\mathfrak{g}_\alpha, \mathfrak{g}_{-\alpha}] \subseteq \mathbf{C}\cdot H_\alpha$. By Corollary 4.3.2, we can find X, Y as above with $\langle X, Y\rangle = 1$; then $[X,Y] = H_\alpha$. This proves (4.3.8).

Since $H_\alpha = [X,Y]$ for suitably chosen $X \in \mathfrak{g}_\alpha$ and $Y \in \mathfrak{g}_{-\alpha}$, ad $H_\alpha =$ [ad X, ad Y] for such X, Y. Then $tr(\text{ad } H_\alpha \mid V) = 0$ for any subspace V of $\mathfrak{g}$ invariant under both ad X and ad Y. This remark leads to the next two lemmas.

Lemma 4.3.5. *Let α, $\beta \in \Delta$. Then there is a rational number $q_{\beta\alpha}$ such that*

$$\langle \beta,\alpha\rangle = q_{\beta\alpha}\langle \alpha,\alpha\rangle. \tag{4.3.9}$$

Moreover, $\langle \alpha,\alpha\rangle$ is a rational number >0.

Proof. Let $V = \sum_{k\in \mathbf{Z}} \mathfrak{g}_{\beta+k\alpha}$ (the sum is finite). By Lemma 4.3.1, $[\mathfrak{g}_\alpha, V] \subseteq V$, $[\mathfrak{g}_{-\alpha}, V] \subseteq V$. Let $d_k = \dim(\mathfrak{g}_{\beta+k\alpha})$. Then, by the remark made above, $tr(\text{ad } H_\alpha \mid V) = 0$. So

$$\sum_k d_k(\langle \beta,\alpha\rangle + k\langle \alpha,\alpha\rangle) = 0.$$

As $\sum_k d_k \geq d_0 > 0$, we can solve for $\langle \beta,\alpha\rangle$ from this relation to get (4.3.9) with $q_{\beta\alpha} = -(\sum_k kd_k)/(\sum_k d_k)$. If $\langle \alpha,\alpha\rangle = 0$ $\langle \beta,\alpha\rangle = 0$ for all $\beta \in \Delta$ by the above result. So $\langle \alpha,\lambda\rangle = 0$ for all $\lambda \in \mathfrak{h}^*$, showing that $\alpha = 0$, a contradiction. Thus $\langle \alpha,\alpha\rangle \neq 0$. By (4.3.4) and (4.3.9)

$$\langle \alpha,\alpha\rangle = \sum_{\beta\in\Delta} \dim(\mathfrak{g}_\beta) q_{\beta\alpha}^2 \langle \alpha,\alpha\rangle^2.$$

Since $\langle \alpha,\alpha\rangle \neq 0$, we can solve for it from this relation to get $\langle \alpha,\alpha\rangle = (\sum_{\beta\in\Delta} q_{\beta\alpha}^2 \dim(\mathfrak{g}_\beta))^{-1}$, which is rational and > 0.

Corollary 4.3.6. *Let*

$$\mathfrak{h}_{\mathbf{R}} = \sum_{\alpha\in\Delta} \mathbf{R}\cdot H_\alpha. \tag{4.3.10}$$

Then $\dim_{\mathbf{R}} \mathfrak{h}_{\mathbf{R}} = l$. Moreover, $\langle \cdot,\cdot\rangle$ is a positive definite scalar product on $\mathfrak{h}_{\mathbf{R}} \times \mathfrak{h}_{\mathbf{R}}$. Each root is real-valued on $\mathfrak{h}_{\mathbf{R}}$.

Proof. By (4.3.9) it is clear that each root is real-valued on $\mathfrak{h}_{\mathbf{R}}$. By (4.3.4),

$$\langle H,H\rangle = \sum_{\beta\in\Delta} \dim(\mathfrak{g}_\beta)\beta(H)^2 \quad (H \in \mathfrak{h}_{\mathbf{R}}).$$

Since $\beta(H)$ is real for $H \in \mathfrak{h}_{\mathbf{R}}$, $\beta \in \Delta$, $\langle H,H\rangle \geq 0$ for $H \in \mathfrak{h}_{\mathbf{R}}$. If $\langle H,H\rangle = 0$ for some $H \in \mathfrak{h}_{\mathbf{R}}$, then $\beta(H) = 0$ for all $\beta \in \Delta$, showing that $H = 0$. So $\langle\cdot,\cdot\rangle$ is a positive definite scalar product on $\mathfrak{h}_{\mathbf{R}} \times \mathfrak{h}_{\mathbf{R}}$. Since $\dim \mathfrak{h} = l$, we can select $\alpha_1, \ldots, \alpha_l \in \Delta$ such that $H_{\alpha_1}, \ldots, H_{\alpha_l}$ span $\mathfrak{h}$ over $\mathbf{C}$. If $H = \sum_{1\leq i\leq l} c_i H_{\alpha_i} \in \mathfrak{h}_{\mathbf{R}}$ $(c_i \in \mathbf{C})$, we have the equations

$$\alpha_j(H) = \sum_{1\leq i\leq l} c_i\langle\alpha_j,\alpha_i\rangle \quad (1 \leq j \leq l).$$

Now the matrix $(\langle\alpha_i,\alpha_j\rangle)_{1\leq i,\,j\leq l}$ is invertible because $\langle\cdot,\cdot\rangle$ is non-singular on $\mathfrak{h}^* \times \mathfrak{h}^*$. Also, its entries are real. Consequently, as the $\alpha_j(H)$ are real, the c_i are also real. In other words, $\mathfrak{h}_{\mathbf{R}}$ is spanned by the H_{α_i} $(1 \leq i \leq l)$ over $\mathbf{R}$. This proves that $\dim_{\mathbf{R}} \mathfrak{h}_{\mathbf{R}} = l$.

Lemma 4.3.7. *Let* $\alpha \in \Delta$. *Then* $\dim(\mathfrak{g}_\alpha) = 1$, *and none of* $\pm 2\alpha, \pm 3\alpha, \ldots,$ *are roots.*

Proof. Select $X \in \mathfrak{g}_\alpha$, $Y \in \mathfrak{g}_{-\alpha}$ such that $H_\alpha = [X,Y]$, and let $V = \mathbf{C}\cdot Y + \mathbf{C}\cdot H_\alpha + \sum_{k\geq 1} \mathfrak{g}_{k\alpha}$. It is clear that V is invariant under both ad X ad Y. Hence $tr(\text{ad } H_\alpha \mid V) = 0$. Write $d_k = \dim(\mathfrak{g}_{k\alpha})$, $k = 1,2, \ldots$. Then

$$\langle\alpha,\alpha\rangle(-1 + d_1 + 2d_2 + \cdots) = 0.$$

Since $\langle\alpha,\alpha\rangle \neq 0$, $d_1 \geq 1$ and $d_i \geq 0$ $(i \geq 2)$, we must have $d_1 = 1$, $d_2 = d_3 = \cdots = 0$.

Since $\langle\alpha,\alpha\rangle$ is a rational number $\neq 0$, we can find a rational number q_α such that $\alpha(q_\alpha H_\alpha) = 2(\alpha \in \Delta)$. Let $\bar{H}_\alpha = q_\alpha H_\alpha$. Thus:

$$\bar{H}_\alpha = \frac{2}{\langle\alpha,\alpha\rangle} H_\alpha \quad (\alpha \in \Delta). \tag{4.3.11}$$

We can find $X_\alpha \in \mathfrak{g}_\alpha$, $X_{-\alpha} \in \mathfrak{g}_{-\alpha}$ such that $[X_\alpha,X_{-\alpha}] = \bar{H}_\alpha$. Then

$$[\bar{H}_\alpha,X_\alpha] = 2X_\alpha, \qquad [\bar{H}_\alpha,X_{-\alpha}] = -2X_{-\alpha}, \qquad [X_\alpha,X_{-\alpha}] = \bar{H}_\alpha. \tag{4.3.12}$$

Write

$$\mathfrak{b}_\alpha = \mathbf{C}\cdot H_\alpha + \mathfrak{g}_\alpha + \mathfrak{g}_{-\alpha}. \tag{4.3.13}$$

Then $\mathfrak{b}_\alpha$ is spanned by $\bar{H}_\alpha, X_\alpha$, and $X_{-\alpha}$ and is a subalgebra of $\mathfrak{g}$ isomorphic to $\mathfrak{sl}(2,\mathbf{C})$. Put

$$\rho_\alpha(X) = \text{ad } X \quad (X \in \mathfrak{b}_\alpha); \tag{4.3.14}$$

ρ_α is then a representation of $\mathfrak{b}_\alpha$ in $\mathfrak{g}$. We now examine this representation using the results of §4.2.

Lemma 4.3.8. *Let $\alpha, \beta \in \Delta$ with $\beta \neq \pm\alpha$. Then there exist two integers $p = p(\alpha,\beta)$ and $q = q(\alpha,\beta)$, both ≥ 0, such that for any integer k, $\beta + k\alpha \in \Delta$ if and only if $-q \leq k \leq p$. The representation ρ_α is irreducible on $\sum_{k\in\mathbf{Z}} \mathfrak{g}_{\beta+k\alpha}$. Moreover, $\beta(\bar{H}_\alpha) = q - p$ is an integer, and $\beta - \beta(\bar{H}_\alpha)\cdot\alpha \in \Delta$.*

Proof. $\beta + k\alpha \neq 0$, $k = 0, \pm 1, \pm 2, \ldots$, so $\dim(\mathfrak{g}_{\beta+k\alpha}) \leq 1$ for all k. Let

$$\mathfrak{g}_{\beta,\alpha} = \sum_{k\in\mathbf{Z}} \mathfrak{g}_{\beta+k\alpha}. \tag{4.3.15}$$

ρ_α leaves $\mathfrak{g}_{\beta,\alpha}$ invariant. If $\mathfrak{g}_{\beta+k\alpha} \neq 0$, its elements are eigenvectors for $\rho_\alpha(\bar{H}_\alpha)$ for the eigenvalue $\beta(\bar{H}_\alpha) + 2k$. So, all eigenvalues of $\rho_\alpha(\bar{H}_\alpha)|\mathfrak{g}_{\beta,\alpha}$ are of multiplicity 1, and any two of them differ by an even integer. By Corollary 4.2.4, ρ_α acts irreducibly on $\mathfrak{g}_{\beta,\alpha}$. Let j be the highest weight of the irreducible representation defined by $\mathfrak{g}_{\beta,\alpha}$. Then since $\mathfrak{g}_\beta \subseteq \mathfrak{g}_{\beta,\alpha}$, we see that $\beta(\bar{H}_\alpha)$ is an integer, while $j - \beta(\bar{H}_\alpha)$ and $\beta(\bar{H}_\alpha) + j$ are nonnegative even integers. Let $p = \frac{1}{2}(j - \beta(\bar{H}_\alpha))$, $q = \frac{1}{2}(j + \beta(\bar{H}_\alpha))$. It is then obvious that

$$\mathfrak{g}_{\beta,\alpha} = \sum_{-q\leq k\leq p} \mathfrak{g}_{\beta+k\alpha}. \tag{4.3.16}$$

Finally, $-\beta(\bar{H}_\alpha)$ must be an eigenvalue of $\rho_\alpha(\bar{H}_\alpha)|\mathfrak{g}_{\beta,\alpha}$ (Corollary 4.2.3), so there is a k with $-q \leq k \leq p$ such that $(\beta + k\alpha)(\bar{H}_\alpha) = -\beta(\bar{H}_\alpha)$. Solving for k, we find that $k = -\beta(\bar{H}_\alpha)$: So

$$\beta - \beta(\bar{H}_\alpha)\cdot\alpha \in \Delta. \tag{4.3.17}$$

Corollary 4.3.9. *Let $\alpha \in \Delta$. If $c \in \mathbf{C}$, then $c\alpha$ is a root if and only if $c = \pm 1$.*

Proof. We may assume that $c \neq 0$. Let $\beta = c\alpha$. Then $2c = \beta(\bar{H}_\alpha)$ is an integer. Similarly, since $\alpha = c^{-1}\beta$, $2c^{-1}$ is an integer too. So $c = \pm\frac{1}{2}, \pm 1$, or ± 2. If $c \neq \pm 1$, then either $\alpha = \pm 2\beta$ or $\beta = \pm 2\alpha$, both of which are impossible by Lemma 4.3.7. So $c = \pm 1$.

Corollary 4.3.10. *Suppose $\alpha, \beta \in \Delta$ and $\beta \neq \alpha$. If $\beta - \alpha$ is not a root, $\langle\alpha,\beta\rangle \leq 0$; moreover, in this case, $q = 0$ and $p = -\beta(\bar{H}_\alpha)$.*

Proof. If $\beta - \alpha$ is not a root, $\mathfrak{g}_{\beta,\alpha} = \mathfrak{g}_\beta + \mathfrak{g}_{\beta+\alpha} + \cdots + \mathfrak{g}_{\beta+p\alpha}$. So $\beta(\bar{H}_\alpha)$ is the lowest eigenvalue of $\rho_\alpha(\bar{H}_\alpha)$ in $\mathfrak{g}_{\beta,\alpha}$, hence $\beta(\bar{H}_\alpha) \leq 0$. This implies that $\langle\alpha,\beta\rangle \leq 0$. The rest is obvious.

Corollary 4.3.11. *Suppose that $\alpha, \beta \in \Delta$ and $\alpha + \beta \in \Delta$. Then $[\mathfrak{g}_\alpha,\mathfrak{g}_\beta] = \mathfrak{g}_{\alpha+\beta}$.*

Proof. Otherwise $[\mathfrak{g}_\alpha,\mathfrak{g}_\beta] = 0$, so $\sum_{-q\leq k\leq 0} \mathfrak{g}_{\beta+k\alpha}$ is invariant under ρ_α. This implies that $p = 0$ or $\beta + \alpha \notin \Delta$, a contradiction.

Corollary 4.3.12. *Let α, $\beta \in \Delta$, with $\beta \neq \pm\alpha$, and let p, q be as in Lemma 4.2.8. Then $p + q \leq 3$. Furthermore, the possible values of $2\langle\alpha,\beta\rangle/\langle\alpha,\alpha\rangle$ are $0, \pm 1, \pm 2, \pm 3$.*

Proof. Let $m = 2\langle\alpha,\beta\rangle/\langle\alpha,\alpha\rangle$ $n = 2\langle\beta,\alpha\rangle/\langle\beta,\beta\rangle$. Then both m and n are integers. Suppose $m \neq 0$. Then $n \neq 0$ also. On the other hand, by Corollary 4.3.6, $\langle\cdot,\cdot\rangle$ is a positive definite scalar product on $\sum_{\gamma\in\Delta} \mathbf{R}\cdot\gamma$; so, using the Cauchy–Schwarz inequality, and remembering that α and β are linearly independent, we get $0 < |m||n| < 4$. So $m = \pm 1, \pm 2$, or ± 3. Replacing β by $\beta + p\alpha$ in this result, we see that the highest weight of the representation ρ_α in $\mathfrak{g}_{\alpha,\beta}$, namely $(\beta + p\alpha)(\bar{H}_\alpha)$, is at most 3. Hence $\dim(\mathfrak{g}_{\beta,\alpha}) \leq 4$. So $p + q \leq 3$.

Corollary 4.3.13. *Let $\alpha,\beta \in \Delta$, with $\beta \neq \pm\alpha$, and let p, q be as in Lemma 4.3.8. Suppose that $Z_\alpha \in \mathfrak{g}_\alpha$ and $Z_{-\alpha} \in \mathfrak{g}_{-\alpha}$, both $\neq 0$. Then for any $X \in \mathfrak{g}_\beta$,*

$$[Z_\alpha,[Z_{-\alpha},X]] = \frac{\langle\alpha,\alpha\rangle}{2}\langle Z_\alpha,Z_{-\alpha}\rangle q(p+1)X. \tag{4.3.18}$$

Proof. Let $X_\alpha \in \mathfrak{g}_\alpha$, $X_{-\alpha} \in \mathfrak{g}_{-\alpha}$ be such that (4.3.12) is satisfied. Since $\bar{H}_\alpha = (2/\langle\alpha,\alpha\rangle)H_\alpha$ and $[X_\alpha,X_{-\alpha}] = \langle X_\alpha,X_{-\alpha}\rangle H_\alpha$ (cf. (4.3.7)), we have

$$\langle X_\alpha,X_{-\alpha}\rangle = \frac{2}{\langle\alpha,\alpha\rangle}. \tag{4.3.19}$$

Let $Z_\alpha = c_\alpha X_\alpha$, $Z_{-\alpha} = c_{-\alpha}X_{-\alpha}$. If j is the highest weight of the representation ρ_α in $\mathfrak{g}_{\beta,\alpha}$, it is clear that $\mathfrak{g}_\beta = (\operatorname{ad} X_{-\alpha}))^p[\mathfrak{g}_{\beta+p\alpha}]$, so we obtain from (4.2.5)

$$\operatorname{ad} X_\alpha \operatorname{ad} X_{-\alpha}\cdot X = (p+1)(j-p)X \quad (X \in \mathfrak{g}_\beta).$$

On the other hand, $\dim(\mathfrak{g}_{\beta,\alpha}) = p + q + 1$, so $j = p + q$. So

$$[X_\alpha,[X_{-\alpha},X]] = q(p+1)X \quad (X \in \mathfrak{g}_\beta). \tag{4.3.20}$$

Now, (4.3.19) implies that $\langle Z_\alpha,Z_{-\alpha}\rangle = c_\alpha c_{-\alpha}\cdot 2/\langle\alpha,\alpha\rangle$, so we get (4.3.18) from (4.3.20).

Consider now $\alpha \in \Delta$. Let

$$\sigma_\alpha = \{H : H \in \mathfrak{h}, \alpha(H) = 0\}. \tag{4.3.21}$$

σ_α is a hyperplane in $\mathfrak{h}$. Since $\alpha(H_\alpha) > 0$, $H_\alpha \notin \sigma_\alpha$. So $\mathfrak{h}$ is the orthogonal direct sum of σ_α and $\mathbf{C}\cdot\bar{H}_\alpha$. We now introduce the "reflection" s_α in the hyperplane σ_α. This is the endomorphism of $\mathfrak{h}$ given by

$$s_\alpha \cdot H = H - \alpha(H)\bar{H}_\alpha. \tag{4.3.22}$$

Note that s_α leaves $\mathfrak{h}_\mathbf{R}$ invariant and $s_\alpha \mid \mathfrak{h}_\mathbf{R}$ is the reflection in the null space of α in $\mathfrak{h}_\mathbf{R}$. It is easy to verify that

$$\begin{aligned} &s_\alpha^2 = 1 \\ &s_\alpha H = H(H \in \sigma_\alpha), \qquad s_\alpha \cdot H_\alpha = -H_\alpha \end{aligned} \tag{4.3.23}$$

Let $\mathfrak{w}$ be the subgroup of $GL(\mathfrak{h})$ generated by the s_α ($\alpha \in \Delta$). It is called the *Weyl group of* $(\mathfrak{g},\mathfrak{h})$, or simply, the *Weyl group*. $\mathfrak{w}$ acts naturally in $\mathfrak{h}^*$ too and the two actions are intertwined by the map $\lambda \mapsto H_\lambda$.

Lemma 4.3.14. *$\mathfrak{w}$ is a finite subgroup of the orthogonal group of $\mathfrak{h}$ with respect to $\langle\cdot,\cdot\rangle$. Each element of $\mathfrak{w}$ induces a permutation of Δ.*

Proof. If $\alpha \in \Delta$, then for any $\beta \in \Delta$, $\beta - \beta(\bar{H}_\alpha)\alpha = \gamma$ is also in Δ. But $H_\gamma = H_\beta - \beta(\bar{H}_\alpha)H_\alpha = H_\beta - \alpha(H_\beta)\bar{H}_\alpha = s_\alpha \cdot H_\beta$. So for the action of $\mathfrak{w}$ in $\mathfrak{h}^*$, $s_\alpha \cdot \beta = \gamma$. Thus $s_\alpha \cdot \Delta \subseteq \Delta$. Since s_α is one-to-one, $s_\alpha \cdot \Delta = \Delta$. Hence $s \cdot \Delta = \Delta$ for any $s \in \mathfrak{w}$. For any $s \in \mathfrak{w}$, let $\bar{s}$ be the permutation of Δ induced by s. Then $s \mapsto \bar{s}$ is a homomorphism of $\mathfrak{w}$ onto a finite group. If $\bar{s}$ is the identity, it is clear from Lemma 4.3.3 that s is the identity. So $s \mapsto \bar{s}$ is an isomorphism. $\mathfrak{w}$ is thus finite. It is obvious from (4.3.23) that s_α is an orthogonal transformation for each $\alpha \in \Delta$. So each $s \in \mathfrak{w}$ is orthogonal.

The essential results in the foregoing analysis may now be summarized as follows. Our notation is as above.

Theorem 4.3.15. *The root spaces are one-dimensional. If $\alpha, \beta \in \Delta$ and $\alpha + \beta \neq 0$, then $[\mathfrak{g}_\alpha,\mathfrak{g}_\beta] = 0$ or $= \mathfrak{g}_{\alpha+\beta}$ according as $\alpha + \beta \notin \Delta$ or $\in \Delta$; if $X \in \mathfrak{g}_\alpha$, $Y \in \mathfrak{g}_{-\alpha}$, $[X,Y] = \langle X,Y\rangle H_\alpha$. $\Delta = -\Delta$; if $\alpha \in \Delta$, then $\pm\alpha$ are the only multiples of α which are roots. If $\alpha, \beta \in \Delta$, $\langle\alpha,\alpha\rangle$ is > 0, both $\langle\alpha,\alpha\rangle$ and $\langle\alpha,\beta\rangle$ are rational numbers, and $2\langle\alpha,\beta\rangle/\langle\alpha,\alpha\rangle$ is an integer whose possible values are $0, \pm1, \pm2, \pm3$. $\mathbf{C}\cdot\Delta = \mathfrak{h}^*, \mathfrak{h}_\mathbf{R} = \sum_{\alpha\in\Delta} \mathbf{R}\cdot H_\alpha$ is of dimension l over $\mathbf{R}$, and $\langle\cdot,\cdot\rangle$ is positive definite on $\mathfrak{h}_\mathbf{R} \times \mathfrak{h}_\mathbf{R}$. For each $\alpha \in \Delta$, $s_\alpha : H \mapsto H - (2\alpha(H)/\langle\alpha,\alpha\rangle)H_\alpha$ is a reflection in $\mathfrak{h}$ in the null space (hyperplane) of α, and the group generated by the s_α is a finite subgroup $\mathfrak{w}$ of the orthogonal group of $\mathfrak{h}$ relative to $\langle\cdot,\cdot\rangle$; moreover, $s \cdot \Delta = \Delta$ for each $s \in \mathfrak{w}$. $\mathfrak{h} \perp \mathfrak{g}_\alpha$ ($\alpha \in \Delta$), and $\mathfrak{g}_\alpha \perp \mathfrak{g}_\beta$ if $\alpha, \beta \in \Delta$ but $\alpha + \beta \neq 0$; $\langle\cdot,\cdot\rangle$ is nonsingular on $\mathfrak{h} \times \mathfrak{h}$ and $\mathfrak{g}_\alpha \times \mathfrak{g}_{-\alpha}$ ($\alpha \in \Delta$).*

2. Positive and simple systems of roots. Weyl group. A set $P \subseteq \Delta$ of roots is called a *positive system* if

$$P \cap (-P) = \varnothing, \qquad P \cup (-P) = \Delta$$
$$\alpha, \beta \in P, \alpha + \beta \in \Delta \Longrightarrow \alpha + \beta \in P \tag{4.3.24}$$

Note that P then contains exactly half the number of elements of Δ. A set $S \subseteq \Delta$ of roots is called a *simple system* if (i) $S = \{\alpha_1, \ldots, \alpha_l\}$ has l elements, and (ii) if $\alpha \in \Delta$, there are integers m_i $(1 \leq i \leq l)$, either all ≥ 0 or all ≤ 0, such that $\alpha = m_1\alpha_1 + \cdots + m_l\alpha_l$. Note that S must be a basis for $\mathfrak{h}^*$, and the m_i corresponding to a given α are uniquely determined. If $S = \{\alpha_1, \ldots, \alpha_l\}$ is a simple system, the set of all roots of the form $m_1\alpha_1 + \cdots + m_l\alpha_l$, where the m_i are all integers ≥ 0, is easily verified to be a positive system; it is called the positive system *corresponding to* S. It is obvious that if P is a positive system, S is a simple system, and $t \in \mathfrak{w}$, then $t \cdot P$ is a positive system, and $t \cdot S$ is a simple system; if P corresponds to S, $t \cdot P$ corresponds to $t \cdot S$.

Before constructing positive and simple systems, we examine some of their properties. Let $S = \{\alpha_1, \ldots, \alpha_l\}$ be a simple system. If P is the corresponding positive system and $\alpha \in P$, then write

$$O(\alpha) = m_1 + \cdots + m_l \quad (\alpha = m_1\alpha_1 + \cdots + m_l\alpha_l). \tag{4.3.25}$$

$O(\alpha)$ is called the *order* of α. $O(\alpha) \geq 1$ for $\alpha \in P$, and $= 1$ if and only if $\alpha \in S$.

Theorem 4.3.16. *Let $S = \{\alpha_1, \ldots, \alpha_l\}$ be a simple system of roots, and let P be the corresponding positive system. Then:*

(i) *If $1 \leq i, j \leq l$ and $i \neq j$, then $\alpha_i - \alpha_j$ is not a root, and*

$$\langle \alpha_i, \alpha_j \rangle \leq 0. \tag{4.3.26}$$

(ii) *If $1 \leq i \leq l$,*

$$s_{\alpha_i} \cdot \alpha_i = -\alpha_i, \; s_{\alpha_i} \cdot (P \setminus \{\alpha_i\}) = P \setminus \{\alpha_i\}; \tag{4.3.27}$$

$\mathfrak{w}$ is generated by $s_{\alpha_1}, \ldots, s_{\alpha_l}$, and

$$\Delta = \bigcup_{1 \leq i \leq l} \mathfrak{w} \cdot \alpha_i. \tag{4.3.28}$$

(iii) *If Q is an arbitrary positive system, $Q = P$ if and only if $\alpha_i \in Q$ for $1 \leq i \leq l$.*

(iv) *$\sum_{\alpha \in P} \mathfrak{g}_\alpha$ (resp. $\sum_{\alpha \in P} \mathfrak{g}_{-\alpha}$) is the subalgebra of $\mathfrak{g}$ generated by $\sum_{1 \leq i \leq l} \mathfrak{g}_{\alpha_i}$ (resp. $\sum_{1 \leq i \leq l} \mathfrak{g}_{-\alpha_i}$), while $\mathfrak{g}$ is the algebra generated by $\sum_{1 \leq i \leq l} \mathfrak{g}_{\alpha_i} + \sum_{1 \leq i \leq l} \mathfrak{g}_{-\alpha_i}$.*

Proof. If $m_1\alpha_1 + \cdots + m_l\alpha_l$ is a root, then we must have either all $m_i \geq 0$ or all $m_i \leq 0$. So $\alpha_i - \alpha_j$ cannot be a root if $i \neq j$; by Corollary 4.3.10, $\langle\alpha_i,\alpha_j\rangle \leq 0$ in this case. Fix i, $1 \leq i \leq l$, and let $\beta \in P$, $\beta \neq \alpha_i$. Then $\beta = m_1\alpha_1 + \cdots + m_l\alpha_l$, where $m_j > 0$ for some $j \neq i$. Now $s_{\alpha_i}\beta = \beta + c\alpha_i$ for some integer c, so $s_{\alpha_i}\beta = m_j\alpha_j + \sum_{p \neq j} m'_p\alpha_p$, where the m'_p are integers. Since $s_{\alpha_i}\beta$ is a root and $m_j > 0$, $s_{\alpha_i}\beta$ must be in P and, in fact, in $P \setminus \{\alpha_i\}$. For if $s_{\alpha_i}\beta = \alpha_i$, then $\beta = s_{\alpha_i}\cdot\alpha_i = -\alpha_i$. This proves (4.3.27).

Let $\mathfrak{w}_0$ be the subgroup of $\mathfrak{w}$ generated by $s_{\alpha_1}, \ldots, s_{\alpha_l}$. We prove first that

$$\Delta = \bigcup_{1 \leq i \leq l} \mathfrak{w}_0 \cdot \alpha_i$$

Since $\mathfrak{w}_0\cdot\alpha_i = \mathfrak{w}_0 s_{\alpha_i}\cdot\alpha_i = -\mathfrak{w}_0\cdot\alpha_i$, it is enough to prove that $P \subseteq \bigcup_{1\leq i\leq l}\mathfrak{w}_0\cdot\alpha_i$. We shall prove by induction on $O(\alpha)$ that $\alpha \in \bigcup_{1\leq i\leq l}\mathfrak{w}_0\cdot\alpha_i$ for all $\alpha \in P$. If $O(\alpha) = 1$, $\alpha = \alpha_i$ for some i, so we may assume that $O(\alpha) > 1$. Then we can find i, $1 \leq i \leq l$, such that $\langle\alpha,\alpha_i\rangle > 0$; for otherwise, $\langle\alpha,\alpha_j\rangle \leq 0$ for all j, so that, writing $\alpha = m_1\alpha_1 + \cdots + m_l\alpha_l$ $(m_j \geq 0)$, we would have $\langle\alpha,\alpha\rangle = \sum_{1\leq i\leq l} m_j\langle\alpha,\alpha_j\rangle \leq 0$. Let $\beta = s_{\alpha_i}\alpha$. Then $\beta = \alpha - c\alpha_i$, where c is an integer > 0. On the other hand, since $\alpha \neq \alpha_i$, $\beta \in P$ by (4.3.27). Since $O(\beta) = O(\alpha) - c < O(\alpha)$, $\beta \in \bigcup_{1\leq i\leq l}\mathfrak{w}_0\cdot\alpha_i$ by the induction hypothesis. This shows that $\alpha = s_{\alpha_i}\beta$ has the required property. In particular, (4.3.28) is proved.

It remains to prove that $\mathfrak{w}_0 = \mathfrak{w}$. It is enough to prove that $s_\alpha \in \mathfrak{w}_0$ for all $\alpha \in \Delta$. Since $s_\alpha = s_{-\alpha}$, we may assume $\alpha \in P$. Let $\alpha \in P$. Then we can find $\tau \in \mathfrak{w}_0$ and i, $1 \leq i \leq l$, such that $\alpha = \tau\cdot\alpha_i$. Since s_α and s_{α_i} are the orthogonal reflections in the null spaces of α and α_i respectively, it is clear that $s_\alpha = \tau\cdot s_{\alpha_i}\cdot\tau^{-1}$. Hence, $s_\alpha \in \mathfrak{w}_0$.

For (iii), Let Q be a positive system with $\alpha_i \in Q$, $1 \leq i \leq l$. We shall prove by induction on $O(\beta)$ that any $\beta \in P$ belongs to Q. If $\beta \in P$ and $O(\beta) = 1$, $\beta = \alpha_i$ for some i, so that $\beta \in Q$. Suppose $\beta \in P$ and $O(\beta) > 1$. As before, we can find i, $1 \leq i \leq l$, for which $\langle\beta,\alpha_i\rangle > 0$. Now, $\beta - \alpha_i \neq 0$, so by Corollary 4.3.10, $\beta - \alpha_i$ must be a root. Since $O(\beta - \alpha_i) = O(\beta) - 1 \geq 0$, this root must be in P. So $\beta - \alpha_i \in Q$ by the induction hypothesis. Hence $\beta = (\beta - \alpha_i) + \alpha_i \in Q$. Thus $P \subseteq Q$. Since both P and Q have the same number of elements, $P = Q$.

We now consider (iv). Let $\mathfrak{p} = \sum_{\alpha\in P}\mathfrak{g}_\alpha$, $\mathfrak{a} = \sum_{1\leq i\leq l}\mathfrak{g}_{\alpha_i}$. Since P is a positive system, $\mathfrak{p}$ is a subalgebra containing $\mathfrak{a}$. Let $\mathfrak{p}'$ be the subalgebra generated by $\mathfrak{a}$. Then $\mathfrak{p}' \subseteq \mathfrak{p}$. We shall prove by induction on $O(\beta)$ that $\mathfrak{g}_\beta \subseteq \mathfrak{p}'$ for each $\beta \in P$. This is obvious if $O(\beta) = 1$. Suppose $\beta \in P$ and $O(\beta) > 1$. As in the proof of (iii), we can find $\gamma \in P$ and i $(1 \leq i \leq l)$ such that $\beta = \gamma + \alpha_i$. Since $O(\gamma) < O(\beta)$, $\mathfrak{g}_\gamma \subseteq \mathfrak{p}'$ by the induction hypothesis. Hence $\mathfrak{g}_\beta = [\mathfrak{g}_\gamma,\mathfrak{g}_{\alpha_i}] \subseteq \mathfrak{p}'$ (Corollary 4.3.11). So $\mathfrak{p} \subseteq \mathfrak{p}'$, proving that $\mathfrak{p} = \mathfrak{p}'$. A similar argument proves that $\sum_{\alpha\in P}\mathfrak{g}_{-\alpha}$ is the subalgebra of $\mathfrak{g}$ generated by

$\sum_{1\leq i\leq l}\mathfrak{g}_{-\alpha_i}$. Since $[\mathfrak{g}_\alpha,\mathfrak{g}_{-\alpha}] = \mathbf{C}\cdot H_\alpha$, it is clear that $\mathfrak{g}$ is the subalgebra generated by $\sum_{1\leq i\leq l}\mathfrak{g}_{\alpha_i} + \sum_{1\leq i\leq l}\mathfrak{g}_{-\alpha_i}$.

We now show that positive and simple systems exist, and determine them. We need a lemma.

Lemma 4.3.17. *Let V be a real vector space with a positive definite scalar product $(\cdot,\cdot)$. Let $v_1,\ldots,v_s$ be elements of V with the following two properties:* (i) $(v_i,v_j)\leq 0$ *for* $i\neq j$, $1\leq i, j\leq s$, *and* (ii) *if $m_1,\ldots,m_s$ are nonnegative numbers with $m_1v_1+\cdots+m_sv_s=0$, then $m_1=\cdots=m_s=0$. Then the v_i are independent.*

Proof. Suppose to the contrary. Let $c_1v_1+\cdots+c_sv_s=0$ be a nontrivial relation between the v_j's. By throwing away the v_i for which $c_i=0$, we may assume that $c_i\neq 0$, $i=1,\ldots,s$. By (ii) the c_i cannot all be of the same sign. So $s\geq 2$, and we may renumber the c_i and v_i in such a manner that for a suitable r with $1\leq r<s$, $c_1,\ldots,c_r$ are >0 and $c_{r+1},\ldots,c_s$ are <0. Write $c_i=-d_i$, $r<i\leq s$. Then $d_i>0$, and we have the relation $c_1v_1+\cdots+c_rv_r=d_{r+1}v_{r+1}+\cdots+d_sv_s=w$ (say). But then $(w,w)=\sum_{1\leq i\leq r<j\leq s}c_id_j\,(v_i,v_j)\leq 0$, by (i). So $w=0$, contradicting (ii).

Let $\mathfrak{h}'_{\mathbf{R}}$ be the set of all regular points in $\mathfrak{h}_{\mathbf{R}}$. $\mathfrak{h}'_{\mathbf{R}}$ is a dense open subset of $\mathfrak{h}_{\mathbf{R}}$ and consists precisely of those points in $\mathfrak{h}_{\mathbf{R}}$ where no root vanishes. Any connected component of $\mathfrak{h}'_{\mathbf{R}}$ is called a *Weyl chamber* in $\mathfrak{h}_{\mathbf{R}}$, or simply a *chamber*. Let C be a chamber. Then each root takes values of the same sign on C, so for each $\alpha\in\Delta$ we can find $\epsilon_C(\alpha)=\pm 1$ such that $\epsilon_C(\alpha)\alpha$ is positive on C. If $C'=\{H: H\in\mathfrak{h}_{\mathbf{R}},\ \epsilon_C(\alpha)\alpha(H)>0 \text{ for all } \alpha\in\Delta\}$, it is clear that $C\subseteq C'\subseteq\mathfrak{h}'_{\mathbf{R}}$. On the other hand, C' is an open convex set, hence connected. So $C=C'$. In particular, C is a convex cone, i.e.,

$$H, H'\in C,\ t,t'\geq 0,\ t+t'>0 \Longrightarrow tH+t'H'\in C. \tag{4.3.29}$$

Since the number of ± 1-valued functions on Δ is finite, there are only finitely many chambers. Let $P(C)$ be the set of all roots which take only positive values on C. It is then easily verified that $P(C)$ is a positive system. Thus positive systems exist. Since C is the set of points of $\mathfrak{h}_{\mathbf{R}}$ where the members of $P(C)$ take positive values, the correspondence $C\mapsto P(C)$ is one-to-one. If $t\in\mathfrak{w}$, $t\cdot\mathfrak{h}'_{\mathbf{R}}=\mathfrak{h}'_{\mathbf{R}}$; hence t permutes the chambers, and $P(t\cdot C)=t\cdot P(C)$.

Theorem 4.3.18. *For any chamber C, let $P(C)$ be the set of roots which are positive on C, and let $S(C)$ be the set of elements of $P(C)$ which cannot be expressed as a sum of two elements of $P(C)$. Then $S(C)$ is a simple system, and $P(C)$ is the corresponding positive system. Moreover, the correspondence*

$C \mapsto S(C)$ (resp. $C \mapsto P(C)$) is a bijection of the set of all chambers onto the set of all simple systems (resp. positive systems); and the Weyl group acts in a simply transitive manner on the set of all chambers (resp. simple systems, positive systems).

Proof. Let C be a chamber and let $P(C)$ be defined as above. Let $S(C)$ be defined as in the theorem. We shall first prove that $S(C)$ is a simple system of roots. To this end, let $H \in C$ be fixed. If $\alpha, \beta \in S(C)$ and $\alpha \neq \beta$, then $\alpha - \beta$ is not a root; for if it is a root, then either $\alpha - \beta$ or $\beta - \alpha$ would be in $P(C)$, and the relations $\alpha = \beta + (\alpha - \beta)$, $\beta = \alpha + (\beta - \alpha)$ would then contradict the definition of $S(C)$. By Corollary 4.3.10, $\langle\alpha,\beta\rangle \leq 0$. On the other hand, let $S(C) = \{\alpha_1, \ldots, \alpha_p\}$, and let m_i be numbers ≥ 0 such that $m_1\alpha_1 + \cdots + m_p\alpha_p = 0$. Then $\sum_{1 \leq i \leq p} m_i\alpha_i(H) = 0$. Since $\alpha_i(H) > 0$ for $1 \leq i \leq p$, we must have $m_1 = \cdots = m_p = 0$. In other words, $\alpha_1, \ldots, \alpha_p$ satisfy the conditions of Lemma 4.3.17. So they are linearly independent. We claim now that if $\beta \in P(C)$, we can find integers $m_1, \ldots, m_p \geq 0$ such that $\beta = m_1\alpha_1 + \cdots + m_p\alpha_p$. Suppose in fact that this is not true. Then we can find a $\beta \in P(C)$, with the smallest possible value of $\beta(H)$, such that β cannot be expressed as a nonnegative integral linear combination $\alpha_1, \ldots, \alpha_p$. Clearly, $\beta \notin S(C)$, so we can write $\beta = \beta_1 + \beta_2$, where $\beta_1, \beta_2 \in P(C)$. Since $\beta_1(H)$ and $\beta_2(H)$ are > 0, $\beta_i(H) < \beta(H)$ for $i = 1, 2$, so we can find integers $m_{ij} \geq 0$ ($i = 1, 2$, $1 \leq j \leq p$) such that $\beta_i = \sum_{1 \leq j \leq p} m_{ij}\alpha_j$. But then $\beta = \sum_{1 \leq j \leq p} (m_{1j} + m_{2j})\alpha_j$, a contradiction. Our claim is thus proved. Since $\Delta = P(C) \cup (-P(C))$ spans $\mathfrak{h}^*$, we see at once that $p = l$ and that $S(C)$ is a simple system. If $S(C) = \{\alpha_1, \ldots, \alpha_l\}$, note that

$$C = \{H : H \in \mathfrak{h}_{\mathbf{R}}, \alpha_i(H) > 0 \text{ for } i = 1, \ldots, l\}. \tag{4.3.30}$$

For the remainder of the proof we fix a chamber C, and write $S(C) = S = \{\alpha_1, \ldots, \alpha_l\}$, $P(C) = P$. S is a simple system, and P is the corresponding positive system. We shall first show that $\mathfrak{w}$ acts transitively on the set of all positive systems. Let Q be a positive system, and let $r(Q)$ be the number of roots in P which are not in Q. We prove by induction on $r(Q)$ that for some $t \in \mathfrak{w}$, $t \cdot Q = P$.

If $r(Q) = 0$, $P \subseteq Q$, so $P = Q$; we may then take $t = 1$. Suppose $r(Q) \geq 1$. Then $P \neq Q$, so by assertion (iii) of Theorem 4.3.16, we can find i, $1 \leq i \leq l$, such that $\alpha_i \notin Q$. Let $Q' = s_{\alpha_i} \cdot Q$. Then Q' is a positive system. If $\beta \in Q \cap P$, then $\beta \in P \setminus \{\alpha_i\}$, so $s_{\alpha_i}\beta \in Q' \cap P$ by (4.3.27); moreover, Q being a positive system, $-\alpha_i \in Q$, so $\alpha_i = s_{\alpha_i}(-\alpha_i) \in Q' \cap P$. So $Q' \cap P$ has more elements than $Q \cap P$, i.e., $r(Q') < r(Q)$. By the induction hypothesis, there is a $t' \in \mathfrak{w}$ such that $t' \cdot Q' = P$. Then $t \cdot Q = P$ if we write $t = t's_{\alpha_i}$.

If $Q = s \cdot P$ for $s \in \mathfrak{w}$, then $Q = P(s \cdot C)$, and Q is the positive system corresponding to the simple system $s \cdot S$. So we conclude that $\mathfrak{w}$ is transitive on the set of chambers and the set of simple systems, and that the correspondences $C \mapsto P(C)$ and $C \mapsto S(C)$, are bijective, from the set of chambers onto the set of positive systems and the set of simple systems.

It remains to show that the action of $\mathfrak{w}$ is simply transitive on these sets. It is enough to consider one of them, say the set of chambers. By the definition of simple transitivity, we must show that if $t \in \mathfrak{w}$ and $t \cdot C = C$, then $t = 1$. Let $t \cdot C = C$. Let $\{H_1, \ldots, H_l\}$ be the basis of $\mathfrak{h}_{\mathbf{R}}$ over $\mathbf{R}$ dual to $\{\alpha_1, \ldots, \alpha_l\}$. Since t leaves C invariant, it must leave $S(C)$ invariant. Hence t permutes the H_i among themselves. Let $H = H_1 + \cdots + H_l$. Then $t \cdot H = H$. On the other hand, $\alpha_i(H) = 1, 1 \leq i \leq l$, so $\alpha(H) > 0$ (resp. < 0) if $\alpha \in P(C)$ (resp. $-P(C)$). In particular, H is regular. $\mathfrak{w}$ being a finite reflection group, the stabilizer of any regular point is trivial, by Lemma 4.15.15 of the appendix. So $t = 1$.

Corollary 4.3.19. *Let P be a positive system, C, the corresponding chamber. Then $-P$ is a positive system with corresponding chamber $-C$, and there is a unique element $s_0 \in \mathfrak{w}$ such that $s_0 P = -P$. Moreover, $s_0^2 = 1$.*

Proof. $-P$ is obviously the positive system corresponding to $-C$. Existence and uniqueness of s_0 are clear. Since $s_0^2 P = P$, $s_0^2 = 1$.

Corollary 4.3.20. *Let P be any positive system, and let S be the set of roots in P which cannot be expressed as a sum of two elements of P. Then S is a simple system, and P is the positive system corresponding to S.*

Proof. Obvious, since the result is valid when $P = P(C)$.

Corollary 4.3.21. *Let w be the order of $\mathfrak{w}$. Then there are exactly w chambers (resp. simple systems, positive systems).*

Proof. Obvious.

It is often useful to have alternative ways of constructing positive systems at our disposal. To introduce these we need to consider the concept of ordering a real vector space.

Let V be a real vector space of dimension $n (1 \leq n < \infty)$, and let V^* be the dual of V. By an *ordering* of V we mean a relation $<$ between pairs of elements of V such that the following properties are satisfied: (i) If $u, v, w \in V$ and $u < v$, $v < w$, then $u < w$. (ii) If $u, v \in V$, then exactly one of the three relations $u < v$, $v < u$, $u = v$ is valid. (iii) If $u, v, w \in V$ and $u < v$, then $u + w < v + w$. (iv) If $u, v \in V$, $u < v$, and $c \neq 0$ is in $\mathbf{R}$, then $cu < cv$ or $cv < cu$ according as $c > 0$ or $c < 0$. It is usual to write $u > v$ instead of $v < u$, and $u \leq v$ (or $v \geq u$) to mean that one of the two relations $u < v$, $u = v$ is valid. It follows from the above assumptions that if $u < v$ and u'

$\leq v'$, then $u + u' < v + v'$, and that if $u_1, \ldots, u_s$ are ≥ 0, $u_1 + \cdots + u_s$ can be zero if and only if all the u_i are 0.

Let $\{v_1, \ldots, v_n\}$ be a basis for V, $\{v_1^*, \ldots, v_n^*\}$ the dual basis for V^*. Given a $v \neq 0$ in V, there is a unique integer r, with $1 \leq r \leq n$, such that $v_r^*(v) \neq 0$ while $v_s^*(v) = 0$ for $1 \leq s < r$; we define $0 < v$ if $v_r^*(v) > 0$. If $u, v \in V$ and $u \neq v$, we define $u < v$ if $0 < v - u$. It is easy to verify that $<$ is an ordering of V. It is said to be the *lexicographic ordering on V induced by the basis* $\{v_1, \ldots, v_n\}$. Similarly, we have the lexicographic ordering induced on V^* by $\{v_1^*, \ldots, v_n^*\}$; if $v^* \in V^*$ and $v^* \neq 0$, then $v^* > 0$ if and only if the first non-zero member of the sequence $v^*(v_1), \ldots, v^*(v_n)$ is positive.

Returning now to $\mathfrak{g}$ and $\mathfrak{h}$, since the roots are real-valued on $\mathfrak{h}_{\mathbf{R}}$, we may regard them as elements of the dual $\mathfrak{h}_{\mathbf{R}}^*$ of the real vector space $\mathfrak{h}_{\mathbf{R}}$. Let $<$ be an ordering in $\mathfrak{h}_{\mathbf{R}}^*$, and let P be the set of all roots which are > 0 in this ordering. It is then clear that P is a positive system of roots (this is the reason for the name). Every positive system can be obtained in this manner. In fact, let P be an arbitrary positive system, and let $S = \{\alpha_1, \ldots, \alpha_l\}$ be the simple system such that P is the positive system corresponding to S; it is then obvious that in the lexicographic ordering of $\mathfrak{h}_{\mathbf{R}}^*$ induced by the basis $\{\alpha_1, \ldots, \alpha_l\}$, P is the set of all roots which are > 0.

3. Weyl basis. For $\alpha, \beta \in \Delta$ let $a_{\alpha,\beta} = 2\langle\alpha,\beta\rangle/\langle\alpha,\alpha\rangle$. We now determine the extent to which the integers $a_{\alpha,\beta}$ determine the semisimple Lie algebra $\mathfrak{g}$. We shall look at this question from a much more significant point of view later; for the present we follow Weyl's discussion of this problem [cf. Weyl, 2, 3, 4].

Select elements $Z_\alpha \in \mathfrak{g}_\alpha$ ($\alpha \in \Delta$) such that

$$\langle Z_\alpha, Z_{-\alpha}\rangle = -1 \quad (\alpha \in \Delta). \tag{4.3.31}$$

By (4.3.7),

$$[Z_\alpha, Z_{-\alpha}] = -H_\alpha \quad (\alpha \in \Delta). \tag{4.3.32}$$

Now define the members $N_{\alpha,\beta}$ ($\alpha, \beta \in \Delta, \alpha + \beta \neq 0$) by

$$\begin{aligned} [Z_\alpha, Z_\beta] &= N_{\alpha,\beta} Z_{\alpha+\beta} \quad \text{if } \alpha, \beta, \alpha+\beta \in \Delta \\ N_{\alpha,\beta} &= 0 \qquad\qquad\quad \text{if } \alpha, \beta \in \Delta, \alpha + \beta \notin \Delta. \end{aligned} \tag{4.3.33}$$

It is clear from Corollary 4.3.11 that $N_{\alpha,\beta} \neq 0$ if $\alpha, \beta, \alpha + \beta \in \Delta$. The following lemma details the relations satisfied by the $N_{\alpha,\beta}$.

Lemma 4.3.22. *Let $N_{\alpha,\beta}$ be defined as above. Then:*

(i) $N_{\alpha,\beta} = -N_{\beta,\alpha}$

(ii) *if* $\alpha, \beta, \gamma \in \Delta$ *and* $\alpha + \beta + \gamma = 0$, *then*

$$N_{\alpha,\beta} = N_{\beta,\gamma} = N_{\gamma,\alpha}$$

(iii) *If* $\alpha, \beta, \gamma, \delta \in \Delta$ *are such that the sum of no two of them is* 0, *and if* $\alpha + \beta + \gamma + \delta = 0$, *then*

$$N_{\alpha,\beta}N_{\gamma,\delta} + N_{\beta,\gamma}N_{\alpha,\delta} + N_{\gamma,\alpha}N_{\beta,\delta} = 0$$

(iv) *Let* $\alpha, \beta \in \Delta$ with $\beta \neq \pm\alpha$, *and let* $p, q \geq 0$ *be defined as in Lemma* 4.3.8. *Then*

$$N_{\alpha,\beta}N_{-\alpha,-\beta} = \tfrac{1}{2}\langle\alpha,\alpha\rangle p(q + 1)$$

Proof. (i) is trivial. For (ii) note that none of $\alpha + \beta$, $\beta + \gamma$, $\gamma + \alpha$ vanishes, so $N_{\alpha,\beta}$ $N_{\beta,\gamma}$ and $N_{\gamma,\alpha}$ are all defined. From the relation $[Z_\alpha,[Z_\beta,Z_\gamma]] + [Z_\beta,[[Z_\gamma,Z_\alpha]] + [Z_\gamma,[Z_\alpha,Z_\beta]] = 0$ we then get, since $\beta + \gamma = -\alpha$, etc.,

$$-N_{\beta,\gamma}H_\alpha - N_{\gamma,\alpha}H_\beta - N_{\alpha,\beta}H_\gamma = 0.$$

Comparing this with the relation $H_\alpha + H_\beta + H_\gamma = 0$, we get

$$N_{\beta,\gamma} = N_{\gamma,\alpha} = N_{\alpha,\beta}.$$

We now take up (iii). We assert that $[Z_\alpha,[Z_\beta,Z_\gamma]] = -N_{\beta,\gamma}N_{\alpha,\delta}Z_{-\delta}$. If $\beta + \gamma \notin \Delta$, then since $\beta + \gamma \neq 0$, $N_{\beta,\gamma} = 0$, and the assertion is trivial. If $\beta + \gamma \in \Delta$, $[Z_\alpha,[Z_\beta,Z_\gamma]] = N_{\beta,\gamma}N_{\alpha,\beta+\gamma}Z_{-\delta}$; and $N_{\alpha,\beta+\gamma} = -N_{\alpha,\delta}$ by (i) and (ii). Similarly we have $[Z_\beta,[Z_\gamma,Z_\alpha]] = -N_{\gamma,\alpha}N_{\beta,\delta}Z_{-\delta}$, and $[Z_\gamma,[Z_\alpha,Z_\beta]] = -N_{\alpha,\beta}N_{\gamma,\delta}Z_{-\delta}$. (iii) now follows from the Jacobi identity. We finally come to (iv). By the definition of p and q, $\beta + k(-\alpha)$ is a root if and only if $-p \leq k \leq q$. So, from (4.3.18), we get

$$[Z_{-\alpha},[Z_\alpha,Z_\beta]] = -\frac{\langle\alpha,\alpha\rangle}{2}p(q + 1)Z_\beta.$$

This implies that when $\alpha + \beta$ is a root,

$$N_{\alpha,\beta}N_{-\alpha,\alpha+\beta} = -\frac{\langle\alpha,\alpha\rangle}{2}p(q + 1).$$

Since $N_{-\alpha,\alpha+\beta} = -N_{-\alpha,-\beta}$ by (i) and (ii), we get (iv), when $\alpha + \beta \in \Delta$; when $\alpha + \beta \notin \Delta$, $N_{\alpha,\beta} = 0$ and $p = 0$, so (iv) is trivially true.

We now formulate and prove the main result towards which this discussion has been directed, namely, that the integers $a_{\alpha,\beta}$ determine $\mathfrak{g}$ up to isomorphism. To this end, let $\bar{\mathfrak{g}}$ be another semisimple Lie algebra over $\mathbf{C}$ with CSA $\bar{\mathfrak{h}}$, and let $\bar{\Delta}$ be the set of roots of $(\bar{\mathfrak{g}},\bar{\mathfrak{h}})$. We put

$$\bar{a}_{\alpha,\beta} = \frac{2\langle\bar{\alpha},\bar{\beta}\rangle}{\langle\bar{\alpha},\bar{\alpha}\rangle} \quad (\bar{\alpha},\bar{\beta} \in \bar{\Delta}), \tag{4.3.34}$$

where $\langle\cdot,\cdot\rangle$ is the bilinear form induced on $\bar{\mathfrak{h}}^* \times \bar{\mathfrak{h}}^*$ by the Cartan–Killing form of $\bar{\mathfrak{g}}$ in the usual way.

Lemma 4.3.23. *Let ξ be a linear isomorphism of $\mathfrak{h}$ onto $\bar{\mathfrak{h}}$ with the property that its dual ξ^* maps $\bar{\Delta}$ onto Δ. Then ξ preserves the respective restrictions of the Cartan–Killing forms, and*

$$\bar{a}_{\alpha,\beta} = a_{\xi^*\alpha,\xi^*\beta} \quad (\bar{\alpha},\bar{\beta} \in \bar{\Delta}).$$

Conversely, let ζ be a bijection of $\bar{\Delta}$ onto Δ such that, for some constant $c \neq 0$,

$$\bar{a}_{\alpha,\beta} = c a_{\zeta\alpha,\zeta\beta} \quad (\bar{\alpha},\bar{\beta} \in \bar{\Delta});$$

then $c = 1$, and there is a unique linear isomorphism ξ of $\mathfrak{h}$ onto $\bar{\mathfrak{h}}$ such that $\xi^*\bar{\alpha} = \zeta\bar{\alpha}$ for $\bar{\alpha} \in \bar{\Delta}$.

Proof. Let ξ be a linear isomorphism of $\mathfrak{h}$ onto $\bar{\mathfrak{h}}$ such that $\xi^*\bar{\Delta} = \Delta$. Then, for any $H, H' \in \mathfrak{h}$,

$$\begin{aligned}\langle \xi H, \xi H' \rangle &= \sum_{\alpha \in \bar{\Delta}} \bar{\alpha}(\xi H)\bar{\alpha}(\xi H') \\ &= \sum_{\alpha \in \bar{\Delta}} (\xi^*\bar{\alpha})(H)(\xi^*\bar{\alpha})(H') \\ &= \sum_{\alpha \in \Delta} \alpha(H)\alpha(H') \\ &= \langle H, H' \rangle.\end{aligned}$$

So ξ preserves the respective restrictions of the Cartan–Killing forms. The relations

$$\bar{a}_{\alpha,\beta} = a_{\xi^*\alpha,\xi^*\beta} \quad (\bar{\alpha},\bar{\beta} \in \bar{\Delta})$$

follow at once from this. For the converse, let ζ be a bijection of $\bar{\Delta}$ onto Δ with the properties described in the lemma. Let $c(\bar{\alpha})$ $(\bar{\alpha} \in \bar{\Delta})$ be constants. We assert that

$$\sum_{\alpha \in \Delta} c(\bar{\alpha})\bar{\alpha} = 0 \Longleftrightarrow \sum_{\alpha \in \Delta} c(\bar{\alpha})\zeta\bar{\alpha} = 0 \tag{4.3.35}$$

In fact,

$$\begin{aligned}\sum_{\alpha \in \bar{\Delta}} c(\bar{\alpha})\bar{\alpha} = \; &\Longleftrightarrow \sum_{\alpha \in \bar{\Delta}} c(\bar{\alpha})\langle \bar{\alpha}, \bar{\beta} \rangle = 0 \quad \forall\, \bar{\beta} \in \bar{\Delta} \\ &\Longleftrightarrow \sum_{\alpha \in \bar{\Delta}} c(\bar{\alpha})\bar{a}_{\beta,\alpha} = 0 \quad \forall\, \bar{\beta} \in \bar{\Delta} \\ &\Longleftrightarrow \sum_{\alpha \in \bar{\Delta}} c(\bar{\alpha})a_{\zeta\beta,\zeta\alpha} = 0 \quad \forall\, \bar{\beta} \in \bar{\Delta} \\ &\Longleftrightarrow \sum_{\alpha \in \bar{\Delta}} c(\bar{\alpha})\langle \zeta\bar{\alpha}, \beta \rangle = 0 \quad \forall\, \beta \in \Delta \\ &\Longleftrightarrow \sum_{\alpha \in \bar{\Delta}} c(\bar{\alpha})\zeta\bar{\alpha} = 0.\end{aligned}$$

It follows from (4.3.35) that there is a unique linear isomorphism η of $\bar{\mathfrak{h}}^*$ onto $\bar{\mathfrak{h}}^*$ such that η coincides with ζ on $\bar{\Delta}$. If we taken ξ to be the linear isomorphism of $\mathfrak{h}$ onto $\bar{\mathfrak{h}}$ such that $\xi^* = \eta$, we are through.

Theorem 4.3.24. *Let $\bar{\mathfrak{g}}$ be a semisimple Lie algebra over* $\mathbf{C}$, *$\bar{\mathfrak{h}}$, a CSA of $\bar{\mathfrak{g}}$, and $\bar{\Delta}$ the set of roots of $(\bar{\mathfrak{g}},\bar{\mathfrak{h}})$. Suppose that ξ is a linear isomorphism of $\mathfrak{h}$ onto $\bar{\mathfrak{h}}$ such that $\xi^*\bar{\Delta} = \Delta$. Then ξ extends to an isomorphism of $\mathfrak{g}$ onto $\bar{\mathfrak{g}}$.*

Proof. We select a lexicographic ordering in $\mathfrak{h}_{\mathbf{R}}^*$. For any $\sigma \in \Delta$ with $\sigma > 0$, write

$$\Delta(\sigma) = \{\alpha : \alpha \in \Delta, -\sigma < \alpha < \sigma\}.$$

For any $\alpha \in \Delta$, let $\bar{\alpha} \in \bar{\Delta}$ be such that $\xi^*\bar{\alpha} = \alpha$. Choose elements $Z_\alpha \in \mathfrak{g}_\alpha$ such that $\langle Z_\alpha, Z_{-\alpha}\rangle = -1$ for all $\alpha \in \Delta$, and define the numbers $N_{\alpha,\beta}$ by (4.3.33). It is enough to construct elements $\bar{Z}_{\bar{\alpha}} \in \bar{\mathfrak{g}}_{\bar{\alpha}}$ ($\alpha \in \Delta$) such that

$$\text{(4.3.36)}\qquad \begin{aligned} &\text{(a)}\ \langle \bar{Z}_{\bar{\alpha}}, \bar{Z}_{-\bar{\alpha}}\rangle = -1 && (\alpha \in \Delta)\\ &\text{(b)}\ [\bar{Z}_{\bar{\alpha}}, \bar{Z}_{\bar{\beta}}] = N_{\alpha,\beta}\bar{Z}_{\bar{\alpha}+\bar{\beta}} && (\alpha, \beta \in \Delta, \alpha + \beta \neq 0) \end{aligned}$$

(in (b) we regard the right side as 0 if $\alpha + \beta \notin \Delta$); the unique linear map η of $\mathfrak{g}$ into $\bar{\mathfrak{g}}$, which is ξ on $\mathfrak{h}$, and which sends Z_α to $\bar{Z}_{\bar{\alpha}}$, is then an isomorphism between the two Lie algebras.

Suppose that $\sigma \in \Delta$ is > 0 and that we have constructed elements $\bar{Z}_{\bar{\alpha}} \in \bar{\mathfrak{g}}_{\bar{\alpha}}$ for all $\alpha \in \Delta(\sigma)$ such that (a) above is satisfied for all $\alpha \in \Delta(\sigma)$ and (b) above is satisfied for all $\alpha, \beta \in \Delta(\sigma)$ for which $\alpha + \beta \neq 0$ and $-\sigma < \alpha + \beta < \sigma$. We now define the elements $\bar{Z}_{\pm\bar{\sigma}}$. If there are no $\alpha, \beta \in \Delta(\sigma)$ such that $\alpha + \beta = \sigma$, we define $\bar{Z}_{\bar{\sigma}} \in \bar{\mathfrak{g}}_{\bar{\sigma}}$ to be any nonzero element therein; otherwise, we choose $\gamma, \delta \in \Delta(\sigma)$ with $\gamma + \delta = \sigma$, and define $\bar{Z}_{\bar{\sigma}}$ by the following requirement (this is possible, since $N_{\gamma,\delta} \neq 0$ and $\bar{\gamma} + \bar{\delta} = \bar{\sigma}$):

$$\text{(4.3.37)}\qquad [\bar{Z}_{\bar{\gamma}}, \bar{Z}_{\bar{\delta}}] = N_{\gamma,\delta}\bar{Z}_{\bar{\sigma}}$$

We then define $\bar{Z}_{-\bar{\sigma}} \in \bar{\mathfrak{g}}_{-\bar{\sigma}}$ by the requirement that $\langle \bar{Z}_{\bar{\sigma}}, \bar{Z}_{\bar{\sigma}}\rangle = -1$. Let ρ be the immediate successor to σ in Δ with respect to the ordering $<$. We prove (a) of (4.3.36) for all $\alpha \in \Delta(\rho)$, and (b) of (4.3.36) for all $\alpha, \beta \in \Delta(\rho)$ for which $\alpha + \beta \neq 0$ and $-\rho < \alpha + \beta < \rho$. This will carry forward the inductive definition of the elements $\bar{Z}_{\bar{\alpha}}$ and complete the proof. For $\alpha, \beta \in \Delta(\rho)$ with $\alpha + \beta \neq 0$ and $-\rho < \alpha + \beta < \rho$, $\bar{N}_{\alpha,\beta}$ is defined as follows: if $\bar{\alpha} + \bar{\beta} \notin \bar{\Delta}$, $\bar{N}_{\alpha,\beta} = 0$; if $\bar{\alpha} + \bar{\beta} \in \bar{\Delta}$, it is defined by

$$[\bar{Z}_{\bar{\alpha}}, \bar{Z}_{\bar{\beta}}] = \bar{N}_{\alpha,\beta}\bar{Z}_{\bar{\alpha}+\bar{\beta}}.$$

It is then enough to prove that $N_{\alpha,\beta} = \bar{N}_{\alpha,\beta}$ whenever α, β, and $\alpha + \beta$ are in $\Delta(\rho)$. Several cases arise.

Case 1: α, β, and $\alpha + \beta$ are in $\Delta(\sigma)$. The result follows from the induction hypothesis.

Case 2: $\alpha, \beta \in \Delta(\rho)$, $\alpha + \beta = \sigma$. Then both α and β must be > 0 and so must belong to $\Delta(\sigma)$. In view of (4.3.37) we may assume that neither of α and β is equal to either of γ and δ. Note that γ and δ are also both > 0. Then $\alpha + \beta + (-\gamma) + (-\delta) = 0$, and no two of $\alpha, \beta, -\gamma, -\delta$ add up to 0. Applying (iii) of Lemma 4.3.22 we get

$$N_{\alpha,\beta}N_{-\gamma,-\delta} = -N_{\beta,-\gamma}N_{\alpha,-\delta} - N_{-\gamma,\alpha}N_{\beta,-\delta}$$
$$\bar{N}_{\alpha,\beta}\bar{N}_{-\gamma,-\delta} = -\bar{N}_{\beta,-\gamma}\bar{N}_{\alpha,-\delta} - \bar{N}_{-\gamma,\alpha}\bar{N}_{\beta,-\delta}$$

Now, $\beta + (-\gamma) \neq 0$ and $-\sigma < \beta, \gamma, \beta + (-\gamma) < \sigma$, so $N_{\beta,-\gamma} = \bar{N}_{\beta,-\gamma}$ by the induction hypothesis. Arguing similarly with the other terms, we conclude that the right sides of the above equations are the same. Hence

$$N_{\alpha,\beta}N_{-\gamma,-\delta} = \bar{N}_{\alpha,\beta}\bar{N}_{-\gamma,-\delta}.$$

We now note that all these numbers are nonzero. On the other hand, since $N_{\gamma,\delta} = \bar{N}_{\gamma,\delta}$, we conclude from (iv) of Lemma 4.3.22 and the isometric nature of ξ^* that $N_{-\gamma,-\delta} = \bar{N}_{-\gamma,-\delta}$. So $N_{\alpha,\beta} = \bar{N}_{\alpha,\beta}$.

Case 3: $\alpha, \beta \in \Delta(\sigma)$ $\alpha + \beta = -\sigma$. Then $-\alpha, -\beta \in \Delta(\rho)$ and $(-\alpha) + (-\beta) = \sigma$. So, by the previous case, $N_{-\alpha,-\beta} = \bar{N}_{-\alpha,-\beta}$. Once again, using (iv) of Lemma 4.3.22 and the isometric nature of ξ^*, we conclude that $N_{\alpha,\beta} = \bar{N}_{\alpha,\beta}$.

Case 4: $\alpha, \beta \in \Delta(\rho)$, $\alpha + \beta \in \Delta(\sigma)$. Unless one of α and β is $\pm\sigma$, we would be in Case 1. If $\alpha = \sigma$, then β must be in $\Delta(\sigma)$. So, since $(\alpha + \beta) + (-\beta) = \sigma$, we are in Case 2. Consequently $N_{\alpha+\beta,-\beta} = \bar{N}_{\alpha+\beta,-\beta}$. From (i) and (ii) of Lemma 4.3.22, we get $N_{-\alpha,-\beta} = \bar{N}_{-\alpha,-\beta}$. As before, this implies that $N_{\alpha,\beta} = \bar{N}_{\alpha,\beta}$. The other alternatives for α and β are handled similarly.

This completes the proof of the theorem.

Corollary 4.3.25. *There exists an automorphism φ of $\mathfrak{g}$ which coincides with -1 on $\mathfrak{h}$. Any such automorphism is involutive, and if φ is one such, φ maps $\mathfrak{g}_\alpha$ onto $\mathfrak{g}_{-\alpha}$ for all $\alpha \in \Delta$.*

Proof. The existence of φ is immediate from the theorem. The fact that φ maps $\mathfrak{g}_\alpha$ onto $\mathfrak{g}_{-\alpha}$ is also straightforward. We now prove that $\varphi^2 = 1$. Select $Z_\alpha \in \mathfrak{g}_\alpha$ such that $\langle Z_\alpha, Z_{-\alpha}\rangle = -1$ for all $\alpha \in \Delta$. Then there are constants $c(\alpha)$ such that $\varphi Z_\alpha = c(\alpha)Z_{-\alpha}$ for all $\alpha \in \Delta$. So $\varphi^2 Z_\alpha = c(\alpha)c(-\alpha)Z_\alpha$ ($\alpha \in \Delta$). On the other hand, since $\langle \cdot, \cdot \rangle$ is invariant under φ (cf. §3.9), we have $\langle \varphi Z_\alpha, \varphi Z_{-\alpha}\rangle = -1$ ($\alpha \in \Delta$), so $c(\alpha)c(-\alpha) = 1$ for all $\alpha \in \Delta$. This proves that φ^2 is the identity.

Let $\{H_1,\ldots,H_l\}$ be a basis for $\mathfrak{h}_{\mathbf{R}}$ over $\mathbf{R}$, and let Z_α be a nonzero element of $\mathfrak{g}_\alpha$ for each $\alpha \in \Delta$. Then the H_i $(1 \leq i \leq l)$ and the Z_α $(\alpha \in \Delta)$ form a basis for $\mathfrak{g}$ over $\mathbf{C}$. We shall call this a *Weyl basis* if (i) $\langle Z_\alpha, Z_{-\alpha}\rangle = -1$ for all $\alpha \in \Delta$, and (ii) if the constants $N_{\alpha,\beta}$ are defined by (4.3.33), then $N_{\alpha,\beta} = N_{-\alpha,-\beta}$ for all $\alpha, \beta \in \Delta$ with $\alpha + \beta \neq 0$.

Theorem 4.3.26. *$\mathfrak{g}$ always admits a Weyl basis. If H_i $(1 \leq i \leq l)$ and $Z_\alpha(\alpha \in \Delta)$ are members of a Weyl basis, the corresponding constants $N_{\alpha,\beta}$* are *real, and*

(4.3.38)
$$\mathfrak{g}_0 = \mathfrak{h}_{\mathbf{R}} + \sum_{\alpha \in \Delta} \mathbf{R}\cdot Z_\alpha$$

is a real form of $\mathfrak{g}$.

Proof. Let P be a positive system of roots and φ an automorphism of $\mathfrak{g}$ such that $\varphi \mid \mathfrak{h} = -1$. It is obvious that we can select, for each $\alpha \in P$, a $Z_\alpha \in \mathfrak{g}_\alpha$ such that $\langle Z_\alpha, \varphi Z_\alpha\rangle = -1$; we define $Z_{-\alpha}$ to be φZ_α. Since φ is involutive, $\langle Z_\alpha, \varphi Z_\alpha\rangle = -1$ for $\alpha \in \Delta$. Now $\varphi Z_\alpha = Z_{-\alpha}$ for all $\alpha \in \Delta$; so, on writing the condition that φ is an automorphism of $\mathfrak{g}$, we see that $N_{\alpha,\beta} = N_{-\alpha,-\beta}$. So the Z_α, together with a basis for $\mathfrak{h}_{\mathbf{R}}$ over $\mathbf{R}$, constitute a Weyl basis. Since

$$N_{\alpha,\beta}^2 = \tfrac{1}{2}\langle \alpha,\alpha\rangle p(q+1) \geq 0$$

by (iv) of Lemma 4.3.22, we conclude that the $N_{\alpha,\beta}$ have to be real numbers. In particular, if we define $\mathfrak{g}_0$ by (4.3.38), $\mathfrak{g}_0$ will be a Lie algebra over $\mathbf{R}$. Obviously, $\dim_{\mathbf{R}} \mathfrak{g}_0 = \dim_{\mathbf{C}} \mathfrak{g}$, so $\mathfrak{g}_0$ is a real form of $\mathfrak{g}$. This proves the theorem.

Cartan matrices and Dynkin diagrams. Let r be an integer ≥ 1, and $A = (a_{ij})$ an $r \times r$ matrix. We say that A is a *Cartan matrix* if the following conditions are satisfied:

(4.3.39)
- (i) a_{ij} is an integer for all i, j; $a_{ij} \leq 0$ if $i \neq j$, $a_{ii} = 2$ for all i; $a_{ij} = 0$ if and only if $a_{ji} = 0$
- (ii) $\det(A) \neq 0$
- (iii) Let V be a vector space over $\mathbf{C}$ with a basis $\{v_1,\ldots,v_r\}$, and let s_i be the endomorphism of V such that $s_i v_j = v_j - a_{ij} v_i$ $(1 \leq i, j \leq r)$. Then $s_1,\ldots,s_r$ generate a finite subgroup of $GL(V)$.

r is said to be the *rank of A*. The finite group in (iii) is called the *Weyl group of A*. Note that if s_i is the endomorphism of V with $s_i v_j = v_j - a_{ij} v_i$ $(1 \leq i, j \leq r)$, then $s_i v_i = -v_i$ and $s_i^2 v_j = v_j$ by direct calculation; so $s_i \in GL(V)$ and $s_i^2 = 1$, $1 \leq i \leq r$. A simple argument shows that $s_i v = -v$ if and only if $v \in \mathbf{C}\cdot v_i$, so each s_i is a reflection in V. Two Cartan matrices $A = (a_{ij})$ and

$A' = (a'_{pq})$ are said to be *equivalent* if they have the same rank (say r), and if there is a permutation $i \mapsto i'$ of $\{1,\ldots,r\}$ such that $a_{ij} = a'_{i'j'}$, $1 \leq i,j \leq r$. A Cartan matrix $A = (a_{ij})$ of rank r is said to be *reducible* if we can find a partition of $\{1,\ldots,r\}$ into two nonempty sets S_1 and S_2 such that $a_{ij} = 0$ for $i \in S_1$ and $j \in S_2$; if there is no such partition, A is said to be *irreducible*. If two Cartan matrices are equivalent, and one of them is irreducible, so is the other.

Let $S = \{\alpha_1,\ldots,\alpha_l\}$ be a simple system of roots of $(\mathfrak{g},\mathfrak{h})$, and let

$$a_{ij} = \frac{2\langle\alpha_i,\alpha_j\rangle}{\langle\alpha_i,\alpha_i\rangle} = \alpha_j(\bar{H}_{\alpha_i}) \quad (1 \leq i,j \leq l). \tag{4.3.40}$$

Then the $l \times l$ matrix $A = (a_{ij})$ is easily verified to be a Cartan matrix of rank l; we remark that if $i \neq j$, $a_{ij} \leq 0$ by (4.3.26), while $a_{ij} = 0 \leftrightarrow a_{ji} = 0 \leftrightarrow \langle\alpha_i,\alpha_j\rangle = 0$, and the group in (iii) of (4.3.39) is the Weyl group of $(\mathfrak{g},\mathfrak{h})$. The construction of A depends on the choice of the CSA and the choice of the simple system S of roots. Since the CSA's of $\mathfrak{g}$ are mutally conjugate under the adjoint group, and since the simple systems of roots of $(\mathfrak{g},\mathfrak{h})$ are mutually conjugate under the Weyl group of $(\mathfrak{g},\mathfrak{h})$, it is clear that different choices of $\mathfrak{h}$ and S will change A only to an equivalent matrix. Thus we can associate with $\mathfrak{g}$ a unique equivalence class of Cartan matrices of rank equal to l. We then have

Theorem 4.3.27. *Two semisimple Lie algebras over* **C** *are isomorphic if and only if the corresponding equivalence classes of Cartan matrices are identical. A Lie algebra over* **C** *is simple if and only if the associated equivalence class of Cartan matrices consists entirely of irreducible elements.*

Proof. Let $\mathfrak{g}$, $\mathfrak{h}$ be as usual. Let $\bar{\mathfrak{g}}$ be a semisimple Lie algebra over **C** with CSA $\bar{\mathfrak{h}}$, and suppose that the two Lie algebras give rise to the same equivalence class of Cartan matrices. Then we can find simple systems $S = \{\alpha_1, \ldots,\alpha_l\}$ and $\bar{S} = \{\bar{\alpha}_1,\ldots,\bar{\alpha}_l\}$ of roots of $(\mathfrak{g},\mathfrak{h})$ and $(\bar{\mathfrak{g}},\bar{\mathfrak{h}})$ respectively such that

$$\frac{2\langle\alpha_i,\alpha_j\rangle}{\langle\alpha_i,\alpha_j\rangle} = \frac{2\langle\bar{\alpha}_i,\bar{\alpha}_j\rangle}{\langle\bar{\alpha}_i,\bar{\alpha}_j\rangle} \quad (1 \leq i,j \leq l). \tag{4.3.41}$$

Let $\bar{\Delta}$ be the set of roots of $(\bar{\mathfrak{g}},\bar{\mathfrak{h}})$ and $\bar{\mathfrak{w}}$ its Weyl group.

We can obviously find a linear isomorphism ξ of $\mathfrak{h}$ onto $\bar{\mathfrak{h}}$ such that its dual ξ^* maps $\bar{\alpha}_i$ onto α_i for $1 \leq i \leq l$. It is then clear from (4.3.41) that $\xi^{*-1}s_{\alpha_i}\xi^* = s_{\bar{\alpha}_i}$ for $1 \leq i \leq l$. Consequently, $\xi^{*-1}\cdot\mathfrak{w}\cdot\xi^* = \bar{\mathfrak{w}}$. Since $\Delta = \bigcup_{1\leq i\leq l}\mathfrak{w}\cdot\alpha_i$ and $\bar{\Delta} = \bigcup_{1\leq i\leq l}\bar{\mathfrak{w}}\cdot\bar{\alpha}_i$, we see at once that $\xi^*\bar{\Delta} = \Delta$. Theorem 4.3.24 now implies that we can extend ξ to an isomorphism of $\mathfrak{g}$ onto $\bar{\mathfrak{g}}$.

We now take up the second assertion. Let $\mathfrak{g}$, $\mathfrak{h}$ be as usual and let $S = \{\alpha_1,\ldots,\alpha_l\}$ be a simple system of roots. Suppose $\mathfrak{g}$ is not simple. Then we can write $\mathfrak{g}$ as the direct sum of two ideals $\mathfrak{g}_1$ and $\mathfrak{g}_2$, both nonzero. Since $[\mathfrak{g}_1,\mathfrak{g}_2] = 0$, ad X ad $Y = 0$ for $X \in \mathfrak{g}_1$ and $Y \in \mathfrak{g}_2$, so $\mathfrak{g}_1 \perp \mathfrak{g}_2$. Since $\mathfrak{g}_1$ and $\mathfrak{g}_2$ are invariant under ad $\mathfrak{h}$, it is clear that for any $\lambda \in \mathfrak{h}^*$, $\mathfrak{g}_\lambda = (\mathfrak{g}_\lambda \cap \mathfrak{g}_1) + (\mathfrak{g}_\lambda \cap \mathfrak{g}_2)$. As the root spaces are one-dimensional, we see that for any $\alpha \in \Delta$, either $\mathfrak{g}_\alpha \subseteq \mathfrak{g}_1$ or $\mathfrak{g}_\alpha \subseteq \mathfrak{g}_2$. Let $S_i = \{j: 1 \leq j \leq l,\ \mathfrak{g}_{\alpha_j} \subseteq \mathfrak{g}_i\}$, $i = 1, 2$. Suppose $S_1 = \varnothing$. Then $\mathfrak{g}_{\alpha_j} \subseteq \mathfrak{g}_2$ for $1 \leq j \leq l$. Hence $\mathfrak{g}_\alpha \subseteq \mathfrak{g}_2$ for all $\alpha \in P$ by (iv) of Theorem 4.3.16. Since $\mathfrak{g}_2$ is an ideal, and since $[\mathfrak{g}_{-\alpha},\mathfrak{g}_\alpha] = \mathbf{C}\cdot H_\alpha, [\mathfrak{g}_{-\alpha},\mathbf{C}\cdot H_\alpha] = \mathfrak{g}_{-\alpha}$, we see that $\mathfrak{g}_2$ contains $\mathfrak{h}$ and $\mathfrak{g}_{-\alpha}$ for all $\alpha \in P$. So $\mathfrak{g}_2 = \mathfrak{g}$ and $\mathfrak{g}_1 = 0$, which is impossible. Similarly, $S_2 \neq \varnothing$. So S is the disjoint union of the nonempty sets S_1 and S_2. If $i \in S_1$ and $j \in S_2$, then the above argument shows that $H_{\alpha_i} \in \mathfrak{g}_1$ and $H_{\alpha_j} \in \mathfrak{g}_2$. Hence $\langle \alpha_i,\alpha_j\rangle = \langle H_{\alpha_i},H_{\alpha_j}\rangle = 0$. This shows that the Cartan matrices corresponding to $\mathfrak{g}$ are reducible.

Conversely, suppose the Cartan matrices associated with $\mathfrak{g}$ are reducible. Then we can find nonempty disjoint subsets S_1 and S_2 of $\{1,\ldots,l\}$, whose union is $\{1,\ldots,l\}$, such that $\langle\alpha_i,\alpha_j\rangle = 0$ for $i \in S_1, j \in S_2$. Let $\mathfrak{w}_r$ be the subgroup of $\mathfrak{w}$ generated by s_{α_i} $(i \in S_r)$, and let $\Delta_r = \bigcup_{i\in S_r} \mathfrak{w}\cdot\alpha_i$ $(r = 1, 2)$. Clearly, $\Delta = \Delta_1 \cup \Delta_2$. For $i \in S_1, j \in S_2$, $s_{\alpha_i}\alpha_j = \alpha_j$, $s_{\alpha_j}\alpha_i = \alpha_i$; from this we see easily that $s_{\alpha_i}s_{\alpha_j} = s_{\alpha_j}s_{\alpha_i}$. Thus the elements of $\mathfrak{w}_1$ commute with those of $\mathfrak{w}_2$. This shows that $\mathfrak{w} = \mathfrak{w}_1\mathfrak{w}_2$ and $\Delta_r = \bigcup_{i\in S_r} \mathfrak{w}_r\cdot\alpha_i$, $r = 1, 2$. The elements of Δ_r are therefore integral linear combinations of only the α_i, with $i \in S_r$. Thus $\Delta_1 \cap \Delta_2 = \varnothing$; moreover, if $s_r \in \mathfrak{w}_r$, $i \in S_1$, $j \in S_2$, then $\langle s_1\alpha_i,s_2\alpha_j\rangle = \langle\alpha_i,\alpha_j\rangle = 0$. We conclude from this that if $\alpha \in \Delta_1$ and $\beta \in \Delta_2$, $\langle\alpha,\beta\rangle = 0$ and $\alpha + \beta$ is neither 0 nor a root. Define $\mathfrak{h}_r$ to be the subspace of $\mathfrak{h}$ spanned by the H_{α_i} $(i \in S_r)$, and let

$$\mathfrak{g}_r = \mathfrak{h}_r + \sum_{\alpha\in\Delta_r} \mathfrak{g}_\alpha \quad (r = 1, 2).$$

It follows from the above observations that $\mathfrak{g}_r$ is a subalgebra of $\mathfrak{g}$ for $r = 1, 2$, $\mathfrak{g}$ is the direct sum of $\mathfrak{g}_1$ and $\mathfrak{g}_2$, and $[\mathfrak{g}_1,\mathfrak{g}_2] = 0$. Consequently, $\mathfrak{g}_1$ and $\mathfrak{g}_2$ are ideals $\neq 0$, and $\mathfrak{g}$ is their direct sum. So $\mathfrak{g}$ is not simple. This completes the proof of the theorem.

The natural question at this stage is whether one can, given an arbitrary equivalence class of Cartan matrices, construct a semisimple Lie algebra which gives rise to this class. In view of Theorem 4.3.27, we may assume the matrices in this class to be irreducible. That this can be done was first shown by E. Cartan. In his famous 1894 thesis [1] he classified the equivalence classes of irreducible Cartan matrices and showed that they form four infinite sequences and five isolated ones; he then proved the theorem by verifying that the four sequences of equivalence classes come from the classical simple Lie algebras, and by explicitly constructing simple Lie algebras corresponding to each of the five isolated classes. The simple Lie algebras corresponding to the five isolated classes of Cartan matrices are usually called *exceptional*.

It is clearly desirable to obtain a general (i.e., not involving case-by-case consideration) proof of this result. This was first done by Chevalley and Harish-Chandra. We shall prove it is §4.8 by following essentially the method of Harish-Chandra.

The above correspondence between Cartan matrices and semisimple Lie algebras can be used to determine all the simple Lie algebras over **C**. The problem is combinatorial, and its solution is effected by the use of the so-called Dynkin diagrams. Let $\mathfrak{g}$, $\mathfrak{h}$ be as usual, $S = \{\alpha_1, \ldots, \alpha_l\}$ a simple system of roots, and $A = (a_{ij})$ the associated Cartan matrix. We then associate with $\mathfrak{g}$, $\mathfrak{h}$ and S a graph consisting of l points $P_1, \ldots, P_l$, wherein P_i is given a weight proportional to $\langle\alpha_i,\alpha_i\rangle$, and P_i and P_j are connected by $a_{ij}a_{ji}$ lines ($i \neq j$). The resulting graph is called the *Dynkin diagram* of $\mathfrak{g}$. Note that $a_{ij}a_{ji}$ is ≥ 0, and is in fact $4\langle\alpha_i,\alpha_j\rangle^2/\langle\alpha_i,\alpha_i\rangle\langle\alpha_j,\alpha_j\rangle$, so that it can take only the values 0, 1, 2, 3. Since (for $i \neq j$)

$$a_{ij} = -(a_{ij}a_{ji})^{1/2}\{\langle\alpha_j,\alpha_j\rangle/\langle\alpha_i,\alpha_i\rangle\}^{1/2}, \tag{4.3.42}$$

it is clear that A is determined by the Dynkin diagram. It is easy to see that A is irreducible if and only if the Dynkin diagram has the following property: given i, j with $i \neq j$, $1 \leq i, j \leq l$, we can find integers $i_0 = i, i_1, \ldots, i_r = j$ ($r \geq 1$, $1 \leq i_0, \ldots, i_r \leq l$, $i_s \neq i_{s+1}$ for $s = 0,1, \ldots, r-1$) such that P_{i_s} and $P_{i_{s+1}}$ are connected for $s = 0,1, \ldots, r-1$. Such a diagram is said to be *connected.*

We shall first examine the classical algebras over **C**, prove that they are all simple, and determine their Dynkin diagrams (§4.4). Then we shall classify all possible connected Dynkin diagrams (§4.5). It turns out that apart from the diagrams coming from the classical Lie algebras, there are only five more possibilities. An elementary calculation reveals that each of these comes from an irreducible Cartan matrix. In view of the correspondence between Cartan matrices and semisimple Lie algebras, we then conclude that each of the five possibilities is the Dynkin diagram of a simple Lie algebra over **C**. Thus the simple Lie algebras over C are all determined (cf. also Dynkin [1]).

4.4. The Classical Lie Algebras

We now examine the structure of the classical Lie algebras. We show that they are semisimple and construct their Cartan matrices and Dynkin diagrams. The fact that they are all simple will follow from the fact that their Dynkin diagrams are connected.

Lemma 4.4.1. *Let* $\mathfrak{g}$ *be a Lie algebra over* **C**. *Let* $\mathfrak{h}$ *be an abelian subalgebra of* $\mathfrak{g}$, Δ *a finite subset of* $\mathfrak{h}^* \setminus \{0\}$. *For each* $\lambda \in \mathfrak{h}^*$ *let*

$$\mathfrak{g}_\lambda = \{X : X \in \mathfrak{g}, [H,X] = \lambda(H)X \textit{ for all } H \in \mathfrak{h}\}.$$

Suppose the following conditions are satisfied:

(i) Δ *spans* $\mathfrak{h}^*$
(ii) $\Delta = -\Delta$, *and* $[\mathfrak{g}_\alpha, \mathfrak{g}_{-\alpha}] \neq 0$ *for each* $\alpha \in \Delta$
(iii) $\mathfrak{g} = \mathfrak{h} + \sum_{\alpha \in \Delta} \mathfrak{g}_\alpha$

Then $\mathfrak{g}$ *is semisimple,* $\mathfrak{h}$ *is a CSA, and (iii) is the root space decomposition of* $\mathfrak{g}$ *with respect to* $\mathfrak{h}$.

Proof. It is obvious that (iii) is a direct sum, $\mathfrak{h} = \mathfrak{g}_0$, and $[\mathfrak{g}_\lambda, \mathfrak{g}_{\lambda'}] \subseteq \mathfrak{g}_{\lambda+\lambda'}$ for all $\lambda, \lambda' \in \mathfrak{h}^*$. In particular, $[\mathfrak{g}_\alpha, \mathfrak{g}_{-\alpha}] \subseteq \mathfrak{h}$ for $\alpha \in \Delta$. It is clear from (ii) that $\mathfrak{g}_\alpha \neq 0$ for $\alpha \in \Delta$.

Fix $\alpha \in \Delta$. In view of (ii), we may select $X'_\alpha \in \mathfrak{g}_\alpha$ and $X'_{-\alpha} \in \mathfrak{g}_{-\alpha}$ such that $H'_\alpha = [X'_\alpha, X'_{-\alpha}] \neq 0$. Now we argue as in Lemma 4.3.5 to conclude that for suitable rational numbers $q_{\beta\alpha}$,

$$\beta(H'_\alpha) = q_{\beta\alpha}\alpha(H'_\alpha) \quad (\beta \in \Delta).$$

If $\alpha(H'_\alpha) = 0$, then $\beta(H'_\alpha) = 0$ for all $\beta \in \Delta$, so $H'_\alpha = 0$ by (i). Consequently,

$$\alpha(H'_\alpha) \neq 0.$$

We may then argue as in Lemma 4.3.7 to conclude that

$$\dim \mathfrak{g}_\alpha = 1 \quad (\alpha \in \Delta).$$

Write $\bar{H}_\alpha = 2\alpha(H'_\alpha)^{-1}H'_\alpha$, $X_\alpha = X'_\alpha$, $X_{-\alpha} = 2\alpha(H'_\alpha)^{-1}X'_{-\alpha}$. Then

$$[\bar{H}_\alpha, X_\alpha] = 2X_\alpha, \qquad [\bar{H}_\alpha, X_{-\alpha}] = -2X_{-\alpha}, \qquad [X_\alpha, X_{-\alpha}] = \bar{H}_\alpha. \tag{4.4.1}$$

Let $\mathfrak{q} = \text{rad}(\mathfrak{g})$. Since $\mathfrak{q}$ is invariant under ad $\mathfrak{h}$, we must have $\mathfrak{q} = \mathfrak{q} \cap \mathfrak{h} + \sum_{\alpha \in \Delta}(\mathfrak{q} \cap \mathfrak{g}_\alpha)$. We claim that $\mathfrak{q} \cap \mathfrak{g}_\alpha = 0$ for all $\alpha \in \Delta$. Suppose otherwise, and let $\alpha \in \Delta$ be such that $\mathfrak{q} \cap \mathfrak{g}_\alpha \neq 0$. Define $\bar{H}_\alpha$, X_α, $X_{-\alpha}$ as above. Since $\dim \mathfrak{g}_\alpha = 1$, $X_\alpha \in \mathfrak{q}$. $\mathfrak{q}$ being an ideal, we conclude from (4.4.1) that $\bar{H}_\alpha$, $X_{-\alpha} \in \mathfrak{q}$ also. Now $\mathfrak{q}$ is solvable, but the space $\mathbf{C}\cdot\bar{H}_\alpha + \mathfrak{g}_\alpha + \mathfrak{g}_{-\alpha}$ is a subalgebra of $\mathfrak{q}$ which is not solvable, a contradiction. So $\mathfrak{q} \subseteq \mathfrak{h}$. If $\mathfrak{q} \neq 0$ and $H \in \mathfrak{q}$ is nonzero, we can choose $\alpha \in \Delta$ such that $\alpha(H) \neq 0$; then $X_\alpha = \alpha(H)^{-1}[H, X_\alpha] \in \mathfrak{q}$, a contradiction. Therefore $\mathfrak{q} = 0$ i.e., $\mathfrak{g}$ is semisimple. It is clear from our assumptions that $\mathfrak{h}$ is maximal abelian and ad H is semisimple for all $H \in \mathfrak{h}$. So $\mathfrak{h}$ is a CSA. This proves the lemma.

The algebras A_l ($l \geq 1$). Let $\mathfrak{g} = \mathfrak{sl}(l+1, \mathbf{C})$, l being an integer ≥ 1. It is customary to write A_l for this Lie algebra. Let $\mathfrak{h}$ be the abelian subalgebra of all diagonal elements of $\mathfrak{g}$; if $a_1, \ldots, a_{l+1} \in \mathbf{C}$, diag$(a_1, \ldots, a_{l+1})$ denotes the diagonal matrix with $a_1, \ldots, a_{l+1}$ as its diagonal entries. We write E_{ij} for the matrix whose ijth entry is 1 and all of whose other entries are 0, $1 \leq i, j \leq l+1$. Obviously, the matrices

$$E_{ii} - E_{i+1,i+1} \quad (1 \leq i \leq l), \qquad E_{ij} \quad (i \neq j, 1 \leq i, j \leq l+1)$$

form a basis for $\mathfrak{g}$.

Let $\lambda_1, \ldots, \lambda_{l+1}$ be the linear functions on $\mathfrak{h}$ defined by

$$\lambda_i: \mathrm{diag}(a_1, \ldots, a_{l+1}) \mapsto a_i.$$

Clearly,

$$\lambda_1 + \cdots + \lambda_{l+1} = 0.$$

Since

$$[\mathrm{diag}(a_1, \ldots, a_{l+1}), E_{ij}] = (a_i - a_j)E_{ij},$$

we have

$$[H, E_{ij}] = (\lambda_i - \lambda_j)(H)E_{ij} \quad (H \in \mathfrak{h}). \tag{4.4.2}$$

Let

$$\Delta = \{\lambda_i - \lambda_j : i \neq j, 1 \leq i, j \leq l+1\}$$

Then

$$\mathfrak{g} = \mathfrak{h} + \sum_{\alpha \in \Delta} \mathfrak{g}_\alpha, \quad \mathfrak{g}_{\lambda_i - \lambda_j} = \mathbf{C} \cdot E_{ij} \quad (i \neq j), \tag{4.4.3}$$

and

$$[E_{ij}, E_{ji}] = E_{ii} - E_{jj} \quad (i \neq j). \tag{4.4.4}$$

It follows from Lemma 4.4.1 that $\mathfrak{g}$ is semisimple, $\mathfrak{h}$ is a CSA, and (4.4.3) is the root-space decomposition. Note also the commutation rules

$$[E_{ij}, E_{jk}] = E_{ik} \quad (i, j, k \text{ distinct}). \tag{4.4.5}$$

We now calculate the Cartan–Killing form. For $H = \mathrm{diag}(a_1, \ldots, a_{l+1})$ and $H' = \mathrm{diag}(a'_1, \ldots, a'_{l+1})$, both in $\mathfrak{h}$, we have

$$\begin{aligned} \langle H, H' \rangle &= \sum_{i \neq j} (a_i - a_j)(a'_i - a'_j) \\ &= 2(l+1) \sum_{1 \leq i \leq l+1} a_i a'_i \end{aligned}$$

It follows from this that for $i \neq j$,

$$H_{\lambda_i - \lambda_j} = \frac{1}{2(l+1)}(E_{ii} - E_{jj}), \tag{4.4.6}$$

so

$$\bar{H}_{\lambda_i-\lambda_j} = E_{ii} - E_{jj}. \tag{4.4.7}$$

Let

$$\alpha_i = \lambda_i - \lambda_{i+1} \quad (1 \leq i \leq l). \tag{4.4.8}$$

Then

$$\Delta = \{\pm(\alpha_i + \alpha_{i+1} + \cdots + \alpha_j) : 1 \leq i < j \leq l\}. \tag{4.4.9}$$

So

$$S = \{\alpha_1, \ldots, \alpha_l\}$$

is a simple system of roots. The corresponding positive system is the set of all $\lambda_i - \lambda_j$ with $i < j$. From (4.4.6) we get

$$\langle \alpha_i, \alpha_i \rangle = \frac{1}{l+1} \quad (1 \leq i \leq l). \tag{4.4.10}$$

The rank of $\mathfrak{g}$ is obviously l.

Let $A = (a_{ij})$ be the Cartan matrix of $\mathfrak{g}$. The off-diagonal entries are now easily seen to be given by

$$a_{ij} = \begin{cases} 0 & \text{if } |j - i| \geq 2 \\ -1 & \text{if } |j - i| = 1 \end{cases} \tag{4.4.11}$$

It is clear from the above discussion that the Dynkin diagram of $\mathfrak{g}$ is

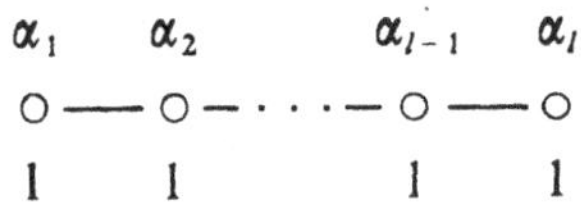

$\mathfrak{g}$ is thus simple.

Let Π_r be the group of permutations of $\{1, \ldots, r\}$. For $s \in \Pi_{l+1}$, we write $\bar{s}$ for the linear transformation of $\mathfrak{h}$ given by

$$\bar{s} : \mathrm{diag}(a_1, \ldots, a_{l+1}) \mapsto \mathrm{diag}(a_{s^{-1}(1)}, \ldots, a_{s^{-1}(l+1)})$$

If $i \neq j$, it is easily seen that the Weyl reflection corresponding to the root $\lambda_i - \lambda_j$ is $\bar{s}_{ij}$, where $s_{ij} \in \Pi_{l+1}$ is the permutation that interchanges i and j, leaving the other elements fixed. Since the permutations of this form generate Π_{l+1}, we have

$$\mathfrak{w} = \{\bar{s} : s \in \Pi_{l+1}\}; \tag{4.4.12}$$

here $\mathfrak{w}$ is the Weyl group of $(\mathfrak{g}, \mathfrak{h})$.

The algebras D_l ($l \geq 2$). Let $l \geq 2$ be an integer and V a vector space over $\mathbf{C}$ of dimension $2l$. Let $(\cdot,\cdot)$ be a non-singular symmetric bilinear form on $V \times V$. We consider the Lie algebra $\mathfrak{g}$ of all endomorphisms L of V such that

$$(Lu,v) + (u,Lv) = 0 \quad (u, v \in V). \tag{4.4.13}$$

It is customary to denote this Lie algebra by D_l.

It is easy to show the existence of a basis $\{u_1,\ldots,u_{2l}\}$ for V such that

$$(u_i,u_j) = \begin{cases} 0 & \text{if } |j-i| \neq l \\ 1 & \text{if } |j-i| = l \end{cases} \tag{4.4.14}$$

For any endomorphism L of V, let $\tilde{L}$ denote its matrix with respect to the basis $\{u_1,\ldots,u_{2l}\}$. If

$$F = \begin{pmatrix} 0 & 1 \\ 1 & 0 \end{pmatrix},$$

where 0 and 1 are the $l \times l$ zero and identity matrices respectively, then it is easy to show that $\mathfrak{g}$ is the set of all endomorphisms L for which

$$\tilde{L}^t F + F\tilde{L} = 0 \tag{4.4.15}$$

(t denoting matrix transposition). If we write, for any endomorphism L,

$$\tilde{L} = \begin{pmatrix} A & B \\ C & D \end{pmatrix},$$

where A, B, C, D are complex $l \times l$ matrices, then $L \in \mathfrak{g}$ if and only if

$$D = -A^t, \qquad B^t = -B, \qquad C^t = -C. \tag{4.4.16}$$

we write (A,B,C) for the element $L \in \mathfrak{g}$ for which

$$\tilde{L} = \begin{pmatrix} A & B \\ C & -A^t \end{pmatrix} \quad (B^t = -B, C^t = -C).$$

A straightforward calculation now shows that

$$[(A_1,B_1,C_1), (A_2,B_2,C_2)] = (A,B,C), \tag{4.4.17}$$

where

$$
\begin{aligned}
A &= [A_1,A_2] + B_1C_2 - B_2C_1 \\
B &= (A_1B_2 - A_2B_1) - (A_1B_2 - A_2B_1)^t \\
C &= (C_1A_2 - C_2A_1) - (C_1A_2 - C_2A_1)^t
\end{aligned}
\tag{4.4.18}
$$

We write E_{ij} for the $l \times l$ matrix whose ijth entry is 1, the other entries being 0. For $1 \leq p < q \leq l$, let $F_{pq} = E_{pq} - E_{qp}$.

Let $\mathfrak{h}$ be the set of all elements of $\mathfrak{g}$ of the form $(A,0,0)$, where A is an arbitrary diagonal matrix. Let λ_i $(1 \leq i \leq l)$ be the linear functions on $\mathfrak{h}$ defined by

$$
\lambda_i : (\mathrm{diag}(a_1, \ldots, a_l),0,0) \mapsto a_i. \tag{4.4.19}
$$

We then obtain the following commutation rules, valid for $1 \leq i,j \leq l$, $1 \leq p < q \leq l$, $H \in \mathfrak{h}$.

$$
\begin{aligned}
[H,(E_{ij},0,0)] &= (\lambda_i - \lambda_j)(H)(E_{ij},0,0) \\
[H,(0,F_{pq},0)] &= (\lambda_p + \lambda_q)(H)(0,F_{pq},0) \\
[H,(0,0,F_{pq})] &= -(\lambda_p + \lambda_q)(H)(0,0,F_{pq}).
\end{aligned}
\tag{4.4.20}
$$

Now,

$$
\mathfrak{g} = \mathfrak{h} + \sum_{i \neq j} \mathbf{C}\cdot(E_{ij},0,0) + \sum_{p<q} \mathbf{C}\cdot(0,F_{pq},0) + \sum_{p<q} \mathbf{C}\cdot(0,0,F_{pq}).
$$

Moreover,

$$
\begin{aligned}
[(E_{ij},0,0),(E_{ji},0,0)] &= (E_{ii} - E_{jj},0,0) \qquad (i \neq j) \\
[(0,F_{pq},0),(0,0,F_{pq})] &= (-E_{pp} - E_{qq},0,0) \quad (p < q).
\end{aligned}
$$

Therefore we conclude from Lemma 4.4.1 that $\mathfrak{g}$ is semisimple of rank l and that $\mathfrak{h}$ is a CSA. Obviously,

$$
\Delta = \{\pm(\lambda_i - \lambda_j) : 1 \leq i < j \leq l\} \cup \{\pm(\lambda_p + \lambda_q) : 1 \leq p < q \leq l\}. \tag{4.4.21}
$$

Let

$$
\alpha_i = \lambda_i - \lambda_{i+1} \quad (1 \leq i \leq l-1), \qquad \alpha_l = \lambda_{l-1} + \lambda_l \tag{4.4.22}
$$

Then $\{\alpha_1, \ldots, \alpha_l\}$ is a simple system of roots. In fact, $\Delta = P \cup -P$, where

$$
P = \{\lambda_i - \lambda_j : 1 \leq i < j \leq l\} \cup \{\lambda_p + \lambda_q : 1 \leq p < q \leq l\},
$$

and the elements of P can be expressed in terms of the α_i as follows:

$$
\begin{aligned}
\lambda_i - \lambda_j &= \alpha_i + \alpha_{i+1} + \cdots + \alpha_{j-1} && (1 \leq i < j \leq l) \\
\lambda_p + \lambda_q &= (\alpha_p + \cdots + \alpha_{l-2}) + (\alpha_q + \cdots + \alpha_l) && (1 \leq p < q \leq l).
\end{aligned}
\tag{4.4.23}
$$

Let $H = (\mathrm{diag}(a_1, \ldots, a_l), 0, 0)$, $H' = (\mathrm{diag}(a'_1, \ldots, a'_l), 0)$ be in $\mathfrak{h}$. It follows from (4.4.21) that

$$\langle H, H' \rangle = 4(l-1) \sum_{1 \le i \le l} a_i a'_i.$$

In particular, we find

$$H_{\alpha_i} = \frac{1}{4(l-1)}(E_{ii} - E_{i+1,i+1}, 0, 0) \quad (1 \le i \le l-1)$$

$$H_{\alpha_l} = \frac{1}{4(l-1)}(E_{l-1,l-1} + E_{ll}, 0, 0)$$

Therefore

$$\langle \alpha_i, \alpha_i \rangle = \frac{1}{2(l-1)} \quad (1 \le i \le l). \tag{4.4.24}$$

Let $A = (a_{ij})$ be the Cartan matrix of $\mathfrak{g}$ with respect to $\mathfrak{h}$ and the α_i. For the off-diagonal elements a_{ij} $(i \ne j)$ we have

$$a_{ij} = \begin{cases} -1 & \text{if } 1 \le i, j \le l-1 \text{ and } |j - i| = 1, \text{ or if one} \\ & \text{of } i \text{ and } j \text{ is } l, \text{ and the other is } l-2 \\ 0 & \text{otherwise} \end{cases} \tag{4.4.25}$$

Thus the Dynkin diagram is

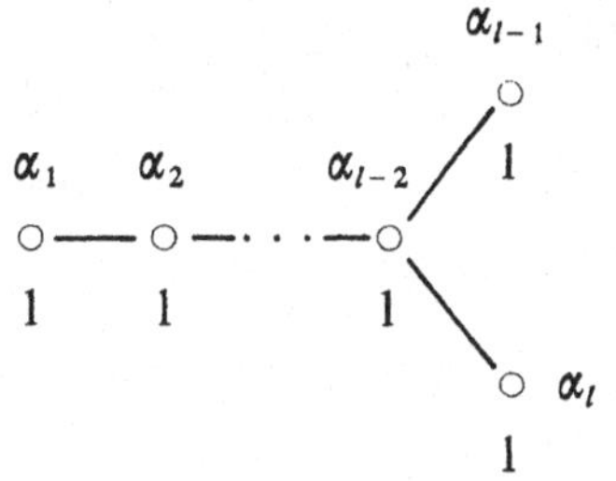

If $l = 2$, the diagram is

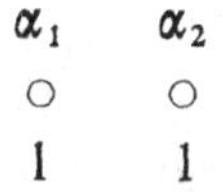

If $l = 3$, the diagram is

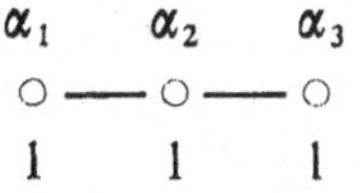

D_2 thus splits as the direct sum of two copies of A_1; D_3 is isomorphic to A_3. For $l \ge 3$, D_l is simple.

Let Π_l be the group of permutations of $\{1,\ldots,l\}$, and let $\mathcal{E}^+$ be the multiplicative group of all l-tuples $(\varepsilon_1,\ldots,\varepsilon_l)$, where the ε_i are all ± 1, $\Pi_{1\leq i\leq l}\,\varepsilon_i = 1$, and multiplication is component-wise. We allow Π_l and $\mathcal{E}^+$ to act on $\mathfrak{h}$ as follows: s: $(\mathrm{diag}(a_1,\ldots,a_l),0,0) \mapsto (\mathrm{diag}(a_{s^{-1}(1)},\ldots,a_{s^{-1}(l)}),0,0)$ and $(\varepsilon_1,\ldots,\varepsilon_l)$: $(\mathrm{diag}(a_1,\ldots,a_l),0,0) \mapsto \mathrm{diag}(\varepsilon_1 a_1,\ldots,\varepsilon_l a_l),0,0)$ $(s \in \Pi_l,\ (\varepsilon_1,\ldots,\varepsilon_l) \in \mathcal{E}^+)$. This gives faithful representations of Π_l and $\mathcal{E}^+$, so we identify Π_l and $\mathcal{E}^+$ with their images in $GL(\mathfrak{h})$. We now observe that if $1 \leq i \leq l-1$ and s_i is the permutation that interchanges i and $i+1$, then s_i is the Weyl reflexion s_{α_i}, while $(1,\ldots,1,-1,-1)$ is the Weyl reflection s_{α_l}. It follows easily from this that $\Pi_l\mathcal{E}^+ = \mathcal{E}^+\Pi_l$ is the Weyl group of $(\mathfrak{g},\mathfrak{h})$ and that it has $l!\ 2^{l-1}$ elements.

The algebras C_l $(l \geq 2)$. Let $l \geq 2$ be an integer, V a vector space of dimension $2l$ over $\mathbf{C}$. Let $(\cdot,\cdot)$ be a non-singular skew-symmetric bilinear form on $V \times V$. We now examine the Lie algebra $\mathfrak{g}$ of all endomorphisms L of V for which

$$(Lu,v) + (u,Lv) = 0 \quad (u, v, \in V) \tag{4.4.26}$$

It is customary to denote this Lie algebra by C_l. It is not difficult[2] to show that there is a basis $\{u_1,\ldots,u_{2l}\}$ for V such that

$$(u_i,u_j) = \begin{cases} 0 & \text{if } |j-i| \neq l \\ 1 & \text{if } j = l+i. \end{cases} \tag{4.4.27}$$

For any endomorphism L of V, let $\tilde{L}$ be its matrix in this basis. Then $L \in \mathfrak{g}$ if and only if

$$\tilde{L}^t F + F\tilde{L} = 0, \tag{4.4.28}$$

where

$$F = \begin{pmatrix} 0 & 1 \\ -1 & 0 \end{pmatrix}; \tag{4.4.29}$$

here 0 and 1 are the zero and identity $l \times l$ matrices. An easy calculation then

[2]This may be seen as follows. For any subspace $W \subseteq V$, let $W^\perp$ be defined as $\{w: (u,w) = 0 \text{ for all } u \in W\}$; then $\dim(W) + \dim(W^\perp) = 2l$. Since $(u,u) = 0\ \forall\ u \in V$, we can select $u_1 \neq 0$ such that $(u_1,u_1) = 0$. Then $u_1 \in (\mathbf{C}u_1)^\perp = W$, and $\dim W = 2l - 1$. Select $u_{l+1} \notin W$ such that $(u_1,u_{l+1}) = 1$. Let $V_0 = \mathbf{C}u_1 + \mathbf{C}u_{l+1}$, and write $V_1 = V_0^\perp$. Then $V = V_0 + V_1$ is a direct sum, $\dim V_1 = 2(l-1)$, and $(\cdot,\cdot)\,|\,V_1 \times V_1$ is skew-symmetric and nonsingular. Our claim follows now by induction on l.

shows that $\mathfrak{g}$ is the set of all L for which

(4.4.30)
$$\tilde{L} = \begin{pmatrix} A & B \\ C & -A^t \end{pmatrix}$$

where A, B, C are complex $l \times l$ matrices and both B and C are symmetric. We shall write $L = (A,B,C)$, if $\tilde{L}$ is as in (4.4.30).

The calculations are similar to the case of the algebras D_l. We use the same notation. In particular, $\mathfrak{h}$ is the abelian subalgebra of all elements of $\mathfrak{g}$ of the form $(A,0,0)$, where A is any $l \times l$ diagonal matrix. Let E_{ij} be as before and let $G_{pq} = E_{pq} + E_{qp}$, $1 \leq p \leq q \leq l$. Let λ_i be defined by (4.4.19). We then have the following commutation rules, valid for $1 \leq i, j \leq l$, $1 \leq p \leq$ $\leq q \leq l$, $H \in \mathfrak{h}$:

(4,4,31)
$$\begin{aligned} [H,(E_{ij},0,0)] &= (\lambda_i - \lambda_j)(H)(E_{ij},0,0) \\ [H,(0,G_{pq},0)] &= (\lambda_p + \lambda_q)(H)(0,G_{pq},0) \\ [H,(0,0,G_{pq})] &= -(\lambda_p + \lambda_q)(H)(0,0,G_{pq}). \end{aligned}$$

Now, $[(E_{ij},0,0), (E_{ji},0,0)] = (E_{ii} - E_{jj},0,0)$, while $[0,G_{pq},0),(0,0,G_{pq})] = (E_{pp} + E_{qq},0,0)$. Further,

$$\mathfrak{g} = \mathfrak{h} + \sum_{i \neq j} \mathbf{C}\cdot(E_{ij},0,0) + \sum_{p \leq q} \mathbf{C}\cdot(0,G_{pq},0) + \sum_{p \leq q} \mathbf{C}\cdot(0,0,G_{pq}),$$

so Lemma 4.4.1 is applicable, and we conclude that $\mathfrak{g}$ is semisimple, of rank l, and that $\mathfrak{h}$ is a CSA. If Δ is the set of roots, then $\Delta = P \cup -P$, where

(4.4.32)
$$P = \{\lambda_i - \lambda_j : 1 \leq i < j \leq l\} \cup \{\lambda_p + \lambda_q : 1 \leq p \leq q \leq l\}.$$

Let

(4.4.33)
$$\alpha_i = \lambda_i - \lambda_{i+1} \quad (1 \leq i \leq l-1), \qquad \alpha_l = 2\lambda_l.$$

Then $\{\alpha_1, \ldots, \alpha_l\}$ is a simple system; for

(4.4.34)
$$\begin{aligned} \lambda_i - \lambda_j &= \alpha_i + \cdots + \alpha_{j-1} && (1 \leq i < j \leq l) \\ \lambda_p + \lambda_q &= (\alpha_p + \cdots + \alpha_{l-1}) + (\alpha_q + \cdots + \alpha_l) && \\ & && (1 \leq p \leq q \leq l) \end{aligned}$$

If $H = (\mathrm{diag}(a_1, \ldots, a_l),0,0)$ and $H' = (\mathrm{diag}(a'_1, \ldots, a'_l),0,0)$ are two elements of $\mathfrak{h}$, then a simple calculation shows that

(4.4.35)
$$\langle H,H' \rangle = 4(l+1) \sum_{1 \leq i \leq l} a_i a'_i.$$

In particular we find

$$H_{\alpha_i} = \frac{1}{4(l+1)}(E_{ii} - E_{i+1,i+1},0,0) \quad (1 \leq i \leq l-1)$$

$$H_{\alpha_l} = \frac{1}{4(l+1)}(2E_{ll},0,0)$$

Thus

$$\langle\alpha_i,\alpha_i\rangle = \begin{cases} \dfrac{1}{2(l+1)} & \text{if } 1 \leq i \leq l-1) \\ \dfrac{1}{l+1} & \text{if } i = l \end{cases} \tag{4.4.36}$$

For the off-diagonal entries a_{ij} $(i \neq j)$ of the Cartan matrix $A = (a_{ij})$ associated with $\mathfrak{g}$, $\mathfrak{h}$, and $\{\alpha_1, \ldots, \alpha_l\}$, we obtain the following formulae:

$$\begin{cases} -1 & \text{if } 1 \leq i,j \leq l-1 \text{ and } |j-i| = 1, \text{ or if } i = l \text{ and } j = l-1 \\ -2 & \text{if } i = l-1 \text{ and } j = l \\ 0 & \text{otherwise} \end{cases} \tag{4.4.37}$$

The Dynkin diagram of $\mathfrak{g}$ is thus

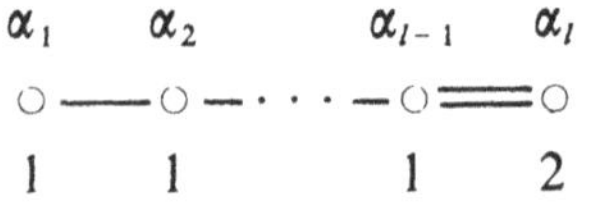

$\mathfrak{g}$ is thus simple.

Let Π be as before the group of permutations of $\{1, \ldots, l\}$, and let $\mathcal{E}$ be the multiplicative group of all l-tuples $(\varepsilon_1, \ldots, \varepsilon_l)$, where each $\varepsilon_l = \pm 1$ and multiplication is component-wise. Π_l and $\mathcal{E}$ act on $\mathfrak{h}$, if we associate with $s \in \Pi_l$ and $(\varepsilon_1, \ldots, \varepsilon_l) \in \mathcal{E}$, the linear transformations of $\mathfrak{h}$ as in the previous discussion. This gives faithful representations of Π_l and $\mathcal{E}$ in $\mathfrak{h}$, so we may identify Π_l and $\mathcal{E}$ with their images in $GL(\mathfrak{h})$. For $1 \leq i \leq l-1$, let s_i be the permutation that interchanges i and $i+1$ while leaving the others fixed. It is then easy to verify that the Weyl reflection s_{α_i} is s_i $(1 \leq i \leq l-1)$ while the reflection s_{α_l} is $(1, \ldots, 1, -1)$. It follows that $\Pi_l\mathcal{E} = \mathcal{E}\Pi_l$ is the Weyl group of $(\mathfrak{g},\mathfrak{h})$ and that it has $l!2^l$ elements.

The algebras B_l $(l \geq 1)$. Let l be an integer ≥ 1, V a vector space of dimension $2l+1$ over $\mathbf{C}$. Let $(\cdot,\cdot)$ be a non-singular symmetric bilinear form on $V \times V$. We write $\mathfrak{g}$ for the Lie algebra of all endomorphisms L of V for which

$$(Lu,v) + (u,Lv) = 0 \quad (u, v \in V). \tag{4.4.38}$$

The algebra is usually denoted by B_l. We now select a basis $\{u_0,u_1,\dots,u_{2l}\}$ for V with

$$(u_i,u_j) = \begin{cases} 1 & \text{if } i,j \geq 1 \text{ and } |j-i| = l \text{ or if } i=j=0 \\ 0 & \text{otherwise} \end{cases} \tag{4.4.39}$$

For any endomorphism L of V, let $\tilde{L}$ be its matrix in this basis. If

$$F = \begin{pmatrix} 1 & 0_{1,l} & 0_{1,l} \\ 0_{l,1} & 0_{l,l} & 1_{l,l} \\ 0_{l,1} & 1_{l,l} & 0_{l,l} \end{pmatrix},$$

where $0_{m,n}$ is the $m \times n$ zero matrix, and $1_{n,n}$ is the $n \times n$ identity matrix, then $L \in \mathfrak{g}$ if and only if

$$\tilde{L}^t F + F\tilde{L} = 0. \tag{4.4.40}$$

A simple calculation shows that this is the case if and only if

$$\tilde{L} = \begin{pmatrix} 0 & a & b \\ -b^t & A & B \\ -a^t & C & -A^t \end{pmatrix} \tag{4.4.41}$$

where A, B, C are $l \times l$ matrices, a, b are $1 \times l$ matrices, and B, C are skew-symmetric. We write

$$L = (a,b : A,B,C).$$

Let the $l \times l$ matrices E_{ij} and F_{pq} be defined as in the discussion of the algebras D_l. Let e_r $(1 \leq r \leq l)$ be the $1 \times l$ matrix whose entries are $\delta_{r1},\dots,\delta_{rl}$ (Kronecker delta).

We now introduce the set $\mathfrak{h}$ of all elements of $\mathfrak{g}$ of the form $(0,0\colon A,0,0)$, where A is an arbitrary diagonal matrix. $\mathfrak{h}$ is clearly an abelian subalgebra of $\mathfrak{g}$. Let λ_r $(1 \leq r \leq l)$ be the linear functions on $\mathfrak{h}$ defined by

$$\lambda_r : (0,0 : \operatorname{diag}(a_1,\dots,a_l),0,0) \mapsto a_r \tag{4.4.42}$$

Then the following commutation rules are valid for $1 \leq i,j \leq l$, $1 \leq p < q \leq l$, $1 \leq r \leq l$ and $H \in \mathfrak{h}$:

$$\begin{aligned} [H,0,0 : E_{ij},0,0)] &= (\lambda_i - \lambda_j)(H)(0,0 : E_{ij}, 0,0) \\ [H,(0,0 : 0,F_{pq},0)] &= (\lambda_p + \lambda_q)(H)(0,0 : 0,F_{pq},0) \\ [H,(0,0 : 0,0,F_{pq})] &= -(\lambda_p + \lambda_q)(H)(0,0 : 0,0,F_{pq}) \\ [H,(0,e_r : 0,0,0)] &= \lambda_r(H)(0,e_r : 0,0,0) \\ [H,(e_r,0 : 0,0,0)] &= -\lambda_r(H)(e_r,0 : 0,0,0) \end{aligned} \tag{4.4.43}$$

The first three relations follow at once from (4.4.20); the last two follow from the relation

$$[(0,0:A,0,0),(a,b:0,0,0)] = (-aA,bA^t:0,0,0).$$

The conditions of Lemma 4.4.1 are now easily seen to be satisfied. In fact, since the elements of $\mathfrak{g}$ of the form $(0,0\colon A,B,C)$ form a subalgebra isomorphic to D_l, this reduces to verifying that $[(0,e_r\colon 0,0,0),(e_r,0\colon 0,0,0)] \neq 0$; but this commutator is $(0,0\colon -E_{rr},0,0)$, so Lemma 4.4.1 may now be used to conclude that $\mathfrak{g}$ is semisimple, is of rank l, and that $\mathfrak{h}$ is a CSA. The set Δ of roots is $P \cup -P$, where

$$P = \{\lambda_i - \lambda_j : 1 \leq i < j \leq l\} \cup \{\lambda_p + \lambda_q : 1 \leq p < q \leq l\} \cup \{\lambda_r : 1 \leq r \leq l\}. \tag{4.4.44}$$

Let

$$\alpha_i = \lambda_i - \lambda_{i+1} \quad (1 \leq i \leq l-1), \qquad \alpha_l = \lambda_l. \tag{4.4.45}$$

Then

$$\begin{aligned} \lambda_i - \lambda_j &= \alpha_i + \cdots + \alpha_{j-1} && (1 \leq i < j \leq l) \\ \lambda_p + \lambda_q &= (\alpha_p + \cdots + \alpha_l) + (\alpha_q + \cdots + \alpha_l) && (1 \leq p < q \leq l) \\ \lambda_r &= \alpha_r + \cdots + \alpha_l && (1 \leq r \leq l) \end{aligned} \tag{4.4.46}$$

These formulae show that $\{\alpha_1, \ldots, \alpha_l\}$ is a simple system of roots.

If $H = (0,0\colon \operatorname{diag}(a_1, \ldots, a_l),0,0)$ and $H' = (0,0\colon \operatorname{diag}(a'_1, \ldots, a'_l),0,0)$ are in $\mathfrak{h}$, a simple calculation shows that

$$\langle H,H' \rangle = 2(2l-1) \sum_{1 \leq i \leq l} a_i a'_i. \tag{4.4.47}$$

In particular, we have

$$\begin{aligned} H_{\alpha_i} &= \frac{1}{2(2l-1)}(0,0 : E_{ii} - E_{i+1,i+1},0,0) \quad (1 \leq i \leq l-1) \\ H_{\alpha_l} &= \frac{1}{2(2l-1)}(0,0 : E_{ll},0,0). \end{aligned} \tag{4.4.48}$$

Thus

$$\langle \alpha_i,\alpha_i \rangle = \begin{cases} 1/2l-1 & 1 \leq i \leq l-1 \\ 1/2(2l-1) & i = l. \end{cases} \tag{4.4.49}$$

The off-diagonal entries a_{ij} $(i \neq j)$ of the Cartan matrix associated with $\mathfrak{g}$, $\mathfrak{h}$, and $\{\alpha_1, \ldots, \alpha_l\}$ are thus given by

$$
(4.4.50) \qquad a_{ij} = \begin{cases} -1 & \text{if } 1 \leq i, j \leq l-1 \text{ and } |j-i| = 1 \text{ or if } i = l-1, j = l \\ -2 & \text{if } i = l, j = l-1 \\ 0 & \text{otherwise} \end{cases}
$$

The Dynkin diagram of $\mathfrak{g}$ is thus

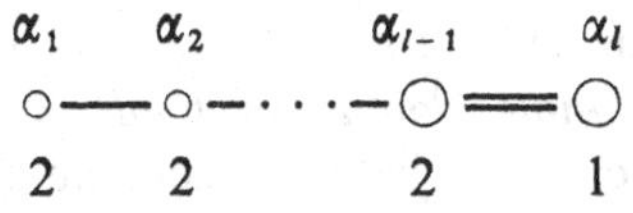

$\mathfrak{g}$ is thus simple. For $l = 1$, B_1 is isomorphic to A_1; for $l = 2$, B_2 is isomorphic to C_2.

Let Π_l and $\mathcal{E}$ have the same meaning as in the discussion of the C_l. Proceeding in an analogous manner, we may define faithful representations of Π_l and $\mathcal{E}$ in $\mathfrak{h}$, using which we can identify Π_l and $\mathcal{E}$ with their images in $GL(\mathfrak{h})$. The form (4.4.45) of the simple roots α_i shows as before that $\Pi_l\mathcal{E} = \mathcal{E}\Pi_l$ is the Weyl group of $(\mathfrak{g},\mathfrak{h})$.

Summarizing, we have shown that the classical algebras A_l $(l \geq 1)$, B_l $(l \geq 2)$, C_l $(l \geq 3)$, and D_l $(l \geq 4)$ are all simple, and have determined the corresponding Dynkin diagrams. It is obvious from the diagrams that these Lie algebras are mutually nonisomorphic.

4.5. Determination of the Simple Lie Algebras over C

We now proceed to the determination of all the simple Lie algebras over **C** (up to isomorphism). Our aim is to prove that in addition to the classical simple Lie algebras A_l $(l \geq 1)$, B_l $(l \geq 2)$, C_l $(l \geq 3)$, and D_l $(l \geq 4)$, there are precisely five exceptional simple Lie algebras: G_2, F_4, E_6, E_7, and E_8 (of respective ranks 2, 4, 6, 7, and 8), and that no two of these are isomorphic.

Let V be a real Euclidean space, the scalar product and norm of which are denoted respectively by $(\cdot,\cdot)$ and $|\cdot|$. By a *scheme* (*in* V) we understand a set $S \subseteq V$ with the following properties:

(4.5.1)
- (i) the elements of S are linearly independent.
- (ii) if $\alpha, \beta \in S$ and $\alpha \neq \beta$, then $a_{\alpha,\beta} = 2(\alpha,\beta)/(\alpha,\alpha)$ is an integer ≤ 0 (by the Schwartz inequality, $a_{\alpha,\beta}a_{\beta,\alpha} \in \{0,1,2,3\}$ in this case).

The number of vectors in S is called the *rank* of S (rk(S)); each vector of S is called a *vertex* of S. It is obvious that each subset of a scheme is a scheme, and these will be referred to as *subschemes*. A scheme S is said to be *connected* if, given $\alpha, \beta \in S$ with $\alpha \neq \beta$, one can find $\gamma_0 = \alpha, \gamma_1, \ldots, \gamma_p = \beta \in S$ such that $(\gamma_i, \gamma_{i+1}) \neq 0$, $0 \leq i \leq p-1$.

Given a scheme S, one can associate with it a diagram consisting of points, with weights and lines connecting pairs of these points, called the *graph* of S. This is done as follows. We represent the vertices α of S by points α with weights proportional to $|\alpha|^2$, and connect distinct vertices α, β by $a_{\alpha,\beta}a_{\beta,\alpha}$ lines. A pair $\{\alpha,\beta\}$ (unordered) of vertices of S is called a *link* if they are connected by at least one line; if $\alpha, \beta \in S' \subseteq S$, the link is said to be *from* S'. A link is said to be *simple*, *double*, or *triple* according as the number of connecting lines is 1, 2, or 3. A *cycle* in a scheme S is a subset $\{\alpha_1, \ldots, \alpha_{n+1}\}$ ($n \geq 1$) such that $\{\alpha_1,\alpha_2\},\{\alpha_2,\alpha_3\}, \ldots, \{\alpha_n,\alpha_{n+1}\},\{\alpha_{n+1},\alpha_1\}$ are all links. A *chain* is a scheme $\{\alpha_1, \ldots, \alpha_{n+1}\}$ ($n \geq 1$) such that $\{\alpha_i,\alpha_{i+1}\}$ ($1 \leq i \leq n$) are precisely all its links. Chains are connected schemes. A chain is called *simple* if all its links are simple. Note that if $\{\alpha,\beta\}$ is a simple link in a scheme S, $a_{\alpha,\beta} = a_{\beta,\alpha} = -1$, so

(4.5.2) $$|\alpha|^2 = |\beta|^2, \qquad (\alpha,\alpha) + 2(\alpha,\beta) = 0.$$

Thus the graph of a simple chain has the form

(4.5.3)

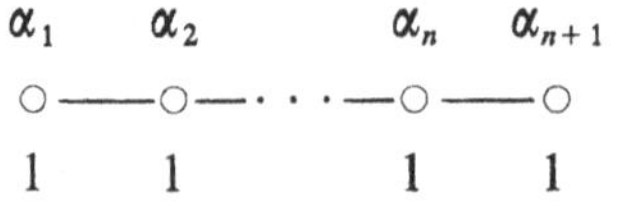

A vertex of a scheme is said to be *simple*, *double*, *triple*, etc. according as the number of lines issuing from it is 1, 2, 3, etc.

Lemma 4.5.1. *Let S be a scheme with n elements. Then there cannot be more than $n-1$ links from S.*

Proof. Let $S = \{\alpha_1, \ldots, \alpha_n\}$, let P be the set of links from S, and let p be the number of members of P. Define $\alpha = \sum_{1 \leq i \leq n} |\alpha_i|^{-1}\alpha_i$. Clearly, $\alpha \neq 0$, so

(4.5.4) $$0 < |\alpha|^2 = n + 2 \sum_{\{\alpha,\beta\} \in P} ((\alpha,\beta)/|\alpha| \cdot |\beta|).$$

But $a_{\alpha,\beta}a_{\beta,\alpha} \geq 1$ for $\{\alpha,\beta\} \in P$, implying that $2(\alpha,\beta)/|\alpha|\cdot|\beta|$ is ≤ -1 for $\{\alpha,\beta\} \in P$. So we find that $n - p > 0$ from (4.5.4).

Corollary 4.5.2. *No scheme can contain a cycle. If S is a scheme, S_1 is a connected subscheme $\neq S$, and $\beta \in S \setminus S_1$, then β cannot be linked to more than one vertex of S_1.*

Proof. Since a cycle with n elements has at least n links, the first assertion is clear. For the second, if $\{\alpha_i,\beta\}$ are links ($\alpha_1, \alpha_2 \in S_1, \alpha_1 \neq \alpha_2$), we can find $\gamma_0 = \alpha_1, \gamma_1, \ldots, \gamma_p = \alpha_2$ in S_1 such that $\{\gamma_i,\gamma_{i+1}\}$ are all links ($0 \leq i \leq p-1$); then $\{\beta,\gamma_0,\ldots,\gamma_p\}$ would be a cycle.

Corollary 4.5.3. *The number of lines issuing from a given vertex in a scheme cannot exceed* 3.

Proof. Let α be a vertex of S, $\beta_1,\ldots,\beta_p$ distinct vertices such that $\{\alpha,\beta_i\}$ are all links, $1 \leq i \leq p$. If γ is the orthogonal projection of α on the linear span of the β_i, $(\gamma,\gamma) < (\alpha,\alpha)$. On the other hand, $(\beta_i,\beta_j) = 0$ for $i \neq j$, for otherwise $\{\alpha,\beta_i,\beta_j\}$ would be a cycle. So $2\gamma = \sum_{1\leq i\leq p} a_{\beta_i,\alpha}\,\beta_i$, and $4(\gamma,\gamma) = (\sum_{1\leq i\leq p} a_{\beta_i,\alpha}\, a_{\alpha,\beta_i})\,(\alpha,\alpha)$. We therefore have $\sum_{1\leq i\leq p} a_{\beta_i,\alpha} a_{\alpha,\beta_i} < 4$.

Lemma 4.5.4. *Let $C = \{\alpha_1,\ldots,\alpha_{n+1}\}$ $(n \geq 1)$ be a simple chain in a scheme S, and let $\alpha = \alpha_1 + \cdots + \alpha_{n+1}$. Then $|\alpha|^2 = |\alpha_i|^2$ $(1 \leq i \leq n+1)$, and $(S \setminus C) \cup \{\alpha\}$ is a scheme whose graph can be obtained from that of S by replacing C by the single vertex α and connecting any $\beta \in S \setminus C$ to α by p lines, where p is the multiplicity with which β is linked to some vertex[3] of C.*

Proof. That $|\alpha|^2 = |\alpha_i|^2$ $(1 \leq i \leq n+1)$ follows at once from (4.5.2). Fix $\beta \in S \setminus C$, and choose i with $1 \leq i \leq n+1$ such that $(\beta,\alpha_j) = 0$ for $j \neq i$. Then $(\beta,\alpha) = (\beta,\alpha_i)$, so $2(\beta,\alpha)/(\alpha,\alpha) = a_{\alpha_i,\beta}$, $2(\beta,\alpha)/(\beta,\beta) = a_{\beta,\alpha_i}$. This implies the assertions of the lemma.

Lemma 4.5.5. *Let C be a chain with at least one double link. Then the graph of C has the form*

(X_{pq}):

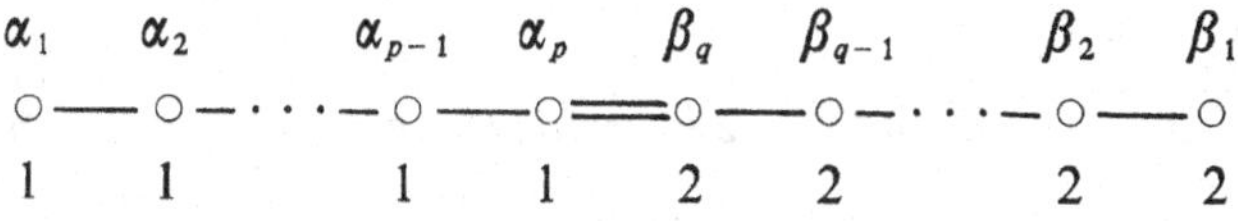

where p, q are integers ≥ 1, and either $\min(p,q) = 1$ or $p = q = 2$.

Proof. We observe first that C does not have more than one double link. If it did there would be a subscheme C' of C whose graph has the form

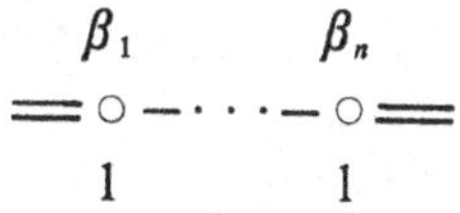

[3] By Corollary 4.5.2, there can be at most one such vertex, so p is uniquely determined; also, $p = 0$ if there is no such vertex.

The double lines at β_1 and β_n indicate that β_1 and β_n are linked to vertices in $C \setminus C'$ with double links; but then, on using Lemma 4.5.4 and replacing C' by the single vector $\beta = \beta_1 + \cdots + \beta_n$, we would obtain a scheme in which 4 lines issue from a vertex, namely β. Thus the graph of C has the form X_{pq} with $p, q \geq 1$.

Now let $\alpha = \sum_{1 \leq i \leq p} i\alpha_i$, $\beta = \sum_{1 \leq j \leq q} j\beta_j$; then α, β are linearly independent, and $(\alpha,\beta) = pq(\alpha_p,\beta_q)$. Further, it follows without difficulty from (4.5.2) that $(\alpha,\alpha) = \frac{1}{2}p(p+1)(\alpha_p,\alpha_p)$ and $(\beta,\beta) = \frac{1}{2}q(q+1)(\beta_q,\beta_q)$. Now, $(\beta_q,\beta_q) = 2(\alpha_p,\alpha_p)$, so $(\alpha_p,\beta_q) = -(\alpha_p,\alpha_p)$. The inequality $(\alpha,\beta)^2 < (\alpha,\alpha)(\beta,\beta)$ then gives us $2pq < (p+1)(q+1)$, or $(p-1)(q-1) < 2$. Since p and q are integers ≥ 1, either $\min(p,q) = 1$ or $p = q = 2$. This proves the lemma.

Lemma 4.5.6. *Let S be a connected scheme whose links are all simple and whose graph has a triple vertex. Then the graph of S has the form*

(Y_{pqr}):

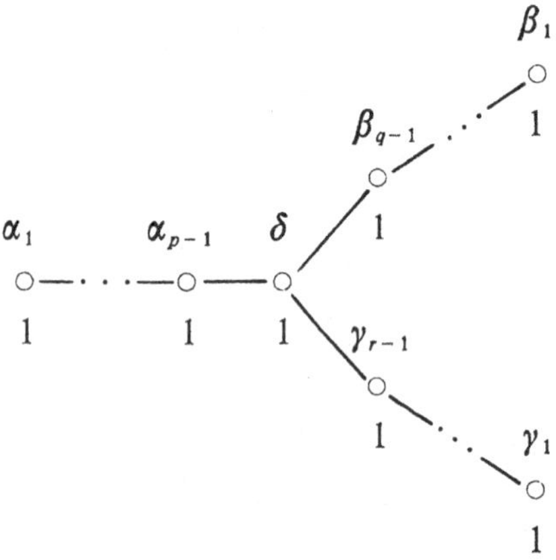

where either $q = r = 2$ and $p \geq 2$ is arbitrary, or $r = 2$, $q = 3$, and $3 \leq p \leq 5$.

Proof. Note first that S does not have more than one triple vertex. If it does, the connectedness of S implies the existence of a scheme $\subseteq S$ whose graph has the form

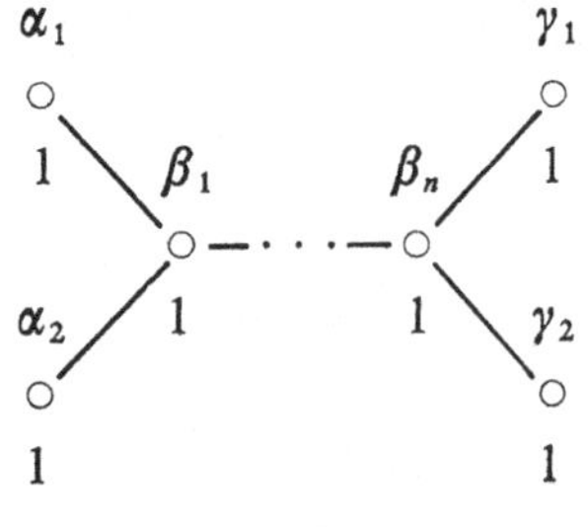

On once again using Lemma 4.5.4 and replacing $\{\beta_1, \ldots, \beta_n\}$ by the single vector $\beta = \beta_1 + \cdots + \beta_n$, we would then obtain a scheme with a vertex (namely β) from which 4 lines issue.

Let δ be the unique triple vertex. Then S has subschemes with graphs of the form Y_{abc} with δ as the triple vertex. Let S' be a maximal such subscheme. We assert that $S' = S$. Otherwise, since S is connected, we can find a link $\{\alpha,\beta\}$ with $\alpha \in S'$ and $\beta \in S \setminus S'$. If α is an extreme vertex of S', $S' \cup \{\beta\}$ would be a scheme with a graph of the form Y_{abc}, contradicting the maximality of S'; $\alpha \neq \delta$, since δ is already a triple vertex of S'; α cannot be one of the remaining vertices of S', for that would mean that S, has more than one triple vertex. We must therefore have $S' = S$. Clearly, we may assume that the graph of S is of the form Y_{pqr} with $p \geq q \geq r \geq 2$.

Let $\alpha = \sum_{1 \leq i \leq p-1} i\alpha_i$, $\beta = \sum_{1 \leq j \leq q-1} j\beta_j$, $\gamma = \sum_{1 \leq k \leq r-1} k\gamma_k$. Let c be the common value of $|v|^2$ for $v \in S$. Then

$$(\alpha,\alpha) = \tfrac{1}{2}p(p-1)c, \qquad (\beta,\beta) = \tfrac{1}{2}q(q-1)c, \qquad (\gamma,\gamma) = \tfrac{1}{2}r(r-1)c.$$

Moreover, α, β, and γ are mutually orthogonal and

$$(\alpha,\delta) = -\tfrac{1}{2}(p-1)c, \qquad (\beta,\delta) = -\tfrac{1}{2}(q-1)c, \qquad (\gamma,\delta) = -\tfrac{1}{2}(r-1)c.$$

The vectors α, β, γ, and δ are linearly independent, so the matrix M of their scalar products is positive definite. A simple calculation shows that $\det(M) = (c^4/16)(p-1)(q-1)(r-1)(pq + qr + rp - pqr)$. So we must have $pq + qr + rp > pqr$ or

$$p^{-1} + q^{-1} + r^{-1} > 1. \tag{4.5.5}$$

If $p \geq q \geq r \geq 3$, then $p^{-1} + q^{-1} + r^{-1} \leq 1$. So $r = 2$ for any integral solution (p,q,r) (with $p \geq q \geq r$) of (4.5.5), and $p^{-1} + q^{-1} > \frac{1}{2}$. If $q \geq 4$, $p^{-1} + q^{-1} \leq \frac{1}{2}$, so $q = 2$ or 3. If $q = 2$, $p \geq 2$, and if $q = 3$, $p < 6$. The integral triple (p,q,r) with $p \geq q \geq r \geq 2$ is thus seen to satisfy (4.5.5) if and only if either $q = r = 2$ and $p \geq 2$ is arbitrary, or $r = 2$, $q = 3$ and $3 \leq p \leq 5$. This completes the proof.

Theorem 4.5.7. *Let S be a connected scheme of rank l. Then the graph of S has one of the following forms:*

(A_l): 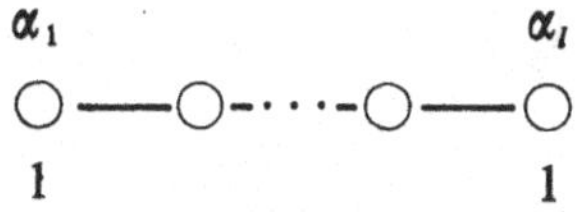

$(l \geq 1)$

(B_l): 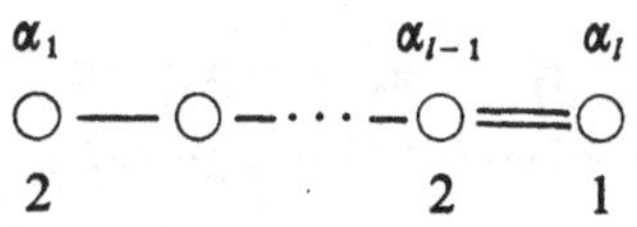

$(l \geq 2)$

$$
(C_l): \quad
\begin{array}{cccccc}
\alpha_1 & \alpha_2 & & \alpha_{l-1} & \alpha_l \\
\bigcirc & \!\!-\!\!\bigcirc\!\!-\cdots-\!\! & \bigcirc & \!\!=\!\!=\!\! & \bigcirc \\
1 & 1 & & 1 & 2
\end{array}
\qquad (l \geq 3)
$$

$$
(D_l): \qquad\qquad\qquad\qquad\qquad\qquad (l \geq 4)
$$

(4.5.6)

$$
(G_2): \quad
\begin{array}{cc}
\alpha_1 & \alpha_2 \\
\bigcirc & \!\!\equiv\!\!\bigcirc \\
1 & 3
\end{array}
$$

$$
(F_4): \quad
\begin{array}{cccc}
\alpha_1 & \alpha_2 & \alpha_3 & \alpha_4 \\
\bigcirc\!\!-\!\!\!\! & \bigcirc\!\!=\!\!=\!\! & \bigcirc\!\!-\!\! & \bigcirc \\
1 & 1 & 2 & 2
\end{array}
$$

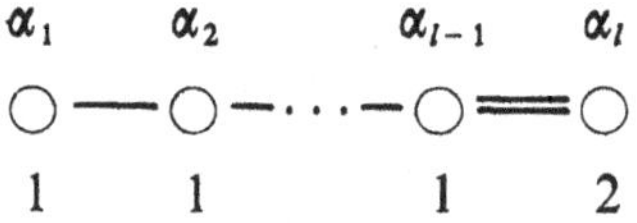

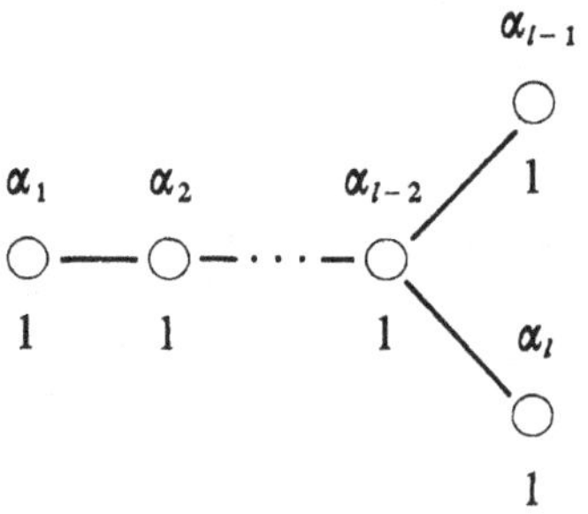

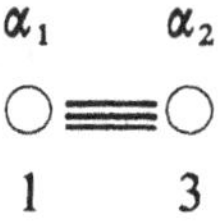

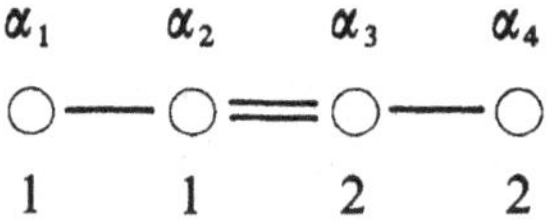

(E_6):

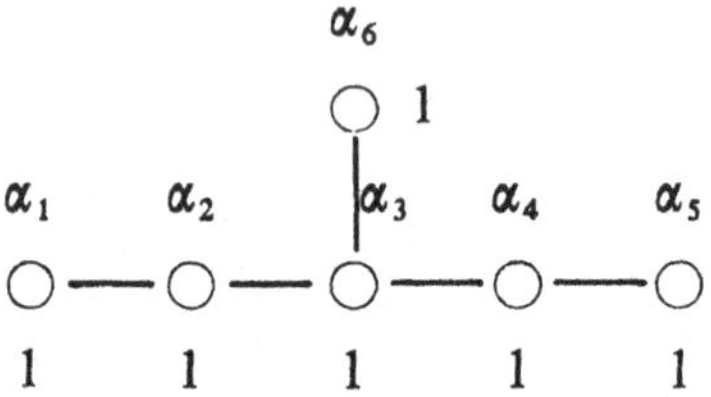

(E_7):

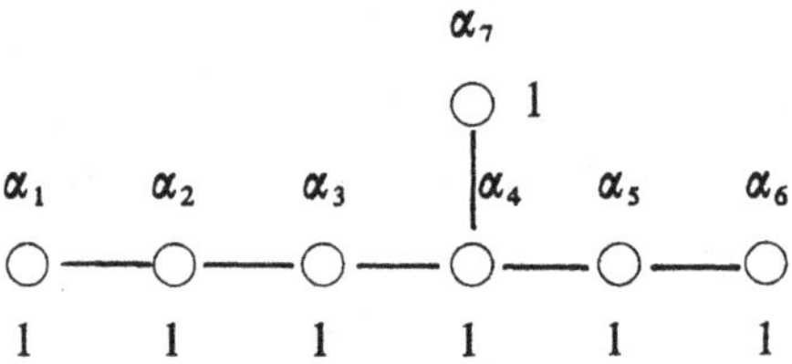

(E_8):

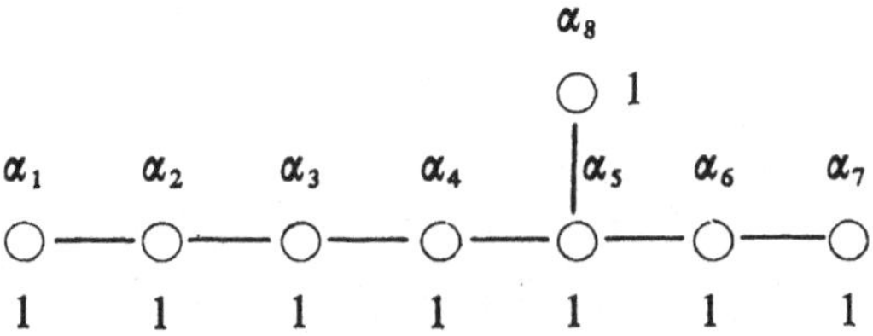

Proof. Let Γ be the graph of S. If $l = 1$, $\Gamma = A_1$. So we may assume that $l \geq 2$. Suppose first that Γ has a triple link $\{\alpha,\beta\}$. By Corollary 4.5.3, neither of α and β can be linked with any other vertex of S. Since S is connected, $S = \{\alpha,\beta\}$, and hence $\Gamma = G_2$. We may assume therefore that Γ has no triple link.

Assume next that all links of Γ are simple. Let C be a maximal chain of S. If $C = S$, $\Gamma = A_l$. Suppose now that $C \neq S$. Let the graph of C be

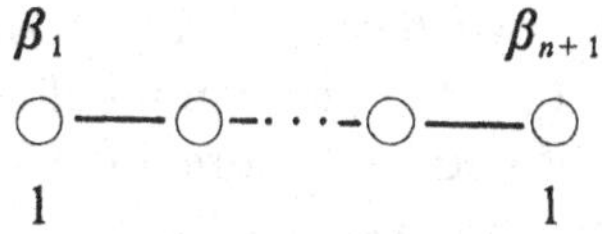

$(n \geq 1)$.

Since S is connected, there would exist a vertex in $S \setminus C$ linked with some β_j. j cannot be 1 or $n + 1$, because C is a maximal chain. So $2 \leq j \leq n$, $l \geq 4$, and S has a triple vertex. Lemma 4.5.6 is now applicable and allows us to conclude that either $\Gamma = D_l$ or Γ is one of E_6, E_7, E_8.

It remains to consider the case when Γ has a double link. Let C be a maximal chain with a double link. By Lemma 4.5.5, the graph of C has the form X_{pq} with either $\min(p,q) = 1$ or $p = q = 2$. We assert that $C = S$. Suppose $C \neq S$. Since S is connected, there exists a vertex β of $S \setminus C$ linked with a vertex α of C. α cannot be either of the extreme vertices of the chain C, because C is maximal; nor can α be one of the vertices forming the double link from C. α is thus one of the remaining vertices of C, so S contains a subscheme S' with graph

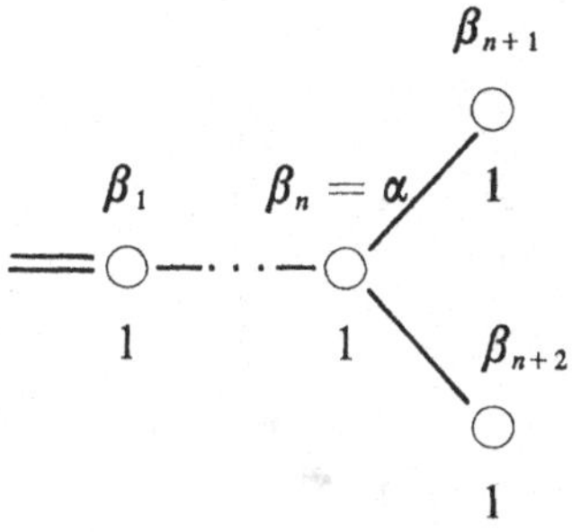

$(n \geq 2)$,

the double line at β_1 indicating that β_1 is in double link with a vertex of S. By Lemma 4.5.4, we may then replace $\{\beta_1, \ldots, \beta_n\}$ by $\bar{\beta} = \beta_1 + \cdots + \beta_n$ and deduce the existence of a scheme having a vertex (namely $\bar{\beta}$) from which 4 lines issue. This contradiction shows that $C = S$. But then Lemma 4.5.5 implies that either $\Gamma = B_l\ (l \geq 2)$, $\Gamma = C_l\ (l \geq 3)$, or $\Gamma = F_4$. This completes the proof of the theorem.

Remark. We have not shown that each of the above diagrams is the graph of a scheme. We shall presently prove a much stronger result. However, an independent proof of this fact is not difficult. It is based on the simple observation that an $l \times l$ matrix is the matrix of scalar products of l linearly independent vectors in a Euclidean space if and only if it is positive definite. We leave it to the reader.

Let $\mathfrak{g}$ be a simple Lie algebra over $\mathbf{C}$, $\mathfrak{h}$ a CSA, and $S = \{\alpha_1, \ldots, \alpha_l\}$ a simple system of roots of $(\mathfrak{g}, \mathfrak{h})$. Then it follows from our work in §4.3 that S is a connected scheme (the Euclidean space in which S is imbedded in the space $\sum_{1 \leq i \leq l} \mathbf{R} \cdot \alpha_i$ equipped with the Cartan–Killing form), and that its graph is none other than what we have called the Dynkin diagrams of $\mathfrak{g}$. We may therefore conclude that the Dynkin diagram of a simple Lie algebra over $\mathbf{C}$ has one of the forms indicated in Theorem 4.5.7. On the other hand, it follows from the calculations of §4.4 that the diagrams (A_l) — (D_l) are the Dynkin diagrams of the classical simple Lie algebras over $\mathbf{C}$. In order to complete the determination of the simple Lie algebras over $\mathbf{C}$, one need therefore only examine which of the remaining graphs are the Dynkin diagrams of simple Lie algebras.

For any scheme $S = \{\alpha_1, \ldots, \alpha_l\}$ let us associate the $l \times l$ matrix $A(S)$ whose ijth element is $2(\alpha_i, \alpha_j)/(\alpha_i, \alpha_i)$ $(1 \leq i, j \leq l)$. In view of the relations that exist between Cartan matrices, Dynkin diagrams and semisimple Lie algebras (cf. §§4.3, 4.8), S will be the Dynkin diagram of a complex simple Lie algebra if and only if S is connected and $A(S)$ is a Cartan matrix. For the "exceptional" graphs of Theorem 4.5.7 the associated matrices $A(S)$ are as follows:

$$A(G_2) = \begin{pmatrix} 2 & -3 \\ -1 & 2 \end{pmatrix} \tag{4.5.7}$$

$$A(F_4) = \begin{pmatrix} 2 & -1 & 0 & 0 \\ -1 & 2 & -2 & 0 \\ 0 & -1 & 2 & -1 \\ 0 & 0 & -1 & 2 \end{pmatrix} \tag{4.5.8}$$

$$A(E_8) = \begin{pmatrix} 2 & -1 & 0 & 0 & 0 & 0 & 0 & 0 \\ -1 & 2 & -1 & 0 & 0 & 0 & 0 & 0 \\ 0 & -1 & 2 & -1 & 0 & 0 & 0 & 0 \\ 0 & 0 & -1 & 2 & -1 & 0 & 0 & 0 \\ 0 & 0 & 0 & -1 & 2 & -1 & 0 & -1 \\ 0 & 0 & 0 & 0 & -1 & 2 & -1 & 0 \\ 0 & 0 & 0 & 0 & 0 & -1 & 2 & 0 \\ 0 & 0 & 0 & 0 & -1 & 0 & 0 & 2 \end{pmatrix} \tag{4.5.9}$$

$A(E_7)$ is the matrix obtained from $A(E_8)$ by deleting the first row and column; $A(E_6)$ is obtained from $A(E_8)$ by deleting the first two rows and columns.

Turning now to the conditions defining a Cartan matrix, it is easy to verify that the five matrices described above satisfy (i) and (ii) of (4.3.39). It can be shown that they also satisfy the crucial finiteness condition (iii) of (4.3.39). The verification of this is quite straightforward, although somewhat tedious; a method of doing this is indicated in the Exercises at the end of this chapter. One may therefore conclude that each of the exceptional diagrams of Theorem 4.5.7 is the Dynkin diagram of a simple Lie algebra over **C**. It is customary to denote these simple algebras by the same symbol as the corresponding Dynkin diagram, and refer to them as the *exceptional Lie algebras*. We have thus obtained the following result, using the results of §4.8:

Theorem 4.5.8. *The Lie algebras* A_l $(l \geq 1)$, B_l $(l \geq 2)$, C_l $(l \geq 3)$, D_l $(l \geq 4)$, G_2, F_4, E_6, E_7, *and* E_8 *are simple. Any simple Lie algebra over* **C** *is isomorphic to exactly one of these.*

We refer the reader to the Exercises for some of the details concerning the dimensions, root structures, and representations of the exceptional Lie algebras.

Corollary 4.5.9. *The connected schemes are precisely the Dynkin diagrams of complex simple Lie algebras.*

4.6. Representations with a Highest Weight

Let $\mathfrak{g}$ be a semisimple Lie algebra over **C**. One of the main problems of the theory is the determination of the irreducible finite-dimensional representations of $\mathfrak{g}$. Let $\mathfrak{h}$ be a CSA, $S = \{\alpha_1, \ldots, \alpha_l\}$ a simple system of roots, and $H_i = \bar{H}_{\alpha_i}$ $(1 \leq i \leq l)$. Let $\mathfrak{h}^*$ be the dual of $\mathfrak{h}$. Then we shall be able to associate with any $\lambda \in \mathfrak{h}^*$ an irreducible representation π_λ of $\mathfrak{g}$ in such a manner that the correspondence sending λ to the equivalence class of π_λ is one-to-one. However, π_λ is in general infinite-dimensional. We shall prove that π_λ is finite-dimensional if and only if $\lambda(H_i)$ is an integer ≥ 0 for $1 \leq i \leq l$, and that corresponding representations π_λ exhaust all the irreducible finite-dimensional representations of $\mathfrak{g}$ up to equivalence.

It was E. Cartan who first proved that the irreducible finite-dimensional representations of $\mathfrak{g}$ are determined by l-tuples of nonnegative integers. His proof was, however, based on the classification of simple Lie algebras over **C** and required detailed calculations in the case of the exceptional Lie algebras [2]. General proofs (i.e., those not based on the classification of simple algebras) were first given by H. Weyl [2, 3, 4]. Weyl's methods were transcendental, involving integrations on compact groups. It was Harish-Chandra who first suceeded in developing a general algebraic theory of the irreducible representations of $\mathfrak{g}$ [4].

The method of Harish-Chandra, also discovered independently by Chevalley [6], goes much further. Let $A = (a_{ij})_{1 \leq i,j \leq l}$ be an arbitrary Cartan matrix of rank l. Let $\mathfrak{g}$ be a Lie algebra over $\mathbf{C}$, *not necessarily of finite dimension*, with the property that $\mathfrak{g}$ is generated by $3l$ linearly independent elements H_i, X_i, Y_i $(1 \leq i \leq l)$ satisfying the following commutation rules, valid for $1 \leq i, j \leq l$:

$$(4.6.1) \qquad \begin{aligned} [H_i,H_j] &= 0, & [X_i,Y_j] &= \delta_{ij}H_i. \\ [H_i,X_j] &= a_{ij}X_j, & [H_i,Y_j] &= -a_{ij}Y_j. \end{aligned}$$

Then the method of Chevalley–Harish-Chandra, as modified by Jacobson [1] and Serre [1], leads to a complete description of the irreducible finite-dimensional representations of $\mathfrak{g}$. Taking for $\mathfrak{g}$ a (*finite-dimensional*) semisimple Lie algebra, one may obtain by this method the main results of the representation theory of such Lie algebras. One the other hand, we may take for $\mathfrak{g}$ the *universal* Lie algebra corresponding to the commutation rules (4.6.1); then one can use the finite-dimensional representations of this algebra to construct a (finite-dimensional) semisimple Lie algebra having A as its Cartan matrix, thereby solving the problem raised in §4.3.

Let $A = (a_{ij})_{1 \leq i,j \leq l}$ be a Cartan matrix of rank l. We work with a Lie algebra $\mathfrak{g}$ over $\mathbf{C}$, generated by $3l$ linearly independent elements H_i, X_i, Y_i $(1 \leq i \leq l)$ which satisfy the commutation rules (4.6.1). *Unless otherwise stated, neither $\mathfrak{g}$ nor its representations are finite-dimensional.* Let $\mathfrak{h}$ be the linear span of $H_1, \ldots, H_l$, and let α_j $(1 \leq j \leq l)$ be the linear functions on $\mathfrak{h}$ defined by

$$(4.6.2) \qquad \alpha_j(H_i) = a_{ij} \quad (1 \leq i \leq l).$$

Note that the α_j are linearly independent because $\det(A) \neq 0$. Put

$$(4.6.3) \qquad \Gamma = \{m_1\alpha_1 + \cdots + m_l\alpha_l : m_1, \ldots, m_l \text{ are integers} \geq 0 \text{ with } m_1 + \cdots + m_l > 0\}.$$

Γ induces a partial ordering $\prec$ on $\mathfrak{h}^*$; elements $\lambda, \lambda' \in \mathfrak{h}^*$ being given, $\lambda \prec \lambda'$ (or $\lambda' \succ \lambda$) if $\lambda' - \lambda \in \Gamma$. $\lambda \in \mathfrak{h}^*$ is called *integral* if $\lambda(H_i)$ is an integer for $1 \leq i \leq l$; it is called *dominant* if $\lambda(H_i)$ is real and ≥ 0 for $1 \leq i \leq l$. The elements of Γ are clearly integral. For $q = m_1\alpha_1 + \cdots + m_l\alpha_l \in \Gamma \cup \{0\}$, let $O(q) = m_1 + \cdots + m_l$.

We shall begin with some lemmas on the algebraic structure of $\mathfrak{g}$.

Lemma 4.6.1. *For any $\lambda \in \mathfrak{h}^*$ let*

$$(4.6.4) \qquad \mathfrak{g}_\lambda = \{X : X \in \mathfrak{g}, [H,X] = \lambda(H)X \text{ for all } H \in \mathfrak{h}\}.$$

Put

$$(4.6.5) \qquad \mathfrak{n}^+ = \sum_{\lambda \in \Gamma} \mathfrak{g}_\lambda, \qquad \mathfrak{n}^- = \sum_{\lambda \in \Gamma} \mathfrak{g}_{-\lambda}$$

Then $\mathfrak{n}^+$ (resp. $\mathfrak{n}^-$) is the subalgebra of $\mathfrak{g}$ generated by $X_1, \ldots, X_l$ (resp. $Y_1, \ldots, Y_l$), and $\mathfrak{g} = \mathfrak{h} + \mathfrak{n}^+ + \mathfrak{n}^-$, the sum being direct. In particular, $\mathfrak{h} = \mathfrak{g}_0$ and is its own normalizer in $\mathfrak{g}$.

Proof. Since $[\mathfrak{g}_\lambda,\mathfrak{g}_{\lambda'}] \subseteq \mathfrak{g}_{\lambda+\lambda'}$ for $\lambda, \lambda' \in \mathfrak{h}^*$, $\mathfrak{n}^+$ and $\mathfrak{n}^-$ are subalgebras of $\mathfrak{g}$. Let $\mathfrak{q}^+$ (resp. $\mathfrak{q}^-$) be the linear span of the X_k and all elements of the form $(\text{ad } X_{i_1} \cdots \text{ad } X_{i_\nu})(X_j)$ $(1 \leq i_1, \ldots, i_\nu, j, k \leq l)$ (resp. Y_k and the $(\text{ad } Y_{i_1} \cdots \text{ad } Y_{i_\nu})(Y_j)$). Clearly, $\mathfrak{q}^\pm \subseteq \mathfrak{n}^\pm$, while $\mathfrak{h}, \mathfrak{n}^+, \mathfrak{n}^-$ are linearly independent. Put $\hat{\mathfrak{g}} = \mathfrak{h} + \mathfrak{q}^+ + \mathfrak{q}^-$. Clearly, it is enough to prove that $\mathfrak{q}^\pm = \mathfrak{n}^\pm$ and $\mathfrak{g} = \mathfrak{h} + \mathfrak{n}^+ + \mathfrak{n}^-$. All of this will be immediate if we prove that $\hat{\mathfrak{g}} = \mathfrak{g}$. Since $\hat{\mathfrak{g}}$ contains H_i, X_i, and Y_i which generate $\mathfrak{g}$, it is enough to prove that $\hat{\mathfrak{g}}$ is a subalgebra of $\mathfrak{g}$. We shall, in fact, prove that $[\mathfrak{g},\hat{\mathfrak{g}}] \subseteq \hat{\mathfrak{g}}$. Since H_i, X_i, Y_i $(1 \leq i \leq l)$ generate $\mathfrak{g}$, it is sufficient to verify that $\text{ad } H_i$, $\text{ad } X_i$, and $\text{ad } Y_i$ leave $\hat{\mathfrak{g}}$ invariant for $1 \leq i \leq l$.

Since $\mathfrak{q}^+$ is spanned by the X_k and the elements $(\text{ad } X_{i_1} \cdots \text{ad } X_{i_\nu})(X_j)$ $(1 \leq i_1, \ldots, i_\nu, j, k \leq l)$, and these are eigenvectors for $\text{ad } H (H \in \mathfrak{h})$, we see that $[\mathfrak{h},\mathfrak{q}^+] \subseteq \mathfrak{q}^+$. Similarly, $[\mathfrak{h},\mathfrak{q}^-] \subseteq \mathfrak{q}^-$, so $[\mathfrak{h},\hat{\mathfrak{g}}] \subseteq \hat{\mathfrak{g}}$. We shall prove that $[X_i,\hat{\mathfrak{g}}] \subseteq \hat{\mathfrak{g}}$, $1 \leq i \leq l$; the argument for Y_i is similar. Fix i, $1 \leq i \leq l$. Clearly, $[X_i,\mathfrak{q}^+] \subseteq \mathfrak{q}^+$ and $[X_i,\mathfrak{h}] \subseteq \mathbf{C}\cdot X_i \subseteq \mathfrak{q}^+$. From (4.6.1) we get $[X_i,Y_j] \in \mathfrak{h}$, $1 \leq j \leq l$. We now prove by induction on $\nu \geq 1$ that $[X_i, Y] \in \mathfrak{q}^-$, where $Y = (\text{ad } Y_{j_1} \cdots \text{ad } Y_{j_\nu})(Y_j)$ $(1 \leq j_1, \ldots, j_\nu, j \leq l)$. For $\nu = 1$, $[X_i, Y] = \delta_{ij_1}[H_i, Y_j] + \delta_{ij}[Y_{j_1},H_i] \in \mathbf{C}\cdot Y_{j_1} + \mathbf{C}\cdot Y_j$. If $\nu > 1$, writing $Z = (\text{ad } Y_{j_2} \cdots \text{ad } Y_{j_\nu})(Y_j)$, we find $[X_i, Y] = [Y_{j_1}[X_i,Z]] + \delta_{ij_1}[H_i,Z] \in \mathfrak{q}^-$ by the induction hypothesis. So $[X_i,\mathfrak{q}^-] \subseteq \mathfrak{q}^-$. Thus ad X_i leaves $\hat{\mathfrak{g}}$ invariant by $1 \leq i \leq l$. As observed earlier, this proves the lemma.

Corollary 4.6.2. *dim $\mathfrak{g}_{\pm\alpha_i} = 1 (1 \leq i \leq l)$, and dim $\mathfrak{g}_{\pm\lambda} < \infty$ $(\lambda \in \Gamma)$. In particular, the dimension of $\mathfrak{g}$ is at most countable.*

Proof. $\mathfrak{n}^+$ is spanned by the X_k and the $(\text{ad } X_{j_1} \cdots \text{ad } X_{j_\nu})(X_j)$ $(1 \leq j_1, \ldots, j_\nu, j \leq l, \nu \geq 1)$, which belong respectively to $\mathfrak{g}_{\alpha_k}$ and $\mathfrak{g}_\lambda$ with $\lambda = \alpha_{j_1} + \cdots + \alpha_{j_\nu} + \alpha_j$. It is clear that if $\nu \geq 1$ and $1 \leq i \leq l$, $\alpha_{j_1} + \cdots + \alpha_{j_\nu} + \alpha_j$ cannot coincide with α_i for any choice of $j_1, \ldots, j_\nu, j$, and moreover that for any $\lambda \in \Gamma$, $\lambda = \alpha_{j_1} + \cdots + \alpha_{j_\nu} + \alpha_j$ for only finitely many choices of $\nu, j_1, \ldots, j_\nu$, and j. It follows at once that dim $\mathfrak{g}_{\alpha_i} = 1$ $(1 \leq i \leq l)$ and dim $\mathfrak{g}_\lambda < \infty$ $(\lambda \in \Gamma)$. The arguments for $\mathfrak{g}_{-\alpha_i}$ and $\mathfrak{g}_{-\lambda}$ are similar.

Lemma 4.6.3 *Define*[4], *for $1 \leq i, j \leq l$ with $i \neq j$,*

$$\theta_{ij}^+ = (\text{ad } X_i)^{-a_{ij}+1}(X_j) \tag{4.6.6}$$
$$\theta_{ij}^- = (\text{ad } Y_i)^{-a_{ij}+1}(Y_j).$$

Then, for $1 \leq k \leq l$,

$$[X_k,\theta_{ij}^-] = 0, \qquad [Y_k,\theta_{ij}^+] = 0. \tag{4.6.7}$$

[4]Note that for $i \neq j$, $a_{ij} \leq 0$.

Proof. Since the proofs for θ_{ij}^+ and θ_{ij}^- are similar, we prove only the relation $[Y_k,\theta_{ij}^+] = 0$. If $k \neq i$ and $k \neq j$, this is immediate, since $[Y_k,X_i] = [Y_k,X_j] = 0$. There are thus two remaining cases.

Case 1: $k = i$. Since H_i, X_i, and Y_i span a subalgebra isomorphic to $\mathfrak{sl}(2,\mathbf{C})$, we have, by (4.2.4)

$$(\text{ad } Y_i)(\text{ad } X_i)^{\mu+1} = (\text{ad } X_i)^{\mu+1}(\text{ad } Y_i) - (\mu + 1)(\text{ad } X_i)^{\mu}(\text{ad } H_i + \mu \cdot 1)$$

for any integer $\mu \geq 0$. In particular, for $\mu = -a_{ij}$,

$$\begin{aligned}[Y_i,\theta_{ij}^+] &= (\text{ad } X_i)^{-a_{ij}+1}(\text{ad } Y_i)(X_j) - (-a_{ij} + 1)(\text{ad } X_i)^{-a_{ij}}(\text{ad } H_i - a_{ij}1)(X_j)\\ &= 0\end{aligned}$$

from (4.6.1).

Case 2: $k = j$. Then ad Y_j commutes with ad X_i, so

$$[Y_j,\theta_{ij}^+] = -(\text{ad } X_i)^{-a_{ij}+1}(H_j).$$

Now, $[X_i,H_j] \in \mathbf{C}\cdot X_i$, so the result follows if $-a_{ij} > 0$. If $a_{ij} = 0$, then $a_{ji} = 0$ (cf. (4.3.39)), so $[Y_j,\theta_{ij}^+] = -[X_i,H_j] = 0$ by (4.6.1).

Let π be a representation of $\mathfrak{g}$ in a vector space V. As we mentioned earlier, we allow V to be infinite-dimensional. Let $\mathfrak{G}$ be the universal enveloping algebra of $\mathfrak{g}$. We extend π to a representation of $\mathfrak{G}$ in V and denote the extension by π too. For any $\lambda \in \mathfrak{h}^*$ let

(4.6.8) $$V_\lambda = \{v : v \in V, \pi(H)v = \lambda(H)v \text{ for all } H \in \mathfrak{h}\}.$$

λ is said to be a *weight* of π if $V_\lambda \neq 0$; V_λ is then called the *weight subspace corresponding to the weight* λ. π is said to be a *representation with weights* if $V = \sum_{\lambda \in \mathfrak{h}^*} V_\lambda$.

Lemma 4.6.4. *Let π be a representation of $\mathfrak{g}$ in a vector space V. Then the V_λ for distinct $\lambda \in \mathfrak{h}^*$ are linearly independent; and for any subspace U invariant under all $\pi(H)$ $(H \in \mathfrak{h})$,*

(4.6.9) $$U \cap \sum_{\lambda \in \mathfrak{h}^*} V_\lambda = \sum_{\lambda \in \mathfrak{h}^*} (U \cap V_\lambda).$$

If π admits a nonzero cyclic vector lying in some V_λ, then π is a representation with weights.

Proof. Let U be a subspace invariant under $\pi(H)$ for all $H \in \mathfrak{h}$. To prove (4.6.9), it is enough to check the inclusion $U \cap \sum_{\lambda \in \mathfrak{h}^*} V_\lambda \subseteq \sum_{\lambda \in \mathfrak{h}^*} (U \cap V_\lambda)$. Let $u \in U \cap \sum_{\lambda \in \mathfrak{h}^*} V_\lambda$. We may assume that for some $s \geq 1$, $u = u_1 + \cdots + u_s$, where $u_i \in V_{\lambda_i}$, $u_i \neq 0$, and $\lambda_1, \ldots, \lambda_s$ are distinct elements of $\mathfrak{h}^*$.

Let $H_0 \in \mathfrak{h}$ be an element of $\mathfrak{h}$ such that the numbers $\lambda_1(H_0), \ldots, \lambda_s(H_0)$ are all distinct. For $1 \leq i \leq s$, we can construct a polynomial p_i in one variable such that $p_i(\lambda_j(H_0)) = \delta_{ij}$ $(1 \leq j \leq s)$. Write L_i for the endomorphism $p_i(\pi(H_0))$ of V. It is then clear that $L_i u = u_i$. Since U is invariant under all $\pi(H)$ $(H \in \mathfrak{h})$, $u_i \in U$ for $1 \leq i \leq s$. This proves that $u \in \sum_{1 \leq i \leq s} (U \cap V_{\lambda_i})$. Taking $U = 0$, we see at once that the V_λ for distinct $\lambda \in \mathfrak{h}^*$ are linearly independent.

To prove the last assertion, we note first that

$$\pi(X)[V_\lambda] \subseteq V_{\lambda+\alpha} \quad (\lambda \in \mathfrak{h}^*, X \in \mathfrak{g}_\alpha, \alpha \in \Gamma \cup -\Gamma) \tag{4.6.10}$$

This shows that $\sum_{\lambda \in \mathfrak{h}^*} V_\lambda = V^0$ is invariant under π. If there is a nonzero vector in V^0 that is cyclic for π, $V^0 = V$. This proves the lemma.

Given a representation π of $\mathfrak{g}$ in V, there can be at most one weight λ with the property that $\lambda' \prec \lambda$ for any weight λ' of π other than λ; if there is such a weight λ, it is called the *highest weight* of π.

Our aim now is to investigate the structure of representations with a highest weight.

Lemma 4.6.5. *Let π be a representation of $\mathfrak{g}$ in a vector space V. Suppose $v \in V$ is a nonzero vector such that*

$$\begin{cases} \text{(i)} & v \in V_\lambda \textit{ for some } \lambda \in \mathfrak{h}^* \\ \text{(ii)} & \pi(X_i)v = 0,\ 1 \leq i \leq l \\ \text{(iii)} & v \textit{ is cyclic for } \pi. \end{cases} \tag{4.6.11}$$

Then π is a representation with weights, and λ is the highest weight of π. Moreover, we have

$$\begin{cases} \text{(i)} & \pi(X)v = 0 \quad (X \in \mathfrak{n}^+) \\ \text{(ii)} & \dim V_\lambda = 1,\ \dim V_\mu < \infty \textit{ for } \mu \in \mathfrak{h}^*, \textit{ and } V_\mu = 0 \textit{ unless } \\ & \lambda - \mu \in \Gamma \cup \{0\}. \\ \text{(iii)} & V = \pi[\mathfrak{N}^-]v, \textit{ where } \mathfrak{N}^- \textit{ is the subalgebra of } \mathfrak{G} \textit{ generated by } \\ & \mathfrak{n}^-. \end{cases} \tag{4.6.12}$$

If, moreover, π is irreducible, then

$$V_\lambda = \{u : u \in V, \pi(X_i)u = 0 \text{ for } 1 \leq i \leq l\}. \tag{4.6.13}$$

Proof. Let v be as above. We have already observed that π is a representation with weights. (i) of (4.6.12) is obvious, since the X_i generate $\mathfrak{n}^+$.

Let us write

$$v_{j_1,\ldots,j_\nu} = \pi(Y_{j_1} \cdots Y_{j_\nu})v \quad (\nu \geq 1, 1 \leq j_1, \ldots, j_\nu \leq l).$$

Obviously, (4.6.10) implies that $v_{j_1,\ldots,j_\nu} \in V_{\lambda-(\alpha_{j_1}+\cdots+\alpha_{j_\nu})}$. Let $\bar{V}$ be the linear span of v and all the $v_{j_1,\ldots,j_\nu}$. $\bar{V}$ is clearly invariant under $\pi(H_i)$ and $\pi(Y_i)$, $1 \leq i \leq l$. We shall prove that it is invariant under $\pi(X_i)$, $1 \leq i \leq l$. Fix i. It is enough to prove that $\pi(X_i)v_{j_1,\ldots,j_\nu} \in \bar{V}$ for all $\nu \geq 1$, $1 \leq j_1, \ldots, j_\nu \leq l$. We do this by induction on ν. If $\nu = 1$, we have, from (4.6.1), $\pi(X_i)v_{j_1} = \pi(Y_{j_1})\pi(X_i)v + \delta_{ij_1}\pi(H_i)v = \delta_{ij_1}\lambda(H_i)v \in \bar{V}$. If $\nu > 1$, then writing $v' = \pi(Y_{j_2} \cdots Y_{j_\nu})v$, we have

$$\pi(X_i)v_{j_1,\ldots,j_\nu} = \pi(Y_{j_1})\pi(X_i)v' + \delta_{ij_1}\pi(H_i)v',$$

which belongs to $\bar{V}$ by the induction hypothesis. So $\bar{V}$ is invariant under $\pi(X_i)$, $1 \leq i \leq l$. Since $\mathfrak{g}$ is generated by H_i, X_i, and Y_i $(1 \leq i \leq l)$, $\bar{V}$ is invariant under π.

Now $v \in \bar{V}$. So by (iii) of (4.6.11), $\bar{V} = V$. This already proves (iii) of (4.6.12). Further, if $\nu \geq 1$, then $v_{j_1,\ldots,j_\nu} \in V_{\lambda'}$, where $\lambda' = \lambda - (\alpha_{j_1} + \cdots + \alpha_{j_\nu}) \prec \lambda$. Consequently, since v and these vectors span V, we conclude that $V_\lambda = \mathbf{C}\cdot v$, that λ is the highest weight of π, and that for any $\mu \prec \lambda$, V_μ is spanned by those $v_{j_1,\ldots,j_\nu}$ for which $\nu, j_1, \ldots, j_\nu$ satisfy the relation $\alpha_{j_1} + \cdots + \alpha_{j_\nu} = \lambda - \mu$. This proves (ii) of (4.6.12).

Let π be irreducible, and let $U = \{u : u \in V, \pi(X_i)u = 0 \text{ for } 1 \leq i \leq l\}$. It is easy to check that U is invariant under all $\pi(H)$, $H \in \mathfrak{h}$. So $U = \sum_{\mu\in\mathfrak{h}^*} (U \cap V_\mu)$ by Lemma 4.6.4. If $u \in U \cap V_\mu$ and $u \neq 0$, the irreducibility of π implies that u is cyclic for π. So by what we have proved just now, μ is the highest weight of π. By uniqueness of highest weights, $\mu = \lambda$. This shows that $U \subseteq V_\lambda$, proving (4.6.13). This completes the proof of the Lemma.

We now examine the problem of constructing irreducible representations with highest weights. Suppose π is an irreducible representation of $\mathfrak{g}$ in a vector space V. Suppose that π has the highest weight λ. Then $\lambda + \alpha_i$ cannot be a weight of π for $1 \leq i \leq l$, so $\pi(X_i)[V_\lambda] = 0$, $1 \leq i \leq l$, by (4.6.10). In particular, any nonzero vector $v \in V_\lambda$ satisfies the conditions (4.6.11). So $\dim V_\lambda = 1$. Now write

$$\mathfrak{M}_\pi = \{a : a \in \mathfrak{G}, \pi(a)[V_\lambda] = 0\}. \tag{4.6.14}$$

Then $\mathfrak{M}_\pi$ is a maximal left ideal of $\mathfrak{G}$ containing X_i and $H_i - \lambda(H_i)1$, $1 \leq i \leq l$. Obviously, $\mathfrak{M}_\pi$ depends only on the equivalence class of π, and π is equivalent to the representation of $\mathfrak{G}$ in $\mathfrak{G}/\mathfrak{M}_\pi$. Conversely, suppose that $\lambda \in \mathfrak{h}^*$ and that $\mathfrak{M}$ is a maximal left ideal of $\mathfrak{G}$ containing X_i and $H_i - \lambda(H_i)1$, $1 \leq i \leq l$. Write $V = \mathfrak{G}/\mathfrak{M}$, let v be the image of 1 in V, and let π be the natural representation of $\mathfrak{G}$ in V. Then $\pi(H_i)v = \lambda(H_i)v$, and $\pi(X_i)v = 0$ for $1 \leq i \leq l$; moreover, π is irreducible. It then follows from Lemma 4.6.5 that π has λ as its highest weight.

The above discussion shows that for a given $\lambda \in \mathfrak{h}^*$, the equivalence classes of irreducible representations of $\mathfrak{g}$ with λ as their highest weight are in natural one-to-one correspondence with maximal left ideals of $\mathfrak{G}$ containing X_i and $H_i - \lambda(H_i)1$, $1 \leq i \leq l$. The next lemma clarifies the situation completely.

Lemma 4.6.6. *Let* $\lambda \in \mathfrak{h}^*$, *and let*

$$\mathfrak{G}_\lambda = \sum_{1 \leq i \leq l} \mathfrak{G}X_i + \sum_{1 \leq i \leq l} \mathfrak{G}(H_i - \lambda(H_i)1). \tag{4.6.15}$$

Then $\mathfrak{G}_\lambda$ *is a left ideal of* $\mathfrak{G}$, $\mathfrak{G}$ *is the direct sum of* $\mathfrak{G}_\lambda$ *and* $\mathfrak{N}^-$, *and there is a unique maximal left ideal of* $\mathfrak{G}$ *containing* $\mathfrak{G}_\lambda$. *In particular, given any* $\lambda \in \mathfrak{h}^*$, *there is exactly one equivalence class of irreducible representations of* $\mathfrak{g}$ *admitting* λ *as highest weight.*

Proof. Fix $\lambda \in \mathfrak{h}^*$. Let $\mathfrak{N}^+$, $\mathfrak{N}^-$, and $\mathfrak{H}$ be the respective subalgebras of $\mathfrak{G}$ generated by $\mathfrak{n}^+$, $\mathfrak{n}^-$, and $\mathfrak{h}$. $\mathfrak{H}$ is clearly isomorphic to the algebra $\mathbf{C}[T_1, \ldots, T_l]$ ($T_1, \ldots, T_l$ indeterminates) under the isomorphism that sends H_i to T_i, $1 \leq i \leq l$. $\mathfrak{n}^+$, $\mathfrak{n}^-$, and $\mathfrak{h}$ being subalgebras of $\mathfrak{g}$, we may canonically identify their respective universal enveloping algebras with $\mathfrak{N}^+$, $\mathfrak{N}^-$, and $\mathfrak{H}$ (cf. Theorem 3.2.5). It is then an easy consequence of the Poincaré–Birkhoff–Witt theorem that there is a linear isomorphism ψ of $\mathfrak{N}^- \otimes \mathfrak{H} \otimes \mathfrak{N}^+$ onto $\mathfrak{G}$ (as vector spaces) that sends $x^- \otimes h \otimes x^+$ to x^-hx^+. In particular, $\mathfrak{G} = \mathfrak{N}^-\mathfrak{H}\mathfrak{N}^+$. From the isomorphism of $\mathfrak{H}$ with $\mathbf{C}[T_1, \ldots, T_l]$ it follows that $\mathfrak{H}$ is the direct sum of $\mathbf{C}\cdot 1$ and $\mathfrak{H}_\lambda$, where

$$\mathfrak{H}_\lambda = \sum_{1 \leq i \leq l} \mathfrak{H}(H_i - \lambda(H_i)1).$$

Moreover, by Corollary 3.2.6, $\mathfrak{N}^+$ is the direct sum of $\mathfrak{N}^+\mathfrak{n}^+$ and $\mathbf{C}\cdot 1$. Therefore, applying ψ, we see that

$$\mathfrak{G} = \mathfrak{G}\mathfrak{n}^+ + \mathfrak{N}^-\mathfrak{H}_\lambda + \mathfrak{N}^-, \tag{4.6.16}$$

the sum being direct. In particular, $\mathfrak{G} = \mathfrak{G}\mathfrak{n}^+ + \mathfrak{N}^-\mathfrak{H}$, the sum being direct.

Now $\mathfrak{n}^+ = \sum_{\alpha \in \Gamma} \mathfrak{g}_\alpha$. If $X \in \mathfrak{g}_\alpha$, $H \in \mathfrak{h}$, then $XH = (H - \alpha(H)\cdot 1)X$. It follows from this that $X\mathfrak{H} \subseteq \mathfrak{G}X$. Consequently, $\mathfrak{G}\mathfrak{n}^+\mathfrak{H} \subseteq \mathfrak{G}\mathfrak{n}^+$. But then $\mathfrak{G}\mathfrak{H}_\lambda = \mathfrak{G}\mathfrak{n}^+\mathfrak{H}_\lambda + \mathfrak{N}^-\mathfrak{H}\mathfrak{H}_\lambda \subseteq \mathfrak{G}\mathfrak{n}^+ + \mathfrak{N}^-\mathfrak{H}_\lambda$. We may thus conclude that $\mathfrak{G}_\lambda \subseteq \mathfrak{G}\mathfrak{n}^+ + \mathfrak{N}^-\mathfrak{H}_\lambda$. On the other hand, since the X_i generate the Lie algebra $\mathfrak{n}^+$, it is obvious that $\sum_{1 \leq i \leq l} \mathfrak{G}X_i = \mathfrak{G}\mathfrak{n}^+$. Hence,

$$\mathfrak{G}_\lambda = \mathfrak{G}\mathfrak{n}^+ + \mathfrak{N}^-\mathfrak{H}_\lambda. \tag{4.6.17}$$

(4.6.16) and (4.6.17) show at once that $\mathfrak{G}$ is the direct sum of $\mathfrak{G}_\lambda$ and $\mathfrak{N}^-$.

The relations (4.6.16) and (4.6.17) show that $\mathfrak{G}_\lambda$ is *properly contained in* $\mathfrak{G}$. So $\mathfrak{G}_\lambda$ is a (proper)[5] left ideal of $\mathfrak{G}$, and we may consider $V = \mathfrak{G}/\mathfrak{G}_\lambda$. Let v

[5] All left ideals are proper by definition.

be the image of 1 in V, and let π be the natural representation of $\mathfrak{G}$ in V. It is obvious that $v \neq 0$ is a cyclic vector for π, and that $\pi(H_i)v = \lambda(H_i)v$, $\pi(X_i)v = 0$, $1 \leq i \leq l$. Lemma 4.6.5 now implies that π is a representation with weights, that λ is its highest weight, and that $\dim V_\lambda = 1$. Write $V^\times = \sum^{\cdot}_{\mu \neq \lambda} V_\mu$.

Suppose U is a subspace of V invariant under π and not containing v. By (4.6.9),

$$U = \sum_{\mu \in \mathfrak{h}^*} (U \cap V_\mu).$$

Since $\dim V_\lambda = 1$ and $v \notin U$, $U \cap V_\lambda = 0$. So $U \subseteq V^\times$. Consequently, if $U_{\max}$ is the linear span of all π-invariant subspaces not containing v, $U_{\max}$ is π-invariant, and $U_{\max} \subseteq V^\times$. In particular, $U_{\max}$ does not contain v. In other words, the π-invariant subspaces of V that do not contain v are all contained in a unique largest such subspace, namely $U_{\max}$. Now, the correspondence which associates with any π-invariant subspace U of V the subset $\mathfrak{M}(U) = \{a : a \in \mathfrak{G}, \pi(a)v \in U\}$ of $\mathfrak{G}$ is a bijection of the set of all π-invariant subspaces of V that do not contain v onto the set of all left ideals of $\mathfrak{G}$ that contain $\mathfrak{G}_\lambda$; moreover, this bijection preserves the inclusion relations. Consequently, if we put $\mathfrak{M}_\lambda = \mathfrak{M}(U_{\max})$, then $\mathfrak{M}_\lambda$ is a left ideal of $\mathfrak{G}$ containing every left ideal of $\mathfrak{G}$ that contains $\mathfrak{G}_\lambda$. Lemma 4.6.6 follows at once.

We write π_λ for any irreducible representation with λ as its highest weight, and $\mathfrak{M}_\lambda$ for the unique maximal left ideal defined above.

The crucial question at this stage is the following: when is the irreducible representation with highest weight λ finite-dimensional? We now examine this problem. We begin with a definition. An endomorphism L of a not necessarily finite-dimensional vector space V is called *locally nilpotent* if for each $v \in V$ there is an integer $n = n_v \geq 1$ such that $L^n v = 0$.

For any i, $1 \leq i \leq l$, let s_i be the transformation of $\mathfrak{h}^*$ into itself given by

$$s_i\mu = \mu - \mu(H_i)\alpha_i \quad (\mu \in \mathfrak{h}^*). \tag{4.6.18}$$

It is easy to see that s_i is a reflexion and that $s_i\alpha_j = \alpha_j - a_{ij}\alpha_i$, $1 \leq i, j \leq l$. From condition (iii) of (4.3.39) we see that $s_1, \ldots, s_l$ generate a finite subgroup of $GL(\mathfrak{h})$. We write $\mathfrak{w}$ for this group.

Lemma 4.6.7. *Let π be a representation of $\mathfrak{g}$ in a vector space V. Assume that:*

(i) *π is a representation with weights;*
(ii) *each weight subspace of V is finite-dimensional;*
(iii) *$\pi(X_i)$ and $\pi(Y_i)$ are locally nilpotent for $1 \leq i \leq l$. Then the weights of π are all integral, and the set of weights of π is invariant under $\mathfrak{w}$.*

Proof. It is enough to prove that if μ is a weight of π and $1 \leq i \leq l$, $\mu(H_i)$ is an integer and $s_i\mu$ is a weight of π. Fix a weight μ of π and an integer i with $1 \leq i \leq l$.

Since $\dim V_\mu < \infty$ and $\pi(Y_iX_i)$ leaves V_μ invariant (cf. (4.6.10)), we can find a $c_0 \in \mathbf{C}$ and a nonzero $v_0 \in V_\mu$ such that $\pi(Y_iX_i)v_0 = c_0v_0$. Note that $\pi(X_iY_i)v_0 = (c_0 + \mu(H_i))v_0$. Write $v_r = \pi(X_i^r)v$ $(r = 0,1,\ldots)$ and $v_{-r} = \pi(Y_i^r)v$ $(r = 0,1,\ldots)$. Clearly, $v_t \in V_{\mu+t\alpha_i}$ $(t \in \mathbf{Z})$. By (iii) above we can find integers $p, q \geq 0$ such that $v_p \neq 0$, $v_{-q} \neq 0$, but $v_{p+1} = v_{-(q+1)} = 0$. Note that $v_t \neq 0$, $-q \leq t \leq p$.

We claim that $U = \sum_{-q\leq t\leq p} \mathbf{C}\cdot v_t$ is invariant under $\pi(H_i)$, $\pi(X_i)$, and $\pi(Y_i)$. The invariance under $\pi(H_i)$ is obvious. The proofs for the other two invariances are similar, so we give the argument only for $\pi(X_i)$. For $t \geq 0$, $\pi(X_i)v_t = v_{t+1}$, and $\pi(X_i)v_p = 0$. So in order to prove the invariance under $\pi(X_i)$, it is enough to prove that $\pi(X_i)v_t \in U$ for $t \leq -1$. We show by induction on $s \geq 1$ that $\pi(X_i)v_{-s} \in \mathbf{C}\cdot v_{-(s-1)}$. For $s = 1$, $\pi(X_i)v_{-1} = \pi(X_iY_i)v_0 = \pi(Y_iX_i)v_0 + \pi(H_i)v_0 = (c_0 + \mu(H_i))v_0$. If $s > 1$, $\pi(X_i)v_{-s} = \pi(Y_i)\pi(X_i)v_{-(s-1)} + \pi(H_i)v_{-(s-1)}$; since $\pi(X_i)v_{-(s-1)} \in \mathbf{C}\cdot v_{-(s-2)}$ by the induction hypothesis, and $\pi(H_i)v_{-(s-1)} = (\mu - (s-1)\alpha_i)(H_i)v_{-(s-1)}$, we find that $\pi(X_i)v_{-s} \in \mathbf{C}\cdot v_{-(s-1)}$.

We thus have a representation of the Lie algebra spanned by H_i, X_i, and Y_i in the finite-dimensional space U. So by Corollary 4.2.3, $\pi(H_i) \mid U$ has only integral eigenvalues, the eigenvalues forming a set symmetric about 0. Since v_t is an eigenvector for $\pi(H_i)$ corresponding to the eigenvalue $\mu(H_i) + 2t$ $(-q \leq t \leq p)$, we see at once that $\mu(H_i)$ is an integer and that for some t_0, with $-q \leq t_0 \leq p$, $\mu(H_i) + 2t_0 = -\mu(H_i)$. This gives $t_0 = -\mu(H_i)$, and implies that $V_{\mu-\mu(H_i)\alpha_i} = V_{s_i\mu} \neq 0$.

Lemma 4.6.8. *Let $\theta_{ij}^{\pm}$ be as in* (4.6.6) *Denote by π_λ the irreducible representation of $\mathfrak{g}$ with $\lambda \in \mathfrak{h}^*$ as its highest weight. Then $\pi_\lambda(\theta_{ij}^{\pm}) = 0$.*

Proof. Let V be the vector space on which π_λ acts. We consider first $\pi_\lambda(\theta_{ij}^+)$. Write $\nu = \alpha_j + (-a_{ij} + 1)\alpha_i$. Then $\theta_{ij}^+ \in \mathfrak{g}_\nu \subseteq \mathfrak{n}^+$, so $\pi_\lambda(\theta_{ij}^+)v = 0$, where v is any nonzero vector in V_λ. Write $U = \{u \colon u \in V, \pi_\lambda(\theta_{ij}^+)u = 0\}$. Since $[\theta_{ij}^+, Y_k] = 0$, $1 \leq k \leq l$, it is obvious that U is invariant under $\pi_\lambda(Y_k)$, $1 \leq k \leq l$. Since $v \in U$, we must have $U = V$ by (iii) of (4.6.12). We now take up $\pi_\lambda(\theta_{ij}^-)v$. $\pi_\lambda(\theta_{ij}^-)v \in V_{\lambda-\nu}$. On the other hand, since $[X_k, \theta_{ij}^-] = 0$, $\pi_\lambda(X_k)\pi_\lambda(\theta_{ij}^-)v = 0$, $1 \leq k \leq l$. So we conclude from (4.6.13) that $\pi_\lambda(\theta_{ij}^-)v = 0$. We prove by induction on $O(q)$ that $\pi_\lambda(\theta_{ij}^-)[V_{\lambda-q}] = 0$, $q \in \mathbf{\Gamma} \cup \{0\}$. Fix $q \in \mathbf{\Gamma}$. Then, since $V_{\lambda-q}$ is spanned by $\pi_\lambda(Y_{i_1} \cdots Y_{i_\nu})v$ with $\alpha_{i_1} + \cdots + \alpha_{i_\nu} = q$, it is enough to prove that $\pi_\lambda(\theta_{ij}^-)(\pi_\lambda(Y_{i_1} \cdots Y_{i_\nu})v) = 0$ for such $\nu, i_1, \ldots, i_\nu$. Consider such a sequence $\nu, i_1, \ldots, i_\nu$, and let $u = \pi_\lambda(Y_{i_1} \cdots Y_{i_\nu})v$.

Then $\pi_\lambda(X_k)\pi_\lambda(\theta_{ij}^-)u = \pi_\lambda(\theta_{ij}^-)\pi_\lambda(X_k)u$, $1 \leq k \leq l$. Now $\pi_\lambda(X_k)u \in V_{\lambda-(q-\alpha_k)}$ so if $q - \alpha_k \in \Gamma \cup \{0\}$, then $\pi_\lambda(\theta_{ij}^-)\pi_\lambda(X_k)u = 0$ by the induction hypothesis; if $q - \alpha_k \notin \Gamma \cup \{0\}$, then $V_{\lambda-(q-\alpha_k)} = 0$, so $\pi_\lambda(X_k)u$ itself is 0. Thus $\pi_\lambda(X_k)\pi_\lambda(\theta_{ij}^-)u = 0$ $(1 \leq k \leq l)$. Once again, (4.6.13) implies that $\pi_\lambda(\theta_{ij}^-) = 0$.

Lemma 4.6.9. *Let π be a representation of $\mathfrak{g}$ in a vector space V. Suppose $\dim V < \infty$. Then π is a representation with weights, all its weights are integral, and for any $\mu \in \mathfrak{h}^*$,*

$$\dim V_\mu = \dim V_{s\mu} \quad (s \in \mathfrak{w}). \tag{4.6.19}$$

Moreover, for any $\lambda \in \Gamma \cup (-\Gamma)$ and $X \in \mathfrak{g}_\lambda$, $\pi(X)$ is nilpotent. Suppose further that π is irreducible. Then π is equivalent to π_λ for a unique dominant integral $\lambda \in \mathfrak{h}^*$.

Proof. Let $\mathfrak{a}_i$ be the linear span of H_i, X_i, Y_i $(1 \leq i \leq l)$. Applying Corollary 4.2.3 to $\pi \mid \mathfrak{a}_i$, we see that $\pi(H_i)$ is a semisimple endomorphism with integral eigenvalues, while $\pi(X_i)$ and $\pi(Y_i)$ are nilpotent. In particular, $\pi \mid \mathfrak{h}$ is a representation of the abelian Lie algebra $\mathfrak{h}$ sending the elements of a basis of $\mathfrak{h}$ into semisimple endomorphisms. This implies that $\pi(H)$ is semisimple for $H \in \mathfrak{h}$ and shows that V is the sum of its weight subspaces. π is thus a representation with weights, and all its weights are integral. Let μ be a weight of π and $U = \sum_{n \in \mathbf{Z}} V_{\mu+n\alpha_i}$. Then U is invariant under the representation $\pi \mid \mathfrak{a}_i$. We now get (4.6.19) in the usual manner. If $\lambda \in \Gamma \cup (-\Gamma)$ and $X \in \mathfrak{g}_\lambda$, then $\pi(X)^n[V_\mu] \subseteq V_{\mu+n\lambda}$ $(n \geq 0)$ by (4.6.10), so the nilpotency of $\pi(X)$ is immediate from the finiteness of the set of weights of π.

Suppose now that π is irreducible. Since the set of weights of π is finite, we can find a weight λ such that π does not have any weight λ' with $\lambda' \succ \lambda$. This implies that $\lambda + \alpha_i$, $1 \leq i \leq l$, cannot be weights of π and hence that $\pi(X_i)[V_\lambda] = 0$, $1 \leq i \leq l$. By Lemma 4.6.5, λ is the highest weight of π. So π is equivalent to π_λ. λ is obviously uniquely determined by π. For $1 \leq i \leq l$, $s_i\lambda = \lambda - \lambda(H_i)\alpha_i$ is a weight of π. So $\lambda - s_i\lambda = \lambda(H_i)\alpha_i \in \Gamma \cup \{0\}$. This shows that $\lambda(H_i) \geq 0$. λ is thus dominant and integral.

Lemma 4.6.10. *Let λ be a dominant integral linear function on $\mathfrak{h}$. Write $\lambda_i = \lambda(H_i)$, $1 \leq i \leq l$, and let*

$$\begin{aligned}\mathfrak{M}_\lambda^0 = &\sum_{1 \leq i \leq l} \mathfrak{G}X_i + \sum_{1 \leq i \leq l} \mathfrak{G}(H_i - \lambda_i \cdot 1)\\ &+ \sum_{1 \leq i, j \leq l} \mathfrak{G}\theta_{ij}^-\mathfrak{G} + \sum_{1 \leq i \leq l} \mathfrak{G}Y_i^{\lambda_i+1}\end{aligned} \tag{4.6.20}$$

Then $\mathfrak{M}_\lambda^0$ is a (proper) left ideal of $\mathfrak{G}$, and $\dim(\mathfrak{G}/\mathfrak{M}_\lambda^0) < \infty$.

Proof. Let λ be dominant integral. Consider first the irreducible representation π_λ corresponding to λ. Let $\bar{V}$ be the vector space on which π_λ acts and $\bar{v}$ a nonzero element in $\bar{V}_\lambda$. The set of all $a \in \mathfrak{G}$ for which $\pi_\lambda(a)\bar{v} = 0$ is then the unique maximal left ideal $\mathfrak{M}_\lambda$ of $\mathfrak{G}$ containing all $X_i, H_i - \lambda_i \cdot 1, 1 \leq i \leq l$. By Lemma 4.6.8, $\mathfrak{G}\theta_{ij}^-\mathfrak{G} \subseteq \mathfrak{M}_\lambda$, $1 \leq i, j \leq l$. Furthermore, by (4.2.4),

$$\begin{aligned}\pi_\lambda(X_i)\pi_\lambda(Y_i)^{\lambda_i+1}\bar{v} &= \pi_\lambda(Y_i)^{\lambda_i+1}\pi_\lambda(X_i)\bar{v} + (\lambda_i + 1)\pi_\lambda(Y_i)^{\lambda_i}(\pi_\lambda(H_i) - \lambda_i)\bar{v} \\ &= 0,\end{aligned}$$

while for $j \neq i$, $\pi_\lambda(X_j)\pi_\lambda(Y_i)^{\lambda_i+1}\bar{v} = \pi_\lambda(Y_i)^{\lambda_i+1}\pi_\lambda(X_j)\bar{v} = 0$. So then, since $\pi_\lambda(Y_i)^{\lambda_i+1}\bar{v} \in \bar{V}_{\lambda-(\lambda_i+1)\alpha_i}$, we may conclude from (4.6.13) that $\pi_\lambda(Y_i^{\lambda_i+1})\bar{v} = 0$. In other words, $Y_i^{\lambda_i+1} \in \mathfrak{M}_\lambda$, $1 \leq i \leq l$. Since $\mathfrak{M}_\lambda$ is a left ideal, we see that $\mathfrak{M}_\lambda^0 \subseteq \mathfrak{M}_\lambda$. In particular, $\mathfrak{M}_\lambda^0$ is a (proper) left ideal.

Let $V = \mathfrak{G}/\mathfrak{M}_\lambda^0$, let v be the image of 1 in V, and let π be the natural representation of $\mathfrak{G}$ in V. We prove first that $\pi(X_i)$ and $\pi(Y_i)$ are locally nilpotent for $1 \leq i \leq l$. Fix i, $1 \leq i \leq l$. If μ is a weight of π, $\lambda - \mu \in \Gamma \cup \{0\}$, so $\lambda - (\mu + n\alpha_i) \notin \Gamma \cup \{0\}$ if $n \geq 1$ is large enough. So $\pi(X_i^n)[V_\mu] = 0$ if $n \geq 1$ is sufficiently large. Now consider $\pi(Y_i)$. Let $U = \{u: \pi(Y_i^s)u = 0$ for some integer $s = s(u) \geq 1\}$. Since $\pi(Y_i^{\lambda_i+1})v = 0$, $v \in U$. We must prove that $U = V$. In view of (iii) of (4.6.12), it is enough to check that U is invariant under $\pi(Y_j)$, $1 \leq j \leq l$. Invariance under $\pi(Y_i)$ is obvious. Fix $j \neq i$, let $u \in U$, and let $c \geq 1$ be an integer such that $\pi(Y_i^c)u = 0$. Now if we write, for any $X \in \mathfrak{g}$, L_X and R_X for the endomorphisms (mutually commuting) $a \mapsto Xa$ and $a \mapsto aX$ of $\mathfrak{G}$, then ad $X = L_X - R_X$, so $Y_i^dY_j = (\text{ad } Y_i + R_{Y_i})^d(Y_j) = \sum_{0 \leq s \leq d}\binom{d}{s}(\text{ad } Y_i)^s(Y_j)Y_i^{d-s}$ for any integer $d \geq 1$. If $d - s \geq c$, $\pi(Y_i)^{d-s}u = 0$. On the other hand, if $s \geq -a_{ij}$, $(\text{ad } Y_i)^s(Y_j)Y_i^{d-s} \in \mathfrak{G}\theta_{ij}^-\mathfrak{G}$; and by definition of $\mathfrak{M}_\lambda^0$, $\pi(\xi) = 0$ if $\xi \in \mathfrak{G}\theta_{ij}^-\mathfrak{G}$. So, for such d, s, $\pi((\text{ad } Y_i)^s(Y_j)Y_i^{d-s})u = 0$. Combining these two observations, we see that $\pi(Y_i^d)\pi(Y_j)u = 0$ if $d \geq -a_{ij} + c$. This proves that $\pi(Y_j)u \in U$.

Lemma 4.6.7 now applies, and we conclude that the set of weights of π is invariant under $\mathfrak{w}$. Since $\mathfrak{w}$ is a finite group, the set $\mathfrak{w} \cdot \lambda$ is finite, so we may select $s_0 \in \mathfrak{w}$ such that $\mu = s_0\lambda$ is a minimal element of the set $\mathfrak{w} \cdot \lambda$ in the partial ordering $\prec$. If $1 \leq i \leq l$, $s_i\mu = \mu - \mu(H_i)\alpha_i$, so by the manimality of μ, $\mu(H_i)$ is ≤ 0. Let u be a nonzero vector in V_μ. We claim that $\pi(Y_i)u = 0$, $1 \leq i \leq l$. Fix i. If $\mu - \alpha_i$ is not a weight of π, this is clear. Suppose $\mu - \alpha_i$ is a weight of π. Then so is $s_0^{-1}(\mu - \alpha_i) = \lambda - s_0^{-1}\alpha_i$. We conclude therefore that $s_0^{-1}\alpha_i \in \Gamma$, and hence that $-s_0^{-1}\alpha_i \notin \Gamma$. Thus $\lambda + s_0^{-1}\alpha_i$ cannot be a weight of π, which implies that $\mu + \alpha_i$ cannot be a weight of π either. This means that $\pi(X_i)u = 0$. Now u is an eigenvector for $\pi(H_i)$ for the eigenvalue $\mu(H_i)$, and $\pi(Y_i)^su = 0$ for $s \geq 1$ sufficiently large. Therefore by Corollary 4.2.5, $\mu(H_i)$ is an integer ≥ 0 and $\pi(Y_i)^{\mu(H_i)+1}u = 0$. Since we know that $\mu(H_i) \leq 0$, $\mu(H_i) = 0$, so $\pi(Y_i)u = 0$. Thus, in all cases,

$\pi(Y_i)u = 0$, $1 \leq i \leq l$. Since π is irreducible, u is cyclic for π. We may now argue exactly as in Lemma 4.6.5 to conclude that V is spanned by u and the vectors $\pi(X_{i_1}) \cdots \pi(X_{i_\nu})u$ ($\nu \geq 1$, $1 \leq i_1, \ldots, i_\nu \leq l$). In particular, for any weight λ' of π, we have $\mu \preccurlyeq \lambda' \preccurlyeq \lambda$.

Write $\mu = \lambda - (m_1\alpha_1 + \cdots + m_l\alpha_l)$, where the m_i are all integers ≥ 0. Then, if n_i are integers ≥ 0, we have $\mu \preccurlyeq \lambda - (n_1\alpha_1 + \cdots + n_l\alpha_l) \preccurlyeq \lambda$ if and only if $0 \leq n_i \leq m_i$ for $i \leq i \leq l$. This shows that the set of weights of π is finite. Each weight subspace of π being finite-dimensional, we see at once that $\dim V < \infty$. This completes the proof of the lemma.

We now have

Theorem 4.6.11. *Let* $\mathfrak{g}$ *be a Lie algebra over* **C** *(possibly infinite-dimensional) generated by* $3l$ *linearly independent elements* H_i, X_i, Y_i $(1 \leq i \leq l)$ *satisfying the commutation rules (4.6.1). Let* $\mathfrak{D}$ *be the set of all dominant integral linear functions on* $\mathfrak{h}$ *(notation as above). Then the correspondence which assigns to any* $\lambda \in \mathfrak{D}$ *the equivalence class of* π_λ *is a bijection of* $\mathfrak{D}$ *onto the set of all equivalence classes of finite-dimensional irreducible representations of* $\mathfrak{g}$.

Proof. For $\lambda \in \mathfrak{D}$, $\dim(\mathfrak{G}/\mathfrak{M}_\lambda) < \infty$ by Lemma 4.6.10, so π_λ is finite-dimensional. Theorem 4.6.11 is now an immediate consequence of the work so far.

Remark. For $\lambda \in \mathfrak{D}$, $\mathfrak{M}_\lambda \supseteq \mathfrak{M}_\lambda^0$, $\mathfrak{M}_\lambda^0$ being defined by (4.6.20). It can be shown that $\mathfrak{M}_\lambda = \mathfrak{M}_\lambda^0$.

The theory of the representations of this algebra may now be used to discuss the two problems we mentioned at the beginning. We now turn to these questions.

4.7. Representations of Semisimple Lie Algebras

Let $\mathfrak{g}$ be a semisimple Lie algebra over **C**. Let $\mathfrak{h}$ be a CSA, Δ the set of roots of $(\mathfrak{g},\mathfrak{h})$, P a positive system of roots, and $S = \{\alpha_1, \ldots, \alpha_l\}$ the corresponding simple system. Write $H_i = \bar{H}_{\alpha_i}$, $1 \leq i \leq l$. $\mathfrak{D}_P$ is the set of all dominant integral linear functions on $\mathfrak{h}$. From Theorem 4.6.11 we get at once

Theorem 4.7.1. *The representations* π_λ $(\lambda \in \mathfrak{D}_P)$ *exhaust the irreducible representations of finite dimension of* $\mathfrak{g}$ *up to equivalence; moreover, the* π_λ *for distinct* $\lambda \in \mathfrak{D}_P$ *are inequivalent.*

We devote this section to some complements to the theory developed in §4.6. Note that if $\lambda \in \mathfrak{h}^*$ is integral, $\lambda(\bar{H}_\alpha)$ is an integer for all $\alpha \in \Delta$. In fact, if $\alpha \in \Delta$ is such that $\lambda(\bar{H}_\alpha)$ is an integer, and $\beta = s_{\alpha_i}\alpha$ $(1 \leq i \leq l)$, then

$\bar{H}_\beta = s_{\alpha_i}\bar{H}_\alpha$, so $\lambda(\bar{H}_\beta) = \lambda(\bar{H}_\alpha) - \alpha_i(\bar{H}_\alpha)\lambda(H_i)$ is also an integer; this shows our claim to be true. In particular, $s\lambda$ is integral for all $s \in \mathfrak{w}$. Also, if $\lambda \in \mathfrak{D}_P$, $\lambda(H_\alpha) \geq 0$ for all $\alpha \in P$, so $\lambda(\bar{H}_\alpha)$ is a nonnegative integer for all $\alpha \in P$.

We also note that

$$\theta_{ij}^+ = 0 \quad (1 \leq i, j \leq l), \tag{4.7.1}$$

where the $\theta_{ij}^\pm$ are defined by (4.6.6). This is seen as follows. Fix $i \neq j$. Then $\alpha_j - \alpha_i$ is not a root, so by Corollary 4.3.10, $\alpha_j + k\alpha_i$ is a root if and only if k is an integer such that $0 \leq k \leq -a_{ij}$. Since $\theta_{ij}^+ \in \mathfrak{g}_\alpha$ for $\alpha = \alpha_j + (-a_{ij} + 1)\alpha_i$, $\theta_{ij}^+ = 0$. Similarly, $\theta_{ij}^- = 0$.

Let $\lambda \in \mathfrak{D}_P$. Let us denote by $\mathfrak{G}$, as usual, the universal enveloping algebra of $\mathfrak{g}$, and let $\mathfrak{M}_\lambda$ be the maximal left ideal of $\mathfrak{G}$ corresponding to π_λ. It turns out that we can describe $\mathfrak{M}_\lambda$ explicitly. We need a lemma.

Lemma 4.7.2. *Let $\lambda \in \mathfrak{D}_P$, and let π be a finite-dimensional representation of $\mathfrak{g}$ such that* (i) *λ is the highest weight of π, and* (ii) *there is a vector of weight λ which is cyclic for π. Then π is equivalent to π_λ.*

Proof. It is enough to prove that π is irreducible. We use the notation and results of §4.6. Let V be the vector space on which π acts, v a nonzero vector in V_λ which is cyclic for π. Since $\lambda + \alpha_i$ is not a weight of π, $\pi(X_i)v = 0$, $1 \leq i \leq l$. By Lemma 4.6.5, $V_\lambda = \mathbf{C}\cdot v$. It is clear from Lemma 4.6.6 that there is a (unique) largest π-invariant subspace of V not containing v. Let U denote this subspace. Since π is semisimple, we can find a π-invariant subspace U' of V complementary to U. If $v \notin U'$, $U' \subseteq U$, which would imply $U = V$. So $v \in U'$, and hence, by the cyclic nature of v, $U' = V$. Thus $U = 0$. In other words, π is already irreducible.

Theorem 4.7.3. *Let $\lambda \in \mathfrak{D}_P$, and let $\mathfrak{M}_\lambda$ be as in* §4.6. *Write $\lambda_i = \lambda(H_i)$, $1 \leq i \leq l$, and use the notation of* §4.6. *Then*

$$\mathfrak{M}_\lambda = \sum_{1 \leq i \leq l} \mathfrak{G}X_i + \sum_{1 \leq i \leq l} \mathfrak{G}(H_i - \lambda_i 1) + \sum_{1 \leq i \leq l} \mathfrak{G}Y_i^{\lambda_i + 1}. \tag{4.7.2}$$

Proof. In view of (4.7.1) and Lemma 4.6.10, it is clear that the right side of (4.7.2) is a left ideal of $\mathfrak{G}$ of positive finite codimension. Let $\mathfrak{M}_\lambda^0$ denote this left ideal. Then the representation of $\mathfrak{G}$ induced in $\mathfrak{G}/\mathfrak{M}_\lambda^0$ is such that Lemma 4.7.2 is applicable to it. We therefore conclude that this representation is irreducible. $\mathfrak{M}_\lambda^0$ is thus maximal. But then $\mathfrak{M}_\lambda^0 = \mathfrak{M}_\lambda$. This proves the theorem.

Theorem 4.7.1, taken with Theorem 4.7.3, gives a complete and precise description of the irreducible finite-dimensional representations of $\mathfrak{g}$.

Fix $\lambda \in \mathfrak{D}_P$. If $s \in \mathfrak{w}$, then it is obvious that $s\cdot\lambda$ is dominant integral with respect to the positive system $s\cdot P$ and that $s\cdot\lambda$ is the highest weight of π_λ in the ordering induced by $s\cdot P$. In particular, if $s_0 \in \mathfrak{w}$ is the element such that $s_0\cdot P = -P$, then $s_0\lambda$ is the highest weight of π_λ in the ordering induced by $-P$, so $s_0\lambda$ is the *lowest* weight of π_λ in the ordering induced by P. In other words,

$$s_0\lambda \preccurlyeq \mu \preccurlyeq \lambda \tag{4.7.3}$$

for any weight μ of π_λ.

For any integral linear function μ on $\mathfrak{h}$ we define dim V_μ to be the *multiplicity of* μ in π_λ. Let $m_\lambda(\mu)$ denote this integer. Clearly,

$$m_\lambda(\lambda) = 1, \qquad m_\lambda(s\cdot\mu) = m_\lambda(\mu) \quad (s \in \mathfrak{w}). \tag{4.7.4}$$

For general combinatorial formulae for $m_\lambda(\mu)$, the reader is referred to Exercise 13.

Lemma 4.7.4.[6] *Let λ be an integral linear function of $\mathfrak{h}$. Then there is a unique element of $\mathfrak{D}_P$ in the orbit $\mathfrak{w}\cdot\lambda$. If $\lambda \in \mathfrak{D}_P$, $s\cdot\lambda \preccurlyeq \lambda$ for all $s \in \mathfrak{w}$. Let*

$$\delta_P = \frac{1}{2}\sum_{\alpha\in P}\alpha. \tag{4.7.5}$$

Then $\delta_P \in \mathfrak{D}_P$, $\delta_P(H_i) = 1$ for $1 \leq i \leq l$, and the transforms $s\cdot\delta_P$ $(s \in \mathfrak{w})$ are all distinct. More generally, if $\lambda \in \mathfrak{D}_P$ and $\lambda(H_i) > 0$ for $1 \leq i \leq l$, the transforms $s\lambda$ $(s \in \mathfrak{w})$ are all distinct.

Proof. Let λ be an integral element of $\mathfrak{h}^*$. Let μ be an element of the orbit $O = \mathfrak{w}\cdot\lambda$ that is maximal with respect to $\prec$. Since $s_{\alpha_i}\mu = \mu - \mu(H_i)\alpha_i$, it is clear from the maximality of μ that $\mu(H_i) \geq 0$, $1 \leq i \leq l$. So $\mu \in \mathfrak{D}_P$. Suppose $\mu' \in O \cap \mathfrak{D}_P$. Then μ' is a weight of π_μ and μ is a weight of $\pi_{\mu'}$, so that $\mu' \preccurlyeq \mu$ and $\mu \preccurlyeq \mu'$. This shows that $\mu = \mu'$. In particular, $O \cap \mathfrak{D}_P$ has a unique element, and $s\cdot\lambda \preccurlyeq \lambda$ for all $s \in \mathfrak{w}$ if $\lambda \in \mathfrak{D}_P$. Let $\delta = \delta_P$ be defined by (4.7.5). Since $s_{\alpha_i}P = \{-\alpha_i\} \cup P \setminus \{\alpha_i\}$, $s_{\alpha_i}\delta = \delta - \alpha_i$ for $1 \leq i \leq l$. This proves that $\delta(H_i) = 1$, $1 \leq i \leq l$, and hence in particular that $\delta \in \mathfrak{D}_P$. It is clear that $\delta(\bar{H}_\alpha) > 0$ for all $\alpha \in P$. Let $\mathfrak{w}' = \{s: s \in \mathfrak{w}, s\cdot\delta = \delta\}$. By Theorem 4.15.17, $\mathfrak{w}'$ is generated by the reflection s_α $(\alpha \in P)$ for which $\langle\delta,\alpha\rangle = 0$. But since $\langle\delta,\alpha\rangle > 0$ for all $\alpha \in P$, $\mathfrak{w}'$ must be trivial. The same argument applies to the $\lambda \in \mathfrak{D}_P$ for which $\lambda(H_i) > 0$ for all i. This proves the lemma.

Let G be a complex analytic simply connected group with Lie algebra $\mathfrak{g}$. Then the representations π_λ lift to irreducible complex analytic representa-

[6]Cf. also Theorem 4.15.13 and Corollary 4.15.14.

tions of G, and we obtain all such representations of G in this manner (up to equivalence). We write π_λ again for these representations of G. It is an important problem to determine the characters of these representations. This was done by H. Weyl. We shall take it up later. It follows from Weyl's work that

$$\dim(\pi_\lambda) = \frac{\prod_{\alpha \in P} \langle \lambda + \delta_P, \alpha \rangle}{\prod_{\alpha \in P} \langle \delta_P, \alpha \rangle} \quad (\lambda \in \mathfrak{D}_P). \tag{4.7.6}$$

Let $\mu_1, \ldots, \mu_l$ be the elements of $\mathfrak{D}_P$ such that

$$\mu_i(H_j) = \delta_{ij} \quad (1 \leq i, j \leq l). \tag{4.7.7}$$

Let V_i be a vector space on which the representation $\pi_i = \pi_{\mu_i}$ acts $(1 \leq i \leq l)$, and let v_i be a nonzero vector of weight μ_i in V_i. Suppose now that $\lambda \in \mathfrak{D}_P$. Then $\lambda = \sum_{1 \leq i \leq l} \lambda_i \mu_i$, where $\lambda_i = \lambda(H_i)$. Write

$$\begin{aligned}
V &= V_1 \otimes \cdots \otimes V_1 \otimes V_2 \otimes \cdots \otimes V_l \otimes \cdots \otimes V_l \\
v &= v_1 \otimes \cdots \otimes v_1 \otimes v_2 \otimes \cdots \otimes v_l \otimes \cdots \otimes v_l \\
\pi &= \pi_1 \otimes \cdots \otimes \pi_1 \otimes \pi_2 \otimes \cdots \otimes \pi_l \otimes \cdots \otimes \pi_l
\end{aligned}$$

wherein there are λ_i factors V_i, v_i, π_i, $1 \leq i \leq l$. The weights of π are of the form $\nu_1 + \cdots + \nu_l$, where ν_i is a weight of π_i, $1 \leq i \leq l$. It follows at once that λ is the highest weight of π and that $V_\lambda = \mathbf{C} \cdot v$. By Lemma 4.7.1, the cyclic π-invariant subspace generated by v is irreducible under π and defines the representation π_λ. In other words, if one can somehow construct the π_i $(1 \leq i \leq l)$, then all finite-dimensional irreducible representations can be very explicitly obtained. This was the method used by E. Cartan to prove that for each $\lambda \in \mathfrak{D}_P$ there is an irreducible representation with highest weight λ.

For any $\lambda \in \mathfrak{h}^*$, the irreducible representation π_λ with highest weight λ is the unique irreducible quotient of the representation of $\mathfrak{G}$ in $\mathfrak{G}/\mathfrak{G}_\lambda$ (cf. (4.6.15)). For a study of the decomposition of the representation of $\mathfrak{G}$ in $\mathfrak{G}/\mathfrak{G}_\lambda$, see Verma [1], Bernšteĭn et al. [1]. The representations π_λ $(\lambda \in \mathfrak{h}^*)$ turn out to be useful in the theory of *unitary* representations of a semisimple Lie group too.

It is of interest to determine explicitly the representations π_i for the classical Lie algebras. In the remarks to follow we use without any explicit mention the notation of §4.4.

The algebras A_l $(l \geq 1)$. Let $V = \mathbf{C}^{l+1}$, and let v_j be the vector whose components are $\delta_{j,1}, \ldots, \delta_{j,l+1}$. $\pi: X \mapsto X (X \in \mathfrak{g} = A_l)$ is a representation of $\mathfrak{g}$. Obviously, $v_j \in V_{\lambda_j}$. Since $\lambda_1 - \lambda_k = \alpha_1 + \cdots + \alpha_{k-1}$, λ_1 is the highest weight of π. Clearly, since $\lambda_1 = \mu_1$, $\pi \simeq \pi_1$.

Let E be the exterior algebra over V, and let $E_k (1 \leq k \leq l)$ be the homogeneous subspace of E of degree k. For $X \in \mathfrak{g}$ let $\bar{X}$ be the derivation of E that extends X. The E_k are then invariant subspaces for these derivations, and we have a representation $\bar{\pi}_k$: $X \mapsto \bar{X} | E_k$ of $\mathfrak{g}$ in E_k. Now

$$\bar{X} \cdot v_{i_1} \wedge \cdots \wedge v_{i_k} = (Xv_{i_1}) \wedge v_{i_2} \wedge \cdots \wedge v_{i_k} + \cdots + v_{i_1} \wedge \cdots \wedge v_{i_{k-1}} \wedge (Xv_{i_k}),$$

so $v_{i_1} \wedge \cdots \wedge v_{i_k}$ is of weight $\lambda_{i_1} + \cdots + \lambda_{i_k}$. This shows that $\lambda_1 + \cdots + \lambda_k$ is the highest weight of $\bar{\pi}_k$. Since $\lambda_{i_1} + \cdots + \lambda_{i_k}$ $(1 \leq i_1 < \cdots < i_k \leq l + 1)$ are all the weights of $\bar{\pi}_k$, and these are all in the $\mathfrak{w}$-orbit of the highest weight $\lambda_1 + \cdots + \lambda_k$ which occurs with multiplicity one, we conclude that π_k is irreducible. It is easy to check that $\lambda_1 + \cdots + \lambda_k = \mu_k$, so $\bar{\pi}_k \simeq \pi_k$ $(1 \leq k \leq l)$.

We now consider the case $\mathfrak{g} = B_l$, C_l, or D_l. In all these cases there is a basic representation, namely the one which realizes $\mathfrak{g}$. Let V denote the vector space on which this representation acts, so that $\mathfrak{g} \subseteq \mathfrak{gl}(V)$. Let T be the tensor algebra over V, T_r $(r \geq 0)$ the space spanned by the homogeneous tensors of degree r. For $X \in \mathfrak{g}$, let $\tilde{X}$ be the derivation of T that extends X. Then $X \mapsto \tilde{X}$ $(X \in \mathfrak{g})$ is a representation of $\mathfrak{g}$ in T, and this representation leaves the T_r invariant. Let $\tilde{\pi}_r$: $X \mapsto \tilde{X} | T_r$ be the representation of $\mathfrak{g}$ in T_r. Since $\mathfrak{g}$ is semisimple, $\tilde{\pi}_r$ decomposes as a direct sum of irreducible representations. It is natural to attempt to construct explicitly the irreducible representations of $\mathfrak{g}$ by splitting the $\tilde{\pi}_r$ $(r \geq 0)$. Note that V has a basis of weight vectors (with respect to some fixed CSA of $\mathfrak{g}$) so that the weights of the irreducible representations of $\mathfrak{g}$, obtained by splitting T_r, are of the form $m_1\mu_1 + \cdots + m_s\mu_s$, where the m_j are all integers and the μ_j are the weights in V. This is the procedure we now use. With Cartan, we restrict ourselves to the decomposition of the exterior algebra (for a deep study of the manner in which T decomposes, see H. Weyl's book [1]).

The algebras C_l $(l \geq 2)$. V is a vector space of dimension $2l$, $(\cdot, \cdot)$ a non-singular skew-symmetric bilinear form on $V \times V$, and $\mathfrak{g}$ the Lie algebra of all endomorphisms of V satisfying (4.4.26). Let π: $X \mapsto X$ be the representation of $\mathfrak{g}$ in V. Let E be the exterior algebra over V, let E_k be the homogeneous subspaces of E of degrees $k = 0, 1, \ldots, 2l$, and let $\bar{\pi}_k$: $X \mapsto \bar{\pi}(X) | E_k$ be the representation of $\mathfrak{g}$ induced in E_k, $\bar{\pi}(X)$ being the derivation of E extending X. The weights of $\bar{\pi}_1 = \pi_1$ are $\pm\lambda_i$, $1 \leq i \leq l$; u_i is of weight λ_i and u_{l+i} of weight $-\lambda_i$ $(1 \leq i \leq l)$. If $1 \leq k \leq l$, it is easy to see that $u_1 \wedge \cdots \wedge u_k$ is of weight $\lambda_1 + \cdots + \lambda_k$ for $\bar{\pi}_k$ and that it is the highest weight. Moreover, $\lambda_1 + \cdots + \lambda_k = \mu_k$, $1 \leq k \leq l$. It is, however, not true that for $k > 1$ the representation $\bar{\pi}_k$ is irreducible. If $E_{k,0}$ is the smallest $\bar{\pi}_k$-invariant subspace of E_k containing $u_1 \wedge \cdots \wedge u_k$, $\bar{\pi}_k$ restricted to $E_{k,0}$ is equivalent to π_k. For the complete decomposition of E_k, see Exercise 24.

The algebras B_l ($l \geq 1$), D_l ($l \geq 2$). For $\mathfrak{g} = B_l$, the representation $\bar{\pi}_1$: $X \mapsto X$, acting on the vector space V of dimension $2l + 1$, has weights $\pm\lambda_i$ ($1 \leq i \leq l$) and 0. So it is clear that λ_1 is its highest weight. Let E be the exterior algebra over V, E_k the subspace spanned by the homogeneous elements of degree k. If $1 \leq k \leq l - 1$, the representation $\bar{\pi}_k$ of $\mathfrak{g}$ induced (as in the preceding cases) on E_k has highest weight $\mu_k = \lambda_1 + \cdots + \lambda_k$, while (4.7.6) gives $\dim(\pi_k) = \binom{2l+1}{k}$; so we conclude that $\bar{\pi}_k$ is irreducible and $\simeq \pi_k$. If $k = l$, a similar argument shows that the representation $\bar{\pi}_l$ of $\mathfrak{g}$ induced in E_l is the irreducible representation with highest weight $\lambda_1 + \cdots + \lambda_l = 2\mu_l$ (cf. Exercises 21, 23). So π_l cannot be obtained by this procedure. It turns out that the representation π_l cannot be obtained by splitting even the tensor algebra over V. In fact, if μ is a weight of any irreducible representation of $\mathfrak{g}$ occurring in the tensor algebra, $\mu = m_1\lambda_1 + \cdots + m_l\lambda_l$, where the $m_1, \ldots, m_l$ are all integers, and a simple calculation shows (cf. (4.4.45)) that $\mu(H_l) = 2m_l \neq 1$. π_l is the so-called *spin representation* of $\mathfrak{g}$.

The situation with respect to the algebras D_l is similar. The representation $\bar{\pi}_1$: $X \mapsto X$ of $\mathfrak{g} = D_l$ in V is irreducible and has highest weight $\mu_1 = \lambda_1$. For $1 \leq k \leq l - 2$, the representation $\bar{\pi}_k$ of $\mathfrak{g}$, induced in the homogeneous subspace E_k (of degree k) of the exterior algebra E over V, has highest weight $\lambda_1 + \cdots + \lambda_k = \mu_k$; since $\dim(\pi_k) = \binom{2l}{k}$ by (4.7.6) (cf. Exercises 21, 23), we conclude that $\bar{\pi}_k \simeq \pi_k$. On the other hand, a simple calculation based on (4.4.22) shows that $\mu_{l-1} = \frac{1}{2}(\lambda_1 + \cdots + \lambda_{l-1} - \lambda_l)$ and $\mu_l = \frac{1}{2}(\lambda_1 + \cdots + \lambda_l)$. Since the weights of any irreducible representation occuring in the decomposition of the tensor algebra over V are of the form $m_1\lambda_1 + \cdots + m_l\lambda_l$, where the m_j are all integers, it follows that π_{l-1} and π_l cannot be obtained from the tensor algebra. These are the so-called *spin representations* of $\mathfrak{g}$.

For the spin representations of B_l and D_l, see Chevalley [4], Weyl [1], Brauer and Weyl [1], and Cartan [4].

4.8. Construction of a Semisimple Lie Algebra from its Cartan Matrix

We now take up the proof of the theorem that for an arbitrary Cartan matrix there is a semisimple Lie algebra having it as its Cartan matrix (cf. Harish-Chandra [4], Jacobson [1], Serre [1]).

Let $A = (a_{ij})_{1 \leq i,j \leq l}$ be an arbitrary Cartan matrix of rank $l \geq 1$. We begin by establishing the existence of the "universal" Lie algebra with generators H_i, X_i, Y_i ($1 \leq i \leq l$) satisfying the commutation rules (4.6.1). To this end, let $\mathfrak{g}'$ be the free[7] Lie algebra over $\mathbf{C}$ generated by elements H_i', X_i', Y_i'

[7]See §3.2.

$(1 \leq i \leq l)$. Let Φ be the set consisting of the following elements of $\mathfrak{g}'$ $(1 \leq i, j \leq l)$:

$$
(4.8.1) \qquad \begin{cases} [H'_i, H'_j] \\ [H'_i, X'_j] - a_{ij}X'_j, [H'_i, Y'_j] + a_{ij}Y'_j \\ [X'_i, Y'_j] - \delta_{ij}H'_i \end{cases}
$$

Lemma 4.8.1. *Let $\mathfrak{p}$ be the ideal in $\mathfrak{g}'$ generated by Φ. Then $\mathfrak{p} \neq \mathfrak{g}'$. More generally, the $3l$ elements H'_i, X'_i, Y'_i $(1 \leq i \leq l)$ are linearly independent modulo $\mathfrak{p}$.*

Proof. Let V be a vector space over $\mathbf{C}$ of dimension l with basis $\{e_1, \ldots, e_l\}$. Let $\mathfrak{J}$ be the tensor algebra over V. We suppress the notation for multiplication in $\mathfrak{J}$, so that we write ab for $a \otimes b$ $(a, b \in \mathfrak{J})$.

Fix $\boldsymbol{\lambda} = (\lambda_1, \ldots, \lambda_l) \in \mathbf{C}^l$. We now define a representation $\pi' = \pi'_\lambda$ of $\mathfrak{g}'$ in $\mathfrak{J}$. Since $\mathfrak{g}'$ is free, we may do this by specifying the endomorphisms $\pi'(X'_i), \pi'(H_i), \pi'(Y'_i)$ $(1 \leq i \leq l)$ in a completely arbitrary fahsion. We define

$$
\begin{aligned}
\pi'(Y'_i)b &= e_i b \quad (b \in \mathfrak{J}) \\
\pi'(H'_i)1 &= \lambda_i 1 \\
\pi'(H'_i)e_{i_1} \cdots e_{i_\nu} &= (\lambda_i - a_{ii_1} - \cdots - a_{ii_\nu})e_{i_1}e_{i_2} \cdots e_{i_\nu} \quad (1 \leq i_1, \ldots, i_\nu \leq l).
\end{aligned}
$$

$\pi'(X'_i)e_{i_1} \cdots e_{i_\nu}$ is defined, by induction on ν, as follows. We put

$$
\pi'(X'_i)1 = 0,
$$

and for $\nu \geq 1$, $1 \leq i_1, \ldots, i_\nu \leq l$,

$$
\pi'(X'_i)e_{i_1} \cdots e_{i_\nu} = e_{i_1}\pi'(X'_i)e_{i_2} \cdots e_{i_\nu} + \delta_{ii_1}\pi'(H'_i)e_{i_2} \cdots e_{i_\nu}
$$

(here, for $\nu = 1$, $e_{i_2} \cdots e_{i_\nu}$ is interpreted as 1).

A simple calculation shows that $\pi'(Z) = 0$ for all $Z \in \Phi$. Hence $\pi'(Z) = 0$ for all $Z \in \mathfrak{p}$. Since $\pi' \neq 0$, $\mathfrak{p} \neq \mathfrak{g}'$. Suppose now that $h_i, x_i, y_i \in \mathbf{C}$ $(1 \leq i \leq l)$ are such that $Z = \sum_{1 \leq i \leq l} (h_i H'_i + x_i X'_i + y_i Y'_i) \in \mathfrak{p}$. Then $\pi'_\lambda(Z) = 0$ for all $\boldsymbol{\lambda} \in \mathbf{C}^l$. The relation $\pi'_\lambda(Z)1 = 0$ gives us

$$
\sum_{1 \leq i \leq l} (\lambda_i h_i \cdot 1 + y_i e_i) = 0
$$

Since $\boldsymbol{\lambda}$ is arbitrary, we have $h_i = 0$, $y_i = 0$, $1 \leq i \leq l$. Then $Z = \sum_{1 \leq i \leq l} x_i X'_i$. The relation $\pi'_\lambda(Z)e_j = 0$ gives

$$
\sum_{1 \leq i \leq l} x_i \delta_{ij} \lambda_i = 0 \quad (1 \leq j \leq l).
$$

This implies that $x_i = 0$, $1 \leq i \leq l$. The lemma is proved.

Let $\mathfrak{g} = \mathfrak{g}'/\mathfrak{p}$, and let H_i, X_i, Y_i $(1 \leq i \leq l)$ be the respective images of H'_i, X'_i, Y'_i in $\mathfrak{g}$. These are linearly independent, generate $\mathfrak{g}$, and satisfy (4.6.1). Moreover, it is clear from its construction that $\mathfrak{g}$ has the following universal property: if $\mathfrak{m}$ is any Lie algebra and $\bar{H}_i, \bar{X}_i, \bar{Y}_i$ $(1 \leq i \leq l)$ are elements of $\mathfrak{m}$ satisfying the commutation rules (4.6.1) (with $\bar{H}_i, \bar{Y}_i, \bar{X}_i$ instead of H_i, X_i, Y_i), there is a unique homomorphism τ of $\mathfrak{g}$ into $\mathfrak{m}$ such that $\tau(H_i) = \bar{H}_i$, $\tau(X_i) = \bar{X}_i$, $\tau(Y_i) = \bar{Y}_i$, $1 \leq i \leq l$. We apply to $\mathfrak{g}$ the notation and results of §4.6

Lemma 4.8.2. *There is exactly one ideal $\mathfrak{q}$ in $\mathfrak{g}$ such that*

$$(4.8.2) \qquad \begin{cases} \text{(i)} & \dim \mathfrak{g}/\mathfrak{q} < \infty \\ \text{(ii)} & \mathfrak{g}/\mathfrak{q} \text{ is semisimple} \\ \text{(iii)} & \mathfrak{q} \cap \mathfrak{h} = 0. \end{cases}$$

Moreover, $\mathfrak{q}$ is the kernel of any finite-dimensional semisimple representation of $\mathfrak{g}$ that is faithful on $\mathfrak{h}$.

Proof. To start with, we observe that in view of (4.6.1) H_i, X_i, Y_i are all in $[\mathfrak{g},\mathfrak{g}]$, $1 \leq i \leq l$. So $\mathfrak{g} = [\mathfrak{g},\mathfrak{g}]$. This shows that if $\mathfrak{g}_1$ is a homomorphic image of $\mathfrak{g}$, $\mathfrak{g}_1 = [\mathfrak{g}_1,\mathfrak{g}_1]$. In particular, such a $\mathfrak{g}_1$ will be semisimple if it is finite-dimensional and reductive (Theorem 3.16.3).

This said, we come to the proof. Let μ_i be the linear functions on $\mathfrak{h}$ with $\mu_i(H_j) = \delta_{ij}$, $1 \leq i, j \leq l$. Since μ_i is dominant and integral, the representation π_{μ_i} is finite-dimensional. Let π be the direct sum of the π_{μ_i}, and let V be the vector space on which π acts. Let $\bar{\mathfrak{g}} = \pi[\mathfrak{g}]$, $\bar{\mathfrak{h}} = \pi[\mathfrak{h}]$, and for $X \in \mathfrak{g}$ let $\bar{X} = \pi(X)$. Since $\bar{\mathfrak{g}}$ has a faithful finite-dimensional semisimple representation, $\bar{\mathfrak{g}}$ is reductive (Theorem 3.16.3). In view of the remark made at the beginning, $\bar{\mathfrak{g}}$ is semisimple. Let $\mathfrak{q}$ be the kernel of π. By very construction, π is faithful on $\mathfrak{h}$. Hence $\mathfrak{q} \cap \mathfrak{h} = 0$. $\mathfrak{q}$ thus satisfies (4.8.2).

Suppose now that $\mathfrak{q}^0$ is an ideal of $\mathfrak{g}$ satisfying (4.8.2). Let $\mathfrak{g}^0 = \mathfrak{g}/\mathfrak{q}^0$, and let $X \mapsto X^0$ be the natural map of $\mathfrak{g}$ onto $\mathfrak{g}^0$. Write $\mathfrak{h}^0$ for the image of $\mathfrak{h}$, and let α_i^0 be the linear functions on $\mathfrak{h}^0$ such that $\alpha_i^0(H_j^0) = a_{ji}$ $(1 \leq i, j \leq l)$. (Note the $H_1^0, \ldots, H_l^0$ are linearly independent by (iii) of (4.8.2.) The linear functions $0, \alpha_1^0, \ldots, \alpha_l^0, -\alpha_1^0, \ldots, -\alpha_l^0$ being distinct, it is clear that the $3l$ elements H_i^0, X_i^0, Y_i^0 $(1 \leq i \leq l)$ are linearly independent. The theory of §4.6 may now be used to study the representations of $\mathfrak{g}^0$. In particular, by Lemma 4.6.1, we have the following: $\mathfrak{g}^0$ is semisimple, $\mathfrak{h}_0$ is a CSA of $\mathfrak{g}^0$, and $S^0 = \{\alpha_1^0, \ldots, \alpha_l^0\}$ is a simple system of roots of $(\mathfrak{g}^0, \mathfrak{h}^0)$. Suppose λ is a dominant integral linear function on $\mathfrak{h}$. Then $\lambda^0 \colon H^0 \mapsto \lambda(H)$ $(H \in \mathfrak{h})$ is a dominant integral linear function on $\mathfrak{h}^0$, so we have the finite-dimensional irreducible representation π_{λ^0} associated with it. The representation $X \mapsto \pi_{\lambda^0}(X^0)$ of $\mathfrak{g}$ is then obviously the irreducible representation of $\mathfrak{g}$ with highest weight λ. In view of Theorem 4.6.11, we conclude that every finite-dimensional irreducible representation of $\mathfrak{g}$ vanishes on $\mathfrak{q}^0$.

It follows from the last statement that if γ is any finite-dimensional semisimple representation of $\mathfrak{g}$, γ is 0 on $\mathfrak{q}^0$. We assert that $\mathfrak{q}^0$ is precisely the kernel of γ provided that γ is faithful on $\mathfrak{h}$. In fact, if γ is faithful on $\mathfrak{h}$ but vanishes on an ideal strictly larger than $\mathfrak{q}^0$, we would obtain (by passage to the quotient) an ideal in $\mathfrak{g}^0$ which is nonzero but is linearly independent of $\mathfrak{h}^0$. But this is impossible, because any nonzero ideal of a semisimple Lie algebra must have a nonzero intersection with any of its CSA's. Thus $\mathfrak{q}^0$ must be precisely the kernel of γ. In particular, $\mathfrak{q}^0$ must be the kernel of the representation π constructed earlier in the proof. So $\mathfrak{q}^0 = \mathfrak{q}$. All statements of the lemma are proved.

Theorem 4.8.3. *Let $A = (a_{ij})_{1 \leq i,j \leq l}$ be a Cartan matrix of rank $l \geq 1$. Then there is a semisimple Lie algebra $\bar{\mathfrak{g}}$ over* **C** *with CSA $\bar{\mathfrak{h}}$ and elements $\bar{H}_i, \bar{X}_i, \bar{Y}_i$ $(1 \leq i \leq l)$ such that (a) $\{\bar{H}_i, \ldots, \bar{H}_l\}$ is a basis for $\bar{\mathfrak{h}}$, (b) the commutation rules (4.6.1) are satisfied with $\bar{H}_i, \bar{X}_i, \bar{Y}_i$ instead of H_i, X_i, Y_i, and (c) $\bar{S} = \{\bar{\alpha}_1, \ldots, \bar{\alpha}_l\}$ is a simple system of roots for $(\bar{\mathfrak{g}}, \bar{\mathfrak{h}})$, where $\bar{\alpha}_j$ is the linear function on $\bar{\mathfrak{h}}$ defined by $\bar{\alpha}_j(\bar{H}_i) = a_{ij}$ $(1 \leq i, j \leq l)$. If $\tilde{\mathfrak{g}}$ is another semisimple Lie algebra with elements $\tilde{H}_i, \tilde{X}_i, \tilde{Y}_i$ $(1 \leq i \leq l)$ and CSA $\tilde{\mathfrak{h}}$ such that the conditions analogous to (a), (b), (c) are satisfied, there is a unique isomorphism τ of $\bar{\mathfrak{g}}$ onto $\tilde{\mathfrak{g}}$ such that $\tau(\bar{H}_i) = \tilde{H}_i$, $\tau(\bar{X}_i) = \tilde{X}_i$, $\tau(\bar{Y}_i) = \tilde{Y}_i$.*

Proof. We define $\bar{\mathfrak{g}} = \mathfrak{g}/\mathfrak{q}$, where $\mathfrak{q}$ is as in the previous lemma. Let $\bar{H}_i$, $\bar{X}_i$, and $\bar{Y}_i$ be the respective images of H_i, X_i, and Y_i in $\bar{\mathfrak{g}}$ $(1 \leq i \leq l)$. In the course of the proof of the preceding lemma we have already seen that the properties (*a*), (*b*), (*c*) of the theorem are satisfied. Suppose now that $\tilde{\mathfrak{g}}$ is another semisimple Lie algebra with elements $\tilde{H}_i, \tilde{X}_i, \tilde{Y}_i$ $(1 \leq i \leq l)$ and CSA $\tilde{\mathfrak{h}}$ satisfying (*a*), (*b*), (*c*). By the universal property of $\mathfrak{g}$ there is a unique homomorphism σ of $\mathfrak{g}$ onto $\tilde{\mathfrak{g}}$ such that $\sigma(H_i) = \tilde{H}_i$, $\sigma(X_i) = \tilde{X}_i$, $\sigma(Y_i) = \tilde{Y}_i$ $(1 \leq i \leq l)$. Now $\tilde{\mathfrak{g}}$, being semisimple, has a faithful finite-dimensional semisimple representation. Let γ be one such. Then $\gamma \circ \sigma$ is a finite-dimensional semisimple representation of $\mathfrak{g}$ that is faithful on $\mathfrak{h}$. So by Lemma 4.8.2, $\mathfrak{q} = \text{kernel } (\gamma \circ \sigma)$. Since γ is faithful, this implies that $\mathfrak{q} = \text{kernel}(\sigma)$. So σ induces an isomorphism τ of $\bar{\mathfrak{g}}$ onto $\tilde{\mathfrak{g}}$. τ obviously has all the required properties. This proves the theorem.

It is clear from Lemma 4.6.8 (cf. also (4.7.1)) that $\theta_{ij}^{\pm} \in \mathfrak{q}$ for $1 \leq i, j \leq l$. It was proved by Serre [1] that $\mathfrak{q}$ is precisely the ideal of $\mathfrak{g}$ generated by the $\theta_{ij}^{\pm}$. It follows from this that given a Cartan matrix $A = (a_{ij})_{1 \leq i,j \leq l}$, the corresponding semisimple Lie algebra can be defined as the Lie algebra generated by the $3l$ elements H_i, X_i, Y_i $(1 \leq i \leq l)$ satisfying the following commutation rules ($1 \leq i,j \leq l$)

$$\begin{aligned} &[H_i,H_j] = 0, \quad [H_i,X_j] - a_{ij}X_j = 0, \quad [H_i,Y_j] + a_{ij}Y_j = 0, \\ &[X_i,Y_j] - \delta_{ij}H_i = 0 \\ &\underbrace{[X_i,[X_i,\ldots[X_i,X_j]\ldots]}_{-a_{ij}+1 \text{ times}} = 0, \quad \underbrace{[Y_i,[Y_i,\ldots[Y_i,Y_j]\ldots]}_{-a_{ij}+1 \text{ times}} = 0, \quad (i \neq j) \end{aligned} \tag{4.8.3}$$

Serre's proof of this is indicated in Exercise 6.

4.9. The Algebra of Invariant Polynomials on a Semisimple Lie Algebra

Let $\mathfrak{g}$ be a semisimple Lie algebra over **C**, G any connected complex analytic group with Lie algebra $\mathfrak{g}$. We denote by $P(\mathfrak{g}) = \mathcal{P}$ (resp. $S(\mathfrak{g}) = \mathcal{S}$) the algebra of all polynomial functions on $\mathfrak{g}$ (resp. the symmetric algebra over $\mathfrak{g}$). The group G acts on $\mathcal{P}$ and $\mathcal{S}$ as a group of automorphisms: if $p \in \mathcal{P}$ and $x \in G$, $p^x(X) = p(X^{x^{-1}})$ ($X \in \mathfrak{g}$), while $u \mapsto u^x$ is the automorphism of $\mathcal{S}$ that extends $\mathrm{Ad}(x)$ from $\mathfrak{g}$. Let I (resp. J) denote the set of all $p \in \mathcal{P}$ (resp. $u \in \mathcal{S}$) such that $p^x = p$ (resp. $u^x = u$) for all $x \in G$. I and J are subalgebras of $\mathcal{P}$ and $\mathcal{S}$, respectively. Note that I and J depend only on $\mathrm{Ad}(G)$. The aim of this section is to prove a theorem of Chevalley which asserts that I (resp. J) is isomorphic to the algebra of all polynomials in l indeterminates, where l = rank of $\mathfrak{g}$.

Since $\langle\cdot,\cdot\rangle$ is a non-singular symmetric bilinear form on $\mathfrak{g} \times \mathfrak{g}$, we have an isomorphism $X \mapsto \tilde{X}$ of $\mathfrak{g}$ with $\mathfrak{g}^*$, the dual of the underlying vector space of $\mathfrak{g}$:

$$\tilde{X}(Y) = \langle X,Y \rangle \quad (Y \in \mathfrak{g}). \tag{4.9.1}$$

We extend this to an isomorphism $u \mapsto \tilde{u}$ of $\mathcal{S}$ with the algebra $\mathcal{P}$. It is easily verified that this isomorphism intertwines the action of G. In particular,

$$\tilde{J} = I. \tag{4.9.2}$$

It is therefore enough to work with I.

We begin with a result on the Weyl group.

Theorem 4.9.1. *Let $\mathfrak{h}$ be a CSA of $\mathfrak{g}$, A the centralizer of $\mathfrak{h}$ in G, and $\tilde{A}$ the normalizer of $\mathfrak{h}$ in G. Then the correspondence $a \mapsto s(a) = \mathrm{Ad}(a)|\mathfrak{h}$ induces an isomorphism of $\tilde{A}/A$ with the Weyl group $\mathfrak{w}$. Moreover, $\mathrm{Ad}[A] = exp\ \mathrm{ad}[\mathfrak{h}]$*

Proof. Note that if $a \in \tilde{A}$, then for any $b \in A$, aba^{-1} is also in A. So A is a normal subgroup of $\tilde{A}$. Also, since $\mathfrak{h}$ is abelian and is its own normalizer in $\mathfrak{g}$, $\exp \mathfrak{h} = A^0$ is the analytic subgroup of G corresponding to $\mathfrak{h}$, and A^0 is the component of the identity of $\tilde{A}$. In particular,

$$A^0 \subseteq A \subseteq \tilde{A}. \tag{4.9.3}$$

Since $\tilde{A}$ is obviously closed in G, A^0 is also closed in G.

Let Δ be the set of roots of $(\mathfrak{g},\mathfrak{h})$, P any positive system. If t is any automorphism of $\mathfrak{g}$ that leaves $\mathfrak{h}$ invariant, t induces a permutation of Δ, and $\mathfrak{g}_\alpha^t = \mathfrak{g}_{\alpha^t}$ for all $\alpha \in \Delta$. We begin by proving that if such a t lies in $\mathrm{Ad}(G)$ and leaves P invariant, then $t|\mathfrak{h}$ is the identity. Let $\tilde{G}$ be the simply connected covering group of G, and let $x \in \tilde{G}$ be such that $\mathrm{Ad}(x) = t$. To prove that $t|\mathfrak{h} =$ identity it is enough to prove that $\lambda^t = \lambda$ for all $\lambda \in \mathfrak{D}_P$. Fix such a λ, and let π_λ be the corresponding irreducible representation of $\mathfrak{g}$. Again, we denote by π_λ the associated representation of $\tilde{G}$. Let V be the vector space on which π_λ acts. Then

$$\pi_\lambda(X^x) = \pi_\lambda(x)\pi_\lambda(X)\pi_\lambda(x)^{-1} \quad (X \in \mathfrak{g}). \tag{4.9.4}$$

Consequently, $\pi_\lambda(x)$ induces a linear bijection of V_μ onto V_{μ^t} for any weight μ of π_λ. Furthermore, since $t \cdot P = P$, $\mathrm{Ad}(x)$ leaves $\sum_{\alpha \in P} \mathfrak{g}_\alpha$ invariant, so we conclude from (4.6.13) that $\pi_\lambda(x)[V_\lambda] = V_\lambda$. Combining these two observations, we see that $\lambda^t = \lambda$. This proves the assertion we made.

Let $\alpha \in P$. Select $X_{\pm\alpha} \in \mathfrak{g}_{\pm\alpha}$ such that the commutation rules (4.3.12) are satisfied. Define the element x_α of G by

$$x_\alpha = \exp X_\alpha \exp(-X_{-\alpha}) \exp X_\alpha. \tag{4.9.5}$$

Then $x_\alpha \in \tilde{A}$, and $s(x_\alpha) = \mathrm{Ad}(x_\alpha)|\mathfrak{h} = s_\alpha$. In fact, a straightforward calculation shows that $H_\alpha^{x_\alpha} = -H_\alpha = H_\alpha^{s_\alpha}$, while for $H \in \mathfrak{h}$ with $\alpha(H) = 0$, $H^{x_\alpha} = H = H^{s_\alpha}$ (since such H commute with $X_{\pm\alpha}$). It follows from this that the range of the map $a \mapsto s(a)$ is a subgroup of $GL(\mathfrak{h})$ containing $\mathfrak{w}$. On the other hand, let $a \in \tilde{A}$. Then it is obvious that $s(a) \cdot P$ is a positive system, so for a suitable $s \in \mathfrak{w}$, $s(a) \cdot P = s \cdot P$. By the earlier result, we can find $b \in \tilde{A}$ such that $s(b) = s$. Then $x = b^{-1} a \in \tilde{A}$, and $s(x) \cdot P = P$. We may then conclude that $s(x)$ is the identity. Thus $s(a) = s \in \mathfrak{w}$. This proves that $a \mapsto s(a)$ induces an isomorphism of $\tilde{A}/A$ with $\mathfrak{w}$.

It remains to prove that if $a \in A$, $\mathrm{Ad}(a) \in \exp \mathrm{ad}[\mathfrak{h}]$. We prove the following more general result:

(4.9.6) Let t be an automorphism of $\mathfrak{g}$ which fixes each element of $\mathfrak{h}$; then there is $H_0 \in \mathfrak{h}$ such that $t = e^{\mathrm{ad} H_0}$; in particular, $t \in \mathrm{Ad}[G]$.

Choose nonzero $X_\alpha \in \mathfrak{g}_\alpha$ $(\alpha \in \Delta)$. Then there are nonzero constants $c(\alpha)$ such that $X_\alpha^t = c(\alpha)X_\alpha$. Let $\alpha_1, \ldots, \alpha_l$ be the simple roots in P. We can find

$H_0 \in \mathfrak{h}$ such that $e^{\alpha_i(H_0)} = c(\alpha_i)$, $1 \leq i \leq l$. Write $t_0 = e^{\operatorname{ad} H_0}$. Then t_0 is an automorphism of $\mathfrak{g}$ and coincides with t on $\mathfrak{h} + \sum_{1 \leq i \leq l} \mathfrak{g}_{\alpha_i} + \sum_{1 \leq i \leq l} \mathfrak{g}_{-\alpha_i}$. Since this subspace generates the Lie algebra $\mathfrak{g}$, $t = t_0$.

From now on it is convenient to assume that G is simply connected. Let $\mathfrak{h}$ be a CSA. Denote by $P(\mathfrak{h})$ the algebra of all polynomial functions on $\mathfrak{h}$, and by $S(\mathfrak{h})$ the symmetric algebra over $\mathfrak{h}$. $\mathfrak{w}$ acts naturally on both of these. Write $I(\mathfrak{h})$ (resp. $J(\mathfrak{h})$) for the set of elements the are $\mathfrak{w}$-invariant in $P(\mathfrak{h})$ (resp. $S(\mathfrak{h})$).

Suppose $p \in I$. Since the elements of $\mathfrak{w}$ are induced from G, it is clear that the restriction of p to $\mathfrak{h}$ is invariant under $\mathfrak{w}$. For any $p \in \mathcal{P}$, let

(4.9.7) $$p_{\mathfrak{h}} = p \,|\, \mathfrak{h}.$$

The map $p \mapsto p_{\mathfrak{h}}$ is thus a homomorphism of I into $I(\mathfrak{h})$. Our first theorem, due to Chevalley, asserts that this is an isomorphism.

Theorem 4.9.2. *The map $p \mapsto p_{\mathfrak{h}}$ is an isomorphism of I onto $I(\mathfrak{h})$.*

Proof. If $p \in I$ and $p_{\mathfrak{h}} = 0$, then $p \,|\, \mathfrak{h}^G = 0$. Since $\mathfrak{h}^G$ contains all the regular points of $\mathfrak{g}$ (cf. §4.1), it is dense in $\mathfrak{g}$. So $p = 0$. $p \mapsto p_{\mathfrak{h}}$ is thus injective. It therefore only remains to prove that this map is surjective. Let R denote the range of this map. For any $q \in P(\mathfrak{h})$ let $\bar{q}$ denote its average over $\mathfrak{w}$:

(4.9.8) $$\bar{q} = \frac{1}{[\mathfrak{w}]} \sum_{s \in \mathfrak{w}} q^s \quad ([\mathfrak{w}] = \text{order of } \mathfrak{w}).$$

Clearly, $I(\mathfrak{h}) = \{\bar{q} : q \in P(\mathfrak{h})\}$. Now it is easy to verify that $P(\mathfrak{h})$ is the linear span of the monomials λ^n ($n = 0,1,2, \ldots, \lambda \in \mathfrak{D}_P$). So on writing

(4.9.9) $$\xi_{n,\lambda} = \frac{1}{[\mathfrak{w}]} \sum_{s \in \mathfrak{w}} (s\lambda)^n,$$

we see that the $\xi_{n,\lambda}$ ($n = 0,1, \ldots, \lambda \in \mathfrak{D}_P$) span $I(\mathfrak{h})$. It is clearly sufficient to prove that $\xi_{n,\lambda} \in R$ for all integers $n \geq 0$ and all $\lambda \in \mathfrak{D}_P$.

Fix $\lambda \in \mathfrak{D}_P$, and consider the representation π_λ of G (and $\mathfrak{g}$). Then $p_{n,\lambda}$: $X \mapsto tr(\pi_\lambda(X)^n)$ is a polynomial function on $\mathfrak{g}$. Further, since $\pi_\lambda(X^x) = \pi_\lambda(x)\pi_\lambda(X)\pi_\lambda(x)^{-1}$ for all $x \in G$ and $X \in \mathfrak{g}$, it is clear that $p_{n,\lambda}$ is invariant under G ($n = 0,1, \ldots$). So $(p_{n,\lambda})_{\mathfrak{h}} \in R$ for $n = 0,1, \ldots$. Clearly,

(4.9.10) $$(p_{n,\lambda})_{\mathfrak{h}} = \sum_{\mu} m(\mu)\mu^n,$$

where the sum is over the weights μ of π_λ and $m(\mu)$ is the multiplicity of μ. Now, if μ is a weight of π_λ, so is $s \cdot \mu$ for any $s \in \mathfrak{w}$, and $m(\mu) = m(s \cdot \mu)$; moreover, in each $\mathfrak{w}$-orbit of a weight there is a unique element of $\mathfrak{D}_P$ (Lemma

4.7.4). So, writing $F_\lambda = \{\mu: \mu \in \mathfrak{D}_P,\ \mu \neq \lambda$, and μ is a weight of $\pi_\lambda\}$, we conclude from (4.9.10) that there are constants $c_\mu > 0$ with the following property:

$$(p_{n,\lambda})_{\mathfrak{h}} = c_\lambda \xi_{n,\lambda} + \sum_{\mu \in F_\lambda} c_\mu \xi_{n,\mu}. \tag{4.9.11}$$

Suppose now that for some integer $n \geq 0$ and some $\lambda \in \mathfrak{D}_P$, $\xi_{n,\lambda} \notin R$. The relation (4.9.11) then shows, since $(p_{n,\lambda})_{\mathfrak{h}} \in R$, that F_λ is nonempty and that $\xi_{n,\mu} \notin R$ for some $\mu \in F_\lambda$. We may now argue in succession and construct a sequence $\lambda_1, \lambda_2, \ldots$ of elements of $\mathfrak{D}_P$ such that $\lambda_{i+1} \neq \lambda_i, \lambda_{i+1}$ is a weight of π_{λ_i}, and $\xi_{n,\lambda_i} \notin R$ $(i = 1, 2, \ldots)$. In particular,

$$\lambda_{i+1} \prec \lambda_i, \quad \lambda_i \in \mathfrak{D}_P \quad (i = 1, 2, \ldots). \tag{4.9.12}$$

We now prove that the *infinite* nature of the sequence $\{\lambda_i\}$ satisfying (4.9.12) already leads to a contradiction. Let s_0 be the element of $\mathfrak{w}$ such that $s_0 \cdot P = -P$. Then for $\nu, \nu' \in \mathfrak{D}_P$, with $\nu' \prec \nu$, one has the inequality $s_0\nu \prec s_0\nu'$, so by Lemma 4.7.4, $s_0\nu \prec s_0\nu' \preccurlyeq \nu' \prec \nu$. Using this, we deduce from (4.9.12) the relations

$$s_0\lambda_1 \prec \lambda_i \prec \lambda_1 \quad (i = 1, 2, \ldots). \tag{4.9.13}$$

This is a contradiction, since there are only finitely many elements μ of $\mathfrak{D}_P$ which can satisfy the inequality $s_0\lambda_1 \prec \mu \prec \lambda_1$.

This completes the proof of the theorem.

Since $\mathfrak{w}$ is a finite group generated by reflexions, we can find (cf. Theorem 4.15.23) homogeneous elements $q_1, \ldots, q_l$ in $I(\mathfrak{h})$ of positive degree such that (i) $q_1, \ldots, q_l$ are algebraically independent, and (ii) $I(\mathfrak{h}) = \mathbf{C}[q_1, \ldots, q_l]$. By the above theorem, there are unique $p_j \in I$ such that $(p_j)_{\mathfrak{h}} = q_j$, $1 \leq j \leq l$. We then have the following consequence of the above theorem, also due to Chevalley:

Theorem 4.9.3. *There exist l homogeneous algebraically independent elements $p_1, \ldots, p_l$ of positive degree such that $I = \mathbf{C}[p_1, \ldots, p_l]$.*

As an example, consider the case $\mathfrak{g} = \mathfrak{sl}(l + 1, \mathbf{C})$. Let $\mathfrak{h}$ be the CSA of all diagonal matrices in $\mathfrak{g}$. For any $X \in \mathfrak{g}$ let $c(X: T)$ be its characteristic polynomial, T being an indeterminate. Then

$$c(X: T) = T^{l+1} + p_1(X)T^{l-1} + \cdots + (-1)^{l+1}p_{l+1}(X).$$

Clearly, p_j $(1 \leq j \leq l)$ are homogeneous elements of I. If $X = \text{diag}(a_1, \ldots, a_{l+1}) \in \mathfrak{h}$, then

$$c(X: T) = \prod_{1 \leq j \leq l+1} (T - a_j),$$

so $(p_1)_{\mathfrak{h}}, \ldots, (p_l)_{\mathfrak{h}}$ are the classical symmetric functions of $a_1, \ldots, a_{l+1}$ of respective degrees $2,3,\ldots,l+1$. By Newton's theorem, these are algebraically independent and generate $I(\mathfrak{h})$. So the p_j $(1 \leq j \leq l)$ are algebraically independent homogeneous generators of I. (For other simple Lie algebras see Exercise 58.)

Let $\mathfrak{q}$ be the orthogonal complement of $\mathfrak{h}$ in $\mathfrak{q}$ with respect to $\langle\cdot,\cdot\rangle$. Then $\mathfrak{S}$ is the direct sum of $S(\mathfrak{h})$ and the ideal $\mathfrak{S}\mathfrak{q}$. In particular, for any $u \in \mathfrak{S}$, there is a unique $u_{\mathfrak{h}} \in S(\mathfrak{h})$ such that

$$u \equiv u_{\mathfrak{h}} \pmod{\mathfrak{S}\mathfrak{q}}. \tag{4.9.14}$$

The map $u \mapsto u_{\mathfrak{h}}$ is a homomorphism of $\mathfrak{S}$ onto $S(\mathfrak{h})$ with kernel $\mathfrak{S}\mathfrak{q}$. On the other hand, since $\langle\cdot,\cdot\rangle | \mathfrak{h} \times \mathfrak{h}$ is also nonsingular, we have an algebra isomorphism $v \mapsto \tilde{v}$ of $S(\mathfrak{h})$ onto $P(\mathfrak{h})$. It is easily verified that

$$(u_{\mathfrak{h}})^{\sim} = (\tilde{u})_{\mathfrak{h}} \quad (u \in \mathfrak{S}). \tag{4.9.15}$$

We then obtain the following consequence of Theorems 4.9.2 and 4.9.3:

Theorem 4.9.4. *There are l algebraically independent homogeneous elements u_j $(1 \leq j \leq l)$ of positive degree such that $J = \mathbf{C}[u_1, \ldots, u_l]$. The map $u \mapsto u_{\mathfrak{h}}$ is an isomorphism of J onto $J(\mathfrak{h})$.*

$\mathfrak{S}$ is obviously a J-module. The group G acts on $\mathfrak{S}$ in a natural fashion, and for each finite-dimensional irreducible representation π of G, we denote by $\mathfrak{S}^{\pi}$ the linear span of all subspaces U of $\mathfrak{S}$ with the following property: U is invariant under G and defines a representation of G equivalent to π. It is then not difficult to show that each $\mathfrak{S}^{\pi}$ is a submodule of $\mathfrak{S}$. A deep study of the structure of $\mathfrak{S}$ and the $\mathfrak{S}^{\pi}$ as J-modules has been made by Kostant [3].

4.10 Infinitesimal Characters

We devote this section to the determination of the infinitesimal characters of the finite-dimensional irreducible representations of a semisimple Lie algebra $\mathfrak{g}$. The results of this section were first obtained by Harish-Chandra [4] and formed the point of departure for his study of infinite-dimensional representations of a semisimple Lie group, from the infinitesimal point of view.

Throughout this section we fix a semisimple Lie algebra $\mathfrak{g}$, a CSA $\mathfrak{h}$, and a positive system P of roots of $(\mathfrak{g},\mathfrak{h})$. $\mathfrak{G}$ is the universal enveloping algebra. For $\lambda \in \mathfrak{h}^*$, write $\mathfrak{G}(\lambda)$ for the weight subspace corresponding to λ for the adjoint representation of $\mathfrak{g}$ in $\mathfrak{G}$, i.e.,

$$\mathfrak{G}(\lambda) = \{a : a \in \mathfrak{G}, Ha - aH = \lambda(H)a \text{ for all } H \in \mathfrak{h}\}. \tag{4.10.1}$$

$\mathfrak{G}(0)$ is an algebra, and $\mathfrak{G}$ is the direct sum of the $\mathfrak{G}(\lambda)$. Let $\mathfrak{H}$ $(\subseteq \mathfrak{G}(0))$ be the subalgebra of $\mathfrak{G}$ generated by $\mathfrak{h}$. Since $\mathfrak{h}$ is abelian, it is obvious that $\mathfrak{H}$ is canonically isomorphic to the symmetric algebra $S(\mathfrak{h})$ over $\mathfrak{h}$. We therefore identify $\mathfrak{H}$ with the algebra of polynomial functions on $\mathfrak{h}^*$.

Lemma 4.10.1. *Let*

$$\mathfrak{P} = \sum_{\alpha \in P} \mathfrak{G}\mathfrak{g}_\alpha. \tag{4.10.2}$$

Then, for each $a \in \mathfrak{G}(0)$, *there is a unique element* $\beta_P(a) \in \mathfrak{H}$ *such that*

$$a \equiv \beta_P(a) \pmod{\mathfrak{P}}; \tag{4.10.3}$$

the map $\beta_P\colon a \mapsto \beta_P(a)$ *is a homomorphism of* $\mathfrak{G}(0)$ *onto* $\mathfrak{H}$.

Proof. Let $P = \{\alpha_1, \ldots, \alpha_t\}$. For any j, let X_j and Y_j be nonzero elements in $\mathfrak{g}_{\alpha_j}$ and $\mathfrak{g}_{-\alpha_j}$, respectively $(1 \leq j \leq t)$. Let $\{H_1, \ldots, H_l\}$ be a basis of $\mathfrak{h}$. For t-tuples $(q) = (q_1, \ldots, q_t)$, $(p) = (p_1, \ldots, p_t)$ and an l-tuple $(b) = (b_1, \ldots, b_l)$ of nonnegative integers, let

$$a((q),(b),(p)) = Y_1^{q_1} \cdots Y_t^{q_t} H_1^{b_1} \cdots H_l^{b_l} X_1^{p_1} \cdots X_t^{p_t}.$$

Then $a((q),(b),(p)) \in \mathfrak{G}(\lambda)$ for $\lambda = \sum_{1 \leq i \leq t} (p_j - q_j)\alpha_j$, and these elements form a basis for $\mathfrak{G}$.

Suppose $a \in \mathfrak{G}(0)$. Then it is clear from the above observation that we can find $h \in \mathfrak{H}$ and constants $c((q),(b),(p))$ such that

$$a = h + \sum c((q),(b),(p))\, a(q),(b),(p)),$$

where the sum is over all (q), (b), (p) with $\sum_{1 \leq j \leq t} (q_j + p_j) > 0$, $\sum_{1 \leq j \leq t} (p_j - q_j)\alpha_j = 0$, and all but finitely many of the $c((q),(b),(p))$ are 0. If the c_j are all ≥ 0, the linear combination $\sum_{1 \leq j \leq t} c_j\alpha_j$ cannot be zero unless $c_1 = \cdots = c_t = 0$. Consequently, $\sum_j q_j > 0$, and $\sum_j p_j > 0$ in every term of the above sum. This shows at once that $a - h \in \mathfrak{P}$.

To prove the uniqueness of h we must prove that

$$\mathfrak{H} \cap \mathfrak{P} = 0. \tag{4.10.4}$$

Suppose $a \in \mathfrak{H} \cap \mathfrak{P}$. For any $\lambda \in \mathfrak{h}^*$, let π_λ be the irreducible representation of $\mathfrak{G}$ with highest weight λ. Let v_λ be a nonzero vector of weight λ. Then $\pi_\lambda(a)v_\lambda = 0$ because $a \in \mathfrak{P}$. On the other hand, since $a \in \mathfrak{H}$, $\pi_\lambda(a)v_\lambda = a(\lambda)v_\lambda$. So $a(\lambda) = 0$. Since λ is arbitrary, $a = 0$.

β_P is thus a well-defined linear map of $\mathfrak{G}(0)$ into $\mathfrak{H}$. If $a \in \mathfrak{H}$, $\beta_P(a) = a$. β_P is thus surjective. For any $a \in \mathfrak{G}(0)$ and any $\lambda \in \mathfrak{h}^*$, $\pi_\lambda(a)v_\lambda = \beta_P(a)(\lambda)v_\lambda$.

So $\beta_P(a_1a_2)(\lambda) = \beta_P(a_1)(\lambda)\beta_P(a_2)(\lambda)$ for $a_1, a_2 \in \mathfrak{G}(0)$ and $\lambda \in \mathfrak{h}^*$. This proves that β_P is a homomorphism.

Lemma 4.10.2. *Let $\lambda \in \mathfrak{h}^*$, and let π be a representation of $\mathfrak{g}$ in a vector space V such that* (i) *λ is the highest weight of π, and* (ii) *there is a nonzero vector v in V_λ which is cyclic for π. Then if $\mathfrak{Z}$ is the center of $\mathfrak{G}$,*

$$\pi(z) = \beta_P(z)(\lambda)1 \quad (z \in \mathfrak{Z}). \tag{4.10.5}$$

Proof. By Lemma 4.6.5, $\dim V_\lambda = 1$. Obviously, $\pi(X)v = 0$ for $X \in \mathfrak{g}_\alpha$, $\alpha \in P$. So $\pi(z)v = \beta_P(z)(\lambda)v$ for $z \in \mathfrak{Z}$. Write $U = \{u : u \in V, \pi(z) = \beta_P(z)(\lambda)u$ for all $z \in \mathfrak{Z}\}$. Then U is a π-invariant subspace containing v, so $U = V$.

It follows from Lemma 4.10.2 that π has an infinitesimal character and that

$$\chi_\lambda : z \mapsto \beta_P(z)(\lambda) \quad (z \in \mathfrak{Z}) \tag{4.10.6}$$

is the infinitesimal character of π. In particular, if $\mathfrak{G}_\lambda$ is defined by (4.6.15), the representation of $\mathfrak{G}$ in $\mathfrak{G}/\mathfrak{G}_\lambda$, as well as the irreducible representation π_λ, have χ_λ for their infinitesimal characters.

We now state and prove the main result of this section. Note that since $\mathfrak{H}$ is canonically isomorphic to $S(\mathfrak{h})$, $\mathfrak{w}$ acts as a group of automorphisms of $\mathfrak{H}$.

Theorem 4.10.3. *Let $\delta_P = \frac{1}{2}\sum_{\alpha \in P} \alpha$. For any $z \in \mathfrak{Z}$, let $\gamma(z)$ be the element of $\mathfrak{H}$ defined by*

$$\gamma(z)(\lambda) = \beta_P(z)(\lambda - \delta_P) \quad (\lambda \in \mathfrak{h}^*). \tag{4.10.7}$$

Then $\gamma(z)$ is independent of the positive system P used to define it. Moreover, $\gamma : z \mapsto \gamma(z)$ is an isomorphism of $\mathfrak{Z}$ onto the subalgebra of all $\mathfrak{w}$-invariant elements of $\mathfrak{H}$.

Proof. First, fix P and let $\alpha_1, \ldots, \alpha_l$ be the simple roots in P. Define $\gamma(z)$ for $z \in \mathfrak{Z}$ by (4.10.7). Fix $z \in \mathfrak{Z}$, and let $1 \leq i \leq l$. Let $\lambda \in \mathfrak{D}_P$, and let π be the representation of $\mathfrak{G}$ in $V = \mathfrak{G}/\mathfrak{G}_\lambda$. Then we know that

$$\pi(z) = \beta_P(z)(\lambda)1. \tag{4.10.8}$$

Select $X_{\pm\alpha} \in \mathfrak{g}_{\pm\alpha}$ ($\alpha \in P$) such that the commutation rules (4.3.12) are satisfied, and write $H_j = \bar{H}_{\alpha_j}, X_j = X_{\alpha_j}, Y_j = X_{-\alpha_j}$ $(1 \leq j \leq l)$. Let v be the image of 1 in V and $u = \pi(Y_i^{\lambda_i+1})v$, where $\lambda_i = \lambda(H_i)$. Then $u \in V_{\lambda-(\lambda_i+1)\alpha_i}$, and since $\mathfrak{N}^- \cap \mathfrak{G}_\lambda = 0$ by Lemma 4.6.6, $u \neq 0$. Moreover, a simple calculation based on (4.6.1) shows (as in the proof of Lemma 4.6.10) that $\pi(X_j)u = 0$, $1 \leq j \leq l$. Therefore, if $U = \pi[\mathfrak{G}]u$, then U is nonzero, π-invariant, and defines a representation of $\mathfrak{G}$ for which $\lambda - (\lambda_i + 1)\alpha_i$ is the highest weight. From Lemma 4.10.2 we obtain

$$\pi(z)u' = \beta_P(z)(\lambda - (\lambda_i + 1)\alpha_i)u' \quad (u' \in U). \tag{4.10.9}$$

From (4.10.8) and (4.10.9) we get

$$\beta_P(z)(\lambda) = \beta_P(z)(\lambda - (\lambda_i + 1)\alpha_i).$$

On the other hand, we have $s_{\alpha_i}\delta_P = \delta_P - \alpha_i$ by Lemma 4.7.4. So

$$\gamma(z)(\mu) = \gamma(z)(s_{\alpha_i}\mu) \quad (\mu = \lambda + \delta_P).$$

Since $\lambda \in \mathfrak{D}_P$ is arbitrary, we conclude[8] from the last equation that $\gamma(z)$ is invariant under s_{α_i}. Since i was arbitrary, $\gamma(z)$ is $\mathfrak{w}$-invariant.

The $\mathfrak{w}$-invariance of $\gamma(z)$ $(z \in \mathfrak{Z})$ implies already that $\gamma(z)$ is independent of the positive system used to define it. In fact, let Q be another positive system, $\delta_Q = \frac{1}{2}\sum_{\alpha \in Q} \alpha$, and let $\gamma'(z)(\lambda) = \beta_Q(z)(\lambda - \delta_Q)$. Then there is an $s \in \mathfrak{w}$ such that $Q = s \cdot P$. Let G be a connected complex Lie group with Lie algebra $\mathfrak{g}$. By Theorem 4.9.1 we can find $x \in G$ such that x normalizes $\mathfrak{h}$ and $s = \mathrm{Ad}(x)\,|\,\mathfrak{h}$. If $z \in \mathfrak{Z}$, then $z^x = z$, so the relation $z \equiv \beta_P(z) \pmod{\sum_{\alpha \in P} \mathfrak{G}X_\alpha}$ implies that $z \equiv \beta_P(z)^s \pmod{\sum_{\alpha \in Q} \mathfrak{G}X_\alpha}$. Consequently, $\beta_Q(z) = \beta_P(z)^s$. But then, for any $\lambda \in \mathfrak{h}^*$, since $\delta_Q = s\delta_P$,

$$\begin{aligned}\gamma'(z)(\lambda) &= \beta_P(z)(s^{-1}(\lambda - s\delta_P) \\ &= \gamma(z)(s^{-1}\lambda) \\ &= \gamma(z)(\lambda).\end{aligned}$$

This proves our assertion.

Let $\bar{J}$ be the algebra of all $\mathfrak{w}$-invariant elements of $\mathfrak{H}$. It remains to prove that γ is an isomorphism of $\mathfrak{Z}$ onto $\bar{J}$. We shall deduce this from Theorem 4.9.4. Let $\mathfrak{S}$ (resp. $S(\mathfrak{h})$) be the symmetric algebra over $\mathfrak{g}$ (resp. over $\mathfrak{h}$), $\mathfrak{S}_d$ (resp. $S_d(\mathfrak{h})$) the homogeneous subspace of degree d. For each $X \in \mathfrak{g}$, let D_X denote the derivation of $\mathfrak{S}$ that extends ad X. Let $\boldsymbol{\lambda}$ ($\mathfrak{S} \mapsto \mathfrak{G}$) denote the usual symmetrizer map. Then we know that $\boldsymbol{\lambda}$ intertwines D_X and ad X for each $X \in \mathfrak{g}$ and is a linear bijection of J onto $\mathfrak{Z}$ as well as of $S(\mathfrak{h})$ onto $\mathfrak{H}$. Write $\mathfrak{H}_d = \boldsymbol{\lambda}[S_d(\mathfrak{h})]$, and $J_d = J \cap \mathfrak{S}_d$. Then obviously, $J = \sum_{d \geq 0} J_d$.

Let $d \geq 0$ and $v \in J_d$. Our entire argument is based on the following relation:

$$\gamma(\boldsymbol{\lambda}v) - \boldsymbol{\lambda}(v_{\mathfrak{h}}) \in \sum_{0 \leq e \leq d-1} \mathfrak{H}_e \tag{4.10.10}$$

[8]We are using here the easily proved principle that if F is a polynomial function on $\mathfrak{h}^*$ and $F(\lambda) = 0$ for all $\lambda \in \mathfrak{D}_P$, then $F = 0$.

We now prove this. We may assume that $d \geq 1$. Let $\{H_1, \ldots, H_l\}$ be a basis for $\mathfrak{h}$, let $P = \{\alpha_1, \ldots, \alpha_t\}$, and let X_j (resp. Y_j) be a nonzero element of $\mathfrak{g}_{\alpha_j}$ (resp. $\mathfrak{g}_{-\alpha_j}$) $(1 \leq j \leq t)$. Then we can write v as $v_{\mathfrak{h}} + \sum_{1 \leq p \leq r} \zeta_p$, where each ζ_p is a monomial in the H_i, X_j, Y_k such that (a) $D_H \zeta_p = 0$ for all $H \in \mathfrak{h}$, $1 \leq p \leq r$, and (b) each ζ_p is not a monomial of the H's only. The conditions (a) and (b) imply, as in the proof of Lemma 4.10.1, that in each ζ_p at least one X_j is present with a strictly positive exponent. This shows that $\boldsymbol{\lambda}(\zeta_p)$ is an element of $\mathfrak{G}(0)$, and differs from an element of $\sum_{1 \leq j \leq t} \mathfrak{G} X_j$ by an element of $\sum_{0 \leq e \leq d-1} \boldsymbol{\lambda}[\mathfrak{S}_e]$. On the other hand, it is clear from the definition of β_P and Poincaré–Birkhoff–Witt theorem that $\deg \beta_P(a) \leq \deg(a)$ for any $a \in \mathfrak{G}(0)$. So for $1 \leq p \leq r$,

$$\beta_P(\boldsymbol{\lambda}(\zeta_p)) \in \sum_{0 \leq e \leq d-1} \mathfrak{H}_e$$

Thus

$$\beta_P(\boldsymbol{\lambda}(v)) - \boldsymbol{\lambda}(v_{\mathfrak{h}}) \in \sum_{0 \leq e \leq d-1} \mathfrak{H}_e.$$

If we now observe that $\beta_P(\boldsymbol{\lambda}(v)) - \gamma(\boldsymbol{\lambda}(v))$ lies in $\sum_{0 \leq e \leq d-1} \mathfrak{H}_e$, we obtain (4.10.10).

Suppose first that γ is not injective. Then there is $z \neq 0$ in $\mathfrak{Z}$ such that $\gamma(z) = 0$. Let $u \in J$ be such that $\boldsymbol{\lambda}(u) = z$. We can write $u = u_0 + \cdots + u_d$, where $u_j \in J_j$ and $u_d \neq 0$. Then (4.10.10) implies at once that

$$\boldsymbol{\lambda}((u_d)_{\mathfrak{h}}) \in \sum_{0 \leq e \leq d-1} \mathfrak{H}_e$$

Since $(u_d)_{\mathfrak{h}} \in S_d(\mathfrak{h})$, $\boldsymbol{\lambda}((u_d)_{\mathfrak{h}}) \in \mathfrak{H}_d$. So we infer that $(u_d)_{\mathfrak{h}} = 0$. But then $u_d = 0$ by Theorem 4.9.4, contradicting our assumption.

It remains to prove that γ is surjective. Let R be the range of γ. For any integer $d \geq 0$, let $\bar{J}_d = \boldsymbol{\lambda}(S_d(\mathfrak{h}) \cap J(\mathfrak{h}))$. It is enough to prove that $\bar{J}_d \subseteq R$ for all $d \geq 0$. For $d = 0$ this is obvious. Let $d > 0$, and let us assume that $\bar{J}_e \subseteq R$ for $0 \leq e \leq d-1$. Let $q \in S_d(\mathfrak{h}) \cap J(\mathfrak{h})$. By Theorem 4.9.4 we can find $v \in J_d$ such that $q = v_{\mathfrak{h}}$. (4.10.10) now implies that

$$q' = \gamma(\boldsymbol{\lambda}(v)) - \boldsymbol{\lambda}(q) \in \sum_{0 \leq e \leq d-1} \mathfrak{H}_e.$$

Since $\boldsymbol{\lambda}(v) \in \mathfrak{Z}$, $\gamma(\boldsymbol{\lambda}(v)) \in \bar{J}$. So $q' \in \bar{J}$, and hence q' belongs to $\sum_{0 \leq e \leq d-1} \bar{J}_e$. By the induction hypothesis, $q' \in R$. The above relation shows at once that $q \in R$. Induction on d proves that $R = \bar{J}$. The proof of the theorem is complete.

We then get the following result as an immediate consequence.

Theorem 4.10.4. *For any* $\boldsymbol{\lambda} \in \mathfrak{h}^*$,

$$\tau_\lambda : z \mapsto \gamma(z)(\lambda) \tag{4.10.11}$$

is a homomorphism of $\mathfrak{Z}$ *into* $\mathbf{C}$. *Every homomorphism of* $\mathfrak{Z}$ *into* $\mathbf{C}$ *is of the form* τ_λ *for some* $\lambda \in \mathfrak{h}^*$. *If* $\lambda, \lambda' \in \mathfrak{h}^*$, $\tau_\lambda = \tau_{\lambda'}$ *if and only if* λ *and* λ' *lie in the same* $\mathfrak{w}$*-orbit.*

Remark. Let G be a simply connected complex analytic group with Lie algebra $\mathfrak{g}$. For any $\lambda \in \mathfrak{D}_P$ let π_λ be the corresponding irreducible representation of G. Let M_λ be the space of matrix elements of π_λ. Then, since $z \mapsto \beta_P(z)(\lambda)$ is the infinitesimal character of π_λ,

$$zf = \gamma(z)(\lambda + \delta_P)f \quad (f \in M_\lambda, z \in \mathfrak{Z}). \tag{4.10.12}$$

In particular, if ψ_λ is the character of π_λ, then ψ_λ is an analytic function on G, is invariant under the inner automorphisms, and satisfies the differential equations:

$$z\psi_\lambda = \gamma(z)(\lambda + \delta_P)\psi_\lambda \quad (z \in \mathfrak{Z}). \tag{4.10.13}$$

4.11. Compact and Complex Semisimple Lie Groups

Our treatment so far of the representation theory of semisimple Lie groups has been completely algebraic. We now wish to develop the transcendental approach of H. Weyl [2, 3, 4] to these questions. Weyl's method is very powerful and illuminates the entire theory from a very unexpected point of view.

The fundamental result in Weyl's theory is his discovery (the "unitarian trick") that any complex semisimple analytic group has a compact real form which is unique, up to conjugacy under an inner automorphism. This compact form is simply connected if and only if the complex group is, and the bijective correspondence that results, between simply connected complex semisimple groups and simply connected compact semisimple groups, reduces the representation theory of the complex groups to that of the compact groups. For a compact semisimple Lie group the method of invariant integration on the group can be used to carry out a thorough investigation of the representations. It was by these methods that Weyl established the semisimplicity of representations of a semisimple Lie group, and obtained his famous formulae for the characters and dimensions of the irreducible representations of semisimple Lie groups.

For example, let $G = SU(l + 1,\mathbf{C})$, $G_c = SL(l + 1,\mathbf{C})$. Then G is compact, G_c is complex analytic, and both groups are semisimple and simply connected. In a sense to be made precise later on, G is a real form of G_c. The unitarian trick of Weyl is then the assertion that if a holomorphic function on G_c vanishes on G, it is identically zero. It follows from this that if π_c is a complex analytic irreducible representation of G_c, then $\pi = \pi_c | G$ is an irreducible representation of G, and the map $\pi_c \mapsto \pi$ induces a bijection between the sets of equivalence classes of irreducible representations of G_c and G.

The aim of this section is to set up and study this correspondence between complex and compact semisimple groups. We begin by proving a theorem which implies Weyl's result on the finiteness of the fundamental group of a compact semisimple group. Our proof is essentially that given by Cartier [1] in *Seminaire Sophus Lie*, exposé n° 22; it treats the problem from the point of view of group cohomology, and for arbitrary compact groups satisfying the second axiom of countability.

Throughout this discussion (Lemmas 4.11.1 through 4.11.4), G is a fixed, connected, locally compact group satisfying the second axiom of countability; C is a discrete central subgroup of G such that $\bar{G} = G/C$ is compact; $x \mapsto \bar{x}$ is the natural map of G onto $\bar{G}$; $d\bar{x}$ is the Haar measure on $\bar{G}$ for which $\int_{\bar{G}} d\bar{x} = 1$; $\mathbf{R}^+$ is the multiplicative group of positive reals; and $\mathbf{Z}$ is the additive group of integers.

Lemma 4.11.1. *Let notation be as above. Then C is finitely generated. In particular, if C is infinite, there exist nontrivial homomorphisms of C into $\mathbf{R}^+$.*

Proof. One can choose a compact set D such that $G = CD^0$, where $D^0 =$ interior D. By enlarging D, we may assume that $1 \in D^0$ and $D = D^{-1}$. Since $D \cdot D^{-1}$ is compact and $\subseteq G = \bigcup_{c \in C} cD^0$, we can find $c_1, \ldots, c_k \in C$ such that $D \cdot D^{-1} \subseteq \bigcup_{i=1}^{k} c_i D^0$. Let C_1 be the subgroup generated by $c_1, \ldots, c_k$. If E is the image of D in G/C_1, then E contains the identity, $E = E^{-1}$, and $E \cdot E^{-1} \subseteq E$; E is therefore a subgroup of G/C_1. Since it contains a neighborhood of the identity (image of D^0), it must be an open and closed subgroup. So, G/C_1 being connected, we must have $G/C_1 = E$. In other words, $G = C_1 D$. Since C is discrete, $D \cap C$ is finite, and an easy argument shows that $C = C_1 \cdot (D \cap C)$. So C is finitely generated.

If C is infinite, $C \approx C_0 \times \mathbf{Z}^q$, where C_0 is finite and $q \geq 1$. The existence of a nontrivial homomorphism of C into $\mathbf{R}^+$ is now obvious.

Lemma 4.11.2. *Let φ be a homomorphism of C into $\mathbf{R}^+$. Then there is a continuous function h on G with positive values such that $h | C = \varphi$ and $h(xc) = h(x)\varphi(c)$ for all $x \in G$, $c \in C$.*

Proof. Select a compact set $D = D^{-1}$ such that $G = CD$, and let g be a continuous function on G with compact support such that (i) $g \geq 0$, and (ii) $g(x) = 1$ for $x \in D$. Let

$$h_1(x) = \sum_{c \in C} g(xc)\varphi(c)^{-1} \quad (x \in G).$$

If $x_0 \in G$ and if U is a compact neighborhood of x_0, then writing $K = \text{supp } g$, we see that $F = U^{-1}K \cap C$ is finite, and

$$h_1(x) = \sum_{c \in F} g(xc)\varphi(c^{-1}) \quad (x \in U)$$

h_1 is thus well defined and continuous on G. If $x \in G$, there is $c \in C$ with $xc \in D$, which implies that $g(xc) > 0$; this shows that $h_1(x) > 0$ for all $x \in G$. Write $h = h_1(1)^{-1}h_1$. Then h has the required properties. Note that $h(1) = 1$.

Lemma 4.11.3. *Let $\bar{H}$ be a continuous real-valued function on $\bar{G} \times \bar{G}$ such that $\bar{H}(\bar{1},\bar{1}) = 0$ and, for all $\bar{x}, \bar{y}, \bar{z} \in G$,*

$$\bar{H}(\bar{x}\bar{y},\bar{z}) + \bar{H}(\bar{x},\bar{y}) = \bar{H}(\bar{x},\bar{y}\bar{z}) + \bar{H}(\bar{y},\bar{z}). \tag{4.11.1}$$

Then there exists a continuous real-valued function $\bar{a}$ on $\bar{G}$ with $\bar{a}(\bar{1}) = 0$ such that for all $\bar{x}, \bar{y} \in \bar{G}$,

$$\bar{H}(\bar{x},\bar{y}) = \bar{a}(\bar{x}\bar{y}) - \bar{a}(\bar{x}) - \bar{a}(\bar{y}). \tag{4.11.2}$$

Proof. Let

$$\bar{a}(\bar{x}) = -\int_{\bar{G}} \bar{H}(\bar{x},\bar{y})\, d\bar{y} \quad (\bar{x} \in \bar{G}).$$

It is then a trivial verification, based on the biinvariant nature of $d\bar{x}$, that $\bar{a}$ has the required properties.

Lemma 4.11.4. *Let φ be a homomorphism of C into $\mathbf{R}^+$. Then there exists a continuous homomorphism χ of G into $\mathbf{R}^+$ such that $\chi | C = \varphi$.*

Proof. Select a continuous function h on G with positive values such that the properties stated in Lemma 4.11.2 are satisfied. Define H by

$$H(x,y) = \log h(xy) - \log h(x) - \log h(y) \quad (x, y \in G).$$

Any easy verification shows that H is constant on the $C \times C$ cosets in $G \times G$. Hence, there is a continuous real-valued function $\bar{H}$ on $\bar{G} \times \bar{G}$ such that for all $x, y \in G$,

$$H(x,y) = \bar{H}(\bar{x},\bar{y}).$$

A simple calculation shows that $\bar{H}(\bar{1},\bar{1}) = 0$ and that $\bar{H}$ satisfies the identities (4.11.1). So by the above lemma, there is a continuous real-valued function $\bar{a}$ with $\bar{a}(\bar{1}) = 0$ such that (4.11.2) is satisfied. Let $a(x) = e^{\bar{a}(\bar{x})}$, $x \in G$. Define

$$\chi(x) = h(x)a(x)^{-1} \quad (x \in G).$$

A straightforward verification shows that χ is a continuous homomorphism of G into $\mathbf{R}^+$ and that $\varphi = \chi | C$.

Theorem 4.11.5. *Let G be a connected locally compact group satisfying the second axiom of countability, C a discrete central subgroup. Suppose that*

(a) *G/C is compact*
(b) *G has no nontrivial continuous homomorphisms into $\mathbf{R}^+$.*

Then G is compact. This is, in particular, the case when the commutator subgroup of G is dense in G.

Proof. If C is infinite, there is a nontrivial homomorphism of C into $\mathbf{R}^+$, and this can be extended to a continuous homomorphism of G into $\mathbf{R}^+$, contradicting (b). If we assume that the commutator subgroup of G is dense in G, then any continuous homomorphism of G into a topological group which is abelian is necessarily trivial, so (b) is satisfied.

Theorem 4.11.5 has Weyl's Theorem as an immediate consequence.

Theorem 4.11.6. *Let $\bar{G}$ be a compact semisimple analytic group. Then its universal covering group is also compact.*

Proof. Let G be the universal covering group of $\bar{G}$. We may then assume that $\bar{G} = G/C$, where C is a discrete central subgroup of G. Clearly G is also semisimple. In order to prove that G is compact it is enough to prove that G satisfies condition (b) of the previous theorem. Let $\mathfrak{g}$ be the Lie algebra of G; $\mathfrak{g}$ is then semisimple. Consequently, $\mathfrak{g} = [\mathfrak{g},\mathfrak{g}]$, and hence all of its one-dimensional representations are trivial. If χ is a continuous homomorphism of G into $\mathbf{R}^+$, χ is analytic (Theorem 2.11.2), and its differential $d\chi$ is a one-dimensional representation of $\mathfrak{g}$. $d\chi$ is therefore 0, showing that χ is trivial. This proves the theorem.

Let $\mathfrak{g}$ be a semisimple Lie algebra over $\mathbf{R}$. We say that $\mathfrak{g}$ is of the *compact type* if its adjoint group is a compact subgroup of $GL(\mathfrak{g})$. Put

$$\omega(X) = tr(ad\, X)^2 \quad (X \in \mathfrak{g}). \tag{4.11.3}$$

ω is the *Casimir polynomial* on $\mathfrak{g}$ and is invariant under all the automorphisms of $\mathfrak{g}$ (cf. §3.9)

Theorem 4.11.7. *Let $\mathfrak{g}$ be a Lie algebra over $\mathbf{R}$, G its adjoint group. Then the following statements are equivalent:*

(i) *$\mathfrak{g}$ is reductive and $\mathfrak{D}\mathfrak{g}$ is of compact type*
(ii) *G is compact*

(iii) *If* $X \in \mathfrak{g}$, *ad* X *is semisimple and has only pure imaginary eigenvalues.*

If $\mathfrak{g}$ *is semisimple, these are all equivalent to*

(iv) $-\omega$ *is a positive definite quadratic form on* $\mathfrak{g}$.

Moreover, in this case, any analytic group whose Lie algebra is isomorphic to $\mathfrak{g}$ *is compact.*

Proof. (i) $\Rightarrow$ (ii) Let $\mathfrak{c} =$ center $\mathfrak{g}$, $\mathfrak{g}_1 = \mathfrak{D}\mathfrak{g}$. Then $Y^y = Y$ for $Y \in \mathfrak{c}$, $y \in G$. So if G_1 is the adjoint group of $\mathfrak{g}_1$, $y \mapsto y|\mathfrak{g}_1$ is an isomorphism of G onto G_1. So G is compact.

(ii) $\Rightarrow$ (iii) Let G be compact. It follows from a classical argument[9] that there is an inner product for $\mathfrak{g}$ which is positive definite and invariant under G. It is easy to see that the ad X ($X \in \mathfrak{g}$) are skew-symmetric with respect to this inner product. A standard result in linear algebra implies that all eigenvalues of ad X are pure imaginary. If $X \in \mathfrak{g}$ and $\mathfrak{m}$ is a subspace of $\mathfrak{g}$ invariant under ad X, then ad X leaves the orthogonal complement (with respect to the above inner product) of $\mathfrak{m}$ in $\mathfrak{g}$ invariant. Hence ad X is semisimple for all $X \in \mathfrak{g}$.

(iii) $\Rightarrow$ (i) The adjoint representation of $\mathfrak{g}$ is semisimple by Theorem 3.16.4, so by Theorem 3.16.3, $\mathfrak{g}$ is reductive. Let $\mathfrak{g}_1 = \mathfrak{D}\mathfrak{g}$, and let ω_1 be the Casimir polynomial of $\mathfrak{g}_1$. If $X \in \mathfrak{g}_1$, all eigenvalues of $(\text{ad } X)^2$ are ≤ 0, so $\omega_1(X) = tr(ad\ X)^2 \leq 0$. Moreover, if $X \in \mathfrak{g}_1$ and $\omega_1(X) = 0$, all eigenvalues of $(\text{ad } X)^2$ are 0. So ad X is nilpotent, showing that ad $X = 0$. Thus $X = 0$. $-\omega_1$ is therefore a positive definite quadratic form on $\mathfrak{g}_1$. Let G_1 be the adjoint group of $\mathfrak{g}_1$. Then by Theorem 3.10.8 G_1 is a closed subgroup of $GL(\mathfrak{g}_1)$. On the other hand, G_1 is contained in the orthogonal group of $\mathfrak{g}_1$ with respect to $-\omega_1$, which is compact. So G_1 itself must be compact. This proves (i).

The equivalence (iv) $\Leftrightarrow$ (i) is proved essentially by the same argument as above. If $\mathfrak{g}$ is semisimple and of compact type, and G is the universal covering group of the adjoint group of $\mathfrak{g}$, $\bar{G}$ is compact by Weyl's Theorem. So any analytic group whose Lie algebra is isomorphic to $\mathfrak{g}$ is necessarily compact.

We now turn to the construction of compact real forms of complex groups. This depends on the existence and properties of Weyl bases of a complex semisimple Lie algebra (cf. §4.3).

[9] More generally, let H be a compact group and π ($h \mapsto \pi(h)$) a continuous representation H on a finite-dimensional Hilbert space V with scalar product $(\cdot, \cdot)$ (over **R** or **C**). If we put $(u,v)_0 = \int_H (\pi(h)u, \pi(h)v)\, dh$ ($u, v \in V$), dh being a Haar measure on H, then $(\cdot, \cdot)_0$ is a scalar product for V invariant under all $\pi(h)$ ($h \in H$). With respect to this scalar product, the $\pi(h)$ ($h \in H$) are all unitary.

Lemma 4.11.8. *Let $\mathfrak{g}$ be a complex semisimple Lie algebra, $\mathfrak{h}$ a CSA, Δ the set of roots of $(\mathfrak{g},\mathfrak{h})$. Let $\mathfrak{h}_{\mathbf{R}} = \sum_{\alpha\in\Delta} \mathbf{R}\cdot H_\alpha$ and $\mathfrak{b} = (-1)^{1/2}\,\mathfrak{h}_{\mathbf{R}}$. Then there exists a real form $\mathfrak{u}$ of $\mathfrak{g}$, of compact type, such that $\mathfrak{b} \subseteq \mathfrak{u}$. If $\bar{\mathfrak{u}}$ is another such, there is $H_0 \in \mathfrak{h}_{\mathbf{R}}$ such that $\bar{\mathfrak{u}} = \mathfrak{u}^{\exp \operatorname{ad} H_0}$.*

Proof. By Theorem 4.3.26 and its proof, we can find a Weyl basis for $\mathfrak{g}$ consisting of a basis for $\mathfrak{h}$, nonzero elements $Z_\alpha \in \mathfrak{g}_\alpha$ $(\alpha \in \Delta)$ with $\langle Z_\alpha, Z_{-\alpha}\rangle = -1$ for all α, and an involutive automorphism φ $(X \mapsto X^\varphi)$ of $\mathfrak{g}$ such that $Z_\alpha^\varphi = Z_{-\alpha}$ for all α; then $H^\varphi = -H$ for all $H \in \mathfrak{h}$. Let $\mathfrak{g}_0$ be defined by (4.3.38). Then $\mathfrak{g}_0$ is a real form of $\mathfrak{g}$. Let σ be the corresponding conjugation of $\mathfrak{g}$, so that $(X + (-1)^{1/2}Y)^\sigma = X - (-1)^{1/2}Y$ $(X, Y \in \mathfrak{g}_0)$. Note also that $Z_\alpha^\sigma = Z_\alpha$ $(\alpha \in \Delta)$, $H^\sigma = H (H \in \mathfrak{h}_{\mathbf{R}})$. Then $\tau = \varphi\sigma$ is also a conjugation of $\mathfrak{g}$ that preserves $[\cdot,\cdot]$, and hence, the set of fixed points of τ will be a real form of $\mathfrak{g}$. Write $\mathfrak{u}$ for this real form. If $X \in \mathfrak{g}$, and we write

$$X = (-1)^{1/2}H + \sum_{\alpha\in\Delta} c(\alpha)Z_\alpha \quad (H \in \mathfrak{h}, c(\alpha) \in \mathbf{C}),$$

then the condition $X^\tau = X$ is equivalent to

$$(4.11.4) \qquad H^\sigma = H, \qquad c(\alpha)^{\mathrm{conj}} = c(-\alpha) \quad (\alpha \in \Delta).$$

Now, $H^\sigma = H$ if and only if $H \in \mathfrak{h}_{\mathbf{R}}$. So for all $X \in \mathfrak{u}$,

$$\begin{aligned}\langle X,X\rangle &= -\langle H,H\rangle - \sum_{\alpha\in\Delta} |c(\alpha)|^2 \\ &< 0.\end{aligned}$$

unless $X = 0$. This proves that $\mathfrak{u}$ is a real form of $\mathfrak{g}$ of compact type. That $\mathfrak{b} \subseteq \mathfrak{u}$ is clear from (4.11.4). Note that $Z_\alpha^\tau = Z_{-\alpha}$ for all $\alpha \in \Delta$ and that

$$(4.11.5) \quad \mathfrak{u} = (-1)^{1/2}\mathfrak{h}_{\mathbf{R}} + \sum_{\alpha\in\Delta} \mathbf{R}\cdot(Z_\alpha + Z_{-\alpha}) + \sum_{\alpha\in\Delta} \mathbf{R}\cdot(-1)^{1/2}(Z_\alpha - Z_{-\alpha});$$

this is also immediate from (4.11.4).

Suppose now that $\bar{\mathfrak{u}}$ is another real form of $\mathfrak{g}$, of compact type, containing $\mathfrak{b}$. Let $\bar{\tau}$ be the conjugation of $\mathfrak{g}$ with respect to $\bar{\mathfrak{u}}$. We show first that

$$(4.11.6) \qquad \langle X, X^{\bar{\tau}}\rangle < 0 \quad (X \in \mathfrak{g}, X \neq 0).$$

In fact, if we write $X = Y + (-1)^{1/2}Z$, with $Y, Z \in \bar{\mathfrak{u}}$, then $\langle X, X^{\bar{\tau}}\rangle = \langle Y,Y\rangle + \langle Z,Z\rangle < 0$ unless $Y = Z = 0$. Moreover, $\bar{\tau}|\mathfrak{b}$ is the identity, so $\bar{\tau}|\mathfrak{h}_{\mathbf{R}} = -$identity.

Let $\zeta = \tau\bar{\tau}$, τ being the conjugation of $\mathfrak{g}$ with respect to $\mathfrak{u}$. Then ζ is an automorphism of $\mathfrak{g}$ which is the identity on $\mathfrak{h}$. So $\mathfrak{g}_\alpha^\zeta = \mathfrak{g}_\alpha$ for all $\alpha \in \Delta$. Consequently, $\mathfrak{g}_\alpha^{\bar{\tau}} = \mathfrak{g}_{-\alpha}$ for all α, and hence there are nonzero $d(\alpha)$ such that

$Z_\alpha^\tau = d(\alpha)Z_{-\alpha}$ for all α. Since $\langle Z_\alpha, Z_\alpha^\tau\rangle = -d(\alpha)$, we see from (4.11.6) that $d(\alpha)$ is real and > 0 for all α. Hence, remembering that $Z_\alpha^\tau = Z_{-\alpha}$ for all α, we see that $Z_\alpha^\zeta = d(\alpha)Z_\alpha$ for all α. Let S be a simple system of roots. Then since $d(\alpha)$ is > 0 for all $\alpha \in S$, we can find $H \in \mathfrak{h}_{\mathbf{R}}$ such that $d(\alpha) = e^{\alpha(H)}$ for all $\alpha \in S$. We may now argue as in the proof of (4.9.6) to conclude that $\zeta =$ exp ad H. In particular, $d(\alpha) = e^{\alpha(H)}$, $\alpha \in \Delta$.

Write $H_0 = -\frac{1}{2}H$. We claim that $\bar{\mathfrak{u}} = \mathfrak{u}^{\exp \operatorname{ad} H_0}$. It is clearly sufficient to prove that $\bar{\tau} = \exp \operatorname{ad} H_0 \circ \tau \circ \exp(-\operatorname{ad} H_0)$. If $\tau_1 = \exp \operatorname{ad} H_0 \circ \tau \circ \exp(-\operatorname{ad} H_0)$, a straightforward calculation shows that $\tau_1 | \mathfrak{b}$ is the identity and that $Z_\alpha^{\tau_1} = d(\alpha)Z_{-\alpha}$ for all α. So $\tau_1 = \bar{\tau}$. This proves the lemma.

Theorem 4.11.9. *Let $\mathfrak{g}$ be a complex semisimple Lie algebra, G its adjoint group. Then $\mathfrak{g}$ admits real forms of compact type. Any two such are conjugate via an element of G.*

Proof. In view of the above lemma it is enough to prove that if $\mathfrak{u}_1$ is a real form of compact type of $\mathfrak{g}$, then there is $x \in G$ such that $\mathfrak{u}_1^x$ contains $\mathfrak{b}$, $\mathfrak{b}$ and $\mathfrak{h}$ being as above. Let $\mathfrak{b}_1$ be a CSA of $\mathfrak{u}_1$. Then $\mathfrak{b}_{1c}$ and $\mathfrak{h}$ are both CSA's of $\mathfrak{g}$, and hence, by Chevalley's theorem, we can find $x \in G$ such that $(\mathfrak{b}_{1c})^x = \mathfrak{h}$. Now, if $X \in \mathfrak{b}_1$, it follows from Theorem 4.11.7 that all the roots of $(\mathfrak{g},\mathfrak{b}_{1c})$ take pure imaginary values at X. On the other hand, as $\mathfrak{b} = (-1)^{1/2}\mathfrak{h}_{\mathbf{R}}$, it is clear that if $Y \in \mathfrak{h}$, $Y \in \mathfrak{b}$ if and only if $\alpha(Y) \in (-1)^{1/2}\mathbf{R}$ for all $\alpha \in \Delta$. So we must have $\mathfrak{b}_1^x \subseteq \mathfrak{b}$. As $\dim \mathfrak{b}_1 = \dim \mathfrak{b} = \operatorname{rk}(\mathfrak{g})$, we must have $\mathfrak{b}_1^x = \mathfrak{b}$. But then $\mathfrak{u}_1^x$ contains $\mathfrak{b}$. This completes the proof.

We are now in a position to obtain the global analogue of this theorem. We need a definition. Let M_c be a complex analytic group with Lie algebra $\mathfrak{m}_c$, M a real analytic subgroup[10] of M_c with corresponding real Lie algebra $\mathfrak{m} \subseteq \mathfrak{m}_c$. M is said to be a *real form* of M_c if $\mathfrak{m}$ is a real form of $\mathfrak{m}_c$, i.e., if $\mathbf{C}\cdot\mathfrak{m} = \mathfrak{m}_c$ and $\dim_{\mathbf{R}} \mathfrak{m} = \dim_{\mathbf{C}} \mathfrak{m}_c$.

Theorem 4.11.10. *Let G be a complex semisimple analytic group, $\mathfrak{g}$ its Lie algebra. Then G admits a compact real form. More precisely, if $\mathfrak{u}$ is a compact type real form of $\mathfrak{g}$ and U is the real analytic subgroup of G defined by $\mathfrak{u}$, then U is a compact real form of G. If U_1 and U_2 are two compact real forms of G, there is $y \in G$ such that $U_2 = yU_1y^{-1}$.*

Proof. If $\mathfrak{u}$, U are as in the statement above, U is compact by Theorem 4.11.7. The theorem follows easily from the previous theorem.

Corollary 4.11.11. *All finite-dimensional representations of a complex semisimple Lie algebra are semisimple.*

[10] We permit ourselves this imprecise way of stating that M is a real analytic subgroup of the real analytic group that underlies M_c.

Proof. Let $\mathfrak{g}$ be a complex semisimple Lie algebra, and let G be a simply connected complex analytic group with Lie algebra $\mathfrak{g}$. Let $\mathfrak{u}$, U be as in the above theorem. Suppose π is a finite-dimensional representation of $\mathfrak{g}$. Then π can be lifted to a representation of G, denoted by $\underline{\pi}$. Let $\rho = \underline{\pi} \mid U$. Then ρ is a representation of U, and its differential is $\pi \mid \mathfrak{u}$. Since any finite-dimensional representation of a compact group is semisimple[11], ρ is semisimple. This shows that $\pi \mid \mathfrak{u}$ is semisimple and implies the semisimplicity of π. The corollary follows at once.

This was the method used by H. Weyl to prove the semisimplicity of representations of complex semisimple Lie algebras. We have already discussed the algebraic method of proving it in Chapter 3.

The connection between a complex analytic semisimple group and its compact real forms is extremely close. We shall examine this connexion when the complex group is simply connected. The general case is treated in the exercises at the end of this chapter. As usual, we use the word "representation" to denote a finite-dimensional representation in a complex vector space.

Lemma 4.11.12. *Let U be a compact semisimple analytic group, $\mathfrak{u}$ its Lie algebra. If every representation of $\mathfrak{u}$ is the differential of a representation of U, then U is simply connected.*

Proof. Let $\tilde{U}$ be the universal covering group of U with covering homomorphism π. By Weyl's theorem, $\tilde{U}$ is compact. Let C be the kernel of π. We identify the Lie algebra of $\tilde{U}$ with $\mathfrak{u}$, so that $d\pi$ is the identity. Suppose there is $c \in C$ different from the identity. By the Peter–Weyl theorem, we can find a representation ρ of $\tilde{U}$ such that $\rho(c) \neq 1$. If $d\rho$ is the differential of ρ, there is a representation ρ' of U such that $d\rho = d\rho'$. Consequently, ρ and $\rho' \circ \pi$ are two representations of $\tilde{U}$ with the same differential. This means that $\rho = \rho' \circ \pi$. In particular, $\rho(c) = 1$, contradiction. So C must be trivial. But then U is simply connected.

Lemma 4.11.13. (*The "unitarian trick"*) *Let G be a complex semisimple analytic group, U a compact real form of G, and N a connected open neighborhood of U in G. If F is a holomorphic function on N such that $F \mid U = 0$, then $F = 0$.*

Proof. Let $\mathfrak{g}$ be the Lie algebra (over **C**) of G, $\mathfrak{u}$ the real subalgebra defined by U. Let $\mathfrak{G}$ be the universal enveloping algebra of $\mathfrak{g}$. As usual, we

[11] In fact, let H be a compact group and γ $(h \mapsto \gamma(h))$ a continuous representation of H in a finite-dimensional vector space V. We can equip V with a Hilbert space structure in such a way that all the $\gamma(h)$ are unitary (cf. footnote 9). If W is a subspace of V invariant under γ, the orthogonal complement of W is also invariant under γ. So γ is semisimple.

regard the elements of $\mathfrak{G}$ as left-invariant holomorphic differential operators on G.

Since the elements of $\mathfrak{u}$ are tangent to U, it follows from the vanishing of F on U that $F(u;X) = 0$ for all $u \in U$, $X \in \mathfrak{u}$. Since we are dealing with holomorphic functions, and since $\mathfrak{g} = \mathbf{C}\cdot\mathfrak{u}$, we have $F(u;X) = 0$ for all $u \in U$, $X \in \mathfrak{g}$. In other words, XF vanishes on U for any $X \in \mathfrak{g}$. Thus XF has the same property as F. A simple induction on $\deg(a)$ now shows that aF vanishes on U for any $a \in \mathfrak{G}$. In particular, $F(1;a) = 0$ for all $a \in \mathfrak{G}$. Since F is holomorphic and N is connected, $F = 0$.

Theorem 4.11.14. *Any compact semisimple analytic group can be imbedded as a real form of some complex semisimple analytic group. Let G be a simply connected, complex, semisimple analytic group. Then its compact real forms are simply connected. Suppose U is one such. Then for any irreducible complex analytic representation π of G, $\pi_U = \pi \mid U$ is irreducible, and the equivalence class of π_U depends only on that of π. Moreover, the resulting correspondence between the respective sets of equivalence classes is a bijection.*

Proof. Let G be a simply connected, complex, semisimple analytic group with Lie algebra $\mathfrak{g}$, $\mathfrak{u}$ a real form of $\mathfrak{g}$ of compact type, and U the corresponding compact real form of G. If π is a representation of $\mathfrak{u}$, we can extend π to a representation of $\mathfrak{g}$ by complexification. Since G is simply connected, there is a complex analytic representation ρ of G such that $d\rho$ is this extension. Clearly, $\rho \mid U$ is a representation of U whose differential is π. Thus U has the property described in Lemma 4.11.12. So U is simply connected. Note that this argument already proves that any representation of U is the restriction to U of a unique complex analytic representation of G.

Suppose now that G and U are as above and that π is a complex analytic representation of G in some vector space V. Let $\pi_U = \pi \mid U$. Denote by $E(\pi)$ (resp. $E(\pi_U)$) the complex linear span of all $\pi(x)$, $x \in G$ (resp. $\pi(u), u \in U$). $E(\pi)$ and $E(\pi_U)$ are then algebras of endomorphisms of V, and $E(\pi_U) \subseteq E(\pi)$. We claim that

$$E(\pi) = E(\pi_U). \tag{4.11.7}$$

If this is not true, there is a linear functional Λ on $E(\pi)$ such that $\Lambda \neq 0$ but $\Lambda \mid E(\pi_U) = 0$. Let $F(x) = \Lambda(\pi(x))$, $x \in G$. Then F is a holomorphic function on G, $\not\equiv 0$, but $\equiv 0$ on U, contradicting Lemma 4.11.13. The relation (4.11.7) already shows that if π is irreducible, so is π_U. If π_1 and π_2 are two complex analytic representations of G, and if S is a linear transformation from the vector space of π_1 into the vector space of π_2, then (4.11.7) implies that $S\pi_1(x)S^{-1} = \pi_2(x)$ for all $x \in G$ if and only if this is true for all $x \in U$. These remarks lead easily to the conclusion that the map $\pi \mapsto \pi_U$ induces a bijection of the set of all equivalence classes of complex analytic irreducible representations of G onto the set of all equivalence classes of irreducible representations of U.

It remains to prove that any compact semisimple analytic group can be imbedded as a real form of a complex semisimple analytic group. Let U be the compact group in question, $\mathfrak{u}$ the Lie algebra of U. Let $\mathfrak{g}$ be the complexification of $\mathfrak{u}$, G a simply connected complex analytic group with Lie algebra $\mathfrak{g}$. Let $\bar{U}$ be the real analytic subgroup of G defined by $\mathfrak{u}$. Then $\bar{U}$ is simply connected, so there is a finite central subgroup C of $\bar{U}$ such that $U \simeq \bar{U}/C$. Clearly C is central in G, so U can be imbedded as a compact real form of G/C. This proves the theorem.

This theorem reduces the problem of constructing all the irreducible representations of a complex semisimple Lie algebra to the corresponding problem for a compact semisimple analytic group. In the following sections we shall examine Weyl's approach to this question.

4.12. Maximal Tori of Compact Semisimple Groups

We now examine the structure of compact semisimple Lie groups somewhat more closely. We wish to prove that such a group has maximal tori any two of which are conjugate under a suitable inner automorphism of it, and that any element of the group belongs to at least one maximal torus. These results enable one to reduce questions involving class functions on such groups to questions concerning functions on a fixed maximal torus.

Lemma 4.12.1. *Let $\mathfrak{g}$ be a semisimple Lie algebra of compact type over* $\mathbf{R}$, *G its adjoint group. Then a subalgebra of $\mathfrak{g}$ is a CSA if and only if it is maximal abelian. Suppose $\mathfrak{b}$ is a CSA and $\mathfrak{z}$ is a subalgebra of $\mathfrak{g}$ such that $\mathfrak{b} \subseteq \mathfrak{z} \subseteq \mathfrak{g}$. Then $\mathfrak{z}$ is reductive,* $\mathrm{rk}(\mathfrak{z}) = \mathrm{rk}(\mathfrak{g})$, *and $\mathfrak{D}\mathfrak{z}$ is a semisimple Lie algebra of compact type. Moreover* $\mathfrak{b} = \mathrm{center}(\mathfrak{z}) + (\mathfrak{b} \cap \mathfrak{D}\mathfrak{z})$, *the sum being direct, and $\mathfrak{b} \cap \mathfrak{D}\mathfrak{z}$ is a CSA of $\mathfrak{D}\mathfrak{z}$. If M is the analytic subgroup of G defined by $\mathfrak{D}\mathfrak{z}$, M is compact, and the map $y \mapsto y|\mathfrak{D}\mathfrak{z}$ maps M onto the adjoint group of $\mathfrak{D}\mathfrak{z}$. Finally, $Y^y = Y$ for $y \in M$ and $Y \in$* $\mathrm{center}(\mathfrak{z})$.

Proof. Theorems 4.1.5 and 4.11.7 imply that a subalgebra of $\mathfrak{g}$ is a CSA if and only if it is maximal abelian. If $\mathfrak{z}$ is a subalgebra of $\mathfrak{g}$ containing a CSA $\mathfrak{b}$, it is obvious that $\mathfrak{b}$ is a CSA of $\mathfrak{z}$; in particular, $\mathrm{rk}(\mathfrak{z}) = \mathrm{rk}(\mathfrak{g})$. Now, for $X \in \mathfrak{z}$, $\mathrm{ad}\, X|\mathfrak{z}$ is semisimple and has only pure imaginary eigenvalues. So by Theorem 4.11.7, $\mathfrak{z}$ is reductive, and $\mathfrak{D}\mathfrak{z}$ is semisimple and of compact type. Let $\mathfrak{c} = \mathrm{center}(\mathfrak{z})$. Then $\mathfrak{z}$ is the direct sum of $\mathfrak{c}$ and $\mathfrak{D}\mathfrak{z}$. On the other hand, since $\mathfrak{b} + \mathfrak{c}$ is abelian, the maximal abelian nature of $\mathfrak{b}$ implies that $\mathfrak{b} + \mathfrak{c} =$

$\mathfrak{b}$. So $\mathfrak{c} \subseteq \mathfrak{b}$, showing that $\mathfrak{b}$ is the direct sum of $\mathfrak{c}$ and $\mathfrak{b} \cap \mathfrak{D}\mathfrak{z}$. Obviously, $\mathfrak{b} \cap \mathfrak{D}\mathfrak{z}$ is maximal abelian in $\mathfrak{D}\mathfrak{z}$. So $\mathfrak{b} \cap \mathfrak{D}\mathfrak{z}$ is a CSA of $\mathfrak{D}\mathfrak{z}$.

Let M be as in the lemma. Then M is compact by theorem 4.11.7. Since M is generated by the exp ad X, $X \in \mathfrak{D}\mathfrak{z}$, it is clear that $y \mapsto y \,|\, \mathfrak{D}\mathfrak{z}$ maps M onto the adjoint group of $\mathfrak{D}\mathfrak{z}$ and that $Y^y = Y$ for $Y \in \mathfrak{c}$, $y \in M$.

Theorem 4.12.2. *Let* $\mathfrak{g}$ *be a semisimple Lie algebra of compact type over* **R**, *G its adjoint group. If* $\mathfrak{b}$ *is any CSA of* $\mathfrak{g}$, *then*

$$\mathfrak{g} = \bigcup_{x \in G} \mathfrak{b}^x \tag{4.12.1}$$

In particular, any two CSA's of $\mathfrak{g}$ *are conjugate under G.*

Proof. We prove (4.12.1) by induction on dim $\mathfrak{g}$. We therefore assume the first statement for all semisimple Lie algebras of compact type over **R** whose dimensions are $<$ dim $\mathfrak{g}$. Let $\mathfrak{b}$ be a CSA of $\mathfrak{g}$ and write, for any subset E of $\mathfrak{g}$, $E^G = \bigcup_{x \in G} E^x$. Since $0^G = 0$, the relation $\mathfrak{b}^G = \mathfrak{g}$ will be proved if we show that $(\mathfrak{b}^\times)^G = \mathfrak{g}^\times$, where $\mathfrak{g}^\times = \mathfrak{g} \setminus \{0\}$ and $\mathfrak{b}^\times = \mathfrak{b} \setminus \{0\}$. Now, since $\mathfrak{g}$ is semisimple, dim $\mathfrak{g} \geq 3$. So $\mathfrak{g}^\times$ is a connected open subset of $\mathfrak{g}$ and $(\mathfrak{b}^\times)^G \subseteq \mathfrak{g}^\times$. As a consequence, in order to prove that $(\mathfrak{b}^\times)^G = \mathfrak{g}^\times$, it is sufficient to show that $(\mathfrak{b}^\times)^G$ is both open and closed in $\mathfrak{g}^\times$.

Since G is compact, we can find a positive definite scalar product on $\mathfrak{g} \times \mathfrak{g}$ that is invariant under G. Let $\|\cdot\|$ be the corresponding norm. Suppose $H_n \in \mathfrak{b}$ and $x_n \in G$ $(n = 1,2, \ldots)$ are such that $H_n^{x_n} \longrightarrow X \in \mathfrak{g}^\times$ as $n \longrightarrow \infty$. Since $\|H_n\| = \|H_n^{x_n}\| \longrightarrow \|X\|$, $\{H_n\}$ is bounded in $\mathfrak{b}$. Therefore, we can find a subsequence $\{n_k\}$, and elements $x \in G$, $H \in \mathfrak{b}$, such that $H_{n_k} \longrightarrow H$ and $x_{n_k} \longrightarrow x$ as $k \longrightarrow \infty$. Clearly, $X = H^x$, so $\|H\| = \|X\| > 0$. Thus $H \in \mathfrak{b}^\times$, showing that $X \in (\mathfrak{b}^\times)^G$. $(\mathfrak{b}^\times)^G$ is thus closed in $\mathfrak{g}^\times$.

We shall now prove that $(\mathfrak{b}^\times)^G$ is open in $\mathfrak{g}$. Since it is G-invariant, it is enough to prove that each element of $\mathfrak{b}^\times$ is an interior point of $(\mathfrak{b}^\times)^G$. Fix $X \in \mathfrak{b}^\times$. Let $\mathfrak{z}$ be the centralizer of X in $\mathfrak{g}$. Note that $\mathfrak{b} \subseteq \mathfrak{z}$. Since $X \neq 0$, $\mathfrak{z} \neq \mathfrak{g}$, so $\dim(\mathfrak{D}\mathfrak{z}) < \dim(\mathfrak{g})$. We may therefore apply Lemma 4.12.1 and the induction hypothesis to deduce the relation $(\mathfrak{b} \cap \mathfrak{D}\mathfrak{z})^M = \mathfrak{D}\mathfrak{z}$, M being as in that lemma. Since $Y^y = Y$ for $Y \in$ center$(\mathfrak{z})$ and $y \in M$, we have $\mathfrak{b}^M = \mathfrak{z}$; in particular, $(\mathfrak{b}^\times)^M = \mathfrak{z}^\times$, where $\mathfrak{z}^\times = \mathfrak{z} \setminus \{0\}$. From this we obtain $(\mathfrak{b}^\times)^G = (\mathfrak{b}^\times)^{GM} = (\mathfrak{z}^\times)^G$. On the other hand, by Corollary 4.1.7, $('\mathfrak{z})^G$ is an open subset of $\mathfrak{g}$ containing X, $'\mathfrak{z}$ being defined by (4.1.15):

$$'\mathfrak{z} = \{Z : Z \in \mathfrak{z}, \det(\text{ad } Z)_{\mathfrak{g}/\mathfrak{z}} \neq 0\}.$$

Obviously, $'\mathfrak{z} \subseteq \mathfrak{z}^\times$, so $X \in ('\mathfrak{z})^G \subseteq (\mathfrak{b}^\times)^G$. This shows that X is an interior point of $(\mathfrak{b}^\times)^G$. $(\mathfrak{b}^\times)^G$ is thus open in $\mathfrak{g}$.

As mentioned earlier, this proves that $(\mathfrak{b}^\times)^G = \mathfrak{g}^\times$. Hence $\mathfrak{b}^G = \mathfrak{g}$. This proves the theorem.

For the rest of this section, G is a compact semisimple analytic group and $\mathfrak{g}$ its Lie algebra. By a *torus* we mean a real analytic group which is compact, connected and abelian. If A is a torus of dimension n, then A is isomorphic to $\mathbf{T} \times \cdots \times \mathbf{T}$ (n factors), where $\mathbf{T}$ is the multiplicative group of complex numbers of modulus 1. Note that if A is a torus and an analytic subgroup of G, the compactness of A implies that it is closed in G.

Theorem 4.12.3. *Let $\mathfrak{b} \subseteq \mathfrak{g}$ be a CSA, B the corresponding analytic subgroup of G. Then B is a maximal torus of G. Every maximal torus can be obtained in this way. If B_1, B_2 are two maximal tori of G, there is an $x \in G$ such that $B_2 = B_1^x \ (= xB_1x^{-1})$.*

Proof. $\mathfrak{b}$ is its own centralizer in $\mathfrak{g}$. So if A is the centralizer of $\mathfrak{b}$ in G, $\mathfrak{b}$ is the subalgebra defined by the closed subgroup A. This shows that B is the component of the identity of A. B is therefore closed. Since it is compact, it is a torus. Since $\mathfrak{b}$ is maximal abelian, B is a maximal torus. Suppose B_1 is any maximal torus and $\mathfrak{b}_1$ is the corresponding subalgebra of $\mathfrak{g}$. $\mathfrak{b}_1$ is abelian, so we can find a maximal abelian subalgebra $\mathfrak{b}_2$ containing $\mathfrak{b}_1$. If B_2 is the analytic subgroup of G defined by $\mathfrak{b}_2$, $B_1 \subseteq B_2$, and B_2 is a torus, so $B_1 = B_2$. $\mathfrak{b}_1$ is thus maximal abelian. Finally, let B_j be a maximal torus, $\mathfrak{b}_j$ the corresponding CSA ($j = 1, 2$). By Theorem 4.12.2 there is $x \in G$ such that $\mathfrak{b}_2 = \mathfrak{b}_1^x$; then $B_2 = B_1^x$. This proves the theorem.

The remainder of this section is devoted to the proof that each element of G lies in some maximal torus and that the maximal tori are centralizers of CSA's. The proof uses induction on dim G and is similar to the proof of Theorem 4.12.2. We begin with some preparation before proving this theorem.

Let $\mathfrak{b} \subseteq \mathfrak{g}$ be a CSA, and let B be the corresponding maximal torus. Fix $b \in B$, and let $\mathfrak{z}$ be the centralizer of b in $\mathfrak{g}$. Then $\mathfrak{z}$ is a subalgebra of $\mathfrak{g}$, and $\mathfrak{b} \subseteq \mathfrak{z} \subseteq \mathfrak{g}$. Lemma 4.12.1 is thus applicable to this context. Let $\mathfrak{c} =$ center($\mathfrak{z}$), and let C, Z_1, Z, B_1 be the respective analytic subgroups of G defined by $\mathfrak{c}$, $\mathfrak{D}\mathfrak{z}, \mathfrak{z}$, and $\mathfrak{b} \cap \mathfrak{D}\mathfrak{z}$. Z_1 is compact, and B_1 is a maximal torus of Z_1. Since Z is the component of the identity of the centralizer of b in G,Z is closed in G, and hence compact. Moreover, C can be characterized as the component of the identity of the center of Z. So C is also closed, and consequently compact. The relations $\mathfrak{z} = \mathfrak{c} + \mathfrak{D}\mathfrak{z}$ and $\mathfrak{b} = \mathfrak{c} + (\mathfrak{b} \cap \mathfrak{D}\mathfrak{z})$ easily imply that $Z = CZ_1$ and $B = CB_1$.

Since B is connected and $\mathfrak{b} \subseteq \mathfrak{z}$, we have $B \subseteq Z$. $\mathfrak{g}$ is a real Hilbert space if we define the norm by

$$||X||^2 = -tr(\text{ad } X)^2 \quad (X \in \mathfrak{g}),$$

and Ad(G) is obviously a subgroup of the orthogonal group of $\mathfrak{g}$. In particular, $\mathrm{Ad}(b)$ is semisimple, so

$$\mathfrak{g} = \mathfrak{z} + \mathfrak{q} \quad \text{(direct sum)}, \tag{4.12.2}$$

where

$$\mathfrak{q} = \text{range of } (\mathrm{Ad}(b)^{-1} - 1) \tag{4.12.3}$$

If $x \in Z$, x commutes with b, so $\mathrm{Ad}(x)$ leaves both $\mathfrak{z}$ and $\mathfrak{q}$ invariant. Let

$$'Z = \{x : x \in Z, \det(\mathrm{Ad}(x)^{-1} - 1)_{\mathfrak{q}} \neq 0\}. \tag{4.12.4}$$

$'Z$ is an open subset of Z containing b.

Lemma 4.12.4. *Let notation be as above. For $x, y \in G$ let $x^y = yxy^{-1}$. Then $('Z)^G$ is open in G.*

Proof. Let ψ be the map of $G \times Z$ into G given by $\psi(y,x) = x^y$, for $y \in G$, $x \in Z$. ψ is clearly analytic. We shall now calculate its differential. We identify as usual the tangent spaces to G and Z at any of their points with $\mathfrak{g}$ and $\mathfrak{z}$ respectively. Fix $y \in G$, $x \in Z$. Then, for $Y \in \mathfrak{g}$ and $X \in \mathfrak{z}$,

$$(d\psi)_{(y,x)}(Y,X) = (d\psi)_{(y,x)}(Y,0) + (d\psi)_{(y,x)}(0,X).$$

We have, for real t,

$$x^{y \exp tY} = x^y \cdot \exp(tY^{yx^{-1}}) \exp(-tY^y).$$

Moreover, as $t \longrightarrow 0$, we have, for X', $Y' \in \mathfrak{g}$,

$$\exp tX' \exp tY' = \exp\{t(X' + Y') + O(t^2)\}$$

by (2.12.10). Hence

$$(d\psi)_{(y,x)}(Y,0) = (Y^{x^{-1}} - Y)^y.$$

Further, since $(x \exp tX)^y = x^y \exp(tX^y)$, we have

$$(d\psi)_{(y,x)}(0,X) = X^y.$$

We thus obtain the formula

$$(d\psi)_{(y,x)} = \mathrm{Ad}(y) \circ L_x, \tag{4.12.5}$$

where L_x is the linear map of $\mathfrak{g} \times \mathfrak{z}$ into $\mathfrak{g}$ given by

$$L_x(Y,X) = (\mathrm{Ad}(x)^{-1} - 1)(Y) + X. \tag{4.12.6}$$

If $x \in {}'Z$, $\mathrm{Ad}(x)^{-1} - 1$ maps $\mathfrak{q}$ onto itself. So we conclude from (4.12.5) and (4.12.6) that $(d\psi)_{(y,x)}$ is surjective for $(y,x) \in G \times {}'Z$. ψ is thus submersive on $G \times {}'Z$. This implies that $({}'Z)^G = \psi[G \times {}'Z]$ is an open subset of G. This proves the lemma.

Theorem 4.12.5. *Let B be a maximal torus of G, $\mathfrak{b}$ the corresponding subalgebra of $\mathfrak{g}$. Then B is the centralizer of $\mathfrak{b}$ in G, and $G = B^G$, i.e.,*

$$G = \bigcup_{x \in G} B^x. \tag{4.12.7}$$

In particular, each element of G lies in some maximal torus. Finally, $G = \exp \mathfrak{g}$.

Proof. We prove (4.12.7) first, by induction on dim G. Let L be the center of G; write $B^\times = B \setminus B \cap L$, $G^\times = G \setminus L$. Since dim $G \geq 3$ and L is finite, $G^\times$ is an open *connected* subset of G. Now, $x^G = x$ for $x \in L$. Consequently, to prove (4.12.7) it is enough to show that $(B^\times)^G = G^\times$, and for this it is sufficient to prove that $(B^\times)^G$ is both open and closed in $G^\times$.

Let $b_n \in B^\times$, $x_n \in G$ be such that $b_n^{x_n} \longrightarrow y \in G^\times$ as $n \longrightarrow \infty$. Since both B and G are compact, we can select a subsequence $\{n_k\}$, and elements $b \in B$, $x \in G$, such that $b_{n_k} \longrightarrow b$ and $x_{n_k} \longrightarrow x$ as $k \longrightarrow \infty$. So $b^x = y$. Since $y \in G^\times$, b cannot belong to L. So $b \in B^\times$, proving that $y \in (B^\times)^G$. $(B^\times)^G$ is therefore closed in $G^\times$.

We now prove that $(B^\times)^G$ is open in G. Since it is G-invariant, it is enough to establish that each element of $B^\times$ is an interior point of $(B^\times)^G$. Fix $b \in B^\times$. We use the notation of Lemma 4.12.4 and the remarks preceding it. Since $b \notin L$, $\mathfrak{z} \neq \mathfrak{g}$, so dim $Z_1 <$ dim G. We may thus apply the induction hypothesis to deduce that $Z_1 = B_1^{Z_1}$. Since $B = CB_1$ and $c^z = c$ for $c \in C$ and $z \in Z_1$, we have $Z = B^{Z_1}$. Let $Z^\times = Z \setminus Z \cap L$. Then we have $(B^\times)^{Z_1} = Z^\times$. Consequently,

$$(B^\times)^G = (Z^\times)^G. \tag{4.12.8}$$

On the other hand, it is obvious from (4.12.4) that ${}'Z \subseteq Z^\times$. So in view of (4.12.8), we have

$$b \in ({}'Z)^G \subseteq (B^\times)^G. \tag{4.12.9}$$

This last relation implies, in view of Lemma 4.12.4, that b is an interior point of $(B^\times)^G$. $(B^\times)^G$ is thus open in G. As mentioned at the beginning, this proves (4.12.7).

Since $B = \exp \mathfrak{b}$ and $G = B^G$, one deduces from the relation $\exp(X^x) = (\exp X)^x$ $(x \in G, X \in \mathfrak{g})$ that $G = (\exp \mathfrak{b})^G = \exp \mathfrak{g}$.

It remains to prove that B is the centralizer of $\mathfrak{b}$ in G. Let A be the centralizer of $\mathfrak{b}$ in G, and let $a \in A$. Let $\mathfrak{g}_c$ be the complexification of $\mathfrak{g}$, Δ the set of roots of $(\mathfrak{g}_c, \mathfrak{b}_c)$. Let S be a simple system of roots. Since $\mathrm{Ad}(a)$ centralizes $\mathfrak{b}$, it leaves each root subspace invariant. So there are nonzero complex numbers $c(\alpha)$ $(\alpha \in \Delta)$, such that

$$X^a = c(\alpha)X \quad (\alpha \in \Delta, X \in (\mathfrak{g}_c)_\alpha).$$

Since $\mathrm{Ad}(a)$ is orthogonal on $\mathfrak{g}$ (with respect to the positive definite quadratic form $-\omega$), its eigenvalues are all of absolute value 1. So $|c(\alpha)| = 1$ $(\alpha \in \Delta)$. Further, since $\mathrm{Ad}(a)$ is an automorphism of $\mathfrak{g}$, and since it fixes $\mathfrak{b}$ elementwise, it follows from the relation $[(\mathfrak{g}_c)_\alpha, (\mathfrak{g}_c)_{-\alpha}] = \mathbf{C}\cdot H_\alpha$ that $c(-\alpha) = c(\alpha)^{-1}$. Now, each root takes only pure imaginary values on $\mathfrak{b}$, so we can find $X \in \mathfrak{b}$ such that

$$c(\alpha) = e^{\alpha(X)} \quad (\alpha \in S).$$

Let $a_0 = \exp X$. Then $a_0 \in B$, and $\mathrm{Ad}(a)$ coincides with $\mathrm{Ad}(a_0)$ on $\mathfrak{b}_c + \sum_{\alpha \in \pm S} (\mathfrak{g}_c)_\alpha$. Since both are automorphisms, we must have $\mathrm{Ad}(a) = \mathrm{Ad}(a_0)$.

In other words, we have proved that $A = BL$. In order to prove that $A = B$, it is thus enough to prove that $L \subseteq B$. Suppose $c \in L$. By (4.12.7), there is $x \in G$ such that $c^x \in B$. But $c^x = c$, so already $c \in B$. This completes the proof.

4.13. An Integral Formula

Throughout this section, G is a compact connected semisimple Lie group with Lie algebra $\mathfrak{g}$, $\mathfrak{b}$ is a CSA, and B is the associated maximal torus. $\mathfrak{g}_c$ is the complexification of $\mathfrak{g}$, $\mathfrak{h} = \mathfrak{b}_c$, Δ is the set of roots of $(\mathfrak{g}_c, \mathfrak{h})$, and P is a positive system of roots. For $\alpha \in \Delta$, $\mathfrak{g}_{c\alpha}$ is the corresponding root subspace. Let $\mathfrak{q}$ be the orthogonal complement of $\mathfrak{b}$ in $\mathfrak{g}$. Then $\mathfrak{q}_c = \sum_{\alpha \in \Delta} \mathfrak{g}_{c\alpha}$.

Theorem 4.13.1. *Let $\tilde{B}$ be the normalizer of B in G. For $x \in \tilde{B}$, let*

$$s(x) = \mathrm{Ad}(x)|\mathfrak{b}_c. \tag{4.13.1}$$

Then $x \mapsto s(x)$ induces an isomorphism of $\tilde{B}/B$ onto the Weyl group $\mathfrak{w}$ of $(\mathfrak{g}_c, \mathfrak{b}_c)$. Moreover, $\mathfrak{w}$ leaves $\mathfrak{b}$ invariant.

Proof. Since $\mathfrak{b} = (-1)^{1/2} \sum_{\alpha \in \Delta} \mathbf{R}\cdot H_\alpha$, it is obvious that $\mathfrak{w}$ leaves $\mathfrak{b}$ invariant. Theorem 4.9.1 implies that $s(x) \in \mathfrak{w}$ for $x \in \tilde{B}$ and that $s(x) = 1$ if and only if x centralizes $\mathfrak{b}$, i.e., $x \in B$. Thus $x \mapsto s(x)$ induces an injection of $\tilde{B}/B$ into $\mathfrak{w}$. To prove that this is an isomorphism, it is enough to prove that for each $\alpha \in \Delta$, s_α arises from $\tilde{B}$. In view of Lemma 4.11.8 and the relation (4.11.5), we may assume that $\mathfrak{g}$ has the form

$$\mathfrak{g} = (-1)^{1/2}\mathfrak{h} + \sum_{\alpha \in \Delta} \mathbf{R}\cdot(Z_\alpha + Z_{-\alpha}) + \sum_{\alpha \in \Delta} \mathbf{R}\cdot(-1)^{1/2}(Z_\alpha - Z_{-\alpha}),$$

where $Z_\alpha \in \mathfrak{g}_{c\alpha}$ and $\langle Z_\alpha, Z_{-\alpha} \rangle = -1 \ \forall \ \alpha \in \Delta$. Fix $\alpha \in \Delta$. Then a simple calculation shows that

$$\left[Z_\alpha + Z_{-\alpha}, H_\alpha \pm i\sqrt{\frac{\langle\alpha,\alpha\rangle}{2}}(Z_\alpha - Z_{-\alpha})\right]$$
$$= \pm i\sqrt{2\langle\alpha,\alpha\rangle}\left(H_\alpha \pm i\sqrt{\frac{\langle\alpha,\alpha\rangle}{2}}(Z_\alpha - Z_{-\alpha})\right),$$

where $i = (-1)^{1/2}$. Let $x(t) = \exp t(Z_\alpha + Z_{-\alpha})$, $t \in \mathbf{R}$. Then $x(t) \in G$ for all $t \in \mathbf{R}$, and if $X_\pm = H_\alpha \pm i\sqrt{\langle\alpha,\alpha\rangle/2}\,(Z_\alpha - Z_{-\alpha})$, then

$$X_\pm^{x(t)} = e^{\pm it\sqrt{2\langle\alpha,\alpha\rangle}} X_\pm.$$

Since $H_\alpha = \frac{1}{2}(X_+ + X_-)$, we see that for $x_0 = x(\pi/\sqrt{2\langle\alpha,\alpha\rangle})$,

$$H_\alpha^{x_0} = -H_\alpha.$$

On the other hand, it is obvious that for any $H \in \mathfrak{h}$ with $\alpha(H) = 0$, $H^{x_0} = H$. So $x_0 \in \tilde{B}$, and $s(x_0) = s_\alpha$. This completes the proof of the theorem.

Let $\hat{B}$ be the group of characters of the compact abelian group B. Since exp is a homomorphism of $\mathfrak{b}$ onto B, it is clear that for any $\chi \in \hat{B}$, there is a unique $\lambda_\chi \in \mathfrak{b}_c^*$ such that

$$\chi(\exp H) = e^{\lambda_\chi(H)} \quad (H \in \mathfrak{b}). \tag{4.13.2}$$

It is obvious that λ_χ takes only pure imaginary values on $\mathfrak{b}$. The map $\chi \mapsto \lambda_\chi$ is an isomorphism of $\hat{B}$ onto a discrete additive subgroup $L(G)$ of $\mathfrak{b}_c^*$. We write $\lambda \mapsto \xi_\lambda$ for the inverse map, so that

$$\xi_\lambda(\exp H) = e^{\lambda(H)} \quad (H \in \mathfrak{b}, \lambda \in L(G)). \tag{4.13.3}$$

Let $L(R)$ denote the additive subgroup of $\mathfrak{b}_c^*$ generated by the roots, and let L be the additive group of all integral linear functions on $\mathfrak{b}_c$. Observe that $\mathfrak{w}$ leaves both $L(R)$ and L invariant. On the other hand, since $\tilde{B}$ normalizes B and $\tilde{B}/B \approx \mathfrak{w}$, it is clear that $\mathfrak{w}$ also leaves $L(G)$ invariant and that we may allow $\mathfrak{w}$ to act naturally on B.

Suppose π is a finite-dimensional representation of G in a complex vector space V. We write π again for the corresponding representation of $\mathfrak{g}_c$. Since B is compact and abelian, there is a basis $\{v_1, \ldots, v_d\}$ for V and $\chi_1, \ldots, \chi_d \in \hat{B}$ such that

$$\pi(b)v_j = \chi_j(b)v_j \quad (1 \leq j \leq d, b \in B);$$

if $\lambda_j \in L(G)$ are such that $\chi_j = \xi_{\lambda_j}$ $(1 \leq j \leq d)$, it is clear that

$$\pi(H)v_j = \lambda_j(H)v_j \quad (1 \leq j \leq d, H \in \mathfrak{b}_c),$$

so $\lambda_1, \ldots, \lambda_d$ are the weights of the representation π of $\mathfrak{g}_c$. We shall often refer to the λ_j as the *weights of the representation* π of G. Clearly, the λ_j are in L. If $\Delta(\pi)$ is the set of these weights, then

$$tr(\pi(b)) = \sum_{\mu \in \Delta(\pi)} m_\pi(\mu)\xi_\mu(b) \quad (b \in B), \tag{4.13.4}$$

where $m_\pi(\mu) = \dim V_\mu$ are integers > 0 and

$$m_\pi(s\mu) = m_\pi(\mu) \quad (s \in \mathfrak{w}, \mu \in \Delta(\pi)). \tag{4.13.5}$$

In other words, *the restriction to* B *of the character of* π *is a finite Fourier series on* B, *whose nonzero Fourier coefficients are all integers* > 0, *and which is invariant under* $\mathfrak{w}$.

Lemma 4.13.2. *We have*

$$L(R) \subseteq L(G) \subseteq L, \tag{4.13.6}$$

and all three are isomorphic as additive groups to $\mathbf{Z}^l$ ($l = \operatorname{rank} \mathfrak{g}$). $L(G)$ *is the set of all those integral linear functions on* $\mathfrak{b}_c$ *which occur as weights of representation of* G.

Proof. Since the roots are the weights of the adjoint representation, we have $L(R) \subseteq L(G)$. By the Frobenius reciprocity theorem (cf. Weil [1]), every character of B occurs in the decomposition with respect to B of some representation of G. So $L(G) \subseteq L$, and $L(G)$ is the set of all those linear functions on $\mathfrak{b}_c$ which occur as weights of representations of G. Since both $L(R)$ and L are isomorphic to $\mathbf{Z}^l$ (as abelian groups), so is $L(G)$.

Corollary 4.13.3. *Let* $\delta = \frac{1}{2}\sum_{\alpha \in P} \alpha$. *Then* $2\delta \in L(G)$. *If* $\lambda \in L$, $s\lambda - \lambda \in L(R)$ *for all* $s \in \mathfrak{w}$.

Proof. Follows on taking Lemma 4.7.4 into account.

We shall see later on that $L(G) = L$ if (and only if) G is simply connected. Let

$$B' = \{b : b \in B, \xi_\alpha(b) \neq 1 \text{ for any } \alpha \in \Delta\}. \tag{4.13.7}$$

If $b \in B$, the centralizer $\mathfrak{z}$ of b in $\mathfrak{g}_c$ is spanned by $\mathfrak{b}_c$ and those root subspaces $\mathfrak{g}_{c\alpha}$ for which $\xi_\alpha(b) = 1$. Consequently $\dim \mathfrak{z} \geq l$, and $\dim \mathfrak{z} = l$ if and only if $b \in B'$. If we now observe that any element of G is conjugate to some element of B, we may conclude that for any $x \in G$, $\mathrm{Ad}(x)$ is semisimple, its centralizer has dimension $\geq l$, and this dimension is $= l$ if and only if $x \in (B')^G$. Let T be an indeterminate. Then the foregoing remarks imply that

$$\det(T - 1 + \mathrm{Ad}(x)) \equiv T^l D(x) + T^{l+1} D_1(x) + \cdots \quad (x \in G), \tag{4.13.8}$$

where $D, D_1, \ldots$ are analytic functions on G and $D \not\equiv 0$. D is invariant under all inner automorphisms, and

$$D(b) = \prod_{\alpha \in \Delta} (\xi_\alpha(b) - 1) \quad (b \in B). \tag{4.13.9}$$

If

$$G' = \{x : x \in G, D(x) \neq 0\}, \tag{4.13.10}$$

then

$$G' = (B')^G. \tag{4.13.11}$$

Elements of G' are said to be *regular*. Since

$$\xi_{-\alpha} = \xi_\alpha^{-1} = \xi_\alpha^{\mathrm{conj}} \quad (\alpha \in \Delta), \tag{4.13.12}$$

we have, on writing

$$D_P(b) = \prod_{\alpha \in P} (\xi_\alpha(b) - 1) \quad (b \in B), \tag{4.13.13}$$

the following relations, valid for $b \in B$:

$$\begin{aligned} D_P(b)^{\mathrm{conj}} &= D_P(b^{-1}) \\ D(b) = D(b^{-1}) &= D_P(b)^{\mathrm{conj}} D_P(b) \end{aligned} \tag{4.13.14}$$

In particular, $D(b) \geq 0$ for $b \in B$. Observe also that

$$D(b) = \det(\mathrm{Ad}(b) - 1)_{\mathfrak{q}} \quad (b \in B). \tag{4.13.15}$$

We now obtain an alternative expression for D_P. For $s \in \mathfrak{w}$ we put

$$\epsilon(s) = \det(s). \tag{4.13.16}$$

$\epsilon(s_\alpha) = -1$ for all $\alpha \in \Delta$. Recall that $\mathfrak{D}_P$ is the set of dominant integral linear functions on $\mathfrak{b}_c$ (relative to P).

Lemma 4.13.4. *Let $\mathfrak{F}$ be the algebra of all finite linear combinations of the exponential e^λ, $\lambda \in L$. Let*

(4.13.17)
$$\mathfrak{D}_P^+ = \{\lambda' : \lambda' \in \mathfrak{D}_P, \lambda'(H_\alpha) > 0 \text{ for all } \alpha \in P\}$$
$$g_\lambda = \sum_{s \in \mathfrak{w}} \epsilon(s) e^{s\lambda} \quad (\lambda \in \mathfrak{D}_P^+).$$

Then $g_\lambda \in \mathfrak{F}$, $g_\lambda^s = \epsilon(s) g_\lambda$ *for* $s \in \mathfrak{w}$. *If* $g \in \mathfrak{F}$ *and* $g^s = \epsilon(s) g$ *for all* $s \in \mathfrak{w}$, *then* g *is a linear combination of the* g_λ, $\lambda \in \mathfrak{D}_P^+$. *Finally, if* $\delta = \frac{1}{2}\sum_{\alpha \in P}\alpha$, *then* $\delta \in \mathfrak{D}_P^+$ *and*

(4.13.18)
$$g_\delta = e^{-\delta} \prod_{\alpha \in P} (e^\alpha - 1)$$

Proof. The properties of the g_λ are obvious. Suppose that $g = \sum_{\mu \in L} c_\mu e^\mu$ is an element of $\mathfrak{F}$ such that $g^s = \epsilon(s) g$ for all $s \in \mathfrak{w}$. Then $c_{s\mu} = \epsilon(s) c_\mu$ for all $\mu \in L$, $s \in \mathfrak{w}$. Suppose $\lambda \in \mathfrak{D}_P \setminus \mathfrak{D}_P^+$. Then there is an $\alpha \in P$ with $\lambda(\bar{H}_\alpha) = 0$, so $s_\alpha\lambda = \lambda$. Since $g^{s_\alpha} = -g$, we must have $c_\lambda = -c_\lambda$. So $c_\lambda = 0$. On the other hand, every $\mathfrak{w}$-orbit in L contains a unique element from $\mathfrak{D}_P$, by Lemma 4.7.4. It is now clear that $g = \sum_{\lambda \in \mathfrak{D}_P^+} c_\lambda g_\lambda$.

Note that if $\lambda \in \mathfrak{D}_P^+$, $\lambda(\bar{H}_\alpha) \geq 1 = \delta(\bar{H}_\alpha)$ for all simple roots α in P. Hence $\lambda - \delta \in \mathfrak{D}_P$. Thus

(4.13.19)
$$\mathfrak{D}_P^+ = \{\Lambda + \delta : \Lambda \in \mathfrak{D}_P\}.$$

Let $g = e^{-\delta}\prod_{\alpha \in P}(e^\alpha - 1)$. If $\alpha \in P$ is a simple root, $s_\alpha\delta = \delta - \alpha$, and $s_\alpha \cdot P = \{-\alpha\} \cup (P \setminus \{\alpha\})$. This implies that $g^{s_\alpha} = -g = \epsilon(s_\alpha) g$. Hence $g^s = \epsilon(s) g$, $s \in \mathfrak{w}$. By the previous result we can find constants c_λ ($\lambda \in \mathfrak{D}_P^+$) such that $g = \sum_{\lambda \in \mathfrak{D}_P^+} c_\lambda g_\lambda$. Suppose that for some $\lambda \in \mathfrak{D}_P^+$ with $\lambda \neq \delta$, $c_\lambda \neq 0$. Write $\lambda = \Lambda + \delta$, where $\Lambda \in \mathfrak{D}_P$ is nonzero. We now expand the product defining g to write g in the form $e^\delta + \sum_{\mu \in \Gamma} a_\mu e^{\delta - \mu}$, where a_μ are constants and Γ is the set of all nonzero sums of elements of P. The condition $c_\lambda \neq 0$ then implies that $-\Lambda \in \Gamma$. Write $-\Lambda = m_1\alpha_1 + \cdots + m_l\alpha_l$, where $\alpha_1, \ldots, \alpha_l$ are the simple roots in P and the m_j are ≥ 0. Since $0 < \langle -\Lambda, -\Lambda \rangle = \sum_{1 \leq i \leq l} m_i \langle -\Lambda, \alpha_i \rangle$, there is some i such that $\langle -\Lambda, \alpha_i \rangle > 0$. This contradicts the relation $\Lambda \in \mathfrak{D}_P$. So $c_\lambda = 0$ unless $\lambda = \delta$. Obviously, $c_\delta = 1$. So $g = g_\delta$.

Corollary 4.13.5. $s\delta + \delta \in L(R)$ for all $s \in \mathfrak{w}$, and

(4.13.20)
$$D_P(b) = \sum_{s \in \mathfrak{w}} \epsilon(s) \xi_{s\delta + \delta}(b) \quad (b \in B).$$

Proof. $s\delta + \delta = s\delta - \delta + 2\delta \in L(R)$ for $s \in \mathfrak{w}$. Now for $H \in \mathfrak{b}$,

$$\begin{aligned} D_P(\exp H) &= \prod_{\alpha \in P} (e^{\alpha(H)} - 1) \\ &= e^{\delta(H)} \cdot e^{-\delta(H)} \prod_{\alpha \in P} (e^{\alpha(H)} - 1) \\ &= \sum_{s \in \mathfrak{w}} \epsilon(s) e^{(s\delta + \delta)(H)} \end{aligned}$$

by the above lemma. This proves (4.13.20).

Our aim now is to obtain the main integral formula on G. Let G^* be the analytic manifold G/B. Put $n = \dim(G)$, $q = \dim(G^*)$. Write ζ $(x \mapsto x^*)$ for the canonical map of G onto G^*. For $x \in G$, let l_x (resp. l_x^*) be the map $y \mapsto xy$ (resp. $y^* \mapsto (xy)^*$) of G onto G (resp. G^* onto G^*). Let g be the left-invariant n-form on G for which the corresponding measure is the Haar measure dx with $\int_G dx = 1$. Let β be a left-invariant l-form on B such that the corresponding measure is the Haar measure db for which $\int_B db = 1$. We now show that there is a G-invariant q-form on G^*. The differential $(d\zeta)_1$ is an isomorphism of $\mathfrak{q}$ onto the tangent space to G^* at 1^*. On the other hand,

$$l_b^*(\exp tX)^* = (\exp tX^b)^* \quad (t \in \mathbf{R}, X \in \mathfrak{q}, b \in B),$$

which implies that $(d\zeta)_1$ intertwines $\mathrm{Ad}(b)_\mathfrak{q}$ and $(dl_b^*)_{1^*}$ $(b \in B)$[12]. Thus $\det \mathrm{Ad}(b)_\mathfrak{q} = \det(dl_b^*)_{1^*} = 1$, $b \in B$. This shows that there is a nonzero q-linear form on $T_{1^*}(G^*) \times \cdots \times T_{1^*}(G^*)$ (q factors) that is invariant under all $(dl_b^*)_{1^*}$, $b \in B$. It follows easily from this that there is an analytic q-form g^* on G^* invariant under all l_x^*, $x \in G$. We assume that for the corresponding measure on G^*, say dx^*, one has $\int_{G^*} dx^* = 1$. Finally, let γ be the n-form on $G^* \times B$ which determines the product measure dx^*db.

If $x \in G$, $b \in B$, $b^x = xbx^{-1}$ depends only on x^*. We write $b^{x^*} = b^x$, and define ψ^* by

$$\psi^*(x^*,b) = b^x. \tag{4.13.21}$$

Then ψ^* is an analytic map of $G^* \times B$ onto G. Moreover, ψ^* maps $G^* \times B'$ onto G', and in fact, $\psi^{*-1}(G') = G^* \times B'$.

Lemma 4.13.6. *$\varphi^* = \psi^* \mid G^* \times B'$ is a covering map of $G^* \times B'$ onto G' and has, everywhere on $G^* \times B'$, a bijective differential. Moreover, there is a constant $c \neq 0$ such that the n-forms $c\tilde{D}\cdot\gamma$ and g correspond under φ^*, $\tilde{D}$ being the function $(x^*, b) \mapsto D(b)$ on $G^* \times B'$.*

Proof. It is obvious that φ^* is a proper map, i.e., that for any compact set $A \subseteq G'$, $\varphi^{*-1}(A)$ is a compact subset of $G^* \times B'$. If $b_0 \in B'$, $\mathrm{Ad}(b_0^{-1}) - 1$ maps $\mathfrak{q}$ onto itself, so by (4.12.6), the map $(x,b) \mapsto b^x$ has surjective differential everywhere on $G \times B'$. It follows from this that $d\varphi^*$ is everywhere surjective on $G^* \times B'$. Since $\dim(G^* \times B') = \dim(G)$, φ^* has, everywhere on $G^* \times B'$, a bijective differential. In particular, it is a local homeomorphism, so since it has already been proved proper, it is a covering map.

[12]For $b \in B$, l_b^* fixes 1^*, so $(dl_b^*)_{1^*}$ is an automorphism of $T_{1^*}(G^*)$.

Let θ be the n-form on $G^* \times B'$ which corresponds to g under φ^*. If $y \in G$ and i_y is the inner automorphism induced by y, it is trivial to check that $\varphi^*(l_y^*(x^*),b) = i_y\varphi^*(x^*,b)$ for $(x^*,b) \in G^* \times B'$. On the other hand, G being compact, dx is invariant under all i_y, so g is invariant under all i_y. So θ is invariant under the diffeomorphisms $(x^*,b) \mapsto l_y^*(x^*),b$ of $G^* \times B'$ $(y \in G)$. In order to prove the required formula for θ, it is therefore enough to prove that for some constant $c \neq 0$,

$$\theta_{(1^*,b)} = cD(b)\gamma_{(1^*,b)} \quad (b \in B'). \tag{4.13.22}$$

Let $\{H_1, \ldots, H_l\}$ be a basis for $\mathfrak{b}$ and $\{X_1, \ldots, X_q\}$ a basis for $\mathfrak{q}$. Write $X_j^* = (d\zeta)_1(X_j)$, $1 \leq j \leq q$. We shall, as usual, identify the tangent spaces to G and B at each of their points with $\mathfrak{g}$ and $\mathfrak{b}$ respectively. Since g and β are left-invariant, $g_y(H_1, \ldots, H_l, X_1, \ldots, X_q)$ and $\beta_b(H_1, \ldots, H_l)$ are nonzero constants $(y \in G, b \in B)$. In particular, $\gamma_{(1^*,b)}(X_1^*, \ldots, X_q^*, H_1, \ldots, H_l) = c_2$ is a nonzero constant for all $b \in B'$. On the other hand, if ψ is the map $(x,b) \mapsto b^x$ of $G \times B'$,

$$(d\psi)_{(1,b)}(X,H) = (d\varphi^*)_{(1^*,b)}((d\zeta)_1(X),H)$$

for $X \in \mathfrak{g}$, $H \in \mathfrak{b}$. Consequently, writing $X_{j,b} = (d\psi)_{(1,b)}((X_j,0))$ $(1 \leq j \leq q)$ and $H_{i,b} = (d\psi)_{(1,b)}((0,H_i))$ $(1 \leq i \leq l)$, we get, for $b \in B'$,

$$\theta_{(1^*,b)}(X_1^*, \ldots, X_q^*, H_1, \ldots, H_l) = g_b(X_{1,b}, \ldots, X_{q,b}, H_{1,b}, \ldots, H_{l,b}).$$

But it follows from (4.12.6) that $X_{j,b} = (\mathrm{Ad}(b^{-1}) - 1)X_j$ and $H_{i,b} = H_i$. Hence, taking into account (4.13.14) and (4.13.15), we find that

$$\theta_{(1^*,b)}(X_1^*, \ldots, X_q^*, H_1, \ldots, H_l) = c_1 D(b) \quad (b \in B'),$$

c_1 being a nonzero constant. This leads to (4.13.22) and proves the lemma.

We are now in a position to obtain the basic integral formula on G. Note that since $G \setminus G'$ has measure zero,[13] we need not make any distinction between the Lebesgue spaces $\mathcal{L}^1(G)$ and $\mathcal{L}^1(G')$. Similarly, we may identify $\mathcal{L}^1(B)$ and $\mathcal{L}^1(B')$. For any continuous function f on G', let

$$\varphi_f(b) = \int_{G^*} f(b^{x^*})\, dx^* \qquad (b \in B'). \tag{4.13.23}$$

φ_f is clearly well defined, and

$$\varphi_f(b) = \int_G f(b^x)\, dx \quad (b \in B'). \tag{4.13.24}$$

[13] More generally, let M be a connected analytic manifold, f a nonzero analytic function on M. Let ω be an analytic m-form on M $(m = \dim M)$ and μ the corresponding measure. Then the set of zeros of f has μ-measure zero.

Since $(x^*,b) \mapsto f(b^{x^*})$ is a continuous function on $G^* \times B'$, it is obvious that φ_f is a continuous function on B'. Moreover,

$$\sup_{b \in B'} |\varphi_f(b)| \leq \sup_{x \in G'} |f(x)|. \tag{4.13.25}$$

If f is invariant under the inner automorphisms of G,

$$\varphi_f = f \,|\, B'. \tag{4.13.26}$$

Theorem 4.13.7. *For any continuous function f on G', $f \in \mathfrak{L}^1(G)$ if and only if $\varphi_{|f|} \cdot (D \,|\, B') \in \mathfrak{L}^1(B)$. In this case,*

$$\int_G f(x)\, dx = [\mathfrak{w}]^{-1} \int_B \varphi_f(b) D(b)\, db, \tag{4.13.27}$$

where $[\mathfrak{w}]$ is the order of $\mathfrak{w}$.

Proof. Since φ^* is a covering map of $G^* \times B'$ onto G', there is an integer $k \geq 1$ such that above any element of G' there are exactly k elements of $G^* \times B'$. From the standard theory of integration with respect to differential forms and Lemma 4.13.6 it follows (with c as in the lemma) that for any continuous function f on G', $f \in \mathfrak{L}^1(G')$ if and only if $\tilde{D}(f \circ \varphi^*) \in \mathfrak{L}^1(G^* \times B')$, and for all such f,

$$\int_{G'} f(x)\, dx = \frac{|c|}{k} \int_{G^* \times B'} f(b^{x^*})\, D(b)\, dx^* db.$$

Let $c' = |c|/k$. Then, invoking the Fubini theorem, we obtain (4.13.27) and the theorem above, except for the identification of c' with $[\mathfrak{w}]^{-1}$. Now, taking $f = 1$, we have, in view of our normalizations,

$$c'^{-1} = \int_B D(b)\, db.$$

But $D = |D_P|^2$ by (4.13.14), so in view of (4.13.20),

$$c'^{-1} = \int_B \Big| \sum_{s \in \mathfrak{w}} \epsilon(s) \xi_{s\delta + \delta}(b) \Big|^2 db$$

On the other hand, by Lemma 4.7.4, the characters $\xi_{s\delta+\delta}$ of B are all distinct. Since there are $[\mathfrak{w}]$ of these, we conclude from the usual orthogonality relations in $\mathfrak{L}^2(B)$ that $c'^{-1} = [\mathfrak{w}]$. This completes the proof.

Corollary 4.13.8. *Let f be a continuous function on G' that is invariant under all the inner automorphisms of G. Then $f \in \mathfrak{L}^1(G)$ if and only if $(Df) \,|\, B' \in \mathfrak{L}^1(B)$, and in this case,*

$$\int_G f(x)\, dx = [\mathfrak{w}]^{-1} \int_B f(b) D(b)\, db. \tag{4.13.28}$$

4.14. The Character Formula of H. Weyl

We are now ready to put together the work of the preceding sections to obtain the famous formulae of H. Weyl [2, 3, 4] for the character and degree of the irreducible representations of compact semisimple groups. G is a fixed compact, connected, semisimple Lie group with Lie algebra $\mathfrak{g}$; the rest of our notation is as in §4.13.

Lemma 4.14.1. *Let f be a continuous function on B' invariant with respect to $\tilde{B}$ (or $\mathfrak{w}$). Then there exists a unique continuous function F on G' such that*

(i) *F is invariant under all inner automorphisms of G*
(ii) *$F|B' = f$.*

Proof. We have $G' = (B')^G$. So we have to set $F(b^x) = f(b)$ for $b \in B'$, $x \in G$. To see that F is well defined, let $b_1, b_2 \in B'$ and $x_1, x_2 \in G$ be such that $b_1^{x_1} = b_2^{x_2}$. Then $b_1^x = b_2$, where $x = x_2^{-1}x_1$. Since $\mathfrak{b}$ is the centralizer of both b_1 and b_2, $\mathfrak{b}^x = \mathfrak{b}$, so $x \in \tilde{B}$. Since f is invariant under $\tilde{B}$, $f(b_1) = f(b_2)$. So F is well defined. It is obviously invariant under the inner automorphisms of G. On the other hand, in the notation of Lemma 4.13.6, $f(b) = (F \circ \varphi^*)(x^*,b)$ for $(x^*,b) \in G^* \times B'$, so $F \circ \varphi^*$ is continuous. This shows, since φ^* is a local homeomorphism, that F is continuous.

Lemma 4.14.2. *For $\lambda \in L(G) \cap \mathfrak{D}_P$, let u_λ be the function on B defined by*[14]

$$u_\lambda = \sum_{s \in \mathfrak{w}} \epsilon(s)\xi_{s\lambda + s\delta + \delta}. \tag{4.14.1}$$

Define v_λ on B' by

$$v_\lambda = u_\lambda D_P^{-1}. \tag{4.14.2}$$

Then there is a unique continuous function F_λ on G', invariant under all inner automorphisms of G, such that $F_\lambda| B' = v_\lambda$. $F_\lambda \in \mathfrak{L}^2(G)$, and for all $\lambda, \lambda' \in L(G) \cap \mathfrak{D}_P$,

$$\int_G F_\lambda F_{\lambda'}^{\text{conj}}\, dx = \delta_{\lambda\lambda'}. \tag{4.14.3}$$

Proof. The existence, uniqueness, and continuity of F_λ will follow from the previous lemma provided we show that v_λ is invariant under $\mathfrak{w}$. If $t \in \mathfrak{w}$, a simple calculation based on (4.14.1) and (4.13.20) reveals that

$$D_P^t = \epsilon(t)D_P\xi_{t\delta-\delta}, \qquad u_\lambda^t = \epsilon(t)u_\lambda\xi_{t\delta-\delta} \tag{4.14.4}$$

This shows that $v_\lambda^t = v_\lambda$. Since $|v_\lambda|^2 \cdot D = |u_\lambda|^2 \cdot |D_P|^{-2} \cdot D = |u_\lambda|^2$ by (4.13.14), Corollary 4.13.8 applies and gives us the relation $F_\lambda \in \mathfrak{L}^2(G)$. Suppose λ,

[14] By Corollary 4.13.5, $\xi_{s(\lambda+\delta)+\delta}$ is well defined for all $s \in \mathfrak{w}$.

$\lambda' \in L(G) \cap \mathfrak{D}_P$. Then $F_\lambda F_{\lambda'}^{\text{conj}} \in \mathfrak{L}^1(G)$, so by (4.13.28),

$$\int_{G'} F_\lambda F_{\lambda'}^{\text{conj}}\, dx = [\mathfrak{w}]^{-1} \int_{B'} v_\lambda(b) v_{\lambda'}(b)^{\text{conj}} D_P(b) D_P(b)^{\text{conj}}\, db$$
$$= [\mathfrak{w}]^{-1} \int_B u_\lambda(b) u_{\lambda'}(b)^{\text{conj}}\, db.$$

If $\lambda \neq \lambda'$, then $\lambda + \delta$ and $\lambda' + \delta$ belong to distinct $\mathfrak{w}$-orbits, so $\int_B u_\lambda u_{\lambda'}^{\text{conj}}\, db = 0$. If $\lambda = \lambda'$, the $[\mathfrak{w}]$ transforms $s(\lambda + \delta)$ $(s \in \mathfrak{w})$ are all distinct by Lemma 4.7.4; so $\int_B |u_\lambda|^2\, db = [\mathfrak{w}]$. This proves (4.14.3).

Theorem 4.14.3. *Let G be a compact connected semisimple Lie group, $\mathfrak{g}$ its Lie algebra, $\mathfrak{b}$ a CSA of $\mathfrak{g}$, and B the associated maximal torus. Let P be a positive system of roots of $(\mathfrak{g}_c, \mathfrak{b}_c)$. Then the irreducible representations of G are precisely those whose corresponding representations of $\mathfrak{g}_c$ have highest weights $\lambda \in L(G) \cap \mathfrak{D}_P$. Denoting these by π_λ $(\lambda \in L(G) \cap \mathfrak{D}_P)$, we have*

$$tr\, \pi_\lambda(b) = \sum_{s \in \mathfrak{w}} \epsilon(s) \xi_{s(\lambda+\delta)+\delta}(b) / D_P(b) \quad (b \in B'). \tag{4.14.4}$$

Here, $\delta = \frac{1}{2} \sum_{\alpha \in P} \alpha$ and $D_P = \prod_{\alpha \in P} (\xi_\alpha - 1)$.

Proof. Let X be the set of irreducible characters of G. By the Schur orthogonality relations, we have

$$\int_G \chi \chi'^{\text{conj}}\, dx = \delta_{\chi\chi'} \quad (\chi, \chi' \in X).$$

Let Δ be the function on $\mathfrak{b}$ given by

$$\Delta(H) = e^{-\delta(H)} \prod_{\alpha \in P} (e^{\alpha(H)} - 1) \quad (H \in \mathfrak{b}).$$

Fix $\chi \in X$, and let χ_B be the restriction of χ to B. We have seen in §4.13 that χ_B is a finite Fourier series on B with positive integral coefficients and that χ_B is $\mathfrak{w}$-invariant. On the other hand, $\Delta^s = \epsilon(s)\Delta$ for all $s \in \mathfrak{w}$, by Lemma 4.13.4. So writing

$$g(H) = \chi(\exp H)\, \Delta(H) \quad (H \in \mathfrak{b}),$$

we see that g is a finite integral linear combination of the exponentials e^μ $(\mu \in L)$ and that $g^s = \epsilon(s) g$ for all $s \in \mathfrak{w}$. So by Lemma 4.13.4 and (4.13.19) we can find a finite subset A of $\mathfrak{D}_P$ and nonzero integers c_λ $(\lambda \in A)$ such that

$$g = \sum_{\lambda \in A} c_\lambda \sum_{s \in \mathfrak{w}} \epsilon(s) e^{s(\lambda+\delta)}$$

Consequently, for $H \in \mathfrak{b}$,

$$(\chi_B \cdot D_P)(\exp H) = \sum_{\lambda \in A} c_\lambda \sum_{s \in \mathfrak{w}} \epsilon(s) e^{(s(\lambda+\delta)+\delta)(H)}.$$

Since $\chi_B \cdot D_P$ is a linear combination of the characters of B and since $s\delta + \delta \in L(G)$ for all $s \in \mathfrak{w}$, this formula shows that $\lambda \in L(G)$ for all $\lambda \in A$, i.e., $A \subseteq L(G)$. So we have, on taking (4.14.1) into account,

$$\chi_B D_P = \sum_{\lambda \in A} c_\lambda u_\lambda.$$

In view of (4.14.2) and Lemma 4.14.2, this gives us

$$\chi = \sum_{\lambda \in A} c_\lambda F_\lambda \quad (\text{on } G').$$

If we now use the Schur orthogonality relations and relation (4.14.3), we get

$$\sum_{\lambda \in A} |c_\lambda|^2 = 1.$$

But each c_λ is an integer $\neq 0$. So A consists of exactly one element, say λ, and $\chi = aF_\lambda$ on G', a being a constant $= \pm 1$.

We now claim that $a = 1$. Let π be an irreducible representation of G with character χ. Let Λ denote the highest weight of the corresponding representation of $\mathfrak{g}_c$. Clearly, $\Lambda \in L(G)$. Let Γ be the subset of $L(R) \setminus \{0\}$ consisting of all elements which are sums of elements of P. For purposes of this argument, we say that a finite Fourier series on B *begins with* ξ_ν ($\nu \in L(G)$) if it is of the form $\xi_\nu + \sum_{\mu \in \Gamma} a_\mu \xi_{\nu-\mu}$ for suitable constants a_μ. Obviously, χ_B begins with ξ_Λ. On the other hand, since $\delta - s\delta \in \Gamma$ for $s \neq 1$ (Lemma 4.7.4), it is clear from (4.13.20) that D_P begins with $\xi_{2\delta}$. So $\chi_B \cdot D_P$ begins with $\xi_{\Lambda+2\delta}$. A similar argument shows that u_λ, as defined by (4.14.1), begins with $\xi_{\lambda+2\delta}$. Since $\chi_B \cdot D_P = au_\lambda$, we see at once that $a = 1$ and $\Lambda = \lambda$. Thus $\chi = F_\lambda$ on G', and π has λ as its highest weight.

Suppose, finally, that for some $\lambda_0 \in L(G) \cap \mathfrak{D}_P$, no irreducible character of G equals F_{λ_0}. By Lemma 4.14.2, F_{λ_0} is an element of $\mathfrak{L}^2(G)$ orthogonal to all irreducible characters of G. Since F_{λ_0} is invariant under all inner automorphisms of G, the Peter–Weyl theorem enables us to conclude that $F_{\lambda_0} = 0$, a contradiction.

This proves the theorem.

Theorem 4.14.4. *Let G be a compact connected semisimple Lie group. Suppose G is simply connected. Then $L(G) = L$, and in particular, $\delta \in L(G)$. If $\lambda \in \mathfrak{D}_P$, the character of the irreducible representation of G with highest weight λ is given by*

$$tr\, \pi_\lambda(b) = \frac{\sum_{s \in \mathfrak{w}} \epsilon(s)\xi_{s(\lambda+\delta)}(b)}{\Delta(b)} \quad (b \in B'), \tag{4.14.5}$$

where

$$\Delta = \sum_{s \in \mathfrak{w}} \epsilon(s)\xi_{s\delta}. \tag{4.14.6}$$

Proof. In view of the work of §4.11 we may assume that G is a real form of a complex analytic simply connected semisimple Lie group. If $\lambda \in \mathfrak{D}_P$, the representation π_λ of $\mathfrak{g}_c$ with highest weight λ lifts to a complex analytic representation of G_c. By restricting it to G we obtain an irreducible representation of G having λ as its highest weight. Theorem 4.11.3 already implies that $L(G) = L$. In particular, $\delta \in L(G)$, and (4.14.5) follows from (4.14.4) since, in this case,

$$\Delta(b) = \xi_{-\delta}(b) \prod_{\alpha \in P} (\xi_\alpha(b) - 1) = \sum_{s \in P} \epsilon(s)\xi_{s\delta}(b) \quad (b \in B)$$

(cf. (4.13.20)). This proves the theorem.

It remains to compute the dimension of the representation π_λ with highest weight λ. We need a lemma.

Lemma 4.14.5. *There is a constant $k \neq 0$ with the following property. If ν is any linear function on $\mathfrak{b}_c$,*

$$\frac{\sum_{s \in \mathfrak{w}} \epsilon(s)(s\nu)^m}{m!} = \begin{cases} 0 & \text{if } 0 \leq m < d \\ k \cdot \prod_{\alpha \in P} \langle \nu, \alpha \rangle \cdot \pi & \text{if } m = d, \end{cases} \tag{4.14.7}$$

where d is the number of elements in P and

$$\pi = \prod_{\alpha \in P} \alpha \tag{4.14.8}$$

Proof. We call a polynomial q on $\mathfrak{b}_c$ *skew* if $q^s = \epsilon(s)q$ for all $s \in \mathfrak{w}$. It is easily seen that π is skew and that if q is any skew polynomial on $\mathfrak{b}_c$, q can be written as $\pi q'$, where q' is a $\mathfrak{w}$-invariant polynomial. In particular, if q is homogeneous and skew, we have $q = 0$ when $\deg(q) < \deg(\pi) = d$, and $q = c\pi$ for a constant c when $\deg(q) = \deg(\pi) = d$. For any $\nu \in \mathfrak{b}_c^*$, the left side of (4.14.7) is a homogeneous skew polynomial of degree m. So it must be 0 when $m < d$ and equal to $k(\nu)\pi$ when $m = d$, $k(\nu)$ being a constant. We now evaluate $k(\nu)$. We have

$$\sum_{s \in \mathfrak{w}} \epsilon(s)(s\nu)^d = k(\nu)d!\pi. \tag{4.14.9}$$

Consider now the element ϖ of the symmetric algebra $S(\mathfrak{b}_c)$ given by

$$\varpi = \prod_{\alpha \in P} H_\alpha. \tag{4.14.10}$$

$\partial(\varpi)$ is a differential operator, and we may apply it to both sides of (4.14.9). Since π is homogeneous of degree d and ϖ is of degree d, $\partial(\varpi)(\pi)$ is a constant, say c. On the other hand, it is clear from the definition of ϖ that

$$\varpi^s = \epsilon(s)\varpi \quad (s \in \mathfrak{w}). \tag{4.14.11}$$

Consequently, for any $s \in \mathfrak{w}$,

$$\begin{aligned}\epsilon(s)\partial(\varpi)((sv)^d) &= \partial(\varpi^s)((sv)^d)\\ &= (\partial(\varpi)(v^d))^s.\end{aligned}$$

But v^d being homogeneous of degree d, $\partial(\varpi)v^d$ is a constant. So, for $s \in \mathfrak{w}$

$$\begin{aligned}\epsilon(s)\partial(\varpi)((sv)^d) &= \partial(\varpi)(v^d)\\ &= \prod_{\alpha\in P}\partial(H_\alpha)(v^d)\\ &= d!\prod_{\alpha\in P}\langle v,\alpha\rangle,\end{aligned}$$

as may be seen by applying the differential operators $\partial(H_\alpha)$ in succession. Hence we get

$$[\mathfrak{w}]\prod_{\alpha\in P}\langle v,\alpha\rangle = k(v)c. \tag{4.14.12}$$

Obviously, there are $v \in \mathfrak{b}_c^*$ for which $\prod_{\alpha\in P}\langle v,\alpha\rangle \neq 0$. If we take v to be such an element in (4.14.12), we may conclude that $c \neq 0$. Moreover, writing $k' = c^{-1}[\mathfrak{w}]$, we have

$$k(v) = k'\prod_{\alpha\in P}\langle v,\alpha\rangle.$$

(4.14.9) now implies (4.14.7) provided we put $k = k'\cdot d!$.

Theorem 4.14.6. *Let notation be as in Theorem* 4.14.3. *Then for $\lambda \in L(G)\cap\mathfrak{D}_P$, the dimension of the corresponding representation π_λ of G is given by*

$$\dim(\pi_\lambda) = \prod_{\alpha\in P}\frac{\langle\lambda+\delta,\alpha\rangle}{\langle\delta,\alpha\rangle}. \tag{4.14.13}$$

Proof. Let $\psi_\lambda(b) = tr\,\pi_\lambda(b)$, $b \in B$. Then $\dim(\pi_\lambda) = \psi_\lambda(1)$. But the formula (4.14.4) becomes indeterminate if we substitute $b = 1$. So we have to calculate its limit when $b \longrightarrow 1$. Obviously, $\dim(\pi_\lambda) = \lim_{H\to 0}\psi_\lambda(\exp H)$. Hence, since $e^{\delta(H)} \longrightarrow 1$ for $H \longrightarrow 0$ ($H \in \mathfrak{b}$), we have

$$\dim(\pi_\lambda) = \lim_{H\to 0}\frac{\sum_{s\in\mathfrak{w}}\epsilon(s)e^{s(\lambda+\delta)(H)}}{\prod_{\alpha\in P}(e^{\alpha(H)}-1)}.$$

Since $(e^{\alpha(H)}-1)/\alpha(H) \longrightarrow 1$ as $H \longrightarrow 0$, we get

$$\dim(\pi_\lambda) = \lim_{H\to 0}\frac{\sum_{s\in\mathfrak{w}}\epsilon(s)e^{s(\lambda+\delta)(H)}}{\varpi(H)}. \tag{4.14.14}$$

Expanding the exponentials and using Lemma 4.14.5, we have

$$\sum_{s \in \mathfrak{w}} \epsilon(s) e^{s(\lambda+\delta)(H)} = \pi(H)\{k \cdot \prod_{\alpha \in P} \langle \lambda + \delta, \alpha \rangle + \sum_{m \geq 1} g_m(H)\}$$

where k is a constant independent of λ and H, while g_m is a homogeneous polynomial on $\mathfrak{h}_c$ of degree m. It follows from this and (4.14.14) that

$$\dim(\pi_\lambda) = k \prod_{\alpha \in P} \langle \lambda + \delta, \alpha \rangle.$$

When $\lambda = 0$, π_λ is the trivial representation, and $\dim(\pi_\lambda) = 1$ in this case. This shows that $k = \prod_{\alpha \in P} \langle \delta, \alpha \rangle^{-1}$. So

$$\dim(\pi_\lambda) = \prod_{\alpha \in P} \frac{\langle \lambda + \delta, \alpha \rangle}{\langle \delta, \alpha \rangle}.$$

This proves the theorem.

4.15. Appendix. Finite Reflection Groups

This appendix is devoted to a discussion of some aspects of the theory of finite linear groups generated by reflections. Some of these results have already been encountered by us when the group in question is the Weyl group of a semisimple Lie algebra over **C**. Our treatment is essentially the same as that of Steinberg [1]. See also Bourbaki [4].

1. Finite reflection groups. Let V be a finite-dimensional vector space of dimension l over a field k of characteristic 0. By a *reflection* in V we mean an element $s \in GL(V)$ such that $s^2 = 1$ and the subspace $H = \{\lambda : \lambda \in V, s\lambda = \lambda\}$ is of dimension $l - 1$. A *finite reflection group* (*frg*) is a finite subgroup of $GL(V)$ that is generated by the reflections it contains. If $k = \mathbf{R}$ and $\mathfrak{w}$ is a finite subgroup of $GL(V)$, it is a classical result that one can choose a positive definite scalar product for V which is invariant under $\mathfrak{w}$. Thus in this case there is no loss of generality in working with finite subgroups of the orthogonal group.

2. Root systems and the associated frg. Let V be a real Hilbert space of finite dimension l. We denote by $(\cdot,\cdot)$ and $|\cdot|$ the scalar product and norm in V. Given any $\alpha \in V$, $\alpha \neq 0$, we denote by s_α the reflection in the hyperplane L_α orthogonal to α; thus s_α is the element of the orthogonal group $O(V)$ of V defined by

$$s_\alpha \lambda = \lambda - \frac{2(\lambda,\alpha)}{(\alpha,\alpha)} \alpha \quad (\lambda \in V). \tag{4.15.1}$$

Note that $s_\alpha \alpha = -\alpha$ and $s_\alpha \lambda = \lambda (\lambda \in L_\alpha)$, and that $s_{c\alpha} = s_\alpha (c \neq 0)$. Also, if $t \in O(V)$,

$$s_{t\alpha} = t s_\alpha t^{-1} \tag{4.15.2}$$

Let Δ be a finite subset of $V \setminus \{0\}$. Δ is called a *root system* (*rs*) if

$$\begin{cases} \text{(i)} & \Delta = -\Delta \\ \text{(ii)} & \alpha \in \Delta, k \in \mathbf{R}, k\alpha \in \Delta \implies k = \pm 1 \\ \text{(iii)} & s_\alpha \Delta = \Delta \text{ for all } \alpha \in \Delta. \end{cases} \tag{4.15.3}$$

The elements of Δ will be called *roots*. We denote by $\mathfrak{w} = \mathfrak{w}(\Delta)$ the subgroup of $O(V)$ generated by the s_α ($\alpha \in \Delta$). If $s \in \mathfrak{w}$, it is obvious that $s\Delta = \Delta$ and that s fixes every element of V that is orthogonal to Δ. This shows that the map which assigns to each $s \in \mathfrak{w}$ the permutation $\alpha \mapsto s\alpha$ of Δ is a faithful homomorphism of $\mathfrak{w}$ into the group of all permutations of Δ. $\mathfrak{w}$ is thus a finite reflection subgroup of $O(V)$; it is said to be *associated* to Δ. Δ is said to be *connected* if it is impossible to write V as the orthogonal direct sum $V_1 \oplus V_2$ such that $\Delta = (\Delta \cap V_1) \cup (\Delta \cap V_2)$; note that $\mathbf{R}\cdot\Delta = V$ in this case. The set of roots of a complex semisimple Lie algebra (with respect to a CSA) is clearly a root system, and the associated *frg* is the Weyl group. This circumstance is the motivation behind our terminology.

3. Positive and simple systems of roots. Chambers. Let Δ, $\mathfrak{w}$ be as above. $P \subseteq \Delta$ will be called a *positive system* (of roots) if P is the set of all roots that are positive in some ordering of the vector space V (cf. §4.3). $S \subseteq \Delta$ will be called a *simple system* (of roots) if

$$\begin{cases} \text{(i)} & \text{The elements of } S \text{ are linearly independent.} \\ \text{(ii)} & \text{If } \alpha \in \Delta, \text{ then } \beta = \sum_{\alpha \in S} c_\alpha \alpha, \text{ wherein either all the } c_\alpha \\ & \text{are } \geq 0 \text{ or all of them are } \leq 0 \text{ (thus}^{15}\ [S] = \dim(\mathbf{R}\cdot\Delta)). \end{cases} \tag{4.15.4}$$

V' is the set of all $\lambda \in V$ such that $(\lambda,\alpha) \neq 0$ for any $\alpha \in \Delta$; elements of V' are called *regular*. It is clear that V' is a dense open subset of V. A *chamber* is a connected component of V'. For any chamber C let $P(C)$ be the set of all roots α such that $(\lambda,\alpha) > 0$ for all $\lambda \in C$. Then $P(C)$ is a positive system[16], and $C = \{\lambda : \lambda \in V, (\lambda,\alpha) > 0 \text{ for all } \alpha \in P(C)\}$; in fact if C' is the set in $\{\cdots\}$, C' is open, convex and $\subseteq V'$, and so is connected and contained in a chamber which has to be C. It is clear from our definition that $\mathfrak{w}$ acts naturally on the positive systems, simple systems, and chambers.

[15] For any set A, $[A]$ is its cardinality.

[16] Let $\{\lambda_1, \ldots, \lambda_l\}$ be a basis of V with $\lambda_1 \in C$. If we define, for $\mu, \nu \in V$ and $\lambda = \mu - \nu$, $\mu > \nu$ whenever the first nonzero member of the sequence $(\lambda,\lambda_1), \ldots, (\lambda_1 \lambda_l)$ is positive, then one obtains an ordering for V, and $P(C)$ is the set of roots > 0 in this ordering.

Lemma 4.15.1. *Any simple system is contained in a unique positive system. Any positive system contains a unique simple system. Let S be any simple system and P, the positive system containing it. Then $(\alpha,\beta) \leq 0$ for distinct elements α, β of S; if $\gamma \in P$, there is $\alpha \in S$ such that $(\gamma,\alpha) > 0$.*

Proof. Let S be a simple system, P a positive system containing S and Q the set of all roots which are nonnegative linear combinations of elements of S. Then $Q \subseteq P$, $\Delta = Q \cup (-Q)$. So since $P \cup (-P) = \Delta$ and $P \cap (-P) = \varnothing$, we must have $Q = P$.

Suppose P is an arbitrary positive system. For any $Q \subseteq P$ let (Q) be the set of all elements of P which are nonnegative linear combinations of elements of Q. Let Ω be the collection of all $Q \subseteq P$ such that $(Q) = P$. Clearly Ω is nonempty; for example, $P \in \Omega$. Let S be an element of Ω such that $[S]$ is minimum. We claim that $\alpha \in S$ if and only if $\alpha \notin (P\backslash\{\alpha\})$. Suppose $\alpha \in P$. Then $\alpha \in (S)$. If $\alpha \notin S$, then $S \subseteq P\backslash\{\alpha\}$, so $\alpha \in (P\backslash\{\alpha\})$. Conversely, suppose $\alpha \in S$ but $\alpha \in (P\backslash\{\alpha\})$ also. Let $S = \{\alpha_1 = \alpha,\alpha_2,\ldots,\alpha_l\}$. We can write $\alpha = c_1\beta_1 + \cdots + c_r\beta_r$, where $c_i > 0$, $\beta_i \in P\backslash\{\alpha\}$ $(1 \leq i \leq r)$. Let $\beta_i = \sum_{1\leq j\leq l} m_{ij}\alpha_j$, $d_j = \sum_{1\leq i\leq r} m_{ij}c_i$ $(1 \leq j \leq l)$, the m_{ij} being all ≥ 0. We then have $d_j \geq 0$ for all j and

$$(1 - d_1)\alpha_1 = d_2\alpha_2 + \cdots + d_l\alpha_l.$$

Since $d_2,\ldots,d_l$ are all ≥ 0 and the α_j's are all >0, we must have $0 \leq d_1 \leq 1$. If $d_1 < 1$, $\alpha_1 \in (\{\alpha_2,\ldots,\alpha_l\})$, so $P = (S\backslash\{\alpha\})$, which contradicts the minimality of $[S]$. So $d_1 = 1$ and $d_2\alpha_2 + \cdots + d_l\alpha_l = 0$. But then $d_2 = \cdots = d_l = 0$, giving us $m_{ij} = 0$ $1 \leq i \leq r$, $2 \leq j \leq l$. So $\beta_i = m_{i1}\alpha$ $(1 \leq i \leq r)$, which implies $\beta_i = \alpha$, a contradiction. This proves that $\alpha \notin (P\backslash\{\alpha\})$.

Using this characterization of S, we shall prove that S is the unique simple system $\subseteq P$. Let α, $\beta \in S$, $\alpha \neq \beta$. We claim that $(\alpha,\beta) \leq 0$. If $(\alpha,\beta) > 0$, then $\gamma = s_\alpha\beta = \beta - c\alpha \in \Delta$, where $c > 0$. If $\beta - c\alpha \in P$, $\beta = (\beta - c\alpha) + c\alpha$, so $\beta \in (P\backslash\{\beta\})$; if $c\alpha - \beta \in P$, $\alpha = c^{-1}\beta + c^{-1}(c\alpha - \beta)$, so $\alpha \in (P\backslash\{\alpha\})$; both of these contradict properties of S. By Lemma 4.3.17, the elements of S are linearly independent. So S is a simple system. Suppose $S' \subseteq P$ is another simple system. Since $\alpha \in P\backslash S' \Rightarrow \alpha \in (P\backslash\{\alpha\})$, we have $S \subseteq S'$. Since elements in $S'\backslash S$ are linear combinations of elements of S, we must have $S = S'$.

Suppose, finally, that $\gamma \in P$. Let $\gamma = \sum_{\alpha\in S} c_\alpha\alpha$, where $c_\alpha \geq 0$ for all α. Since $0 < (\gamma,\gamma) = \sum_{\alpha\in S} c_\alpha\,(\gamma,\alpha)$, (γ,α) must be >0 for at least one $\alpha \in S$. This proves everything.

As in the theory of the Weyl group, the fundamental lemma is the following.

Lemma 4.15.2. *Let S be a simple system, P the positive system containing S. Then for $\alpha \in S$,*

(4.15.5) $$s_\alpha\alpha = -\alpha, \qquad s_\alpha\cdot(P\setminus\{\alpha\}) = P\setminus\{\alpha\}.$$

Proof. If $\beta \in P\setminus\{\alpha\}$, then $\exists\ \delta \neq \alpha$ in S and $c > 0$ such that $\beta = c\delta + \sum_{\delta\neq\gamma\in S} c_\gamma\gamma$. Now $s_\alpha\beta$ is of the form $\beta - a\alpha$ for some constant a. So $\alpha \neq s_\alpha\beta = c\delta + \sum_{\delta\neq\gamma\in S} c'_\gamma$. Since the coefficient of δ in the above expression is >0 and since $s_\alpha\beta$ is a root, all the coefficents are ≥ 0 and $s_\alpha\beta \in P$. Thus $s_\alpha\beta \in P\setminus\{\alpha\}$.

Theorem 4.15.3. *The correspondence $C \mapsto P(C)$ is a bijection of the set of all chambers onto the set of all positive systems. The group $\mathfrak{w}$ is transitive on the sets of positive systems, simple systems, and chambers.*

Proof. Let P be a positive system, S the simple system contained in P. For any positive system Q, let $r(Q) = [(-Q)\cap P]$. We prove by induction on $r(Q)$ that Q is conjugate to P under $\mathfrak{w}$. If $r(Q) = 0$, $Q = P$. Let $r(Q) \geq 1$. Then Q contains $-\alpha$ for some $\alpha \in S$. Let $Q' = s_\alpha\cdot Q$, $P' = s_\alpha\cdot P$. Then $r(Q') = [(-s_\alpha\cdot Q)\cap P] = [-Q\cap P']$. But $[-Q\cap P'] = r(Q) - 1$ by the above lemma. So $r(Q') < r(Q)$, and hence there is $s' \in \mathfrak{w}$ such that $Q' = s'\cdot P$. This gives $Q = s_\alpha s'\cdot P$. Since we have already seen that the correspondence $C \mapsto P(C)$ is one-to-one, the remaining assertions of the theorem are immediate.

Theorem 4.15.4. *Let S be a simple system of roots. Then $\mathfrak{w}$ is generated by the s_α ($\alpha \in S$), and $\Delta = \mathfrak{w}\cdot S$.*

Proof. Let P be the positive system containing S. For $\beta \in P$ we define the *order* $O(\beta)$ of β by

(4.15.6) $$O(\beta) = \sum_{\alpha\in S} c_\alpha \quad \left(\beta = \sum_{\alpha\in S} c_\alpha\cdot\alpha\right).$$

Let $\mathfrak{w}_0$ be the subgroup of $\mathfrak{w}$ generated by the $s_\alpha(\alpha \in S)$.

Let $\beta \in P$, and let $\gamma \in (\mathfrak{w}_0\cdot\beta)\cap P$ be chosen such that $O(\gamma) = \min\{O(\beta'): \beta' \in (\mathfrak{w}_0\beta)\cap P\}$. We claim that $\gamma \in S$. Suppose this is not true. By Lemma 4.15.1 we can select $\alpha \in S$ such that $(\gamma,\alpha) > 0$. Let $\beta' = s_\alpha\gamma$. Clearly, $\beta' = \gamma - c\alpha$, where $c > 0$. Since $\gamma \neq \alpha$, $\beta' \in P$ by (4.15.5), so we can conclude from the equation $\beta' + c\alpha = \gamma$ that $O(\beta') + c = O(\gamma)$, or $O(\beta') < O(\gamma)$. Since $\beta' \in (\mathfrak{w}_0\cdot\beta)\cap P$, we have a contradiction.

The above argument shows that for any $\beta \in P$, $\mathfrak{w}_0\cdot\beta$ meets S. So $P \subseteq \mathfrak{w}_0\cdot S$. Since $s_\alpha\alpha = -\alpha$, $-S \subseteq \mathfrak{w}_0\cdot S$, so $-P \subseteq \mathfrak{w}_0\cdot S$ too. Hence $\mathfrak{w}_0\cdot S = \Delta$.

It remains to prove that $\mathfrak{w} = \mathfrak{w}_0$. It is enough to prove that $s_\alpha \in \mathfrak{w}_0$ for $\alpha \in P$. Let $\alpha \in P$. Then there is $\beta \in S$ and $t \in \mathfrak{w}_0$ such that $\alpha = t\beta$. But this means (cf. (4.15.2)) that $s_\alpha = ts_\beta t^{-1} \in \mathfrak{w}_0$.

It may be of interest to note that the following result has been obtained in the course of the above proof.

Corollary 4.15.5. *Let S, P be as above. Then for $\beta \in P$, $O(\beta) \geq 1$. $O(\beta) = 1$ if and only if $\beta \in S$.*

4. The function N. Let Δ, $\mathfrak{w}$ be as before. Choose an ordering ($<$) for V and let P be the set of positive roots in this ordering. Let S be the simple system in P. We put[15]

$$P(t) = P \cap t^{-1}(-P) \qquad N(t) = [P(t)] \quad (t \in \mathfrak{w}). \tag{4.15.7}$$

Thus $N(t)$ is the number of roots $\alpha > 0$ for which $t\alpha < 0$. Given $t \in \mathfrak{w}$, we can write $t = s_{\alpha_1} \cdots s_{\alpha_m}$ ($\alpha_i \in S$); this expression of t is said to be *minimal* if m has the smallest possible value among all such representations of t as a product of reflections corresponding to the simple roots.

Lemma 4.15.6. *Let $\alpha \in S$, $t \in \mathfrak{w}$. Then $N(t) = N(t^{-1})$ and*

$$\begin{aligned} N(ts_\alpha) &= N(t) \pm 1 \quad \textit{according as} \quad t\alpha \gtrless 0 \\ N(s_\alpha t) &= N(t) \pm 1 \quad \textit{according as} \quad t^{-1}\alpha \gtrless 0. \end{aligned} \tag{4.15.8}$$

Proof. $N(t) = [-P(t)] = [-t \cdot P(t)] = N(t^{-1})$, proving the first assertion. We now take up (4.15.8). By (4.15.5) it is easily seen that

$$\begin{aligned} P(ts_\alpha) &= P(t) \cup \{\alpha\} \quad \text{if } t\alpha > 0 \\ P(ts_\alpha) &= P(t) \setminus \{\alpha\} \quad \text{if } t\alpha < 0. \end{aligned} \tag{4.15.9}$$

These relations imply the first identity in (4.15.8). The second follows from the first on replacing t by t^{-1} and observing that $N(t) = N(t^{-1})$ and $N(s_\alpha t) = N(t^{-1}s_\alpha)$.

Corollary 4.15.7. *$N(tt') \equiv N(t) + N(t')$ (mod 2) for $t, t' \in \mathfrak{w}$. In particular, if $t = s_{\alpha_1} \cdots s_{\alpha_m}$ ($\alpha_i \in S$), $N(t) \equiv m$ (mod 2).*

Proof. From (4.15.8) we find $N(ts_\alpha) \equiv N(t) + N(s_\alpha)$(mod 2), for $t \in \mathfrak{w}$, $\alpha \in S$. This leads quickly to the first assertion. The second follows trivially from the first.

Lemma 4.15.8. *$N(tt') \leq N(t) + N(t')$ for $t, t' \in \mathfrak{w}$. In particular, if $t = s_{\alpha_1} \cdots s_{\alpha_m}$ ($\alpha_i \in S$), $N(t) \leq m$.*

Proof. It is easily seen that for $t, t' \in \mathfrak{w}$

(4.15.10) $P(tt') = \{((P \setminus P(t'))) \cap t'^{-1}(P(t))\} \cup \{P(t)' \cap t'^{-1}(-(P \setminus P(t)))\}$,

from which we get $N(tt') \leq N(t) + N(t')$. The second assertion follows trivially from the first.

The crucial result concerning N is the following lemma.

Lemma 4.15.9. *Let $t = s_1 \cdots s_n$, $s_i = s_{\alpha_i}$ where $\alpha_i \in S$ $(1 \leq i \leq n)$. Then the following assertions are equivalent:*

(i) $N(t) < n$.
(ii) *for some j $(1 \leq j \leq n-1)$, $s_1 s_2 \cdots s_j \alpha_{j+1} < 0$.*
(iii) *for some i, j $(1 \leq i \leq j \leq n-1)$, $\alpha_i = s_{i+1} \cdots s_j \alpha_{j+1}$.*
(iv) *$s_{i+1} s_{i+2} \cdots s_{j+1} = s_i s_{i+1} \cdots s_j$ for some i, j $(1 \leq i \leq j \leq n-1)$.*
(v) *$t = s_1 \cdots \hat{s}_i \cdots \hat{s}_{j+1} \cdots s_n$ for some i, j $(1 \leq i \leq j \leq n-1)$ (the $\hat{}$ over s_k means s_k is omitted).*

Proof. If (iii) is true for some i, j, then we get from (4.15.2) the relation $s_i = s_{i+1} \cdots s_j s_{j+1} (s_{i+1} \cdots s_j)^{-1}$, leading to (iv) (for the same i, j). If (iv) is assumed, (v) follows (for the same i, j) on replacing $s_{i+1} s_{i+2} \cdots s_{j+1}$ by $s_i s_{i+1} \cdots s_j$ in the formula $t = s_1 \cdots s_n$. (v) implies (i) trivially.

It therefore remains to prove the implications (i) $\Rightarrow$ (ii) and (ii) $\Rightarrow$ (iii). Assume now that $N(t) < n$. By (4.15.8), if $1 \leq k \leq n-1$, we have $N(s_1 \cdots s_{k+1}) = N(s_1 \cdots s_k) + 1$ provided $s_1 \cdots s_k \alpha_{k+1} > 0$. So we cannot have $s_1 \cdots s_k \alpha_{k+1} > 0$ for all such k, since it would mean $N(t) = n$. Thus for some j, $1 \leq j \leq n-1$, $s_1 \cdots s_j \alpha_{j+1} < 0$, proving (ii). Suppose (ii) is true for some j, $1 \leq j \leq n-1$; then for some i with $1 \leq i \leq j$, $s_{i+1} s_{i+2} \cdots s_j \alpha_{j+1} > 0$ and $s_i s_{i+1} \cdots s_j \alpha_{j+1} < 0$. Since $\alpha_i \in S$, (4.15.5) implies that $s_{i+1} \cdots s_j \alpha_{j+1} = \alpha_i$. This gives (iii).

Theorem 4.15.10. *For $t \in \mathfrak{w}$, $N(t)$ is the number of terms in any minimal expression of t as a product of reflections corresponding to simple roots. If $t = s_{\alpha_1} \cdots s_{\alpha_n}$ is such a minimal expression, $s_{\alpha_1} \cdots s_{\alpha_{j-1}} \alpha_j$ $(1 \leq j \leq n)$ are precisely the n positive roots which are mapped into negative ones by t^{-1}.*

Proof. Let $t = s_1 \cdots s_n$ $(s_i = s_{\alpha_i}, \alpha_i \in S)$ be a minimal expression of t. By Lemma 4.15.8, $N(t) \leq n$. If $N(t) < n$, (v) of Lemma 4.15.9 would contradict the minimality of the expression for t. So $N(t) = n$. To prove the second assertion, note first that since $N(t) = n$, (ii) of the preceding lemma implies that $s_1 \cdots s_{j-1} \alpha_j = \beta_j$ $(1 \leq j \leq n)$ are all > 0. It is therefore sufficient to prove that $\beta_1, \ldots, \beta_n$ are all distinct and that $t^{-1}\beta_j < 0$ for $1 \leq j \leq n$. Suppose $1 \leq i \leq j \leq n-1$ and $\beta_i = \beta_{j+1}$. Then $\alpha_i = s_i s_{i+1} \cdots s_j \alpha_{j+1}$, or $-\alpha_i = s_{i+1} \cdots s_j \alpha_{j+1}$. Since $s_{-\alpha_i} = s_i$, we get (iv) of Lemma 4.15.9 on taking (4.15.2) into account. Thus $N(t) < n$, a contradiction. Since $N(t^{-1}) = n$ and $t^{-1} = s_n \cdots s_1$, (ii) of the preceding lemma implies that $t^{-1}\beta_j = -s_n s_{n-1} \cdots s_{j+1} \alpha_j < 0$ for $1 \leq j \leq n$. This proves the theorem.

Corollary 4.15.11. *There exists $s_0 \in \mathfrak{w}$ such that $s_0 \cdot P = -P$. If $s_0 = s_1 \cdots s_n$ ($s_i = s_{\alpha_i}$, $\alpha_i \in P$) is a minimal expression for s_0, then $n = [P]$ and $s_1 \cdots s_{i-1}\alpha_i$ $(1 \leq i \leq n)$ are precisely all the positive roots.*

Proof. Existence of s_0 follows from Theorem 4.15.3, since $-P$ is also a positive system. Clearly, $N(s_0) = n$ and $P(s_0^{-1}) = P$, so the second statement follows from the above theorem.

Theorem 4.15.12. *$\mathfrak{w}$ acts simply transitively on the set of positive systems (resp. simple systems, chambers).*

Proof. It is enough to consider the action of $\mathfrak{w}$ on the positive systems. We must therefore prove that if $t \in \mathfrak{w}$ and $t \cdot P = P$, then $t = 1$. If $t \neq 1$, then the minimal expression for t contains at least one term, so $N(t) \geq 1$ by Theorem 4.15.10. This is a contradiction, since $N(t) = [P \cap t^{-1}(-P)] = 0$.

Remark. The element $s_0 \in \mathfrak{w}$ of Corollary 4.15.11 is thus uniquely determined, and $s_0^2 = 1$.

5. Fundamental domain. Stabilizers. Let Δ, $\mathfrak{w}$, S, P be as before. Let[16] C be the chamber such that $P = P(C)$. Then

$$C = \{\lambda : \lambda \in V, (\lambda,\alpha) > 0 \text{ for all } \alpha \in S\}. \tag{4.15.11}$$

If $\lambda \in V$ and $(\lambda,\alpha) \geq 0$ for all $\alpha \in S$, then $\lambda + \epsilon v \in C$ for all $\epsilon > 0$ and all $v \in C$. So (Cl denoting closure)

$$Cl(C) = \{\lambda : \lambda \in V, (\lambda,\alpha) \geq 0 \text{ for all } \alpha \in S\}. \tag{4.15.12}$$

Theorem 4.15.13. *Every element of V is conjugate to exactly one element of $Cl(C)$ under $\mathfrak{w}$.*

Proof. Let $\lambda \in V$. Choose a sequence $\{\lambda_n\}$ from V' such that $\lambda_n \longrightarrow \lambda$ as $n \longrightarrow \infty$. Since there are only finitely many chambers, we may assume that all the λ_n belong to a fixed chamber C_1. Choose $s \in \mathfrak{w}$ such that $s \cdot C_1 = C$. Then $\mu = \lim_n s\lambda_n = s\lambda \in Cl(C)$.

Suppose now that $\lambda, \mu \in Cl(C)$ and that for some $t \in \mathfrak{w}$, $t\lambda = \mu$. We prove by induction on $N(t)$ that $\lambda = \mu$. If $N(t) = 0$, $t = 1$ and there is nothing to prove. Let $N(t) \geq 1$. Then $t \cdot P \neq P$, so we can find an $\alpha \in S$ such that $t\alpha < 0$. Since $0 \leq (\lambda,\alpha) = (\mu,t\alpha) \leq 0$, $(\lambda,\alpha) = 0$, so $s_\alpha\lambda = \lambda$. Write $t' = ts_\alpha$. Then $t'\lambda = \mu$ and $N(t') = N(t) - 1$ by (4.15.8); so the induction hypothesis is now applicable and proves that $\lambda = \mu$.

The above theorem shows that $Cl(C)$ is a fundamental domain for the action of $\mathfrak{w}$.

Corollary 4.15.14. *If $\lambda \in Cl(C)$, $\lambda - s\lambda$ is a linear combination of elements of S with coefficients which are all ≥ 0.*

Proof. Given $\mu, \nu \in V$, we write $\mu \succeq \nu$ $(\nu \preceq \mu)$ if $\mu - \nu = \sum_{\gamma \in S} c_\gamma \gamma$, where the c_γ are all ≥ 0. $\preceq$ is a partial order in V. Let $\nu \in \mathfrak{w} \cdot \lambda$. Among the elements of $\mathfrak{w} \cdot \lambda$ which are $\succeq \nu$ select one, say μ, which is maximal with respect to the partial ordering defined above. If $(\mu,\alpha) < 0$ for some $\alpha \in S$, $\mu \neq s_\alpha \mu$ and $\nu \preceq \mu \preceq s_\alpha \mu$, contradicting the maximality of μ. Thus $\mu \in Cl(C)$. By the above theorem, $\lambda = \mu \succeq \nu$.

For any $\lambda \in V_c$ let

$$\mathfrak{w}(\lambda) = \{t : t \in \mathfrak{w}, t\lambda = \lambda\}. \tag{4.15.13}$$

More generally, if Φ is any subset of V_c,

$$\mathfrak{w}(\Phi) = \bigcap_{\lambda \in \Phi} \mathfrak{w}(\lambda) \tag{4.15.14}$$

Lemma 4.15.15. *Let $\lambda \in V$. Then $\mathfrak{w}(\lambda)$ is the subgroup of $\mathfrak{w}$ generated by those s_α $(\alpha \in P)$ for which $(\lambda,\alpha) = 0$. In particular, $\mathfrak{w}(\lambda)$ is trivial if and only if $\lambda \in V'$.*

Proof. Select $t_0 \in \mathfrak{w}$ such that $\mu = t_0\lambda \in Cl(C)$. Then $\mathfrak{w}(\mu) = t_0 \mathfrak{w}(\lambda) t_0^{-1}$, and it is clearly sufficient to prove the lemma with μ instead of λ. Let $\tilde{\mathfrak{w}}(\mu)$ be the subgroup of $\mathfrak{w}$ generated by those s_α $(\alpha \in P)$ for which $(\mu,\alpha) = 0$. Clearly, $\tilde{\mathfrak{w}}(\mu) \subseteq \mathfrak{w}(\mu)$, so we need to prove the inclusion $\mathfrak{w}(\mu) \subseteq \tilde{\mathfrak{w}}(\mu)$. Let $t \in \mathfrak{w}(\mu)$. We shall prove that $t \in \tilde{\mathfrak{w}}(\mu)$ by induction on $N(t)$. If $N(t) = 0$, $t = 1$. Let $N(t) \geq 1$. Then $t \cdot P \neq P$, and hence we can find $\alpha \in S$ such that $t\alpha < 0$. Since $0 \leq (\mu,\alpha) = (\mu,t\alpha) \leq 0$, $(\mu,\alpha) = 0$. Write $t' = ts_\alpha$. Then $t' \in \mathfrak{w}(\mu)$, and $N(t') = N(t) - 1 < N(t)$ by (4.15.8). Consequently, $t' \in \tilde{\mathfrak{w}}(\mu)$ by the induction hypothesis. Since $s_\alpha \in \tilde{\mathfrak{w}}(\mu)$ and $t = t's_\alpha$, we have $t \in \tilde{\mathfrak{w}}(\mu)$.

Corollary 4.15.16. *The s_α $(\alpha \in P)$ are the only reflections in $\mathfrak{w}$.*

Proof. Suppose $t \in \mathfrak{w}$ is a reflection which is not any one of the s_α. Let L be the hyperplane of points fixed by t. Then no root α is orthogonal to L. So we can find $\lambda \in L$ such that $(\lambda,\alpha) \neq 0$ for any $\alpha \in \Delta$. By the above lemma, $\mathfrak{w}(\lambda)$ is trivial. This is a contradiction, since $t \in \mathfrak{w}(\lambda)$.

Remark. If $\Delta(\lambda)$ is the set of all $\alpha \in \Delta$ orthogonal to λ, it is clear that $\Delta(\lambda)$ is a root system and that $\mathfrak{w}(\lambda)$ is its associated *frg*.

For the next theorem, V_c is the complexification of V $(V \subseteq V_c)$. We extend $(\cdot,\cdot)$ to a Hermitian scalar product for V_c and identify $GL(V)$ with its natural image in $GL(V_c)$.

Theorem 4.15.17. *Let Φ be any subset of V_c. Then $\mathfrak{w}(\Phi)$ is generated by those s_α ($\alpha \in P$) for which α is orthogonal to Φ. If $\Phi \subseteq Cl(C)$, $\mathfrak{w}(\Phi)$ is generated by those s_α ($\alpha \in S$) for which α is orthogonal to Φ.*

Proof. We may replace Φ by the set consisting of the real and imaginary parts[17] of its members without changing $\mathfrak{w}(\Phi)$. We may therefore assume $\Phi \subseteq V$. Since we may replace Φ by any basis of the **R**-linear subspace of V spanned by Φ, we may also assume that Φ is finite. We use induction on $[\Phi]$. If $[\Phi] = 1$, Lemma 4.15.15 gives the required result. Suppose $[\Phi] > 1$, $\lambda \in \Phi$, and $\Psi = \Phi \setminus \{\lambda\}$. Clearly, $\mathfrak{w}(\Phi)$ is the stabilizer of Ψ in $\mathfrak{w}(\lambda)$. But $\mathfrak{w}(\lambda)$ is the frg of the root system $\Delta(\lambda)$ and $[\Psi] = [\Phi] - 1$, so by the induction hypothesis, this stabilizer is the group generated by those s_α ($\alpha \in \Delta(\lambda)$) for which α is orthogonal to Ψ. However, these are precisely the s_α ($\alpha \in P$) for which α is orthogonal to Φ. Note that $\mathfrak{w}(\Phi)$ is the frg associated with $\Delta(\Phi) =$ the set of all $\alpha \in \Delta$ which are orthogonal to Φ, which is a root system.

Suppose $\Phi \subseteq Cl(C)$. If $\beta \in P$ and $\beta = \sum_{\alpha \in S} c_\alpha \alpha$, then the nonnegativity of the c_α implies that β is orthogonal to Φ if and only if $c_\alpha = 0$ for every $\alpha \in S$ which is not orthogonal to Φ. So $S \cap \Delta(\Phi)$ is a simple system for $\Delta(\Phi)$. This shows that $\mathfrak{w}(\Phi)$ is generated by those s_α ($\alpha \in S$) for which α is orthogonal to Φ.

Corollary 4.15.18. *Let $\lambda \in V_c$. In order that the orbit $\mathfrak{w} \cdot \lambda$ have $[\mathfrak{w}]$ elements it is necessary and sufficient that $\prod_{\alpha \in P} (\lambda,\alpha) \neq 0$.*

For α, $\beta \in S$, we define $n(\alpha,\beta)$ to be 1 if $\alpha = \beta$, and to be the order of $s_\alpha s_\beta$ in $\mathfrak{w}$ if $\alpha \neq \beta$. It can then be proved that $\mathfrak{w}$ is defined by the generators s_α ($\alpha \in S$) subject to the relations $(s_\alpha s_\beta)^{n(\alpha,\beta)} = 1$ (α, $\beta \in S$).

6. Integral root systems (irs). A root system $\Delta \subseteq V$ is said to be *integral* if $2(\alpha,\beta)/(\alpha,\alpha)$ is an integer for each α, $\beta \in S$. The root systems of complex semisimple Lie algebras are clearly integral. If we add the (normalizing) requirement that $\mathbf{R} \cdot \Delta = V$, these are the only integral root systems. In fact, let $\mathfrak{w}$ be an integral root system such that $\mathbf{R} \cdot \Delta = V$. Let $S = \{\alpha_1, \ldots, \alpha_l\}$ ($l = \dim V$) be a simple system. Define A to be the $l \times l$ matrix $(a_{ij})_{1 \leq i, j \leq l}$ whose ijth entry is $2(\alpha_i,\alpha_j)/(\alpha_i,\alpha_i)$. Note that if $i \neq j$, $a_{ij} \leq 0$ by Lemma 4.15.1 and that $a_{ij} = 0 \Leftrightarrow a_{ji} = 0$. It is then immediate that A is a Cartan matrix. By the results of §4.8 we then have a semisimple Lie algebra $\mathfrak{g}$ over **C**, a CSA

[17]Any $v \in V_c$ can be written uniquely as $\mu + i\mu'$, where $i^2 = -1$, μ, $\mu' \in V$. μ (resp. μ') is called the real part (resp. imaginary part) of v. If $s \in \mathfrak{w}$ (resp. $\alpha \in V$), $sv = v$ (resp. $(\alpha,v) = 0$) if and only if $s\mu = \mu$, $s\mu' = \mu'$ (resp. $(\alpha,\mu) = (\alpha,\mu') = 0$).

$\mathfrak{h}$, and a simple system $\{\bar{\alpha}_1, \ldots, \bar{\alpha}_l\}$ of roots of $(\mathfrak{g},\mathfrak{h})$ whose associated Cartan matrix is precisely A. Let $\bar{\Delta}$ be the set of roots of $(\mathfrak{g},\mathfrak{h})$, $\mathfrak{w}$ the Weyl group of $(\mathfrak{g},\mathfrak{h})$, and $\bar{V} = \sum_{1\leq i\leq l} \mathbf{R}\cdot\bar{\alpha}_i$. It is then easy to see that the correspondence $\bar{\alpha}_i \mapsto \alpha_i$ $(1 \leq i \leq l)$ extends to a linear isomorphism ζ of $\bar{V}$ onto V that maps $\bar{\Delta}$ onto Δ and $\bar{\mathfrak{w}}$ onto $\mathfrak{w}$. It is also easy to show that Δ is connected $\Longleftrightarrow$ $\mathfrak{g}$ is simple, and that in this case there is a constant $c > 0$ such that $c\cdot\zeta$ is an isometry.[18] The connected integral root systems Δ with $\mathbf{R}\cdot\Delta = V$ are therefore precisely the root systems of the simple Lie algebras over $\mathbf{C}$ (up to isomorphism). We leave the proofs of these statements to the reader.

It follows from these remarks that if Δ is an integral root system, P a positive system, and S the simple system in P, the elements of P are nonnegative *integral* linear combinations of the members of S. Further, if $\lambda \in V$ and $2(\lambda,\alpha)/(\alpha,\alpha)$ is an integer ≥ 0 for each $\alpha \in S$, then $\lambda - s\lambda$ is a nonnegative integral linear combination of members of S. Both these assertions can also be proved directly. We leave the proofs to the reader.

7. Extended integral roots systems. By an *extended integral root system* we mean a subset Δ of $V \setminus \{0\}$ such that

$$(4.15.15)\quad \begin{cases} \text{(i)} & \Delta = -\Delta \\ \text{(ii)} & \alpha, \beta \in \Delta \Rightarrow 2(\alpha,\beta)/(\alpha,\alpha) \in \mathbf{Z} \\ \text{(iii)} & \text{If } \alpha \in \Delta \text{ and } s_\alpha \text{ is the corresponding reflection, then} \\ & s_\alpha\Delta = \Delta. \end{cases}$$

Note that we do not require Δ to satisfy (ii) of (4.15.3) (hence the qualification "extended"). Δ is said to be *normalized* if $\mathbf{R}\cdot\Delta = V$. Extended integral root systems occur in the theory of symmetric spaces and the Iwasawa decomposition of real semisimple Lie algebras. The elements of Δ will be called *roots*. $\mathfrak{w} = \mathfrak{w}(\Delta)$ is the subgroup of $O(V)$ generated by the s_α $(\alpha \in \Delta)$. It is clearly finite. Δ is said to be *connected* if it is impossible to write V as a nontrivial orthogonal direct sum $V_1 \oplus V_2$ such that $\Delta = (\Delta \cap V_1) \cup (\Delta \cap V_2)$. Connected systems are obviously normalized. Any irs is an extended irs, but not necessarily conversely (as will be seen later on).

Let Δ be an extended irs. Suppose $\alpha, \beta \in \Delta$ and $\beta = c\alpha$ $(c \in \mathbf{R})$. Then (ii) of (4.15.15) implies that $2c$ and $2c^{-1}$ are both integers. Thus $c = \pm\frac{1}{2}$, ± 1, ± 2. $\frac{1}{2}\alpha$ and 2α cannot both be in Δ, since $2\alpha = 4(\frac{1}{2}\alpha)$. So $\mathbf{R}\cdot\alpha \cap \Delta$ is one of the three sets $\{\pm\alpha\}$, $\{\pm\alpha,\pm 2\alpha\}$, $\{\pm\frac{1}{2}\alpha,\pm\alpha\}$. $\alpha \in \Delta$ is said to be a *short* root if $\frac{1}{2}\alpha \notin \Delta$.

Let Δ_s be the set of short roots. $\Delta_s = -\Delta_s$, and if $\alpha \in \Delta$, $s_\alpha\Delta_s = \Delta_s$. It follows from these observations that Δ_s is an integral root system and that $\mathfrak{w} = \mathfrak{w}(\Delta_s)$. This device permits one to reduce the proofs of many results concerning Δ to the proofs of similar results concerning root systems. In

[18]The scalar product in $\bar{V}$ is, of course, the one coming from the Cartan–Killing form.

particular, since $\mathfrak{w}$ is the frg associated with Δ_s, the entire theory described earlier is applicable to $\mathfrak{w}$.

We now consider positive and simple systems. With applications in mind, we define them in a somewhat different way this time. A subset $P \subseteq \Delta$ is said to be a *positive system* if

$$
\begin{aligned}
&P \cap (-P) = \varnothing, \qquad P \cup (-P) = \Delta, \\
&\alpha, \beta \in P, \alpha + \beta \in \Delta \implies \alpha + \beta \in P.
\end{aligned} \tag{4.15.16}
$$

A subset $S \subseteq \Delta$ is called a *simple system* if

$$
\left\{
\begin{array}{ll}
\text{(i)} & \text{the elements of } S \text{ are linearly independent} \\
\text{(ii)} & \text{if } \beta \in \Delta, \text{ then } \beta = \sum_{\alpha \in S} c_\alpha \alpha, \text{ where the } c_\alpha \text{ are integers} \\
& \text{which are either all } \geq 0 \text{ or all } \leq 0.
\end{array}
\right. \tag{4.15.17}
$$

If Δ is actually a root system, these definitions are equivalent to our earlier ones. To see this we may assume Δ to be normalized; Δ is then the root system of a semisimple Lie algebra over **C**, and our observation is a consequence of the work in §4.3 (cf. Theorem 4.3.18). In the general case of an arbitrary extended irs Δ and a positive system P, we have

$$
\beta \in P, c > 0, c\beta \in \Delta \implies c\beta \in P. \tag{4.15.18}
$$

This is clear if $c = 2$; if $c = \frac{1}{2}$ and $\frac{1}{2}\beta \notin P$, then $-\beta = (-\frac{1}{2}\beta) + (-\frac{1}{2}\beta) \in P$, which is impossible.

The set of all roots which are > 0 in some ordering of the vector space V is a positive system. For more intrinsic constructions of positive systems we proceed as follows. Let V' be the set of all $\lambda \in V$ such that $(\lambda,\alpha) \neq 0$ for each $\alpha \in \Delta$. If C is any connected component of V' and $P(C)$ is the set of all $\alpha \in \Delta$ such that $(\lambda,\alpha) > 0$ for all $\lambda \in C$, then $P(C)$ is a positive system. As before, C will be called a *chamber*. Note that the chambers do not change if we pass from Δ to Δ_s.

Suppose Q is an arbitrary positive system of Δ_s. Let C be the corresponding chamber so that $\lambda \in C \iff (\lambda,\alpha) > 0$ for all $\alpha \in Q$. It is then immediate that $P(C) = Q \cup (2Q \cap \Delta)$ and $Q = P(C) \cap \Delta_s$. Moreover, if S is the simple system in Q, the above expression for $P(C)$ shows that S is a simple system for Δ. S is also the only simple system of Δ contained in $P(C)$. For if S' is any simple system $\subseteq P(C)$, it follows from (ii) of (4.15.17) that $S' \subseteq \Delta_s$, i.e., $S' \subseteq Q$; thus $S' = S$.

Conversely, let P be any positive system of Δ. Then $P_s = P \cap \Delta_s$ satisfies (4.15.16), with Δ_s instead of Δ. It is thus a positive system of Δ_s, and $P = P_s \cup (2P_s \cap \Delta)$. P is thus obtained in the manner described in the preceding paragraph. In particular, the simple system of P_s is the unique simple system in P. It is clear from the theory of semisimple Lie algebras that the members of the system of P are precisely those roots in P which cannot be written as the sum of two roots in P.

It is not difficult to determine all connected (hence normalized) extended irs. Let Δ be one such. We may assume that $\Delta \neq \Delta_s$. Let S be a simple system for Δ_s (and Δ). We write $S = \{\alpha_1, \ldots, \alpha_l\}$ and assume that $2\alpha_s$ $(r \leq s \leq l)$ are precisely all the roots in $2S \cap \Delta$. Let A be the $l \times l$ matrix (a_{ij}) where $a_{ij} = 2(\alpha_i,\alpha_j)/(\alpha_i,\alpha_i)$. Since Δ_s is also connected, A is the Cartan matrix of a simple Lie algebra, whose Dynkin diagram we shall denote by $\mathfrak{D}$. From (ii) of (4.15.15) and the relations $2\alpha_s \in \Delta$ $(r \leq s \leq l)$, it is immediate that a_{sj} is an *even* integer for $r \leq s \leq l$, $1 \leq j \leq l$. This implies that if s and j are as above and if the vertices (in $\mathfrak{D}$) corresponding to α_s and α_j are linked, then the link is double and $(\alpha_s,\alpha_s) = \frac{1}{2}(\alpha_j,\alpha_j)$. A glance at the diagrams (4.5.6) shows that $\mathfrak{D}$ is then B_l. Now we can (cf. §4.4) choose an orthonormal basis $\{\lambda_1, \ldots, \lambda_l\}$ in V such that

$$\Delta_s = \{\pm(\lambda_i \pm \lambda_j), 1 \leq i < j \leq l, \pm\lambda_i, 1 \leq i \leq l\},$$

with $\alpha_1 = \lambda_1 - \lambda_2, \ldots, \alpha_{l-1} = \lambda_{l-1} - \lambda_l, \alpha_l = \lambda_l$, $\mathfrak{w}$ being the group of permutations of the λ's followed by arbitrary sign changes. It follows easily from these remarks that Δ is the extended irs[19] BC_l defined by

$$BC_l = \{\pm(\lambda_i \pm \lambda_j), 1 \leq i < j \leq l, \pm\lambda_i, \pm 2\lambda_i, 1 \leq i \leq l\}. \tag{4.15.19}$$

We have thus obtained the following theorem.

Theorem 4.15.19. *Let Δ be an extended irs, Δ_s the irs of all short roots in Δ, and $\mathfrak{w} = \mathfrak{w}(\Delta)$. Then the map $Q \mapsto P(Q) = Q \cup (2Q \cap \Delta)$ (resp. $C \mapsto P(C)$) is a bijection of the set of all positive systems of Δ_s (resp. chambers) onto the set of all positive systems of Δ, and $Q = P(Q) \cap \Delta_s$. If P is any positive system of Δ and S is the set of roots in P which cannot be written as the sum of two roots in P, then S is a simple system of both P and $P \cap \Delta_s$ and is the unique simple system $\subseteq P$. Finally, if $l = \dim V$, there is (up to isomorphism) exactly one connected extended irs which is not a root system, and it is the system BC_l.*

8. Representatations in polynomial rings. Invariants. Let V be a real Hilbert space of finite dimension l, V_c its complexification. $\mathcal{P}$ is the algebra of all complex-valued polynomial functions on V_c. For any integer $d \geq 0$ we denote by $\mathcal{P}_d$ the subspace of $\mathcal{P}$ of homogeneous elements of degree d. Throughout this discussion we shall fix a finite subgroup $\mathfrak{w}$ of $O(V)$. $\mathfrak{w}$ acts

[19] It is easy to check that BC_l is an extended integral root system.

naturally on $\mathcal{P}$ and leaves $\mathcal{P}_d$ invariant. For $p \in \mathcal{P}$ and $s \in \mathfrak{w}$, p^s is the transform of p by $s: p^s(\lambda) = p(s^{-1}\lambda)$ $(\lambda \in V)$. Let

$$I = \{p : p \in \mathcal{P}, p^s = p \;\forall\; s \in \mathfrak{w}\}. \tag{4.15.20}$$

I is the direct sum of the $I_d = I \cap \mathcal{P}_d$ $(d \geq 0)$. I^+ is defined as the linear span of the I_d $(d > 0)$. I^+ is an ideal in I. Let

$$\mathfrak{F} = \mathcal{P}I^+ \tag{4.15.21}$$

$\mathfrak{F}$ is an ideal in $\mathcal{P}$ contained in $\mathcal{P}^+ = \sum_{d>0} \mathcal{P}_d$.

A vector space A over $\mathbf{C}$ (of finite or infinite dimension) is said to be *graded* if there are finite-dimensional subspaces A_i $(i = 0,1,2,\ldots)$ which are linearly independent and span A; the *Poincaré Series* $P_A(t)$ of A is then defined as the formal power series $\sum_{i\geq 0} \dim(A_i)t^i$ in the indeterminate t. If $B \subseteq A$ is a subspace, it is said to be *graded by* A if $B = \sum_{i\geq 0} (B \cap A_i)$. When $\mathcal{P}$ is graded in the usual way, I becomes a graded subalgebra of $\mathcal{P}$. The Poincaré Series of $\mathcal{P}$ is $(1 - t)^{-l}$. More generally, let $\mathfrak{A}$ be a graded subalgebra of $\mathcal{P}$ (containing 1), and let $\mathfrak{A} = \mathbf{C}[q_1, \ldots, q_m]$, where $q_1, \ldots, q_m$ are algebraically independent homogeneous elements of respective positive degrees $v_1, \ldots, v_m$ $(m \leq l$ necessarily); it is then easily checked that the Poincaré series of $\mathfrak{A}$ is $\prod_{1\leq i\leq l} (1 - t^{v_i})^{-1}$.

Lemma 4.15.20. *There are finite sets M of homogeneous elements of I^+ such that $\mathfrak{F}$ is the ideal generated by M. If $L = \{p_1, \ldots, p_m\}$ is one such, $I = \mathbf{C}[p_1, \ldots, p_m]$.*

Proof. Let $\mathfrak{M}$ be the collection of all finite sets of homogeneous elements in I^+; for $M \in \mathfrak{M}$ let $\mathfrak{F}_M$ be the ideal in $\mathcal{P}$ generated by M. By the Hilbert basis theorem there is a finite set $N \subseteq \mathfrak{F}$ such that $\mathfrak{F} = \sum_{n\in N} \mathcal{P}n$. Since $\mathfrak{F} = \sum_{M\in\mathfrak{M}} \mathfrak{F}_M$, we can find $M_1, \ldots, M_k \in \mathfrak{M}$ such that $N \subseteq \sum_{1\leq i\leq k} \mathfrak{F}_{M_i}$. If $L = \bigcup_{1\leq i\leq k} M_i$, $\mathfrak{F} = \mathfrak{F}_L$.

Fix such an $L = \{p_1, \ldots, p_m\}$. Let $I' = \mathbf{C}[p_1, \ldots, p_m]$. We prove by induction on d that $I_d \subseteq I'$ for all $d \geq 0$. We may assume $d \geq 1$. Let $p \in I_d$. Then $p \in \mathfrak{F}$, so there are $q_i \in \mathcal{P}$ such that $p = q_1p_1 + \cdots + q_mp_m$. We may assume that the q_i are homogeneous and that $\deg(q_i) + \deg(p_i) = d$ if $q_i \neq 0$. Averaging over $\mathfrak{w}$, we find $p = \bar{q}_1p_1 + \cdots + \bar{q}_mp_m$, where $\bar{q}_i = [\mathfrak{w}]^{-1} \sum_{s\in\mathfrak{w}} q_i^s$. By the induction hypothesis, $\bar{q}_i \in I'$ for all i. So $p \in I'$.

For any $p \in \mathcal{P}$ let

$$\bar{p} = [\mathfrak{w}]^{-1} \sum_{s\in\mathfrak{w}} p^s. \tag{4.15.22}$$

It seems reasonable to expect that the p_i will be algebraically independent if L above is a minimal ideal basis. This is actually true if $\mathfrak{w}$ is a frg. The key step in the argument is the following lemma.

Lemma 4.15.21. *Let $\mathfrak{w}$ be a frg. Let $p_1, \ldots, p_m \in I$ be such that $p_1 \notin \sum_{2 \le j \le m} Ip_j$.* If $q_1, \ldots, q_m$ *are homogeneous elements of $\mathcal{P}$ with*

$$q_1 p_1 + \cdots + q_m p_m = 0, \tag{4.15.23}$$

then $q_1 \in \mathfrak{F}$.

Proof. We argue by induction on $\deg(q_1)$. From (4.15.23) we get $\bar{q}_1 p_1 + \cdots + \bar{q}_m p_m = 0$. If q_1 is constant, this constant is 0; for otherwise $p_1 = -q_1^{-1} \sum_{2 \le i \le m} \bar{q}_i p_i$, a contradiction. Suppose $\deg(q_1) \ge 1$. let $s \in \mathfrak{w}$ be a reflection and f a linear function whose zeros are precisely the fixed points of s. From (4.15.23) we have $\sum_{1 \le i \le m}(q_i - q_i^s) p_i = 0$. But there are homogeneous $r_i \in \mathcal{P}$ such that $q_i - q_i^s = r_i f (1 \le i \le m)$. So $\sum_{1 \le i \le m} r_i p_i = 0$. By the induction hypothesis, $r_1 \in \mathfrak{F}$. In other words, $q_1 \equiv q_1^s \pmod{\mathfrak{F}}$. Since s was arbitrary and $\mathfrak{w}$ is a frg, $q_1 \equiv q_1^s \pmod{\mathfrak{F}}$ for each $s \in \mathfrak{w}$. This implies that $q_1 \equiv \bar{q}_1 \pmod{\mathfrak{F}}$. As $\deg(\bar{q}_1) = \deg(q_1) > 0$, $\bar{q}_1 \in \mathfrak{F}$. Thus $q_1 \in \mathfrak{F}$.

Lemma 4.15.22. *Let $\mathfrak{w}$ be a frg. Let $p_1, \ldots, p_m \in I^+$ be such that* (i) *the p_i are homogeneous and $\mathfrak{F} = \sum_{1 \le i \le m} \mathfrak{F} p_i$, and* (ii) $p_i \notin \sum_{j \ne i} Ip_j$ $(1 \le i \le m)$. *Then the p_i are algebraically independent.*

Proof. Let $d_i = \deg(p_i)$. If the p_i are not algebraically independent we can select a nonzero $H \in \mathbf{C}[T_1, \ldots, T_m]$ (T_i are indeterminates) of minimal degree such that $H(p_1, \ldots, p_m) = 0$. We may assume that there is an integer $h > 0$ such that H contains only those monomials $T_1^{k_1} \cdots T_m^{k_m}$ for which

$$k_1 d_1 + \cdots + k_m d_m = h. \tag{4.15.24}$$

Let $H_i = \partial H(T_1, \ldots, T_m)/\partial T_i$ and $q_i = H_i(p_1, \ldots, p_m) (1 \le i \le m)$. Clearly, $q_i \in I$, $q_i = 0 \Leftrightarrow H_i = 0$, and there are i for which $q_i \ne 0$. We may therefore renumber the indices to assume that for some integers s, n with $1 \le s \le n \le m$ the following are true: $q_j \ne 0$ if only if $j \le n$, all these q_j belong to $\sum_{1 \le i \le s} Iq_i$, and for $1 \le i \le s$, q_i does not belong to the ideal in I generated by the remaining q_j. Notice that in view of (4.15.24), q_j is homogeneous of degree $h - d_j$ $(1 \le j \le n)$. By our assumptions we can choose $v_{ji} \in I$ such that

$$q_{s+j} = \sum_{1 \le i \le s} v_{ji} q_i \quad (1 \le j \le n - s). \tag{4.15.25}$$

We may assume that the v_{ji} are homogeneous and that if $v_{ji} \ne 0$, $\deg(v_{ji}) + \deg(q_i) = h - d_{s+j}$.

Let x_k $(1 \leq k \leq l)$ be linear coordinates on V. Differentiating the relation $H(p_1, \ldots, p_m) = 0$ with respect to x_k, we find that

$$\sum_{1 \leq j \leq n} q_j \frac{\partial p_j}{\partial x_k} = 0 \quad (1 \leq k \leq l).$$

So, using (4.15.25), we have

$$\sum_{1 \leq i \leq s} \left(\frac{\partial p_i}{\partial x_k} + \sum_{1 \leq j \leq n-s} v_{ji} \frac{\partial p_{s+j}}{\partial x_k} \right) q_i = 0 \quad (1 \leq k \leq l).$$

The terms within the parentheses are homogeneous, so we may apply the preceding lemma. We thus conclude that

$$\frac{\partial p_i}{\partial x_k} + \sum_{1 \leq j \leq n-s} v_{ji} \frac{\partial p_{s+j}}{\partial x_k} \in \mathfrak{F} \quad (1 \leq k \leq l, 1 \leq i \leq s).$$

Multiplying by x_k and adding the resulting relations, we get

$$d_i p_i + \sum_{1 \leq j \leq n-s} d_{s+j} p_{s+j} v_{ji} \in \mathcal{P}^+ I^+ \quad (1 \leq i \leq s).$$

So we can find $g_{ir} \in \mathcal{P}^+$ $(1 \leq i \leq s, 1 \leq r \leq m)$ such that

$$d_i p_i + \sum_{1 \leq j \leq n-s} d_{s+j} p_{s+j} v_{ji} = \sum_{1 \leq r \leq m} g_{ir} p_r \quad (1 \leq i \leq s). \tag{4.15.26}$$

We now observe that the left side of this relation is homogeneous of degree d_i. So we may assume that $g_{ir} = 0$ unless $\deg(p_r) < d_i$. (4.15.26) then implies that for suitable $f_j \in \mathcal{P}$, $p_i = \sum_{j \neq i} f_j p_j$. Thus $p_i = \sum_{j \neq i} \bar{f}_j p_j$, contradicting the assumption concerning the p_i. This completes the proof of the lemma.

We can now prove the following theorem. It was first proved by Chevalley [7]. Our proof is exactly his.

Theorem 4.15.23. *Let $\mathfrak{w}$ be a finite reflection subgroup of $O(V)$. Then there are l algebraically independent homogeneous elements $p_1, \ldots, p_l$ of positive degree such that $I = \mathbf{C}[p_1, \ldots, p_l]$.*

Proof. By Lemma 4.15.21 and 4.15.22, we can find homogeneous p_i $(1 \leq i \leq m)$ which are algebraically independent and of positive degree such that $I = \mathbf{C}[p_1, \ldots, p_m]$. It only remains to prove that $m = l$. Clearly, $m \leq l$. Let $K = \mathbf{C}(p_1, \ldots, p_m)$ be the quotient field of I and $\mathcal{Q}$ the quotient field of $\mathcal{P}$. To prove that $m = l$ it is enough[20] to prove that $\mathcal{Q}$ is algebraic over K. This will

[20]For then the transcendence degrees over $\mathbf{C}$ of $\mathcal{Q}$ and K will be the same.

be done if we can prove that each element of $\mathcal{P}$ is algebraic over K. Let $f \in \mathcal{P}$. Let T be an indeterminate, and let $\prod_{s \in \mathfrak{w}} (T - f^s) \equiv T^w + g_1 T^{w-1} + \cdots + g_w$ $(w = [\mathfrak{w}])$. Clearly, $g_1, \ldots, g_w$ are in I, and f satisfies the equation

$$f^w + g_1 f^{w-1} + \cdots + g_w = 0,$$

so f is actually integral over I.

Corollary 4.15.24. *Let p_i be as above, $d_i = deg(p_i)$. Then the Poincaré series of I is given by*

$$P_I(t) = \prod_{1 \leq j \leq l} (1 - t^{d_i})^{-1}. \tag{4.15.27}$$

We now prove Theorem 4.15.26, which is the converse, due to Shephard and Todd [1], of Chevalley's theorem. We need a lemma.

Lemma 4.15.25. *Let $\mathfrak{w}$ be any finite subgroup of $O(V)$. Then the Poincaré series of I is given by*

$$P_I(t) = [\mathfrak{w}]^{-1} \sum_{s \in \mathfrak{w}} (\det(1 - ts))^{-1}. \tag{4.15.28}$$

Proof. For any linear automorphism L of V_c and any integer $d \geq 0$, let $L^{(d)}$ be the corresponding induced linear transformation of $\mathcal{P}_d$. Then

$$(\det(1 - tL))^{-1} = \sum_{d \geq 0} tr(L^{(d)}) t^d. \tag{4.15.29}$$

This formula is immediate if L is diagonalizable; the general case follows from this, since the set of such L is a dense open subset of $GL(V_c)$. On the other hand, it is an easy consequence of the theory of characters of $\mathfrak{w}$ that

$$[\mathfrak{w}]^{-1} \sum_{s \in \mathfrak{w}} tr(s^{(d)}) = \dim(I_d) \quad (d = 0, 1, \ldots). \tag{4.15.30}$$

These two relations lead at once to (4.15.28).

Theorem 4.15.26. *Let $\mathfrak{w}$ be a finite subgroup of $O(V)$. Suppose there are algebraically independent homogeneous elements $p_1, \ldots, p_m$ of positive degree such that $I = \mathbf{C}[p_1, \ldots, p_m]$. Then $m = l$, and $\mathfrak{w}$ is a frg.*

Proof. The proof that $m = l$ is the same as in Chevalley's theorem. Let $d_i = \deg(p_i)$. Then the Poincaré series of I is $\prod_{1 \leq i \leq l} (1 - t^{d_i})^{-1}$, so by (4.15.28) we have, writing $w = [\mathfrak{w}]$,

$$w \prod_{1 \leq i \leq l} (1 - t^{d_i})^{-1} = \sum_{s \in \mathfrak{w}} \det(1 - ts)^{-1}; \tag{4.15.31}$$

here and in what follows we shall treat t as a complex variable. If $\lambda_{s1}, \ldots, \lambda_{sl}$ are the eigenvalues of $s \in \mathfrak{w}$ (repeated according to multiplicity), (4.15.31) gives us the following identity on multiplying both sides by $(1-t)^l$:

$$w \prod_{1 \leq i \leq l} (1 + t + \cdots + t^{d_i - 1})^{-1} = \sum_{s \in \mathfrak{w}} \prod_{1 \leq j \leq l} \left(\frac{1-t}{1 - t\lambda_{sj}}\right) \tag{4.15.32}$$

We now set $t = 1$ in (4.15.32). If $s \neq 1$, the term on the right side vanishes for $t = 1$. So we find that

$$w = d_1 \cdots d_l. \tag{4.15.33}$$

We now differentiate both sides of (4.15.32) with respect to t at $t = 1$. An easy argument shows that the term corresponding to $s \in \mathfrak{w}$ contributes 0 unless the multiplicity of 1 as an eigenvalue of s is $l - 1$. But if 1 is an eigenvalue of s of multiplicity $l - 1$, s is *necessarily a reflection*, and the corresponding contribution is $-\frac{1}{2}$. We therefore obtain, on taking (4.15.33) into account.

$$\sum_{1 \leq i \leq l} (d_i - 1) = \text{number of reflections in } \mathfrak{w}. \tag{4.15.34}$$

Let $\mathfrak{w}'$ be the subgroup of $\mathfrak{w}$ generated by the reflections in $\mathfrak{w}$, and let I' be the algebra of $\mathfrak{w}'$-invariant elements of $\mathcal{P}$. By Chevalley's theorem, we can find algebraically independent, homogeneous elements $p'_1, \ldots, p'_l$ of respective degrees $d'_1, \ldots, d'_l$ (>0) such that $I' = \mathbf{C}[p'_1, \ldots, p'_l]$. Let $w' = [\mathfrak{w}']$. Clearly, the arguments of the preceding paragraph are equally applicable to $\mathfrak{w}'$. So $d'_1, \ldots, d'_l = w'$; moreover, since $\mathfrak{w}'$ contains all the reflections in $\mathfrak{w}$, (4.15.34) gives us

$$\sum_{1 \leq i \leq l} d'_i = \sum_{1 \leq i \leq l} d_i. \tag{4.15.35}$$

Now $p_i \in I'$ for all i. So we can find polynomials $A_i \in \mathbf{C}[T_1, \ldots, T_l]$ (T_i being indeterminates) such that

$$p_i = A_i(p'_1, \ldots, p'_l) \quad (1 \leq i \leq l).$$

Since $p_1, \ldots, p_l$ are algebraically independent, the Jacobian $\partial(A_1, \ldots, A_l)/\partial(T_1, \ldots, T_l)$ cannot be zero. Consequently, we can find a permutation $(i_1, \ldots, i_l)$ of $(1, \ldots, l)$ such that

$$\frac{\partial A_{i_1}}{\partial T_1} \cdots \frac{\partial A_{i_l}}{\partial T_l} \neq 0.$$

This implies at once the inequalities

$$d_{i_r} \geq d'_r \quad (1 \leq r \leq l).$$

So by (4.15.35) we have $d_{i_r} = d'_r$, $1 \leq r \leq l$. In particular, $d'_1 \cdots d'_l = d_1 \cdots d_l$, i.e., $[\mathfrak{w}] = [\mathfrak{w}']$, thereby proving that $\mathfrak{w} = \mathfrak{w}'$. This proves the theorem.

Corollary 4.15.27. *Let $\mathfrak{w}$ be a finite reflection subgroup of $O(V)$ and let p_i $(1 \leq i \leq l)$ be as in Theorem 4.15.23. Let $d_i = deg(p_i)$. Then $[\mathfrak{w}] = d_1 \cdots d_l$, and $\sum_{1 \leq i \leq l} (d_i - 1)$ is the number of reflections in $\mathfrak{w}$.*

Remark. It is possible to relax the condition $\mathfrak{w} \subseteq O(V)$ in the two theorems above and to obtain suitable generalizations for finite subgroups of the unitary group $U(V_c)$ of V_c. This was done by Shephard and Todd [1]. They have proved that in order that the algebra of polynomials invariant under a finite subgroup $\mathfrak{w}$ of $U(V_c)$ be isomorphic to $\mathbf{C}[T_1, \ldots, T_l]$ (T_i indeterminates) it is necessary and sufficient that $\mathfrak{w}$ be generated by the elements s with the following property: $\exists$ an integer $m \geq 1$ such that $s^m = 1$ and that the set of fixed points of s has dimension $l - 1$.

We conclude this section with the following theorem, due also to Chevalley [7], on the structure of $\mathcal{P}$ as an I-module.

Theorem 4.15.28. *Let $\mathfrak{w}$ be a finite reflection subgroup of $O(V)$, and let notation be as above. Write $w = [\mathfrak{w}]$. Then $\mathcal{P}$ is a free I-module of rank w. More precisely, let H be a graded subspace of $\mathcal{P}$ such that $\mathcal{P} = \mathfrak{F} + H$ is a direct sum; then $dim(H) = w$, and the map $p \otimes u \mapsto pu$ $(p \in I, u \in H)$ extends to a linear isomorphism of $I \otimes H$ onto $\mathcal{P}$. Finally, the Poincaré Series of H is given by*

$$P_H(t) = \prod_{1 \leq i \leq l} (1 + t + \cdots + t^{d_i - 1}); \tag{4.15.36}$$

here $d_i = deg(p_i)$, where the p_i are as in Chevalley's theorem.

Proof. It is obvious that there are graded subspaces H of $\mathcal{P}$ such that $\mathcal{P} = \mathfrak{F} + H$ is a direct sum. Choose and fix one such. Then the map $p,u \mapsto pu$ $(p \in I, u \in H)$ "extends" to a linear map τ of $I \otimes H$ into $\mathcal{P}$ such that $\tau(p \otimes u) = pu$ for $p \in I$, $u \in H$.

We show first that τ is surjective. Let $\mathcal{R}$ be the range of τ. Clearly, $H \subseteq \mathcal{R}$ and $I\mathcal{R} \subseteq \mathcal{R}$. We shall prove by induction on d that $\mathcal{P}_d \subseteq \mathcal{R}$ for all $d \geq 0$. For $d = 0$ there is nothing to prove. Let $d \geq 1$, and let $f \in \mathcal{P}_d$. There is $v \in H \cap \mathcal{P}_d$ such that $f - v \in \mathfrak{F}$. So we can write $f = v + \sum_{1 \leq i \leq l} g_i p_i$, where $g_i \in \mathcal{P}$ for all i. For reasons of homogeneity we may assume that the g_i are homogeneous and that $\deg(g_i) + \deg(p_i) = d$ whenever $g_i \neq 0$. But then $g_i \in \mathcal{R}$, $1 \leq i \leq l$, by the induction hypothesis. So $f \in \mathcal{R}$ too.

We prove next that τ is injective. Since H is spanned by homogeneous elements, this will be done if we prove the following: if $\varphi_1, \ldots, \varphi_m$ are homogeneous linearly independent elements of H, then

$$f_1, \ldots, f_m \in I, \qquad \sum_{1 \leq i \leq m} f_i \varphi_i = 0 \Longrightarrow f_1 = \cdots = f_m = 0.$$

Suppose the φ_i and f_j are as above. If some $f_i \neq 0$, we may renumber the indices so that f_1 does not belong to the ideal in I generated by $f_2, \ldots, f_m$. But then, by Lemma 4.15.21 $\varphi_1 \in \mathfrak{F}$, a contradiction.

τ is thus a linear isomorphism. If $P_I(t)$, $P_H(t)$, and $P(t)$ are the respective Poincaré Series of I, H, and $\mathcal{P}$, one has $P_I(t)P_H(t) = P(t) = (1 - t)^{-l}$. In view of (4.15.27), we then obtain

$$p_H(t) = \prod_{1 \leq i \leq l} (1 + t + \cdots + t^{d_i - 1}). \tag{4.15.37}$$

Since $P_H(t)$ is a *polynomial* in t, $\dim(H) < \infty$. If we take $t = 1$ in the above relation, we find $\dim(H) = d_1 \cdots d_l$. Corollary 4.15.27 now shows that $\dim(H) = [\mathfrak{w}]$.

This proves the theorem.

It can be shown that the representation of $\mathfrak{w}$ in H is equivalent to its regular representation (cf. Exercise 70).

EXERCISES

1. Determine all CSA's of $\mathfrak{g} = \mathfrak{sl}(n,\mathbf{R})$

2. Let $G = SO(1,n)$ $(n \geq 2)$ and let $r(G)$ denote the maximum number of mutually nonconjugate CSA's of $\mathfrak{g}$. Prove that $r(G) = 1$ or 2 according as n is odd or even. Determine also the number of connected components of $\mathfrak{g}'$.

3. Let $\mathfrak{g} = A_l$ and $X = E_{12} + E_{23} + \cdots + E_{l,l+1}$ (notation as in §4.4). Determine the centralizer of X in $\mathfrak{g}$ and verify that it is a maximal abelian subalgebra of $\mathfrak{g}$ of dimension l consisting entirely of nilpotent elements.

4. Let π_j $(j = 0,1,2, \ldots)$ be the representations of $SL(2,\mathbf{C})$ constructed in §4.2. Let j, j' be two integers, $0 \leq j \leq j'$. Prove that $\pi_j \otimes \pi_{j'}$ is the direct sum of π_r, $r = j' - j + 2k$, $k = 0,1, \ldots, j$ (Clebsch–Gordon series).

5. Let notation and assumptions be as in Lemma 4.6.7. Prove that in this case, for any weight μ of π, $\dim V_{s\mu} = \dim V_\mu$ for all $s \in \mathfrak{w}$. (Hint: Fix i, select $c_{j0} \in \mathbf{C}$, $v_{j0} \in V_\mu$ $(1 \leq j \leq m$, $v_{0,0} = 0)$ such that the v_{j0} form a basis for V_μ and $\pi(Y_i X_i)v_{j0} \equiv c_{j0}v_{j0} \pmod{v_{10}, v_{20}, \ldots, v_{j-1,0}}$ $(1 \leq j \leq m)$. Proceed as in Lemma 4.6.7.)

The next exercise gives an outline of Serre's approach to Theorem 4.8.3 (cf. remarks at the end of §4.8). $\mathfrak{g}$ is as in §4.8. The rest of the notation is as in §§4.6 and 4.8.

6. (a) Let $\mathfrak{n}^{\pm}$ be as in Lemma 4.6.1 and let $\mathfrak{q}^+$ (resp. $\mathfrak{q}^-$) be the ideal in $\mathfrak{n}^+$ generated by all the θ_{ij}^+ (resp. ideal in $\mathfrak{n}^-$ generated by all the θ_{ij}^-). Prove that $\mathfrak{q}^+$ and $\mathfrak{q}^-$ are ideals in $\mathfrak{g}$. Deduce that $\mathfrak{q} = \mathfrak{q}^+ + \mathfrak{q}^-$ is the ideal in $\mathfrak{g}$ generated by all the $\theta_{ij}^{\pm}$. (Hint: To prove that $\mathfrak{q}^+$ is an ideal, show by induction on $s \geq 1$ that $[Y_k, \operatorname{ad} X_{k_1} \cdots \operatorname{ad} X_{k_s}(\theta_{ij}^+)] \in \mathfrak{q}^+$ for $1 \leq k, k_1, \ldots, k_s \leq l$.)

(b) Let $\mathfrak{u}$ be an ideal of $\mathfrak{g}$, $\hat{\mathfrak{g}} = \mathfrak{g}/\mathfrak{u}$, $X \mapsto \hat{X}$ the natural map of $\mathfrak{g}$ onto $\hat{\mathfrak{g}}$. Assume that $\mathfrak{u} \cap \mathfrak{h} = 0$ and that $\operatorname{ad} \hat{X}_i$ and $\operatorname{ad} \hat{Y}_i$ are locally nilpotent endomorphisms of $\hat{\mathfrak{g}}$ for $1 \leq i \leq l$. For $\lambda \in \mathfrak{h}^*$, let $\hat{\mathfrak{g}}_\lambda = \{X : X \in \hat{\mathfrak{g}}, [\hat{H}, X] = \lambda(H)X \text{ for all } H \in \mathfrak{h}\}$. Let $\hat{\mathfrak{h}}$ be the image of $\mathfrak{h}$ in $\hat{\mathfrak{g}}$, and let $\hat{\mathfrak{w}}$ be the subgroup of $GL(\hat{\mathfrak{h}})$ that corresponds to $\mathfrak{w}$ under the mapping $H \mapsto \hat{H}$ $(H \in \mathfrak{h})$. Prove that given any $s \in \hat{\mathfrak{w}}$, there is an "inner" automorphism $x(s)$ of $\hat{\mathfrak{g}}$ such that $x(s)$ leaves $\hat{\mathfrak{h}}$ invariant and $x(s)|\hat{\mathfrak{h}} = s$. Deduce that $\dim \hat{\mathfrak{g}}_{s\lambda} = \dim \hat{\mathfrak{g}}_\lambda$ for $\lambda \in \mathfrak{h}^*$, $s \in \mathfrak{w}$. (Hint: For $1 \leq i \leq l$, consider $x(s_i) = \exp(\operatorname{ad} \hat{X}_i) \exp(-\operatorname{ad} \hat{Y}_i) \exp(\operatorname{ad} \hat{X}_i)$.)

(c) Let $\Delta = \bigcup_{1 \leq i \leq l} \mathfrak{w} \cdot \alpha_i$. For $\lambda \in \Delta$, prove that $\dim \hat{\mathfrak{g}}_{c\lambda} = 1$ or 0 according as $c = \pm 1$ or $c \neq \pm 1$. (Hint: $\mathfrak{n}^+$ is spanned by the X_j and the elements $\operatorname{ad} X_{k_1} \cdots \operatorname{ad} X_{k_s}(X_k)$. Hence $\dim \mathfrak{g}_{c\alpha_j} = 1$ or 0 according as $c = \pm 1$ or $c \neq \pm 1$. Now use (b).)

(d) Let $P = \Delta \cap \Gamma$. Prove that $\Delta = P \cup (-P)$ and that if $1 \leq i \leq l$, $s_i[P \setminus \{\alpha_i\}] = P \setminus \{\alpha_i\}$. (Hint: Use (c).)

(e) Let $\mathfrak{h}_0$ be the linear span of the H_i over $\mathbf{R}$, $\mathfrak{h}_0'$ the set where no element of Δ vanishes, and $\mathfrak{h}_0^+$ the set where all the α_i take positive values. Prove that $\mathfrak{h}_0^+$ is a connected component of $\mathfrak{h}_0'$ and that $\mathfrak{w}$ acts transitively on the connected components of $\mathfrak{h}_0'$.

(f) Suppose $\lambda \neq 0$ lies in $\mathfrak{h}^* \setminus \Delta$. Prove that $\hat{\mathfrak{g}}_\lambda = 0$. Deduce that $\dim(\hat{\mathfrak{g}}) < \infty$. (Hint: One may assume that λ is not proportional to any member of Δ. Select $H \in \mathfrak{h}_0'$ such that $\lambda(H) = 0$ and an $s \in \mathfrak{w}$ such that $sH \in \mathfrak{h}_0^+$. Then $(s\lambda)(sH) = 0$, so $s\lambda \notin \Gamma \cup (-\Gamma)$.)

(g) Prove that $\hat{\mathfrak{g}}$ is semisimple. (Hint: Let $\hat{\mathfrak{q}} = \operatorname{rad}(\hat{\mathfrak{g}})$. By (b) $\hat{\mathfrak{w}}$ comes from the adjoint group of $\hat{\mathfrak{g}}$, so if $\hat{\mathfrak{q}} \cap (\sum_{\lambda \in \Delta} \hat{\mathfrak{g}}_\lambda) \neq 0$, there is some i such that $X_i, H_i, Y_i \in \hat{\mathfrak{q}}$. So $\hat{\mathfrak{q}} \subseteq \hat{\mathfrak{h}}$. But then all the α_j vanish on $\hat{\mathfrak{q}}$.)

(h) Prove that $\mathfrak{q}$ is the unique ideal of $\mathfrak{g}$ such that $\mathfrak{q} \cap \mathfrak{h} = 0$ and $\dim(\mathfrak{g}/\mathfrak{q}) < \infty$, and hence that it is the unique ideal described in Lemma 4.8.2.

7. Let $\mathfrak{g} = \mathfrak{sl}(2,\mathbf{C})$, $\mathfrak{G}$ the universal enveloping algebra of $\mathfrak{g}$, and H, X, Y as in (4.2.2). For $\lambda \in \mathbf{C}$ let $\bar{\pi}_\lambda$ be the natural representation of $\mathfrak{G}$ in $\mathfrak{G}/\mathfrak{G}_\lambda$, where $\mathfrak{G}_\lambda = \mathfrak{G}X + \mathfrak{G}(H - \lambda \cdot 1)$. Prove that $\bar{\pi}_\lambda$ is irreducible if and only if λ is not a nonnegative integer.

8. Let V be a real Hilbert space of finite dimension, V_c the complexification of V. We regard V_c as a complex Hilbert space in the natural fashion. Let $\Delta \subseteq V$ be an integral root system and $\mathfrak{w}$ the associated finite reflection group. Let D be the additive subgroup of V generated by Δ. For $\lambda \in V_c$, let $\mathfrak{w}(\lambda) = \{s : s \in \mathfrak{w}, \lambda - s\lambda \in D\}$ and $\Delta(\lambda) = \{\alpha : \alpha \in \Delta, 2(\lambda,\alpha)/(\alpha,\alpha) \in \mathbf{Z}\}$. Let P be a positive system $\subseteq \Delta$, and let Γ be the set of all finite sums of elements of P.

(a) Prove that $\Delta(\lambda)$ is an integral root system.

(b) Prove that $\mathfrak{w}(\lambda)$ is the finite reflection group generated by the s_α $(\alpha \in \Delta(\lambda))$.

(c) Let $\mathbf{Z}^+$ be the set of positive integers, and suppose that $2(\lambda,\alpha)/(\alpha,\alpha) \notin \mathbf{Z}^+$

for each $\alpha \in P$. If $\lambda - s\lambda \in D \setminus \{0\}$ for some $s \in \mathfrak{w}$, prove that $\lambda - s\lambda \in -\Gamma$. (Hint: $P(\lambda) = P \cap \Delta(\lambda)$ is a positive system for $\Delta(\lambda)$, $s \in \mathfrak{w}(\lambda)$, and $(\lambda,\alpha) \leq 0$ for all $\alpha \in P(\lambda)$. Now use Corollary 4.15.14.)

9. Let $\mathfrak{g}$ be a semisimple Lie algebra over **C**, and let other notation be as in §4.7. Fix $\lambda \in \mathfrak{h}^*$, and let $\mathfrak{G}_\lambda = \sum_{\alpha \in P} \mathfrak{G}\mathfrak{g}_\alpha + \sum_{H \in \mathfrak{h}} \mathfrak{G}(H - \lambda(H)1)$. Denote by $\bar{\pi}_\lambda$ the natural representation of $\mathfrak{G}$ in $V = \mathfrak{G}/\mathfrak{G}_\lambda$. Let $\Omega(\lambda)$ be the set of all $\mu \in \mathfrak{h}^*$ such that $\mu \prec \lambda$ and $\mu + \delta \in \mathfrak{w} \cdot (\lambda + \delta)$.

(a) Let U_1 and U_2 be two subspaces of V invariant under $\bar{\pi}_\lambda$ with $U_1 \subsetneqq U_2$. Let $W = U_2/U_1$, and let π_W be the representation induced in W. Prove that there exists $\mu \in \mathfrak{h}^*$ and a nonzero $u \in W_\mu$ such that $\pi_W[\mathfrak{g}_\alpha]u = 0$ for all $\alpha \in P$, and that any such μ belongs to $\Omega(\lambda)$.

(b) Take $U_1 = 0$, $U_2 = V$ in (a), let μ and u be as above, and put $V' = \bar{\pi}_\lambda[\mathfrak{G}]u$. Prove that the subrepresentation of $\bar{\pi}_\lambda$ defined by V' is equivalent to $\bar{\pi}_\mu$. (Hint: Observe that the restriction of $\bar{\pi}_\lambda$ to $\mathfrak{N}$ is equivalent to the left regular representation of $\mathfrak{N}$.)

(c) Prove that any strictly monotonic sequence of invariant subspaces of V is finite. Deduce that $\bar{\pi}_\lambda$ has a finite Jordan series and that the irreducible constituents of the series are of the form π_μ, where $\mu \in \Omega(\lambda)$.

(d) Suppose $(\lambda + \delta)(\bar{H}_\alpha) \in \mathbf{C} \setminus \mathbf{Z}^+$ for all $\alpha \in P$. Prove that $\bar{\pi}_\lambda$ is irreducible and hence equivalent to π_λ. (Hint: Use Exercise 8(c).)

(e) Suppose λ is such that for some simple root α, $\lambda(\bar{H}_\alpha)$ is an integer ≥ 0. Let $\mu = \lambda - (\lambda(\bar{H}_\alpha) + 1)\alpha$. Prove that $\bar{\pi}_\mu$ occurs as a subrepresentation of $\bar{\pi}_\lambda$. (Hint: Let $\alpha_1 = \alpha, \alpha_2, \ldots, \alpha_l$ be the simple roots. Let v be a nonzero vector of highest weight in V. Prove that $\bar{\pi}_\lambda(X_{\alpha_j})(\bar{\pi}_\lambda(X_{-\alpha_1})^{\lambda_1+1}v) = 0$ for $1 \leq j \leq l$, where $\lambda_1 = \lambda(\bar{H}_\alpha)$.)

(f) Suppose λ is dominant integral. Prove that for any $s \in \mathfrak{w}$, $\bar{\pi}_{s(\lambda+\delta)-\delta}$ occurs as a subrepresentation of $\bar{\pi}_\lambda$. (Hint: Let $s = s_{i_q}s_{i_{q-1}} \cdots s_{i_1}$ be a minimal expression for s, where s_{i_r} is the reflection corresponding to the simple root α_{i_r}. Then $\alpha_{i_1}, s_{i_1}\alpha_{i_2}, s_{i_1}s_{i_2}\alpha_{i_3}, \ldots, s_{i_1} \cdots s_{i_{q-1}}\alpha_{i_q}$ are all positive roots. So, writing $\bar{H}_i$ for $\bar{H}_{\alpha_i}$, $(\lambda + \delta)(\bar{H}_{i_1})$, $(s_{i_1}(\lambda + \delta))(H_{i_2})$, $(s_{i_1}s_{i_2}(\lambda + \delta))(H_{i_3}), \ldots$ are all > 0. Now use (e).)

10. We continue with the above setup. The partition function **P** is defined on $\mathfrak{h}^*$ in the following way: $\mathbf{P}(\mu) = 0$ if $\mu \notin \bar{\Gamma} = \Gamma \cup \{0\}$; if $\mu \in \bar{\Gamma}$, $\mathbf{P}(\mu)$ is the number of distinct functions $\alpha \mapsto k(\alpha)$ defined on P such that $k(\alpha)$ is a nonnegative integer for all $\alpha \in P$ and $\mu = \sum_{\alpha \in P} k(\alpha)\alpha$.

(a) If $\lambda \in \mathfrak{h}^*$, the multiplicity of the weight μ in $\bar{\pi}_\lambda$ is $\mathbf{P}(\lambda - \mu)$.

(b) Prove the existence of a polynomial function p on $\mathfrak{h}^*$ such that $|\mathbf{P}(\mu)| \leq |p(\mu)|$ for all μ.

(c) Let $c_1, \ldots, c_p$ be constants ≥ 0 and $\nu_1, \ldots, \nu_p$ fixed elements of Γ. Suppose $\mathbf{P}(\nu) \geq \sum_i c_i\mathbf{P}(\nu - \nu_i)$ for all $\nu \in \Gamma$. Then show that $\sum_i c_i \leq 1$. (Hint: Let $c = \sum_i c_i$, $\mu = \sum_i \nu_i$. Prove that $\mathbf{P}(n\mu) \geq c^n$ for $n = 1,2,\ldots$ and use (b).)

(d) Prove that each irreducible constituent of a Jordan series of the representation $\bar{\pi}_\lambda$ occurs with multiplicity 1. (Hint: If for some $\mu \in \Omega(\lambda)$, π_μ occurs more than once, we would have $\mathbf{P}(\lambda - \mu + \nu) \geq 2\mathbf{P}(\nu)$ for all $\nu \in \Gamma$. This would contradict (c).)

For these results see Verma [1] and Bernstein et al [1], where a deep study of the representations $\bar{\pi}_\lambda$ has been carried out.

The next exercise develops an algebraic proof of the Weyl character formula. Notation is as in §4.7.

11. Let $\mathfrak{F}$ be the complex vector space of all formal sums of exponentials $f = \sum_{\nu \in \mathfrak{h}^*} c_\nu e^\nu$ ($c_\nu \in \mathbf{C}$); we write $c_\nu = c_\nu(f)$ and denote by supp(f) the set of all ν such that $c_\nu(f) \neq 0$. We denote by $\mathcal{E}$ the subset of all $f \in \mathfrak{F}$ for which supp(f) has the following property: there exists a finite subset $\Phi = \Phi_f \subseteq \mathfrak{h}^*$ such that $\text{supp}(f) \subseteq \bigcup_{\mu \in \Phi} (\mu - \bar{\Gamma})$.

(a) Prove that $\mathcal{E}$ is a linear subspace of $\mathfrak{F}$.

(b) Given $f, g \in \mathcal{E}$, prove the existence of a unique element $h \in \mathcal{E}$ such that $c_\nu(h) = \sum_{\nu' + \nu'' = \nu} c_{\nu'}(f) c_{\nu''}(g)$ for all $\nu \in \mathfrak{h}^*$. Writing $h = f \cdot g$, prove that $\mathcal{E}$ becomes an associative and commutative algebra with unit under this definition of multiplication.

(c) Let $\Delta = e^\delta \prod_{\alpha \in P} (1 - e^{-\alpha})$. Prove that Δ is an invertible element of $\mathcal{E}$ and that $\Delta^{-1} = \sum_{\mu \in \bar{\Gamma}} \mathbf{P}(\mu) e^{-(\mu + \delta)}$.

(d) For any representation π of $\mathfrak{G}$ with weights such that all its weights are of finite multiplicity, we define the formal character $\theta(\pi)$ to be the element $\sum_\mu m(\pi : \mu) e^\mu$ of $\mathfrak{F}$ where $m(\pi : \mu)$ is the multiplicity of μ in π. If there is a finite subset $\Phi \subseteq \mathfrak{h}^*$ such that all the weights of π are of the form $\mu - \nu$ with $\mu \in \Phi$, $\nu \in \bar{\Gamma}$, prove that $\theta(\pi) \in \mathcal{E}$.

(e) For any $\lambda \in \mathfrak{h}^*$, prove that $\theta(\bar{\pi}_\lambda) = e^{\lambda + \delta} \cdot \Delta^{-1}$.

(f) Fix $\lambda \in \mathfrak{h}^*$ and let $\mathcal{E}(\lambda)$ be the linear span of all $\theta(\pi)$ where π is an irreducible representation which has a highest weight and whose infinitesimal character is the same as π_λ. Prove that $\{\theta(\bar{\pi}_\mu) : \mu + \delta \in \mathfrak{w} \cdot (\lambda + \delta)\}$ is a basis for $\mathcal{E}(\lambda)$. (Hint: Since $\bar{\pi}_\mu$ has a finite Jordan series, $\theta(\bar{\pi}_\mu) \in \mathcal{E}(\lambda)$ if $\mu + \delta \in \mathfrak{w} \cdot (\lambda + \delta)$. By (e), $\{\theta(\bar{\pi}_\mu) : \mu + \delta \in \mathfrak{w} \cdot (\lambda + \delta)\}$ is a linearly independent set. On the other hand, $\mathcal{E}(\lambda)$ is spanned by the $\theta(\pi_\mu)$ with $\mu + \delta \in \mathfrak{w} \cdot (\lambda + \delta)$ and so $\dim \mathcal{E}(\lambda) \leq [\mathfrak{w} \cdot (\lambda + \delta)]$.)

(g) Suppose λ is dominant integral. Prove that $\theta(\pi_\lambda) = \sum_{s \in \mathfrak{w}} \epsilon(s)^{s(\lambda + \delta)} \cdot \Delta^{-1}$. (Hint: By (f), $\theta(\pi_\lambda) = \sum_{s \in \mathfrak{w}} c_s e^{s(\lambda + \delta)} \cdot \Delta^{-1}$ for suitable constants c_s. Observe now that $\theta(\pi_\lambda)$ is a finite sum of exponentials invariant under $\mathfrak{w}$, and deduce that $c_s = \varepsilon(s)$.)

12. Obtain the following expressions for the Cartan–Killing forms of the classical Lie algebras:

$$\mathfrak{g} = \mathfrak{sl}(l + 1, \mathbf{C}) : \langle X, Y \rangle = 2(l + 1)\, tr(XY)$$

$$\mathfrak{g} = \mathfrak{o}(2l, \mathbf{C}) : \langle X, Y \rangle = 2(l - 1)\, tr(XY)$$

$$\mathfrak{g} = \mathfrak{o}(2l + 1, \mathbf{C}) : \langle X, Y \rangle = (2l - 1)\, tr(XY)$$

$$\mathfrak{g} = \mathfrak{sp}(l, \mathbf{C}) : \langle X, Y \rangle = 2(l + 1)\, tr(XY).$$

13. We use the notation of §4.7.

(a) Let $\lambda \in \mathfrak{D}_P$, and for any integral linear function ν on $\mathfrak{h}$ let $m_\lambda(\nu)$ be the multiplicity of the weight ν in π_λ. Prove that

$$m_\lambda(\nu) = \sum_{s \in \mathfrak{w}} \epsilon(s) \mathbf{P}(s(\lambda + \delta) - (\nu + \delta)).$$

(Hint: Observe that $\prod_{\alpha \in P}(1 - e^{-\alpha})^{-1} = \sum \mathbf{P}(\mu)e^{-\mu}$, and use Weyl's character formula.)

(b) Let $\lambda_1 \lambda_2 \in \mathfrak{D}_P$, and for $\Lambda \in \mathfrak{D}_P$, let $M(\Lambda)$ be the multiplicity of π_Λ in $\pi_{\lambda_1} \otimes \pi_{\lambda_2}$. Prove that

$$M(\Lambda) = \sum_{s,t \in \mathfrak{w}} \epsilon(s)\epsilon(t)\mathbf{P}(s(\lambda_1 + \delta) + t(\lambda_2 + \delta) - (\Lambda + 2\delta)).$$

(Hint: Observe that $\sum m_{\lambda_1}(\nu)e^\nu$. $\sum_{s \in \mathfrak{w}} \epsilon(s)e^{s(\lambda_2 + \delta)} = \sum_\Lambda M(\Lambda) \cdot \sum_{t \in \mathfrak{w}} \epsilon(t)e^{t(\Lambda+\delta)}$. Now use the result of (a), and identify coefficients on both sides.)

The formula in (a) is due to Kostant [2]; that in (b) to Steinberg (cf. Jacobson [1]).

14. Let $\mathfrak{g} = A_3$; $S = \{\alpha_1, \alpha_2\}$ a simple system of roots. For integers $m_1, m_2 \geq 0$ let π_{m_1, m_2} denote the irreducible representation π_λ where $\lambda(\bar{H}_{\alpha_i}) = m_i$ $(1 = 1,2)$. Prove the decomposition formula

$$\pi_{m,0} \otimes \pi_{0,m} \simeq \pi_{0,0} \oplus \pi_{1,1} \oplus \cdots \oplus \pi_{m,m}.$$

15. (a) Let $\mathfrak{g}$ be arbitrary semisimple, $\mathfrak{h}$ a CSA, and S a simple system of roots. Suppose $\lambda \in \mathfrak{h}^*$ is such that $\langle \lambda, \alpha \rangle \geq 0$ (resp. > 0) for all $\alpha \in S$. Then prove that $\lambda = \sum_{\alpha \in S} m(\alpha)\alpha$ where the $m(\alpha)$ are all ≥ 0 (resp. > 0). If λ is integral, prove that the $m(\alpha)$ are rational. (Hint: Let $S = \{\alpha_1, \ldots, \alpha_l\}$, $\lambda = \sum_{1 \leq k < r} m_k \alpha_k - \sum_{r \leq k \leq l} m_k \alpha_k = \mu - \nu$ say, where $r \geq 1$, $m_k \geq 0$, with $m_k > 0$ for $r \leq k \leq l$. Consider $\langle \nu, \nu \rangle = \langle \mu, \nu \rangle - \langle \lambda, \nu \rangle$.)

(b) Let A be the Cartan matrix $(\alpha_j(\bar{H}_{\alpha_i}))$. Show that all the entries of the matrix A^{-1} are ≥ 0.

(c) Let $\mathfrak{h}_\mathbf{R} = \sum_{1 \leq i \leq l} \mathbf{R} \cdot H_{\alpha_i}$ and let $\mathfrak{h}_\mathbf{R}^+$ be the set of all $H \in \mathfrak{h}_\mathbf{R}$ such that $\alpha_i(H) > 0$ for $1 \leq i \leq l$. Prove that $\langle H, H' \rangle \geq 0$ for all $H, H' \in Cl(\mathfrak{h}_\mathbf{R}^+)$.

16. Let $\mathfrak{g}$, $\mathfrak{h}$ and $S = \{\alpha_1, \ldots, \alpha_l\}$ be as above. Let Γ denote, as usual, the set of all $\mu \in \mathfrak{h}^* \setminus \{0\}$ of the form $m_1\alpha_1 + \cdots + m_l\alpha_l$ where the m_i are all integers ≥ 0. Prove that for any $\lambda \in \mathfrak{D}_P$, the irreducible representation π_λ has 0 as a weight if and only if $\lambda \in \Gamma \cup \{0\}$. (Hint: Let Δ_λ be the set of weights of π_λ. Let ν be a minimal element of $\Delta_\lambda \cap (\Gamma \cup \{0\})$ in the partial ordering $\prec$. Suppose $\nu = \sum_{1 \leq i \leq l} m_i \alpha_i \neq 0$. If $m_i > 0$, $\pi_\lambda[\mathfrak{g}^{-\alpha_i}]$ annihilates the weight space of weight ν, and so $\langle \nu, \alpha_i \rangle \leq 0$. Thus $\langle \nu, \nu \rangle \leq 0$.)

17. Let notation be as above. $\mathfrak{w}$ is the Weyl group and Δ the set of roots.

(a) Suppose $\mathfrak{h}$ is the direct sum of subspaces $\mathfrak{h}_1, \ldots, \mathfrak{h}_r$ which are $\mathfrak{w}$-invariant. Let $\Delta_j = \{\alpha : \alpha \in \Delta, H_\alpha \in \mathfrak{h}_j\}$. Prove that $\Delta = \bigcup_{1 \leq j \leq r} \Delta_j$ and that $[\mathfrak{g}_\alpha, \mathfrak{g}_\beta] = 0$ whenever $\alpha \in \Delta_j$, $\beta \in \Delta_k$, and $j \neq k$. Deduce that $\mathfrak{g}_j = \mathfrak{h}_j + \sum_{\alpha \in \Delta_j} \mathfrak{g}_\alpha$ are ideals of $\mathfrak{g}$ and $\mathfrak{g}$ is their direct sum. (Hint: If $\alpha \in \Delta$, s_α leaves $\mathfrak{h}_j$ invariant and so $\mathfrak{h}_j$ is the direct sum of $\mathfrak{h}_j \cap (\mathbf{C} \cdot H_\alpha)$ and $\mathfrak{h}_j \cap \sigma_\alpha$ where σ_α is the nullspace of α. Deduce from this that $\alpha \in \bigcup_{1 \leq j \leq r} \Delta_j$, and hence that the $\mathfrak{h}_j$ are mutually orthogonal. Use the relation $\Delta = \bigcup_{1 \leq j \leq r} \Delta_j$ and the mutual orthogonality of the $\mathfrak{h}_j$ to prove that if $\alpha \in \Delta_j$, $\beta \in \Delta_k$, and $j \neq k$, then $\alpha \pm \beta$ are not roots.)

(b) Prove that $\mathfrak{g}$ is simple if and only if $\mathfrak{w}$ acts irreducibly on $\mathfrak{h}$, and that this is equivalent to the condition that $\mathfrak{w}$ acts irreducibly on $\mathfrak{h}_\mathbf{R}$.

(c) Suppose $\mathfrak{g}$ is simple. If α and β are two roots, then, in order that there should exist an $s \in \mathfrak{w}$ such that $s\alpha = \beta$ it is necessary and sufficient that $\langle\alpha,\alpha\rangle = \langle\beta,\beta\rangle$. (Hint: Assume $\langle\alpha,\alpha\rangle = \langle\beta,\beta\rangle$. By (b), $\langle t\alpha,\beta\rangle \neq 0$ for some $t \in \mathfrak{w}$. So one may assume $\langle\alpha,\beta\rangle < 0$. Then $2\langle\alpha,\beta\rangle/\langle\alpha,\alpha\rangle = 2\langle\alpha,\beta\rangle/\langle\beta,\beta\rangle = -1$ and $s_\beta s_\alpha(\beta) = \alpha$.)

(d) Let n be the number of $\mathfrak{w}$-orbits in Δ. Prove that $n = 1$ if $\mathfrak{g} = A_l$ $(l \geq 1)$, D_l $(l \geq 3)$, E_l $(l = 6,7,8)$ while $n = 2$ if $\mathfrak{g} = B_l$ $(l \geq 2)$, C_l $(l \geq 3)$, G_2, F_4.

18. Let $\mathfrak{g}$, $\mathfrak{h}$, P be as in §4.7. Let $\mathfrak{G}$ be the universal enveloping algebra of $\mathfrak{g}$. Let $S = \{\alpha_1, \ldots, \alpha_l\}$ be the simple system of roots in P, $H_i = \bar{H}_{\alpha_i}$ $(1 \leq i \leq l)$. Let $\mathfrak{N}$ be the subalgebra of $\mathfrak{G}$ generated by 1 and the $\mathfrak{g}_{-\alpha_i}$, $1 \leq i \leq l$. Prove that for any $\lambda \in \mathfrak{D}_P$

$$\mathfrak{N} \cap \mathfrak{M}_\lambda = \sum_{1 \leq i \leq l} \mathfrak{N} Y_i^{\lambda_i + 1};$$

here $\lambda_i = \lambda(H_i)$, Y_i is a nonzero element of $\mathfrak{g}_{-\alpha_i}$, and $\mathfrak{M}_\lambda$ is the maximal left ideal in $\mathfrak{G}$ corresponding to π_λ. (Hint: Prove that $\sum_{1 \leq i \leq l} \mathfrak{G}(H_i - \lambda_i 1) + \sum_{1 \leq i \leq l} \mathfrak{G} X_i + \sum_{1 \leq i \leq l} \mathfrak{N} Y_i^{\lambda_i+1}$ $(0 \neq X_i \in \mathfrak{g}_{\alpha_i})$ is a left ideal of $\mathfrak{G}$ and use (4.7.2); cf. Parthasarathy, Ranga Rao, and Varadarajan [1]).

19. We continue in the above context. Let $s_0 \in \mathfrak{w}$ be such that $s_0 \cdot P = -P$. For $\lambda \in \mathfrak{h}^*$ let us write $\lambda^* = -s_0\lambda$. Let X_i, Y_i $(1 \leq i \leq l)$ be as in the previous exercise. Given $\mu, \nu \in \mathfrak{D}_P$, we define, for any integral linear function γ on $\mathfrak{h}$, $V^+(\mu : \gamma : \nu) = \{v : v \text{ in the space of } \pi_\mu,\ v \text{ of weight } \gamma, \text{ and } \pi_\mu(X_i)^{\nu_i+1}v = 0 \text{ for } 1 \leq i \leq l\}$; $V^-(\mu : \gamma : \nu)$ is defined analogously, with Y_i replacing X_i $(\nu_i = \nu(H_i)$ as usual).

(a) Prove that for $\mu, \nu \in \mathfrak{D}_P$ and γ integral, $\dim V^+(\mu : \gamma : \nu) = \dim V^-(\mu : -\gamma^* : \nu^*)$. (Hint: If x is an element of the simply connected group corresponding to $\mathfrak{g}$ inducing s_0, show that $\pi_\mu(x) \cdot V^+(\mu : \gamma : \nu) = V^-(\mu : -\gamma^* : \nu^*)$.)

(b) Let $\lambda_1, \lambda_2 \in \mathfrak{D}_P$, W_1, W_2 the spaces on which $\pi_{\lambda_1^*}$ and π_{λ_2} act, V the vector space of linear maps of W_1 into W_2. For $X \in \mathfrak{g}$ let $\pi(X)$ be the endomorphism of V defined by $\pi(X)L = \pi_{\lambda_2}(X)L - L\pi_{\lambda_1^*}(X)$, $L \in V$. Prove that π is a representation of $\mathfrak{g}$ equivalent to $\pi_{\lambda_1} \otimes \pi_{\lambda_2}$. (Hint: Note that $\pi_{\lambda_1^*}$ and π_{λ_1} are mutually contragredient.)

(c) With notation as in (b), let $U = \{L : L \in V, \pi(Y_i)L = 0, 1 \leq i \leq l\}$. Let w be a nonzero vector of weight λ_1^* in W_1. Prove that the map $\xi : L \mapsto Lw$ is a linear isomorphism of U onto the subspace U' of W_2 given by $U' = \{v : v \in W_2,\ \pi_{\lambda_2}(Y_i)^{\lambda_{1i}^*+1}v = 0,\ 1 \leq i \leq l\}$, where $\lambda_{1i}^* = \lambda_1^*(H_i)$. (Hint: For the injectivity observe that $\pi_{\lambda_1^*}[\mathfrak{N}]w = W_1$. If $v \in U'$ use Exercise 18 to prove that the map $\pi_{\lambda_1^*}(y)w \mapsto \pi_{\lambda_2}(y)v$, $y \in \mathfrak{N}$ is well defined and belongs to U. Deduce surjectivity.)

(d) Deduce from (c) that for $\gamma \in \mathfrak{D}_P$, the multiplicity of π_γ in $\pi_{\lambda_1} \otimes \pi_{\lambda_2}$ is precisely $\dim V^-(\lambda_2 : -\gamma^* + \lambda_1^* : \lambda_1^*)$. (Hint: U is invariant under $\pi[\mathfrak{h}]$ and is the direct sum of weight subspaces $U_1, \ldots, U_k$, U_i being of weight $-\gamma_i^*$ for $\gamma_i \in \mathfrak{D}_P$. Show that $\pi \simeq \sum_i \dim(U_i)\pi_{\gamma_i}$. Use result of (c) to show that ξ is an isomorphism of U_i onto $V^-(\lambda_2 : -\gamma_i^* + \lambda_1^* : \lambda_1^*)$.)

(e) Show that the multiplicity of π_γ in $\pi_{\lambda_1} \otimes \pi_{\lambda_2}$ is also equal to the dimension of $V^-(\gamma : \lambda_1 - \lambda_2^* : \lambda_1)$ as well as to the dimension of $V^+(\gamma : \lambda_2 - \lambda_1^* : \lambda_1^*)$. (Hint: This multiplicity is also the multiplicity of π_{λ_2} in $\pi_{\lambda_1^*} \otimes \pi_\gamma$. Now use (d) and (a).)

(f) Let ν be the unique element of $\mathfrak{D}_P$ in $\mathfrak{w} \cdot (\lambda_1 - \lambda_2^*)$. Prove that π_ν occurs in $\pi_{\lambda_1} \otimes \pi_{\lambda_2}$ with multiplicity 1. Prove also that ν is a weight of every π_γ ($\gamma \in \mathfrak{D}_P$) that enters the decomposition of $\pi_{\lambda_1} \otimes \pi_{\lambda_2}$.

(g) For fixed $\gamma \in \mathfrak{D}_P$, prove that the multiplicity of π_γ in $\pi_{\lambda_1} \otimes \pi_{\lambda_2}$ is the multiplicity of the weight $\lambda_1 - \lambda_2^*$ in π_γ whenever $\lambda_1(H_i) \geq \dim(\pi_\gamma) - 1$ for $1 \leq i \leq l$, in particular if all the $\lambda_1(H_i)$ are sufficiently large.

(h) Prove that the number of irreducible constituents of $\pi_{\lambda_1} \otimes \pi_{\lambda_2}$ cannot exceed $\min(\dim(\pi_{\lambda_1}), \dim(\pi_{\lambda_2}))$.

In this connection, see Kostant [2], Parthasarathy et al [1].

20. Let notation be as in Exercises 18 and 19.

(a) Let τ be the adjoint representation of $\mathfrak{g}$ in $\mathfrak{G}$. Let $\mu \in \mathfrak{D}_P$; let π_μ be the corresponding irreducible representation of $\mathfrak{G}$ acting on a vector space V^μ; and let E^μ be the algebra of endomorphisms of V^μ. For $X \in \mathfrak{g}$, let $\tau^\mu(X)v = [\pi_\mu(X), v]$ ($v \in E^\mu$). Show that the map $a \mapsto \pi_\mu(a)$ ($a \in \mathfrak{G}$) intertwines the representations τ and τ^μ of $\mathfrak{g}$.

(b) Let $\lambda \in \mathfrak{D}_P$ be such that π_λ has 0 as a weight; write d_λ for the multiplicity of 0 in π_λ. Prove that for suitable μ, π_λ occurs as an irreducible constituent of τ^μ with multiplicity d_λ. (Hint: Use Exercise 19(g).)

(c) Let $S(\mathfrak{g})$ be the symmetric algebra over $\mathfrak{g}$. For each $X \in \mathfrak{g}$ let $\sigma(X)$ denote the derivation of $S(\mathfrak{g})$ that extends ad X. Prove, using (a) and (b) that π_λ ($\lambda \in \mathfrak{D}_P$) occurs as an irreducible constituent of σ if and only if 0 is weight of π_λ, i.e., if and only if $\lambda \in \Gamma \cup \{0\}$. (Hint: Consider the map $\boldsymbol{\lambda} : S(\mathfrak{g}) \longrightarrow \mathfrak{G}$ intertwining σ and τ).

21. Let $\mathfrak{g}$ be one of A_l ($l \geq 1$), B_l ($l \geq 1$), C_l ($l \geq 2$), D_l ($l \geq 2$). Let notation be as in 4.4. Define the $\mu_i \in \mathfrak{D}_P$ by $\mu_i(\bar{H}_{\alpha_j}) = \delta_{ij}$, $1 \leq i, j \leq l$. Write $d_i = \dim(\pi_{\mu_i})$, $1 \leq i \leq l$.

(a) Let $\mathfrak{g} = A_l$; then $d_i = \binom{l+1}{i}$.

(b) Let $\mathfrak{g} = B_l$; then $d_i = \binom{2l+1}{i}$ for $1 \leq i \leq l-1$, $d_l = 2^l$. Verify also that $\dim(\pi_{2\mu_l}) = \binom{2l+1}{l}$.

(c) Let $\mathfrak{g} = C_l$; then $d_i = \binom{2l}{i} - \binom{2l}{i-2}$ for $1 \leq i \leq l$. (For $r < 0$, $\binom{n}{r}$ is defined to be 0).

(d) Let $\mathfrak{g} = D_l$; then $d_i = \binom{2l}{i}$ for $1 \leq i \leq l-2$, $d_{l-1} = d_l = 2^{l-1}$. Verify also that the linear forms $\lambda_1 + \cdots + \lambda_{l-1}$, $\lambda_1 + \cdots + \lambda_l$, $\lambda_1 + \cdots + \lambda_{l-1} - \lambda_l$ are in $\mathfrak{D}_P$ and that the dimensions of the corresponding representations are respectively $\binom{2l}{l-1}$, $\frac{1}{2}\binom{2l}{l}$ and $\frac{1}{2}\binom{2l}{l}$.

22. Let notation be as in Exercise 21. Let s_0 be the element of the Weyl group that sends positive roots to negative ones.

(a) If $\mathfrak{g} = A_l$, $s_0\lambda_i = \lambda_{l+2-i}$ $(1 \leq i \leq l+1)$
(b) If $\mathfrak{g} = B_l$ or C_l, $s_0 = -$identity.
(c) If $\mathfrak{g} = D_l$, l even, then $s_0 = -$identity; if l is odd, $s_0\lambda_j = -\lambda_j$ for $1 \leq j \leq l-1$ and $s_0\lambda_l = \lambda_l$.
(d) Deduce that all representations of B_l and C_l are self-contragredient.

23. Let $\mathfrak{g} = B_l$ $(l \geq 1)$ or D_l $(l \geq 2)$; other notation as in 4.4. Let $\bar{\pi}_1$ be the representation of $\mathfrak{g}$ in V; $\bar{\pi}_k$, the representation induced by $\bar{\pi}_1$ in E_k, the subspace of elements of degree k of the exterior algebra $E = E(V)$ over V.
(a) Let $\mathfrak{g} = B_l$. Then the $\bar{\pi}_k$ $(0 \leq k \leq 2l+1)$ are all irreducible, $\bar{\pi}_k \simeq \bar{\pi}_{2l+1-k}$, $\bar{\pi}_0$ is the trivial representation, $\bar{\pi}_i \simeq \pi_{\mu_i}$ for $1 \leq i \leq l-1$, and $\bar{\pi}_l \simeq \pi_{2\mu_l}$.
(b) Let $\mathfrak{g} = D_l$. Then $\bar{\pi}_k \simeq \pi_{2l-k}$ $(0 \leq k \leq 2l)$, and $\bar{\pi}_0$ is the trivial representation; for $1 \leq k \leq l-1$, $\bar{\pi}_k$ is irreducible, while $\bar{\pi}_l$ splits as a direct sum of two inequivalent irreducible representations; $\bar{\pi}_k \simeq \pi_{\mu_k}$ $(1 \leq k \leq l-2)$, $\bar{\pi}_{l-1} \simeq \pi_{\mu_{l-1}+\mu_l}$, and $\bar{\pi}_l$ is the direct sum of $\pi_{2\mu_{l-1}}$ and $\pi_{2\mu_l}$. (For the last equivalence, observe that $\lambda_1 + \cdots + \lambda_{l-1} \pm \lambda_l$ are weights of $\bar{\pi}_l$, but $(\lambda_1 + \cdots + \lambda_{l-1} \pm \lambda_l) + \alpha_i$ is not a weight of $\bar{\pi}_l$ for $1 \leq i \leq l$; for the rest, use Exercises 21 and 22.)

24. Let $\mathfrak{g} = C_l$ $(l \geq 2)$, $\bar{\pi}_1$ the representation of $\mathfrak{g}$ in V, $\bar{\pi}_k$ the corresponding representation in the subspace E_k of the exterior algebra E over V; these and other notation as in §4.4.
(a) Prove that $\bar{\pi}_k \simeq \bar{\pi}_{2l-k}$ $(0 \leq k \leq 2l)$ and that $\bar{\pi}_{2l}$ is the trivial representation.
(b) Let $E_{k,0}$ $(1 \leq k \leq l)$ be the smallest $\bar{\pi}_k$-invariant subspace of E_k containing $u_1 \wedge \cdots \wedge u_k$, and let $\bar{\pi}_{k,0}$ be the representation defined by $E_{k,0}$. Prove that $\bar{\pi}_{1,0} = \bar{\pi}_1 \simeq \pi_{\mu_1}$ and $\bar{\pi}_{k,0} \simeq \pi_{\mu_k}$ $(1 \leq k \leq l)$.
(c) Let $\varphi = u_1 \wedge u_{l+1} + u_2 \wedge u_{l+2} + \cdots + u_l \wedge u_{2l}$. Prove that $\mathbf{C}\cdot\varphi$ is invariant under π_2 and defines the trivial representation of $\mathfrak{g}$, and that E_2 is the direct sum of $E_{2,0}$ and $\mathbf{C}\cdot\varphi$.
(d) Let $2 \leq k \leq l$, and let $E_{k,s} = \underbrace{\varphi \wedge \varphi \wedge \cdots \wedge \varphi}_{s \text{ factors}} \wedge E_{k-2s,0}$, $1 \leq s \leq \frac{1}{2}k$.
Prove that the $E_{k,s}$ $(0 \leq s \leq \frac{1}{2}k)$ are all nonzero and linearly independent and span E_k. Deduce that $\bar{\pi}_k$ is equivalent to the direct sum of π_{μ_i} $(i = k, k-2, \ldots,$ the sequence continuing as long as $i \geq 0$; $\mu_0 = 0)$. (Use Exercise 21.)

25. Suppose G is a semisimple real analytic group whose Lie algebra is simple. Prove that if G is not compact, G has no nontrivial (finite-dimensional) unitary representation.

26. (a) Let G be a real analytic group with Lie algebra $\mathfrak{g}$, $H \subseteq G$ a closed subgroup, and $\mathfrak{h} \subseteq \mathfrak{g}$ the corresponding subalgebra. Let $\mathfrak{g}_c$ be the complexification of $\mathfrak{g}$, $X \mapsto X^{\text{conj}}$ the associated conjugation of $\mathfrak{g}_c$, and $\mathfrak{h}_c = \mathbf{C}\cdot\mathfrak{h}$. Prove that G/H has a G-invariant complex structure if and only if there are subalgebras $\mathfrak{p}^+$ and $\mathfrak{p}^-$ of $\mathfrak{g}_c$ with the following properties: (i) $\mathfrak{p}^+ \cap \mathfrak{p}^- = \mathfrak{h}_c$, $\mathfrak{p}^+ + \mathfrak{p}^- = \mathfrak{g}_c$, (ii) $(\mathfrak{p}^+)^{\text{conj}} = \mathfrak{p}^-$, and (iii) $\mathfrak{p}^+$ and $\mathfrak{p}^-$ invariant under $\text{Ad}(H)$.

(b) Let G be a compact semisimple analytic group, B a maximal torus. Prove that G/B has a unique G-invariant complex structure.

(c) Let $G = SL(2,\mathbf{R})$, $H = SO(2,\mathbf{R})$. Determine all the G-invariant complex structures on G/H.

Exercises 27–30 lead (among other things) to the proofs of existence of the exceptional simple Lie algebras. In these, $A = (a_{ij})_{1 \leq i, j \leq l}$ is one of the matrices corresponding to the Dynkin diagrams G_2, F_4, E_p ($p = 6,7,8$) (cf. (4.5.7)–(4.5.9)); V is a vector space over $\mathbf{C}$ with basis $\alpha_1, \ldots, \alpha_l$; s_i ($1 \leq i \leq l$) are the reflexions given by $s_i\alpha_j = \alpha_j - a_{ij}\alpha_i$ ($1 \leq i, j \leq l$); $\mathfrak{w}(A)$ is the subgroup of $GL(V)$ generated by the s_i; $\Delta = \bigcup_{1 \leq i \leq l} \mathfrak{w}(A)\cdot\alpha_i$; μ_i are the basic dominant integral linear forms.

27. Let $l = 2$, $A = A(G_2)$.

(a) Write $\lambda_1 = \alpha_1$, $\lambda_2 = \alpha_1 + \alpha_2$, $\lambda_0 = -(\lambda_1 + \lambda_2)$. Verify that $s_1\lambda_1 = -\lambda_1$, $s_1\lambda_2 = -\lambda_0$, $s_2\lambda_1 = \lambda_2$, $s_2\lambda_2 = \lambda_1$, and deduce that $\mathfrak{w}(A)$ leaves the set $\{\pm\lambda_i\ (i = 0, 1, 2),\ \pm(\lambda_i - \lambda_j)\ (0 \leq i < j \leq 2)\}$ invariant.

(b) Use (a) to show that $\mathfrak{w}(A)$ is finite, and deduce the existence of a simple Lie algebra (also denoted by G_2) whose Cartan matrices are equivalent to $A(G_2)$.

(c) Prove that Δ is the set $\{\pm\lambda_i\ (i = 0, 1, 2),\ \pm(\lambda_i - \lambda_j)\ (0 \leq i < j \leq 2)\}$, and hence verify that $\dim(G_2) = 14$.

(d) Let Γ_1 be the group of permutations of $\{0,1,2\}$ acting naturally on $\mathbf{C}^3 = \{(x_0,x_1,x_2)\}$, and let Γ be the group generated by Γ_1 and $-$identity. Prove that Γ leaves the plane $x_0 + x_1 + x_2 = 0$ invariant and that there is an isomorphism of this plane with V that transforms the action of Γ into that of $\mathfrak{w}(A)$. Deduce that $[\mathfrak{w}(G_2)] = 12$.

(e) Prove that $\mu_1 = 2\alpha_1 + \alpha_2$ and $\mu_2 = 3\alpha_1 + 2\alpha_2$, that $\delta = 5\alpha_1 + 3\alpha_2$, and that μ_2 is the highest root.

(f) Verify that $\dim(\pi_{\mu_1}) = 7$, $\dim(\pi_{\mu_2}) = 14$, and that π_{μ_2} is the adjoint representation.

(g) Prove that every representation of G_2 is self-contragredient and contains 0 as a weight.

28. Let $l = 4$, $A = A(F_4)$.

(a) Write $\lambda_1 = 2\alpha_1 + 3\alpha_2 + 2\alpha_3 + \alpha_4$, $\lambda_2 = \alpha_2$, $\lambda_3 = \alpha_2 + \alpha_3$, $\lambda_4 = \alpha_2 + \alpha_3 + \alpha_4$. Verify that s_2, s_3, s_4 fix λ_1 and that $s_1\lambda_i = \frac{1}{2}(\lambda_1 + \epsilon_{i2}\lambda_2 + \epsilon_{i3}\lambda_3 + \epsilon_{i4}\lambda_4)$, where $\epsilon_{ij} = +1$ or -1 according as $j = i$ or $j \neq i$ ($i = 2, 3, 4$), and $\epsilon_{ij} = +1$ for all j, if $i = 1$. Determine the action of s_2, s_3, and s_4 on λ_2, λ_3, and λ_4. Prove that $\mathfrak{w}(A)$ leaves invariant the set $\{\pm\lambda_i, \pm\lambda_i \pm \lambda_j, \frac{1}{2}(\pm\lambda_1 \pm \lambda_2 \pm \lambda_3 \pm \lambda_4)\}$ ($i < j$, $i, j = 1, 2, 3, 4$).

(b) Use (a) to prove that there is a simple Lie algebra (denoted again by F_4) for which $A(F_4)$ is a Cartan matrix.

(c) Prove that Δ is the set described in (a) and $\dim(F_4) = 52$.

(d) Verify that

$$\mu_1 = 2\alpha_1 + 3\alpha_2 + 2\alpha_3 + \alpha_4,$$

$$\mu_2 = 3\alpha_1 + 6\alpha_2 + 4\alpha_3 + 2\alpha_4,$$

$$\mu_3 = 4\alpha_1 + 8\alpha_2 + 6\alpha_3 + 3\alpha_4,$$

$$\mu_4 = 2\alpha_1 + 4\alpha_2 + 3\alpha_3 + 2\alpha_4,$$
$$\delta = 11\alpha_1 + 21\alpha_2 + 15\alpha_3 + 8\alpha_4,$$

and that μ_4 is the highest root of F_4.

(e) Verify that the dimensions of the representations π_{μ_i} $(i \leq i \leq 4)$ are 26, 273, 1274, 52 respectively and that π_{μ_4} is the adjoint representation.

(f) Show that every representation of F_4 is self-contragredient and contains 0 as a weight.

(g) Show that $[\mathfrak{w}(F_4)] = 2^7 \cdot 3^2 = 1152$. (Hint: First check that the orbit of the root μ_4 has 24 elements. The stabilizer of μ_4 is isomorphic to $\mathfrak{w}(C_3)$ and so has $2^3 \cdot 3! = 48$ elements.)

29. Let $l = 8$, $A = A(E_8)$.

(a) Define the elements λ_j $(1 \leq i \leq 8)$ of V as follows:

$$3\lambda_i = 3(\alpha_i + \alpha_{i+1} + \cdots + \alpha_5) + 2\alpha_6 + \alpha_7 + \alpha_8 \quad (1 \leq i \leq 5),$$
$$3\lambda_6 = 2\alpha_6 + \alpha_7 + \alpha_8,$$
$$3\lambda_7 = -\alpha_6 + \alpha_7 + \alpha_8,$$
$$3\lambda_8 = -\alpha_6 - 2\alpha_7 + \alpha_8.$$

Verify that the λ_i form a basis for V, and that s_i permutes λ_i and λ_{i+1} leaving the others fixed $(1 \leq i \leq 7)$. Also verify that

$$s_8\lambda_i = \lambda_i + \frac{1}{3}(\lambda_6 + \lambda_7 + \lambda_8) \quad (1 \leq i \leq 5),$$
$$s_8\lambda_6 = \frac{1}{3}(\lambda_6 - 2\lambda_7 - 2\lambda_8)$$
$$s_8\lambda_7 = \frac{1}{3}(-2\lambda_6 + \lambda_7 - 2\lambda_8),$$
$$s_8\lambda_8 = \frac{1}{3}(-2\lambda_6 - 2\lambda_7 + \lambda_8).$$

(b) Use the results of (a) to show that $\mathfrak{w}(A)$ leaves the set

$$\{\pm(\lambda_i - \lambda_j), \pm(\lambda_i + \lambda_j + \lambda_k), \pm(\lambda_i + \lambda_j + \lambda_k + \lambda_p + \lambda_q + \lambda_r), \pm(\lambda_j + \lambda_1 + \cdots + \lambda_8)\}$$

invariant ($i < j < k < p < q < r$ are from $(1, \ldots, 8)$). Deduce the existence of a simple Lie algebra (denoted by E_8) having $A = A(E_8)$ as a Cartan matrix.

(c) Prove that Δ is the set described in (b). Deduce that $\dim(E_8) = 248$.

(d) Let $s_0 \in \mathfrak{w}$ be such that $s_0 \cdot P = -P$. Prove that $s_0 = -id$. Deduce that all representations of E_8 are self-contragredient.

(e) Show that $\det(A(E_8)) = 1$. Deduce that the additive group of integral linear forms is already generated by the roots. Hence conclude that all representations of E_8 contain 0 as a weight.

(f) Verify that $\mu_1 = 2\alpha_1 + 3\alpha_2 + 4\alpha_3 + 5\alpha_4 + 6\alpha_5 + 4\alpha_6 + 2\alpha_7 + 3\alpha_8$ is the highest root.

30. (a) Let $\mathfrak{g}$ be a semisimple Lie algebra over $\mathbf{C}$, $\mathfrak{h}$ a CSA, $S = \{\bar{\alpha}_1, \ldots, \bar{\alpha}_l\}$ a simple system of roots, and F a subset of $\{1, \ldots, l\}$. Let Δ_F be the set of all those roots of $(\mathfrak{g},\mathfrak{h})$ which are linear combinations of the α_i with $i \in F$. Prove that $\mathfrak{g}_F = \sum_{i \in F} \mathbf{C} \cdot \bar{H}_{\alpha_i} + \sum_{\alpha \in \Delta_F} \mathfrak{g}_\alpha$ is a semisimple Lie algebra, with CSA $\mathfrak{h}_F = \sum_{i \in F} \mathbf{C} \cdot \bar{H}_{\alpha_i}$, and Cartan matrix $(\bar{a}_{ij})_{i,j \in F}$, $(\bar{a}_{ij})_{1 \le i,j \le l}$ being the Cartan matrix of $(\mathfrak{g},\mathfrak{h})$.

(b) Use (a) to prove the existence of simple Lie algebras E_6 and E_7 with respective Cartan matrices $A(E_6)$ and $A(E_7)$.

(c) Show that $\dim(E_6) = 78$ and $\dim(E_7) = 133$. (Hint: Use (a) above and Exercise 29(b).)

(d) For both E_6 and E_7 show that $-id$ is the element of the Weyl group that sends positive roots to negative ones. Deduce that all representations of E_6 and E_7 are self-contragredient.

(e) Show that $\det(A(E_6)) = 3$ and $\det(A(E_7)) = 2$.

(f) Let V_6 be spanned by α_i $(1 \le i \le 6)$. Define λ_i $(1 \le i \le 6)$ by $3\lambda_i = 3(\alpha_i + \cdots + \alpha_3) + 2\alpha_4 + \alpha_5 + \alpha_6$ $(1 \le i \le 3)$, $3\lambda_4 = 2\alpha_4 + \alpha_5 + \alpha_6$, $3\lambda_5 = -\alpha_4 + \alpha_5$, $3\lambda_6 = -\alpha_4 - 2\alpha_5 + \alpha_6$. Show that the roots of E_6 are

$$\pm(\lambda_i - \lambda_j) \pm (\lambda_i + \lambda_j + \lambda_k), \qquad \pm(\lambda_1 + \cdots + \lambda_6)$$
$$(1 \le i < j < k \le 6).$$

Deduce that $\mu_6 = \alpha_1 + 2\alpha_2 + 3\alpha_3 + 2\alpha_4 + \alpha_5 + 2\alpha_6$ is the highest root. (Hint: Use (a) above and Exercise 29(b).)

(g) Let V_7 be spanned by α_i $(1 \le i \le 7)$. Define λ_i $(1 \le i \le 7)$ by $3\lambda_i = 3(\alpha_i + \cdots + \alpha_4) + 2\alpha_5 + \alpha_6 + \alpha_7$ $(1 \le i \le 4)$, $3\lambda_5 = 2\alpha_5 + \alpha_6 + \alpha_7$, $3\lambda_6 = -\alpha_5 + \alpha_6 + \alpha_7$, $3\lambda_7 = -\alpha_5 - 2\alpha_6 + \alpha_7$. Show that the roots of E_7 are

$$\pm(\lambda_i - \lambda_j), \qquad \pm(\lambda_i + \lambda_j + \lambda_k), \qquad \pm(\lambda_i + \lambda_j + \lambda_k + \lambda_p + \lambda_q + \lambda_r),$$

where $1 \le i < j < k < p < q < r \le 7$. Deduce that $\mu_6 = \alpha_1 + 2\alpha_2 + 3\alpha_3 + 4\alpha_4 + 3\alpha_5 + 2\alpha_6 + 2\alpha_7$ is the highest root.

(h) Show that $\mathfrak{w}(E_6) = 2^7 \cdot 3^4 \cdot 5$, $[\mathfrak{w}(E_7)] = 2^{10} \cdot 3^4 \cdot 5 \cdot 7$, and $[\mathfrak{w}(E_8)] = 2^{14} \cdot 3^5 \cdot 5^2 \cdot 7$. (Hint: By (f), the $\mathfrak{w}(E_6)$-orbit of μ_6 has 72 elements; the stabilizer of μ_6 in $\mathfrak{w}(E_6)$ is isomorphic to $\mathfrak{w}(A_5)$. By (g), the $\mathfrak{w}(E_7)$-orbit of μ_6 has 126 elements, while its stabilizer is $\approx \mathfrak{w}(D_6)$. Finally, by (f) of Exercise 29, the $\mathfrak{w}(E_8)$-orbit of μ_1 has 240 elements, and its stabilizer is $\approx \mathfrak{w}(E_7)$.)

For explicit realizations of the exceptional Lie algebras, see Jacobson [1].

31. Let $G = U(n,\mathbf{C})$, B the subgroup of all diagonal matrices of G. We identify B with $\mathbf{T}^n$ via the map $\mathrm{diag}(a_1, \ldots, a_n) \mapsto (a_1, \ldots, a_n)$. If $h_1 \ge \cdots \ge h_n \ge 0$ are integers, $[h_1, \ldots, h_n]$ denotes the function $(a_1, \ldots, a_n) \mapsto \det((a_i^{h_j})_{1 \le i,j \le n})$ on B.

(a) Let $h_1 \geq \cdots \geq h_n \geq 0$ be integers. Prove that $[n-1, \ldots, 0] = \prod_{1 \leq i < j \leq n} (a_i - a_j)$ and that $[h_1, \ldots, h_n]$ is divisible by $[n-1, \ldots, 0]$ in the ring of polynomials in $a_1, \ldots, a_n$. Given integers $f_1 \geq \cdots \geq f_n \geq 0$, let $h_i = f_i + n - i$ $(1 \leq i \leq n)$, and let $\langle f_1, \ldots, f_n \rangle$ be the quotient $[h_1, \ldots, h_n]/[n-1, \ldots, 0]$. Prove that $\langle f_1, \ldots, f_n \rangle$ is a homogeneous polynomial in $a_1, \ldots, a_n$ of degree $f_1 + \cdots + f_n$ with the following properties: (i) $a_1^{f_1} \cdots a_n^{f_n}$ occurs in it with coefficient 1, and (ii) if $a_1^{g_1} \cdots a_n^{g_n}$ occurs with a nonzero coefficient, then $(g_1, \ldots, g_n) \leqq (f_1, \ldots, f_n)$ in the usual lexicographic ordering on $\mathbf{R}^n$ ($(x_1, \ldots, x_n) < (y_1, \ldots, y_n)$ if for some i, $1 \leq i \leq n$, $x_i < y_i$ and $x_j = y_j$ for $j < i$).

(b) Prove that there is a unique irreducible character $\chi(f_1, \ldots, f_n)$ of G such that $\chi(f_1, \ldots, f_n) | B = \langle f_1, \ldots, f_n \rangle$. Prove that $\chi^0(f_1, \ldots, f_n) = \chi(f_1, \ldots, f_n) | SU(n, \mathbf{C})$ is an irreducible character of $SU(n, \mathbf{C})$, that every irreducible character of $SU(n, \mathbf{C})$ is of this form, and that $\chi^0(f_1, \ldots, f_n) = \chi^0(f'_1, \ldots, f'_n)$ if and only if $f_i - f_{i+1} = f'_i - f'_{i+1}$ $(1 \leq i \leq n-1)$. (Hint: Use Weyl's formula to construct $\chi^0(f_1, \ldots, f_n)$ first. Let π^0 be the corresponding representation of $SU(n, \mathbf{C})$. Prove that $ax \mapsto a^{f_1 + \cdots + f_n} \pi^0(x)$ $(a \in \mathbf{C}, |a| = 1, x \in SU(n, \mathbf{C}))$ is an irreducible representation of G with character $\chi(f_1, \ldots, f_n)$.)

(c) Let $\varphi(x) = \det(x)$ $(x \in G)$. Prove that for any integer s, $\varphi^s \chi(f_1, \ldots, f_n)$ is an irreducible character of G. Prove, further, that all irreducible characters of G are of this form. Prove, finally, that $\varphi^s \chi(f_1, \ldots, f_n) = \varphi^{s'} \chi(f'_1, \ldots, f'_n)$ if and only if $f_i + s = f'_i + s'$ $(1 \leq i \leq n)$.

(e) (Branching law). Identify $U(n-1, \mathbf{C})$ with the subgroup of G of all elements of the form $\begin{pmatrix} A & 0 \\ 0 & 1 \end{pmatrix}$ $(n \geq 2)$. Denote by $\pi(f_1, \ldots, f_n)$ the representation of G with character $\chi(f_1, \ldots, f_n)$. Prove that $\pi(f_1, \ldots, f_n)$ maps $a \cdot 1$ into the scalar $a^{f_1 + \cdots + f_n} \cdot 1$. Prove, further, that the irreducible constituents of the restriction of $\pi(f_1, \ldots, f_n)$ to $U(n-1, \mathbf{C})$ are precisely all the representations $\pi(f'_1, \ldots, f'_{n-1})$, where $f'_1, \ldots, f'_{n-1}$ are integers such that $f_1 \geq f'_1 \geq f_2 \geq f'_2 \geq \cdots \geq f_{n-1} \geq f'_{n-1} \geq f_n \geq 0$; and that each of these occurs with multiplicity 1. (Hint: Put $a_n = 1$ in $\chi(f_1, \ldots, f_n)$, and simplify the determinants by subtracting the $(i+1)$th column from the ith, $1 \leq i \leq n-1$).

(f) Let $D(z_1, \ldots, z_n) = \prod_{1 \leq i < j \leq n} (z_i - z_j)$ $(z_1, \ldots, z_n \in \mathbf{C})$. Prove that

$$\dim(\pi(f_1, \ldots, f_n)) = \frac{D(h_1, \ldots, h_n)}{D(n-1, \ldots, 0)} \quad (h_i = f_i + n - i).$$

(g) Let $V = \mathbf{C}^n$, and let $\{e_1, \ldots, e_n\}$ be the canonical basis of V. Let $\mathfrak{J}$ be the tensor algebra over V, $\mathfrak{J}_f$ $(f \geq 0)$ the homogeneous subspace of $\mathfrak{J}$ of degree f. Denote by λ_f the natural representation of G in $\mathfrak{J}_f$. If $f, g \geq 0$, $t \in \mathfrak{J}_f$, $t' \in \mathfrak{J}_g$, we write $t \wedge t'$ for $(1/(f+g)!) \sum_{s \in \Pi_{f+g}} \epsilon(s)\, t \otimes t'$, where Π_{f+g} is the group of permutations of $\{1, 2, \ldots, f+g\}$. Prove that if $g_i = f_i - f_{i+1}$ $(1 \leq i \leq n-1)$ and $g_n = f_n$, the vector

$$e_1 \otimes \cdots \otimes e_1 \otimes (e_1 \wedge e_2) \otimes \cdots \otimes (e_1 \wedge e_2) \otimes \cdots$$
$$\otimes (e_1 \wedge \cdots \wedge e_n) \otimes \cdots \otimes (e_1 \wedge \cdots \wedge e_n),$$

in which $e_1 \wedge \cdots \wedge e_k$ occurs g_k times, belongs to a subspace of $\mathfrak{J}_f$ that is irreducibly invariant under λ_f and defines the representation $\pi(f_1, \ldots, f_n)$.

(h) Prove that the $\pi(f_1, \ldots, f_n)$ $(f_1 + \cdots + f_n = f)$ are precisely all the irreducible constituents of λ_f.

32. Let G be a complex analytic group with Lie algebra $\mathfrak{g}$, G_0 the **R**-analytic group underlying G, and $\mathfrak{g}_0$ the real Lie algebra underlying $\mathfrak{g}$. Let $\mathfrak{g}_c$ be the complexification of $\mathfrak{g}_0$, φ the canonical imbedding of $\mathfrak{g}_0$ in $\mathfrak{g}_c$.

(a) Let J be the endomorphism of $\mathfrak{g}_c$ such that $\varphi(iX) = J\varphi(X)$ for all $X \in \mathfrak{g}_c$. Prove that $\pm i$ are the only eigenvalues of J and that the corresponding eigenspaces $\mathfrak{g}_c^{\pm}$ are ideals of $\mathfrak{g}_c$ having $\mathfrak{g}_c$ as their direct sum.

(b) Let $\beta^{\pm}(X) = \frac{1}{2}(\varphi(X) \mp iJ\varphi(X))$ $(X \in \mathfrak{g}_0)$. Prove that β^+ (resp. β^-) is a Lie algebra isomorphism of $\mathfrak{g}$ (resp. the complex conjugate of $\mathfrak{g}$) with $\mathfrak{g}_c^+$ (resp. $\mathfrak{g}_c^-$).

(c) Let A and B be associative algebras over an algebraically closed field of characteristic 0, and let $C = A \otimes B$. Prove that the irreducible representations of C are exactly those of the form $\rho_A \otimes \rho_B$, where ρ_A (resp. ρ_B) is an irreducible representation of A (resp. B). (Hint: Let ρ be an irreducible representation of C in V. Let $U \subseteq V$ be a subspace invariant and irreducible under the representation ρ'_A $(a \mapsto \rho(a \otimes 1_B))$ of A. For $b \in B$, let $\rho'_B(b) = \rho(1_A \otimes b)$, and let $U_b = \rho'_B(b)[U]$; if $U_b \neq 0$, it is invariant under ρ'_A, and the subrepresentations defined by U and U_b are equivalent. Observe now that $V = \sum_{b \in B} U_b$, and conclude that $\rho'_A \cong \rho_A \otimes 1$ for some irreducible representation ρ_A of A.)

(d) Assume that G is simply connected. Prove that the irreducible representations of G_0 in complex vector spaces are precisely all representations of the form $\pi_1 \otimes \pi_2^{\text{conj}}$, where π_1 and π_2 are irreducible complex analytic representations of G and π_2^{conj} is the complex conjugate of the representation π_2. (Hint: Let $\mathfrak{G}_c$ be the universal enveloping algebra of $\mathfrak{g}_c$, $\mathfrak{G}_c^{\pm}$ the subalgebras generated by $\mathfrak{g}_c^{\pm}$. Then $\mathfrak{G}_c \simeq \mathfrak{G}_c^+ \otimes \mathfrak{G}_c^-$. Now apply (c) to conclude that the irreducible representations of $\mathfrak{g}_c \simeq \mathfrak{g}_c^+ \times \mathfrak{g}_c^-$ are precisely those of the form $\rho^+ \otimes \rho^-$, where $\rho^{\pm}$ are irreducible representations of $\mathfrak{g}_c^{\pm}$, and pull back $\rho^{\pm}$ to $\mathfrak{g}$ through $\beta^{\pm}$.)

For details, see Cartier, Exposé n° 22, Séminaire Sophus Lie [1].

33. Let G be a complex semisimple analytic group with Lie algebra $\mathfrak{g}$. Let G_0 (resp. $\mathfrak{g}_0$) denote the real analytic group (resp. Lie algebra over **R**) underlying G (resp. $\mathfrak{g}$). Let $U \subseteq G_0$ be a compact real form of G and $\mathfrak{u} \subseteq \mathfrak{g}_0$ the corresponding subalgebra. Write $\mathfrak{p} = (-1)^{1/2}\mathfrak{u}$.

(a) Prove that $[\mathfrak{u},\mathfrak{p}] \subseteq \mathfrak{p}$ and $[\mathfrak{p},\mathfrak{p}] \subseteq \mathfrak{u}$.

(b) Prove that for any $X \in \mathfrak{p}$, ad X is semisimple and has only real eigenvalues, and that $(\text{ad } X)^2$ leaves $\mathfrak{p}$ invariant.

(c) Let σ $(Z \mapsto Z^{\sigma})$ be the conjugation of $\mathfrak{g}$ with respect to $\mathfrak{u}$. For $X, Y \in \mathfrak{g}$, let $(X,Y) = -\langle X,Y^{\sigma}\rangle$. Prove that $(\cdot,\cdot)$ is a positive definite scalar product for $\mathfrak{g}$ and that for $X \in \mathfrak{u}$ (resp. $X \in \mathfrak{p}$) ad X is a skew-Hermitian (resp. Hermitian) endomorphism of $\mathfrak{g}$ with respect to this scalar product.

(d) Prove that $\exp[\mathfrak{p}]$ is closed in G. (Hint: Let $X_n \in \mathfrak{p}$ and $\exp X_n =$

$p_n \longrightarrow p \in G$ as $n \longrightarrow \infty$. By (c) above and Exercise 12(c) of Chapter 2, there exists a Hermitian $L \in \mathfrak{gl}(\mathfrak{g})$ such that ad $X_n \longrightarrow L$. Since ad is faithful, $X_n \longrightarrow X \in \mathfrak{p}$.)

(e) Let ψ be the map $(k,X) \mapsto k \exp X$ of $U \times \mathfrak{p}$ into G_0. Let f, g be the entire functions on **C** defined by $f(z) = (\cosh z - 1)/z = \sum_{n\geq 0} z^{2n+1}/(2n+2)!$, $g(z) = (\sinh z)/z = \sum_{n\geq 0} z^{2n}/(2n+1)!$ ($z \in \mathbf{C}$). Prove that, with appropriate identifications of Lie algebras, $(d\psi)_{(k,X)}$ ($k \in U$, $X \in \mathfrak{p}$) is the linear map $\mathrm{Ad}(\exp X) \circ L$ of $\mathfrak{u} \times \mathfrak{p}$ into $\mathfrak{g}_0$, where $L : \mathfrak{u} \times \mathfrak{p} \longrightarrow \mathfrak{g}_0$ is the map given by

$$L(Z,Y) = (Z + f(\mathrm{ad}\,X)(Y)) + g(\mathrm{ad}\,X)(Y) \quad (Z \in \mathfrak{u},\ Y \in \mathfrak{p}).$$

(f) Prove that ψ is an analytic diffeomorphism of $U \times \mathfrak{p}$ onto G_0. (Hint: By (b), $g(\mathrm{ad}\,X)$ leaves $\mathfrak{p}$ invariant and is invertible on $\mathfrak{p}$. Conclude from (e) that L is bijective. By (d), $U \exp[\mathfrak{p}]$ is both open and closed in G_0, so $G_0 = U \exp[\mathfrak{p}]$. To prove that ψ is one-to-one, use the uniqueness of polar decomposition in $GL(\mathfrak{g})$.)

(g) Deduce from (f) that U and G have the same fundamental group.

(h) Deduce from (g) that center (G) is finite and coincides with center (U).

(i) Deduce from (h) that U is a maximal compact subgroup of G. (Otherwise there would exist nonzero $X \in \mathfrak{p}$ such that all eigenvalues of $\mathrm{Ad}(\exp X)$ $= e^{\mathrm{ad}\,X}$ are of absolute value 1.)

34. (a) Let G_i ($i = 1, 2$) be a complex semisimple analytic groups, and let U_i be a compact real form of G_i. Prove that any real analytic homomorphism of U_1 onto U_2 can be extended uniquely to a complex analytic homomorphism of G_1 onto G_2. (Use (h) of Exercise 33.)

(b) Deduce from (a) that the complex analytic group containing a given compact connected semisimple group as a real form is determined up to a complex analytic isomorphism.

(c) Let G be a complex analytic semisimple group, U a compact real form of G. Prove that the restriction map $\pi \mapsto \pi | U$ induces a bijection of the set of all equivalence classes of irreducible complex analytic representations of G onto the set of all equivalence classes of irreducible representations of U.

35. Let G be a compact analytic group, $\mathfrak{g}$ its Lie algebra.

(a) Prove that G is isomorphic to $(C \times H_1 \times \cdots \times H_s)/F$, where C is a torus, the H_i are compact, semisimple, and simply connected and have simple Lie algebras, and F is a finite normal subgroup.

(b) Deduce from (a) that $\pi_1(G)$ is finite if and only if G is semisimple.

(c) Prove that $G = \exp[\mathfrak{g}]$.

(d) Use (a) to generalize the results involving maximal tori of compact semisimple groups to the case of arbitrary compact connected Lie groups.

Exercises 36–40 deal with universal complexifications of compact analytic groups (cf. Hochschild [1] for full details). The main result is that these are precisely the complex reductive groups—namely complex analytic groups that possess a faithful complex analytic representation and all of whose complex analytic representations are semisimple.

36. (a) Let G be a compact topological group satisfying the second axiom of countability. Prove that given any open neighborhood N of the identity, there is a closed normal subgroup H of G and an integer $n \geq 1$ such that $H \subseteq N$ and G/H is isomorphic to a compact subgroup of $U(n,\mathbf{C})$. (Hint: By the Peter–Weyl theorem, for each $x \in G \setminus N$ there is a representation ρ_x of G such that $\rho_x(x) \neq id$. Select a neighborhood N_x of x such that $\rho_x(y) \neq 1$ for $y \in N_x$, choose $x_1, \ldots, x_k \in G \setminus N$ such that $G \setminus N \subseteq \bigcup_{1 \leq i \leq k} N_{x_i}$, and consider the direct sum of the ρ_{x_i}.)

(b) Suppose G is a compact Lie group. Prove that G has a faithful representation. (Hint: Use Exercise 39 of Chapter 2.)

(c) Suppose G_c is a complex analytic semisimple group and that G is a compact real form of G_c. If π is a complex analytic representation of G_c whose kernel F is a discrete subgroup of G_c, prove that $F \subseteq$ center (G). Deduce that π is faithful if and only if $\pi | G$ is a faithful representation of G.

(d) Prove that any complex analytic semisimple group has a faithful complex analytic representation. (Hint: Apply (c) to the universal covering group G_c of the given complex analytic group.)

37. (a) Let G be a compact analytic group, (G_c,γ) its universal complexification. Prove that γ is an isomorphism of G onto a compact real form of G_c.

(b) Let G_c be a complex analytic semisimple group, G a compact real form of G_c, and γ the inclusion map of G into G_c. Prove that (G_c,γ) is a universal complexification of G.

38. (a) Let $G = \mathbf{T}^n$, $G_c = \mathbf{C}^{\times n}$, and γ the inclusion map of G in G_c. Prove that G_c is reductive, that G is the largest compact subgroup of G_c, and that (G_c,γ) is a universal complexification of G. Prove also that if π is a complex analytic representation of G_c such that kernel $(\pi) \cap G$ is discrete, then kernel $(\pi) \subseteq G$. (Hint: For the last assertion, let $S \subseteq \mathbf{Z}^n$ be such that the characters $(z_1, \ldots, z_n) \longrightarrow z_1^{a_1} \cdots z_n^{a_n}$ $((a_1, \ldots, a_n) \in S)$ are all the irreducible constituents of π. Then S generates a subgroup of finite index in $\mathbf{Z}^n$ and so S spans $\mathbf{R}^n$. If $(z_1, \ldots, z_n) \in$ kernel (π), $\sum_j a_j \log |z_j| = 0$ for all $(a_1, \ldots, a_n) \in S$, and hence $\log |z_j| = 0$ for all j.)

(b) Let A_c be a complex analytic abelian group of dimension n. Prove the equivalence of the following statements:

(i) A_c is isomorphic to $\mathbf{C}^{\times n}$ as a complex analytic group.

(ii) A_c has a faithful complex analytic representation, and there exists a finite subgroup F of A_c such that all complex analytic representations of A_c/F are semisimple.

(iii) A_c is reductive.

(iv) There exists a real form A of A_c which is a maximal compact subgroup of A_c.

(v) Let γ be the identity map of A_c. Then there exists a compact real form A of A_c such that (A_c,γ) is a universal complexification of A.

(vi) There exists a compact real form A of A_c with the following property: every character A extends to a complex analytic homomorphism of A_c into $\mathbf{C}^{\times}$.

(Hint: To prove (iii) $\Rightarrow$ (i), observe that we can write the reductive group A_c in the form $\mathbf{C}^{\times n}/D$, where D is a discrete subgroup with $D \cap \mathbf{T}^n$

$= \{1\}$. Now use (a). To prove (ii) $\Rightarrow$ (iii), let π be a complex analytic representation of A_n such that $\pi | F$ is a character of F, and let m be the order of F. Then $\underbrace{\pi \otimes \cdots \otimes \pi}_{m \text{ factors}} = \pi^{\otimes m}$ is semisimple, so $\pi(x)^{\otimes m}$ is semisimple for all $x \in A_c$. This implies that $\pi(x)$ is semisimple for all $x \in A_c$.)

39. Let G_c be a complex analytic group with Lie algebra $\mathfrak{g}_c$. Assume that $\mathfrak{g}_c$ is reductive, and let $\mathfrak{h}_c = [\mathfrak{g}_c,\mathfrak{g}_c]$, $\mathfrak{a}_c = \text{center}(\mathfrak{g}_c)$. Denote by H_c and A_c the complex analytic subgroups of G_c defined respectively by $\mathfrak{h}_c$ and $\mathfrak{a}_c$.
(a) Prove that H_c and A_c are closed and $F = H_c \cap A_c$ is finite.
(b) Prove that G_c is reductive if and only if A_c is reductive. (Hint: If G_c is reductive, so is A_c by the implication (ii) $\Rightarrow$ (iii) of Exercise 38(b). For the converse, if A_c is reductive, use the results of §3.16 to prove the semisimplicity of complex analytic representations of G_c. To construct a faithful complex analytic representation of G_c, observe that any complex analytic representation of H_c may be extended to one of G_c.)
(c) Suppose G_c is reductive. Prove that G_c has compact real forms, that these are precisely the maximal compact subgroups of G_c, and that any two such are conjugate in G_c. (Hint: First prove these results when $G_c \approx H_c \times A_c$.)

40. Let G_c be a complex analytic group, G a compact real form of it. Denote by γ the inclusion map of G into G_c. Prove the equivalence of the following statements:
(i) G_c is reductive.
(ii) G is a maximal compact subgroup of G_c.
(iii) (G_c,γ) is a universal complexification of G.
(Hint: Write $G = HA$, where H is a compact real form of H_c and A is a compact real form of A_c. Assume (ii), and let B be a maximal compact subgroup of A_c. Then $A \subseteq B$ and $HB = G$, so $B \subseteq (G \cap A_c)^0 = A$. Now use Exercises 38 and 39 to get (i). To prove (i) $\Rightarrow$ (iii), observe that $H \cap A = H_c \cap A_c$ under (i). To prove (iii) $\Rightarrow$ (i), prove that any character φ of A extends to a complex analytic homomorphism of A_c into $\mathbf{C}^\times$ by constructing a representation ρ of G such that $\rho | A = \varphi \cdot id$, and using (iii).)

41. Let G be a compact analytic group.
(a) Let $x \in G$, and let G_x be the centralizer of x in G. Prove that $x \in G_x^0$, the component of identity of G_x. (Note that x belongs to a torus.)
(b) Let $A \subseteq G$ be a torus, and let $x \in G$ centralize A. Prove that there is a maximal torus of G containing A and x. (Let T be a maximal torus of G_x^0 containing A. Use (a) to prove that $x \in T$ and that T is a maximal torus of G.)
(c) Deduce from (b) that centralizers of tori in G are connected.
(d) Let $G = SO(3,\mathbf{R})$ and $x = \text{diag}(1,-1,-1)$. Verify that the centralizer of x in G is not connected.

42. Let G be a compact semisimple analytic group, $\mathfrak{g}$ its Lie algebra, B a maximal torus of G, and $\mathfrak{b}$ the corresponding subalgebra of $\mathfrak{g}$. G' is the set of regular points of G; $B' = G \cap B'$. $\tilde{B}$ is the normalizer of B in G. Identify $\tilde{B}/B$ with the Weyl group $\mathfrak{w}$. B^+ is a connected component of B'. $G^* = G/B$, $x \mapsto x^*$ is the natural map of G onto G^*, and $\psi^*(x^*,b) = b^x$ $(b \in B, x \in G)$.
(a) G' is connected and $\pi_1(G) = \pi_1(G')$. (The main point here is that

$\dim(G \setminus G') = \dim(G) - 3$. This remarkable fact was noticed by Weyl and was exploited by both Weyl and Cartan. See Helgason [1], Chapter 7, for a very careful treatment of this result.)

(b) Prove that ψ^* is a covering map of $G^* \times B^+$ onto G' and that $\psi^*(x_1^*,b_1) = \psi^*(x_2^*,b_2)$ ($b_i \in B^+$, $x_i \in G$) if and only if there is an element $y \in \tilde{B}$ such that y leaves B^+ invariant and $b_2 = b_1^y$, $x_2 = x_1 y^{-1}$. (First prove the corresponding result with B' instead of B^+. Then note that $\psi^*[G^* \times B^+]$ is both open and closed in G'.)

(c) Deduce from (b) that $\mathfrak{w}$ acts transitively on the set of all connected components of B'.

(d) Suppose G is simply connected. Prove that both B^+ and G^* are simply connected and that ψ^* is an analytic diffeomorphism of $G^* \times B^+$ onto G'. Deduce that in this case $\mathfrak{w}$ acts simply transitively on the set of all connected components of B'.

(e) Let G be simply connected. Prove that the centralizers in G of regular elements are maximal tori. (Let $x \in B^+$, and let $yxy^{-1} = x$. Then $y \in \tilde{B}$ and $(B^+)^y = B^+$, so y centralizes $\mathfrak{b}$ by (d). Compare this with (d) of exercise 41.)

43. Let G be a compact analytic group which is semisimple, $\mathfrak{g}$ its Lie algebra, B a maximal torus, and $\mathfrak{b}$ the corresponding subalgebra of $\mathfrak{g}$. Let $L(R)$, $L(G)$, L and ξ_λ be as in §4.13.

(a) Prove that the map $\lambda \mapsto \xi_\lambda$ induces an isomorphism of $L(R)$ with the group of all characters of B which are trivial on center(G). (Use Exercise 46 below to prove that the weights of the representations of $AD[G]$ belong to $L(R)$.)

(b) Deduce from (a) that $L(G)/L(R)$ is canonically isomorphic to the character group of center(G).

(c) Prove by similar reasoning that $L/L(G)$ is canonically isomorphic to the character group of $\pi_1(G)$.

(d) Deduce the isomorphisms (non canonical) center(G) $\approx L(G)/L(R)$, $\pi_1(G) \approx L/L(G)$.

(e) Suppose that G is simply connected. Let $A = (a_{ij})$ be the Cartan matrix of $(\mathfrak{g}_c,\mathfrak{b}_c)$ with respect to some simple system of roots. Prove that center(G) is isomorphic to the abelian group which has l generators ($l = \text{rk}(G)$) $\xi_1, \ldots, \xi_l$ subject to the relations $\sum_{1 \le i \le l} a_{ij}\xi_i = 0$ $(1 \le j \le l)$. Hence show that the order of center(G) is $|\det(A)|$.

(f) For any simple Lie algebra $\mathfrak{g}$ over $\mathbf{C}$, let $C(\mathfrak{g})$ denote the center of the corresponding simply connected complex (or compact) group. Obtain from (d) the following isomorphisms (here, for any integer $p \geq 2$, $\mathbf{Z}_p$ is the group $\mathbf{Z}/p\mathbf{Z}$ and 0 is the group having only the identity):

$$C(A_l) \approx \mathbf{Z}_{l+1}, \qquad C(B_l) \approx \mathbf{Z}_2, \qquad C(C_l) \approx \mathbf{Z}_2, \qquad C(D_l) \approx \mathbf{Z}_4$$

(l odd) and $C(D_l) \approx \mathbf{Z}_2 \oplus \mathbf{Z}_2$ (l even), $C(E_6) \approx \mathbf{Z}_3$, $C(E_7) \approx \mathbf{Z}_2$, while $C(E_8)$, $C(G_2)$, and $C(F_4)$ are all ≈ 0.

(g) Deduce the simple connectedness of $SU(n,\mathbf{C})$ $(n \geq 2)$ and $Sp(n)$ $(n \geq 1)$ and the relations $\pi_1(SO(n,\mathbf{R})) = \mathbf{Z}_2$ $(n \geq 3)$.

44. Let G be a compact analytic group, B a maximal torus, E and F two subsets of B. Suppose there is a $y \in G$ such that $E^y = F$. Prove that there is $z \in G$ such that z normalizes B and $E^z = F$. (Hint: Let H be the component of the identity of the centralizer of F in G. Note that B and B^y are two maximal tori of H.)

45. $\mathfrak{g}$ is a semisimple Lie algebra over $\mathbf{C}$, $\mathfrak{h}$ a CSA, G a complex analytic group with Lie algebra $\mathfrak{g}$.

(a) Prove that $\exp[\mathfrak{h}]$ is the centralizer of $\mathfrak{h}$ in G. (Use Theorems 4.9.1 and 4.12.5 and Exercise 33(h).)

(b) Let $\mathfrak{l}$ be a subspace of $\mathfrak{h}$, $\mathfrak{z}$ the centralizer of $\mathfrak{l}$ in $\mathfrak{g}$. Determine $\mathfrak{z}$ in terms of the root space decomposition of $(\mathfrak{g},\mathfrak{h})$, and prove that $\mathfrak{z}$ is reductive. Prove also that the center of $\mathfrak{z}$ consists of all $H \in \mathfrak{h}$ with the property that $\alpha(H) = 0$ for every root α of $(\mathfrak{g},\mathfrak{h})$ which vanishes identically on $\mathfrak{l}$. Deduce that the adjoint representation of $\mathfrak{g}$ remains semisimple on restriction to $\mathfrak{z}$.

(c) Let Z be the complex analytic subgroup of G defined by $\mathfrak{z}$. Prove that Z is the centralizer of $\mathfrak{l}$ in G. (Let $y \in G$ centralize $\mathfrak{l}$. By Theorem 4.1.3, for some $z \in Z$, $x = zy$ centralizes $\mathfrak{l}$ and $\mathfrak{h}^x = \mathfrak{h}$. By (4.9.5) and Theorem 4.15.17 of the appendix, for some $z' \in Z$, $z'x$ centralizes $\mathfrak{h}$. Now use (a).)

46. Let G be a compact Lie group. Suppose $\mathcal{R}$ is a set of irreducible representations of G such that (i) the trivial representation belongs to $\mathcal{R}$, (ii) if $\pi \in \mathcal{R}$, then $\mathcal{R}$ contains a representation equivalent to the contragredient of π, and (iii) if $x \in G$ and $x \neq 1$, there is a $\pi \in \mathcal{R}$ such that $\pi(x) \neq 1$. Prove that every irreducible representation of G occurs as a constituent of some tensor product $\pi_1 \otimes \cdots \otimes \pi_k$ for suitable $\pi_1, \ldots, \pi_k \in \mathcal{R}$.

47. (a) Let G be a semisimple real analytic group with Lie algebra $\mathfrak{g}$, G_c a complex analytic group with Lie algebra $\mathfrak{g}_c$, γ an $\mathbf{R}$-analytic homomorphism of G into G_c such that $\mathbf{C} \cdot d\gamma[\mathfrak{g}] = \mathfrak{g}_c$. Prove that the following statements are equivalent:

(i) (G_c,γ) is a universal complexification of G.

(ii) If φ is any representation of G in a complex vector space V, there exists a complex analytic representation π of G_c in V such that $\varphi = \pi \circ \gamma$.

(iii) If H_c is a complex semisimple group and φ is an $\mathbf{R}$-analytic homomorphism of G into H_c, there exists a complex analytic homomorphism π of G_c into H_c such that $\varphi = \pi \circ \gamma$.

(Hint: To prove that (iii) $\Longrightarrow$ (i), let A_c be a complex analytic group with Lie algebra $\mathfrak{a}_c$ and $\psi : G \longrightarrow A_c$ an $\mathbf{R}$-analytic homomorphism. Prove the existence of a Levi subalgebra $\mathfrak{m}_c \subseteq \mathfrak{a}_c$ such that $d\psi[\mathfrak{g}] \subseteq \mathfrak{m}_c$.)

(b) Let G be as in (a), and let (G_c,γ) be a universal complexification of G. Suppose D is the kernel of γ. Prove that D is a discrete central subgroup of G.

(c) Show that G/D has a faithful representation and that D is the intersection of the kernels of all representations of G.

(d) Give an example of a complex analytic semisimple group G_c and a real form G of it such that G_c, together with the inclusion map of G into it, is not a universal complexification of G.

(Observe the contrast with compact real forms. For another treatment of complexifications of real semisimple groups, see Harish-Chandra [3].)

48. Let $\mathfrak{g}$ be a semisimple Lie algebra over **C**, $\mathfrak{G}$ its universal enveloping algebra, $\omega \in \mathfrak{G}$ the Casimir element.

(a) Let π be any representation of $\mathfrak{G}$. Show that $tr\, \pi(\omega)$ is a rational number ≥ 0.

(b) Let $\mu \in \mathbf{C}$ and let $\mathfrak{F}_\mu = \mathfrak{G}(\omega - \mu \cdot 1)\mathfrak{G}$. Prove that $\mathfrak{F}_\mu$ is a proper two-sided ideal in $\mathfrak{G}$. (Hint: Observe that $\mathfrak{F}_\mu = \mathfrak{G}(\omega - \mu \cdot 1)$, and show that all elements of $\mathfrak{F}_\mu$ have degree ≥ 2.)

(c) Let $\mathfrak{A}_\mu = \mathfrak{G}/\mathfrak{F}_\mu$. Prove that $\mathfrak{A}_\mu$ is a finitely generated algebra and that if μ is not a nonnegative rational number, $\mathfrak{A}_\mu$ has no finite-dimensional representation. (See Harish-Chandra [1].)

Exercises 49–51 deal with the famous reciprocity between representations of the permutation and unitary groups. For full details, see Weyl [1, 5].

49. Let Π be a finite group, **A** its group algebra of all formal sums $\mathbf{a} = \sum_{s \in \Pi} a(s)s$ ($a(s) \in \mathbf{C}$). For $\mathbf{a} = \sum_s a(s)s \in \mathbf{A}$, $\hat{\mathbf{a}} = \sum_s a(s^{-1})^{\text{conj}}s$. We denote by r the right regular representation of A. U is a finite-dimensional vector space, and π is a representation of Π in U.

(a) Prove that $\pi : \mathbf{a} = \sum_s a(s)s \mapsto \sum_s a(s)\pi(s)$ is a representation of **A** in U. Prove further that there is a unique two-sided ideal $\mathbf{A}_0 \subseteq A$ such that π is a bijection of $\mathbf{A}_0$ onto $\pi[\mathbf{A}]$.

(b) Let $\mathbf{B} = \pi[\mathbf{A}]'$ (= the algebra of endomorphisms of U commuting with $\pi[\mathbf{A}]$). Prove that the action of **B** on U is semisimple and $\mathbf{B}' = \pi[\mathbf{A}]$.

(c) For any $\mathbf{a} \in \mathbf{A}$, let $U(\mathbf{a})$ be the range of $\pi(\mathbf{a})$. Prove that $U(\mathbf{a})$ is invariant under **B**. Conversely, let U' be a subspace of U that is invariant under **B**. Prove that there is an idempotent $\mathbf{e} \in \mathbf{A}_0$ ($\mathbf{e}^2 = \mathbf{e}$) such that $U' = U(\mathbf{e})$. (Hint: Let U'' be a **B**-invariant subspace complementary to U', and let L be the projection $U \longrightarrow U'$ mod U''. Then $L \in \mathbf{B}'$.)

(d) Let $\mathbf{a}, \mathbf{b} \in \mathbf{A}_0$. Prove that $U(\mathbf{a}) \subseteq U(\mathbf{b})$ if and only if $\mathbf{aA} \subseteq \mathbf{bA}$. Deduce that $\rho : \mathbf{aA} \mapsto U(\mathbf{a})$ is a well-defined inclusion preserving bijection of the lattice $L(\mathbf{A}_0)$ of all right ideals of **A** that are contained in $\mathbf{A}_0$ onto the lattice $L_\mathbf{B}(U)$ of all B-invariant subspaces of U and that for any $\mathbf{E} \in L(\mathbf{A}_0)$, $\rho(\mathbf{E}) = \sum_{\mathbf{a} \in \mathbf{E}} U(\mathbf{a})$. (Hint: Let $U(\mathbf{a}) \subseteq U(\mathbf{b})$, let N = null space of $\pi(\mathbf{b})$, and let U'' be a **B**-invariant subspace complementary to N. If $u \in U$, $\exists$ unique $u'' = Lu \in U''$ such that $\pi(\mathbf{a})u = \pi(\mathbf{b})u''$. Show that $L = \pi(\mathbf{c})$ for some $\mathbf{c} \in \mathbf{A}_0$ and $\mathbf{a} = \mathbf{bc}$.)

(e) For any $\mathbf{E} \in L(\mathbf{A}_0)$ (resp. $V \in L_\mathbf{B}(U)$), let $r(\mathbf{E})$ (resp. $\beta(V)$) be the subrepresentation of the right regular representation of **A** (resp. the representation of **B** in U) defined by **E** (resp. V). Suppose $\mathbf{E}, \mathbf{F} \in L(\mathbf{A}_0)$, $V = \rho(\mathbf{E})$, $W = \rho(\mathbf{F})$. Prove that the following statements are equivalent:

(i) For some $\mathbf{c} \in \mathbf{A}_0$, $\mathbf{cE} = \mathbf{F}$.

(ii) $\beta(W)$ is equivalent to a subrepresentation of $\beta(V)$.

(iii) $r(\mathbf{F})$ is equivalent to a subrepresentation of $r(\mathbf{E})$.

(f) Let $\mathbf{e}_0$ be the idempotent such that $\mathbf{e}_0\mathbf{A}_0 = \mathbf{A}_0\mathbf{e}_0 = \mathbf{A}_0$, and let $\mathbf{e}_0 = \sum_i \mathbf{e}_i$, where the $\mathbf{e}_i$ are primitive idempotents. Prove that the $U(\mathbf{e}_i)$ are irreducible, $U = \sum_i U(\mathbf{e}_i)$ is a direct sum, and that the subrepresentations defined by $U(\mathbf{e}_i)$ and $U(\mathbf{e}_{i'})$ are equivalent if and only if the irreducible representations of Π defined on $\mathbf{e}_i\mathbf{A}$ and $\mathbf{e}_{i'}\mathbf{A}$ by right translations are equivalent. Let $\mathcal{E}(\mathbf{B})$ be the set of equivalence classes of irreducible representations of $\mathbf{B}$ occuring in U, and for each $\mathfrak{v} \in \mathcal{E}(\mathbf{B})$ let $U_\mathfrak{v}$ be the linear span of all invariant subspaces of U that are irreducible and transform according to $\mathfrak{v}$. Further, let $\mathbf{e}_\alpha$ $(\alpha \in I)$ be the irreducible characters of Π such that $\mathbf{A}_0 = \sum_{\alpha \in I} \mathbf{e}_\alpha \mathbf{A}$ is the decomposition of $\mathbf{A}_0$ into minimal two-sided ideals. Prove that there is a bijection $\alpha \mapsto \mathfrak{v}_\alpha$ of I onto $\mathcal{E}(\mathbf{B})$ such that $U(\mathbf{e}_\alpha) = U_{\mathfrak{v}_\alpha}$ for all $\alpha \in I$.

(g) Convert $\mathbf{A}$ into a Hilbert space by defining $(\mathbf{a},\mathbf{b}) = \sum_{s \in \Pi} a(s)b(s)^{\text{conj}}$. For any $\mathbf{E} \in L(\mathbf{A}_0)$ let $\mathbf{E}' = \mathbf{A}_0 \cap \mathbf{E}^\perp$. Prove that $\mathbf{E}' \in L(\mathbf{A}_0)$ too. Supposing U to be a Hilbert space and π to be a unitary representation, prove that for any $\mathbf{E} \in L(\mathbf{A}_0)$, $\rho(\mathbf{E}')$ is the orthogonal complement of $\rho(\mathbf{E})$ in U, and further that

$$\rho(\mathbf{E}) = \{u : u \in U, \pi(\hat{a})u = 0 \text{ for all } a \in \mathbf{E}'\}.$$

Hence deduce the result that the $\mathbf{B}$-invariant subspaces of U are precisely those that are given by "symmetry conditions" of the form $\sum_{s \in \Pi} c_s\pi(s)u = 0$ $(c_s \in \mathbf{C})$.

50. Let $V = \mathbf{C}^n$, let $\{e_1, \ldots, e_n\}$ be the canonical basis of V, and let $G = U(n,\mathbf{C})$. B is the subgroup of diagonal matrices of G which we identify with $\mathbf{T}^n$ via the map $\operatorname{diag}(a_1, \ldots, a_n) \mapsto (a_1, \ldots, a_n)$. Let $f \geq 1$ be an integer, $U = V \otimes \cdots \otimes V$ (f factors). For any endomorphism L of V, $L^{(f)} = L \otimes \cdots \otimes L$ (f factors). Π is the permutation group of $\{1, \ldots, f\}$, and $\pi : s \mapsto \pi(s)$ the natural representation of Π in U. Let notation be as in Exercise 49.

(a) Prove that $\mathbf{A}_0 = \mathbf{A}$ (i.e., π is a faithful representation of $\mathbf{A}$) if and only if $n \geq f$.

(b) Prove that $\mathbf{B}$ is the linear span of the endomorphisms $x^{(f)}$ $(x \in G)$. (Hint: By the unitarian trick, the $x^{(f)}$ $(x \in G)$ have the same linear span as the $L^{(f)}$ when L runs through the algebra of endomorphisms of V. Prove now that if E is an endomorphism of U such that $tr(UL^{(f)}) = 0$ for all L, then $tr(Uy) = 0$ for all $y \in \mathbf{B}$.)

(c) Let $n \geq f$. For integers $f_1 \geq \cdots \geq f_n \geq 0$, let $\pi(f_1, \ldots, f_n)$ be the irreducible representation of G described in Exercise 31. Use the results of the preceding exercise to set up the one-to-one correspondence between the irreducible characters of Π and the irreducible representations $\pi(f_1, \ldots, f_n)$ with $f_1 + \cdots + f_n = f$.

(d) Prove that the spaces of symmetric and skew-symmetric tensors in U are invariant and irreducible under the action of $G^{(f)}$, the corresponding representations of G being respectively $\pi(f,0, \ldots, 0)$ and $\pi(\underbrace{1, \ldots, 1}_{f \text{ terms}},0, \ldots, 0)$.

(e) For any sequence $(g_1, \ldots, g_n)$ of integers $g_i \geq 0$ with $g_1 + \cdots + g_n = f$, let $\Pi(g_1, \ldots, g_n)$ denote the subgroup of Π of all permutations which permute the first g_1 numerals of $\{1, \ldots, f\}$ among themselves, the next g_2 numerals among themselves, and so on. For any conjugacy class $\mathfrak{k}$

of Π let $c_{g_1,\dots,g_n}(\mathfrak{k})$ be the number of elements on $\mathfrak{k} \cap \Pi(g_1, \dots, g_n)$. Suppose X is the character of an irreducible representation of G occuring in U and ξ is the corresponding character of Π; then prove that for any diagonal matrix $a = (a_1, \dots, a_n)$,

$$X(a_1, \dots, a_n) = \sum_{\mathfrak{k}} \xi(\mathfrak{k}) \sum_{g_1,\dots,g_n} \frac{c_{g_1,\dots,g_n}(\mathfrak{k})}{g_1! \cdots g_n!} a_1^{g_1} \cdots a_n^{g_n}.$$

(Hint: Let $\mathbf{e} = \sum_{s \in \Pi} e(s)s$ be a primitive idempotent corresponding to X. For $i_1, \dots, i_f$, let $\Gamma(i_1, \dots, i_f)$ be the stabilizer in Π of $e_{i_1} \otimes \cdots \otimes e_{i_f}$. Then

$$\begin{aligned} X(a) &= \sum_{s \in \Pi} e(s)\, tr(\pi(s) a^{(f)}) \\ &= \sum_{i_1,\dots,i_f} a_{i_1} \cdots a_{i_f} \sum_{s \in \Gamma(i_1,\dots,i_f)} e(s) \\ &= \sum_{g_1,\dots,g_n} \left\{ \left(\frac{a_1^{g_1} \cdots a_n^{g_n}}{g_1! \cdots g_n!} \right) \sum_{\substack{s \in \Pi(g_1,\dots,g_n) \\ t \in \Pi}} e(tst^{-1}) \right\} \end{aligned}$$

Now ξ is the character of the subrepresentation defined by $\mathbf{eA}$ of the right regular representation of Π, so $\xi(s) = \sum_{t \in \Pi} e(tst^{-1})$.)

51. Let notation be as above.

(a) Prove that any $s \in \Pi$ can be written as a product of cyclic permutations no two of which involve a common numeral and that this representation is unique.

(b) Let $i_1, i_2, \dots$ be integers ≥ 0 such that $i_1 + 2i_2 + \cdots = f$. Prove that the set $[i_1 i_2 \dots]$ of all $s \in \Pi$ in whose decomposition there are i_1, cycles with 1 numeral, i_2 cycles with 2 numerals, etc., is a conjugacy class of Π and that every conjugacy class of Π may be obtained this way. Deduce that if $\mathfrak{k}$ is any conjugacy class, $\mathfrak{k} = \mathfrak{k}^{-1}$. Prove also that the number of elements in $[i_1 i_2 \dots]$ is $f!/1^{i_1} i_1! 2^{i_2} i_2! 3^{i_3} i_3! \dots$.

(c) Let $g_1, \dots, g_n \geq 0$ be integers with $g_1 + \cdots + g_n = f$. Prove that the number of elements in $\Pi(g_1, \dots, g_n)$ belonging to the conjugacy class $[i_1 i_2 \cdots]$ is

$$\frac{1}{1^{i_1} 2^{i_2} \cdots} \sum_{(i)} \left\{ \frac{\prod_\alpha g_\alpha!}{i_{\alpha 1}! i_{\alpha 2}! \cdots} \right\};$$

here α runs from 1 to n, and the sum is extended over all possible sequences $i_{11}, \dots, i_{21}, \dots, \dots, i_{n1}, \dots$ of integers ≥ 0 satisfying the equations

$$\sum_\alpha i_{\alpha 1} = i_1, \qquad \sum_\alpha i_{\alpha 2} = i_2, \dots$$
$$\sum_\nu \nu i_{1\nu} = g_1, \qquad \sum_\nu \nu i_{2\nu} = g_2, \dots, \qquad \sum_\nu \nu i_{n\nu} = g_n$$

(d) Let $\sigma_r = a_1^r + a_2^r + \cdots + a_n^r$ $(r = 1,2, \dots)$. Prove that the characters X of G and the character ξ of Π of Exercise 50(e) are related by the formula

$$X(a_1, \dots, a_n) = \sum_{[i_1 i_2 \cdots]} \left\{ \left(\frac{\xi([i_1 i_2 \cdots])}{1^{i_1} 2^{i_2} \cdots i_1! i_2! \cdots} \right) \sigma_1^{i_1} \sigma_2^{i_2} \cdots \right\}.$$

(e) Let $n(\mathfrak{k})$ be the number of elements in the conjugacy class $\mathfrak{k}$ of Π. Prove that

$$\sum_{\xi} \xi(\mathfrak{k}')\xi(\mathfrak{k}) = \delta_{\mathfrak{k}\mathfrak{k}'}\frac{f!}{n(\mathfrak{k})},$$

the sum being extended over all irreducible characters of Π. (Hint: Use the orthogonality relations.)

(f) Deduce from (e) the following formula:

$$\sigma_1^{i_1}\sigma_2^{i_2}\cdots = \sum_{\xi} \xi([i_1,i_2\cdots])X(a_1,\ldots,a_2).$$

(g) Let $f_1 \geq \cdots \geq f_n \geq 0$ be integers with $f_1 + \cdots + f_n = f$, X the character of the representation $\pi(f_1,\ldots,f_n)$ of G (cf. exercise 31), and ξ the corresponding character of Π. Let $D(a_1,\ldots,a_n) = \prod_{i<j}(a_i - a_j)$, $h_i = f_i + n - i\ (1 \leq i \leq n)$. Show that $\xi([i_1 i_2 \cdots])$ is the coefficient of $a_1^{h_1}\cdots a_n^{h_n}$ in $D(a_1,\ldots,a_n)\sigma_1^{i_1}\sigma_2^{i_2}\ldots$. Deduce that the characters of Π are integer-valued functions on Π.

In Exercises 52–55, for any C^∞ manifold N of dimension n, $H(N)$ denotes its De Rham cohomology algebra (cf. Exercises 29–34 of Chapter 2); $H^k(N)$ is the subspace of $H(N)$ spanned by elements of degree k $(0 \leq k \leq n)$; $P_N(t) = \sum_{0\leq k\leq n} \dim H^k(N)t^k$ is the Poincaré polynomial of N.

52. Let G be a compact connected Lie group of dimension g acting smoothly and transitively on a C^∞ manifold M of dimension m. We assume that (G,M) is a symmetric pair (cf. Exercise 33, Chapter 2). Let $y_0 \in M$, let H be the stabilizer of y_0 in G, and let ρ $(h \mapsto \rho(h))$ be the representation of H induced on the tangent space $T_{y_0}(M)$ to M at y_0. For $1 \leq k \leq m$, let ρ_k be the representation of H in $\Lambda_k(T_{y_0}(M))$ obtained from ρ. dh is the Haar measure on H such that $\int_H dh = 1$.

(a) Prove that $\dim H^k(M) = \int_H tr\,\rho_k(h)\,dh$ $(1 \leq k \leq m)$. (The right side is the dimension of the subspace of ρ_k-invariant elements in $\Lambda_k(T_{y_0}(M))$; now use Exercise 33 of Chapter 2.)

(b) Deduce from (a) that $P_M(t) = \int_H \det(1 + t_\rho(h))\,dh$

$$(\det(1 + t_\rho(h)) \equiv 1 + \sum_{1\leq k\leq m} tr(\rho_k(h))\cdot t^k).$$

(c) Prove that M is orientable if and only if $\det\rho(h) \equiv 1$ $(h \in H)$.

(d) Use (a) and (c) to prove that if M is orientable, $\dim H^k(M) = \dim H^{m-k}(M)$ $(0 \leq k \leq m)$. (Hint: If $\sigma(h)$ is orthogonal with respect to a scalar product in $T_{y_0}(M)$, $tr\,\sigma_k(h) = tr\,\sigma_{m-k}(h)$, $1 \leq k \leq m$, since $\det\sigma(h) = 1$.)

53. Let G be a compact, semisimple, connected Lie group; B, a maximal torus of G; $l = rk(G)$

(a) Prove that $P_G(t) \equiv [\mathfrak{w}]^{-1}(1 + t)^l \int_B D(b)\prod_{\alpha\in\Delta}(1 + t\xi_\alpha(b))\,db$, where D is defined by (4.13.9) and $[\mathfrak{w}]$ is the order of the Weyl group $\mathfrak{w}$.

(b) Deduce from (a) that $\sum_{0 \leq k \leq \dim(G)} \dim H^k(G) = 2^l$. (Taking $t = 1$ in (a), this reduces to proving that $\int_B \prod_\alpha (1 - \xi_\alpha^2(b))\, db = [\text{1U}]$. Observe now that the Fourier expansions of $\prod_\alpha (1 - \xi_{2\alpha})$ and $\prod_\alpha (1 - \xi_\alpha)$ have the same constant term.)

54. (a) Let $H = U(n,\mathbf{C}) \times U(1,\mathbf{C})$, and let ρ be the representation of H in $\mathbf{C}^n$ given by $\rho(x,\zeta) = (1/\zeta)x$ $(x \in U(n,\mathbf{C}), |\zeta| = 1)$. Prove that the representations ρ_k induced by H in $\Lambda_k(\mathbf{C}^n)$ $(0 \leq k \leq n$; ρ_0 is the trivial representation) are all irreducible and mutually inequivalent. (Use the fact that the representations of the Lie algebra A_{n-1} in $\Lambda_k(\mathbf{C}^n)$ $(1 \leq k \leq n)$ are irreducible and mutually inequivalent.)

(b) Deduce from (a) that $\int_H |\det(1 + t\rho(h))|^2\, dh \equiv \sum_{0 \leq k \leq n} |t|^{2k}$ for $t \in \mathbf{C}$, dh being the Haar measure on H such that $\int_H dh = 1$.

(c) Let $M = \mathbf{P}_n(\mathbf{C})$ be the complex n-dimensional projective space (= space of one-dimensional linear subspaces of $\mathbf{C}^{n+1}$; cf. Exercise 38 of Chapter 2). Prove that $P_M(t) \equiv 1 + t^2 + \cdots + t^{2n}$. (Let $y_0 \in M$ be the one-dimensional subspace of $\mathbf{C}^{n+1}$ generated by $(0, \ldots, 0,1)$, H_0, the stabilizer of y_0 in $U(n + 1,\mathbf{C})$ which acts transitively on M. Identify H_0 with H in such a way that the representation of H_0 in $T_{y_0}(M)$ becomes equivalent to ρ defined in (a), but considered as a representation of H in the *real vector space underlying* $\mathbf{C}^n$.)

(d) Let S be the unit sphere in $\mathbf{C}^{n+1}$ given by $z_1\bar{z}_1 + \cdots + z_{n+1}\bar{z}_{n+1} = 1$, and let $\pi : S \longrightarrow M$ be the map which sends each point of S into the one-dimensional subspace generated by it. Prove that there is a 2-form Ω on M such that $\pi^*\Omega = dz_1 \wedge d\bar{z}_1 + \cdots + dz_{n+1} \wedge d\bar{z}_{n+1}$ on S. Verify that Ω is invariant under $U(n + 1,\mathbf{C})$.

(e) Prove that $1, [\Omega], [\Omega \wedge \Omega], \ldots, [\Omega \wedge \cdots \wedge \Omega]$ (n factors) are linearly independent and span $H(M)$ (here, for any exterior differential form ω on M with $d\omega = 0$, $[\omega]$ is the De Rham class in $H(M)$ that contains ω).

(f) Let $1 \leq k \leq n$, and let M_k be the subset of M consisting of all one-dimensional subspaces lying in the subspace of $\mathbf{C}^{n+1}$ defined by $z_{k+2} = \cdots = z_{n+1} = 0$. Verify that M_k is a compact submanifold of M, and evaluate $\int_{M_k} \Omega \wedge \cdots \wedge \Omega$ (k factors) (after orienting M_k).

(g) Let M_R be the subset of M consisting of all one-dimensional subspaces of $\mathbf{C}^{n+1}$ spanned by elements of $\mathbf{R}^{n+1}$. Prove that M_R is an analytic compact submanifold of M of (real) dimension n. Deduce from (c)–(e) that $\int_{M_R} \omega = 0$ for any closed n-form on M. (If n is odd, this is clear because $\dim H^n(M) = 0$ by (c). If $n = 2k$, note that $\Omega \wedge \cdots \wedge \Omega$ (k factors) is identically 0 on M_R.)

55. Let M be the space of projective lines in $\mathbf{P}_3(\mathbf{C})$ (= space of 2-dimensional linear subspaces of $\mathbf{C}^4$; cf. Exercise 38 of Chapter 2). Prove that $P_M(t) = 1 + t^2 + 2t^4 + t^6 + t^8$. (Let $U = U(4,\mathbf{C})$. Then U acts transitively on M, and (U,M) is a symmetric pair.)

For these results cf. Cartan [3]. See Weyl [1] for calculations leading to the determination of the P_G when G is compact classical. See also Samelson [1], Borel [1].

56. Let $\mathfrak{g}$ be a semisimple Lie algebra over $\mathbf{C}$, I the algebra of polynomial functions on $\mathfrak{g}$ invariant under the adjoint group of $\mathfrak{g}$.

(a) Let $p_1, \ldots, p_l$ be homogeneous elements of I of degrees $d_1, \ldots, d_l$ respectively ($d_j > 0$) which are algebraically independent and which generate I (together with 1). Prove that the coefficient of t^k in the formal expansion of $\prod_{1 \le i \le l}(1 - t^{d_i})^{-1}$ in powers of the indeterminate t is precisely the dimension of the subspace of homogeneous elements of degree k in I.

(b) Deduce from (a) that the integers $d_1, \ldots, d_l$ are uniquely determined (up to a permutation) by I (the d_i are called the *primitive degrees* and the p_i are said to *generate I freely*).

(c) Let $q_1, \ldots, q_l$ be homogeneous algebraically independent elements of I such that $\deg(q_i) = \deg(p_i)$, $1 \le i \le l$. Prove that the q_i freely generate I.

(d) Let $\mathfrak{g}_0$ be a real form of $\mathfrak{g}$. Prove the existence of homogeneous elements $p_1, \ldots, p_l$ of positive degree freely generating I such that each p_i is real-valued on $\mathfrak{g}_0$.

57. Let k be a field of characteristic 0, l an integer ≥ 1, and Π_l the group of permutations of $\{1, \ldots, l\}$ acting naturally on k^l. $\mathcal{E} = \{\varepsilon = (\varepsilon_1, \ldots, \varepsilon_l), \varepsilon_i = \pm 1$ for all $i\}$; $\mathcal{E}^+ = \{\varepsilon : \varepsilon_1 \cdots \varepsilon_l = +1\}$. $\mathcal{E}$ is considered an abelian group under componentwise multiplication and is allowed to act on k^l as follows: $\varepsilon : (a_1, \ldots, a_l) \mapsto (\varepsilon_1 a_1, \ldots, \varepsilon_l a_l)$. $\mathfrak{D}$ and $\mathfrak{D}^+$ are the subgroups of $GL(l,k)$ defined by $\mathfrak{D} = \mathcal{E}\,\Pi_l$, $\mathfrak{D}^+ = \mathcal{E}^+\,\Pi_l$. x_i are the linear forms $(a_1, \ldots, a_l) \mapsto a_i$ on k^l. J, I, I^+ are the respective algebras of polynomials in the x_i invariant under Π_l, $\mathfrak{D}$, $\mathfrak{D}^+$. t is an indeterminate.

(a) Prove that coefficient of the polynomial $\prod_{1 \le i \le l}(t + x_i)$ freely generate J.

(b) Prove that the coefficients of $\prod_{1 \le i \le l}(t + x_i^2)$ freely generate I.

(c) Let p_j be the coefficient of t^j in $\prod_{1 \le i \le l}(t + x_i^2)$ $(1 \le j \le l - 1)$. Let $p_l = x_1 \cdots x_l$. Prove that $p_1, \ldots, p_l$ are algebraically independent and, together with 1, generate I^+. (Hint: Write $\sigma(\varepsilon) = \varepsilon_1 \cdots \varepsilon_l$, and call a polynomial f in the x_i *skew* if $f^\varepsilon = \sigma(\varepsilon)f$ for all $\varepsilon \in \mathcal{E}$. First prove that f is skew if and only if $f = p_l g$ for some polynomial g which is invariant under $\mathcal{E}$. Use this to prove that $I^+ = I + p_l I$, the sum being direct. Now use (b).)

58. Let $\mathfrak{g}$ be a classical simple Lie algebra over $\mathbf{C}$ and π, the basic representation of $\mathfrak{g}$, namely, the one which defines it. Let $F(t : X) \equiv \det(t + \pi(X))$ $(X \in \mathfrak{g}$, t an indeterminate). I is as in Exercise 56.

(a) Let $\mathfrak{g} = A_l$. Then $F(t : X) \equiv t^{l+1} + \sum_{1 \le \nu \le l} p_\nu(X)^{\nu-1}$, and the p_ν $(1 \le \nu \le l)$ freely generate I.

(b) Let $\mathfrak{g} = B_l$. Then $F(t : X) \equiv t^{2l+1} + \sum_{1 \le \nu \le l} p_\nu(X)t^{2\nu-1}$, and the p_ν $(1 \le \nu \le l)$ freely generate I.

(c) Let $\mathfrak{g} = C_l$. Then $F(t : X) \equiv t^{2l} + \sum_{1 \le \nu \le l} p_\nu(X)t^{2(\nu-1)}$, and the p_ν $(1 \le \nu \le l)$ freely generate I.

(Hint: For these results, use Exercise 57, the calculations of §4.4, and Chevalley's theorem.)

(d) Let $\mathfrak{g} = D_l$. Prove the existence of an element $p_l \in I$ such that $p_l(X)^2 = \det \pi(X)$ for all $X \in \mathfrak{g}$. Verify that p_l is homogeneous of degree l.

(In the notation of §4.4, $\det \pi(X) = (-1)^l \lambda_1(X)^2 \cdots \lambda_l(X)^2$ if $X = (\text{diag}(a_1, \ldots, a_l), 0, 0)$. Observe now that $\lambda_1 \cdots \lambda_l$ is invariant under the Weyl group, so $\exists\, p_l \in I_P$ such that $p_l | \mathfrak{h} = (-1)^{l/2} \lambda_1 \cdots \lambda_l$.)

(e) Let $\mathfrak{g} = D_l$. Then $F(t : X) \equiv t^{2l} + p_l(X)^2 + \sum_{1 \le \nu \le l-1} p_\nu(X) t^{2\nu}$, and the p_ν $(1 \le \nu \le l)$ freely generate I. (Hint: Use Exercise 57.)

(f) Obtain the primitive degrees of the classical simple Lie algebras.

59. Prove that the primitive degrees of $\mathfrak{g} = G_2$ are 2 and 6, and that those of F_4 are 2, 6, 8, and 12.

60. (a) Determine for which of the simple Lie algebras $\mathfrak{g}$ over **C** it is true that the coefficients of the characteristic polynomial of ad X generate I.

(b) Let $\mathfrak{g} = A_l$, π an irreducible representation of $\mathfrak{g}$ with highest weight λ. Determine necessary and sufficient conditions on λ which ensure that the coefficients of the characteristic polynomial of $\pi(X)$ generate I.

61. Let k be a field of characteristic 0, $\mathfrak{g} = \mathfrak{sl}(2,k)$, and H, X, and Y as in (4.2.1). ρ is a representation $\mathfrak{g}$ in a (finite-dimensional) vector space V over k.

(a) Prove that the formulae (4.2.5) define an irreducible representation π_j of $\mathfrak{g}$, that these are all the irreducible representations of $\mathfrak{g}$, and that they stay irreducible in any extension field of k.

(b) Let d_k be the multiplicity with which π_k occurs in ρ. Let $V_t = \{v : v \in V, \rho(H)v = tv\}$ $(t \in \mathbf{Z})$. If W is the null space of $\rho(X)$, prove that W is invariant under $\rho(H)$, that $\rho(H)|W$ has only nonnegative integral eigenvalues, and that $d_k = \dim(V_k \cap W)$ $(k = 0,1,2, \ldots)$.

(c) Prove that V is the direct sum of the range of $\rho(X)$ and the null space of $\rho(Y)$.

(d) Prove that $\rho(X)$ maps V_k onto V_{k+2} $(k = 0,1,2, \ldots)$, while $\rho(Y)$ maps V_k onto V_{k-2} $(k = 0,-1,-2, \ldots)$.

(e) Prove that $\sum_{k \ge 0} d_{2k} = \dim(V_0)$.

(f) Prove that $d_k = \dim(V_k) - \dim(V_{k+2})$ $(k = 0,1,2, \ldots)$.

62. (a) (Theorem of Jacobson–Morozov) Let $\mathfrak{g}$ be a semisimple Lie algebra over a field k of characteristic 0, $X \ne 0$ a nilpotent element of $\mathfrak{g}$. Prove the existence of elements $H, Y \in \mathfrak{g}$ such that $[H,X] = 2X$, $[H,Y] = -2Y$, $[X,Y] = H$. (See Jacobson [1] and also Kostant [1]. $\{H,X,Y\}$ is called an *S-triple.*)

(b) Let $\{H,X,Y\}$ be an S-triple, $\mathfrak{a} = k \cdot H + k \cdot X + k \cdot Y$, $\mathfrak{g}_X$ the centralizer of X. Prove that X and Y are nilpotent, H is semisimple, and ad H has only integral eigenvalues. If $n^0(\mathfrak{a})$ (resp. $n^e(\mathfrak{a})$) is the number of irreducible constituents in the decomposition of the adjoint representation of $\mathfrak{g}$ restricted to $\mathfrak{a}$ with odd dimension (resp. even dimension), prove that $n^0(\mathfrak{a}) \ge l$ $(= rk\, \mathfrak{g})$ and that $n(\mathfrak{a}) = n^0(\mathfrak{a}) + n^e(\mathfrak{a}) = \dim \mathfrak{g}_X \ge l$. Deduce that the following statements are equivalent: (i) $n(\mathfrak{a}) = l$, (ii) $n^e(\mathfrak{a}) = 0$ and H is regular, and (iii) $\dim \mathfrak{g}_X = l$.

(c) Let $\mathfrak{g}_H$ be the centralizer of H and $\mathfrak{g}_2 = \{Z : Z \in \mathfrak{g}, [H,Z] = 2Z\}$. Prove that $[\mathfrak{g}_2, \mathfrak{g}_H] \subseteq \mathfrak{g}_2$. If $\mathfrak{g}_2' = \{Z : Z \in \mathfrak{g}_2, [Z, \mathfrak{g}_H] = \mathfrak{g}_2\}$, prove that $X \in \mathfrak{g}_2'$. (Hint: Use representation theory of $\mathfrak{a}$ and Exercise 61.)

(d) Let $k = \mathbf{R}$ or $\mathbf{C}$, G the adjoint group of $\mathfrak{g}$, and G_H the analytic subgroup of G defined by $\mathfrak{g}_H$. If $k = \mathbf{C}$ and $X' \in \mathfrak{g}_2$, prove that there exists $Y' \in \mathfrak{g}$

such that $\{H,X',Y'\}$ is an S-triple if and only if $X' \in \mathfrak{g}_2'$. Prove also that in both the real and complex cases G_H acts naturally on the set $\mathfrak{S}_H$ of all S-triples containing H and that $\mathfrak{S}_H$ splits into finitely many G_H orbits. Prove, finally, that $\mathfrak{S}_H = G_H \cdot \{H,X,Y\}$ if $k = \mathbf{C}$. (For the first assertion, $X' \in \mathfrak{g}_2'$ is necessary by (c). Prove, by a differential calculation, that $G_H \cdot X'$ is an open subset of $\mathfrak{g}_2'$ for each $X' \in \mathfrak{g}_2'$. Deduce from Whitney's theorem that $\mathfrak{g}_2'$ splits into finitely many orbits, since $\mathfrak{g}_2 \setminus \mathfrak{g}_2'$ is an algebraic set. If $k = \mathbf{C}$, $\mathfrak{g}_2'$ is connected, so there is only one orbit.)

Exercises 63–68 study the orbit structure of the adjoint representation of a complex semisimple Lie algebra. As an outcome we also obtain the values of the primitive degrees as defined in Exercise 56. (See Kostant [1]; also, Varadarajan [1].) $\mathfrak{g}$ is a fixed semisimple Lie algebra over $\mathbf{C}$, $\mathfrak{N}$ is the set of nilpotent elements of $\mathfrak{g}$, and G is the adjoint group of $\mathfrak{g}$. $\mathfrak{h}$ is a CSA, P is a positive system of roots of $(\mathfrak{g},\mathfrak{h})$, $\alpha_1, \ldots, \alpha_l$ are the simple roots in P, and $\mathfrak{n} = \sum_{\alpha \in P} \mathfrak{g}_\alpha$.

63. Let $0 \neq X \in \mathfrak{g}$. Denote by $\mathfrak{z}$ the centralizer of X in $\mathfrak{g}$.

(a) Prove that $(Y,Z) \mapsto \langle Y,[X,Z]\rangle$ $(Y, Z \in \mathfrak{g})$ induces a nonsingular skew-symmetric bilinear form on $\mathfrak{g}/\mathfrak{z} \times \mathfrak{g}/\mathfrak{z}$.

(b) Deduce from (a) that $\dim([X,\mathfrak{g}]) = \dim(\mathfrak{g}/\mathfrak{z})$ is even.

(c) Let G_X be the stabilizer of X in G. Prove that the map $xG_X \mapsto X^x$ is an imbedding of G/G_X into $\mathfrak{g}$. Deduce that the orbit X^G of X in $\mathfrak{g}$ is an analytic submanifold of $\mathfrak{g}$ of even dimension.

64. (a) Suppose that $\{H,X,Y\}$ is an S-triple with $H \in \mathfrak{h}$ and that $\alpha_i(H) \geq 0$ for $1 \leq i \leq l$. Prove that $\alpha_i(H) = 0, 1$, or 2 for each $i = 1, \ldots, l$. (Let $Y_i \in \mathfrak{g}_{-\alpha_i}$ be nonzero, $Z = [X,Y_i]$. Then $Z \in \mathfrak{h} + \mathfrak{n}$. But $[H,Z] = (2 - \alpha_i(H))Z$, so $2 - \alpha_i(H) \geq 0$.)

(b) Let X be nilpotent. Prove that $X^x \in \mathfrak{n}$ for some $x \in G$. (Take an S-triple $\{H,X,Y\}$ and find $x \in G$ such that $H^x \in \mathfrak{h}$ and $\alpha(H^x) \geq 0$ for all $\alpha \in P$.)

(c) Prove that the set of nilpotent elements of a semisimple Lie algebra over $\mathbf{R}$ or $\mathbf{C}$ splits into finitely many orbits under its adjoint group. (By (a), $\exists$ only finitely many H in a given CSA which form the first member of an S-triple. Now use (d) of Exercise 62, and note that any semisimple element can be moved by the adjoint group into one of a fixed finite set of CSA's.)

65. A nilpotent $X \in \mathfrak{g}$ is called *principal* if the dimension of its centralizer is $l\,(= rk(\mathfrak{g}))$.

(a) Let $X \in \mathfrak{N}$, $\{H,X,Y\}$ an S-triple. Prove that X is principal if and only if H is regular and all eigenvalues of ad H are even integers.

(b) Let $H_i \in \mathfrak{h}$ be such that $\alpha_j(H_i) = 2\delta_{ij}$, $H = H_1 + \cdots + H_l$. Let $X_i \in \mathfrak{g}_{\alpha_i}$ be nonzero, and let m_i be such that $H = m_1 H_{\alpha_1} + \cdots + m_l H_{\alpha_l}$. Prove that $m_i > 0$ for all i. If $Y_i \in \mathfrak{g}_{\alpha_i}$ is such that $\langle X_i,Y_i\rangle = 1$, $X = X_1 + \cdots + X_l$, and $Y = m_1 Y_1 + \cdots + m_l Y_l$, prove that $\{H,X,Y\}$ is an S-triple and X is principal. ($m_i > 0$ by Exercise 15(a); X is principal by (a) above.)

(c) The principal nilpotents in $\mathfrak{n}$ form a Zariski-open nonempty subset of $\mathfrak{n}$. (Hint: $X' \in \mathfrak{n}$ is principal nilpotent if and only if ad X' has rank $n - l$, where $n = \dim \mathfrak{g}$.)

(d) Prove that the set of principal nilpotents is an open, dense, connected subset of $\mathfrak{N}$ and forms a single orbit under G. Deduce that it is a regular complex submanifold of $\mathfrak{g}$ of dimension $n - l$ ($n = \dim \mathfrak{g}$). (If X' is a principal nilpotent, for suitable $H, Y, x \in G$, $\{H, X = X'^x, Y\}$ is an S-triple, with $H \in \mathfrak{h}$, and $\alpha_i(H) \geq 0$ for all i. Then $H = H_1 + \cdots + H_l$ by Exercise 64(a) and (a) above. Now use Exercise 62(d). For denseness, use (c) above and Exercise 64(b). X'^G is thus locally compact; by Exercise 63(c), the imbedding of $X^{G'}$ in $\mathfrak{g}$ is regular.)

(e) Let X be a principal nilpotent of $\mathfrak{g}$, $\{H,X,Y\}$ an S-triple. Prove that Y is also a principal nilpotent. Prove further that ad H leaves $\mathfrak{g}_X$ and $\mathfrak{g}_Y$ invariant and that we can select bases $\{X_1, \ldots, X_l\}$ for $\mathfrak{g}_X$, $\{Y_1, \ldots, Y_l\}$ for $\mathfrak{g}_Y$, and even integers $\lambda_1, \ldots, \lambda_l > 0$ such that $[H,X_i] = \lambda_i X_i$, $[H,Y_i] = -\lambda_i Y_i$, $1 \leq i \leq l$. Deduce that $\mathfrak{g}_X$ and $\mathfrak{g}_Y$ consist entirely of nilpotent elements. (By Exercise 62(d) and (d) above, we may assume that H, X, Y are as in (b). If $Z \in \mathfrak{g}_X$ and $[H,Z] = 0$, $Z \in \mathfrak{g}_X \cap \mathfrak{h} = 0$. So all eigenvalues of ad H in $\mathfrak{g}_X$ are > 0. If $\mathfrak{g}_s = \{Z : Z \in \mathfrak{g}, [H,Z] = sZ\}$ and $X' \in \mathfrak{g}_X$, then $(\text{ad } X')^q(Z) \in \sum_{t \geq 2q+s} \mathfrak{g}_t$ for $Z \in \mathfrak{g}_s$, $q \geq 1$; thus ad X' is nilpotent.)

66. Let notation be as above. Let p_i ($1 \leq i \leq l$) be homogeneous elements of I (cf. Exercise 56) generating I. Let $\mathbf{p}$ be the map $X \mapsto (p_1(X), \ldots, p_l(X))$ of $\mathfrak{g}$ into $\mathbf{C}^l$. For $\xi \in \mathbf{C}^l$, V_ξ is the set $\mathbf{p}^{-1}(\{\xi\})$. If $X \in \mathfrak{g}$ is semisimple, nilpotent, etc., the orbit X^G is said to be semisimple, nilpotent, etc.

(a) Let $\bar{\mathbf{p}} = \mathbf{p}|\mathfrak{h}$. Prove that $\bar{\mathbf{p}}$ is proper and $d\bar{\mathbf{p}}$ is nonsingular at all regular points of $\mathfrak{h}$. (If $\{\omega_i\}$ is a basis of $\mathfrak{h}^*$, z_i the coordinates on $\mathbf{C}^l$, and π the product of all $\alpha \in P$, then $\exists$ a constant $c \neq 0$ such that $\bar{\mathbf{p}}^*(dz_1 \wedge \cdots \wedge dz_l) = c\pi \cdot \omega_1 \wedge \cdots \wedge \omega_l$. $\bar{\mathbf{p}}$ is proper because if $\mathbf{p}(X)$ is bounded, the coefficients of the characteristic polynomial of ad X are bounded, so $\alpha(X)$ is bounded for each root α ($X \in \mathfrak{h}$).)

(b) Deduce from (a) that $\bar{\mathbf{p}}$ maps $\mathfrak{h}$ onto $\mathbf{C}^l$, and hence prove that $\mathbf{p}$ induces a bijection of the set of all semisimple orbits onto $\mathbf{C}^l$. Deduce that for each $\xi \in \mathbf{C}^l$, V_ξ is nonempty and contains a unique semisimple orbit.
($\bar{\mathbf{p}}[\mathfrak{h}]$ contains a Zariski-open subset of $\mathbf{C}^l$ by a standard result—cf. Jacobson [1] Chapter 9; $\bar{\mathbf{p}}[\mathfrak{h}]$ is closed since $\bar{\mathbf{p}}$ is proper. So $\bar{\mathbf{p}}[\mathfrak{h}] = \mathbf{C}^l$. For the second result, note that if $X, X' \in \mathfrak{h}$, $\mathbf{p}(X) = \mathbf{p}(X')$ if and only if X and X' lie in the same $\mathfrak{w}$-orbit, by Chevalley's theorem. Now use Theorem 4.9.1.)

(c) Let $S \in \mathfrak{g}$ be a semisimple element; $\mathfrak{z}$ the centralizer of S; $\mathfrak{z}_1 = \mathfrak{D}\mathfrak{z}$ (cf. Exercise 45). Prove that if $X \in \mathfrak{z}_1$, X is nilpotent in $\mathfrak{z}_1$ if and only if it is nilpotent in $\mathfrak{g}$ and that $\mathfrak{z} \cap \mathfrak{N} \subseteq \mathfrak{z}_1$. Prove also that if $\{H,X,Y\}$ is an S-triple in $\mathfrak{z}_1$, $(s + X)^{\exp tH} \longrightarrow S$ as $t \longrightarrow 0$. Deduce that the closure of any orbit contains a semisimple orbit. (For the last result, use Jordan decomposition.)

(d) Let $\xi \in \mathbf{C}^l$, and let $S \in V_\xi$ be semisimple. Let $\mathfrak{z}$ and $\mathfrak{z}_1$ be as in (c). Prove that $V_\xi = \{(S + N)^x : x \in G, N \in \mathfrak{z}_1 \cap \mathfrak{N}\}$. Prove, further, that if $N, N' \in \mathfrak{z}_1 \cap \mathfrak{N}$, $S + N$ and $S + N'$ are conjugate under G if and only if N and N' are conjugate under the adjoint group of $\mathfrak{z}_1$. (If $x \in G$ and $(S + N)^x = S + N'$, then x centralizes S and so belongs to the *analytic subgroup* of G defined by $\mathfrak{z}$, by Exercise 45. But then $x|\mathfrak{z}_1$ lies in the adjoint group of $\mathfrak{z}_1$.)

(e) Prove that each V_ξ splits into finitely many orbits.

(f) Prove that the semisimple orbits are precisely the closed ones. Prove also that if $X \in \mathfrak{g}$ is regular and $\xi = \mathbf{p}(X)$, then $X^G = V_\xi$. (Each orbit is σ-compact, so by (e) and the Baire category theorem, V_ξ contains an orbit open in it. Using (e) repeatedly, we can write $V_\xi = \mathbf{\Omega}_1 \cup \cdots \cup \mathbf{\Omega}_k$, where each $\mathbf{\Omega}_i$ is an orbit, and is open in $\mathbf{\Omega}_i \cup \mathbf{\Omega}_{i+1} \cup \cdots \cup \mathbf{\Omega}_k$, which is closed. So V_ξ has a closed orbit, namely $\mathbf{\Omega}_k$. By (c) this must be semisimple.)

(g) For any $X \in \mathfrak{g}$, prove that X^G is open in its closure, hence that it is locally compact. Deduce that X^G is a regular complex submanifold of $\mathfrak{g}$ of dimension $\dim(\mathfrak{g}) - \dim(\mathfrak{g}_X)$, $\mathfrak{g}_X$ being the centralizer of X.

67. An element $X \in \mathfrak{g}$ is called *principal* if $\dim \mathfrak{g}_X = l$ where $\mathfrak{g}_X$ is the centralizer of X.

(a) Prove that $X \in \mathfrak{g}$ is principal if and only if the orbit X^G has maximal dimension (among all orbits) and that this dimension is $n - l$.

(b) Let $X \in \mathfrak{g}$, and let $X = S + N$ be its Jordan decomposition. Prove that X is principal if and only if N is a principal nilpotent in the derived algebra of the centralizer of S. (Let $\mathfrak{z}$ be as in Exercise 66(c). Then $\mathrm{rk}(\mathfrak{z}) = \mathrm{rk}(\mathfrak{g})$ and $\mathfrak{g}_X = \mathfrak{z} \cap \mathfrak{g}_N$.)

(c) Let $\xi \in \mathbf{C}^l$. Prove that V_ξ contains a unique principal orbit which is open and dense in V_ξ. (Hint: Use Exercises 66(d) and 65(c).)

(d) Deduce from (c) that $\mathbf{p}$ induces a bijection of the set of all principal orbits onto $\mathbf{C}^l$.

(e) Let $X \in \mathfrak{g}$ be principal. Prove that $\mathfrak{g}_X$ is abelian. (Take X_n regular, $n = 1,2, \ldots, X_n \longrightarrow X$. $\mathfrak{g}_{X_k}$ are CSA's and $\longrightarrow \mathfrak{g}_X$ in the Grassman manifold of l-planes of $\mathfrak{g}$. Use a compactness argument to show that $\mathfrak{g}_X$ is abelian; see Kostant's paper [3].)

68. Let $p_1, \ldots, p_l$ be homogeneous generators of I, $d_i = \deg(p_i)$. We assume that $d_1 \leq \cdots \leq d_l$. $X \in \mathfrak{g}$ is a principal nilpotent; $\{H,X,Y\}$ an S-triple; $\mathfrak{g}_Y$, $Y_1, \ldots, Y_l$, $\lambda_1, \ldots, \lambda_l$ have the same meaning as in Exercise 67(f). We assume $\lambda_1 \leq \cdots \leq \lambda_l$. Put $\nu_j = 1 + \frac{1}{2}\lambda_j$. E is the Euler vector field on $\mathfrak{g}$ which assigns to $Z \in \mathfrak{g}$ the tangent vector Z at Z. As usual, we identify the tangent spaces to $G \times \mathbf{C}^l$ (resp. $\mathfrak{g}$) at each point with $\mathfrak{g} \times \mathbf{C}^l$ (resp. $\mathfrak{g}$).

(a) Prove that $[X,\mathfrak{g}] + \mathfrak{g}_Y = \mathfrak{g}$ is a direct sum. (Hint: See Exercise 61(c).)

(b) Let ψ be the map $(x, (u_1, \ldots, u_l)) \mapsto (X + u_1 Y_1 + \cdots + u_l Y_l)^x$ of $G \times \mathbf{C}^l$ into $\mathfrak{g}$. Deduce from (a) the existence of a connected open neighborhood N of $(0, \ldots, 0)$ in $\mathbf{C}^l$ such that $(d\psi)_{(x,(u_1,\ldots,u_l))}$ is surjective for all $x \in G$, $(u_1, \ldots, u_l) \in N$.

(c) Let $Z = \frac{1}{2}H$, $v_j = \nu_j u_j$ $(1 \leq j \leq l,\ u_j \in \mathbf{C})$. Verify that

$$(d\psi)_{(1,(u_1,\ldots,u_l))}(Z,(v_1, \ldots, v_l)) = X + u_1 Y_1 + \cdots + u_l Y_l.$$

(d) Let $\mathbf{\Omega}_N = \{(X + u_1 Y_1 + \cdots + u_n Y_n)^x : (u_1, \ldots, u_l) \in N,\ x \in G\}$. Let $\mathfrak{A}$ be the algebra of all G-invariant holomorphic functions on $\mathbf{\Omega}_N$ and for

$f \in \mathfrak{a}$, let $\tilde{f}$ be the holomorphic function on N defined by $\tilde{f}(u_1, \ldots, u_l) = f(X + u_1 Y_1 + \cdots + u_l Y_l)$, $(u_1, \ldots, u_l) \in N$. Prove that $f \mapsto \tilde{f}$ is an injection of $\mathfrak{a}$ into the algebra of holomorphic functions on N and that if $\tilde{E}$ is the differential operator $\sum_{1 \le j \le l} v_j u_j (\partial/\partial u_j)$ on N, $\widetilde{Ef} = \tilde{E}\tilde{f}$ $(f \in \mathfrak{a})$.

(e) Let $1 \le j \le l$. Prove that $\tilde{p}_j$ is a linear combination of those monomials $u_1^{m_1} \cdots u_l^{m_l}$ for which $\sum_{1 \le k \le l} v_k m_k = d_j$.

$$(\tilde{E}\tilde{p}_j = d_j \tilde{p}_j \text{ and } \tilde{E}(u_1^{m_1} \cdots u_l^{m_l}) = (\textstyle\sum v_k m_k) \cdot (u_1^{m_1} \cdots u_l^{m_l})).$$

(f) Prove that $v_j = d_j$ for $1 \le j \le l$. (If $v_j > d_j$ for some j, then $d_s < v_t$ for $s \le j \le t$, so by (e), for $s \le j$, $\tilde{p}_s$ is a polynomial of only $u_1, \ldots, u_{j-1}$. $\tilde{p}_1, \ldots, \tilde{p}_j$ are thus algebraically dependent, contradicting (d). (Further, $\sum_{1 \le j \le l} v_j = \sum_{1 \le j \le l} d_j = \frac{1}{2}(l + \dim(\mathfrak{g}))$, by Chevalley's theorem and the results of the appendix.)

(g) For any $\alpha = m_1\alpha_1 + \cdots + m_l\alpha_l \in P$, let $O(\alpha) = m_1 + \cdots + m_l$. For any integer $k \ge 1$, let b_k be the number of $\alpha \in P$ with $O(\alpha) = k$. Prove that $b_1 \ge b_2 \ge \cdots$ and that b_k is the number of the d_j's which are equal to $k + 1$. (Let H, X, Y be as in Exercise 65(b). Then $[H, X_\alpha] = 2O(\alpha)X_\alpha$ if $\alpha \in P$, $X_\alpha \in \mathfrak{g}_\alpha$. By representation theory of $\mathbf{C} \cdot H + \mathbf{C} \cdot X + \mathbf{C} \cdot Y$, b_k is the number of the λ_j's which are $\ge 2k$.)

69. V, V_c, Δ, $\mathfrak{w}$ are as in 2–5 of the appendix to Chapter 4. $\mathcal{P}$ (resp. $\mathcal{S}$) is the graded polynomial algebra (resp. symmetric algebra) over V_c; $p \mapsto \tilde{p}$ is the algebra isomorphism of $\mathcal{P}$ onto $\mathcal{S}$ that extends the linear isomorphism of V^* with V induced by the scalar product of V. $p \mapsto p^{\text{conj}}$ (resp. $u \mapsto u^{\text{conj}}$) is the conjugation of $\mathcal{P}$ (resp. $\mathcal{S}$) corresponding to the conjugation of V_c^* (resp. V_c) induced by V^* (resp. V).

(a) Prove that $p, q \mapsto \langle p, q \rangle = (\partial(\tilde{q}^{\text{conj}})p)(0)$ is a positive definite Hermitian bilinear form on $\mathcal{P} \times \mathcal{P}$. (Hint: Let $\{e_1, \ldots, e_l\}$ be an orthonormal basis of V^* (over $\mathbf{R}$). Then $\langle e_1^{r_1} \cdots e_l^{r_l}, e_1^{s_1} \cdots e_l^{s_l} \rangle = \delta_{r_1 s_1} \cdots \delta_{r_l s_l} r_1! \cdots r_l!$).

(b) Let I (resp. J) be the algebra of $\mathfrak{w}$-invariant elements of $\mathcal{P}$ (resp. $\mathcal{S}$), I^+ (resp. J^+) the linear span in I (resp. J) of the homogeneous elements of positive degree. Let $H = \{p : p \in \mathcal{P}, \partial(u)p = 0 \text{ for all } u \in J^+\}$. Prove that H is a graded subspace of $\mathcal{P}$ and that $\mathcal{P} = \mathcal{P}I^+ + H$ is an orthogonal direct sum. (Hint: $\mathcal{S}J^+ = (\mathcal{P}I^+)^\sim$ and $p \in H \Leftrightarrow (\partial(u)p)(0) = 0 \;\forall\; u \in \mathcal{S}J^+$. The elements of H are called *harmonic*.)

(c) Deduce from (b) that $\dim H = [\mathfrak{w}]$.

(d) Let P be a positive system. Let $\pi \in \mathcal{P}$ be such that $\tilde{\pi}$ is the element $\prod_{\alpha \in P} \alpha$ of $\mathcal{S}$. For $s \in \mathfrak{w}$ let $\epsilon(s) = \det(s)$. Prove that π is skew-symmetric, i.e., $\pi^s = \epsilon(s)\pi \;\forall\; s \in \mathfrak{w}$, and that $I\pi$ is the space of all skew-symmetric elements of $\mathcal{P}$.

(e) Prove that $\pi \in H$. (If $u \in J^+$, $\partial(u)\pi$ is skew-symmetric and of lower degree.)

(f) Prove that $\deg(P) \le \deg(\pi) \;\forall\; p \in H$ and that $\deg(p) = \deg(\pi)$ if and only if $p = c\pi$ for some nonzero constant c. (Use the Poincaré Series for H.)

70. Let notation be as above, with $w = [\mathfrak{w}]$; $u_1 = 1, u_2, \ldots, u_w$, a basis of homogeneous elements of H. $V_c' = \{\lambda : \lambda \in V_c, \pi(\lambda) \ne 0\}$.

(a) Prove that $\sum_{1 \leq i \leq w} \deg(u_i) = \frac{1}{2}w[P]$. (If $P(t)$ is the Poincaré Series for H,

$$\sum_{1 \leq i \leq w} \deg(u_i) = \left(\frac{d}{dt}P(t)\right)_{t=1}.)$$

(b) For $\xi \in V_c$, let $\mathcal{P}_\xi$ (resp. I_ξ) be the set of all p in $\mathcal{P}$ (resp. I) vanishing at ξ. Prove that if $\xi \in V_c'$, $\mathcal{P}_\xi = \mathcal{P}I_\xi$ and that $\mathcal{P} = \mathcal{P}_\xi + H$ is a direct sum. (Since $\mathcal{P} \simeq I \otimes H$, $\mathcal{P} = \mathcal{P}I_\xi + H$ is a direct sum; observe now that $[\mathfrak{w} \cdot \xi] = w$.)

(c) Prove that if $\xi \in V_c'$, the restriction map $u \mapsto u|\mathfrak{w} \cdot \xi$ is a linear isomorphism of H onto the complex vector space of all functions on $\mathfrak{w} \cdot \xi$. Deduce that the representation of $\mathfrak{w}$ in H is equivalent to its regular representation.

(d) Let U be the $w \times w$ matrix $(u_{js})_{1 \leq j \leq w, s \in \mathfrak{w}}$, where $u_{js} = u_j^s$. Prove that $\det(U) = c\pi^{(1/2)w}$, where $c \neq 0$ is a constant. (Let $\alpha \in \widetilde{P}$. Subtracting column $s_\alpha s$ from column s, prove that $\alpha^{(1/2)w}$ divides $\det(U)$. Hence $\det(U) = c\pi^{(1/2)w}$, where $c \in \mathcal{P}$. c is a nonzero constant by (a) and (c).)

(e) Prove that there are unique rational functions $v_1, \ldots, v_w$ such that $\sum_{s \in \mathfrak{w}} u_j^s v_k^s = \delta_{jk}$ $(1 \leq j,k \leq w)$. (Let $Q = U^{-1} = (q_{tk})_{t \in \mathfrak{w}, 1 \leq k \leq w}$; write $q_{1k} = v_k$. Then $QU = 1 \Rightarrow \sum_{1 \leq k \leq w} v_k u_k^s = \delta_{1s}$ $(s \in \mathfrak{w}) \Rightarrow \sum_{1 \leq k \leq w} v_k^t u_k^{t'} = \delta_{tt'}$ $(t, t' \in \mathfrak{w})$. So $q_{tk} = v_k^t$. Now use $UQ = 1$.)

(f) Prove that v_k is well defined on V_c' and that $\pi v_k \in \mathcal{P}$ $(1 \leq k \leq w)$. (Argue as in (d) to prove that $\pi^{(1/2)w-1}$ divides the cofactors in $\det(U)$.) (The results are due to Harish-Chandra [7, 8].)

71. Let notation be as above. For $p \in \mathcal{P}$, Let $z_{ij:p}$ be the unique elements of I such that $pu_j = \sum_{1 \leq i \leq w} z_{ij:p} u_i$ $(1 \leq j \leq w)$. Let $\Gamma(p)$ be the $w \times w$ matrix $(z_{ji:p})$. For $\xi \in V_c$ let $\Gamma(p:\xi)$ be the $w \times w$ matrix $(z_{ji:p}(\xi))$.

(a) Prove that $p \mapsto \Gamma(p:\xi)$ is a representation of $\mathcal{P}$ by $w \times w$ matrices.

(b) For $\xi \in V_c'$ and $s \in \mathfrak{w}$, let $\psi_s(\xi)$ be the vector in $\mathbf{C}^w$ whose components are $v_1^s(\xi), \ldots, v_w^s(\xi)$. Prove that the $\psi_s(\xi)$ $(s \in \mathfrak{w})$ form a basis for $\mathbf{C}^w$ and that $\Gamma(p:\xi)\psi_s(\xi) = p^s(\xi)\psi_s(\xi)$ $(s \in \mathfrak{w})$. Deduce that the characteristic polynomial of $\Gamma(p:\xi)$ is $\prod_{s \in \mathfrak{w}} (T - p^s(\xi))$ $(\xi \in V_c)$ and that $\Gamma(p:\xi)$ is semisimple if $\xi \in V_c'$, $p \in \mathcal{P}$.

(c) Let M be the matrix $(u_j v_k)_{1 \leq j,k \leq w}$ be order w. Prove that if $\xi \in V_c'$, then $M^s(\xi)$ $(s \in \mathfrak{w})$ are the spectral projections of $\Gamma(p:\xi)$, i.e., the projections $\mathbf{C}^w \longrightarrow \mathbf{C} \cdot \psi_s(\xi)$ corresponding to the direct sum $\mathbf{C}^w = \sum_{s \in \mathfrak{w}} \mathbf{C} \cdot \psi_s(\xi)$. (See Harish-Chandra [7, 8].)

BIBLIOGRAPHY

ADO, I. D.

[1] Über die Darstellung der endlichen kontinuierlichen Gruppen durch lineare substitutionen, *Bull. Soc. Physico-Mathématique Kazan*, **7**, 1934–35, 3–43.

[2] The representation of Lie algebras by matrices. *Uspekhi Mat. Nauk* (N.S.) **2**, No. 6 (22) 1947, 159–173. Amer. Math. Soc. Transl. No. 2 (1949).

BAKER, H. F.

[1] Alternants and continuous groups, *Proceedings of the London Math. Soc.*, Second Series **3** (1905), 24–47.

BERNSTEIN, I. N., GELFAND, I. M., and GELFAND, S. I.

[1] Structure of representations with highest weights (in Russian), *Journal of Functional Analysis and its Applications* **5** (1) (1971), 1–9.

BISHOP, R. L., and R. J. CRITTENDEN

[1] *Geometry of Manifolds*, Academic Press, New York and London, 1964.

BOREL, A.

[1] Topology of Lie groups and characteristic classes, *Bull. Amer. Math. Soc.* **61** (1955), 397–432.

BOURBAKI, N.

[1] *Éléments de Mathématique*, Livre II, *Algèbre*, Chapitre 2: Algèbre Lineaire, Hermann, Paris, 1962.

[2] *Éléments de Mathématique*, Livre II, *Algèbre*, Chapitre 7: Modules sur les Anneaux Principaux, Hermann, Paris, 1964.

[3] *Éléments de Mathématique*, *Groupes et algèbres de Lie*, Chapitre 1: Algèbres de Lie, Hermann, Paris, 1960.

[4] *Éléments de Mathématique*, *Groupes et algèbres de Lie*, Chapitre 4: Groupes de Coxeter et systèmes de Tits. Chapitre 5: Groupes engendrés par des réflexions. Chapitre 6: Systèmes de racines, Hermann, Paris, 1968.

[5] *Éléments de Mathématique*, *Groupes et algèbres de Lie*, Chapitre 2: Algèbres de Lie libres. Chapitre 3: Groupes de Lie, Hermann, Paris, 1972.

BRAUER, R., and H. WEYL

[1] Spinors in n dimensions, *Amer. J. Math.* **57** (1935), 425–449.

CARTAN, E.

[1] *Sur la structure des groupes de transformations finis et continus*, Thèse, Paris, Nony, 1894 (Oeuvres Completes, Partie I, Vol. 1, pp. 137–287).

[2] Les groupes projectifs qui ne laissent invariante aucune multiplicité plane, *Bulletin de la Société mathématique de France*, **41** (1913), 53–96 (Oeuvres Complètes, Partie I, Vol. 1, pp. 355–398).

[3] Sur les invariants intégraux de certains espaces homogènes clos et les propriétés topologiques de ces espaces, *Ann. Soc. Pol. Math.*, **8** (1929), 181–225 (Oeuvres Complètes, Partie I, Vol. 2, pp. 1081–1125).

[4] *Lecons sur la theorie des spineurs*, Hermann, Paris, 1938.

[5] Les représentations linéaires des groups de Lie, *J. Math. pures et appliquées*, **17** (1938), 1–12.

CHEVALLEY, C.

[1] *Theory of Lie Groups*, I, Princeton University Press, Princeton, 1946.

[2] *Théorie des Groupes de Lie*, II: *Groupes Algébriques*, Hermann, Paris, 1951.

[3] *Théorie des Groupes de Lie*, III: *Théorèmes Généraux sur les algèbres de Lie*, Hermann, Paris, 1955.

[4] *The Algebraic Theory of Spinors*, Columbia University Press, New York, 1954.

[5] A new kind of relationship between matrices, *Amer. J. Math.* **65** (1943), 521–531.

[6] Sur la classification des algèbres de Lie simples et de leurs representations, *C. R. Acad. Sci.*, Paris **227** (1948), 1136–1138.

[7] Invariants of finite groups generated by reflexions, *Amer. J. Math.* **77** (1955), 778–782.

DYNKIN, E. B.

[1] The structure of semisimple Lie algebras, *Uspekhi Mat. Nauk* (*N.S.*) **2** (1947), 59–127 (*Amer. Math. Soc. Translations*. No. 17 (1950)).

[2] Normed Lie algebras and analytic groups, *Amer. Math. Soc. Transl.* (1) 9, 470–534.

HARISH-CHANDRA

[1] On representations of Lie algebras, *Ann. Math.* **50** (1949), 900–915.

[2] Faithful representations of Lie algebras, *Ann. Math.* **50** (1949), 68–76.

[3] Faithful representations of Lie groups, *Proc. Amer. Math. Soc.* **1** (1950), 205–210.

[4] On some applications of the universal enveloping algebra of a semi-simple Lie algebra, *Trans. Amer. Math. Soc.* **70** (1951), 28–96.

[5] Representations of a semi-simple Lie group on a Banach space, I, *Trans. Amer. Math. Soc.* **75** (1953), 185–243.

[6] A formula for semi-simple Lie groups, *Amer. J. Math.* **79** (1957), 733–760.

[7] Spherical functions on a semi-simple Lie group, I, *Amer. J. Math.* **80** (1958), 241–310.

[8] Spherical functions on a semi-simple Lie group, II, *Amer. J. Math.* **80** (1958), 553–613.

HAUSDORFF, F.

[1] *Die symbolische Exponential Formel in der Gruppen theorie*, Berichte Über die Verhandlungen, Leipzig 1906, 19–48.

HELGASON, S.

[1] *Differential Geometry and Symmetric Spaces*, Academic Press, New York and London, 1962.

HOCHSCHILD, G.

[1] *The Structure of Lie Groups*, Holden-Day Inc., 1965.

JACOBSON, N.

[1] *Lie Algebras*, Interscience Publishers, New York and London, 1962.

KOBAYASHI, S., and K. NOMIZU

[1] *Foundations of Differential Geometry*, Interscience Publishers, New York and London, 1963.

KOSTANT, B.

[1] The principal three-dimensional subgroups and the Betti numbers of a complex simple Lie group, *Amer. J. Math.* **81** (1959), 973–1032.

[2] A formula for the multiplicity of a weight, *Trans. Amer. Math. Soc.* **93** (1959), 53–73.

[3] Lie group representations on polynomial rings, *Amer. J. Math.* **85** (1963), 327–404.

KOSZUL, J. L.

[1] Homologie et cohomologie des algèbres de Lie, *Bull. de la Société mathématique de France* **78** (1950), 65–127.

LAZARD, M.

[1] Théorie des répliques. Critère de Cartan. *Séminarie Sophus Lie: Théorie des Algèbres de Lie, Topologie des groupes de Lie* (1955), 6°, 1–9.

MAL'ČEV, A.

[1] On the simple connectedness of invariant subgroups of Lie groups, *Doklady Akademii Nauk SSSR* **34** (1942), 10–13.

MONTGOMERY, D., and L. ZIPPIN

[1] *Topological Transformation Groups*, Interscience Publishers, New York and London, 1955.

NARASIMHAN, R.

[1] *Analysis on Real and Complex Manifolds*, North Holland Publishing Company, Amsterdam, 1968.

PALAIS, R. S.

[1] On the existence of slices for actions of noncompact Lie groups, *Ann. Math.* **72** (1961), 295–323.

[2] A global formulation of the Lie theory of transformation groups. *Memoirs of the American Mathematical Society*, Number 22, 1957.

PARTHASARATHY, K. R., R. RANGA RAO, and V. S. VARADARAJAN
[1] Representations of complex semi-simple Lie groups and Lie algebras, *Ann. of Math.* **85** (1967), 383–429.

PONTRYAGIN, L. S.
[1] *Topological Groups*, 2nd ed., Gordon and Breach, New York, 1966.

SAMELSON, H.
[1] Topology of Lie groups, *Bull. Amer. Math. Soc.* **58** (1952), 2–37.

SEMINAIRE SOPHUS LIE
[1] *Théorie des Algèbres de Lie, Topologie des groupes de Lie*, Paris, 1955.

SERRE, J. P.
[1] *Algèbres de Lie Semi-Simple Complexes*, W. A. Benjamin, Inc., New York, 1966.

SHEPHARD, G. C., and I. A. TODD
[1] Finite unitary reflexion groups, *Canadian J. Math.* **6** (1954), 274–304.

SPANIER, E. H.
[1] *Algebraic Topology*, McGraw-Hill Book Company, New York, Toronto, London, 1966.

STEINBERG. R.
[1] *Finite reflexion groups* (Unpublished notes).

VARADARAJAN, V. S.
[1] On the ring of invariant polynomials on a semi-simple Lie algebra, *Amer. J. Math.* **90** (1968), 308–317.

VERMA, DAY-NAND
[1] Structure of certain induced representations of complex semisimple Lie algebras, *Bull. Amer. Math. Soc.* **74** (1968), 160–166.

WEIL, A.
[1] *L'integration dans les groups topologiques et se applications*, Hermann, Paris, 1940.

WEYL, H.
[1] *The Classical Groups*, Princeton University Press, Princeton, 1946.

[2] Theorie der Darstellung kontinuerlicher halb-einfacher Gruppen durch lineare transformationen, Teil I, *Mathematische Zeitschrift* **23** (1925), 271–309.

[3] Theorie der Darstellung kontinuerlicher halb-einfacher Gruppen durch lineare transformationen, Teil II, *Mathematische Zeitschrift* **24** (1926), 328–376.

[4] Theorie der Darstellung kontinuerlicher halb-einfacher Gruppen durch lineare transformationen, Teil III, *Mathematische Zeitschrift* **24** (1926), 377–395.

[5] *The theory of groups and quantum mechanics*, Dover Press

WHITNEY, H.
[1] Elementary structure of real algebraic varieties, *Ann. of Math.* **66** (1957), 545–556.

INDEX

H

I

J

L

M

N

O

P

Q

R

U

V

W

Graduate Texts in Mathematics

(continued from page ii)

62 KARGAPOLOV/MERLZJAKOV. Fundamentals of the Theory of Groups.
63 BOLLOBAS. Graph Theory.
64 EDWARDS. Fourier Series. Vol. I 2nd ed.
65 WELLS. Differential Analysis on Complex Manifolds. 2nd ed.
66 WATERHOUSE. Introduction to Affine Group Schemes.
67 SERRE. Local Fields.
68 WEIDMANN. Linear Operators in Hilbert Spaces.
69 LANG. Cyclotomic Fields II.
70 MASSEY. Singular Homology Theory.
71 FARKAS/KRA. Riemann Surfaces. 2nd ed.
72 STILLWELL. Classical Topology and Combinatorial Group Theory. 2nd ed.
73 HUNGERFORD. Algebra.
74 DAVENPORT. Multiplicative Number Theory. 2nd ed.
75 HOCHSCHILD. Basic Theory of Algebraic Groups and Lie Algebras.
76 IITAKA. Algebraic Geometry.
77 HECKE. Lectures on the Theory of Algebraic Numbers.
78 BURRIS/SANKAPPANAVAR. A Course in Universal Algebra.
79 WALTERS. An Introduction to Ergodic Theory.
80 ROBINSON. A Course in the Theory of Groups. 2nd ed.
81 FORSTER. Lectures on Riemann Surfaces.
82 BOTT/TU. Differential Forms in Algebraic Topology.
83 WASHINGTON. Introduction to Cyclotomic Fields. 2nd ed.
84 IRELAND/ROSEN. A Classical Introduction to Modern Number Theory. 2nd ed.
85 EDWARDS. Fourier Series. Vol. II. 2nd ed.
86 VAN LINT. Introduction to Coding Theory. 2nd ed.
87 BROWN. Cohomology of Groups.
88 PIERCE. Associative Algebras.
89 LANG. Introduction to Algebraic and Abelian Functions. 2nd ed.
90 BRØNDSTED. An Introduction to Convex Polytopes.
91 BEARDON. On the Geometry of Discrete Groups.
92 DIESTEL. Sequences and Series in Banach Spaces.
93 DUBROVIN/FOMENKO/NOVIKOV. Modern Geometry—Methods and Applications. Part I. 2nd ed.
94 WARNER. Foundations of Differentiable Manifolds and Lie Groups.
95 SHIRYAEV. Probability. 2nd ed.
96 CONWAY. A Course in Functional Analysis. 2nd ed.
97 KOBLITZ. Introduction to Elliptic Curves and Modular Forms. 2nd ed.
98 BRÖCKER/TOM DIECK. Representations of Compact Lie Groups.
99 GROVE/BENSON. Finite Reflection Groups. 2nd ed.
100 BERG/CHRISTENSEN/RESSEL. Harmonic Analysis on Semigroups: Theory of Positive Definite and Related Functions.
101 EDWARDS. Galois Theory.
102 VARADARAJAN. Lie Groups, Lie Algebras and Their Representations.
103 LANG. Complex Analysis. 3rd ed.
104 DUBROVIN/FOMENKO/NOVIKOV. Modern Geometry—Methods and Applications. Part II.
105 LANG. $SL_2(\mathbf{R})$.
106 SILVERMAN. The Arithmetic of Elliptic Curves.
107 OLVER. Applications of Lie Groups to Differential Equations. 2nd ed.
108 RANGE. Holomorphic Functions and Integral Representations in Several Complex Variables.
109 LEHTO. Univalent Functions and Teichmüller Spaces.
110 LANG. Algebraic Number Theory.
111 HUSEMÖLLER. Elliptic Curves.
112 LANG. Elliptic Functions.
113 KARATZAS/SHREVE. Brownian Motion and Stochastic Calculus. 2nd ed.
114 KOBLITZ. A Course in Number Theory and Cryptography. 2nd ed.
115 BERGER/GOSTIAUX. Differential Geometry: Manifolds, Curves, and Surfaces.
116 KELLEY/SRINIVASAN. Measure and Integral. Vol. I.
117 SERRE. Algebraic Groups and Class Fields.
118 PEDERSEN. Analysis Now.
119 ROTMAN. An Introduction to Algebraic Topology.

120 ZIEMER. Weakly Differentiable Functions: Sobolev Spaces and Functions of Bounded Variation.
121 LANG. Cyclotomic Fields I and II. Combined 2nd ed.
122 REMMERT. Theory of Complex Functions. *Readings in Mathematics*
123 EBBINGHAUS/HERMES et al. Numbers. *Readings in Mathematics*
124 DUBROVIN/FOMENKO/NOVIKOV. Modern Geometry—Methods and Applications. Part III.
125 BERENSTEIN/GAY. Complex Variables: An Introduction.
126 BOREL. Linear Algebraic Groups. 2nd ed.
127 MASSEY. A Basic Course in Algebraic Topology.
128 RAUCH. Partial Differential Equations.
129 FULTON/HARRIS. Representation Theory: A First Course. *Readings in Mathematics*
130 DODSON/POSTON. Tensor Geometry.
131 LAM. A First Course in Noncommutative Rings.
132 BEARDON. Iteration of Rational Functions.
133 HARRIS. Algebraic Geometry: A First Course.
134 ROMAN. Coding and Information Theory.
135 ROMAN. Advanced Linear Algebra.
136 ADKINS/WEINTRAUB. Algebra: An Approach via Module Theory.
137 AXLER/BOURDON/RAMEY. Harmonic Function Theory.
138 COHEN. A Course in Computational Algebraic Number Theory.
139 BREDON. Topology and Geometry.
140 AUBIN. Optima and Equilibria. An Introduction to Nonlinear Analysis.
141 BECKER/WEISPFENNING/KREDEL. Gröbner Bases. A Computational Approach to Commutative Algebra.
142 LANG. Real and Functional Analysis. 3rd ed.
143 DOOB. Measure Theory.
144 DENNIS/FARB. Noncommutative Algebra.
145 VICK. Homology Theory. An Introduction to Algebraic Topology. 2nd ed.
146 BRIDGES. Computability: A Mathematical Sketchbook.
147 ROSENBERG. Algebraic *K*-Theory and Its Applications.
148 ROTMAN. An Introduction to the Theory of Groups. 4th ed.
149 RATCLIFFE. Foundations of Hyperbolic Manifolds.
150 EISENBUD. Commutative Algebra with a View Toward Algebraic Geometry.
151 SILVERMAN. Advanced Topics in the Arithmetic of Elliptic Curves.
152 ZIEGLER. Lectures on Polytopes.
153 FULTON. Algebraic Topology: A First Course.
154 BROWN/PEARCY. An Introduction to Analysis.
155 KASSEL. Quantum Groups.
156 KECHRIS. Classical Descriptive Set Theory.
157 MALLIAVIN. Integration and Probability.
158 ROMAN. Field Theory.
159 CONWAY. Functions of One Complex Variable II.
160 LANG. Differential and Riemannian Manifolds.
161 BORWEIN/ERDÉLYI. Polynomials and Polynomial Inequalities.
162 ALPERIN/BELL. Groups and Representations.
163 DIXON/MORTIMER. Permutation Groups.
164 NATHANSON. Additive Number Theory: The Classical Bases.
165 NATHANSON. Additive Number Theory: Inverse Problems and the Geometry of Sumsets.
166 SHARPE. Differential Geometry: Cartan's Generalization of Klein's Erlangen Program.
167 MORANDI. Field and Galois Theory.
168 EWALD. Combinatorial Convexity and Algebraic Geometry.
169 BHATIA. Matrix Analysis.
170 BREDON. Sheaf Theory. 2nd ed.
171 PETERSEN. Riemannian Geometry.
172 REMMERT. Classical Topics in Complex Function Theory.
173 DIESTEL. Graph Theory.
174 BRIDGES. Foundations of Real and Abstract Analysis.
175 LICKORISH. An Introduction to Knot Theory.
176 LEE. Riemannian Manifolds.
177 NEWMAN. Analytic Number Theory.
178 CLARKE/LEDYAEV/STERN/WOLENSKI. Nonsmooth Analysis and Control Theory.
179 DOUGLAS. Banach Algebra Techniques in Operator Theory. 2nd ed.

180 SRIVASTAVA. A Course on Borel Sets.
181 KRESS. Numerical Analysis.
182 WALTER. Ordinary Differential Equations.
183 MEGGINSON. An Introduction to Banach Space Theory.
184 BOLLOBAS. Modern Graph Theory.
185 COX/LITTLE/O'SHEA. Using Algebraic Geometry.
186 RAMAKRISHNAN/VALENZA. Fourier Analysis on Number Fields.
187 HARRIS/MORRISON. Moduli of Curves.
188 GOLDBLATT. Lectures on the Hyperreals: An Introduction to Nonstandard Analysis.
189 LAM. Lectures on Modules and Rings.
190 ESMONDE/MURTY. Problems in Algebraic Number Theory.
191 LANG. Fundamentals of Differential Geometry.
192 HIRSCH/LACOMBE. Elements of Fuctional Analysis.

Printed in the United States
By Bookmasters